AF879533

Geometry, Topology and Physics for Raoul Bott

Geometry, Topology and Physics for Raoul Bott

Shing-Tung Yau
editor

International Press
www.intlpress.com

Geometry, Topology and Physics for Raoul Bott

Editor:
Shing-Tung Yau, Harvard University

ISBN 978-1-57146-261-9

Paperback reissue 2013.
Previously published in 1995 under ISBN 1-57146-024-1 (clothbound).

Printed in the United States of America.

Table of Contents

Preface

In 1993, we held a conference dedicated to Raoul Bott on his 70th birthday. Many of his friends and students came from all over the world to celebrate this special occasion.

Raoul Bott is a valued member of the Department of Mathematics, Harvard University, and a great mathematician. The conference in his honor attracted a large audience over the course of three days. The lectures given during that conference are presented in this volume in addition to other important mathematical contributions that have been dedicated to Raoul Bott.

We are very grateful to all the people who participated in this conference and to the authors who contributed to this volume. Their involvement was one of the basic reasons for the success of this conference.

Along with support from Harvard, we had support from the Journal of Differential Geometry, which provided the funding for the speakers. I am very grateful to the editors of the journal: C. C. Hsiung, Robert Bryant, Simon Donaldson, Blaine Lawson, and Rick Schoen. I also wish to thank Wilfried Schmid, the Department Chair who made resources and lecture halls available, Ruby Aguirre, the Department Administrator, Sarah Jaffe, and Svetlana Nasonova-Szenes, for their work on the manuscripts.

S.-T. Yau
Harvard University

Toward a Geometry of Differential Equations

Robert Bryant
Duke University
Durham, NC 27708

Phillip Griffiths
Institute for Advanced Study
Princeton, NJ 08540

Lucas Hsu
Institute for Advanced Study
Princeton, NJ 08540

In this expository paper we will discuss the geometry of differential equations. On the one hand this subject has very classical origins going back to the earliest systematic studies of differential equations. On the other hand, in much of this half-century, its main thrust has been more towards structural and foundational matters within the theory than towards applications to particular equations. However, recently this balance has been changing and there are indications that this type of geometric understanding of differential equations is of value in understanding geometric and analytic problems.

In this paper the emphasis will be on illustrating the general theory rather than explaining it. Thus we will attempt to illustrate how the geometry of differential equations may be used to gain insight into selected particular equations and classes of equations. The examples we have chosen will for the most part be elementary. The presentation is designed to show how one may look at familiar materials in PDE theory in a somewhat more geometric fashion.

What do we mean by a "*geometry*"? In the first approximation, this will mean a pair

$$(B_G \to M,\, \phi)$$

where M is an n-manifold, $G \subset \mathrm{GL}(n,\mathbb{R})$ is a Lie subgroup and $B_G \to M$ is a G-structure on M, and ϕ is a connection on $B_G \to M$. (Actually, B_G will sometimes turn out to be a sub-bundle of a higher order coframe bundle and ϕ will be a pseudo-connection, but these refinements will not be insisted upon here.) To

This research was supported in part by NSF Grant DMS 9205222 (Bryant), and the Institute for Advanced Study (Griffiths and Hsu).

Geometry, Topology, and Physics for Raoul Bott

the geometry $(B_G \to M, \phi)$ we may invariantly attach other objects, such as the *curvature* and *geodesics* of the connection ϕ.

The main general observation is that to a differential equation (ordinary or partial) one may canonically attach a geometry in the above sense. More precisely, given a differential equation E *and* a group Γ of admissable transformations (for example gauge, point, or contact transformations, as explained below), there is, under mild regularity assumptions, an intrinsically associated geometry. For general PDE systems, including many interesting special cases, the pseudo-connection ϕ will be unique. However, the construction

$$(E, \Gamma) \rightsquigarrow (B_G \to M, \phi)$$

is somewhat subtle, and in fact the general result that the construction is a finite process has (to our knowledge) never really been proved. In this paper we will usually think of this process as a "black box"; in the literature it is more formally known as the *equivalence problem.* The consequences of the construction will be discussed in the text for a number of examples. In the Appendix we have worked through the "black box" in these examples. Here we emphasize the fundamental point that *there is a natural geometry associated to a differential equation.* Thus to a differential equation one may invariantly attach its *curvatures* (classically known as differential invariants) and the notion of *completeness* (which pertains to existence of global solutions).

Although the association of a geometry to a differential equation may not be generally familiar, certain aspects of this construction—"partial geometries", so to speak—are, of course, well known. For example, many equations are written in "almost canonical" coordinates that do have physical significance, so that coordinate calculations have at least partially invariant meaning. Or again, the *symbol* (and sometimes *sub-principal symbol*) of a partial differential equation, which is usually discussed in a non-invariant way, turns out to have geometric meaning. Our point is that consideration of the full geometry associated to a differential equation can sometimes supplement these existing methods.

Before turning to a discussion of the contents of this paper, we would like to suggest what the objectives of a "geometry of differential equations" should be. First and foremost, the theory should provide new information and perspectives on solutions[1] to differential equations. In fact, the classical theory was originally developed in order to provide explicit solutions to interesting differential equations, and then later it sought to provide explicit methods (integral formulas, etc.) for constructing the solutions (see *[17]*). A second objective of the theory is to treat $(B_G \to M, \phi)$ as a geometric object of interest in its own right. Thus, as we shall illustrate below, certain interesting differential equations may be uniquely characterized in terms of their associated geometry.

Perhaps one may think here of the subject of algebraic geometry. Its original objective was to solve algebraic equations. Later the geometric objects these

[1]In fact, as we shall illustrate below, the theory once again raises the question "What is meant by a *solution* to a differential equation?"

equations define—the algebraic varieties—became an object of interesting study in their own right. In algebraic geometry the above two objectives are now basically coincident. As to what will happen with the geometric theory of differential equations, it has been said that it is easier to invent the future than to predict it.

We now proceed to give an outline of the contents of this paper.

In Section 0 we give a brief discussion of differential equations and their equivalence under various natural groups of transformations. In addition, we give a sketch of how one may canonically associate an exterior differential system to a differential equation. We do not belabor the technicalities of this construction as these may be found in the literature (see for instance, *[2]*) and, in any case, they will be explained in the examples.

In Section 1 we study two elementary examples—the first being the geometry associated to the second order ODE

$$\frac{d^2y}{dx^2} = f\left(x, y, \frac{dy}{dx}\right) \tag{1}$$

with the group being induced by coordinate transformations in the xy-plane. The geometry associated to this simple example is extremely rich and, with hindsight, may be seen to portend the double fibration construction of twistor theory. The geodesics of the canonical connection associated to (1) turn out to give solution curves to the equation together with a distinguished choice of parameter defined up to a linear fractional transformation; this leads to an intrinsic notion of completeness for solutions of (1). An interesting observation is that the second order ODE above has a natural "dual equation". It turns out that the "curvature" of (1) has two principal components—the vanishing of one (or the other) being equivalent to the condition that the equation (or its dual) is the geodesic equation of a projective connection. É. Cartan observed that in the latter case, generically, equation (1) has a complete set of first integrals expressed rationally in terms of f and its derivatives.

In the Appendix we have worked through the equivalence problem for (1). In fact, this simple example is somewhat subtle and illustrates many aspects of the procedure of associating a geometry to a differential equation.

In our second example, we study the geometry associated to the non-linear scalar PDE

$$u_t + g(x,t,u)u_x = h(x,t,u), \qquad g_u \neq 0. \tag{2}$$

It is observed that this geometry is equivalent to the geometry associated to (1). This leads to an intrinsic notion of completeness for solutions of (2), and of global integral surfaces of the exterior differential system associated to (2). We will show that the sign of the curvature associated to the geometry governs the development of singularities of classical smooth solutions $u(x,t)$ to (2).

Next, in Section 2, we turn to the geometry associated to the non-linear scalar conservation law

$$u_t + \partial_x(F(x,t,u)) = 0, \qquad F_{uu} \neq 0 \tag{3}$$

where ∂_x denotes the total derivative with respect to the x variable. This geometry is a refinement of that associated to (2) since two distinct classes of scalar conservation laws may be equivalent as PDEs. There is again a notion of completeness and of global integral surfaces of the exterior differential system associated to (3). For example, any translation-invariant conservation law

$$u_t + f(u)u_x = 0, \qquad f'(u) \neq 0 \tag{4}$$

where $f : \mathbb{R} \to \mathbb{R}$ is a smooth function defined on all of $\mathbb{R}$, is complete. Again we interpret classical results on the existence or non-existence of global smooth solutions $u(x,t)$ in terms of the sign of the curvature of (3). Finally, for equations such as (4) the classical results on global existence and uniqueness of "shock solutions" (see *[20]*) to the equation are discussed in terms of the geometry of the global, smooth integral surface of the associated exterior differential system.

In the Appendix we discuss the equivalence problem associated to the scalar conservation law (3).

In Section 3.1 we study the exterior differential system associated to a hyperbolic PDE system

$$\begin{aligned} u_y + au_x + bv_x + f &= 0 \\ v_y + cu_x + ev_x + g &= 0 \end{aligned} \tag{5}$$

where the coefficients functions are functions of x, y, u and v. This exterior differential system is given by the data $(M; \Omega_1, \Omega_2)$ consisting of a 4-manifold M together with a pair of transverse, decomposable 2-forms Ω_1, Ω_2. This relatively simple structure turns out to have a very rich geometry, the basic aspects of which we explain in the Appendix. For example, many interesting non-linear PDEs (5) are explicitly linearizable as exterior differential systems—i.e., they may be linearized by a suitable contact transformation. In the Appendix we give necessary and sufficient conditions on the torsion and curvature of the geometry associated to (5) that its associated exterior differential system be linearizable.

In Section 1 we discuss conditions on the curvature of (1) that imply that there be a complete set of first integrals of (1). For PDEs one analog of having a complete set of first intergals is that there should exist explicit formulae for the general solution of (5) of the form

$$\begin{aligned} u &= U\big(x, y, \alpha(x), \alpha'(x), \dots, \alpha^{(k)}(x), \beta(y), \beta'(y), \dots, \beta^{(k)}(y)\big) \\ v &= V\big(x, y, \alpha(x), \alpha'(x), \dots, \alpha^{(k)}(x), \beta(y), \beta'(y), \dots, \beta^{(k)}(y)\big) \end{aligned} \tag{6}$$

where α and β are arbitrary functions of one variable. This is closely related to the concept of *Darboux integrability* of the exterior differential system associated to (5), and in Section 3.2 we will express the necessary and sufficient conditions for Darboux integrability at level one (roughly, this is equivalent to the existence of (6) with $k = 1$) in terms of the vanishing of suitable components of the torsion and curvature of (5). This result is then illustrated by a number of examples.

In Section 4.1 we begin a discussion of translation-invariant hyperbolic systems of the form

$$\begin{aligned} u_y &+ a(u,v)u_x + b(u,v)v_x = 0 \\ v_y &+ c(u,v)u_x + e(u,v)v_x = 0. \end{aligned} \tag{7}$$

The exterior differential systems associated to (7) are always linearizable. For a general hyperbolic system (5) we may intrinsically formulate the condition (which we call *characteristic completeness*) that the non-characteristic initial value problem of the associated exterior differential problem admit global, smooth integral surfaces. Then we show that a wide class of equations (7) satisfy the condition to be characteristically complete, and we illustrate this result in several examples.

In Section 4.2 we tie in the classical concept of genuine non-linearity of (7) with the discussion of the elementary equation (2). Then we derive, from our perspective, the well-known result that genuinely non-linear systems (7) do not admit global smooth solutions with compactly supported initial data. On the other hand we will show that in this case there exists a unique, global smooth integral surface for the initial value problem of the associated exterior differential system.

Finally, in Section 4.3 we discuss the case where (7) is given as a hyperbolic system of conservation laws

$$\begin{aligned} u_y &+ \partial_x\big(f(u,v)\big) = 0 \\ v_y &+ \partial_x\big(g(u,v)\big) = 0. \end{aligned} \tag{8}$$

As mentioned above, for smooth non-characteristic initial data there is a unique, global smooth integral surface S for the associated exterior differential system. For $\mathbb{R}^2_+ = \{(x,y) : y \geq 0\}$ the mapping

$$\pi : S \longrightarrow \mathbb{R}^2_+ \tag{9}$$

is shown to be proper. We may think of S as the "geometric" solution of (8).

On the other hand, it is known *[20]* that under suitable conditions on the initial data and with the assumption that (8) is genuinely non-linear, there is a unique global shock solution. It is of interest to compare the geometric and shock solutions.

For a scalar conservation law, the shock solution arises by taking a suitable cross-section of (9). We may say that the geometric solution "captures" the shock solution.

For a system of conservation laws, the situation is more complex. It turns out that on the uv-plane there are two naturally defined pairs of families of curves. The first pair consists of the two foliations ρ_λ, ρ_μ given by the level sets of the Riemann invariants. The other is the two families δ_λ, δ_μ of jump curves obtained by imposing the Rankine-Hugonoit (or jump) conditions. At each point of the uv-plane the δ-curves and ρ-curves osculate to second, but generally not to third, order.

In general, the two families of δ-curves do *not* define a pair of (local) foliations of the uv-plane—the relation defined by a δ-curve is symmetric and reflexive but not necessarily transitive. The condition that δ_λ and δ_μ define a pair of foliations is seen to be given by a pair of second-order PDEs for the defining functions $f(u,v)$ and $g(u,v)$ of the conservation laws (8). (These equations may be explicitly solved.) We shall call such hyperbolic systems of conservation laws *special.* For special systems, the δ-curves and ρ-curves coincide since two foliations which are everywhere tangent must coincide.

We formulate and study the *Riemann problem* for the exterior differential system associated to (8). For this case, it will be shown that the geometric solution "captures" the shock solution if, and only if, the system is special. Special systems form a remarkable class of hyperbolic conservation laws—for arbitrary smooth initial data, there appears to be a unique shock solution to each of these systems arising as a cross-section of (9); in particular, the shock solution should be piecewise smooth if the initial data is piecewise smooth. On the other hand, for non-special systems, it seems necessary to perform some sort of "PDE surgery" on S in order to obtain an integral surface that captures the shock solution. We conclude the paper with a brief discussion of this point.

0 Preliminaries

Recall that a first order PDE system may be written in coordinates as

$$F^\mu(x^i,\, u^\alpha(x),\, \partial u^\alpha(x)/\partial x^i) = 0, \qquad \mu = 1,\ldots,m \tag{1}$$

where $x = (x^1,\ldots,x^n) \in X \subset \mathbb{R}^n$ and $u = (u^1,\ldots,u^s) \in U \subset \mathbb{R}^s$ are the independent and dependent variables respectively. Introducing variables p_i^α to stand for the derivatives $\partial u^\alpha/\partial x^i$, we consider the locus in xup-space defined by

$$M = \{F^\mu(x^i,\, u^\alpha,\, p_i^\alpha) = 0, \quad \mu = 1,\ldots,m\}.$$

We assume M to be a smooth manifold. Clearly, a graph mapping $X \longrightarrow M$ given by $x \mapsto (x, u(x), p(x))$ is a solution to (1) if, and only if, the differential forms

$$\theta^\alpha = du^\alpha - p_i^\alpha dx^i$$

restrict to be zero on M. This suggests that we more generally consider immersions

$$f : N \longrightarrow M$$

of a manifold N into M which satisfy

$$f^*\theta^\alpha = 0, \qquad \alpha = 1,\ldots,s. \tag{2}$$

Since

$$f^* : \Omega^*(M) \longrightarrow \Omega^*(N)$$

is a mapping of differential graded algebras, i.e., f^* commutes with exterior derivative and wedge products, we should consider the *differential ideal $\mathcal{I}$* in $\Omega^*(M)$ generated by the θ^α. Immersions satisfying

$$f^*\theta = 0, \qquad \theta \in \mathcal{I}$$

are the same as those satisfying (2) and are called *integral manifolds* of the differential ideal $\mathcal{I}$. The pair $(M, \mathcal{I})$ consisting of a manifold M together with a differential ideal $\mathcal{I} \subset \Omega^*(M)$ is by definition an *exterior differential system.*

This construction of associating an exterior differential system to a PDE system given in coordinates may be extended to higher order systems, either directly using higher order jet coordinates and the associated contact forms (see, for instance, [2]) or by rewriting the higher order system as a (larger) first order system.

How invariant is the above coordinate construction of an exterior differential system associated with a PDE system? In an increasing order of generality there are four natural types of change of variables associated to (1):

Classical transformations: These are induced by a change $\bar{x} = \bar{x}(x)$, $\bar{u} = \bar{u}(u)$ of independent and dependent variables separately. Geometrically we are viewing (1) as imposing conditions on the differential of a mapping

$$\begin{array}{ccc} X & \longrightarrow & U \\ \cap & & \cap \\ \mathbb{R}^n & & \mathbb{R}^s. \end{array}$$

Gauge transformations: These are induced by a change of variables of the form $\bar{x} = \bar{x}(x)$, $\bar{u} = \bar{u}(x, u)$. Here we are viewing (1) as imposing differential conditions on cross sections of a fibration

$$\begin{array}{c} U \subset \mathbb{R}^{n+s} \\ \downarrow \\ X \subset \mathbb{R}^n. \end{array}$$

Point transformations: These are induced by the following change of variables $\bar{x} = \bar{x}(x, u)$, $\bar{u} = \bar{u}(x, u)$. Now we are viewing (1) as imposing differential conditions on an immersion

$$\begin{array}{ccc} X & \longrightarrow & \mathbb{R}^{n+s} \\ \cap & & \\ \mathbb{R}^n & & \end{array}$$

where we have happened to write the immersion in (1) locally in the form of a graph $(x^i) \mapsto (x^i, u^\alpha(x))$.

In all the above transformations, the change $\bar{p} = \bar{p}(x, u, p)$ is induced by differentiating the change among the x, u variables. The last type of transformation we shall consider is of a different character.

Contact transformations: These are induced by a change of variables $\bar{x} = \bar{x}(x, u, p)$, $\bar{u} = \bar{u}(x, u, p)$, and $\bar{p} = \bar{p}(x, u, p)$ which map M to itself and preserve the differential ideal generated by the θ^α.

More or less by construction, the association of an exterior differential system to a PDE system is contact invariant. For a given problem we may therefore choose the group Γ to be generated by the classical, gauge, point or contact transformations of (1). Under point transformations there is no notion of "independent" and "dependent" variables. More generally, under contact transformations there is not even a notion of "base" variables. To illustrate the difference, we note that under contact transformations the following equations for a function $u(x, y)$:

$$u_{xx} - u_{yy} = 0 \qquad \text{and} \qquad u_{xx}u_{yy} - u_{xy}^2 = -1$$

are globally equivalent even though there is no change of variables in xyu-space which will convert one of these equations into the other. Special contact transformations (Legendre transformation, hodographic method) have of course been classically used to study differential equations (see, for instance, *[12]*). But the systematic utilization of the invariants of a differential equation under any of the above groups (as in Klein's Erlangen Program) has not been so much an aspect of the modern theory of differential equations. In the following we shall illustrate the use of each of the above groups.

1 Two Elementary Examples and their Geometries

In this section, with the help of two elementary examples, we shall show how the geometry associated with a differential equation might be used to deduce interesting properties of the solutions of the equation. For both examples, the geometry of the defining equation have been worked out in the Appendix.

EXAMPLE 1: *The second-order ODE.* Consider the differential equation

$$\frac{d^2y}{dx^2} = f\left(x, y, \frac{dy}{dx}\right). \tag{1}$$

Of course, such equations have been objects of study for over three centuries, with a primary goal being to describe the solutions in a reasonably explicit way. The classical theory generally concentrates on constructing "first integrals" of the equation, with the most common technique being that of assuming some sort of symmetry, so that some version of Noether's Theorem can be applied to yield a conserved quantity, i.e., a first integal.

We shall study (1) under the equivalence relation given by point transformations. Thus, we shall think of (1) as a second-order ODE for curves on a surface S, where x, y are local coordinates on S and (1) is the ODE for those curves $(x, y(x))$ which are graphs over the x-axis.

To study (1) geometrically we consider the exterior differential system $(\mathcal{I}, \underline{\alpha})$ in xyp-space, henceforth denoted by M, generated by the 1-forms

$$\underline{\theta} = dy - p\,dx \qquad \text{and} \qquad \underline{\omega} = dp - f(x, y, p)\,dx$$

with independence condition given by the 1-form $\underline{\alpha} = dx \neq 0$. Integral curves $c : \mathbb{R} \to M$ of the differential system

$$\underline{\theta} = \underline{\omega} = 0$$

satisfying the transversality condition $c^*(\underline{\alpha}) \neq 0$ are in one-to-one correspondence with solution curves of (1).

In intrinsic terms, $M = \mathbb{P}(TS)$ is the projectived tangent bundle of S and x, y, p are local coordinates in the open set of tangent directions ξ for which $dx(\xi) \neq 0$. The equations $\underline{\theta} = \underline{\omega} = 0$ define a field of line elements on M whose integral curves project to S to give solutions to (1).

By construction, M is a contact manifold with contact structure defined by the 1-form $\underline{\theta}$. Diffeomorphisms of M which preserve the differential ideal generated by $\underline{\theta}$ are classically called *contact transformations*. Under such transformations, a century-old theorem of Sophus Lie states that every ODE of the form (1) is locally equivalent to the "flat" model

$$\frac{d^2y}{dx^2} = 0.$$

Although this is an interesting geometric fact, it is well known that finding a contact transformation taking (1) into the flat model involves solving a PDE for the generating function of the contact transformation. Thus Lie's result is generally of little help in explicitly solving (1).

In the following, we shall instead study the geometry of (1) with respect to the smaller pseudo-group Γ of those contact transformations of M which preserve the differential ideal $\mathcal{J}$ generated by the 1-forms

$$\underline{\alpha} = dx \qquad \text{and} \qquad \underline{\theta} = dy - p\,dx.$$

Clearly this is the same as the ideal generated by dx and dy and so such contact transformations are in fact induced by point transformations. Following Cartan, we will explain the consequences of a geometric construction that singles out a large class of second-order ODEs that may be (globally) explicitly solved, in spite of there being no assumption of symmetry. This will illustrate how a geometric study of a differential equation can provide new perspectives on solutions to the equation.

The starting point is to observe that the ideals $\mathcal{I}$ and $\mathcal{J}$ on M determine a canonical double fibration

$$\begin{array}{ccc} & M & \\ {}^{\lambda}\swarrow & & \searrow^{\rho} \\ S & & \Sigma \end{array}$$

where $\ker \lambda_* = \{\underline{\alpha}, \underline{\theta}\}^\perp$ and $\ker \rho_* = \{\underline{\theta}, \underline{\omega}\}^\perp$ define a pair of line fields on M with S and Σ the locally defined quotient surfaces. Geometrically, viewing (1) as defining a 2-parameter family Σ of paths on a surface S—each point $s \in S$ and each tangent vector $\xi \in T_sS$ determining a unique path $c \in \Sigma$—the quotient surface Σ may be viewed as the "space of solutions" of the second-order ODE (1) on S.

In local coordinates x, y on S and σ, τ on Σ, the solutions of (1) may be written as

$$F(x, y, \sigma, \tau) = 0 \tag{2}$$

where for each fixed σ, τ the above relation gives a solution curve to (1)—i.e., σ, τ are constants of integration. Explicitly, the equation $F = 0$ describes the image of the mapping $(\lambda, \rho) : M \to S \times \Sigma$.

Note that this geometric picture is symmetric with respect to S and Σ. In other words, while each point of Σ represents a curve in S, dually, each point $p \in S$ represents a curve $\rho(\lambda^{-1}(p))$ in Σ. It is easy to see that the 2-parameter family of curves in Σ defined by (2) are the solutions to a certain second-order ODE in the $\sigma\tau$-plane (i.e., Σ)

$$\frac{d^2\sigma}{d\tau^2} = \varphi\left(\sigma, \tau, \frac{d\sigma}{d\tau}\right). \tag{3}$$

The relation between (1) and (3) is symmetric. In the classical literature, these two second order ODEs were said to be "dual" equations. By construction, functions on Σ are first integrals of (1), and dually, functions on S are first integrals of (3).

The goal of Cartan's approach was to find an explicit procedure for finding functions on M which are the pull back of (non-constant) functions on Σ. Clearly, the level sets of such functions implicitly define the solution curves of (1). More exactly, Cartan wanted to describe classes of equations (1) for which such procedures existed.

Cartan's method is quite general, but here we want to describe one particular such class of equations. Recall that the geodesics of a Riemannian metric on S are locally given by an ODE (1) where f is a cubic polynomial in dy/dx. More generally, Cartan *[6]* has defined the concept of a *normal projective connection*, and shown that the geodesics of such a connection satisfy an equation of the form (1) where

$$\frac{\partial^4 f}{\partial p^4} = 0. \tag{4}$$

Conversely, every such equation can be realized as the geodesic equation of a unique normal projective connection on S. Curvature invariants of this connection then produce functions on S which are generally non-constant.

Cartan *[7]* showed that the condition on (1) that its dual equation (3) describe the geodesics of a projective connection on Σ is that the function f satisfy the differential equation

$$\frac{d^2}{dx^2}(f_{pp}) - 4\frac{d}{dx}(f_{py}) + f_p\left(4f_{py} - \frac{d}{dx}(f_{pp})\right) - 3f_y f_{pp} + 6f_{yy} = 0 \tag{5}$$

where d/dx denotes total derivative with respect to x. Moreover, he showed that, in this case, starting with only the knowledge of the ideals $\mathcal{I}$ and $\mathcal{J}$ on M, which, as we have seen, are completely determined by equation (1), there is an algorthmic procedure to compute the curvatures of the projective connection on Σ, thus yielding first integrals of the original equation (1). It follows that the class of equations (1) satisfying (5) should be *geometrically integrable* in the sense that a complete set of first integrals (or conservation laws) may be determined explicitly from f and its derivatives.[2] Thus, Cartan *[7]* was able to state the following result:

> *Every second-order ODE (1) satisfying (5) may be integrated by a process involving only algebraic operations, differentiation, and quadratures.*

Remark: Strictly speaking, Cartan's statement is not quite correct. It turns out that, up to equivalence, there is a 1-parameter family of such equations whose integration by Cartan's method requires solving a Ricatti equation, which is well-known not to be solvable by quadrature. (This is caused by the fact that the projective curvatures on Σ turn out to be constant.) This expectional family of equations is uniquely characterized by the condition that its symmetry group is $SL(2, \mathbb{R})$ (see *[5]* for more details).

The actual computation of the first integrals of the above equations requires a detailed study of the G-structure associated with the above double fibration. (The G-structure construction for (1) is described in the Appendix.) Associated to the G-structure is an intrinsic connection ϕ whose curvature Φ has two principal components, which we denote by K_1 and K_2. These components have the property that K_2 vanishes if, and only if, the ODE (1) is the geodesic equation of a projective connection ϕ_S on S and K_1 vanishes if, and only if, the ODE (3) is the geodesic equation of a projective connection ϕ_Σ on Σ. Furthermore, the curvatures of ϕ_Σ may be computed in terms of Φ and therefore in terms of the original equation (1). In fact, they may be expressed rationally in terms of f and its derivatives. The functions on M constructed from the curvatures of ϕ_Σ together with their covariant derivatives will then generate the desired set of first integrals of (1).

Cartan's approach to integrating differential equations is rather general. It has been extended to third order ODEs by Chern *[10]* and to fourth order equations by Bryant *[1]* in connection with his work on exceptional holonomies. In fact, Chern's paper directly anticipates several aspects of twistor theory while Bryant's makes explicit reference to it. To paraphrase a famous remark, one may see in the double fibration picture a "piece of twistor theory that fell into the early 20th century".

EXAMPLE 2: *Scalar first order PDE.* We shall now discuss the geometry associated to a first order quasi-linear PDE

$$u_t + g(x, t, u)u_x = h(x, t, u). \tag{6}$$

This equation is usually introduced at the beginning of introductory textbooks on PDE theory and one may well wonder what is new and interesting that the geometry

---

[2] In recent years there has been much interest in "integrable systems". The Cartan approach provides a complementary, perhaps more directly geometric approach to integrating a differential equation.

associated to (6) can tell us? We shall show that when (6) is *genuinely non-linear* (a condition defined in the PDE literature on this equation to mean that $g_u \neq 0$), the geometry associated to (6) under equivalence by gauge transformations is equivalent to the geometry associated to the ODE (1) above under equivalence by point transformations. Having already worked out the latter geometry we shall then be able to apply the results to draw conclusions for (6).

To explain this identification we return briefly to EXAMPLE 1. We may equivalently give the above data $(M; \mathcal{I}, \mathcal{J})$ of the 3-manifold M with Pfaffian systems $\mathcal{I}, \mathcal{J}$ as the data $(M; \Phi, \Omega)$ consisting of the 3-manifold M together with a pair of everywhere linearly independent 2-forms Φ, Ω, defined up to non-zero multiples and satisfying the non-degeneracy condition that the field of 2-planes $\Phi^\perp \wedge \Omega^\perp$ be a contact structure. To explain this we observe that $\Phi^\perp$ and $\Omega^\perp$ are linearly independent line fields on M, and therefore span a field $\Phi^\perp \wedge \Omega^\perp$ of 2-plane elements. Our non-degeneracy condition is that this rank 2 distribution be non-integrable. Given $\mathcal{I}$ and $\mathcal{J}$ as in EXAMPLE 1 where locally $\mathcal{I}$ is generated by 1-forms θ, ω and $\mathcal{J}$ is generated by 1-forms α, θ, we may set

$$\Phi = \theta \wedge \omega \qquad \text{and} \qquad \Omega = \alpha \wedge \theta.$$

Then $\Phi^\perp \wedge \Omega^\perp$ is the field of 2-planes defined by $\theta = 0$ and the non-degeneracy condition is $\theta \wedge d\theta \neq 0$. The construction

$$(M; \mathcal{I}, \mathcal{J}) \to (M; \Phi, \Omega)$$

is easily seen to be well-defined and reversible.

We now proceed to show that the PDE (6) gives rise to a $(M; \Phi, \Omega)$ structure. Letting M denote xtu-space, we consider the 2-forms on M defined by

$$\Phi = (du - h\,dt) \wedge (dx - g\,dt) \qquad \text{and} \qquad \Omega = dx \wedge dt.$$

The motivation for introducing Φ is that on graph surfaces $(x, t) \mapsto (x, t, u(x, t))$ the exterior equation

$$\Phi = 0$$

is equivalent to the PDE (6). Thus the exterior differential system $\Phi = 0$ with independence condition $\Omega \neq 0$ models (6).

Clearly the 2-forms Φ and Ω are well-defined up to non-zero factors under gauge transformations. Furthermore, since $\Omega = (dx - g\,dt) \wedge dt$, we see that Φ and Ω have the common linear factor

$$\theta = dx - g\,dt.$$

Our assumption of genuine non-linearity now implies that

$$\theta \wedge d\theta = g_u\,du \wedge dx \wedge dt \neq 0$$

and so θ defines a contact structure on M. Thus, to the gauge-equivalence class of non-linear PDEs (6) is associated a $(M; \Phi, \Omega)$ structure, and conversely.

Now, in view of the above identification of the two geometries associated with (1) and (6), the G-structure construction for (1) now associates to the gauge-equivalence class of (6) a principal G-bundle $(B_G \longrightarrow M, \phi)$ with connection ϕ. In the following we shall discuss how the geometry $(B_G \longrightarrow M, \phi)$ determines the global behavior of solutions of (6).

Here, *global* has the following intrinsic meaning: Associated to the connection ϕ are its geodesics, which are a special class of curves in B_G equipped with a distinguished parameter τ (see the Appendix for details). Two geodesics in B_G that project to the same curve γ in M have their parameters related by a linear fractional transformation

$$\tilde{\tau} = \frac{a\,\tau + b}{c\,\tau + d}, \qquad ad - bc = 1.$$

We shall say that γ is complete in case its parameter τ takes values in all of $\mathbb{R}$ viewed as a subset of the projective line $\mathbb{P}$. The geometry $(B_G \longrightarrow M,\, \phi)$ is then defined to be *complete* in case all such curves γ are complete.

In the case of the ODE (1) above, the curves γ are the canonical lifts to $\mathbb{P}(TS)$ of solution curves to (1). In the case of the non-linear PDE (6), the curves γ are the characteristic curves (to be explained momentarily). In both cases the completeness of the curves γ gives an intrinsic meaning to the concept of a *global solution* to (1) or (6). In examples this concept turns out to agree with what one usually thinks of as a global solution to a differential equation.

We now formulate the initial value problem of the PDE (6). Recall that associated to (6) is the *characteristic vector field*

$$X = \partial_t + g\,\partial_x + h\,\partial_u$$

on M. Invariantly, X is the vector field uniquely defined up to scaling by the relation $X \lrcorner \Phi = 0$. We note that X depends only on the exterior differential system $\Phi = 0$ and not on the independence condition Ω.

Non-characteristic initial data is given by an immersed curve Γ

$$s \longmapsto \big(x(s),\, t(s),\, u(s)\big)$$

which is nowhere tangent to X. Flowing Γ along integral curves of X then generates the general solution surface N to the exterior differential system $\Phi = 0$. At points of N where $\Omega \neq 0$ the mapping

$$N \longrightarrow xt\text{-plane}$$

is locally one-to-one. Thus, near such points, N is locally given as a graph

$$(x,\, t) \longmapsto \big(x,\, t,\, u(x,\, t)\big)$$

where $u(x,\, t)$ is a solution to (6). This is the method of characteristics, here illustrated in its simplest form.

Now what does the geometry $(B_G \to M,\, \phi)$ tells us about (6)? First, in the flat case, where the curvature of ϕ is zero, working through the identification of the geometries associated to (1) and (6) shows that the PDE (6) is gauge equivalent to the (inviscid) Burgers' equation

$$u_t + uu_x = 0. \tag{7}$$

If, in this case, we consider the corresponding exterior differential system

$$\Phi = du \wedge (dx - u\,dt) = 0 \tag{8}$$

on M, then there are global solutions to (8) as follows: Given an initial curve

$$s \mapsto \big(s,\, 0,\, u_0(s)\big), \qquad s \in \mathbb{R},$$

we may uniquely extend this to a mapping of $\mathbb{R}^2$ giving an integral surface of (8) by

$$(s,\, t) \mapsto \big(s + tu_0(s),\, t,\, u_0(s)\big).$$

This solution may be said to be global for the geometric reason that the characteristic curves $s =$ const are geodesics for the connection ϕ that are *complete* in the sense described above. The question of whether the projection $N \to \mathbb{R}^2$ is one-to-one or onto will be taken up in the next section when we discuss the "shock" behavior of (7).

What does this perfectly global solution to (8) say about classical solutions to (7)? We let $\mathbb{R}^2_+$ and $\mathbb{R}^3_+$ denote the half-space $t \geq 0$ in $\mathbb{R}^2$ and $\mathbb{R}^3$ respectively, and we have the picture

$$\begin{array}{ccccc} \Gamma \subset & N & \subset \mathbb{R}^3_+ & \ni & (x,\, t,\, u) \\ {\scriptstyle \delta_0}\uparrow & \downarrow{\scriptstyle \pi} & & & \downarrow \\ \mathbb{R} \subset & \mathbb{R}^2_+ & & \ni & (x,\, t) \end{array}$$

where $\mathbb{R}$ is the x-axis $t = 0$ and the initial data is given by δ_0, and we seek to extend δ_0 to as large a neighborhood of the initial curve $t = 0$ as possible. To see that we may in general expect difficulties we note that up on N

$$\pi^*(dx \wedge dt) = \big(1 + tu_0'(s)\big)\, ds \wedge dt. \tag{9}$$

Thus, if there are points on the initial curve where $u_0'(s) < 0$ (which will certainly happen if $u_0(s)$ is compactly supported and not identically zero) then δ_0 cannot be extended to a smooth map δ defined on all of $\mathbb{R}^2_+$. This is also evident from the explicit formula

$$\delta(x,\, t) = \big(x,\, t,\, u(x,\, t)\big)$$

where $u(x,\, t)$ is defined implicitly by the equation

$$u = u_0(x - tu). \tag{10}$$

Returning to the general discussion we ask what is the meaning of the vanishing of the curvatures K_1, K_2 of ϕ? The condition $K_1 = 0$ is sufficient to guarantee that the characteristic equation associated to $\Phi = 0$ has a complete set of first integrals. Thus the characteristic curves of the PDE (6) may be found explicitly. In this case we may then write down an implicit equation for u which solves the initial value problem, as was done in (10) for Burgers' equation.

Perhaps more interesting is the condition $K_2 = 0$, which is the case when the second order ODE (1) associated to (6) is the equation of geodesics of a projective connection on a surface. For example, suppose that (6) is

$$u_t + f(u)u_x = h(x, t, u).$$

Our non-linearity assumption is $f'(u) \neq 0$, and so we may write

$$h(x, t, u) = H(x, t, u)/f'(u)$$

for some function H. The condition $K_2 = 0$ then implies that

$$H(x, t, u) = A + Bf(u) + Cf(u)^2 + Df(u)^3.$$

Geometrically the ODE associated to (6) arises as follows: In the picture

$$\begin{array}{ccc} \mathbb{R}^3 \ni & (x, t, u) \\ \downarrow \pi & \downarrow \\ \mathbb{R}^2 \ni & (x, t) \end{array}$$

the projection to $\mathbb{R}^2$ of the characteristic curves of this equation will, by our non-linearity assumption, generate a 2-parameter family of curves which are the solutions of an ODE of the form (1) in $\mathbb{R}^2$. For instance, in the case $K_2 = 0$ as above the ODE is

$$\frac{d^2x}{dt^2} = A + B\left(\frac{dx}{dt}\right) + C\left(\frac{dx}{dt}\right)^2 + D\left(\frac{dx}{dt}\right)^3. \tag{11}$$

The initial value problem for (6) then has the following meaning: Along an initial curve γ

$$s \mapsto \big(x(s), t(s)\big)$$

in the xt-plane we prescribe the initial values $(dt/dx)(s)$. Then we take the integral curves of (1) emanating from γ with the given initial values. As long as these curves don't cross or focus, we will have a classical solution to (6). In fact, *it is exactly the focusing of geodesics that gives classical development of singularities*. This phenomenon is well-known in the study of wave fronts and the Hamilton-Jacobi equation.

Thus for example, suppose that (11) is the geodesic equation for a complete Riemannian metric in the xt-plane whose Gauss curvature $K \geq 0$. (For Burgers' equation (7) this is the situation, with $K = 0$.) Suppose also that γ is a geodesic and that there is a sub-interval of γ along which the angle of the initial data is decreasing. (This is the geometric form of the assumption $u_0' < 0$ in Burgers' equation.) Then the geodesics which represent the projections of the characteristic curves on the surface N will focus in the xt-plane and thus singularities will develop. If, however, the Gauss curvature K is strictly negative and the initial data is appropriately slowly varying, then the spreading of geodesics as given by the Jacobi equation will imply that there is no focussing and global solutions to the PDE will exist. Pictorially, moderate ocean waves will not break along a negatively curved beach.

2 The Scalar Conservation Law

In EXAMPLE 2 above we considered the geometry associated to the scalar first order PDE

$$u_t + g(x,\, t,\, u)u_x = h(x,\, t,\, u). \tag{1}$$

We assumed that $g_u \neq 0$ (i.e., genuine non-linearity) and saw that the geometry of (1) was equivalent to that of a second order ODE

$$\frac{d^2y}{dx^2} = f\left(x,\, y,\, \frac{dy}{dx}\right). \tag{2}$$

The initial value problem and development of singularities of solutions to (1) then had interpretations in terms of the "path geometry" defined by (2).

In this section we will look more deeply into singularity development and so-called "shock solutions" in terms of a geometry associated to a special subclass of equations (1)—the ones which can be written as a scalar conservation law

$$u_t + \partial_x\big(F(x,t,u)\big) = 0.$$

To motivate the introduction of this class of equations, we assume given a PDE (1) whose associated geometry is complete. Given a non-characteristic initial curve Γ, we sweep out a surface S by flowing Γ along the characteristic vector field X. This constructs S as an integral surface of the exterior differential system $\Phi = 0$ associated to (1). If Γ is given in coordinates by $s \mapsto \big(s,\, 0,\, u_0(s)\big)$ and we set $\mathbb{R}^2_+ = \{(x,\, t) : t \geq 0\}$, then suitable conditions on the initial data (see below) plus the completeness of X will insure that

$$\pi : S \to \mathbb{R}^2_+ \tag{3}$$

is surjective. Singularities of a solution $u(x,\, t)$ to (1), defined say for $0 \leq t < t_0$, will then develop in relation to the projection in the xt-plane of the *fold locus* $\mathcal{F}$ where the differential π_* drops rank. We may seek to understand this situation by applying singularity theory to the projection π.[3]

[3]The idea of multi-valued solutions to PDEs is of course classical. The issue is to show in examples that these exist and have interesting properties. In this regard we would like to call attention to the recent interesting paper *[9]* by Caflish et al which studies the branching behavior of multi-valued solutions to certain non-linear PDEs.

Typically $\mathcal{F}$ will be a union of curves, with the typical one looking something like a classical "cusp" (Figure 2.1). Over the lines $t = t_-$, $t = t_0$, $t = t_+$ the graphs of the projection will resemble Figure 2.2. More generally the fold locus will project to a curve that might contain several such cusps (Figure 2.3) with the following cross-sections of S over the lines $t = t_i$ (Figure 2.4).

So far, the whole picture is gauge invariant. However, if, as is commonly done in the PDE literature, we seek to define a "generalized solution" to (1) as a cross-section (Figure 2.5) satisfying certain local conditions (which are typically *not* invariant under gauge transformations), then we need more information than that given by the gauge equivalence class of (1). To determine the cross-section u, we need to know the *breaking curve* along which u jumps from one sheet of S to another, and we also need to know to which sheet u will jump as it crosses this curve. This data cannot be determined from the gauge equivalence class of (1), so we will consider a special subclass for which a more restrictive geometry can be defined.

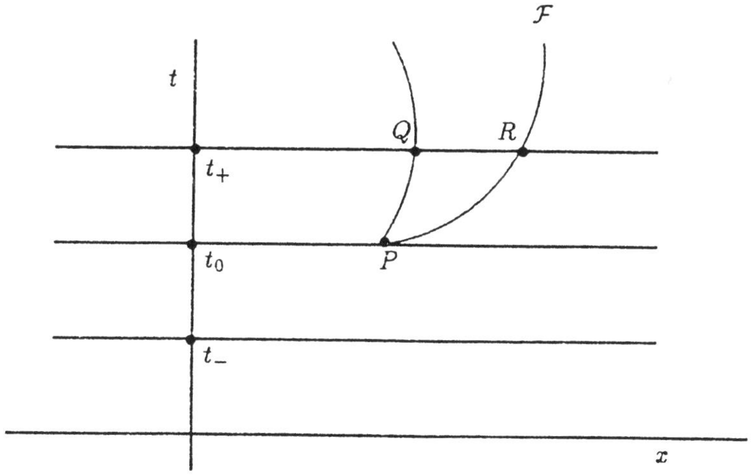

Figure 2.1

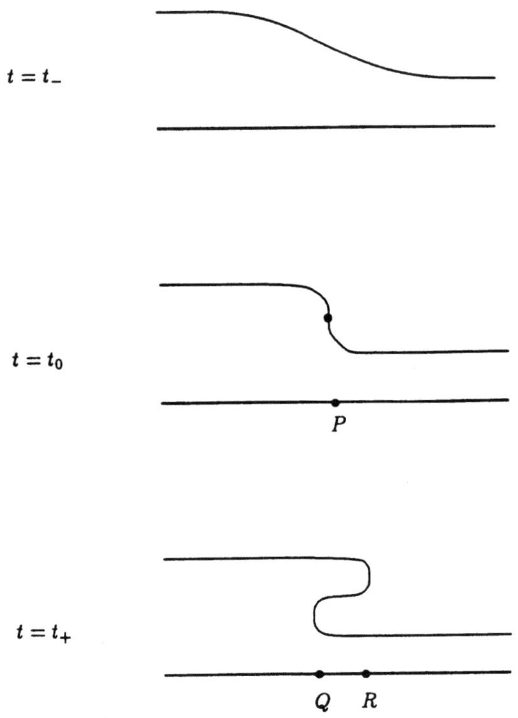

Figure 2.2

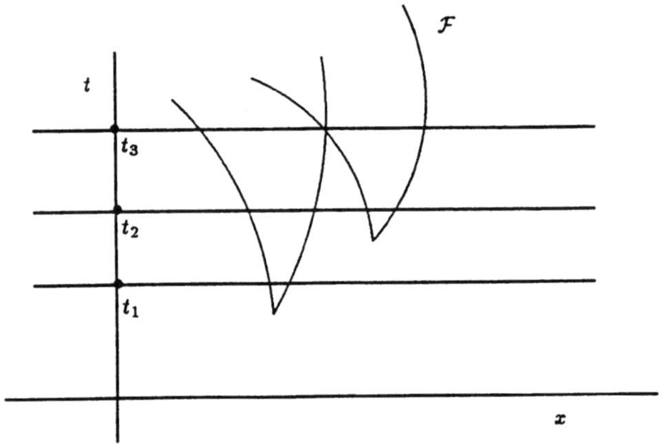

Figure 2.3

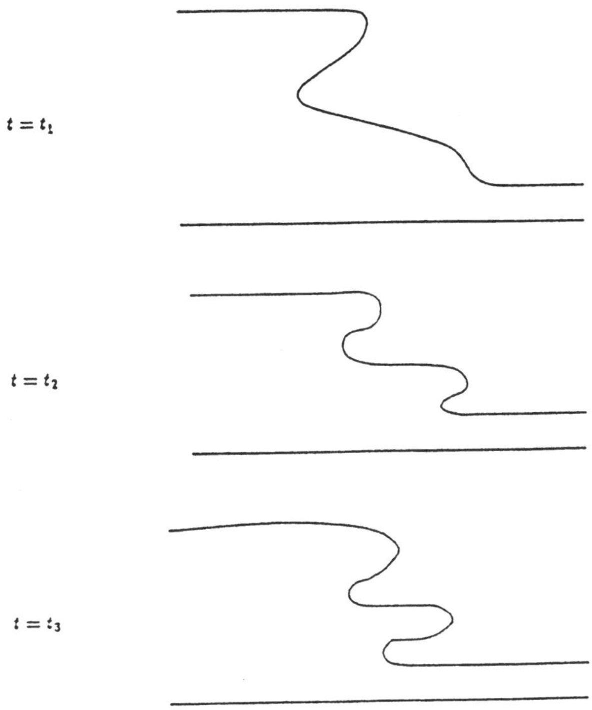

Figure 2.4

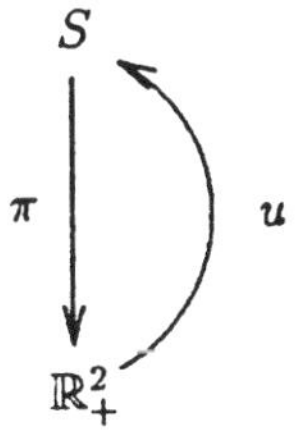

Figure 2.5

EXAMPLE 3: *The scalar conservation law.* We now discuss the geometry associated to a scalar conservation law

$$u_t + \partial_x\big(F(x,\,t,\,u)\big) = 0 \tag{4}$$

where ∂_x denotes total derivative with respect to x. This is an equation of the form (1) where $g = F_u$ and $h = -F_x$. The associated exterior differential system is

$$0 = \Phi = (du + F_x\,dt) \wedge (dx - F_u\,dt) = d\big(u\,dx - F(x,t,u)\,dt\big).$$

Note that Φ is an *exact* 2-form.[4]

This suggests that we try to construct a geometry from the data $(M;\,\Phi,\,\Omega)$ consisting of a 3-manifold M endowed with two linearly independent 2-forms Φ (which is closed) and Ω which is invariant under diffeomorphisms which preserve Φ and preserve Ω up to a non-vanishing factor. We shall always make the non-degeneracy assumption on our data that the common linear factor of Φ and Ω define a contact structure on M. For simplicity of exposition, we assume that M is connected and simply connected.

In the Appendix we have worked out the corresponding G-structure problem for $(M;\,\Phi,\,\Omega)$. It turns out that there is a principal relative invariant K, and when $K = 0$ there is a secondary relative invariant L. The geometry of $(M;\,\Phi,\,\Omega)$ thus divides into three cases:

(i) $K \neq 0$; (ii) $K = 0, L \neq 0$; (iii) $K = L = 0$.

In all three cases we will see that there is a canonical *affine* connection on M such that the characteristic curves of (4) are geodesics, with a canonical parameter τ defined up to an affine transformation $\tau \longrightarrow a\tau + b$. We shall then assume that this geometry is *complete*. An integral surface S of the differential system $\Phi = 0$ with initial curve Γ as above may then be said to be *global* in the intrinsic sense that each characteristic curve in S is defined for $\tau \in [0,\,\infty)$, where $\tau = 0$ gives a point on Γ.

Definition: Let S be an integral surface for $\Phi = 0$. We will say that S has a *singularity* at $p \in S$ in case Ω pulled back to S vanishes at p.

The points of S where $\Omega(p) = 0$ are precisely where the differential of π in (3) drops rank. Thus, when the surface S constructed from Γ has singularities in this sense, a global classical solution to (4) with the given initial data will not exist. We will now establish the following result:

Let Γ be a non-characteristic initial data for $(M;\,\Phi,\,\Omega)$ and let $\lambda_0(s)$ be the real-valued function on Γ constructed from Γ and $(M;\,\Phi,\,\Omega)$ below. There is a curvature function Q associated with the geometry of $(M;\,\Phi,\,\Omega)$ such that if

$$Q \leq 0, \qquad \textit{and} \qquad \lambda_0(\underline{s}) > 0$$

[4]More generally we may consider equations $\partial_t\big(F(x,\,t,\,u)\big) - \partial_x\big(G(x,\,t,\,u)\big) = 0$, expressing the condition that the 1-form $\varphi = F\,dx + G\,dt$ be closed on solutions.

for some $\underline{s}$, then the global integral surface of the exterior system $\Phi = 0$ with initial data Γ develops singularities. On the other hand, assuming that the curvatures of $(M;\ \Phi,\ \Omega)$ are constants, if the curvature function $Q > 0$ and the initial data $\lambda_0(s)$ is small relative to Q, then there exists a global solution to the PDE which $(M; \Phi, \Omega)$ models.

These results are similar to known statements in PDE theory (although there may not be an exact analog of the existence result); our main point is that they may be intrinsically formulated in terms of the geometry associated to the PDE.

Intuitively, the function λ_0 measures the tangent of the "angle" of the unique integral element at each point of Γ relative to Ω. Thus $\lambda_0(p) = \infty$ is equivalent to the condition $\Omega(p) = 0$.

A special case occurs when (4) is a translation invariant conservation law

$$u_t + f(u)u_x = 0, \qquad f'(u) > 0. \tag{5}$$

In this case, the curvature Q vanishes identically and the condition on the initial data for blowup is the classical one

$$u_0'(\underline{s}) < 0.$$

We will now give a proof of the above result in the cases (i) and (ii) above. Using the results in the Appendix, the argument will apply to case (iii) as well.

Since M is simply connected, in each of the cases (i)—(iii) there will be a coframing $\alpha,\ \theta,\ \omega$ of M such that

$$\Phi = \theta \wedge \omega \qquad \text{and} \qquad \Omega = \alpha \wedge \theta.$$

The characteristic vector field $X = \partial_\alpha$ is then dual to α, and we will assume it to be complete. Although the structure equations will be a little different in each of these cases, they all have in common the equation

$$d\theta \equiv \alpha \wedge \omega \mod \theta \tag{6}$$

reflecting the non-linearity of (4).

An *integral element* of the exterior differential system $\Phi = 0$ with the independence condition $\Omega \neq 0$ is by definition given by a point $p \in M$ and 2-plane $E \subset T_pM$ such that

$$\Phi|_E = 0 \qquad \text{and} \qquad \Omega|_E \neq 0.$$

Integral elements are thus defined by a linear equation in the tangent space

$$\omega - \lambda\,\theta = 0, \qquad \lambda \in \mathbb{R} \tag{7}$$

and so the set of all such integral elements forms the manifold

$$M^{(1)} = M \times \mathbb{R}$$

underlying the *first prolongation* $\mathcal{I}^{(1)}$ of the differential ideal $\mathcal{I}$ generated by Φ. On $M^{(1)}$ we consider the exterior differential system $\mathcal{I}^{(1)}$ generated by the 1-form

$$\sigma = \omega - \lambda\,\theta;$$

thus by definition $d\sigma$ lies in $\mathcal{I}^{(1)}$ as does $\Phi = \theta\wedge\sigma$. Every integral surface $S \subset M$ of $\Phi = 0$ satisfying $\Omega \neq 0$ lifts to a unique integral surface $S^{(1)} \subset M^{(1)}$ of $\sigma = 0$ satisfying $\Omega \neq 0$ and moreover, every integral surface of $\sigma = 0$ satisfying $\Omega \neq 0$ is the lifting $S^{(1)} \subset M^{(1)}$ of an integral surface $S \subset M$ of $\Phi = 0$ satisfying $\Omega \neq 0$. The structure equations in both case (i) and case (ii) will give that

$$d\lambda \wedge \theta \equiv (\lambda^2 - Q)\,\alpha \wedge \theta \mod \mathcal{I}^{(1)} \tag{8}$$

where Q is a curvature of the system.

Now let $\Gamma \subset M$ be a non-characteristic initial curve with parameter s along which $\theta \neq 0$. We let S be the flow of Γ under X for $0 \leq t < \infty$. Then S has coordinates $s,\ t$ where $t \geq 0$ and is a global integral surface of $\Phi = 0$.

The open subset $U \subset S$ where $\Omega(p) \neq 0$ will contain a neighborhood of Γ. In U the condition $\sigma = 0$ defines a function $\lambda = \lambda(s,\ t)$. We write $d\lambda \equiv \dot{\lambda}\,\alpha \mod \theta$ where $\dot{\lambda} = \mathcal{L}_X\lambda$ is the Lie derivative of λ along the characteristic vector field X. By (8)

$$\dot{\lambda} = \lambda^2 - Q. \tag{9}$$

Thus, for each s the function $\lambda(s,\ t)$ satisfies a Ricatti equation in t with initial value $\lambda(s,\ 0)$. Under the conditions $Q \leq 0$ and $\lambda(\underline{s},\ 0) > 0$ for some $\underline{s}$, the usual Ricatti analysis then implies that we will have finite time blowup. This is the proof of the first part of the above result.

We now formulate and prove a converse statement in the special case where the structure equations have the form

$$d\alpha = 0, \qquad d\theta = \alpha \wedge \omega, \qquad d\omega = -Q\,\alpha \wedge \theta \tag{10}$$

where Q is a constant.

Setting $\Phi = \theta\wedge\omega$ and $\Omega = \alpha\wedge\theta$, it is straightforward to verify that the system $(M;\ \Phi, \Omega)$ models the PDE

$$u_t + uu_x - Qx = 0. \tag{11}$$

In fact, in $M = \mathbb{R}^3$ with coordinates $x,\ t,\ u$ we may take

$$\alpha = dt, \qquad \theta = dx - u\,dt, \qquad \omega = du - Qx\,dt, \tag{12}$$

and then (10) is satisfied. Moreover, the vector fields $\partial_\theta = \partial_x$, $\partial_\omega = \partial_u$, and $\partial_\alpha = \partial_t + u\,\partial_x + Qx\,\partial_u$ are complete, and thus the 1-forms (12) are the Maurer-Cartan forms on the unique simply connected Lie group with the structure equations (10). We will show that

> *For $Q > 0$ with initial data $u_0(x)$ satisfying $|u_0'(x)| < \sqrt{Q}$, the solution to (11) exists for all time.*

To see why this result should be true (but not yet to give a proof), we write $Q = q^2$ and examine the ODE (9)

$$\lambda = \lambda^2 - q^2$$

with initial condition $\lambda(0) = \lambda_0$. For $|\lambda_0| < q$ this equation has a solution for $0 \leq t < \infty$. In fact, $\lambda(t)$ is obtained by solving for λ in the linear fractional equation

$$\frac{q-\lambda}{q+\lambda} = \left(\frac{q-\lambda_0}{q+\lambda_0}\right) e^{2qt}.$$

In case $(M;\, \Phi,\, \Omega)$ models the PDE (11) we have

$$\lambda_0(s) = u_0'(s).$$

Proof: For notational simplicity we specialize to $Q = 1$ so that the characteristic vector field becomes

$$X = \partial_t + u\,\partial_x + x\,\partial_u.$$

With a given initial curve

$$s \mapsto \big(s,\, 0,\, u_0(s)\big)$$

in xtu-space the global integral surface of $\Phi = 0$ is

$$(s,t) \mapsto \left(\tfrac{1}{2}\big(s-u_0(s)\big)e^t + \tfrac{1}{2}\big(s-u_0(s)\big)e^{-t}, t, \tfrac{1}{2}\big(s+u_0(s)\big)e^t - \tfrac{1}{2}\big(s-u_0(s)\big)e^{-t}\right).$$

The projection $\pi : S \longrightarrow \mathbb{R}^2_+$ is given explicitly by

$$2x = \big(s + u_0(s)\big)e^t + \big(s - u_0(s)\big)e^{-t} \qquad \text{and} \qquad t = t.$$

If $\pi(s,\, t) = \pi(\bar{s},\, \bar{t})$ then $t = \bar{t}$ and, assuming $s \geq \bar{s}$ and writing

$$u_0(s) - u_0(\bar{s}) = u_0'(\xi)(s - \bar{s}), \qquad \bar{s} \leq \xi \leq s$$

we have

$$(s - \bar{s})\left[\big(1 + u_0'(\xi)\big)e^t + \big(1 - u_0'(\xi)\big)e^{-t}\right] = 0.$$

With the assumption $|u_0'| < 1$ this implies that $s = \bar{s}$ and $\pi : S \longrightarrow \mathbb{R}^2_+$ is one-to-one. To show that π is onto we note that

$$2\partial_s x = \big(1 + u_0'(s)\big)e^t + \big(1 - u_0'(s)\big)e^{-t} > 0$$

so that for each fixed t

$$x(-\infty,\, t) = -\infty \qquad \text{and} \qquad x(+\infty,\, t) = +\infty$$

and so the mapping $\pi : S \longrightarrow \mathbb{R}^2_+$ is onto. □

We now return to our discussion of equation (5) where we make the genuine non-linearity assumption $f'(u) > 0$. We assume initial data given in the form

$$x \mapsto \big(x,\, 0,\, u_0(x)\big) \tag{13}$$

where $u_0(x)$ is constant outside of a compact set, say for $|x| > R$, but where $u_0'(x_0) \neq 0$ for some x_0. Then the situation is this:

(i) *the geometry of the equation is complete, so that there is a unique global, smooth integral surface S of the associated exterior differential system*

$$\Phi = du \wedge \big(dx - f(u)\,dt\big) = 0$$

with the initial curve Γ corresponding to (13);

(ii) *the mapping $\pi : S \to \mathbb{R}^2_+$ is surjective but develops a singularity.*

We have proved everything except for the surjectivity of π. In fact, π is *proper*. To see this we think of S as the half-plane $\{(s, t) : -\infty < s < \infty, t \geq 0\}$. The line $t = 0$ maps bijectively to Γ. For each fixed s_0 the half-line $s = s_0$, $t \geq 0$ maps bijectively to the half-line $\big(s_0 + tu(x_0), t\big), t \geq 0$. For $x_0 > R$ we therefore have a family of parallel lines, and similarly for $x_0 < -R$. The picture is something like Figure 2.6 and from this it follows that π is proper.

Now in view of (ii) above a natural question is whether or not the scalar conservation (5) admits some sort of global "generalized solution" $u(x, t)$. Since (5) is a non-linear differential equation care must be taken in defining a generalized solution. The key observation is that, writing $f(u)u_x = \partial_x\big(F(u)\big)$ for a smooth function $F(u)$ with $F'(u) = f(u)$, a locally bounded, measurable function $u(x, t)$ may be defined to be a weak solution to (5) if the equation holds in the distribution sense, i.e., if

$$\int\!\!\int \big(\alpha_t u + \alpha_x F(u)\big)\,dx\,dt = 0 \tag{14}$$

holds for all smooth functions α with compact support in $\mathbb{R}^2_+ = \{(x, t) : t > 0\}$. We will be primarily interested in the situation where u is locally bounded and of class C^1 outside a set of *break curves* across which u has a jump discontinuity. These break curves will consist of piecewise C^1 arcs, and if $\gamma = \big(x(\sigma), t(\sigma)\big)$ is one such arc then (14) is easily seen to imply the *Rankine-Hugonoit* (or jump) condition

$$[F(u)] = s[u] \tag{15}$$

where $s = x'(\sigma)/t'(\sigma)$ is the propagation speed of the discontinuity and [ ] represents the magnitude of the jump across γ (thus $[u] = u_- - u_+$ where u_-

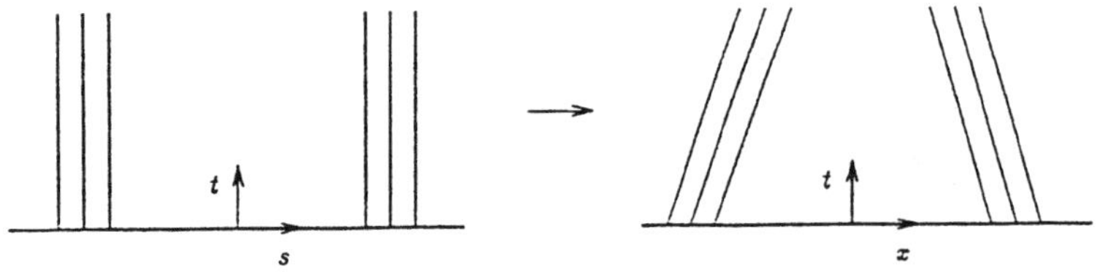

Figure 2.6

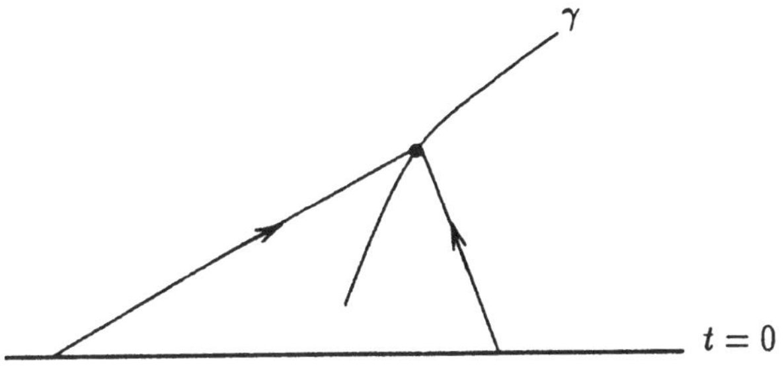

Figure 2.7

and u_+ are respectively the left and right hand limits of $u(x, t)$ at a point of γ). In general the jump condition (15) are not sufficient to uniquely specify a weak solution to (5), and so one traditionally adds the *entropy* condition

$$f'(u_+) < s < f'(u_-). \tag{16}$$

Geometrically this mean that at a point of the break curve γ the two characteristics should impinge (Figure 2.7). With our assumption that $f' > 0$ the entropy condition means that $[u] > 0$, i.e., u should "jump down" across γ.

If one now defines a *shock solution* to (5) to be a weak solution satisfying the entropy condition, then under rather general assumptions on the initial data (for instance that it be smooth with compact support) there is a unique, global shock solution to the initial value problem for (5) (for details, see *[20]*).

The question now arises: *Does the shock solution arise by taking a suitable cross-section of the proper mapping $\pi : S \to \mathbb{R}^2_+$?* In particular for the shock solution u, does $\big(x, t, u(x, t)\big)$ lie on the geometric solution surface S for all $(x, t) \in \mathbb{R}^+_2$? We shall argue that this is indeed the case, at least when the singularities of π are generic.

Referring to Figure 2.1 and Figure 2.2 above, there will be a break curve γ corresponding to each component of the fold locus $\mathcal{F}$. Inside $\mathcal{F}$ the jump will be known since we must jump from the top sheet to the bottom sheet (jumping to the intermediate sheet will run us into a singularity on $\mathcal{F}$). Thus, inside the cusp defined by $\mathcal{F}$ we may view (15) as a differential equation for the break curve γ. More precisely, in this region, equation (15) defines a vector field X. The genuine non-linearity assumption $F'' > 0$ implies that, aside from the cusp points themselves, at each point along $\mathcal{F}$, the vector field X will be transverse to $\mathcal{F}$. Thus there is a unique integral curve of X which emanates from the cusp point of $\mathcal{F}$, represented in Figure 2.8 by the dashed line. This is the desired break curve.

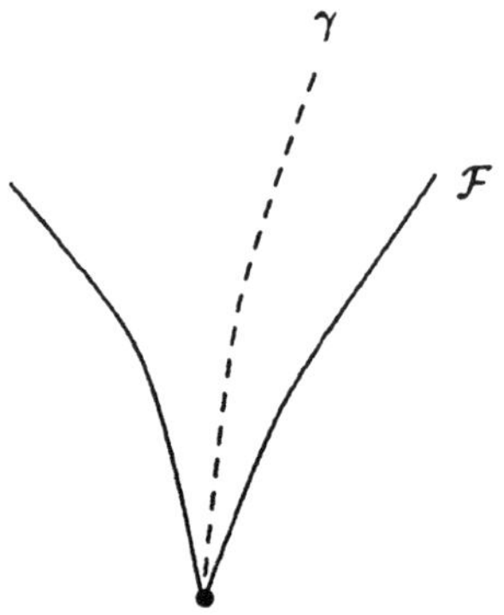

Figure 2.8

In summary, under the assumption that the singularities of the projection π are generic, we have given heuristic reasoning to explain the well-known existence and uniqueness of shock solutions to (5). The point is that the exterior differential system perspective naturally leads to the "picture" of the shock solution.

In the following sections, we will see that the situation for hyperbolic systems of conservation laws (as opposed to a single conservation law) is both more complex and also more interesting.

3 Geometry of Hyperbolic Systems

3.1 *Hyperbolic Exterior Differential Systems.* In the preceeding sections we have studied the geometry associated to a second-order ODE, a first-order non-linear scalar PDE and a non-linear scalar conservation law. In each case the geometry is given by a suitable G-structure $B_G \longrightarrow M$ with an intrinsic connection ϕ and the geodesics of ϕ gives rise to solution curves in the case of the ODE, and characteristic curves in the other two cases. Moreover, each geodesic γ has a natural parameter τ defined up to an affine transformation $\tau \longrightarrow a\tau + b$ once we fix a point of γ, and we may then speak of what it means for γ to be *complete.* With this concept in hand we could define what it means for an integral surface $S \subset M$ passing through a given initial curve to be *global.*

The curvatures of ϕ had several interpretations. One is that the vanishing of certain components implied that there were a complete set of first integrals for the solution curves of the ODE. Another is that the sign of suitable curvature components could be interpreted as a spreading or focusing of geodesics, thereby relating to the existence or non-existence of global classical solutions to the non-linear scalar PDE. Finally, the exterior differential system point of view led to a picture of the classical existence and uniqueness of shock solutions to the non-linear conservation law in terms of the geometry of the proper map $\pi : S \longrightarrow \mathbb{R}^2_+$.

When we seek to study the geometry of less elementary differential equations the situation is more interesting and correspondingly more complex. We shall now discuss certain aspects of the geometry associated to a hyperbolic PDE system. For further details on this geometry, consult the Appendix and *[4]*.

EXAMPLE 4: Beyond the scalar equation already discussed, the simplest non-linear PDE is a first order quasi-linear system for two unknowns over a domain $U \subset \mathbb{R}^2$. We assume that the system is hyperbolic. Then, after a suitable change of dependent and independent variables, the system will assume the form

$$\begin{aligned} u_y + a\,u_x + b\,v_x + f &= 0 \\ v_y + c\,u_x + e\,v_x + g &= 0 \end{aligned} \tag{1}$$

where the coefficients $a,\ b,\ \ldots$ etc. are functions of $x,\ y,\ u,\ v$. Hyperbolicity of (1) amounts to the condition that the matrix

$$A = \begin{pmatrix} a & b \\ c & e \end{pmatrix}$$

have everywhere distinct real eigenvalues. Examples of such equations include hyperbolic systems of conservation laws for two unknowns in $(1+1)$-dimensional spacetime.

To write (1) as an exterior differential system, we set

$$\begin{aligned} \Phi &= -du{\wedge}dx + (a\,du + b\,dv){\wedge}dy + f\,dx{\wedge}dy \\ \Psi &= -dv{\wedge}dx + (c\,du + e\,dv){\wedge}dy + g\,dx{\wedge}dy. \end{aligned}$$

Integral surfaces of the exterior differential system $\Phi = \Psi = 0$ on which $dx{\wedge}dy \neq 0$ are locally in one-to-one correspondence with solutions to (1). A little computation shows that

$$(\Phi + \xi\Psi)^2 = 2(b + (e-a)\xi - c\xi^2)\,du \wedge dv \wedge dx \wedge dy.$$

The discriminant of the quadratic polynomial in ξ is

$$(e-a)^2 + 4bc = (\operatorname{tr} A)^2 - 4\det A.$$

Thus, by hyperbolicity, the equation

$$(\Phi + \xi\Psi)^2 = 0$$

has two distinct real roots $\xi_1 < \xi_2$, giving rise to two linearly independent 2-forms $\Omega_1 = \Phi + \xi_1\Psi$ and $\Omega_2 = \Phi + \xi_2\Psi$ satisfying

$$\begin{cases} \Omega_1{\wedge}\Omega_1 = 0 = \Omega_2{\wedge}\Omega_2 \\ \Omega_1{\wedge}\Omega_2 \neq 0. \end{cases} \tag{2}$$

Definition: A *hyperbolic exterior differential system* (of class $s = 0$) on a 4-manifold M is a differential system which can be generated locally by a pair Ω_1 and Ω_2 of 2-forms satisfying (2).

As we have seen, the quasi-linear system (1) above generates a hyperbolic exterior differential system. A number of well-known examples of system (1) will be given below.

Another interesting source of examples are the hyperbolic Monge-Ampere equations of the form

$$E\,(z_{xx}z_{yy} - z_{xy}^2) + A\,z_{xx} + 2B\,z_{xy} + C\,z_{yy} + D = 0 \tag{3}$$

where the coefficient functions A, B, C, D, and E are functions of x, y, $p = z_x$ and $q = z_y$ alone, i.e., they have no explicit dependence on z. In this case, the exterior differential system in $xypq$-space generated by the pair of 2-forms

$$\begin{aligned}\Phi &= \qquad\qquad\qquad\qquad dp{\wedge}dx + dq{\wedge}dy\\ \Psi &= E\,dp{\wedge}dq + A\,dp{\wedge}dy + B\,(dq{\wedge}dy + dx{\wedge}dp) + C\,dx{\wedge}dq + D\,dx{\wedge}dy\end{aligned}$$

is a hyperbolic exterior differential system. Again, solution surfaces of $\Phi = \Psi = 0$ on which $dx{\wedge}dy \neq 0$ are locally in one-to-one correspondence with solutions to (3). The hyperbolicity of (3) is equivalent to the equation $(\Phi + \xi\Psi)^2 = 0$ having distinct real roots. Among the many well-known hyperbolic Monge-Ampere equation of this type, we cite only the classical hyperbolic Monge-Ampere equation

$$z_{xx}z_{yy} - z_{xy}^2 = -1$$

and the analog, for timelike surfaces in Minkowski 3-space, of the classical minimal surface equation

$$z_y^2\,z_{xx} - 2(1 + z_x z_y)z_{xy} + z_x^2\,z_{yy} = 0. \tag{4}$$

(This is the Euler-Lagrange equation for the area functional for surfaces of the form $\big(x, y, z(x,y)\big)$ in 3-space with metric $dz^2 + 2\,dx \circ dy$.)

Before going further in the study of hyperbolic exterior differential systems, we want to impose some non-degeneracy conditions. Recall that a non-zero exterior 2-form Ω is *decomposable* in the sense that Ω can be locally written in the the form $\Omega = \alpha{\wedge}\beta$ for a pair of (linearly independent) 1-forms α and β if, and only if, $\Omega{\wedge}\Omega = 0$. We say that such an Ω is *integrable* if it can be locally written in the form $\Omega = f\,dx{\wedge}dy$ for some local functions x, y, and f. By an elementary application of the Frobenius theorem, the integrability of Ω is equivalent to the condition that there exist a 1-form ω so that $d\Omega = \omega{\wedge}\Omega$. In this case, the integral manifolds of Ω can locally be described by elementary methods. In particular, in a region where we can write $\Omega = f\,dx{\wedge}dy$, the integral manifolds are just given as hypersurfaces cut out by a non-trivial equation of the form $F(x, y) = 0$.

For a hyperbolic exterior differential system $(M; \Omega_1, \Omega_2)$, the case where either of Ω_i is integrable may thus be analyzed by essentially elementary methods and will not be further discussed here. Thus, we shall assume without further mention that the system is *non-degenerate* in the sense that neither Ω_1 nor Ω_2 is integrable. In this case one may seek to determine the geometry associated to

$(M;\Omega_1,\Omega_2)$ with the non-degeneracy assumption built in from the outset, in a manner similar to the use of the non-linearity assumption in EXAMPLE 2 and EXAMPLE 3 above. We shall discuss some of the specific implications of the equivalence problem in this case.

Before doing this, a remark on the group Γ of allowable coordinate transformations is in order. In the case of EXAMPLE 1 we took Γ to be the group of point transformations while in EXAMPLE 2 we used for Γ the gauge group, since under the larger group of contact transformations any equations of either EXAMPLE 1 or EXAMPLE 2 are locally equivalent to the corresponding trivial model. In EXAMPLE 3 we used a subgroup of gauge transformations which preserve a closed 2-form, since it is only under this smaller group that the "jump conditions" for weak solutions are invariant. For hyperbolic systems, however, we shall use for Γ the full group of contact transformations. Here the geometry is already quite rich.

For instance, suppose we define a hyperbolic system $(M;\Omega_1,\Omega_2)$ to be *linear* if it is locally contact equivalent to the exterior differential system arising from a linear hyperbolic PDE system (1). Linear hyperbolic systems include many arising from non-linear PDEs, such as the equation $z_{xx}z_{yy}-z_{xy}^2=-1$. It turns out (see the Appendix and *[4]*) that linear hyperbolic exterior differential systems may be characterized by the vanishing of certain "curvature" components in the geometry associated to $(M;\Omega_1,\Omega_2)$ and that in turn this geometry induces on solution surfaces an intrinsic pseudo-Riemannian metric whose own geometry (constancy of Gauss curvature, etc.) has meaning for solutions to the original PDE.

In the Appendix we study the equivalence problem associated with the non-degenerate hyperbolic system $(M;\Omega_1,\Omega_2)$ and derive its geometry. A natural question to ask is: What can that geometry tell us about the original PDE? By analogy with EXAMPLE 1 and EXAMPLE 2 in §1 above we could ask

(i) for conditions on the curvatures that would allow us to explicitly "solve" the PDE;

(ii) for a notion of completeness that would guarantee that the exterior differential system has "global" integral surfaces which project onto the half-plane $\mathbb{R}^2_+=\{(x,y):y\geq 0\}$ to provide global multi-valued solutions to the PDE system; and

(iii) for curvature-type conditions that will imply that there are no global smooth classical solutions to (1).

In this section we will take up (i); in later sections there will be a discussion of (ii) and (iii).

3.2 *Remarks on Integration by the Method of Darboux.* For a hyperbolic system given by (1) or (3) what does it mean to explicitly "solve" the PDE? One notion is that a general solution is given by an explicit formula

$$\begin{aligned}
x &= X\big(s,t,\ \alpha(s),\alpha'(s),\dots,\alpha^{(k)}(s);\ \beta(t),\beta'(t),\dots,\beta^{(k)}(t)\big)\\
y &= Y\big(s,t,\ \alpha(s),\alpha'(s),\dots,\alpha^{(k)}(s);\ \beta(t),\beta'(t),\dots,\beta^{(k)}(t)\big)\\
u &= U\big(s,t,\ \alpha(s),\alpha'(s),\dots,\alpha^{(k)}(s);\ \beta(t),\beta'(t),\dots,\beta^{(k)}(t)\big)\\
v &= V\big(s,t,\ \alpha(s),\alpha'(s),\dots,\alpha^{(k)}(s);\ \beta(t),\beta'(t),\dots,\beta^{(k)}(t)\big)
\end{aligned} \tag{5}$$

where α and β are arbitrary functions of one variable. Note that the "function count" is correct, since, in particular, initial data for a hyperbolic system of type (1) or (3) is given by two arbitrary functions of one variable. A more refined notion of "explicitly solving the PDE" is that there should be explicit expressions of the general form (5) in terms of the initial data.

The existence of formulae of the above form for the general integral surface of a hyperbolic exterior exterior differential system is part of what constitutes the phenomenon of *integrability by the method of Darboux.* For a more complete discussion of this method, see *[13, 17]* and *[4]*, but intuitively this corresponds to the equations being solvable in the above form.

If a hyperbolic system is Darboux-integrable, then the values of a solution at (x, y) depend only on the initial data at the two points p and q where the characteristics enamating from (x, y) meet the curve along which the initial data is posed. This is not true for the general equation (1) where, even in the linear case, the solution usually depends on the initial data along the whole segment between p and q on the initial curve. Whether this might be an analytic characterization of Darboux integrability is an interesting question.

We now provide some examples of Darboux-integrable equations:

EXAMPLE: Consider a surface endowed with a Lorentzian metric (i.e., a metric of signature $(1, 1)$) of constant curvature -1. A natural coordinate representation of the equation

$$\Box z = -k(1-k)z, \qquad k \in \mathbb{R}^*$$

(here $\Box$ denotes the Lorentzian Laplacian of the metric) is given by

$$z_{xy} = k(1-k)(x-y)^{-2}z$$

where the domain is taken to be the half-plane $x - y > 0$. This equation admits a general solution of the form (5) exactly when k is a positive integer, and in that case the formula is

$$z = (x-y)^k \frac{\partial^{2k-2}}{\partial x^{k-1}\partial y^{k-1}}\left(\frac{f(x)-g(y)}{x-y}\right)$$

where f and g are arbitrary functions of one variable.

EXAMPLE: Equation (4) for a "maximal" time-like surface in Minkowski 3-space is Darboux-integrable, with a formula of the form (5) with $k = 2$. This formula for the general solution is the analog of the well-known Weierstrass representation for the minimal surface equation in Euclidean 3-space.

In fact every time-like surface $\big(x,\, y,\, z(x,\, y)\big)$ in $\mathbb{R}^3$ where z satisfies (4) is seen to be a surface of translation, where the two translation curves are null curves of the quadratic form $dz^2 + 2dx \circ dy$. This allows us to write the general solution in the form

$$\begin{aligned} x(s,\, t) &= \alpha''(s) + \tfrac{1}{2}t^2\beta''(t) - t\beta'(t) + \beta(t) \\ y(s,\, t) &= \beta''(t) + \tfrac{1}{2}s^2\alpha''(s) - s\alpha'(s) + \alpha(s) \end{aligned}$$

and

$$z(x,\, t) = s\alpha''(s) - \alpha'(s) + t\beta''(t) - \beta'(t)$$

where α and β are arbitrary functions of one variable. (Note that, in this case, we cannot eliminate the "characteristic" parameters s and t to get a formula for z directly as an expression in x and y and two arbitrary functions of them.)

For hyperbolic systems our assumption of non-degeneracy rules out the possibility of a "level zero" formula

$$\begin{aligned} u &= U\big(\alpha(x),\, \beta(y)\big) \\ v &= V\big(\alpha(x),\, \beta(y)\big) \end{aligned}$$

(i.e., one for which no derivatives of the arbitrary functions are needed) such as one obtains for the $s = 0$ classical wave system

$$\begin{aligned} u_x &= 0 \\ v_y &= 0 \end{aligned}$$

this being the model system when both Ω_1 and Ω_2 are integrable. We therefore turn to the question of the existence of a "level one" formula, this being a representation (5) when $k = 1$. It may be shown that the necessary and sufficient conditions for such a formula are

$$p_3 = p_4 = q_1 = q_2 = 0; \qquad (k_{13}-1) = (k_{31}-1) = k_{14} = k_{32} = 0 \tag{6}$$

where the p_i and q_i are components of the torsion and the k_{ij} are suitable components of the curvature derived from the geometry associated to the hyperbolic system (cf. the Appendix). This gives in practice an algorithm to test if a hyperbolic system has an explicit level one general solution of the form (5). Perhaps more interestingly, it leads to the result that there are exactly two contact equivalence classes of hyperbolic systems for which this is true.

EXAMPLE: If (6) holds together with

$$p_1 = q_3 = 0,$$

then (cf. again the Appendix) the hyperbolic system is linear. Moreover, the induced metric on solution surfaces has constant Gauss curvature $K = 1$ and invariant $F = 0$ (for this terminology, see the Appendix). From this it follows that it is locally equivalent as an exterior differential system to the system in $xyuv$-space generated by

$$\Omega_1 = \left(du - \frac{v\,dy}{\cos(x+y)}\right) \wedge dx$$
$$\Omega_2 = \left(dv - \frac{u\,dx}{\cos(x+y)}\right) \wedge dy$$

where $|x+y| < \pi/2$. This exterior differential system models the hyperbolic PDE system

$$u_y = \frac{v}{\cos(x+y)}$$
$$v_x = \frac{u}{\cos(x+y)}.$$

The general solution to this system is provided by the formulas

$$u(x,\, y) = g(y)\sec(x+y) + f(x)\tan(x+y) + f'(x)$$
$$v(x,\, y) = f(x)\sec(x+y) + g(y)\tan(x+y) + g'(y)$$

where f and g are arbitrary functions of one variable.

If $p_1q_3 \neq 0$ then it can be shown that the hyperbolic exterior differential system is equivalent to one modeling the $s = 0$ Liouville system

$$u_y = e^v$$
$$v_x = e^u.$$

The representation (5) of a general solution is

$$e^u = \frac{-a'(x)}{a(x)+b(y)}$$
$$e^v = \frac{-b'(y)}{a(x)+b(y)}.$$

4 Translation Invariant Hyperbolic Systems

4.1 *Characteristic Completeness and Global Integral Surfaces.* We will now discuss the geometry of translation invariant, quasi-linear hyperbolic systems

$$\begin{aligned} u_y + a(u,\, v)u_x + b(u,\, v)v_x &= 0 \\ v_y + c(u,\, v)u_x + e(u,\, v)v_x &= 0. \end{aligned} \tag{1}$$

For such systems the natural group of equivalences to consider are classical transformations which change the independent and dependent variables separately. It will be seen that the exterior differential system $(M;\, \Omega_1,\, \Omega_2)$ associated to (1) is *linear* in the appropriate sense, and in this section we will use that observation to establish a global existence and uniqueness result for integral surfaces of the exterior differential system. In subsequent sections we will discuss the geometry of these surfaces as they pertain to classical solutions of (1).

Before doing this we want to comment on the initial value problem for integral surfaces of a hyperbolic system. Initial data is given by a mapping $\phi : [0, 1] \to M$ whose image is an immersed, non-characteristic curve, where, by *non-characteristic*, we mean that $\phi'(t) \lrcorner \Omega_i \neq 0$ for $i = 1$ or 2. We consider the unit square $\Sigma = \{(s, t) \in \mathbb{R}^2 : 0 \leq s, t \leq 1\}$ and set Δ to be the diagonal in Σ. The *initial value problem* seeks to extend the initial data ϕ to a mapping

$$f_\phi : U_\phi \to M \tag{2}$$

of a neighborhood $U_\phi \subset \Sigma$ of the diagonal, which is an integral surface of the hyperbolic system $(M; \Omega_1, \Omega_2)$, with the additional property that the characteristic foliations pull-back to be the $s =$ const and $t =$ const lines. This last requirement removes the reparametrization ambiguity usually associated with integral surfaces of differential systems since any diffeomorphism of the unit square fixing the diagonal pointwise and the foliations $s =$ const and $t =$ const must be the identity.

Now, the standard existence and uniqueness theorems for hyperbolic PDEs imply that there is a neighborhood U of the diagonal such that the desired extension (2) exists and is unique. We shall say that the solution is *characteristically complete* in case we may take U to be the whole unit square Σ. That is, the solution is characteristically complete if we are able to extend the mapping ϕ to include all points lying on the characteristics emanating from the initial curve to the point where they meet. Pictorially, we have the following situation (see Figure 4.1). As stated above, the initial data, prescribed on the diagonal Δ, may always be extended to give a solution surface in a neighborhood U of Δ. However, if an extension could not be further extended beyond the U depicted, the two characteristic curves drawn could not be extended until they met in U, so such a solution would be incomplete. In fact, unless we can extend the mapping f_ϕ to the whole square Σ, this condition of characteristic completeness fails to be satisfied. The question of existence of complete solutions with arbitrary non-characteristic initial data is obviously a global one. We shall give a large class of hyperbolic systems for which such complete integral surfaces exist with arbitrary initial data. By illustration we shall also show that in some examples one may expect characteristic completeness for some but not all initial data.

We now return to the system (1), which we write in vector form as

$$\mathbf{u}_y + \mathbf{a}(\mathbf{u})\mathbf{u}_x = 0 \tag{3}$$

where $\mathbf{u} = {}^t(u, v)$ and $\mathbf{a}(\mathbf{u})$ is a 2-by-2 matrix which has distinct real eigenvalues $\lambda = \lambda(\mathbf{u})$ and $\mu = \mu(\mathbf{u})$ at all points under consideration. The exterior differential system associated to (3) is given by

$$\mathbf{\Phi} = d\mathbf{u} \wedge dx - \mathbf{a}(\mathbf{u})\, d\mathbf{u} \wedge dy = 0 \tag{4}$$

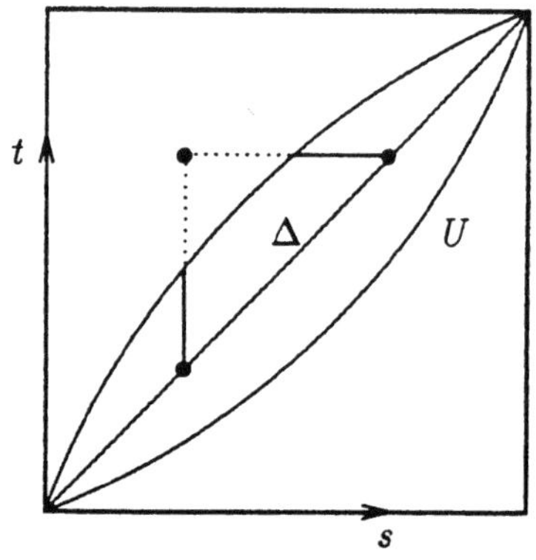

Figure 4.1

where Φ is an $\mathbb{R}^2$-valued 2-form. Let $\mathbf{c}(\mathbf{u})$ be a 2-by-2 matrix such that

$$\mathbf{c}(\mathbf{u})\mathbf{a}(\mathbf{u})\mathbf{c}(\mathbf{u})^{-1} = \Lambda(\mathbf{u}) = \begin{pmatrix} \lambda & 0 \\ 0 & \mu \end{pmatrix}$$

Setting $\Pi = {}^t(\pi_1, \pi_2) = \mathbf{c}(\mathbf{u})\, d\mathbf{u}$, the hyperbolic system (4) becomes

$$\Pi \wedge dx - \Lambda(\mathbf{u})\, \Pi \wedge dy = 0$$

or equivalently, in components

$$\begin{aligned} \pi_1 \wedge (dx - \lambda\, dy) &= 0 \\ \pi_2 \wedge (dx - \mu\, dy) &= 0. \end{aligned} \tag{5}$$

In the following, we shall define a notion of completeness for the PDE system (3). We shall show that for such systems the corresponding exterior differential system has a unique, complete smooth integral surface for arbitrary non-characteristic initial data.

We begin by noticing that for the PDE system (3), the 1-forms π_1 and π_2 defined above are integrable in the uv-plane and hence there are locally defined functions $p = p(u, v)$ and $q = q(u, v)$ such that π_1 is a multiple of dp and π_2 is a multiple of dq. These functions $p(u, v)$ and $q(u, v)$ are the well-known *Riemann invariants* associated with hyperbolic PDE systems.

A hyperbolic system (3) is said to be *complete* on a domain D in the uv-plane if we can choose the Riemann invariants p and q to be global coordinates on D in such a way that the image under the mapping $(p, q) : D \longrightarrow \mathbb{R}^2$ is a coordinate box in the pq-plane. If (3) is complete on $D = \mathbb{R}^2$, then we simply say that (3) is complete. We will say that non-characteristic initial data

$$s \mapsto (x(s), y(s), u(s), v(s))$$

is complete if the image of the data in the uv-plane lies in a domain D in which the equation (3) is complete.

We can now establish the following result:

For any hyperbolic system (3), any complete non-characteristic initial curve extends to a unique, smooth characteristically complete integral surface of the associated exterior differential system. Furthermore, the integral surface is an immersion in a neighborhood of the initial curve.

Before proceeding to the proof of this result, we consider an example:

EXAMPLE: Recall that the Fermi-Pasta-Ulam (FPU) equation is given by

$$z_{yy} - k^2(z_x)z_{xx} = 0, \tag{6}$$

where k is a smooth positive function. This is a special case of the Monge-Ampere equation (3) introduced in the preceeding section. The exterior differential system corresponding to (6) is generated by

$$\begin{aligned} \Phi_1 &= du\wedge dx + \quad dv\wedge dy \\ \Phi_2 &= dv\wedge dx + \big(k(u)\big)^2\, du\wedge dy. \end{aligned}$$

Integral surfaces of the system $\Phi_1 = \Phi_2 = 0$ on which $dx\wedge dy \neq 0$ are (up to an additive constant) locally in one-to-one correspondence with solutions to the FPU equation. Setting

$$\begin{aligned} \pi_1 &= dv + k(u)\,du, \qquad & \omega^1 &= dx + k(u)\,dy \\ \pi_2 &= dv - k(u)\,du, \qquad & \omega^2 &= dx - k(u)\,dy \end{aligned}$$

and

$$\Omega_1 = \pi_1 \wedge \omega^1, \qquad \Omega_2 = \pi_2 \wedge \omega^2$$

we note that span $\{\Phi_1, \Phi_2\}$ = span $\{\Omega_1, \Omega_2\}$. Thus we obtain an exterior differential system of the form (5) where π_1, π_2 are exact. In fact, letting $K(u)$ be an anti-derivative of the function $k(u)$ we have

$$\begin{aligned} dp &= dv + k(u)\,du = d\big(v + K(u)\big) \\ dq &= dv - k(u)\,du = d\big(v - K(u)\big). \end{aligned}$$

From this it follows that the condition for the FPU system to be complete is that the one-to-one mapping $K : \mathbb{R} \to \mathbb{R}$ should be surjective. More generally, if the image of K is some proper subinterval (a, b) of the line (where either $a = -\infty$ or $b = \infty$ is allowed, but not both), then initial data of the form

$$\phi(s) = \big(x_0(s),\, y_0(s),\, u_0(s),\, v_0(s)\big)$$

is complete if and only if, for all s and t in the domain of ϕ, we have

$$a < \tfrac{1}{2}\big(K(u_0(s)) + K(u_0(t)) + v_0(s) - v_0(t)\big) < b$$

(note that this inequality clearly holds when s and t are sufficiently close, since the value in the middle will then be close to $K(u_0(s))$).

We now turn to the proof of the above claim. Let the initial data be given by a complete non-characteristic curve written in suitable $xypq$-coordinates in the form

$$\phi(s) = \big(x_0(s),\, y_0(s),\, p_0(s),\, q_0(s)\big)$$

where $0 \le s \le 1$. We seek a mapping $f : \Sigma \to M = \mathbb{R}^2 \times P \times Q$, where P and Q are the image intervals in $\mathbb{R}$ of the functions p_0 and q_0 respectively, of the form

$$(s,\, t) \mapsto \big(x(s,\, t),\, y(s,\, t),\, p(s,\, t),\, q(s,\, t)\big) \tag{7}$$

and satisfying the initial conditions

$$x(s,\, s) = x_0(s), \quad y(s,\, s) = y_0(s), \quad p(s,\, s) = p_0(s), \quad q(s,\, s) = q_0(s) \tag{8}$$

as well as the exterior equations

$$\begin{aligned} dp(s,\, t) \wedge ds &= 0, \qquad & \big(dx(s,\, t) - \lambda(s,\, t)\, dy(s,\, t)\big) \wedge ds &= 0 \\ dq(s,\, t) \wedge dt &= 0, \qquad & \big(dx(s,\, t) - \mu(s,\, t)\, dy(s,\, t)\big) \wedge dt &= 0 \end{aligned} \tag{9}$$

where $\lambda(s,\, t) = \lambda\big(p(s,\, t),\, q(s,\, t)\big)$ and $\mu(s,\, t) = \mu\big(p(s,\, t),\, q(s,\, t)\big)$. In fact, these equations imply not only that Ω_1 and Ω_2 pull back to zero under the mapping (7), but also that the two linear factors of Ω_1 will each pull back to define the characteristic foliation $s = \text{const}$, and similarly the linear factors of Ω_2 will define the curves $t = \text{const}$.

The first set of equations in (9) above together with the initial conditions (8) imply that

$$p(s,\, t) = p_0(s), \qquad\qquad q(s,\, t) = q_0(t).$$

(Note that this does well-define p and q as functions of s and t since the allowable values of (p, q) form a coordinate box in the pq-plane.) In particular, $\lambda(s,\, t)$ and $\mu(s,\, t)$ are determined from the initial data. The second set of equations in (9) are equivalent to

$$\begin{aligned} x_t - \lambda(s, t) y_t &= 0, \\ x_s - \mu(s, t) y_s &= 0. \end{aligned} \tag{10}$$

This is a *linear* hyperbolic PDE system posed in the unit square $\Sigma = [0,\, 1] \times [0,\, 1]$, with characteristics given by the line $s = \text{const}$ and $t = \text{const}$ and with initial data given on the diagonal. By standard existence and uniqueness theorems there is a unique solution $x(s,\, t)$, $y(s,\, t)$ to the system (10).

The condition that the initial data be non-characteristic may be easily seen to be equivalent to (7) being an immersion along the diagonal, and hence in a neighborhood of the diagonal. □

In the next section we will show that, under certain restrictions on the initial data, the mapping (7) is an immersion outside a compact set in the st-plane and that it is everywhere an immersion for "generic" initial data. It may be that (7) is always an immersion everywhere for all initial data but we have not been able to establish this in general.

We now consider equations which are not complete—many interesting equations, such as those arising from gas dynamics, belong to this class. For such equations, the above proof may still allow us to establish a global existence result for some (but not all) initial data. We illustrate this with the following

EXAMPLE: Consider the coupled Burgers' equations

$$\begin{aligned} u_y + v v_x &= 0 \\ v_y + u u_x &= 0. \end{aligned} \tag{11}$$

Note that this system is hyperbolic whenever $uv > 0$, and so in what follows we shall restrict to the region $\mathbb{R}^4_+ = \{(x, y, u, v) \in \mathbb{R}^4 : u > 0, v > 0\}$. The corresponding exterior differential system on $\mathbb{R}^4_+$ has the form

$$\begin{aligned} \Phi &= -du \wedge dx + v\, dv \wedge dy = 0 \\ \Psi &= -dv \wedge dx + u\, du \wedge dy = 0. \end{aligned}$$

From the relation

$$(\Phi + \xi\Psi)^2 = 2(v - \xi^2 u)(du \wedge dv \wedge dx \wedge dy)$$

we see that $\xi = \pm\sqrt{v/u}$ gives the decomposable linear combinations of Φ and Ψ. A little computation shows that (up to a multiple) these decomposables are

$$\begin{aligned} \Omega_1 &= \tfrac{3}{2}(\sqrt{u}\, du + \sqrt{v}\, dv) \wedge (dx - \sqrt{uv}\, dy), \\ \Omega_2 &= \tfrac{3}{2}(\sqrt{u}\, du - \sqrt{v}\, dv) \wedge (dx + \sqrt{uv}\, dy). \end{aligned}$$

In what follows, we shall find it convenient to introduce the following change of variables

$$p = \tfrac{1}{2}\left(u^{3/2} + v^{3/2}\right), \qquad q = \tfrac{1}{2}\left(u^{3/2} - v^{3/2}\right).$$

Now, the exterior differential system on $M = \{(x, y, p, q) \in \mathbb{R}^4 : p > |q|\}$ is given by

$$\begin{aligned} \Omega_1 &= dp \wedge \left(dx - (p^2 - q^2)^{1/3}\, dy\right) = 0, \\ \Omega_2 &= dq \wedge \left(dx + (p^2 - q^2)^{1/3}\, dy\right) = 0. \end{aligned}$$

Notice that this system is *not* complete on its domain of hyperbolicity.

We now seek solutions $f : \Sigma \to M$ of the above hyperbolic system. As in the proof above, we have

$$p(s,t) = p_0(s), \qquad q(s,t) = q_0(t). \tag{12}$$

In order to have $p(s,t) > |q(s,t)|$ for all $(s,t) \in \Sigma$, the initial data must clearly satisfy

$$\max_s |q_0(s)| < \min_s p_0(s)$$

and these are necessary and sufficient restrictions on the initial data to have a characteristically complete integral surface.

4.2 *The Geometry of Global Integral Surfaces.* We continue our discussion of integral surfaces of the exterior differential system on $M \subset \mathbb{R}^4$ associated to the PDE system

$$\mathbf{u}_y + \mathbf{a}(\mathbf{u})\mathbf{u}_x = 0. \tag{1}$$

As was established in the previous section we may locally write this exterior differential system as

$$\begin{aligned} \Omega_1 &= dp \wedge (dx - \lambda\, dy) = 0 \\ \Omega_2 &= dq \wedge (dx - \mu\, dy) = 0 \end{aligned} \tag{2}$$

where p and q are functions of u and v for which $dp \wedge dq \neq 0$. Henceforth, we shall assume that the system (1) above is complete and so we may choose p, q to be global coordinates on the uv-plane. In particular, λ and μ are then well-defined functions of p and q.

We now consider special solution surfaces of the above exterior differential system given by a mapping $f : U \to M$ where the domain U is an open subset of the st-plane. These solution surfaces are characterized by the dimension of their image under the projection to the uv-plane.

TYPE 0: These are solution surfaces $S \subset M$ on which

$$\pi_1 = \pi_2 = 0 \tag{3}$$

or equivalently $p = \text{const}$ and $q = \text{const}$. The image of such a solution surface is thus a point in the uv-plane.

In a region of the st-plane where (3) is valid, x and y satisfy a linear hyperbolic PDE system with constant coefficients (cf. equation (10) of §4.1).

TYPE 1: These solutions $S \subset M$ are defined by the condition that

$$\pi_1 \wedge \pi_2 = 0, \tag{4}$$

and in addition are not of TYPE 0. Thus the image of a TYPE 1 solution surface in the uv-plane is a curve. These solutions are classically known as *simple waves.*

Note that for simple waves, the hyperbolic system $(M;\, \Omega_1,\, \Omega_2)$ reduces to the system $(N_1;\, \Omega_1)$ or $(N_2;\, \Omega_2)$ where N_1 and N_2 are the hypersurfaces in M defined respectively by $q = q_0$ and $p = p_0$. In what follows we shall have occasion to study classical solutions of (1) which then naturally requires us to introduce the independence 2-form $\Omega = dx \wedge dy$. Now recall that the condition for genuine non-linearity of $(N_A;\, \Omega_A,\, \Omega)$ is that the common linear factor θ_A of Ω_A and Ω should be a contact form on N_A.

From (2) above we have that

$$\begin{aligned} \theta_1 &= dx - \lambda(p, q_0)\, dy, \\ \theta_2 &= dx - \mu(p_0, q)\, dy, \end{aligned}$$

and hence

$$\theta_1 \wedge d\theta_1 = \lambda_p \, dp \wedge dx \wedge dy,$$
$$\theta_2 \wedge d\theta_2 = \mu_q \, dq \wedge dx \wedge dy.$$

This leads to the classical definition of *genuine non-linearity* for the PDE system (1), namely, that it should satisfy the conditions

$$\lambda_p \neq 0 \qquad \text{and} \qquad \mu_q \neq 0. \tag{5}$$

This concept is intrinsic to the data $(M;\, \Omega_1,\, \Omega_2;\, \Omega)$ consisting of the hyperbolic exterior differential system $(M;\, \Omega_1,\, \Omega_2)$ together with the independence condition Ω.

TYPE 2: These are solutions on which

$$\pi_1 \wedge \pi_2 \neq 0.$$

Clearly, the image of such a solution surface is an open subset of the uv-plane.

We now proceed to develop a "picture" of the solutions to the system (2). For this we assume given classical initial data

$$x \mapsto (x,\, 0,\, u_0(x),\, v_0(x)) \tag{6}$$

where $u_0(x)$ and $v_0(x)$ are constant outside a compact set, say for $|x| > R$. We then use coordinates

$$\xi = \tfrac{1}{2}(s+t) \qquad \text{and} \qquad \eta = \tfrac{1}{2}(s-t)$$

in the st-plane and assume, as we may, that the initial data (6) is given by

$$\xi \mapsto (\xi,\, 0,\, p_0(\xi),\, q_0(\xi))$$

where p_0 and q_0 are constant for $|\xi| > R$. We also recall that

$$\begin{aligned} p(s,\, t) &= p(s,\, s) = p_0(s) \\ q(s,\, t) &= q(t,\, t) \;= q_0(t) \end{aligned} \tag{7}$$

are determined by the initial curve. This suggests we consider the following picture in the $\xi\eta$-plane

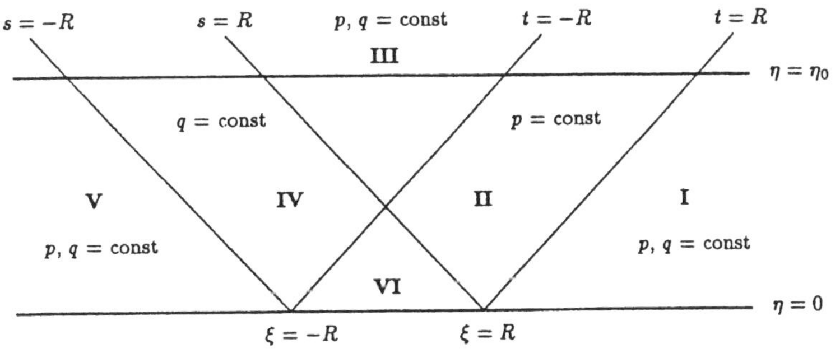

Figure 4.2

From (7) it follows that p and q are constant in regions **I**, **III**, **V**, so that the solution there is of TYPE 0. Moreover, p is constant in region **II** and q is constant in region **IV**, so that the solution there is of TYPE 1. For fixed $\eta = \eta_0 > R$ the solution along the ξ-axis

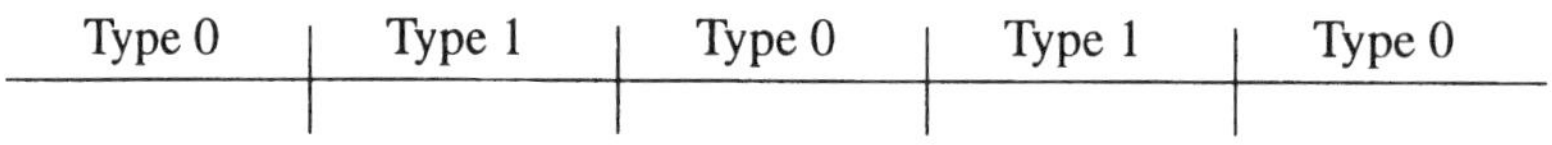

Figure 4.3

is of a wave that is alternately constant and simple with the simple wave components separating as η increases. We also see from the above picture that the boundary of a region where we have a solution of TYPE 0 consists of characteristic curves and in the adjacent region the solution is of TYPE 1. Furthermore, outside of region **VI** and its mirror image in the half-plane $\eta < 0$, the mapping giving the complete integral surface with initial data (6) is an immersion.

We shall use the above picture to deduce the following well-known fact (see *[19]*):

> *If the PDE system (1) is genuinely non-linear then any classical solution* $\mathbf{u}(x, y)$ *with non-constant compactly supported initial data will develop a singularity in finite time.*

Proof: The proof is by contradiction. Setting

$$\omega^1 = dx(\xi, \eta) - \lambda(\xi, \eta)\, dy(\xi, \eta),$$
$$\omega^2 = dx(\xi, \eta) - \mu(\xi, \eta)\, dy(\xi, \eta),$$

where $\lambda(\xi, \eta)$, $\mu(\xi, \eta)$ are given with the non-linearity property (5), we will show that $\Omega = \omega^1 \wedge \omega^2$ vanishes somewhere in the $\xi\eta$-plane.

Assuming this is not the case there will exist smooth functions $\rho(\xi, \eta)$ and $\sigma(\xi, \eta)$ such that

$$\pi^1 - \rho\,\omega^1 = 0 \qquad \text{and} \qquad \pi^2 - \sigma\,\omega^2 = 0.$$

The exterior derivative of the first of these equations gives

$$(\lambda - \mu)\, d\rho + \rho(\lambda_p \rho + \lambda_q \sigma)\, \omega^2 \equiv 0 \mod \omega^1.$$

Recall that for genuinely non-linear systems $\lambda_p \neq 0$. Now, because we want to illustrate the geometric approach rather than derive the most general result, we will assume that λ_p is bounded away from zero, say $\lambda_p \geq \lambda_0 > 0$.

Note that for a fixed $s = s_0$ characteristic line we may assume that the parametrization is chosen so that $\omega^2 = (\mu - \lambda)\, dt$. Along this characteristic we then have

$$\rho_t = \rho(\lambda_p \rho + \lambda_q \sigma). \tag{8}$$

By assumption the initial data vanishes outside a compact set, say $|\xi| > R$. Thus we have

$$\left.\begin{aligned} p(\xi, 0) &= 0 \\ q(\xi, 0) &= 0 \end{aligned}\right\} \qquad \text{for } |\xi| > R.$$

Now notice that along the initial curve $\eta = 0$ we have

$$\rho(\xi, 0) = \alpha(\xi) p_\xi(\xi, 0) \qquad \text{and} \qquad \sigma(\xi, 0) = \beta(\xi) q_\xi(\xi, 0) \tag{9}$$

where $\alpha(\xi)$ and $\beta(\xi)$ are nowhere vanishing functions. Thus we have that on the set $|\xi| > R$

$$\rho(\xi, 0) = \sigma(\xi, 0) = 0.$$

Pictorially we have Figure 4.4. From equation (8), we have that $\rho = 0$ along $s =$ const characteristics that begin to the right of $\xi = R$. And by assumption there will be a point P on the line $\eta = 0$ where $p_\xi(P, 0)$ vanishes, and changes sign as we cross from one side of P to the other. Again by (8) ρ will vanish along the characteristic represented by the dashed line, while we may assume that ρ is positive at Q. Finally, along the characteristic emanating from Q we will have $\sigma = 0$ and hence along this characteristic equation (8) gives

$$\rho_t = \lambda_p \rho^2 \geq \lambda_0 \rho^2$$

with

$$\lambda_0 > 0 \qquad \text{and} \qquad \rho|_Q > 0$$

from which we infer that ρ blows up in finite time. □

Recall that for a complete hyperbolic system (1), smooth initial data

$$s \mapsto (s, 0, u_0(s), v_0(s))$$

defined along the x-axis extends to a unique, smooth complete integral surface given by

$$f : \mathbb{R}^2_{(s,t)} \longrightarrow \mathbb{R}^4$$

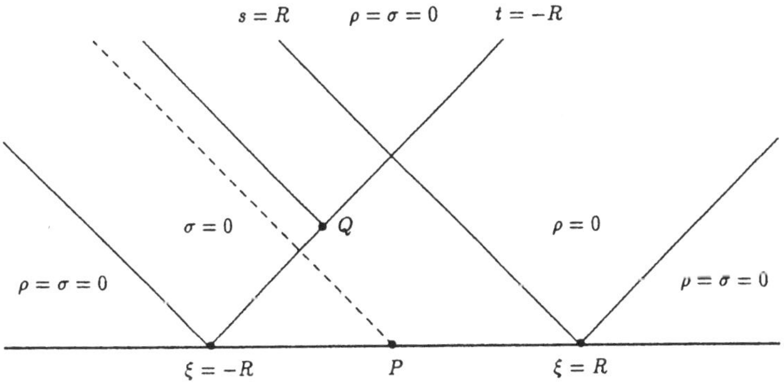

Figure 4.4

where $\mathbb{R}^2_{(s,t)}$ denotes the st-plane. Let us consider the mapping

$$\pi : \mathbb{R}^2_{(s,t)} \rightarrow \mathbb{R}^2_{(x,y)} \tag{10}$$

where π is the composition of f with the projection onto the xy-plane. By the above result we see that for compactly supported initial data the differential of π must drop rank somewhere. In fact, we have the relation (cf. equation (10) of §4.1)

$$\det \begin{pmatrix} x_s & x_t \\ y_s & y_t \end{pmatrix} = (\mu - \lambda) y_s y_t$$

which shows that the curve where π_* drops rank is given by

$$y_s y_t = 0.$$

Thus it has two "branches", one where $y_s = 0$ (and thus $x_s = 0$) and the other $y_t = 0$ (and thus $x_t = 0$). These branches intersect at a point where $x_s = x_t = y_s = y_t = 0$, which is a cusp for the mapping π. Along the branch $y_s = 0$, $y_t \neq 0$ the kernel of π_* is spanned by ∂_s, i.e., it is a characteristic direction.

We now address the question of whether the mapping (10) is surjective; i.e., does our construction give a global albeit "multi-valued" solution to (1)? In the following we shall show that

> *If the initial data is constant outside of a compact set, then the mapping (10) is proper. In particular, it is surjective.*

Proof: Explicitly, the mapping (10) is given by

$$\pi(s, t) = \big(x(s, t), y(s, t)\big)$$

where $x(s, t)$ and $y(s, t)$ satisfy the linear PDE (10) of the preceding section. To show that π is proper we need to understand the behavior of the characteristic curves, which are the images of the straight lines parallel to the coordinate axes in the st-plane. Thus we consider Figure 4.5. The characteristic curves in the xy-plane are the integral curves of

$$\begin{aligned} dx - \lambda(p, q)\, dy &= 0 \\ dx - \mu(p, q)\, dy &= 0. \end{aligned}$$

Recall that $p = p(s, t) = p(s)$ and $q = q(s, t) = q(t)$ by the argument in the preceding section. It follows that

$$\begin{aligned} p &= \text{const} \quad \text{in regions } \mathbf{I},\ \mathbf{III},\ \mathbf{V} \text{ and } \mathbf{II}, \\ q &= \text{const} \quad \text{in regions } \mathbf{I},\ \mathbf{III},\ \mathbf{V} \text{ and } \mathbf{IV}. \end{aligned}$$

Thus, the characteristic curves in the images of regions **I**, **III** and **V** are two families of parallel lines.

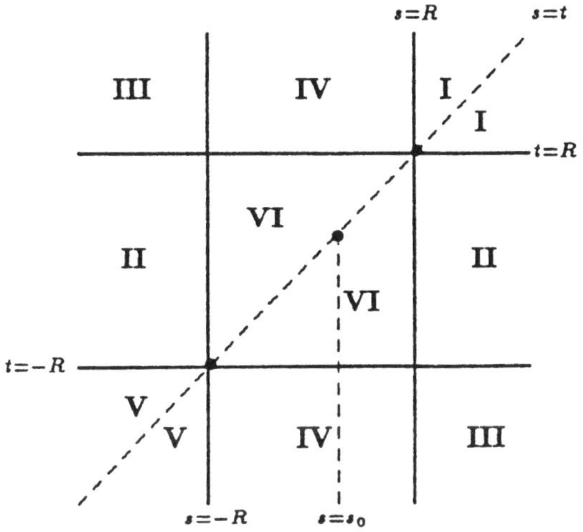

Figure 4.5

We will now examine what happens to the images of the lines $s = s_0$ as s_0 moves from R to $-R$. In region **IV**, q is a constant q_0 and along the dotted vertical line $p = p(s_0)$ is also constant. Thus the image of the $t \leq -R$ part of the dotted line in region **IV** will be a line in the xy-plane since it satisfies

$$dx - \lambda\big(p(s_0),\, q_0\big)dy = 0.$$

As s_0 moves from right to left from R to $-R$ these lines will not be parallel but will have slope $\lambda\big(p(s_0),\, q_0\big)$ that changes continuously for $-R \leq s_0 \leq R$. (Thus in the xy-plane they may focus, causing folds for π in region **IV**). However, we see from this picture that the mapping will be proper in region **IV** as s tends to ∞. Similar arguments apply to the other region **IV** and to the two regions **II**. From this we may conclude that the mapping $\pi : \mathbb{R}^2_{(s,t)} \to \mathbb{R}^2_{(x,y)}$ is proper. □

4.3 *The Riemann Problem for Hyperbolic Exterior Differential Systems.* In §2 we have considered the geometry associated to the non-linear, scalar conservation law

$$u_y + f'(u)u_x = 0 \tag{1}$$

where the non-linearity is expressed by the condition $f''(u) \neq 0$. Setting $\Phi = du\wedge\big(dx - f'(u)\,dy\big) = d\varphi$ for some 1-form $\varphi = u\,dx - f(u)\,dy$, the associated exterior differential system is given by $\Phi = 0$ on the manifold $M = \mathbb{R}^3_+ = \{(x,\, y,\, u) : y \geq 0\}$. Initial data given by $\big(x,\, 0,\, u_0(x)\big)$ where $u_0(x)$ is smooth and constant for $|x| > R$ gives rise to an initial curve $\gamma \subset M$ which then extends to a unique, smooth global integral surface $S \subset M$ such that the projection

$$\pi : S \to \mathbb{R}^2_+ = \{(x,\, y) : y \geq 0\} \tag{2}$$

is proper. (Here global means that S is characteristically complete in the sense that the characteristic curves are complete geodesics in the intrinsic geometry associated to (1).) Furthermore, it is well known that there is a unique, global shock solution $u(x, y)$ of (1) associated to the above initial data, and in §2 we gave heuristic reasoning to the effect that $u(x, y)$ arises by taking a suitable (possibly discontinuous) cross section

$$(x, y) \mapsto (x, y, u(x, y)) \in S$$

of the projection (2).

We now turn to a hyperbolic system of conservation laws

$$\begin{cases} u_y + \partial_x (f(u, v)) = 0 \\ v_y + \partial_x (g(u, v)) = 0 \end{cases} \tag{3}$$

which we also write as

$$\mathbf{u}_y + \partial_x (\mathbf{f}(\mathbf{u})) = 0$$

where $\mathbf{u} = {}^t(u, v)$ and $\mathbf{f} = {}^t(f, g)$. This gives rise to the hyperbolic exterior differential system $\Phi = \Psi = 0$ on $M = \mathbb{R}^4_+ = \{(x, y, u, v) : y \geq 0\}$ where $\Phi = d\phi$ and $\Psi = d\psi$ for 1-forms ϕ, ψ defined by

$$\begin{cases} \phi = u\,dx - f(u, v)\,dy \\ \psi = v\,dx - g(u, v)\,dy. \end{cases} \tag{4}$$

With an appropriate assumption on the completeness of (3) we shall show that smooth initial data

$$x \mapsto (x, 0, \mathbf{u}_0(x))$$

satisfying $\mathbf{u}_0(x) = \text{const}$ for $|x| > R$ gives a unique, smoothly parametrized global integral surface $S \subset M$ such that the projection

$$\pi : S \to \mathbb{R}^2_+ = \{(x, y) : y \geq 0\} \tag{5}$$

is proper. (Here, global means that S should be characteristically complete as explained above.)

In PDE theory (see *[20]* for a general reference and *[14]* for recent developments) it is proved that if (3) is genuinely non-linear and the initial data has small mean oscillation, then there is a unique shock solution $\mathbf{u}(x, y)$ to (3) with the given initial data.

A natural question to ask is whether this shock solution arises by taking a suitable cross-section of (5)? The situation is much more subtle than in the scalar case, and as we shall see when we discuss the Riemann problem below the answer appears to be no in general. However, for a special class of hyperbolic systems which we shall describe in the following, this construction is indeed possible.

Briefly, the problem is that the Rankine-Hugonoit conditions for a weak solution—the analog of the jump condition (14) in §2—are now the pair of equations

$$[f(u, v)] = s[u] \qquad \text{and} \qquad [g(u, v)] = s[v]. \tag{6}$$

On a solution surface S, this is an over-determined set of equations for the shock speed s. We shall see that, at least for the Riemann problem, these conditions can be satisfied on S only when the defining functions $f(u, v)$ and $g(u, v)$ of the hyperbolic system (3) satisfy a pair of second-order PDEs (to be derived below). We shall call such hyperbolic systems of conservation laws *special*.

We now proceed to study the analogue for exterior differential systems of the classical Riemann problem for (1) and (3). Recall that this is the PDE initial value problem for the initial data

$$u_0(x) = \begin{cases} u_- & \text{for } x < 0, \\ u_+ & \text{for } x > 0, \end{cases} \tag{7a}$$

in the case of equation (1) and

$$\mathbf{u}_0(x) = \begin{cases} \mathbf{u}_- & \text{for } x < 0, \\ \mathbf{u}_+ & \text{for } x > 0, \end{cases} \tag{7b}$$

in the case of equation (3).

Notice that the initial data (7) gives a disconnected curve γ_0 in M. In the following we shall show how γ_0 may be uniquely completed to a connected, continuous and piecewise smooth curve γ such that there is a unique, global weak solution surface S to the corresponding exterior differential system with the additional property that S consists only of TYPE 0 or TYPE 1 solutions in the sense of the preceeding section. (The reason that we restrict S to consists of TYPE 0 and TYPE 1 solutions is that the graph of the classical shock solutions have this property. Furthermore, this is a natural geometric condition to impose on a solution surface.) In the scalar case the shock solution to the Riemann problem arises as a suitable cross-section of

$$\pi : S \longrightarrow \mathbb{R}^2_+,$$

but this is not in general true for (3). As stated above, this will be valid only when the system (3) is special.

We begin with a brief discussion of the non-linear scalar conservation law (1). In the following we shall see that there are two basic cases:

(i) $u_- \geq u_+$ — breaking waves; (ii) $u_- \leq u_+$ — rarefaction waves.

We will show that "upstairs" on the geometric solution surface S these two cases are essentially the same, which is not the case "downstairs" for the classical shock solution $\mathbf{u}(x, y)$.

We prescribed the initial data by a mapping

$$s \mapsto \gamma(s) = \big(x(s),\, 0,\, u_0(s)\big), \qquad s \in \mathbb{R}$$

as follows: For $a \leq b$ with $b - a = |f'(u_+) - f'(u_-)|$ we set

$$\gamma(s) = \begin{cases} (s - a,\, 0,\, u_-) & \text{for } s \leq a \\ (s - b,\, 0,\, u_+) & \text{for } s \geq b \end{cases}$$

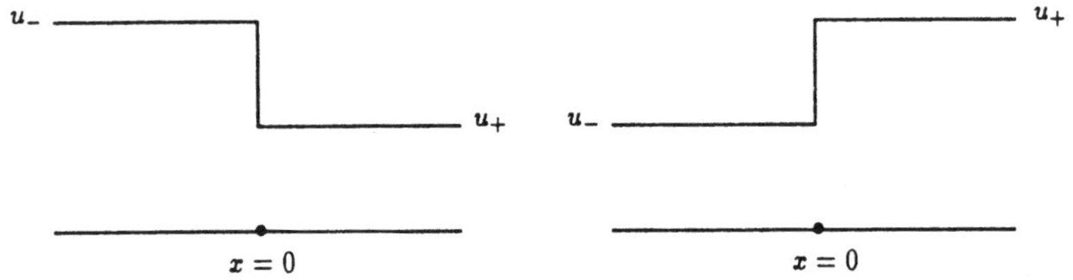

Figure 4.6

while for $a \leq s \leq b$, the data $\gamma(s)$ is defined by

$$x(s) = 0 \qquad \text{and} \qquad f'\left(u_0(s)\right) = \pm s$$

where $u_0(a) = u_-$ and $u_0(b) = u_+$. Here, $\pm$ is the sign of $u_+ - u_-$ (thus $+$ for rarefaction and $-$ for breaking waves). The picture is as follows (see Figure 4.6). From the relation

$$f''\left(u_0(s)\right)u_0'(s) = \pm 1$$

together with the non-linearity assumption $f'' > 0$, we see that the sign is determined by whether $u_0(s)$ is increasing or decreasing on the vertical segment.

The characteristic vector field is

$$X = \partial_y + f'(u)\,\partial_x$$

and we define S to be the image under the mapping Γ of $\mathbb{R}^2_+ = \{(s,t) : t \geq 0\}$ obtained by flowing γ along the integral curves of X. Explicitly

$$\Gamma(s,t) = \left(x(s) + tf'(u_0(s)),\ t,\ u_0(s)\right). \tag{8}$$

This is a continuous, piecewise smooth mapping with (finite) jumps in the differential Γ_* arising from

$$f''\left(u_0(s)\right)u_0'(s) = \pm 1, \qquad a \leq s \leq b,$$

which gives the jump at $s = a$

$$[u_0'] = \pm\frac{1}{f''(u_-)},$$

and similarly at $s = b$. We note that in $a \leq s \leq b$ we have

$$u_0(s) = (f')^{-1}(\pm s)$$

according to whether $u_- \leq u_+$ or $u_- \geq u_+$. Thus, upstairs a shock wave looks like a "backwards" rarefaction wave.

The mapping (8) is a *weak solution* to the exterior differential system

$$d\varphi = 0 \tag{9}$$

in the following sense: First, $\Gamma^*(\varphi)$ is a continuous 1-form in $\mathbb{R}^2_+$. Secondly, for any function α with compact support in $\mathbb{R}^3_+ = \{(x, y, u) : y > 0\}$, $\Gamma^*(\alpha)$ is a compactly supported piecewise smooth function in $\mathbb{R}^2_+ = \{(s, t) : t > 0\}$ and thus the equation

$$\int d\Gamma^*(\alpha) \wedge \Gamma^*(\varphi) = 0 \tag{10}$$

makes sense, and is satisfied for all α.

For $S_0 = S - \gamma$ the mapping $\pi : S_0 \longrightarrow \tilde{\mathbb{R}}^2_+ = \{(x, y) : y > 0\}$ is one-to-one in the rarefraction case, while in the shock case the image in the xy-plane of the region $a \leq s \leq b$, $t > 0$ is covered three times by the mapping π. The entropy condition (see §2) tells us that along the line

$$\frac{x}{y} = \frac{f(u_-) - f(u_+)}{u_- - u_+}$$

we should jump from the top to the bottom sheet of $\pi^{-1}(x, y)$ to obtain the shock solution.

Turning to the Riemann problem for the exterior differential system arising from (3), it turns out that through each point of the uv-plane there are two pairs of curves, one arising from the level sets of Riemann invariants and the other arising from the solution curves to the jump conditions (6). In order to understand the Riemann problems "upstairs" and "downstairs", as well as their relationship, we shall need to understand the local geometry of these curves and how they relate to each other.

Recall that a *Riemann invariant* is a function $r(u, v)$ that is constant on one of the families of characteristics on solutions $\big(u(x, y), v(x, y)\big)$ of (3). The characteristics are integral curves of

$$\text{(i) } dx - \lambda\, dy = 0 \qquad \text{or} \qquad \text{(ii) } dx - \mu\, dy = 0.$$

We shall denote by r_λ, r_μ the Riemann invariants corresponding to (i) and (ii) respectively.

Next, we observe that the form of the equation (3) remains invariant under the affine linear transformation

$$\tilde{\mathbf{u}} = \mathbf{a}_0\mathbf{u} + \mathbf{v}_0 \tag{11}$$

where $\mathbf{a}_0$ is an invertible 2-by-2 matrix and $\mathbf{v}_0$ is a vector. Thus, we may intrinsically think of (u, v) as coordinates in an affine vector space. For each point $\mathbf{u}$ in the affine space we denote by $\mathbf{e}_\lambda(\mathbf{u}), \mathbf{e}_\mu(\mathbf{u})$ eigenvectors of $\mathbf{f}'(\mathbf{u})$ corresponding to $\lambda(\mathbf{u}), \mu(\mathbf{u})$ respectively. We may identify the tangent space at each point $\mathbf{u}$ with the affine space itself and so in this way we may think of $\mathbf{e}_\lambda, \mathbf{e}_\mu$ as a pair of direction fields.

Lemma 1: *The level sets $r_\lambda = const$ and $r_\mu = const$ are the integral curves of $\mathbf{e}_\lambda$ and $\mathbf{e}_\mu$ respectively.*

Proof: We recall that a simple wave for (3) is a solution $\mathbf{u}(x, y)$ whose image in the uv-plane is a 1-dimensional curve. We shall show that the level sets of simple waves are also level sets of Riemann invariants as well as integral curves of the above direction fields.

For a simple wave the level sets $\mathbf{u}(x, y) = \text{const}$ are a family of curves giving a foliation $\xi(x, y) = \text{const}$ in an open set in the xy-plane. This implies that $\mathbf{u} = \mathbf{u}(\xi)$, a function of 1-variable. The PDE (3) gives

$$(\xi_y + \xi_x \mathbf{f}'(\mathbf{u}))\mathbf{u}' = 0$$

which means that $\lambda = -\xi_y/\xi_x$ is an eigenvalue of $\mathbf{f}'(\mathbf{u})$ with eigenvector $\mathbf{e}_\lambda = \mathbf{u}'$. Thus we have

$$\text{(i) } \xi_y + \lambda(\mathbf{u}(\xi))\xi_x = 0; \qquad \text{(ii) } \mathbf{u}'(\xi) = \mathbf{e}_\lambda(\mathbf{u}(\xi)).$$

Rewriting equation (i) as

$$(\partial_y + \lambda(\mathbf{u}(\xi))\partial_x)\xi = 0$$

and using the identity

$$\langle dx - \lambda\, dy,\, \partial_y + \lambda\, \partial_x \rangle = 0$$

we see that the curve $\xi \mapsto \mathbf{u}(\xi)$ in the uv-plane is a level set of the Riemann invariant r_λ.

Now, from equation (ii) we have

$$0 = dr_\lambda(\mathbf{u}(\xi)) = \langle dr_\lambda, \mathbf{e}_\lambda \rangle (\mathbf{u}(\xi))$$

and so $\langle dr_\lambda, \mathbf{e}_\lambda \rangle = 0$ as require. □

To study the geometry of the various curves in a neighborhood of a point $\mathbf{u}_0$, we make an affine transformation (11) so that $\mathbf{u}_0 = (\mathbf{0})$ is the origin and $\mathbf{e}_\lambda(0) = \partial_u, \mathbf{e}_\mu(0) = \partial_v$. Then we have the series expansions

$$\begin{aligned} f(u, v) &= \lambda_0 u + \tfrac{1}{2} f_{uu} u^2 + f_{uv} uv + \tfrac{1}{2} f_{vv} v^2 + \mathrm{O}(3) \\ g(u, v) &= \mu_0 v + \tfrac{1}{2} g_{uu} u^2 + g_{uv} uv + \tfrac{1}{2} g_{vv} v^2 + \mathrm{O}(3) \end{aligned} \tag{12}$$

where $\mathrm{O}(k)$ denotes terms of order k.

Lemma 2: *In a neighborhood of the origin $\mathbf{u} = (\mathbf{0})$, we have*

$$\begin{aligned} \lambda(u, v) &= \lambda_0 + f_{uu} u + f_{uv} v + \mathrm{O}(2), \\ \mu(u, v) &= \mu_0 + g_{vv} v + g_{uv} u + \mathrm{O}(2). \end{aligned} \tag{i}$$

Futhermore, we may choose the eigenvectors and Riemann invariants such that

$$\begin{aligned} \mathbf{e}_\lambda(\mathbf{u}, \mathbf{v}) &= \partial_u + (\lambda_0 - \mu_0)^{-1}(g_{uu} u + g_{uv} v)\partial_v + \mathrm{O}(2), \\ \mathbf{e}_\mu(\mathbf{u}, \mathbf{v}) &= \partial_v + (\mu_0 - \lambda_0)^{-1}(f_{vv} v + f_{uv} u)\partial_u + \mathrm{O}(2), \end{aligned} \tag{ii}$$

and

$$\begin{aligned} r_\lambda(u,v) &= v - (\lambda_0-\mu_0)^{-1}(\tfrac{1}{2}g_{uu}u^2 + g_{uv}uv + \tfrac{1}{2}g_{vv}v^2) + \mathrm{O}(3), \\ r_\mu(u,v) &= u - (\mu_0-\lambda_0)^{-1}(\tfrac{1}{2}f_{uu}u^2 + f_{uv}uv + \tfrac{1}{2}f_{vv}v^2) + \mathrm{O}(3). \end{aligned} \qquad \text{(iii)}$$

We remark that $\mathbf{e}_\lambda$, $\mathbf{e}_\mu$ and r_λ, r_μ are not unique—i.e., the eigenvectors are defined up to a non-zero factor while any non-constant function of the Riemann invariants will still be a Riemann invariant. The above formula should be interpreted as giving expressions for representatives of equivalence classes.

Proof: The linear terms in λ and μ are determined from the equations

$$\begin{pmatrix} 1+\alpha & \beta \\ \gamma & 1+\delta \end{pmatrix}\begin{pmatrix} \lambda & 0 \\ 0 & \mu \end{pmatrix}\begin{pmatrix} 1-\alpha & -\beta \\ -\gamma & 1-\delta \end{pmatrix} = \begin{pmatrix} \lambda_0+f_{uu}u+f_{uv}v & f_{uv}u+f_{vv}v \\ g_{uu}u+g_{uv}v & \mu_0+g_{uv}u+g_{vv}v \end{pmatrix}$$

where α, β, γ, δ are homogeneous linear forms in u, v. Expanding this out gives (i).

For (ii) we write $\mathbf{e}_\lambda = {}^t(1, \eta) = \partial_u + \eta\,\partial_v$ where η is homogeneous and linear in u, v, and then we solve for η in

$$\begin{pmatrix} \lambda_0+f_{uu}u+f_{uv}v & f_{uv}u+f_{vv}v \\ g_{uu}u+g_{uv}v & \mu_0+g_{uv}u+g_{vv}v \end{pmatrix}\begin{pmatrix} 1 \\ \eta \end{pmatrix} = (\lambda_0+f_{uu}u+f_{vv}v)\begin{pmatrix} 1 \\ \eta \end{pmatrix}$$

to obtain

$$\eta = (\lambda_0-\mu_0)^{-1}(g_{uu}u + g_{uv}v) + \mathrm{O}(2).$$

A similar argument gives the expression for $\mathbf{e}_\mu$.

Finally, (iii) follows from the previous lemma and

$$\begin{aligned} \langle dr_\lambda, \mathbf{e}_\lambda\rangle &\equiv 0 \mod \mathrm{O}(2) \\ \langle dr_\mu, \mathbf{e}_\mu\rangle &\equiv 0 \mod \mathrm{O}(2). \end{aligned}$$

□

From equation (iii) above we see that the level sets of the Riemann invariants, passing through the origin, are given respectively by

$$v = \tfrac{1}{2}(\lambda_0-\mu_0)^{-1}g_{uu}u^2 + \mathrm{O}(3) \qquad \text{(14a)}$$

and

$$u = \tfrac{1}{2}(\mu_0-\lambda_0)^{-1}f_{vv}v^2 + \mathrm{O}(3). \qquad \text{(14b)}$$

We now consider the solution curves of the jump equations (6) where we assume (without loss of generality) that the starting point is again $\mathbf{u} = (\mathbf{0})$. These equations have the form

$$\begin{aligned} (s-\lambda_0)u &= P(u,v) \\ (s-\mu_0)v &= Q(u,v) \end{aligned} \qquad \text{(15)}$$

where

$$P(u, v) = \tfrac{1}{2} f_{uu} u^2 + f_{uv} uv + \tfrac{1}{2} f_{vv} v^2 + \mathrm{O}(3)$$
$$Q(u, v) = \tfrac{1}{2} g_{uu} u^2 + g_{uv} uv + \tfrac{1}{2} g_{vv} v^2 + \mathrm{O}(3).$$

Eliminating s gives

$$(\lambda_0 - \mu_0)uv + vP - uQ = 0.$$

This equation has two smooth branches, with tangent vectors ∂_u and ∂_v, given respectively by

$$v = \tfrac{1}{2}(\lambda_0 - \mu_0)^{-1} g_{uu} u^2 + \mathrm{O}(3) \tag{16a}$$

and

$$u = \tfrac{1}{2}(\mu_0 - \lambda_0)^{-1} f_{vv} v^2 + \mathrm{O}(3). \tag{16b}$$

Comparing with (14) we see that:

Through each point of the uv-plane, the solutions with $s \neq 0$ of the jump equations (6) form a pair of curves which osculate to second, but not in general to third, order to the pair of level curves of the Riemann invariants.

Clearly the level sets of the Riemann invariants r_λ and r_μ define a pair of local foliations, henceforth denoted by ρ_λ and ρ_μ respectively, of the uv-plane. Somewhat more subtle is the fact that the two families of jump curves, now denoted by δ_λ and δ_μ, *need not* necessarily form a pair of local foliations.

The problem is that the relation defined by, say, the δ_λ curves is symmetric but not necessarily transitive. Consider two nearby points P, Q in the uv-plane. We say that Q is λ-related to P in case Q lies on the λ-jump curve of P, denoted by $\delta_\lambda(P)$. Since the jump equation (6) is symmetric, it is clear that P is also λ-related to Q, i.e., P lies on the λ-jump curve of Q, $\delta_\lambda(Q)$, but in general $\delta_\lambda(Q)$ will be different from $\delta_\lambda(P)$ (see Figure 4.7). For the point R as depicted above we have that Q and R are λ-related but in general P and R will *not* be λ-related. This situation is clarified by the following

Lemma 3: *The two families of jump curves form a pair of local foliations if and only if they coincide with the two Riemann foliations, and furthermore are local foliations of the uv-plane by straight lines.*

An obvious case when the δ and ρ foliations coincide is when the system (3) *uncouples*, in the sense that there is an affine linear change of variables such that (3) becomes

$$u_y + \partial_x(f(u)) = 0$$
$$v_y + \partial_x(g(u)) = 0.$$

In this case the δ-curves are two families of *parallel* lines. In general, however, the conditions of the lemma may be satisfied by two families of non-parallel lines (although, clearly, if the foliations are *global* in the uv-plane then uncoupling will occur).

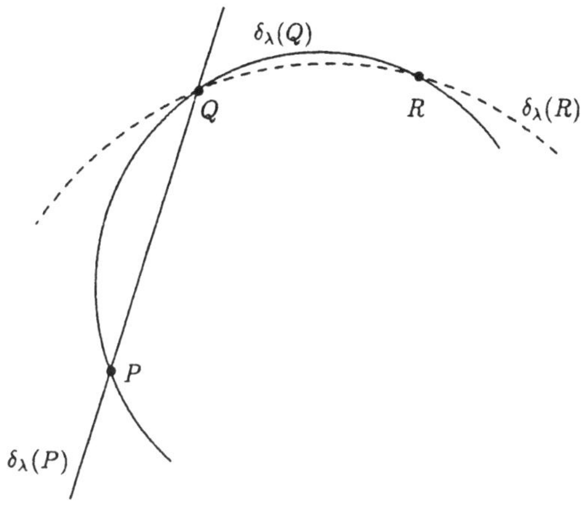

Figure 4.7

Proof: For **v** close to **u** we consider the locus Σ defined by the equations

$$\mathbf{f}(\mathbf{v}) - \mathbf{f}(\mathbf{u}) = s(\mathbf{v} - \mathbf{u})$$

in $(\mathbf{u}, \mathbf{v}, s) \in \mathbb{R}^2 \times \mathbb{R}^2 \times \mathbb{R}$. From the previous lemma, we may infer that

$$\Sigma = \Sigma_\Delta \cup \Sigma_\lambda \cup \Sigma_\mu$$

where Σ_Δ, Σ_λ, Σ_μ are smooth 3-manifolds with

$$\Sigma_\lambda = \{(\mathbf{u}, \mathbf{v}, s) \in \mathbb{R}^5 : \ \mathbf{v} \text{ is } \lambda\text{-related to } \mathbf{u}\},$$
$$\Sigma_\mu = \{(\mathbf{u}, \mathbf{v}, s) \in \mathbb{R}^5 : \ \mathbf{v} \text{ is } \mu\text{-related to } \mathbf{u}\},$$

and

$$\Sigma_\Delta = \{(\mathbf{u}, \mathbf{v}, s) \in \mathbb{R}^5 : \ \mathbf{u} = \mathbf{w} \text{ and } \mathbf{v} = \mathbf{w}\}.$$

Thus, $(\mathbf{u}, \mathbf{v}, s) \in \Sigma_\lambda$ means that $\mathbf{v} \in \delta_\lambda(\mathbf{u})$ and that if $\mathbf{v} \neq \mathbf{u}$ then s is uniquely determined by the jump equations above. We note that, as **v** approaches **u** along $\delta_\lambda(\mathbf{u})$, s approaches $\lambda(\mathbf{u})$. This suggests that we take $(\mathbf{u}, s)$ as local coordinates on Σ_λ and use $s - \lambda(\mathbf{u})$ as parameter along $\delta_\lambda(\mathbf{u})$. We then write the Taylor's series for **v** along $\delta_\lambda(\mathbf{u})$ as

$$\mathbf{v}_\lambda(\mathbf{u}, s) = \mathbf{u} + \big(s - \lambda(\mathbf{u})\big)\mathbf{e}_\lambda(\mathbf{u}) + O\big((s - \lambda(\mathbf{u}))^2\big).$$

We now determine the conditions that the δ_λ-curves define a transitive relation. Thus suppose that $\mathbf{v}_1 = \mathbf{v}(\mathbf{u}, r)$ and $\mathbf{v}_2 = \mathbf{v}(\mathbf{u}, s)$ are λ-related to $\mathbf{u}$, i.e.,

$$\begin{aligned}\mathbf{f}\big(\mathbf{v}_\lambda(\mathbf{u}, r)\big) - \mathbf{f}(\mathbf{u}) &= r\big(\mathbf{v}_\lambda(\mathbf{u}, r) - \mathbf{u}\big)\\ \mathbf{f}\big(\mathbf{v}_\lambda(\mathbf{u}, s)\big) - \mathbf{f}(\mathbf{u}) &= s\big(\mathbf{v}_\lambda(\mathbf{u}, s) - \mathbf{u}\big).\end{aligned}$$

Transitivity now implies that $\mathbf{v}_2$ is λ-related to $\mathbf{v}_1$ and so there a $t = t_\lambda(\mathbf{u}, s, r)$ such that

$$\mathbf{f}\big(\mathbf{v}_\lambda(\mathbf{u}, r)\big) - \mathbf{f}\big(\mathbf{v}_\lambda(\mathbf{u}, s)\big) = t_\lambda(\mathbf{u}, s, r)\big(\mathbf{v}_\lambda(\mathbf{u}, r) - \mathbf{v}_\lambda(\mathbf{u}, s)\big).$$

This equation uniquely determines $t_\lambda(\mathbf{u}, s, r)$ as a continuous function of $(\mathbf{u}, s, r)$ for s and r distinct and close to $\lambda(\mathbf{u})$. Moreover, setting $t_\lambda(\mathbf{u}, s, s) = \lambda\big(\mathbf{v}_\lambda(\mathbf{u}, s)\big)$ defines $t_\lambda(\mathbf{u}, s, r)$ continuously for all $(\mathbf{u}, s, r)$ with s and r close to $\lambda(\mathbf{u})$.

The above equations give

$$r\big(\mathbf{v}_\lambda(\mathbf{u}, r) - \mathbf{u}\big) - s\big(\mathbf{v}_\lambda(\mathbf{u}, s) - \mathbf{u}\big) = t_\lambda(\mathbf{u}, s, r)\big(\mathbf{v}_\lambda(\mathbf{u}, r) - \mathbf{v}_\lambda(\mathbf{u}, s)\big).$$

Now, fix $\mathbf{u}$ and consider the following curve in the uv-plane

$$\gamma(s) = \mathbf{v}_\lambda(\mathbf{u}, s) - \mathbf{u}.$$

The previous equation now gives

$$0 = \big(r\gamma(r) - s\gamma(s)\big) \wedge \big(\gamma(r) - \gamma(s)\big) = (s - r)\,\gamma(r) \wedge \gamma(s).$$

Thus $\gamma(s)$ lies on a line through the origin, as was to be shown.

Conversely, if the δ-curves are straight lines then transitivity is clear. □.

Definition: We will say that (3) is a *special* hyperbolic system of conservation laws in case the level sets of the Riemann invariants are straight lines.

We note that this is an intrinsic condition on a hyperbolic system of conservation laws (since only affine linear transformations of the uv-plane will preserve the form of (3)). We now proceed to describe all such special systems.

For an eigen-direction $\mathbf{e}(\mathbf{u}, \mathbf{v})$ of $\mathbf{f}'(\mathbf{u})$ we define

$$\dot{\mathbf{e}} = \nabla_{\mathbf{e}}\mathbf{e}.$$

Then $\dot{\mathbf{e}}$ is well-defined modulo $\mathbf{e}$ and

$$\mathbf{e} \wedge \dot{\mathbf{e}} = 0$$

is the condition that the integral curves of $\mathbf{e}$ be straight lines. This equation is a second-order condition on $\mathbf{f}(\mathbf{u})$ at each point of the uv-plane—taking into account both eigen-directions, we see that special systems of hyperbolic conservation laws are given by (3) with $\mathbf{f}(\mathbf{u}) = {}^t\big(f(u, v), g(u, v)\big)$ satisfying a pair of second-order PDEs. We may suspect that locally the solutions to this system depend on 4 arbitrary functions of 1 variable. We will see that this is indeed the case by the following geometric construction.

Each point $\mathbf{u}_0$ has a neighborhood consisting of a quadrilateral made up of lines in the two foliations. By an affine transformation we may assume that two edges of the quadrilateral are the u and v axes and that the λ and μ foliations are described by lines of the form

$$\begin{aligned} u &= \alpha(p)v + p, \qquad & 0 \leq p < p_0,\ \alpha(0) = 0, \\ v &= \beta(q)u + q, \qquad & 0 \leq q < q_0,\ \beta(0) = 0, \end{aligned}$$

respectively. We may solve for u and v to obtain

$$\begin{pmatrix} u \\ v \end{pmatrix} = \big(1 - \alpha(p)\beta(q)\big)^{-1} \begin{pmatrix} p + q\alpha(p) \\ q + p\beta(q) \end{pmatrix}.$$

As long as $\alpha(p)\beta(q) < 1$ this gives a valid formula and furthermore, if $\big(1+\alpha'(p)v\big)\big(1+\beta'(q)u\big) \neq 0$ it will be smoothly solvable for p and q as functions of u and v.

Now, we can construct another map to a $\tilde{u}\tilde{v}$-plane by considering the equations

$$\begin{aligned} \tilde{u} &= \alpha(p)\tilde{v} + \varphi(p) \\ \tilde{v} &= \beta(q)\tilde{u} + \psi(q) \end{aligned}$$

which solves to give

$$\begin{pmatrix} \tilde{u} \\ \tilde{v} \end{pmatrix} = \big(1 - \alpha(p)\beta(q)\big)^{-1} \begin{pmatrix} \varphi(p) + \alpha(p)\psi(q) \\ \psi(q) + \beta(q)\varphi(p) \end{pmatrix}.$$

It is not hard to see that, as long as φ and ψ satisfy an open condition, the induced mapping $\mathbf{f}(u, v) = (\tilde{u}, \tilde{v})$ will be hyperbolic and genuinely non-linear. Moreover, this mapping carries each p-line to a p-line of the same slope and each q-line to a q-line of the same slope. Thus, it has the desired properties, and consequently, the solutions to our problem depend on four functions of one variable.

Note that if α and β are not constant, this system will not uncouple.

We now return to our general discussion where the δ and ρ curves osculate to second order but need not coincide. Assuming that (3) is genuinely non-linear we shall show that each of the δ and ρ curves has an *orientation*. These orientations will be constructed from the entropy conditions used in the construction of shock solutions to the Riemann problem.

Recall that the conditions for genuine non-linearity are

$$\langle d\lambda, \mathbf{e}_\lambda \rangle \neq 0 \qquad \text{and} \qquad \langle d\mu, \mathbf{e}_\mu \rangle \neq 0.$$

From Lemma 2 we have that $f_{uu}g_{vv} \neq 0$. For definiteness we will assume that

$$f_{uu} > 0 \qquad \text{and} \qquad g_{vv} > 0.$$

Now, a shock solution $\mathbf{u}(x, y)$ to (3) is given for $y > 0$ by (see *[20]*)

$$\mathbf{u} = \begin{cases} \mathbf{u}_- & \text{for } x/y < s, \\ \mathbf{u}_+ & \text{for } x/y > s. \end{cases}$$

In addition, one of the following *entropy conditions*

$$s < \lambda(\mathbf{u}_-) \qquad \text{and} \qquad \lambda(\mathbf{u}_+) < s < \mu(\mathbf{u}_+)$$

or

$$\mu(\mathbf{u}_+) < s \qquad \text{and} \qquad \lambda(\mathbf{u}_-) < s < \mu(\mathbf{u}_-)$$

must be satisfied. A shock solution satisfying the first set of entropy conditions above is called a 1-*shock* while a solution satisfying the second set of conditions is called a 2-*shock*.

Now, let us set $\mathbf{u}_- = (\mathbf{0})$ and assume that $\mathbf{u}_+$ is in a neighborhood of the origin. The jump conditions specify that $\mathbf{u}_+$ must lie on the $\delta_\lambda(\mathbf{0})$ or $\delta_\mu(\mathbf{0})$ curve. We shall show that the condition for a 1-shock solution is that

$$\mathbf{u}_+ \in \delta_\lambda^-(\mathbf{0}) = \{(u, p(u)) \in \delta_\lambda(\mathbf{0}) : u < 0\}.$$

Proof: Solving for s in (15) along the arc $\delta_\lambda(\mathbf{0})$ gives

$$s = \lambda_0 + \tfrac{1}{2} f_{uu} u + \mathrm{O}(2).$$

The conditions for a 1-shock are

$$\lambda_0 + \tfrac{1}{2} f_{uu} u + \mathrm{O}(2) < \lambda_0$$

and

$$\lambda_0 + \tfrac{1}{2} f_{uu} u + \mathrm{O}(2) < \lambda_0 + \tfrac{1}{2} f_{uu} u + \mathrm{O}(2) < \mu_0 + \tfrac{1}{2} g_{uv} u + \mathrm{O}(2)$$

and these are satisfied only for $u < 0$, $|u| \leq \epsilon$. Similar calculations show that the conditions for a 2-shock cannot be satisfied anywhere along $\delta_\lambda(\mathbf{0})$. □

The orientation on δ_λ and δ_μ is now clear: The positive direction is from δ_λ^- to δ_λ^+, and similarly for δ_μ. Since the ρ_λ, ρ_μ curves are everywhere tangent to the δ_λ, δ_μ curves, the ρ_λ, ρ_μ curves also inherit an orientation.

We are now ready to proceed with the construction of the initial curve γ upstairs to show the existence of a unique "weak" integral surface S emanating from γ, and to contrast S with the graph of the classical shock solution to the Riemann problem.

Let us assume that the initial data $\mathbf{u}_-$ and $\mathbf{u}_+$ for the Riemann problem (7b) is sufficiently close so that $\mathbf{u}_-$ and $\mathbf{u}_+$ are related as in Figure 4.8. We prescribed the initial data by a mapping

$$s \mapsto \gamma(s) = (x(s), 0, \mathbf{u}_0(s)), \qquad s \in \mathbb{R}$$

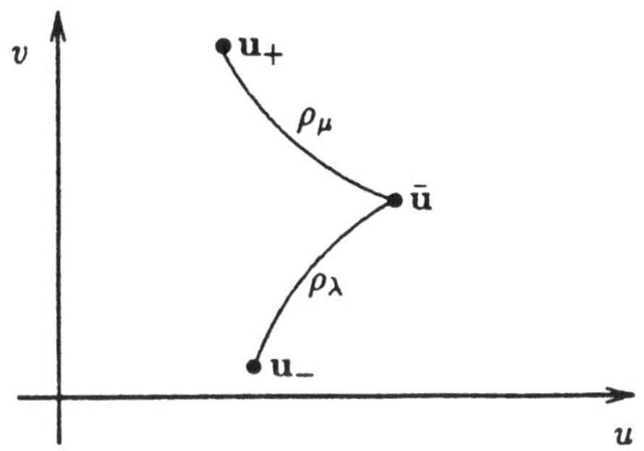

Figure 4.8

as follows: For an interval $[a, c] \subset \mathbb{R}$ to be determined below, we set

$$\gamma(s) = \begin{cases} (s-a,\, 0,\, \mathbf{u}_-) & \text{for } s \le a \\ (s-c,\, 0,\, \mathbf{u}_+) & \text{for } s \ge c \end{cases}$$

and furthermore we set

$$x(s) = 0, \qquad \mathbf{u}_0(s) \in \rho_\lambda, \qquad \text{for } s \in [a, b],$$

and

$$x(s) = 0, \qquad \mathbf{u}_0(s) \in \rho_\mu, \qquad \text{for } s \in [b, c].$$

Pictorially we have Figure 4.9, where the uv-plane lies over $x = 0$. The speed at which $\mathbf{u}_0(s)$ transverses ρ_λ and ρ_μ is specified in the following. Before doing so, we notice that the condition of genuine non-linearity implies that λ and μ are strictly monotone functions on ρ_λ and ρ_μ respectively. In what follows, we shall asume that λ and μ are increasing functions along ρ_λ and ρ_μ (a similar construction as below would work for the other cases as well).

For ξ in the interval with endpoints $\lambda(\mathbf{u}_-)$, $\lambda(\bar{\mathbf{u}})$ we solve the ODE

$$\frac{d}{d\xi}\mathbf{u}_\lambda(\xi) = \mathbf{e}_\lambda\big(\mathbf{u}_\lambda(\xi)\big) \qquad \text{with} \qquad \mathbf{u}_\lambda\big(\lambda(\mathbf{u}_-)\big) = \mathbf{u}_-,\ \mathbf{u}_\lambda\big(\lambda(\bar{\mathbf{u}})\big) = \bar{\mathbf{u}}.$$

For η in the interval with endpoints $\mu(\bar{\mathbf{u}})$, $\mu(\mathbf{u}_+)$ we solve the ODE

$$\frac{d}{d\eta}\mathbf{u}_\mu(\eta) = \mathbf{e}_\mu\big(\mathbf{u}_\mu(\eta)\big) \qquad \text{with} \qquad \mathbf{u}_\mu\big(\mu(\mathbf{u}_+)\big) = \mathbf{u}_+,\ \mathbf{u}_\mu\big(\mu(\bar{\mathbf{u}})\big) = \bar{\mathbf{u}}.$$

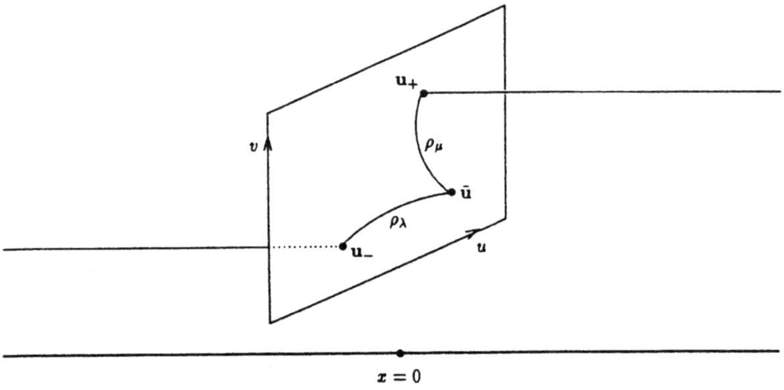

Figure 4.9

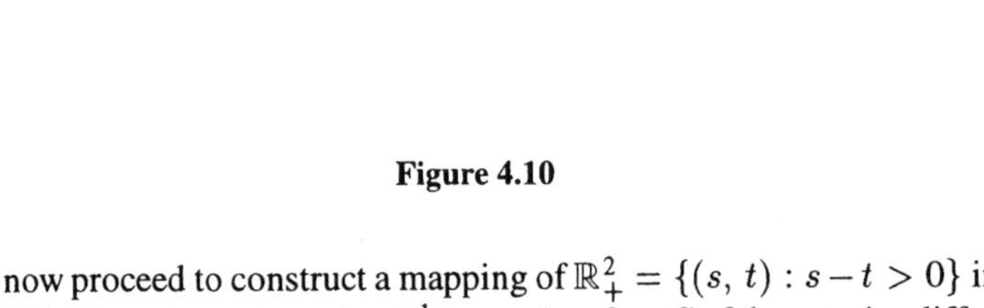

Figure 4.10

We now proceed to construct a mapping of $\mathbb{R}^2_+ = \{(s,\, t) : s - t > 0\}$ into M which will be the unique piecewise C^1 integral surface S of the exterior differential system associated to (3) consisting only of TYPE 0 and TYPE 1 solutions. We begin by dividing $\mathbb{R}^2_+$ into 5 regions as in Figure 4.10. The mapping for regions **I** and **V** are given respectively by

$$(s,\, t) \mapsto (\lambda(\mathbf{u}_-)t + s - a,\, t,\, \mathbf{u}_-),$$

and

$$(s,\,t) \mapsto \big(\mu(\mathbf{u}_+)t + s - c,\, t,\, \mathbf{u}_+\big).$$

Thus the solution in each of these regions is of TYPE 0.

We now consider region **II**. As mentioned above, we are assuming that $\lambda(\mathbf{u})$ is increasing along ρ_λ as we pass from $\mathbf{u}_-$ to $\bar{\mathbf{u}}$. Now we set $a = \lambda(\mathbf{u}_-)$ and $b = \lambda(\bar{\mathbf{u}})$. Along the lines $s = \xi$, the mapping is given by

$$(s,\,t) \mapsto \big(\lambda(\mathbf{u}_\lambda(s))t,\, t,\, \mathbf{u}_\lambda(s)\big)$$

i.e., the lines $s = \xi$ map to the xy-plane to the fan of lines with x/y-slopes increasing from $\lambda(\mathbf{u}_-)$ to $\lambda(\bar{\mathbf{u}})$. Symmetrically, region **IV** is mapped into M using the lines $t = \eta$ in place of $s = \xi$, and where the fan in the xy-plane consists of lines with slopes running between $\mu(\bar{\mathbf{u}})$ and $\mu(\mathbf{u}_+)$. The solution in each of these region is of TYPE 1.

Finally, the boundary of region **III** is mapped into M as indicated in Figure 4.12. Now, let T be the unique linear mapping of the st-plane to the xy-plane taking ∂_t to $\frac{1}{\sqrt{2}}(-\partial_x + \partial_y)$ and ∂_s to $\frac{1}{\sqrt{2}}(\partial_x + \partial_y)$. In region **III**, we set

$$(s,\,t) \mapsto \big(T(s,\,t), \bar{\mathbf{u}}\big).$$

Clearly, the solution in this region is of TYPE 0.

It is clear that S is the image of a continuous, piecewise C^1 mapping. It is also clear that it is a solution surface, in the usual sense, to the exterior differential system $\Phi = \Psi = 0$ in each region. Somewhat less clear is that accross the

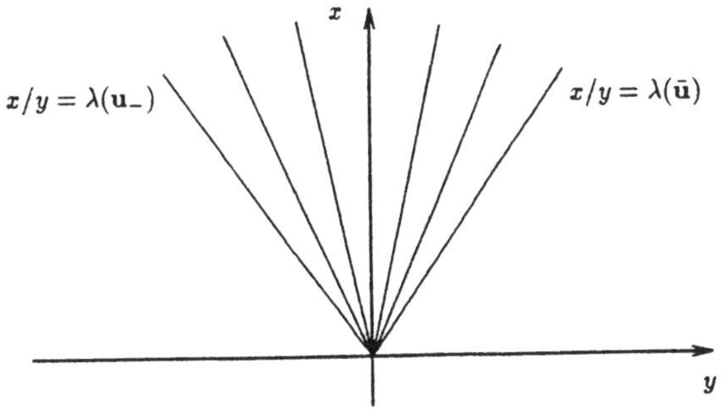

Figure 4.11

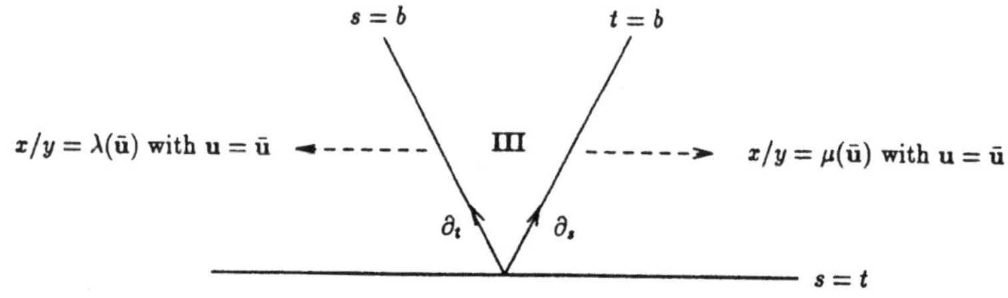

Figure 4.12

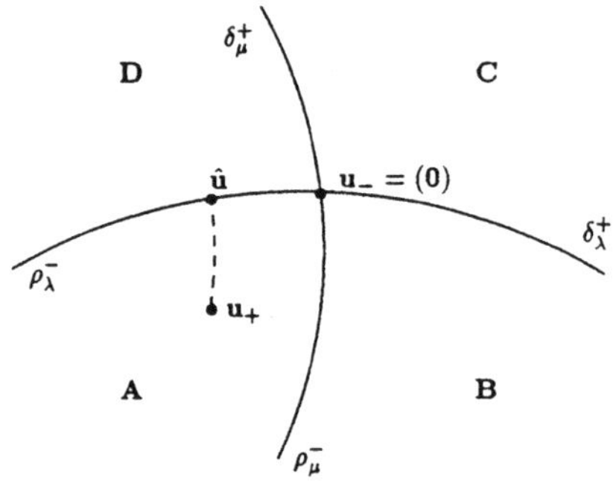

Figure 4.13

boundaries it is a weak solution to $d\varphi = d\psi = 0$. But by our construction the calculation here is essentially the same as in the scalar case.

It remains to compare S with the graph of the classical shock solution $\mathbf{u}(x, y)$. Recall that in a neighborhood of the origin in the uv-plane we have Figure 4.13. These curves are of class C^2 but not generally of class C^3 at the origin. Without loss of generality we set $\mathbf{u}_- = (\mathbf{0})$, and for $\mathbf{u}_+$ lying in each of regions **A**, **B**, **C**, **D**, the solution $\mathbf{u}(x, y)$ consists respectively of a 1-shock followed by a 2-shock, a 1-rarefaction wave followed by a 2-shock, a 1-rarefaction wave followed by a 2-rarefaction wave, and a 1-shock followed by a 2-rarefaction wave. For example, if $\mathbf{u}_+$ lies in region **A**, then there is a unique $\hat{\mathbf{u}}$ as depicted above, and the picture

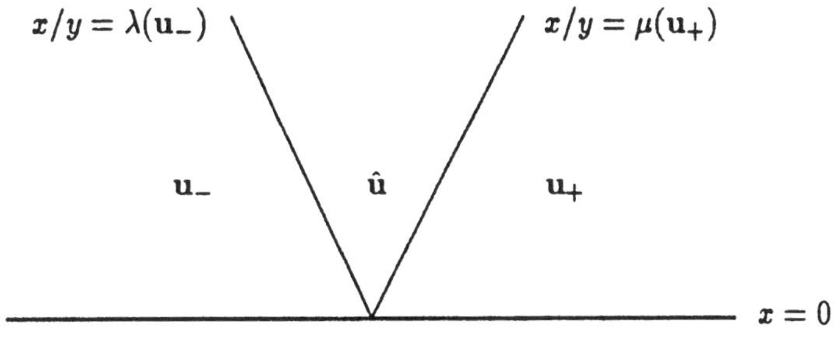

Figure 4.14

of the shock solution $\mathbf{u}(x, y)$ is shown in Figure 4.14.

In general ρ_λ^- will not coincide with δ_λ^-, in which case $\hat{\mathbf{u}} \neq \bar{\mathbf{u}}$ and so the graph of $\mathbf{u}(x, y)$ will in general not lie on the geometric surface S. However, as we have seen, an exception occurs in case (3) is a special hyperbolic system of conservation laws. For such systems we have the following result:

For the Riemann problem associated to a special hyperbolic system of conservation laws, the shock solution arises by taking a cross-section of the geometric solution.

For smooth initial data with suitable bounds on the oscillation there is the Glimm scheme *[16]* leading to a shock solution. Since this scheme is based on iteratively solving Riemann problems one may suspect that here also the shock solution $\mathbf{u}(x, y)$ arise by taking a cross-section of the geometric solution. In this case the singularities of $\mathbf{u}(x, y)$ would reflect the singularities of the map. In particular, $\mathbf{u}(x, y)$ would be piecewise smooth if the initial data is piecewise smooth.

In summary, for Riemann problems we have the following results:

(i) *For special hyperbolic systems of conservation laws, the geometric solution captures the shock solution*;

(ii) *For non-special systems, we must "cut" the geometric solution along suitable break curves and glue in other pieces of integral surfaces to obtain a surface upstairs that captures the shock solution.*

One may reasonably conjecture that both of these statements remains true for arbitrary smooth compactly supported initial data.

Appendix: The Equivalence Problem

In the preceeding sections we have made extensive use of the geometry associated to a number of differential equations to deduce information about the behavior of their solutions. In this appendix we shall explain how to associate a geometry to each of these equations. The main tool is É. Cartan's theory of G-structures—commonly known in the literature as the Method of Equivalence. Aside from Cartan's own exposition *[8]*, there are several other sources for this material, notably Chern *[11]*, Gardner *[15]*, and Bryant-Griffiths *[3]*, which the reader may consult for more examples and information.

A.1 *The equivalence problem for second-order ODE.* We seek to attach an intrinsic geometry to a second-order ordinary differential equation of the form

$$\frac{d^2y}{dx^2} = f\left(x,\, y,\, \frac{dy}{dx}\right). \tag{1}$$

Thinking of x and y as local coordinates on a surface S, the solutions of equation (1) form a 2-parameter family Σ of curves on S in the domain of the coordinates x and y.

Conversely, suppose that we are given a surface S together with a 2-parameter family Σ of curves on S with the property that, through each point p of S, the curves in Σ which pass through p form a smooth curve $\Sigma_p \subset \Sigma$ smoothly parametrized by the assignment $C \mapsto T_pC \in \mathbb{P}(T_pS)$. Such a family Σ is called a *path geometry* on S. Two path geometries, Σ_1 on S_1 and Σ_2 on S_2, are to be regarded as equivalent if there exists a diffeomorphism $\phi : S_1 \to S_2$ which carries the curves of Σ_1 onto the curves of Σ_2.

In any local coordinate system $(x, y)\colon U \to \mathbb{R}^2$ on an open set $U \subset S$, the curves of a path geometry Σ are the solutions of a differential equation of the form (1). We say that two such equations are equivalent if they induce equivalent path geometries on their domains of definition.

Given a path geometry Σ on S, we can define an incidence correspondence

$$M = \{(p, C) \in S \times \Sigma \,|\, p \in C\}.$$

By hypothesis, the assignment $(p, C) \mapsto T_pC$ defined for all $(p, C) \in M$ induces a smooth embedding of M as a subset of $\mathbb{P}(TS)$, the projectivized tangent bundle of S. For all intents and purposes, we may therefore regard M as an open subset of $\mathbb{P}(TS)$, which we do from now on.

As we remarked in §1, this picture is symmetric with respect to interchanging the roles of S and Σ, since, for each fixed p, the set of curves in Σ which contain p is itself a curve Σ_p in Σ. Thus, S defines a path geometry on Σ which is called the *dual* path geometry, with a corresponding local representation as an ordinary differential equation called the *dual equation* to (1).

Now, for a fixed $C \in \Sigma$, as p varies on C, the assignment $p \mapsto T_C\Sigma_p$ then immerses C *as a curve* into the 1-dimensional projective space $\mathbb{P}(T_C\Sigma)$. Thus, every curve in a path geometry inherits a canonical projective structure and hence a parameter well-defined up to a linear fractional change of parameter. Already, one can see that there is quite a bit of local and global geometry attached to a second order ODE. One could say, for example, that an equation of the form (1) (or more generally, a path geometry) is *projectively complete* if the canonical projective structure on its curves is projectively complete. (Note that projective completeness of a projective structure on a curve does not imply that the curve is closed.)

As we discussed in §1, this leads to the double fibration

$$\begin{array}{ccc} & M & \\ {}^{\lambda}\swarrow & & \searrow^{\rho} \\ S & & \Sigma \end{array}$$

where $\lambda(p, C) = p$ and $\rho(p, C) = C$ and each of λ and ρ is a smooth submersion. It follows that M can be covered by open sets U on which there exist coframings $(\underline{\alpha},\, \underline{\theta},\, \underline{\omega})$ so that on U, we have

$$\ker \lambda_* = \{\underline{\alpha},\, \underline{\theta}\}^{\perp} \qquad \text{and} \qquad \ker \rho_* = \{\underline{\theta},\, \underline{\omega}\}^{\perp}. \tag{2}$$

We say that a coframing $(\alpha,\, \theta,\, \omega)$ is *0-adapted* to the double fibration if the relations in (2) are satisfied. Note that the 1-form $\underline{\theta}$ is well-defined up to a scalar multiple and that, as explained in §1, it is a contact form, i.e., $\underline{\theta} \wedge d\underline{\theta}$ is nowhere vanishing.

Now, any other 0-adapted coframing in U is seen to be of the form

$$\begin{pmatrix} \alpha \\ \theta \\ \omega \end{pmatrix} = \begin{pmatrix} a_1 & b_1 & 0 \\ 0 & a_3 & 0 \\ 0 & b_2 & a_2 \end{pmatrix}^{-1} \begin{pmatrix} \underline{\alpha} \\ \underline{\theta} \\ \underline{\omega} \end{pmatrix} \tag{3}$$

where the "transition matrix" takes values in the obvious 5-dimensional subgroup G_0 of $\mathrm{GL}(3, \mathbb{R})$. Thus the local 0-adapted coframings of the double fibration are the local sections of a principal G_0-subbundle $B_{G_0} \to M$ of the bundle of general coframes of M. In other words, B_{G_0} is a G_0-structure on M.

It is important to note that the induced G_0-structure on $M \subset \mathbb{P}(TS)$ can be computed *without* knowing the solutions of (1) in advance. For example, using coordinates x, y, and p $(= dy/dx)$ on $\mathbb{R}^3 \subset \mathbb{P}(T\mathbb{R}^2)$, we can use

$$\begin{pmatrix} \underline{\alpha} \\ \underline{\theta} \\ \underline{\omega} \end{pmatrix} = \begin{pmatrix} dx \\ dy - p\,dx \\ dp - f(x, y, p)\,dx \end{pmatrix}$$

as a 0-adapted coframing on M, even though we may not know explicit functions σ and τ on M whose differentials have the same span as $\{\underline{\theta}, \underline{\omega}\}$, i.e., a pair of independent first integrals of (1). Indeed, one of the reasons for studying the G_0-structure B_{G_0} in the first place is to find ways to compute such functions σ and τ.

Following Cartan, we now apply the equivalence method to this G_0-structure to compute its invariants. Differentiating (3), and collecting terms as much as possible, we see that there exist 1-forms φ_1, φ_2, φ_3, μ_1, and μ_2 and a function t on B_{G_0} so that

$$d\begin{pmatrix}\alpha\\ \theta\\ \omega\end{pmatrix} = -\begin{pmatrix}\varphi_1 & \mu_1 & 0\\ 0 & \varphi_3 & 0\\ 0 & \mu_2 & \varphi_2\end{pmatrix} \wedge \begin{pmatrix}\alpha\\ \theta\\ \omega\end{pmatrix} + \begin{pmatrix}0\\ t\,\alpha\wedge\omega\\ 0\end{pmatrix}$$

In the terminology of the equivalence method, the $\mathfrak{g}_0$-valued 1-form

$$\begin{pmatrix}\varphi_1 & \mu_1 & 0\\ 0 & \varphi_3 & 0\\ 0 & \mu_2 & \varphi_2\end{pmatrix}$$

is called the *pseudo-connection matrix* and the term $t\,\alpha\wedge\omega$ constitutes the *torsion* of the structure. The pseudo-connection 1-forms are not uniquely determined by the structure equation and we will return to this point below.

Now, the contact condition $\theta\wedge d\theta \neq 0$ implies that $t \neq 0$, which suggests that we may restrict to coframings which satisfy $t = 1$. Thus, we say that a 0-adapted local coframing $(\underline{\alpha}, \underline{\theta}, \underline{\omega})$ is *1-adapted* if, in addition to being 0-adapted, it satisfies $d\underline{\theta} \equiv \underline{\alpha}\wedge\underline{\omega} \mod \theta$. The local coframings which are 1-adapted define a G_1-structure on M which is a sub-structure of the given G_0-structure. Its structure group is the 4-dimensional subgroup

$$G_1 = \left\{ \begin{pmatrix}a_1 & b_1 & 0\\ 0 & a_1a_2 & 0\\ 0 & b_2 & a_2\end{pmatrix} \,\middle|\, a_1, a_2 \neq 0 \right\} \subset G_0\,.$$

The structure equation of B_{G_1} now assumes the form

$$d\begin{pmatrix}\alpha\\ \theta\\ \omega\end{pmatrix} = -\begin{pmatrix}\underline{\varphi}_1 & \underline{\mu}_1 & 0\\ 0 & \underline{\varphi}_1 + \underline{\varphi}_2 & 0\\ 0 & \underline{\mu}_2 & \underline{\varphi}_2\end{pmatrix} \wedge \begin{pmatrix}\alpha\\ \theta\\ \omega\end{pmatrix} + \begin{pmatrix}0\\ \alpha\wedge\omega\\ 0\end{pmatrix}$$

for some 1-forms $\underline{\varphi}_1$, $\underline{\varphi}_2$, $\underline{\mu}_1$, and $\underline{\mu}_2$ which are, however, not uniquely determined by this equation.

It is easy to see that if φ_1, φ_2, μ_1, and μ_2 are any 1-forms on B_{G_1} which satisfy

$$d\begin{pmatrix}\alpha\\ \theta\\ \omega\end{pmatrix} = -\begin{pmatrix}\varphi_1 & \mu_1 & 0\\ 0 & \varphi_1 + \varphi_2 & 0\\ 0 & \mu_2 & \varphi_2\end{pmatrix} \wedge \begin{pmatrix}\alpha\\ \theta\\ \omega\end{pmatrix} + \begin{pmatrix}0\\ \alpha\wedge\omega\\ 0\end{pmatrix}, \tag{4}$$

then there exist functions p_1, p_2, q_1, and q_2 so that

$$\varphi_1 = \underline{\varphi}_1 + p_1\,\theta, \qquad \mu_1 = \underline{\mu}_1 + q_1\,\theta + p_1\,\alpha,$$
$$\varphi_2 = \underline{\varphi}_2 + p_2\,\theta, \qquad \text{and} \qquad \mu_2 = \underline{\mu}_2 + q_2\,\theta + p_2\,\omega,$$

and conversely that, for any such functions on B_{G_1} the 1-forms defined by the above equations satisfy (4). For such a set of 1-forms satisfying (4), we will say that the collection of 1-forms $(\alpha, \theta, \omega; \varphi_1, \varphi_2, \mu_1, \mu_2)$ define a $(1,1)$-adapted coframing of the 7-manifold B_{G_1}. It is clear that the $(1,1)$-adapted coframings are the sections of a $G_1^{(1)}$-structure $B_{G_1^{(1)}}$ on B_{G_1} where $G_1^{(1)} \subset \mathrm{GL}(7, \mathbb{R})$ is a certain 4-dimensional Lie group. Note that $B_{G_1^{(1)}}$ is a sub-bundle of the second-order coframe bundle of M and defines what is often called a second-order G-structure on M.

We now want to apply the equivalence method to the $G_1^{(1)}$-structure $B_{G_1^{(1)}}$. The structure equation of $B_{G_1^{(1)}}$ consists of two parts—the first is equation (4). The second will consist of the formulae for $d\varphi_1$, $d\varphi_2$, $d\mu_1$, and $d\mu_2$. We can get information on these by differentiating (4), yielding

$$(d\varphi_1 + \mu_1{\wedge}\omega){\wedge}\alpha + (d\mu_1 - \varphi_2{\wedge}\mu_1){\wedge}\theta = 0$$
$$(d\varphi_2 - \mu_2{\wedge}\alpha){\wedge}\omega + (d\mu_2 - \varphi_1{\wedge}\mu_2){\wedge}\theta = 0$$

and

$$d(\varphi_1 + \varphi_2) - \mu_1 \wedge \omega + \mu_2 \wedge \alpha \equiv 0 \quad \mathrm{mod}\, \theta.$$

From these relations, it follows that there are 1-forms π_1, π_2 and ψ_1, ψ_2 and a function t_0 on $B_{G_1^{(1)}}$ such that

$$d\varphi_1 = -\pi_1{\wedge}\theta - \mu_1{\wedge}\omega - 2\mu_2{\wedge}\alpha + t_0\, \alpha{\wedge}\omega$$
$$d\varphi_2 = -\pi_2{\wedge}\theta + \mu_2{\wedge}\alpha + 2\mu_1{\wedge}\omega + t_0\, \omega{\wedge}\alpha$$

and

$$d\mu_1 = -\psi_1{\wedge}\theta - \pi_1{\wedge}\alpha + \varphi_2{\wedge}\mu_1$$
$$d\mu_2 = -\psi_2{\wedge}\theta - \pi_2{\wedge}\omega + \varphi_1{\wedge}\mu_2.$$

Note that $(\pi_1, \pi_2, \psi_1, \psi_2)$ are the pseudo-connection 1-forms of the $G_1^{(1)}$-structure.

We now compute how the function t_0 (which is the only non-constant component of the torsion) varies on the fibers of $B_{G_1^{(1)}} \to B_{G_1}$. Differentiating the above structure equations yields

$$dt_0 \equiv t_0\,(\varphi_1 + \varphi_2) + 2(\pi_1 - \pi_2) \quad \mathrm{mod}\, \alpha, \theta, \omega.$$

Thus, the equation $t_0 = 0$ defines a submanifold $B_{G_2} \subset B_{G_1^{(1)}}$ which is a G_2-structure where G_2 is a 3-dimensional subgroup of $G_1^{(1)}$. Because of the formula for dt_0, we see that, on B_{G_2}, there must exist a 1-form σ and functions s_1 and s_2 so that

$$\left.\begin{aligned} \pi_1 &\equiv -\sigma + s_1\,\alpha + s_2\,\omega \\ \pi_2 &\equiv -\sigma - s_1\,\alpha - s_2\,\omega \end{aligned}\right\} \quad \mathrm{mod}\, \theta.$$

Thus, by modifying the forms ψ_1 and ψ_2, the structure equations can be reduced to the form

$$\begin{aligned} d\varphi_1 &= \sigma\wedge\theta - \mu_1\wedge\omega - 2\mu_2\wedge\alpha - (s_1\,\alpha + s_2\,\omega)\wedge\theta \\ d\varphi_2 &= \sigma\wedge\theta + \mu_2\wedge\alpha + 2\mu_1\wedge\omega + (s_1\,\alpha + s_2\,\omega)\wedge\theta \end{aligned}$$

and

$$\begin{aligned} d\mu_1 &= -\psi_1\wedge\theta + \sigma\wedge\alpha + \varphi_2\wedge\mu_1 + s_2\,\alpha\wedge\omega \\ d\mu_2 &= -\psi_2\wedge\theta + \sigma\wedge\omega + \varphi_1\wedge\mu_2 + s_1\,\alpha\wedge\omega. \end{aligned}$$

The functions s_1 and s_2 constitute the non-constant terms in the torsion of this G_2-structure.

To calculate how s_1 and s_2 vary on the fibers of $B_{G_2} \longrightarrow B_{G_1}$, we differentiate the above structure equations, yielding

$$\left.\begin{aligned} ds_1 &\equiv s_1\,(2\varphi_1 + \varphi_2) - 3\psi_2 \\ ds_2 &\equiv s_2\,(2\varphi_2 + \varphi_1) - 3\psi_1 \end{aligned}\right\} \quad \operatorname{mod} \alpha,\ \theta,\ \omega.$$

Thus, the equations $s_1 = s_2 = 0$ define a submanifold $B_{G_3} \subset B_{G_2}$ which is a G_3-structure on B_{G_1} where the structure group G_3 is now 1-dimensional. On B_{G_3}, we can now differentiate the identity

$$\tfrac{1}{3}d(\varphi_2 - \varphi_1) = \mu_1 \wedge \omega + \mu_2 \wedge \alpha$$

and use the structure equations found so far to conclude that

$$\psi_1 \wedge \omega + \psi_2 \wedge \alpha \equiv 0 \quad \operatorname{mod} \theta,$$

which implies that there exist functions K_1, K_2 and S so that

$$\begin{pmatrix} \psi_1 \\ \psi_2 \end{pmatrix} \equiv \begin{pmatrix} K_2 & S \\ S & K_1 \end{pmatrix} \begin{pmatrix} \omega \\ \alpha \end{pmatrix} \quad \operatorname{mod} \theta.$$

It follows that by adding appropriate multiple of θ to each of ψ_1, ψ_2, and σ, the structure equations reduce to the form

$$\begin{aligned} d\varphi_1 &= \sigma\wedge\theta - \mu_1\wedge\omega - 2\mu_2\wedge\alpha \\ d\varphi_2 &= \sigma\wedge\theta + \mu_2\wedge\alpha + 2\mu_1\wedge\omega \\ d\mu_1 &= \sigma\wedge\alpha + \varphi_2\wedge\mu_1 + K_2\,\theta\wedge\omega \\ d\mu_2 &= \sigma\wedge\omega + \varphi_1\wedge\mu_2 + K_1\,\theta\wedge\alpha. \end{aligned}$$

Now, the remaining pseudo-connection 1-form σ is uniquely defined by the above structure equations. Differentiating these equations yields

$$d\sigma = (\varphi_1 + \varphi_2)\wedge\sigma - \mu_1\wedge\mu_2 + \theta\wedge(L_1\,\alpha + L_2\,\omega)$$

for some functions L_1 and L_2 on B_{G_3}.

At this stage we have determined a canonical parallelism on B_{G_3} with structure equations given by

$$d\begin{pmatrix}\alpha\\ \theta\\ \omega\end{pmatrix} = -\begin{pmatrix}\varphi_1 & \mu_1 & 0\\ 0 & \varphi_1+\varphi_2 & 0\\ 0 & \mu_2 & \varphi_2\end{pmatrix}\wedge\begin{pmatrix}\alpha\\ \theta\\ \omega\end{pmatrix} + \begin{pmatrix}0\\ \alpha\wedge\omega\\ 0\end{pmatrix}$$

where

$$\begin{aligned}
d\varphi_1 &= \sigma\wedge\theta \; -\mu_1\wedge\omega \; -2\mu_2\wedge\alpha\\
d\varphi_2 &= \sigma\wedge\theta \; +\mu_2\wedge\alpha \; +2\mu_1\wedge\omega\\
d\mu_1 &= \sigma\wedge\alpha \; +\varphi_2\wedge\mu_1 \; +K_2\,\theta\wedge\omega\\
d\mu_2 &= \sigma\wedge\omega \; +\varphi_1\wedge\mu_2 \; +K_1\,\theta\wedge\alpha
\end{aligned}$$

and

$$d\sigma = (\varphi_1+\varphi_2)\wedge\sigma - \mu_1\wedge\mu_2 + \theta\wedge(L_1\,\alpha + L_2\,\omega).$$

Moreover, as the method of equivalence shows, any self-equivalence of S with itself preserving the paths geometry Σ lifts to a unique diffeomorphism of B_{G_3} with itself which preserves the eight 1-forms in this coframe. In particular, the space of automorphisms of the path geometry Σ on S is a Lie group of dimension at most 8. In fact, the dimension of this group can be 8 only if the functions K_i and L_i are constants.

These structure equations can be collected into a more coherent form as follows: Define an $\mathfrak{sl}(3,\mathbb{R})$-valued matrix ϕ by

$$\phi = \begin{pmatrix} -\frac{1}{3}(2\varphi_1+\varphi_2) & -\mu_2 & \sigma\\ \alpha & \frac{1}{3}(\varphi_1-\varphi_2) & \mu_1\\ \theta & \omega & \frac{1}{3}(\varphi_1+2\varphi_2)\end{pmatrix}.$$

Then the structure equations above take the form

$$d\phi + \phi\wedge\phi = \begin{pmatrix} 0 & K_1\alpha\wedge\theta & \theta\wedge(L_1\alpha+L_2\omega)\\ 0 & 0 & K_2\theta\wedge\omega\\ 0 & 0 & 0\end{pmatrix}.$$

Setting $\Phi = d\phi + \phi\wedge\phi$, the Bianchi identity $d\Phi = \Phi\wedge\phi - \phi\wedge\Phi$ then gives rise to the equations

$$\begin{aligned}
dK_1 &\equiv (3\varphi_1+\varphi_2)K_1 \; -L_1\,\omega && \mod \theta,\alpha\\
dK_2 &\equiv (3\varphi_2+\varphi_1)K_2 \; -L_2\,\alpha && \mod \theta,\omega\\
dL_1 &\equiv (3\varphi_1+2\varphi_2)L_1 - K_1\mu_1 + J\omega && \mod \theta,\alpha\\
dL_2 &\equiv (3\varphi_2+2\varphi_1)L_2 + K_2\mu_2 + J\alpha && \mod \theta,\omega
\end{aligned}$$

where J is some function on B_{G_3}.

From these structure equations, it follows, for example, that the expressions

$$\mathcal{K}_1 = (K_1\,\alpha - L_1\theta) \otimes (\alpha{\wedge}\theta)$$
$$\mathcal{K}_2 = (K_2\,\omega + L_2\theta) \otimes (\theta{\wedge}\omega)$$

are well-defined sections of $T^* \otimes \Lambda^2 T^*$ on M. Thus, these define tensors which are the simplest invariants of a path geometry or second order ODE up to point transformations. In the classical terminology, the functions K_1 and K_2 are what is known as *relative invariants*. Furthermore, a little computation shows that J is a relative invariant and that the cubic form $\mathcal{J} = J\,\theta^3$ is well-defined on M and hence is another invariant of the path geometry.

Note however, that these invariants are defined on M, and not on either S or Σ. As an example of an invariant function on M, the ratio $J^4/(K_1K_2)^3$ is seen to be constant on the fibers of $B_{G_3} \to M$ and hence is a well-defined function on M.

Let us consider the meaning of the vanishing of the first two invariants, i.e., the case where K_1 and K_2 vanish identically. (Note that, by the Bianchi identities, this is the only way that K_1 and K_2 can be constants anyway.) By the Bianchi identities, this forces L_1 and L_2 to vanish identically as well, so that Ω vanishes identically. Then, since $d\phi = -\phi{\wedge}\phi$, it follows that there exists a local diffeomorphism $B_{G_3} \to \mathrm{SL}(3,\mathbb{R})$ which identifies ϕ with the canonical left-invariant 1-form on $\mathrm{SL}(3,\mathbb{R})$. This identification induces an identification of S (which can be thought of as the leaves in B_{G_3} of the integrable distribution $\alpha = \theta = 0$) with the left cosets in $\mathrm{SL}(3,\mathbb{R})$ of the subgroup P whose Lie algebra is of the form

$$\begin{pmatrix} * & * & * \\ 0 & * & * \\ 0 & * & * \end{pmatrix}.$$

In other words, S is is locally identified with $\mathrm{SL}(3,\mathbb{R})/P = \mathbb{RP}^2$. Moreover, we also get an identification of Σ (which can be thought of as the leaves in B_{G_3} of the integrable distribution $\omega = \theta = 0$) with the left cosets in $\mathrm{SL}(3,\mathbb{R})$ of the subgroup Π whose Lie algebra is of the form

$$\begin{pmatrix} * & * & * \\ * & * & * \\ 0 & 0 & * \end{pmatrix}.$$

Of course, the homogeneous space $\mathrm{SL}(3,\mathbb{R})/\Pi$ is also diffeomorphic to $\mathbb{RP}^2$, but its points can be regarded as the lines in the first $\mathbb{RP}^2$. Thus, the path geometry in this case is just the classical path geometry of straight lines in the plane. Moreover, the intrinsic geometric structure B_{G_3} is identified with the group of symmetries of this path geometry, i.e., the projective transformations of $\mathbb{RP}^2$. Note that a consequence of our calculations is that this is the only path geometry (up to local equivalence) for which the group of symmetries is actually of dimension 8.

Now let us consider the meaning of the vanishing of each of K_1 or K_2 separately. In terms of the ODE (1), Cartan showed that K_2 vanishes if and only if the function f satisfies $\partial^4 f/\partial p^4 = 0$, i.e., if and only if f is a cubic polynomial in p. On the other hand K_1 vanishes if and only if f satisfies equation (5) of §1.

Now, in the first case, when K_2 vanishes, it follows that the ODE (1) is the local equation for the geodesics of a projective connection on S in the usual sense of differential geometry. Since a projective connection on a surface has a curvature tensor which is well-defined on that surface, it should not be surprising that when K_2 vanishes we can create expressions on M which are well-defined on S. In fact, the vanishing of K_2 implies directly that the tensor $\mathcal{K}_1$ (which is clearly semi-basic for the map $\lambda : M \to S$) is actually well-defined on S. Computing higher derivatives of K_1 as functions on B_{G_3} and taking the right combinations yields functions which are actually well-defined on S. Thus, they are first integrals of the dual ODE on Σ.

Perhaps more interesting is the other side. When K_1 vanishes, it follows that the dual equation to (1) (which we may not know explicitly) is actually the local equation of the geodesics of a projective connection. Again, the same arguments show that by taking the right combinations of the (iterated) derivatives of K_2 (which we can compute from f *without having to solve any ODE*), we can arrive at functions on B_{G_3} which are constant on the fibers of the map $B_{G_3} \to \Sigma$. Hence, these will be functions on M which are first integrals of the equation (1). Thus, equations (1) satisfying the curvature condition $K_1 = 0$ can usually be integrated by differentiation alone! (The reason we have to say "usually" is that there are certain special cases where this procedure won't lead to a pair of independent first integrals. However, these cases are very rare and can only happens when the path geometry has a non-trivial symmetry group, so that other methods, due to Lie, can be applied.)

Although Cartan did not pursue this, there are higher order conditions like K_1 vanishing which will guarantee that the equations which satisfy them can be integrated by an algorithm (see *[5]*). Unfortunately, we do not have time to pursue this topic here.

A.2 *The equivalence problem for scalar conservation laws.* In §2 we have shown that a scalar conservation law determines the data $(M;\, \Phi,\, \Omega)$ consisting of a 3-manifold M endowed with two linearly independent 2-forms Φ (which is closed) and Ω (which is nowhere vanishing and well-defined up to a multiple). In addition the non-linearity of the conservation law is expressed by the condition that the common linear factor of Φ and Ω defines a contact structure on M. We shall study the geometry associated with the above $(M;\, \Phi,\, \Omega)$ structure under diffeomorphisms of M which preserve Φ and preserve Ω up to a non-vanishing factor.

Let $(\underline{\alpha},\, \underline{\theta},\, \underline{\omega})$ be a local coframing of M such that

$$\Phi = \underline{\theta} \wedge \underline{\omega} \qquad \text{and} \qquad \Omega = \underline{\alpha} \wedge \underline{\theta} \tag{1}$$

with $\underline{\theta}$ defining a contact structure on M, i.e.,

$$\underline{\theta} \wedge d\underline{\theta} \neq 0. \tag{2}$$

We shall say that a coframing $(\alpha,\,\theta,\,\omega)$ is 0-*adapted* to the $(M;\,\Phi,\,\Omega)$-structure if the relations in (1) and (2) are satisfied with Φ preserved and Ω preserved up to a non-vanishing multiple. It is easy to see that any two 0-adapted coframes are related as follows

$$\begin{pmatrix}\bar\alpha\\ \bar\theta\\ \bar\omega\end{pmatrix} = \begin{pmatrix}* & * & 0\\ 0 & a^{-1} & 0\\ 0 & * & a\end{pmatrix}^{-1}\begin{pmatrix}\alpha\\ \theta\\ \omega\end{pmatrix} \tag{3}$$

where the "transition matrices" take values in a 4-dimensional subgroup G_0 of $\mathrm{GL}(3,\mathbb{R})$. Thus these local 0-adapted coframings define a G_0-structure B_{G_0} on M.

We now apply the equivalence method to compute the invariants of the above G_0-structure. Differentiating (3) and using the condition $d\Phi = 0$, we see that on B_{G_0} we have the following structure equation

$$d\begin{pmatrix}\alpha\\ \theta\\ \omega\end{pmatrix} = -\begin{pmatrix}\varphi_1 & \mu_1 & 0\\ 0 & -\varphi_2 & 0\\ 0 & \mu_2 & \varphi_2\end{pmatrix}\wedge\begin{pmatrix}\alpha\\ \theta\\ \omega\end{pmatrix} + \begin{pmatrix}0\\ t\,\alpha\wedge\omega\\ 0\end{pmatrix} \tag{4}$$

where t is a nowhere vanishing function on B_{G_0}. The G_0-action on t can be deduced by differentiating (4) which yields the relation

$$dt \equiv t\,(\varphi_1 + 2\varphi_2) \quad \mathrm{mod}\,\alpha,\,\theta,\,\omega.$$

This suggests that we may restrict to local coframings on M which satisfy $t = 1$. We call such coframings 1-*adapted*. They are the sections of a G_1-structure B_{G_1} on which the structure equation has the form

$$d\begin{pmatrix}\alpha\\ \theta\\ \omega\end{pmatrix} = -\begin{pmatrix}2\varphi & \mu_1 & 0\\ 0 & \varphi & 0\\ 0 & \mu_2 & -\varphi\end{pmatrix}\wedge\begin{pmatrix}\alpha\\ \theta\\ \omega\end{pmatrix} + \begin{pmatrix}t_0\,\alpha\wedge\omega\\ \alpha\wedge\omega\\ 0\end{pmatrix}$$

for a smooth function t_0 on B_{G_1}. The structure group G_1 is now 3-dimensional.

Again, we deduce the G_1-action on the function t_0 by differentiating the above equation which yields

$$dt_0 \equiv -t_0\,\varphi - 3\mu_1 \quad \mathrm{mod}\,\alpha,\,\theta,\,\omega.$$

This suggests that we further restrict to the submanifold $B_{G_2} \subset B_{G_1}$ defined by $t_0 = 0$. On B_{G_2} we have

$$d\begin{pmatrix}\alpha\\ \theta\\ \omega\end{pmatrix} = -\begin{pmatrix}2\varphi & 0 & 0\\ 0 & \varphi & 0\\ 0 & \mu & -\varphi\end{pmatrix}\wedge\begin{pmatrix}\alpha\\ \theta\\ \omega\end{pmatrix} + \begin{pmatrix}K\,\theta\wedge\omega\\ \alpha\wedge\omega\\ 0\end{pmatrix}$$

where K is a smooth function on B_{G_2}. Note that the structure group $G_2 \subset G_3$ is now 2-dimensional.

At this stage we notice that the pseudo-connection form φ is uniquely determined by the above structure equation. Thus the 1-form α is now well-defined up to scale. Furthermore, from

$$\alpha \wedge d\alpha = K\, \alpha \wedge \theta \wedge \omega$$

we see that the invariant K vanishes identically if and only if α is integrable.

Now, the G_2-action on the function K is expressed by the relation

$$dK \equiv -2K\, \varphi \quad \operatorname{mod} \alpha,\, \theta,\, \omega.$$

The equivalence problem now branches into two cases according to whether K vanishes or not.

• $K = 0$—In this case the structure equation reduces to

$$d\begin{pmatrix}\alpha\\ \theta\\ \omega\end{pmatrix} = -\begin{pmatrix}2\varphi & 0 & 0\\ 0 & \varphi & 0\\ 0 & \mu & -\varphi\end{pmatrix} \wedge \begin{pmatrix}\alpha\\ \theta\\ \omega\end{pmatrix} + \begin{pmatrix}0\\ \alpha\wedge\omega\\ 0\end{pmatrix}.$$

Since φ is now canonically defined, we can differentiate the above equation to get

$$d\varphi = -\mu \wedge \alpha$$

which in turn (together with the above structure equation) uniquely determines the pseudo-connection form μ. Further differentiation yields

$$d\mu = -2\mu \wedge \varphi + L\, \alpha \wedge \theta$$

for some function L on B_{G_2}.

At this stage that the 1-forms $(\alpha,\, \theta,\, \omega;\, \varphi,\, \mu)$ determines a canonical parallelism on B_{G_2} with structure equations given as above. In the case $L = 0$, it is easy to show that B_{G_2} is a Lie group with $(\alpha,\, \theta,\, \omega;\, \varphi,\, \mu)$ giving a basis for the left-invariant 1-forms. On the other hand, if $L \neq 0$, then a further differentiation of the structure equations above gives

$$dL \equiv 5L\, \varphi \quad \operatorname{mod} \alpha,\, \theta.$$

Thus we can further restrict to the submanifold $B_{G_3} \subset B_{G_2}$ defined by $L = 1$. On B_{G_3} we can set

$$\varphi = P_0\, \alpha + \tfrac{1}{2} P\, \theta$$

and hence from the above structure equations we obtain

$$dP_0 \equiv -\mu + \tfrac{1}{2} P\, \omega \quad \operatorname{mod} \alpha,\, \theta.$$

Now restricting to the submanifold defined by $P_0 = 0$, we obtain an e-structure on M, i.e., a canonical parallelism on M with structure equations

$$d\alpha = P\, \alpha \wedge \theta, \qquad d\theta = \alpha \wedge \omega, \qquad d\omega = (Q\, \alpha - P\, \omega) \wedge \theta.$$

Notice that P and Q are well-defined functions on M. In what follows, they will be refered to as the curvatures of the associated conservation law.

• $K \neq 0$—In this case, assuming, say, that $K > 0$, we may further restrict to the submanifold $B_{G_3} \subset B_{G_2}$ defined by $K = 1$. On B_{G_3} the structure equation has the form

$$d\begin{pmatrix}\alpha\\ \theta\\ \omega\end{pmatrix} = -\begin{pmatrix}0&0&0\\0&0&0\\0&\mu&0\end{pmatrix}\wedge\begin{pmatrix}\alpha\\ \theta\\ \omega\end{pmatrix} + \begin{pmatrix}-2\varphi\wedge\alpha+\theta\wedge\omega\\ -\ \varphi\wedge\theta\ +\alpha\wedge\omega\\ \varphi\wedge\omega\end{pmatrix}$$

where

$$\varphi \equiv 0 \quad \mathrm{mod}\, \alpha,\ \theta,\ \omega.$$

An easy computation shows that we can further reduce to an e-structure on M, i.e., we have a canonical parallelism on M with structure equations given by

$$d\alpha = -2\varphi\wedge\alpha + \theta\wedge\omega, \qquad d\theta = -\varphi\wedge\theta + \alpha\wedge\omega, \qquad d\omega = -\mu\wedge\theta$$

where

$$\varphi \equiv 0 \quad \mathrm{mod}\, \theta,\ \omega \qquad \text{and} \qquad \mu \equiv 0 \quad \mathrm{mod}\, \alpha,\ \theta,\ \omega.$$

Expanding out these congruences and substituting them into the structure equations above produce a collection of well-defined functions on M which have interpretations as curvature functions of the associated conservation law.

A.3 *The equivalence problem for hyperbolic systems.* In the following we will show how to associate a geometry to a (non-degenerate) hyperbolic exterior differential system $(M;\ \Omega_1,\ \Omega_2)$.

Recall that a hyperbolic system is given by a transverse pair of decomposable 2-forms Ω_1, Ω_2 on a 4-manifold M. This implies that there are local coframings $(\omega^1,\ \omega^2,\ \omega^3,\ \omega^4)$ on M such that

$$\Omega_1 = \omega^1\wedge\omega^2, \qquad \Omega_2 = \omega^3\wedge\omega^4. \tag{1}$$

Clearly such coframings are determined up to the group $\mathrm{GL}(2,\mathbb{R}) \times \mathrm{GL}(2,\mathbb{R})$. Thus the Pfaffian systems Ξ_1 and Ξ_2 generated respectively by ω^1, ω^2 and ω^3, ω^4 are well-defined on M. Furthermore on any solution surface $S \subset M$ to the hyperbolic system $\Omega_1 = \Omega_2 = 0$, each of Ξ_1 and Ξ_2 will have rank exactly equal to one and will therefore induce two foliations on S by curves. These are the classical *characteristic curves* and for this reason we shall call Ξ_1 and Ξ_2 the *characteristic systems.* The interaction of the geometry with the characteristic systems and their prolongations is probably the deepest aspect of our subject. In the following we shall only consider hyperbolic systems for which the behavior of the characteristic systems—such as the ranks of their derived flags—is symmetric.

For a non-degenerate hyperbolic system the first derived system of each characteristic system has rank 1. This suggests that we restrict to local coframings satisfying (1) and the conditions

$$d\omega^1 \equiv 0 \quad \mathrm{mod}\,\omega^1,\ \omega^2, \qquad d\omega^3 \equiv 0 \quad \mathrm{mod}\,\omega^3,\ \omega^4. \tag{2}$$

Thus, ω^1 and ω^3 span the first derived systems of the two characteristic systems.

Now, for any coframing satisfying (1) and (2), there must exist functions A and C so that

$$d\omega^2 \equiv A\,\omega^3 \wedge \omega^4 \mod \omega^1, \omega^2 \qquad d\omega^4 \equiv C\,\omega^1 \wedge \omega^2 \mod \omega^3, \omega^4. \tag{3}$$

By assumption, Ξ_1 and Ξ_2 are non-integrable. It follows from this and (2) that neither A nor C can vanish and thus we can further restrict to coframings which satisfy $A = C = 1$. We shall say that a coframing is 1-adapted to the hyperbolic system if it satisfies the conditions (1), (2), and (3) with $A = C = 1$.

If $\omega = (\omega^1, \omega^2, \omega^3, \omega^4)$ is a 1-adapted coframing on a domain $U \subset M$, then any other coframing on U, say $\bar\omega = (\bar\omega^1, \bar\omega^2, \bar\omega^3, \bar\omega^4)$ is seen to be 1-adapted if and only if there exist functions a^2_1, a^4_3, $a^2_2 \neq 0$, and $a^4_4 \neq 0$ on U so that

$$\begin{pmatrix} \bar\omega^1 \\ \bar\omega^2 \\ \bar\omega^3 \\ \bar\omega^4 \end{pmatrix} = \begin{pmatrix} a^4_4/a^2_2 & 0 & 0 & 0 \\ a^2_1 & a^2_2 & 0 & 0 \\ 0 & 0 & a^2_2/a^4_4 & 0 \\ 0 & 0 & a^4_3 & a^4_4 \end{pmatrix} \begin{pmatrix} \omega^1 \\ \omega^2 \\ \omega^3 \\ \omega^4 \end{pmatrix}.$$

The "transition matrices" take values in a certain 4-dimensional lower triangular subgroup of $\mathrm{GL}(4,\mathbb{R})$ which we shall henceforth denote G. Thus, the local coframings which are 1-adapted to the hyperbolic system are the local sections of a principal G-bundle $B_G \to M$ which is a subbundle of the bundle of all coframes of M. In other words, B_G is a G-structure on M in the usual sense. We will refer to B_G as the G-structure *associated to* (or *determined by*) the non-degenerate hyperbolic exterior differential system $(M; \Omega_1, \Omega_2)$.

Now we shall apply the equivalence method to the G-structure B_G in order to understand its invariants. Accordingly, we write the structure equations on B_G in the form

$$d\begin{pmatrix} \omega^1 \\ \omega^2 \\ \omega^3 \\ \omega^4 \end{pmatrix} = -\begin{pmatrix} \phi_{44}-\phi_{22} & 0 & 0 & 0 \\ \phi_{21} & \phi_{22} & 0 & 0 \\ 0 & 0 & \phi_{22}-\phi_{44} & 0 \\ 0 & 0 & \phi_{43} & \phi_{44} \end{pmatrix} \wedge \begin{pmatrix} \omega^1 \\ \omega^2 \\ \omega^3 \\ \omega^4 \end{pmatrix} + \begin{pmatrix} T^1 \\ T^2 \\ T^3 \\ T^4 \end{pmatrix} \tag{4}$$

where, in the terminology of the equivalence method, the ϕ_{ij} are the *pseudo-connection forms* and the T^i are the *torsion* terms (which are semi-basic[5]). These forms are not uniquely determined by these equations, and, following the usual method of equivalence, we now want to understand how modifications of the pseudo-connection forms can be employed to simplify the torsion terms.

Now, by the defining properties of the G-structure B_G, we have

$$\begin{aligned} d\omega^1 &\equiv 0 && \mod \omega^1, \omega^2 \\ d\omega^2 &\equiv \omega^3 \wedge \omega^4 && \mod \omega^1, \omega^2 \\ d\omega^3 &\equiv 0 && \mod \omega^3, \omega^4 \\ d\omega^4 &\equiv \omega^1 \wedge \omega^2 && \mod \omega^3, \omega^4 \end{aligned} \qquad \text{so} \qquad \begin{aligned} T^1 &\equiv 0 && \mod \omega^1, \omega^2 \\ T^2 &\equiv \omega^3 \wedge \omega^4 && \mod \omega^1, \omega^2 \\ T^3 &\equiv 0 && \mod \omega^3, \omega^4 \\ T^4 &\equiv \omega^1 \wedge \omega^2 && \mod \omega^3, \omega^4. \end{aligned}$$

[5] I.e., these terms have the form $T^i = T^i_{jk}\,\omega^j \wedge \omega^k$ for some functions $T^i_{jk} = -T^i_{kj}$ on B.

It follows that there exist 1-forms χ_1, χ_2, χ_3, and χ_4 which are linear combinations of the ω^i so that

$$T^2 = \omega^3{\wedge}\omega^4 + \chi_1{\wedge}\omega^1 + \chi_2{\wedge}\omega^2,$$
$$T^4 = \omega^1{\wedge}\omega^2 + \chi_3{\wedge}\omega^3 + \chi_4{\wedge}\omega^4.$$

The equations for $d\omega^2$ and $d\omega^4$ can therefore be written in the form

$$\begin{aligned} d\omega^2 &= -(\phi_{21} - \chi_1){\wedge}\omega^1 - (\phi_{22} - \chi_2){\wedge}\omega^2 + \omega^3{\wedge}\omega^4 \\ d\omega^4 &= -(\phi_{43} - \chi_3){\wedge}\omega^3 - (\phi_{44} - \chi_4){\wedge}\omega^4 + \omega^1{\wedge}\omega^2. \end{aligned}$$

It follows that we may assume that the ϕ_{ij} have been chosen so that

$$T^2 = \omega^3 \wedge \omega^4 \qquad \text{and} \qquad T^4 = \omega^1 \wedge \omega^2,$$

so we assume this from now on. This condition still does not determine the ϕ_{ij} since making the replacements

$$\begin{pmatrix} \phi_{21} \\ \phi_{22} \\ \phi_{43} \\ \phi_{44} \end{pmatrix} \mapsto \begin{pmatrix} \phi_{21} + a_1\,\omega^1 + a_2\,\omega^2 \\ \phi_{22} + a_2\,\omega^1 + a_3\,\omega^2 \\ \phi_{43} + c_3\,\omega^3 + c_4\,\omega^4 \\ \phi_{44} + c_4\,\omega^3 + c_1\,\omega^4 \end{pmatrix}$$

in the above equations will clearly not affect T^2 or T^4. However, the above congruences on T^1 and T^3 imply that

$$\begin{aligned} T^1 &\equiv T^1_{13}\,\omega^1{\wedge}\omega^3 + T^1_{14}\,\omega^1{\wedge}\omega^4 \quad \mod \omega^2 \\ T^3 &\equiv T^3_{31}\,\omega^3{\wedge}\omega^1 + T^3_{32}\,\omega^3{\wedge}\omega^2 \quad \mod \omega^4 \end{aligned}$$

and the above replacements can be chosen so that $T^1_{13} = T^1_{14} = T^3_{31} = T^3_{32} = 0$. Note that the only replacements of the above form which preserve these latter conditions are ones with $a_2 = a_3 = c_4 = c_1 = 0$.

The upshot of this discussion is that, for the G-structure we have associated to a non-degenerate hyperbolic system, there is a choice of pseudo-connection so that the torsion takes the form

$$\begin{pmatrix} T^1 \\ T^2 \\ T^3 \\ T^4 \end{pmatrix} = \begin{pmatrix} \omega^2{\wedge}(p_1\,\omega^1 + p_3\,\omega^3 + p_4\,\omega^4) \\ \omega^3{\wedge}\omega^4 \\ \omega^4{\wedge}(q_3\,\omega^3 + q_1\,\omega^1 + q_2\,\omega^2) \\ \omega^1{\wedge}\omega^2 \end{pmatrix} \tag{5}$$

Moreover, with the structure equations in this form, the 1-forms ϕ_{22} and ϕ_{44} are unique, the form ϕ_{21} is determined up to the addition of a multiple of ω^1, and the form ϕ_{43} is determined up to the addition of a multiple of ω^3. At this stage, no further reduction of this G-structure can be made without making some non-vanishing assumptions on the invariants.

To complete the discussion of the structure equations, it will be necessary to compute their "Bianchi identities" by differentiating the equations in (5). We will not give the details of the calculations (which are straightforward, if tedious), but shall describe the results. First of all, differentiation of the equations (5) and reduction of the results modulo various combinations of the ω^i shows that there are relations of the form

$$\left.\begin{aligned} dp_1 &\equiv p_1\,\phi_{22} && -q_2\,\phi_{43} \\ dp_3 &\equiv p_3\,(3\phi_{22}-2\phi_{44}) && +p_4\,\phi_{43} \\ dp_4 &\equiv p_4\,(2\phi_{22}) && \\ dq_3 &\equiv q_3\,\phi_{44} && -p_4\,\phi_{21} \\ dq_1 &\equiv q_1\,(3\phi_{44}-2\phi_{22}) && +q_2\,\phi_{21} \\ dq_2 &\equiv q_2\,(2\phi_{44}) && \end{aligned}\right\} \mod \omega^1,\omega^2,\omega^3,\omega^4. \tag{6}$$

We shall use the notation ∇p_4 to mean $dp_4 - 2p_4\,\phi_{22}$, i.e., the semi-basic part of the exterior derivative of p_4, and similarly for the other quantities.

If we now introduce "curvature" 2-forms Φ_{22}, Φ_{44}, Φ_{21}, and Φ_{43} by the equations

$$\begin{aligned} d\phi_{22} &= -\phi_{21}\wedge(p_3\,\omega^3 + p_4\,\omega^4) + q_3\,\omega^1\wedge\omega^2 + \tfrac{1}{2}p_3q_1\,\omega^2\wedge\omega^4 + \Phi_{22} \\ d\phi_{44} &= -\phi_{43}\wedge(q_1\,\omega^1 + q_2\,\omega^2) + p_1\,\omega^3\wedge\omega^4 + \tfrac{1}{2}q_1p_3\,\omega^4\wedge\omega^2 + \Phi_{44} \\ d\phi_{21} &= -\phi_{21}\wedge(\phi_{44} - 2\phi_{22} - p_1\,\omega^2) + \Phi_{21} \\ d\phi_{43} &= -\phi_{43}\wedge(\phi_{22} - 2\phi_{44} - q_3\,\omega^4) + \Phi_{43}, \end{aligned} \tag{7}$$

the exterior derivatives of the equations (5) become

$$\begin{aligned} 0 &= (\Phi_{22}-\Phi_{44})\wedge\omega^1 - \left(\nabla p_1\wedge\omega^1+\nabla p_3\wedge\omega^3+\nabla p_4\wedge\omega^4-q_3p_3\,\omega^3\wedge\omega^4\right)\wedge\omega^2 \\ 0 &= -\Phi_{21}\wedge\omega^1-\Phi_{22}\wedge\omega^2-\omega^1\wedge\omega^2\wedge\omega^3 \\ 0 &= (\Phi_{44}-\Phi_{22})\wedge\omega^3 - \left(\nabla q_3\wedge\omega^3+\nabla q_1\wedge\omega^1+\nabla q_2\wedge\omega^2-p_1q_1\,\omega^1\wedge\omega^2\right)\wedge\omega^4 \\ 0 &= -\Phi_{43}\wedge\omega^3-\Phi_{44}\wedge\omega^4-\omega^3\wedge\omega^4\wedge\omega^1. \end{aligned} \tag{8}$$

(Note that because ϕ_{21} and ϕ_{43} are not canonical, the expression ∇p_3 is actually only well-defined modulo ω^3. However, since this term only occurs wedged with ω^3, the resulting term is well defined. A similar comment applies to the other ambiguities caused by the ambiguity in the pseudo-connection.) The identities (8) now give relations among the coefficients of the derivatives of the primary invariants (i.e., the torsion coefficients) and the curvature coefficients. It is not useful to write these out here; the form (8) will suffice for our purposes.

A little exterior algebra shows that the relations (8) imply that Φ_{22} and Φ_{44} are semi-basic 2-forms, i.e., they are quadratic expressions in the ω^i. In fact,

$$\Phi_{22} = -\kappa_1\wedge\omega^1 - \kappa_2\wedge\omega^2 \qquad \text{and} \qquad \Phi_{44} = -\kappa_3\wedge\omega^3 - \kappa_4\wedge\omega^4 \tag{9}$$

where

$$\kappa_i = k_{ij}\omega^j$$

for functions k_{ij}, suitably skew-symmetrized so as to be well-defined. Using this plus some more exterior algebra, it follows that there are 1-forms ϕ_{211} and ϕ_{433} so that

$$\begin{aligned}\Phi_{21} &= -(\kappa_1 - \omega^3)\wedge\omega^2 - \phi_{211}\wedge\omega^1 \\ \Phi_{43} &= -(\kappa_3 - \omega^1)\wedge\omega^4 - \phi_{433}\wedge\omega^3.\end{aligned} \tag{10}$$

We now want to interpret the vanishing of the torsion coefficients in (5) in terms of integrability of various bundles intrinsically associated to the original hyperbolic system and use this to derive (local) normal forms in various special cases. We will then use the structure equations to develop a test for 'linearizability' of non-degenerate hyperbolic systems.

We begin our first interpretation by noting that the rank 2 Pfaffian system $\Theta = \{\omega^1, \omega^3\}$ spanned by the first derived systems of Ξ_1 and Ξ_2 is well-defined. Indeed, from the structure equations, the 2-form

$$\Omega = \omega^1 \wedge \omega^3$$

itself is well-defined, since the scalings of ω^1 and ω^3 cancel. The integrability of Ω has the following interpretation:

For any non-degenerate hyperbolic system, the system $\Theta = \{\omega^1, \omega^3\}$ is Frobenius if and only if the torsion coefficients satisfy $p_4 = q_2 = 0$. Moreover, the rank one systems $\{\omega^1\}$ and $\{\omega^3\}$ are integrable if and only if we have, in addition to the above, the relations $p_3 = q_1 = 0$.

In these special cases we have the following normal forms result:

Proposition: *Let $(M; \Omega_1, \Omega_2)$ be a non-degenerate hyperbolic system with $p_4 = q_2 = 0$. If $p_3 q_1 \neq 0$, then the hyperbolic system is locally generated by the PDE system*

$$\begin{aligned} u_y + uu_x &= C, \qquad & C_v &\neq 0, \\ v_x + vv_y &= D, \qquad & D_u &\neq 0 \end{aligned}$$

where C and D are functions of x, y, u and v. On the other hand, if $p_3 = q_1 = 0$, then the local model is given by the following (wave equations)

$$\begin{aligned} u_y &= C, \qquad & C_v &\neq 0, \\ v_x &= D, \qquad & D_u &\neq 0. \end{aligned}$$

Remark: Non-degeneracy of the hyperbolic system is expressed by the coupling condition $C_v D_u \neq 0$.

As another example of the use of the invariants of B_G to understand normal forms, we shall give a characterization of linear systems of PDE in terms of these invariants.

Proposition: *A non-degenerate hyperbolic system $(M; \Omega_1, \Omega_2)$ satisfies $p_4 = q_2 = p_3 = q_1 = p_1 = q_3 = 0$ and $\Phi_{22} + \Phi_{44} = F\omega^1\wedge\omega^3$ for some function F if and only if it is locally the hyperbolic system associated to a linear first order hyperbolic PDE system.*

Remark: In this case, a little algebra shows that

$$(\Phi_{22} - \Phi_{44}) = K\omega^1 \wedge \omega^3.$$

Furthermore, the quadratic differential form

$$ds^2 = \omega^1 \circ \omega^3$$

is well-defined and induces on solution surfaces an intrinsic pseudo-Riemannian metric whose null-geodesics are the characteristic curves. The Gauss curvature of this metric may then be verified to be the function K above. In this way the geometry associated to a PDE induces, in an intrinsic manner, a geometry on solution surfaces.

References

[1] R. Bryant, Two exotic holonomies in dimension four, path geometries and twistor theory, in: *Complex Geometry and Lie Theory, Proc. Sym. Pure Math.* **53** (1991), 33–88.

[2] R. Bryant, S.-S. Chern, R. Gardner, H. Goldschmidt, and P. Griffiths, Exterior differential systems, MSRI Publications #18, Springer-Verlag (1991).

[3] R. Bryant and P. Griffiths, Characteristic cohomology of differential systems (I): General theory, IAS preprint (1993); (II): Conservation Laws for a Class of Parabolic Equations, IAS preprint (1993).

[4] R. Bryant, P. Griffiths and L. Hsu, Hyperbolic exterior differential systems and their conservation laws, IAS preprint (1994).

[5] R. Bryant and L. Hsu, Geometric integrability and Cartan's class (C) equations, (in preparation).

[6] É. Cartan, Sur les variétés à connexion projective, Bull. Soc. Math. France **52** (1924), 205–241.

[7] É. Cartan, Les espaces généralisés et l'intégration de certaines classes d'équations différentielles, C. R. Acad. Sci. **206** (1938), 1689–1693.

[8] É. Cartan, Les problèmes d'équivalence, in Œuvres Complètes, Partie II, Éditions du CNRS, Paris (1984).

[9] R. E. Caflisch, N. Ercolani, T. Y. Hou and Y. Landis, Multi-valued solutions and branch point singularities for non-linear hyperbolic or elliptic systems, Comm. Pure and Applied Math. **46** (1993), 453–499.

[10] S.-S. Chern, The geometry of the differential equation $y''' = F(x, y, y', y'')$, Sci. Rep. Tsing Hua Univ. **4** (1940), 97–111.

[11] S.-S. Chern, Geometry of G-structures, Bulletin of the AMS **72** (1966), 167–219.

[12] R. Courant and D. Hilbert, Methods of Mathematical Physics, Volume I, Interscience, New York (1953), Volume II, Interscience, New York (1962).

[13] G. Darboux, Leçons sur la theorie general des surfaces, Tome II, Hermann, Paris (1915).

[14] Ph. Le Floch and Z. Xin, Uniqueness via the adoint problems for systems of conservation laws, Comm. Pure and Applied Math. **46** (1993), 1499–1533.

[15] R. B. Gardner, The Method of Equivalence and Its Applications, CBMS–NSF Regional Conference Series in Applied Math. **58**, SIAM, Philadelphia (1989).

[16] J. Glimm, Solutions in the large for non-linear hyperbolic systems of equations, Comm. Pure and Applied Math. **18** (1965), 697–715.

[17] E. Goursat, Leçons sur l'intégration des équations aux dérivées partielles du second ordre à deux variables indépendantes, Tome 1, Hermann, Paris (1896), Tome 2, Hermann, Paris (1898).

[18] J. Hurtubise and N. Kamran, Projective connections, double fibrations, and formal neighbourhoods of lines, Math. Ann. **292** (1992), 383–409.

[19] P. Lax, Development of Singularities of Solutions of Nonlinear Hyperbolic Partial Differential Equations, J. Math. Phys. **5** (1964), 611–613.

[20] J. Smoller, Shock Waves and Reaction-Diffusion Equations, Springer, New York (1983).

Operators on Noncommutative Differential Forms and Cyclic Homology

Joachim Cuntz

Mathematisches Institut
Universität Heidelberg
In Neuenheimer Feld 288
D-69120 Heidelberg, Germany

Daniel Quillen

Mathematical Institute
University of Oxford
24 St. Giles'
Oxford OX1 3LB, ENGLAND

Let A be an associative unital algebra over the complex numbers and let ΩA be the differential graded algebra of its noncommutative differential forms (*[3]*). The first topic of this paper is a family of canonical operators b, d, κ, B, P, G on ΩA arising in the following way.

On ΩA we have two differentials

$$\Omega^0 A \ \underset{d}{\overset{b}{\leftrightarrows}} \ \Omega^1 A \ \underset{d}{\overset{b}{\leftrightarrows}} \ \Omega^2 A \ \underset{d}{\overset{b}{\leftrightarrows}}$$

where d is the differential associated to the DG algebra structure, and where b, the Hochschild differential, is such that $(\Omega A, b)$ is the standard normalized complex calculating the Hochschild homology $HH_n(A)$.

Let us now view ΩA as analogous to the space of differential forms on a Riemannian manifold, with b playing the role of the adjoint of d. The 'Laplacian' $bd + db$ then has the form $1 - \kappa$, where κ is an operator introduced by Karoubi (*[6]*, 2.12). The Karoubi operator κ satisfies a polynomial identity in each degree, hence ΩA decomposes into generalized eigenspaces for the Laplacian. Thus we have the 'harmonic decomposition'

$$\Omega A \ = \ P\Omega A \oplus P^{\perp}\Omega A \qquad (1)$$

where P is the spectral projection onto the generalized nullspace for the Laplacian and $P^{\perp} = 1 - P$. The 'Green's operator' G is then defined to be zero on $P\Omega A$ and the inverse of the Laplacian on $P^{\perp}\Omega A$. Finally, the Connes boundary operator B

Geometry, Topology, and Physics for Raoul Bott

Cambridge, MA 02138 USA

on normalized chains may be written $B = NPd$, where N is multiplication by n on $\Omega^n A$.

The mixed complex given by ΩA equipped with the operators b, B is the smallest of three 'standard' mixed complexes (2) giving rise to the cyclic homology $HC_n(A)$. Since the harmonic decomposition (1) respects the mixed complex structure and $P^\perp \Omega A$ is a contractible mixed complex with contraction Gd, it follows that $P\Omega A$ also yields the cyclic homology of A.

Thus, like the case of a Riemannian manifold, the harmonic subspace $P\Omega A$ contains the pertinent homology information. In contrast however, the Laplacian on $P\Omega A$ is nonzero in general, although its square is zero. Furthermore, $P\Omega A$ is more than just a graded vector space, as it has a natural mixed complex structure.

The main application of these canonical operators on ΩA can be found in *[4]*. In the present paper we shall explore their relation to Connes' work (*[1, 2]*). This is not completely straightforward, because for example, Connes uses instead of ΩA the universal *nonunital* DG algebra generated by A, even in the case of a unital algebra.

To organize matters we introduce the following acyclicity property for mixed complexes, which is abstracted from Connes' fundamental lemma *[1]*, p. 337, 36. Let us say that a mixed complex M has *Connes' property* when the B-homology $\operatorname{Ker} B/\operatorname{Im} B$ is acyclic with respect to b. This is a vanishing condition on the E^2 term of one of the spectral sequences associated to the b, B bicomplex of M. It implies that the cyclic homology of M is computed by the quotient complex $M/\operatorname{Ker} B$ of (M, b), and also that the homology of M as a supercomplex with differential $b + B$ vanishes.

It turns out that a mixed complex has Connes' property iff it is the direct sum of a B-acyclic mixed complex and a b-acyclic one on which $B = 0$. It follows that a mixed complex has Connes' property iff it is homotopy equivalent to a B-acyclic mixed complex. Moreover, a quasi-isomorphism of mixed complexes with Connes' property is necessarily a homotopy equivalence.

Now ΩA unfortunately does not have Connes' property. So for certain purposes, e.g. if one wants the cyclic homology to be given as in *[1]* by a quotient complex of a complex computing the Hochschild homology, one uses other standard mixed complexes which do have Connes' property.

First, there is the reduced space $\bar{\Omega} A = \Omega A/\mathbb{C}$ which yields the reduced cyclic homology $\bar{HC}_n(A)$. The harmonic decomposition in this case splits $\bar{\Omega} A$ into a B-acyclic mixed complex $P\bar{\Omega} A$ and a b-acyclic mixed complex $P^\perp \bar{\Omega} A$ killed by B. Thus $\bar{\Omega} A$ has Connes' property.

Next, let $\mathcal{A}$ be a (possibly) nonunital algebra and let $\tilde{\mathcal{A}}$ be the augmented algebra obtained by adjoining an identity to $\mathcal{A}$. The reduced space $\bar{\Omega}\tilde{\mathcal{A}}$ can then be canonically identified with the augmentation ideal of $\Omega\tilde{\mathcal{A}}$, hence it has a natural structure as nonunital DG algebra. This is the DG algebra of noncommutative differential forms associated to $\mathcal{A}$ used by Connes. As an instance of $\bar{\Omega} A$ we know $\bar{\Omega}\tilde{\mathcal{A}}$ is also naturally a mixed complex with Connes' property. It gives rise to the cyclic homology $HC_n(\mathcal{A}) = \bar{HC}_n(\tilde{\mathcal{A}})$ of the nonunital algebra $\mathcal{A}$.

Finally, for A unital there is besides ΩA and $\bar{\Omega}\tilde{A}$ another mixed complex yielding the cyclic homology of A. This is Connes' $C(A)$ consisting of (unnormalized) chains equipped with the Hochschild differential b and his boundary operator B, and it has Connes' property by his fundamental lemma. These three mixed complexes are linked by surjective quasi-isomorphisms

$$\bar{\Omega}\tilde{A} \xrightarrow{j} C(A) \xrightarrow{p} \Omega A \tag{2}$$

where j comes from a construction of Kassel (*[9]*, §5), and p is the obvious map from chains to normalized chains.

If we view the mixed complexes with Connes' property we have mentioned as functors defined on the categories of unital or nonunital algebras, then their relation can be summarized as follows. The functor $\bar{\Omega}A$ on unital algebras when pulled back by $\mathcal{A} \to \tilde{\mathcal{A}}$ yields $\bar{\Omega}\tilde{\mathcal{A}}$ on nonunital algebras. This restricts to the functor $\bar{\Omega}\tilde{A}$ on unital algebras, and Connes' $C(A)$ is a simplification of $\bar{\Omega}\tilde{A}$.

In the first two sections we discuss the canonical operators on ΩA and various consequences of this structure. The results about Connes' property we have mentioned are derived in §3 and the first part of §6. We also study B-acyclic mixed complexes in §3, and we identify them up to isomorphism with mixed complexes coming from S-modules in the sense of the canonical adjoint functors between S-modules and mixed complexes (*[5]*, §1).

The fourth section is devoted to applications in the case of $\bar{\Omega}A$, beginning with the proof that it has Connes' property. This implies immediately that the reduced cyclic homology is the homology of the reduced cyclic complex (*[12]*, 4.4), and moreover that the $b+B$ homology of $\bar{\Omega}A$ is zero. We also give an efficient proof of the Connes-Karoubi theorem relating noncommutative de Rham homology to cyclic homology, which proceeds by reducing to a general formula pertaining to B-acyclic mixed complexes.

In §5 we describe $\bar{\Omega}\tilde{\mathcal{A}}$ and its canonical operators in terms of the graded vector space $C(\mathcal{A})$ of chains on $\mathcal{A}$ and its canonical operators b, b', λ, etc. This allows us to identify $\bar{\Omega}\tilde{\mathcal{A}}/\mathrm{Ker}\, B$ with the cyclic complex $C^{\lambda}(\mathcal{A})$, and the S-module associated to $\bar{\Omega}\tilde{\mathcal{A}}$ with the total complex of the cyclic bicomplex of $\mathcal{A}$. Our description of the harmonic space $P\bar{\Omega}\tilde{\mathcal{A}}$ leads to an explicit S operator on the cyclic complex, which we show agrees with the one constructed by Kassel (*[9]*, §4), and also corresponds by duality to Connes' explicit S operator on cyclic cochains (*[1]*, p. 322, 11).

The mixed complex $C(A)$ is studied in §6. The B operator and map j in (2) can be defined more generally for any H-unital algebra starting from a choice of contraction for the complex of chains with differential b'. Connes' boundary operator B on $C(A)$ is obtained from the standard contraction which puts 1 at the right with the appropriate sign. For this choice we prove Connes' lemma that $C(A)$ has Connes' property. It follows then that j is actually a homotopy equivalence of mixed complexes, although we do not know an explicit formula for the homotopy inverse.

In §7 we show the map p in (2) is a quasi-isomorphism by using a suitable Karoubi operator on $C(A)$ instead of the simplicial normalization theorem.

The last section contains an account of Connes' infinite dimensional cycles (*[2]*). We show supertraces on $QA = A * A$ are essentially equivalent to linear functionals $f \in (\Omega A)^*$ which are cocycles with respect to $b+B$ and κ-invariant.

From this we deduce Connes' version that supertraces on the nonunital free product of A with itself are essentially equivalent to $b+B$ cocycles $\psi \in C(A)^*$ satisfying his normalization condition.

This paper has evolved considerably from a preprint with the same title dating from the spring of 1992. The investigation of Connes' property and the details of the relation to Connes' framework are substantial improvements on the old version.

1 Operators on Differential Forms

In this paper we use the notations and assumptions of *[3]*. Our groundfield is the complex numbers $\mathbb{C}$, and algebras are assumed to be associative and unital unless stated otherwise.

Let $\Omega A = \bigoplus_n \Omega^n A$ be the DG algebra of (noncommutative) differential forms over the algebra A. We have the basic identification *[CQ1 §1]*

$$\begin{aligned} \Omega^n &= A \otimes \bar{A}^{\otimes n} \\ a_0 da_1 \dots da_n &\leftrightarrow (a_0, .., a_n) \end{aligned} \tag{1}$$

where $\bar{A} = A/\mathbb{C}$ and the symbol $(a_0, .., a_n)$ stands for the image of $a_0 \otimes \cdots \otimes a_n$. In this way an n-form is identified with what might be called a *normalized n-chain* on A.

Since A will be fixed, we abbreviate $\Omega^n A$, ΩA to Ω^n, Ω.

The operator d

With respect to the two notations we are using the differential d is given by

$$\begin{aligned} d(a_0 da_1 \dots da_n) &= da_0 \dots da_n \\ d(a_0, .., a_n) &= (1, a_0, .., a_n) \end{aligned}$$

Hence we have

$$\begin{aligned} d\Omega^n &\cong \bar{A}^{\otimes n+1} \\ da_0 \dots da_n &\leftrightarrow (a_0, .., a_n) \end{aligned} \tag{2}$$

for $n \geq 0$. We assume from now on that A is not the zero algebra. We then have an exact sequence

$$0 \to \bar{A}^{\otimes n} \xrightarrow{s} A \otimes \bar{A}^{\otimes n} \longrightarrow \bar{A}^{\otimes n+1} \to 0$$

where $s(x) = 1 \otimes x$, hence an exact sequence

$$0 \to d\Omega^{n-1} \longrightarrow \Omega^n \xrightarrow{d} d\Omega^n \to 0 \tag{3}$$

for $n \geq 1$, and a surjection $d : A \to dA \cong \bar{A}$ with kernel $\mathbb{C}$ for $n = 0$. Consequently the d-homology of Ω is

$$H_n(\Omega, d) = \begin{cases} \mathbb{C} & n = 0 \\ 0 & n \neq 0 \end{cases} \tag{4}$$

Hochschild homology and the operator b

The *Hochschild homology* of A is $HH_n(A) = H_n(A, A)$, where $H_n(A, M)$, the Hochschild homology of degree n with the bimodule M as coefficients, is the n-th left derived functor of the commutator quotient space functor

$$M \mapsto M \otimes_A = M/[A, M] \tag{5}$$

Thus the Hochschild homology $HH_n(A)$ can be computed from any projective bimodule resolution of A by applying the commutator quotient space functor and taking homology.

Let us consider the standard normalized resolution of A by free A-bimodules

$$\xrightarrow{b'} A \otimes \bar{A}^{\otimes 2} \otimes A \xrightarrow{b'} A \otimes \bar{A} \otimes A \xrightarrow{b'} A \otimes A \xrightarrow{b'} A \to 0 \tag{6}$$

$$b'(a_0, .., a_{n+1}) = \sum_{i=0}^{n} (-1)^i (a_0, .., a_i a_{i+1}, .., a_{n+1}) \tag{7}$$

Using the identification (1) this can be written *[*CQ1, §3*]*

$$\xrightarrow{b'} \Omega^2 \otimes A \xrightarrow{b'} \Omega^1 \otimes A \xrightarrow{b'} A \otimes A \xrightarrow{b'} A \to 0 \tag{8}$$

$$\begin{aligned} b'(\omega da \otimes a') &= (-1)^{|\omega|}(\omega a \otimes a' - \omega \otimes a a') \\ b'(a \otimes a') &= a a' \end{aligned} \tag{9}$$

where $|\omega|$ denotes the degree of ω.

Applying the commutator quotient space functor to the complex $\Omega \otimes A$ with differential b', and using the canonical isomorphism

$$(X \otimes A) \otimes_A \xrightarrow{\sim} X \qquad x \otimes a \mapsto ax$$

for any left A-module X, we obtain the complex $\Omega = \bigoplus_n \Omega^n$ with the differential b given by

$$b(a_0, .., a_n) = \sum_{j=0}^{n-1} (-1)^j (a_0, .., a_j a_{j+1}, .., a_n) + (-1)^n (a_n a_1, .., a_{n-1}) \tag{10}$$

$$\begin{aligned} b(\omega da) &= (-1)^{|\omega|}(\omega a - a\omega) = (-1)^{|\omega|}[\omega, a] \\ b(a) &= 0 \end{aligned} \tag{11}$$

in the two notations we are using.

Thus the Hochschild homology is the b-homology of Ω:

$$H_n(\Omega A, b) = HH_n(A) \tag{12}$$

The Karoubi operator κ

The *Karoubi operator [6, 2.12]* is the degree zero operator on Ω given in positive degrees by

$$\kappa(\omega da) = (-1)^{|\omega|} da\,\omega \tag{13}$$

and by the identity in degrees ≤ 0. We have

$$\begin{aligned}(bd+db)(\omega da) &= b(d\omega da) + (-1)^{|\omega|} d[\omega, a] \\ &= (-1)^{|\omega|+1}[d\omega, a] + (-1)^{|\omega|} d[\omega, a] \\ &= [\omega, da] \quad = \quad \omega da - (-1)^{|\omega|} da\,\omega\end{aligned}$$

giving the important identity

$$bd + db = 1 - \kappa \tag{14}$$

This shows that κ is homotopic to the identity with respect to either of the differentials b, d. In particular, it commutes with them:

$$b\kappa = \kappa b \qquad\qquad d\kappa = \kappa d \tag{15}$$

We have

$$\begin{aligned}\kappa(a_0 da_1 \cdots da_n) &= (-1)^{n-1} da_n a_0 da_1 \cdots da_{n-1} \\ &= (-1)^n a_n da_0 \cdots da_{n-1} + (-1)^{n-1} d(a_n a_0) da_1 \cdots da_{n-1}\end{aligned}$$

which in chain notation is

$$\kappa(a_0, .., a_n) = (-1)^n (a_n, a_0, .., a_{n-1}) + (-1)^{n-1}(1, a_n a_0, .., a_{n-1}) \tag{16}$$

Notice that the first term on the right is the familiar cyclic permutation operator with sign

$$\lambda(a_0, .., a_n) = (-1)^n (a_n, a_0, .., a_{n-1}) \tag{17}$$

on $A^{\otimes n+1}$, and the second term is a correction such that the sum descends to $A \otimes \bar{A}^{\otimes n}$. In this way κ appears as a substitute for λ on normalized chains.

Now

$$\kappa(da_0 \cdots da_n) = (-1)^n da_n da_0 \cdots da_{n-1} \tag{18}$$

showing that κ on $d\Omega^n$ corresponds to λ with respect to the identification (2). In particular $\kappa^{n+1} d = d$ on Ω^n.

We have

$$\kappa^j(a_0 da_1 \dots da_n) = (-1)^{j(n-1)} da_{n-j+1} \dots da_n a_0 da_1 \dots da_{n-j}$$

for $0 \leq j \leq n$. Hence

$$\begin{aligned}\kappa^n(a_0 da_1 \dots da_n) &= da_1 \dots da_n a_0 \\ &= a_0 da_1 \dots da_n + [da_1 \dots da_n, a_0] \\ &= a_0 da_1 \dots da_n + (-1)^n b(da_1 \dots da_n da_0)\end{aligned}$$

yielding

$$\kappa^n = 1 + b\kappa^n d \tag{19}$$

on Ω^n. Then $\kappa^{n+1} = \kappa + b\kappa^{n+1}d = \kappa + bd$, so using (14) we have

$$\kappa^{n+1} = 1 - db \tag{20}$$

From (19), (20) we find that κ on Ω^n satisfies the polynomial relation

$$(\kappa^n - 1)(\kappa^{n+1} - 1) = 0 \tag{21}$$

This implies κ is invertible, since the polynomial has constant term 1. One can understand this relation from the exact sequence (3), namely, $\kappa^{n+1} - 1$ kills the quotient space $d\Omega^n$, hence it maps Ω^n into the subspace $d\Omega^{n-1}$, which is killed by $\kappa^n - 1$.

We recall that the operator B of Connes is the operator of degree $+1$ on Ω given on Ω^n by

$$\begin{aligned} B &= \sum_{j=0}^{n} \kappa^j d \\ B(a_0, .., a_n) &= \sum_{i=0}^{n} (-1)^{in}(1, a_i, .., a_n, a_0, .., a_{i-1}) \end{aligned} \tag{22}$$

We have

$$dB = Bd = B^2 = 0 \tag{23}$$

$$B\kappa = \kappa B = B \tag{24}$$

since $\kappa^{n+1}d = d$ on Ω^n.

We next compute $\kappa^{n(n+1)} - 1$ in two ways. First, using (19) we have

$$\kappa^{n(n+1)} - 1 = \sum_{j=0}^{n} \kappa^{nj}(\kappa^n - 1) = \sum_{j=0}^{n} b\kappa^{nj+n}d = bB$$

On the other hand, using (20) we have

$$\kappa^{n(n+1)} - 1 = \sum_{j=0}^{n-1} \kappa^{(n+1)j}(\kappa^{n+1} - 1) = -\sum_{j=0}^{n-1} \kappa^{nj+j}db = -Bb$$

since $\kappa^n d = d$ on Ω^{n-1}. Thus we obtain

$$\kappa^{n(n+1)} = 1 + bB = 1 - Bb \tag{25}$$

$$bB + Bb = 0 \tag{26}$$

Remark: As $Bb = -bB$ has square zero, the operator $\kappa^{n(n+1)} = 1 - Bb$ on Ω^n is unipotent, and hence κ is a *quasi-unipotent* operator in the terminology of the theory of monodromy. When $Bb \neq 0$ on Ω^n, $1 - Bb$ has infinite order, and hence so does κ. For example, when $n = 1$ we have $Bb(a_0 da_1) = d[a_0, a_1]$, so $Bb \neq 0$ unless every commutator is a scalar.

2 Harmonic Decomposition

Let us regard

$$\Omega^0 \underset{d}{\overset{b}{\leftrightarrows}} \Omega^1 \underset{d}{\overset{b}{\leftrightarrows}} \Omega^2 \underset{d}{\overset{b}{\leftrightarrows}}$$

as analogous to the de Rham complex of a Riemannian manifold. Thus we pretend as if b and d are adjoints of each other and $bd + db = 1 - \kappa$ is the Laplacian. The natural thing to do then is to examine the spectrum of $1 - \kappa$ and focus upon the zero eigenvalue. The corresponding subspace of Ω is the analogue of the space of harmonic forms, and it should contain all the cohomological information.

Spectral decomposition with respect to κ

We have seen that on Ω^n the operator κ satisfies the polynomial relation

$$(\kappa^n - 1)(\kappa^{n+1} - 1) = 0$$

We recall that an operator satisfying a polynomial relation gives rise to a direct sum decomposition into generalized eigenspaces corresponding to the distinct roots of the polynomial.

Now the roots of $(x^n - 1)(x^{n+1} - 1)$ are the n different n-th roots of unity and the $n + 1$ different roots of unity of order dividing $n + 1$. Since n and $n + 1$ are relatively prime, these two sets of roots have only $x = 1$ in common. Thus 1 is a double root and all the other roots are simple.

Consequently Ω^n decomposes into the direct sum of the generalized eigenspace $\mathrm{Ker}\,(1 - \kappa)^2$ corresponding to the eigenvalue $\zeta = 1$ and the ordinary eigenspaces $\mathrm{Ker}\,(\kappa - \zeta)$ for each root of unity $\zeta \neq 1$ of order dividing n or $n + 1$.

Combining this for all n we obtain the following spectral decomposition with respect to κ

$$\Omega = \mathrm{Ker}\,(\kappa - 1)^2 \oplus \bigoplus_{\zeta} \mathrm{Ker}\,(\kappa - \zeta) \tag{1}$$

where ζ runs over all roots of unity $\neq 1$. Each of these generalized eigenspaces is stable under any operator commuting with κ.

The harmonic projection P and Green's operator G

Lumping the eigenvalues $\zeta \neq 1$ together, we have

$$\Omega = \mathrm{Ker}\,(1 - \kappa)^2 \oplus \mathrm{Im}\,(1 - \kappa)^2 \tag{2}$$

which is the unique decomposition stable under κ such that $1 - \kappa$ is nilpotent on the first factor and invertible on the second. The generalized nullspace $\mathrm{Ker}\,(1 - \kappa)^2$ is analogous to the space of harmonic forms. We note, however, that in contrast to the Riemannian manifold situation, the Laplacian $1 - \kappa$ and the operators b, d are in general nonzero on this space.

We call (2) the *harmonic decomposition* of Ω. Define the *harmonic projection* P to be the projection operator which is one on the first factor and zero on the second; it is the *spectral projection* for κ associated to the eigenvalue 1. The harmonic decomposition can then be written

$$\Omega = P\Omega \oplus P^{\perp}\Omega \tag{3}$$

where $P^{\perp} = 1 - P$ is the spectral projection for κ associated to the set of eigenvalues $\neq 1$. This decomposition is stable under any operator commuting with κ, for instance b, d, B.

On the complementary space $P^{\perp}\Omega$ the operator $1 - \kappa = bd + db$ is both invertible and homotopic to zero with respect to either differential b or d, hence the homology with respect to either of these differentials vanishes. We can make this explicit by introducing the *Green's operator* G for $1 - \kappa$, which is defined to be the inverse of $1 - \kappa$ on $P^{\perp}\Omega$ and 0 on $P\Omega$.

The operators P, G clearly commute with any operator commuting with κ, hence P and G commute with κ, b, d, B and each other. One has $PG = GP = 0$ and

$$P^{\perp} = G(1 - \kappa) = G(bd + db) \tag{4}$$

As in the theory of harmonic forms the complementary space $P^{\perp}\Omega$ to the harmonic forms then splits into subspaces of 'exact' and 'coexact' forms.

Proposition 2.1: *One has $P^{\perp}\Omega = dP^{\perp}\Omega \oplus bP^{\perp}\Omega$, where d maps $bP^{\perp}\Omega$ isomorphically onto $dP^{\perp}\Omega$ with inverse Gb, and where b maps $dP^{\perp}\Omega$ isomorphically onto $bP^{\perp}\Omega$ with inverse Gd. Consequently $H_*(P^{\perp}\Omega, b) = H_*(P^{\perp}\Omega, d) = 0$, and $P\Omega$ and Ω have the same b-homology and d-homology.*

Proof: This is a formal consequence of the identity $1 = G(bd + db)$ on $P^{\perp}\Omega$ and the fact that G commutes with b and d. Thus $(Gdb)d = G(bd + db)d = d$ implies Gdb is a projector with image $dP^{\perp}\Omega$. Similarly Gbd is a projector with image $bP^{\perp}\Omega$, and as these projectors sum to 1, we obtain the desired decomposition. As $(Gd)b = Gdb$ and $b(Gd) = Gbd$, we see b and Gd are inverse isomorphisms between $dP^{\perp}\Omega$ and $bP^{\perp}\Omega$; similarly for the other case.

We next exploit the fact that κ is of finite order on $d\Omega$ in each degree by §1 (18). The general principle is that on any κ-invariant subspace (more generally subquotient space) on which κ has finite order, the projection P is given by averaging over the cyclic group generated by κ.

We use this first to derive a useful characterization of elements of $P\Omega$.

Proposition 2.2: *One has $\omega \in P\Omega$ iff $d\omega$ and $db\omega$ are κ-invariant.*

Proof: As κ has finite order on $d\Omega$ in each degree, the image of P on $d\Omega$ is the fixpoint subspace for κ. Thus the implication $\Rightarrow$ is clear. Conversely, if $d\omega$ and $db\omega$ are fixed under κ, then they are in the image of P, hence killed by G, so $\omega - P\omega = bGd\omega + Gdb\omega = 0$.

We now would like to describe P and G in a more explicit way using the fact that κ has finite order on $d\Omega$ in each degree.

Let us consider more generally an operator T on a vector space V satisfying $T^m = 1$, where $m \geq 1$. Set

$$P_T = \frac{1}{m}\sum_{i=0}^{m-1} T^i \qquad G_T = \frac{1}{m}\sum_{i=0}^{m-1}\left(\frac{m-1}{2} - i\right) T^i \tag{5}$$

Then T, P_T, G_T commute, and we have the identities

$$\begin{aligned} TP_T &= P_T & P_T^2 &= P_T \\ (1-T)G_T &= 1 - P_T & P_T G_T &= 0 \end{aligned} \tag{6}$$

It follows that P_T is a projection operator whose image is the fixpoint subspace $V^T = \operatorname{Ker}(1-T)$ and whose kernel is $(1-T)V$. Thus $V = V^T \oplus (1-T)V$. Also G_T is zero on V^T and the inverse of $1-T$ on $(1-T)V$. Thus P_T is the spectral projection for T associated to eigenvalue 1, and G_T is the Green's operator for the operator $1-T$. This implies in particular that the sums in (5) give the same operators for any multiple m of the order of T.

Applying this to κ on $d\Omega^n$, we obtain from §1 (2) and §1 (18) the identifications

$$\begin{gathered} d\Omega^n \cong \bar{A}^{\otimes n+1} \\ da_0 \cdots da_n \leftrightarrow (a_0, \ldots, a_n) \\ \kappa \leftrightarrow \lambda \\ P \leftrightarrow P_\lambda \\ G \leftrightarrow G_\lambda \end{gathered} \tag{7}$$

where P_λ, G_λ are given by (5) with $m = n+1$. Consequently we have

$$\begin{aligned} Pd\Omega^n &\cong P_\lambda \bar{A}^{\otimes n+1} = (\bar{A}^{\otimes n+1})^\lambda \\ P^\perp d\Omega^n &\cong P_\lambda^\perp \bar{A}^{\otimes n+1} = (1-\lambda)\bar{A}^{\otimes n+1} \end{aligned} \tag{8}$$

and on Ω^n we have

$$Pd = \frac{1}{n+1}\sum_{i=0}^{n} \kappa^i d = \frac{1}{n+1} B \tag{9}$$

$$Gd = \frac{1}{n+1}\sum_{i=0}^{n}\left(\frac{n}{2} - i\right)\kappa^i d \tag{10}$$

The first formula may be written

$$B = NPd \tag{11}$$

where N is the degree operator on Ω, i.e. multiplication by n on Ω^n. The second formula combined with $P^\perp = (Gd)b + b(Gd)$ yields an explicit formula for P on Ω^n. A similar sort of formula for G can be obtained using $G = (G^2d)b + b(G^2d)$.

From the splitting 2.1 in degree n we obtain a canonical isomorphism

$$\begin{aligned} P^{\perp}\Omega^{n} &= dP^{\perp}\Omega^{n-1} \oplus bP^{\perp}\Omega^{n+1} \\ &= dP^{\perp}\Omega^{n-1} \oplus b(dP^{\perp}\Omega^{n}) \\ &\cong (1-\lambda)\bar{A}^{\otimes n} \oplus (1-\lambda)\bar{A}^{\otimes n+1} \end{aligned}$$

Another way to understand this splitting is via eigenvalues. The first (resp. second) summand on the right is the sum of the κ-eigenspaces for the eigenvalues $\zeta \neq 1$ satisfying $\zeta^n = 1$ (resp. $\zeta^{n+1} = 1$).

We can also describe the eigenspace Ω_ζ of κ on Ω with eigenvalue ζ, where ζ is a root of unity $\neq 1$. If $m \geq 2$ is the order of ζ, then it is clear that $\Omega_\zeta^n \neq 0$ only in degrees $n \equiv 0, -1 \bmod m$. Moreover b is an isomorphism from degree km to degree $km-1$, and d is an isomorphism in the other direction, so that Ω_ζ has the form

$$\underset{0}{\overset{0}{\leftrightarrows}}\ \Omega_\zeta^{km-1}\ \underset{d}{\overset{b}{\leftrightarrows}}\ \Omega_\zeta^{km}\ \underset{0}{\overset{0}{\leftrightarrows}}$$

where bd and db here are both equal to multiplication by $1-\zeta$. For instance, on the eigenspace Ω_{-1} where $\kappa = -1$, the operator b is an isomorphism from even to odd with inverse given by $\frac{1}{2}d$.

Since the Laplacian $1-\kappa = [b,d]$ is nonzero in general on the harmonic subspace $P\Omega$, the following may be of interest. If we replace d by the rescaled operator Nd, then the corresponding Laplacian $L = [b, Nd]$ is zero on $P\Omega$ by (11), as $[b, B] = 0$. Moreover, $L = km(1-\zeta)$ on Ω_ζ^{km-1}, Ω_ζ^{km} above, so that L is invertible on $P^{\perp}\Omega$. Thus one has $P\Omega = \operatorname{Ker} L$, $P^{\perp}\Omega = \operatorname{Im} L$. In fact Ω splits into ordinary eigenspaces for L, and this decomposition refines the spectral decomposition associated to κ.

The reduced cyclic complex and B on $P\Omega$

We next derive the consequences of the formula $B = NPd$. In the first place this implies that

$$B(P^{\perp}\Omega) = 0 \tag{12}$$

Secondly, because B and d agree up to nonzero scalars on $P\Omega$, we have

$$H_n(P\Omega, B) = H_n(P\Omega, d) = H_n(\Omega, d) = \begin{cases} \mathbb{C} & n = 0 \\ 0 & n \neq 0 \end{cases} \tag{13}$$

by §1 (4) and 2.1.

We now consider Ω as a complex with differential b and denote by $\operatorname{Ker} B$, $\operatorname{Im} B$ the kernel and image of B on Ω. These are subcomplexes of Ω as b, B anticommute.

From (12) and (13) we obtain

$$\operatorname{Ker} B/\operatorname{Im} B = \mathbb{C} \oplus P^{\perp}\Omega \tag{14}$$

where $\mathbb{C}$ here means the complex with $\mathbb{C}$ in degree zero and 0 elsewhere. Since $P^{\perp}\Omega$ is b-acyclic this yields

$$H_n(\operatorname{Ker} B/\operatorname{Im} B, b) = \begin{cases} \mathbb{C} & n = 0 \\ 0 & n \neq 0 \end{cases} \tag{15}$$

Let us define the *reduced cyclic complex* $\bar{C}^{\lambda} = \bar{C}^{\lambda}(A)$ to be the quotient complex

$$\bar{C}^{\lambda} = \Omega/\operatorname{Ker} B \tag{16}$$

Thus $\bar{C}^{\lambda}_n$ is the quotient $\bar{A}^{\otimes n+1}/(1-\lambda)\bar{A}^{\otimes n+1}$ of $\Omega^n = A \otimes \bar{A}^{\otimes n}$.

From (14) we get $\operatorname{Ker} B = \mathbb{C} \oplus \operatorname{Im} B \oplus P^{\perp}\Omega$, hence $\bar{C}^{\lambda} = P\Omega/(\mathbb{C} \oplus \operatorname{Im} B)$. On the other hand, B induces an isomorphism of complexes $\bar{C}^{\lambda}[1] \tilde{\to} \operatorname{Im} B$ by (16), where the suspension (§3 (1)) occurs because B has degree $+1$.

It follows that we have an exact sequence of complexes

$$0 \longrightarrow \mathbb{C} \oplus \bar{C}^{\lambda}[1] \xrightarrow{i} P\Omega \xrightarrow{j} \bar{C}^{\lambda} \longrightarrow 0 \tag{17}$$

where j is the canonical surjection, i on $\mathbb{C}$ is the obvious inclusion, and i on $\bar{C}^{\lambda}[1]$ is induced by B in the sense that $ij = B$.

3 Mixed Complexes

By a complex we shall mean a complex of vector spaces $X = \bigoplus_{n \in \mathbb{Z}} X_n$ with differential of degree -1. Recall that the *k-th suspension* $X[k]$ of X is defined by

$$X[k]_n = X_{n-k} \qquad d_{X[k]} = (-1)^k d_X \tag{1}$$

A *mixed complex* ([7]) is a complex M with differential usually denoted b, which is bounded below and equipped with an operator B of degree $+1$ such that $B^2 = 0$ and $[b, B] = 0$. A (DG) *S-module* ([8]) is a complex, which is bounded below and equipped with an operator S of degree -2 commuting with the differential.

Given a mixed complex M there is an associated S-module $\mathcal{B}(M)$, which is the total complex of the double complex given by M_{q-p} in bidegree (p, q) for $p \geq 0$, where the vertical and horizontal differentials are b and B respectively. Elements of $\mathcal{B}(M)_n$ may be identified with polynomials $\sum_{p \geq 0} u^p m_{n-2p}$ where $m_{n-2p} \in M_{n-2p}$. With this notation S is given by $S(u^0 m_k) = 0$ and $S(u^p m_k) = u^{p-1} m_k$ for $p \geq 1$, and the differential is $d = b + SB$.

One has an exact sequence of complexes

$$0 \to M \xrightarrow{I} \mathcal{B}(M) \xrightarrow{S} \mathcal{B}(M)[2] \to 0 \tag{2}$$

with I the inclusion of the column $p = 0$. Defining the *Hochschild and cyclic homology* of M by

$$HH_n(M) = H_n(M) \qquad HC_n(M) = H_n(\mathcal{B}(M)) \tag{3}$$

respectively, we obtain the Connes exact sequence

$$\to HH_{n+2}(M) \xrightarrow{I} HC_{n+2}(M) \xrightarrow{S} HC_n(M) \xrightarrow{B} HH_{n+1}(M) \to \quad (4)$$

where the arrow B here, i.e. the connecting map, is induced by $\sum_{p\geq 0} u^p m_{n-2p} \mapsto Bm_n$.

Using the Connes exact sequence the *de Rham homology* of M is defined by

$$\begin{aligned} HD_n(M) &= \mathrm{Im}\{S : HC_{n+2}(M) \to HC_n(M)\} \\ &= \mathrm{Ker}\{B : HC_n(M) \to HH_{n+1}(M)\} \end{aligned} \quad (5)$$

This terminology is suggested by the Connes-Karoubi theorem (cf. 4.2).

Besides these types of homology we also want to consider the $b+B$ homology $H_\nu(M, b+B)$, $\nu \in \mathbb{Z}/2$, where M is considered as a supercomplex with the even-odd grading and the differential $b+B$. In general this homology is different from the *periodic cyclic homology*

$$HP_\nu(M) \;=\; H_\nu\Big(\prod_n M_n, b+B\Big) \quad (6)$$

We next examine mixed complexes which are B-*acyclic*, i.e. $\mathrm{Ker}\, B = \mathrm{Im}\, B$, and those satisfying a weaker condition playing an important role in the work of Connes. We will say that a mixed complex has *Connes' property* when $\mathrm{Ker}\, B/\mathrm{Im}\, B$ is b-acyclic. When formulated dually his key lemma (*[1]*, p. 337, 36) says that the mixed complex $C(A)$ of chains on A has this property (cf. §6). On the other hand ΩA does not have Connes' property by §2 (15).

The next result generalizes specific applications of this lemma.

Proposition 3.1: *When M has Connes' property one has canonical isomorphisms*:
1) $HC_n(M) = H_n(M/\mathrm{Im}\, B, b) = H_n(M/\mathrm{Ker}\, B, b)$.
2) $H_\nu(M, b+B) = 0$.

Proof: The increasing row filtration of the b, B bicomplex gives rise to a spectral sequence abutting to $H_n(\mathcal{B}(M)) = HC_n(M)$, whose E^2 term is the vertical homology of the horizontal homology. By hypothesis E^2 vanishes except along the 0-th column, so we conclude that the edge map $\mathcal{B}(M) \to M/\mathrm{Im}\, B$ is a quasi-isomorphism of complexes. The hypothesis also implies that the obvious surjection $M/\mathrm{Im}\, B \to M/\mathrm{Ker}\, B$ is a quasi-isomorphism. This proves 1).

Consider next the periodic b, B bicomplex with M_{q-p} in bidegree (p, q) for all p, q. The spectral sequence arising from the increasing row filtration abuts to the homology of the uncompleted total complex, i.e. consisting of elements of finite support. By periodicity the homology of this total complex in degree n is $H_{n+2\mathbb{Z}}(M, b+B)$. As $E^2 = 0$ by the hypothesis, this homology vanishes, proving 2).

Proposition 3.2: *When M is B-acyclic one has*
1) $HC_n(M) = H_n(M/BM, b)$.

2) $H_\nu(M, b+B) = 0$.
3) $HD_n(M) = H_n(M/bM, B)$.

Proof: As the first two formulas are included in 3.1, it remains to prove the third. The hypothesis implies there is an exact sequence of complexes

$$0 \to (M/BM)[1] \xrightarrow{i} M \xrightarrow{j} M/BM \to 0 \tag{7}$$

where j is the canonical surjection and i is induced by B in the sense that $ij = B$. The corresponding long exact sequence in homology is the Connes exact sequence with the opposite sign on the S arrow. More precisely, letting ∂ be the connecting map on homology arising from this exact sequence, we have isomorphisms

$$\begin{array}{ccccccc} H_n(M) & \xrightarrow{I} & H_n(\mathcal{B}(M)) & \xrightarrow{S} & H_{n-2}(\mathcal{B}(M)) & \xrightarrow{B} & H_{n-1}(M) \\ \| & (+) & \downarrow \cong & (-) & \downarrow \cong & (+) & \| \\ H_n(M) & \xrightarrow{j} & H_n(M/BM) & \xrightarrow{\partial} & H_{n-2}(M/BM) & \xrightarrow{i} & H_{n-1}(M) \end{array} \tag{8}$$

such that the middle square anticommutes and the others commute.

It follows that we have

$$HD_n(M) = \operatorname{Ker}\{H_n(M/BM, b) \xrightarrow{i} H_{n+1}(M, b)\} \tag{9}$$

where i is induced by the degree one map of complexes $B : M/BM \to M$.

Consequently we have a diagram of exact sequences

$$\begin{array}{ccccccc} & & 0 & & 0 & & \\ & & \downarrow & & \downarrow & & \\ & & HD_n(M) & & H_n(M/bM, B) & & 0 \\ & & \downarrow & & \downarrow & & \downarrow \\ 0 & \to & H_n(M/BM, b) & \to & M_n \big/ BM_{n-1} + bM_{n+1} & \xrightarrow{b} & M_{n-1}/BM_{n-2} \\ & & \downarrow i & & \downarrow B & & \downarrow -B \\ 0 & \to & H_{n+1}(M, b) & \to & M_{n+1}/bM_{n+2} & \xrightarrow{b} & M_n \end{array}$$

such that the arrow $-B$ at the right is injective. The third formula is then clear.

We next use the exact sequence (7) to describe B-acyclic mixed complexes in terms of S-modules.

Let Λ be the DG algebra $\mathbb{C} \oplus \mathbb{C}B$, where B has degree one, boundary zero and square zero. Notice that a mixed complex is the same as a DG module over Λ.

By *[5]*, 1.2 the functor $M \mapsto \mathcal{B}(M)$ from mixed complexes to S-modules has a left adjoint $Q \mapsto \Lambda \tilde{\otimes} Q$. The mixed complex $\Lambda \tilde{\otimes} Q$ is the free graded Λ-module $\Lambda \otimes Q$ equipped with a twisted differential, specifically

$$\begin{aligned} \Lambda \tilde{\otimes} Q &= 1 \otimes Q \, \oplus \, B \otimes Q \\ b(1 \otimes x + B \otimes y) &= 1 \otimes dx \, - \, B \otimes (Sx + dy) \\ B(1 \otimes x + B \otimes y) &= B \otimes x \end{aligned} \tag{10}$$

where d is the differential in Q.

It is clear that $\Lambda \tilde{\otimes} Q$ is B-acyclic. Conversely, given a B-acyclic mixed complex M, we can choose a splitting of (7) which is compatible with the grading but not necessarily with the differential b. Let $l : M/BM \to M$ be the section of j corresponding to the splitting. Then the formula

$$-iS \; = \; [b, l] \tag{11}$$

defines an operator S of degree -2 on M/BM commuting with b, thereby making M/BM an S-module. By the anticommuting square in (8), this operator S induces the Connes S operator on the cyclic homology.

Moreover we have an isomorphism

$$\Lambda \tilde{\otimes} (M/BM) \; \tilde{\to} \; M \tag{12}$$

given by $1 \otimes x \mapsto l(x)$, $B \otimes x \mapsto i(x)$. Using (11) and the fact that $Bl(x) = i(x)$ by definition of i, one checks easily that this isomorphism is compatible with the mixed complex structure.

Another lifting for j has the form $l' = l + iv$, where v can be any operator of degree -1 on M/BM. Changing l to l' changes the operator S on M/BM to $S' = S - [b, v]$, so we see S' can be any operator of degree -2 which is homotopic to S. But over a field the homotopy class of a map of complexes is determined by the induced map on homology, so we have proved

Proposition 3.3: *A B-acyclic mixed complex M is isomorphic to $\Lambda \tilde{\otimes} Q$, where Q is the S-module given by the complex M/BM equipped with any degree -2 operator inducing the Connes S-operator on $HC_*(M) = H_*(M/BM)$.*

It follows that a B-acyclic mixed complex is described up to isomorphism by a complex equipped with a homotopy class of endomorphisms of degree -2.

We conclude this section with some comments about homotopy equivalence and quasi-isomorphism for mixed complexes.

Two maps of mixed complexes $f, g : M \to M'$ (i.e. of degree 0 commuting with b, B) are said to be homotopic when $f - g = [b, h]$ for some linear $h : M \to M'$ of degree 1 anticommuting with B. Using this homotopy relation on maps one defines homotopy equivalent mixed complexes and contractible mixed complex in the obvious way.

For example, in the harmonic decomposition $\Omega = P\Omega \oplus P^{\perp}\Omega$ the mixed complex $P^{\perp}\bar{\Omega}$ is contractible, since the operator Gd on it satisfies $[b, Gd] = 1$ and $BGd = GdB = 0$. Consequently the mixed complexes Ω and $P\Omega$ are homotopy equivalent. In fact Gd is a special contraction on $P^{\perp}\bar{\Omega}$ (i.e. $(Gd)^2 = 0$), and the operators P, Gd give a special deformation retraction (*[9]*, 1.1) of Ω onto $P\Omega$.

Homotopic maps of mixed complexes induce homotopic maps between the associated S-modules, hence they induce the same map on all cyclic type homology and cohomology, e.g. Hochschild, cyclic, de Rham, and periodic cyclic homology, and also bivariant cyclic cohomology (which includes negative cyclic homology and cyclic cohomology). They also induce homotopic maps on the associated complexes $\operatorname{Ker} B$, $\operatorname{Im} B$, $M/\operatorname{Ker} B$, etc. as well as homotopic maps on the supercomplexes with differential $b + B$.

Thus homology such as $H_n(M/\operatorname{Ker} B, b)$, $H_n(\operatorname{Ker} B/\operatorname{Im} B, b)$, and $H_\nu(M, b+B)$ is homotopy invariant. In particular a mixed complex which is homotopy equivalent to one with Connes' property also has this property. In fact, it turns out (cf. 6.2) that a mixed complex has Connes' property iff it is homotopy equivalent to a B-acyclic mixed complex.

A map f of mixed complexes is called a *quasi-isomorphism* when the map of underlying complexes is a quasi-isomorphism, i.e. when f induces an isomorphism on Hochschild homology. Since we are over a field f is a quasi-isomorphism iff it is a homotopy equivalence of complexes ignoring B.

A quasi-isomorphism of mixed complexes induces isomorphisms on all cyclic type homology and cohomology. For cyclic type homology this is easily seen from the definitions via bicomplexes, and for bivariant cyclic cohomology it is *[5]*, 4.6.

On the other hand, a quasi-isomorphism need not induce isomorphisms on homology of the form $H_n(M/\operatorname{Ker} B, b)$ and $H_\nu(M, b+B)$. To see this we use the fact that for any mixed complex M there is a surjective quasi-isomorphism $L \to M$ such that L is B-acyclic; one can take L to be the bar resolution $\Lambda \tilde{\otimes} \mathcal{B}(M)$, where the map $L \to M$ is the canonical adjunction map (*[5]*, 1.2). In the case where M is $\mathbb{C}$ in degree zero and zero elsewhere, the homology in question is different for L and M.

4 $\bar{\Omega}A$ and Reduced Cyclic Homology

We next give some applications of the ideas discussed so far. Because ΩA does not have Connes' property (§2 (15)), we consider the reduced space $\bar{\Omega}A$, which is the same as ΩA except in degree zero, where $\bar{\Omega}^0 A = \bar{A}$. It may also be described as the cokernel of the map $\mathbb{C} = \Omega\mathbb{C} \subset \Omega A$ induced by the inclusion of scalars $\mathbb{C} \subset A$, so the canonical operators d, b, κ, etc. on ΩA descend to $\bar{\Omega}A$. In particular, $\bar{\Omega}A$ equipped with b, B is a mixed complex, whose Hochschild and cyclic homology are by definition (*[12]*, §4) the *reduced Hochschild homology* $\bar{H}H_n(A)$ and *reduced cyclic homology* $\bar{H}C_n(A)$ of the algebra A.

Abbreviating as before, we put $\Omega = \Omega A$, $\bar{\Omega} = \bar{\Omega}A$.

As in the case of Ω we have the harmonic decomposition

$$\bar{\Omega} = P\bar{\Omega} \oplus P^{\perp}\bar{\Omega} \tag{1}$$

compatible with the operators b, d, etc. Note that $P\bar{\Omega} = P\Omega/\mathbb{C}$ and $P^{\perp}\bar{\Omega} = P^{\perp}\Omega$, since $\kappa = 1$ in degree zero.

The advantage of $\bar{\Omega}$ over Ω is the fact that the d-homology of $\bar{\Omega}$ is zero. Consequently the formulas §2 (13)-(17) pertaining to Ω also hold in the case of $\bar{\Omega}$ with $\mathbb{C}$ replaced by zero. We thus obtain

Proposition 4.1: 1) *On $P^{\perp}\bar{\Omega}$ one has $B = 0$ and $[b, Gd] = 1$. Thus $P^{\perp}\bar{\Omega}$ is a contractible mixed complex.*
2) *The mixed complex $P\bar{\Omega}$ is B-acyclic, and $P\bar{\Omega}/B(P\bar{\Omega})$ equipped with differential b is canonically isomorphic to the reduced cyclic complex $\bar{C}^{\lambda}$. Hence one has an exact sequence of complexes*

$$0 \to \bar{C}^{\lambda}[1] \xrightarrow{i} P\bar{\Omega} \xrightarrow{j} \bar{C}^{\lambda} \to 0 \tag{2}$$

such that $ij = B$.
3) *The mixed complex $\bar{\Omega}$ has Connes' property*: $H_*(\mathrm{Ker}\, B/\mathrm{Im}\, B, b) = 0$.

We now give three applications.

First we deduce the fact (*[12]*, 4.4) that reduced cyclic homology is computed by the reduced cyclic complex:

$$\bar{HC}_n(A) = H_n(\bar{C}^{\lambda}) \tag{3}$$

This follows from 3.1 1) and the fact that $\bar{\Omega}/\mathrm{Ker}\, B = \bar{C}^{\lambda}$.

Next we compute some $b+B$ homology. Applying 3.1 2) to $\bar{\Omega}$ and $P\bar{\Omega}$ we get

$$H_{\nu}(\bar{\Omega}, b+B) = H_{\nu}(P\bar{\Omega}, b+B) = 0 \tag{4}$$

Then using the long exact sequences arising from $\Omega/\mathbb{C} = \bar{\Omega}$ and $P\Omega/\mathbb{C} = P\bar{\Omega}$ we get

$$H_{\nu}(\Omega, b+B) = H_{\nu}(P\Omega, b+B) = \begin{cases} \mathbb{C} & \nu \text{ even} \\ 0 & \nu \text{ odd} \end{cases} \tag{5}$$

Finally we derive the Connes-Karoubi theorem reducing noncommutative de Rham homology to cyclic homology (*[1]*, p. 334, Theorem 33; *[6]*, 2.15).

Let

$$\bar{\Omega}_{\natural} = \Omega \big/ \mathbb{C} + [\Omega, \Omega]$$

be the reduced supercommutator quotient space of Ω. This is a complex with differential d, whose homology is by definition the reduced noncommutative de Rham homology of A.

The next result is the reduced version of the Connes-Karoubi theorem. By applying it to the algebra obtained by adjoining an identity one obtains the unreduced version §5 (10).

Proposition 4.2: *One has* $H_n(\bar{\Omega}_\natural, d) = \mathrm{Ker}\{B : \bar{H}C_n(A) \to \bar{H}H_{n+1}(A)\}$.

Proof: Since Ω is generated by A and dA, the supercommutator subspace is

$$[\Omega, \Omega] \;=\; [\Omega, A] + [\Omega, dA] \;=\; b\Omega + (1 - \kappa)\Omega$$

where $[\Omega, A] = b\Omega$ and $[\Omega, dA] = (1 - \kappa)\Omega$ by §1 (11) and §1 (16). Thus

$$\Omega \big/ \mathbb{C} + [\Omega, \Omega] \;=\; \bar{\Omega} \big/ b\bar{\Omega} + (1 - \kappa)\bar{\Omega}$$

As $\kappa^n = 1 + b\kappa^n d$ on Ω^n, we see that κ has order dividing n on $\bar{\Omega}/b\bar{\Omega}$ in degree n, hence dividing this space by the image of $1 - \kappa$ is the same as applying P. Thus we have

$$\bar{\Omega}_\natural \;=\; P(\bar{\Omega}/b\bar{\Omega}) \;=\; P\bar{\Omega}/bP\bar{\Omega}$$

Since the differential d on $P\bar{\Omega}$ agrees with B up to nonzero numerical factors, we have

$$H_n(\bar{\Omega}_\natural, d) \;=\; H_n(P\bar{\Omega}/bP\bar{\Omega}, B)$$

and the result follows from 3.2 3).

5 Nonunital Algebras

So far we have considered a unital algebra A, and our constructions ΩA, $\bar{\Omega} A$, as well as the associated operators and homology are functorial over the category of unital algebras. We now carry over our investigations to nonunital algebras $\mathcal{A}$ using the augmented algebra $\tilde{\mathcal{A}} = \mathbb{C} \oplus \mathcal{A}$ obtained by adjoining an identity.

An augmented algebra is by definition a unital algebra A equipped with a unital algebra homomorphism $\epsilon : A \to \mathbb{C}$ called the augmentation. In this situation we may identify $\bar{A} = A/\mathbb{C}$ with the augmentation ideal $\mathrm{Ker}(\epsilon)$, whence $\bar{A}$ becomes a nonunital algebra in a natural way such that

$$A \;=\; \bar{A}^{\sim} \;=\; \mathbb{C} \oplus \bar{A}$$

The functors $\mathcal{A} \mapsto \tilde{\mathcal{A}}$, $A \mapsto \bar{A}$ give an equivalence between the categories of nonunital algebras and augmented algebras.

Let $\mathcal{A}$ be a fixed nonunital algebra. The augmentation $\tilde{\mathcal{A}} \to \mathbb{C}$ induces an augmentation $\Omega\tilde{\mathcal{A}} \to \Omega\mathbb{C} = \mathbb{C}$, making $\Omega\tilde{\mathcal{A}}$ into an augmented DG algebra whose augmentation ideal we write $\bar{\Omega}\tilde{\mathcal{A}}$. Thus $\bar{\Omega}\tilde{\mathcal{A}}$ is the nonunital DG algebra such that

$$\Omega\tilde{\mathcal{A}} \;=\; \left(\bar{\Omega}\tilde{\mathcal{A}}\right)^{\sim} \;=\; \mathbb{C} \oplus \bar{\Omega}\tilde{\mathcal{A}} \qquad (1)$$

and it is closed under the operators b, d, κ, B, P, G on $\Omega\tilde{\mathcal{A}}$.

From the universal property of $\Omega\tilde{\mathcal{A}}$ one sees easily that $\bar{\Omega}\tilde{\mathcal{A}}$ is the universal nonunital DG algebra generated by $\mathcal{A}$ in the obvious sense. Thus $\bar{\Omega}\tilde{\mathcal{A}}$ is the universal algebra $\Omega(\mathcal{A})$ used by Connes [C1]; it is the differential envelope of $\mathcal{A}$ in the terminology of Kastler (*[10]*).

Let $C = C(\mathcal{A})$ be the graded vector space of (unnormalized) chains on $\mathcal{A}$ defined by $C_n(\mathcal{A}) = \mathcal{A}^{\otimes n+1}$ for $n \geq 0$ and $C_n(\mathcal{A}) = 0$ for $n < 0$. Since $\Omega^n \tilde{\mathcal{A}} = \tilde{\mathcal{A}} \otimes \mathcal{A}^{\otimes n}$ there are isomorphisms

$$\begin{array}{ccccc} C_n & \oplus & C_{n-1} & \xrightarrow{\sim} & \bar{\Omega}^n \tilde{\mathcal{A}} \\ (a_0, \ldots, a_n) & & & \mapsto & a_0 da_1 \cdots da_n \\ & & (a_1, \ldots, a_n) & \mapsto & da_1 \cdots da_n \end{array} \tag{2}$$

for all n, which allow us to identify elements of $\bar{\Omega}^n \tilde{\mathcal{A}}$ with column vectors $\begin{pmatrix} x \\ y \end{pmatrix}$ such that $x \in C_n$, $y \in C_{n-1}$. In terms of this identification operators on $\bar{\Omega}\tilde{\mathcal{A}}$ are described by 2×2 matrices of operators on C.

We next give the matrices corresponding to the operators b, d, κ, B, P, G. Among the possible operators on $\bar{\Omega}\tilde{\mathcal{A}}$ one might consider, these have the property of making sense more generally on $\bar{\Omega}A$ for any unital algebra A, not just an augmented algebra. However, the extra structure given by the augmentation on $\tilde{\mathcal{A}}$ leads to other interesting operators, for instance those related to the explicit S operator on the cyclic complex of $\mathcal{A}$. We refer to Kastler *[10]* for an extensive study of operators on $\bar{\Omega}\tilde{\mathcal{A}}$[1].

On C we have the operators b', b, c, λ, where b', b, λ are given by the formulas (7), (10), (17) of §1 with $a_i \in \mathcal{A}$, and where $c = b - b'$ is the crossover term. Associated with λ we have the operators P_λ, G_λ as in §2 (5), and N_λ defined by

$$N_\lambda = \sum_0^n \lambda^i = (n+1) P_\lambda \quad \text{on} \quad \mathcal{A}^{\otimes n+1} \tag{3}$$

Let us write $\tilde{b}, \tilde{B}$ for the operators b, B on $\bar{\Omega}\tilde{\mathcal{A}}$ in order to distinguish them from the b operator on C and the B operator on C to be defined below in §6.

Then with this notation it is straightforward to obtain the following formulas

[1]Note added in proof. The operator κ in the case of $\bar{\Omega}\tilde{\mathcal{A}}$ appears already in Kastler's paper *[15]*.

from the definitions.

$$
\begin{aligned}
d &= \begin{pmatrix} 0 & 0 \\ 1 & 0 \end{pmatrix} & \tilde{b} &= \begin{pmatrix} b & 1-\lambda \\ 0 & -b' \end{pmatrix} \\
\kappa &= \begin{pmatrix} \lambda & 0 \\ -c & \lambda \end{pmatrix} & \kappa^j d &= \begin{pmatrix} 0 & 0 \\ \lambda^j & 0 \end{pmatrix} \\
\tilde{B} &= \begin{pmatrix} 0 & 0 \\ N_\lambda & 0 \end{pmatrix} & Gd &= \begin{pmatrix} 0 & 0 \\ G_\lambda & 0 \end{pmatrix} \\
\tilde{b}Gd &= \begin{pmatrix} P_\lambda^\perp & 0 \\ -b'G_\lambda & 0 \end{pmatrix} & Gd\tilde{b} &= \begin{pmatrix} 0 & 0 \\ G_\lambda b & P_\lambda^\perp \end{pmatrix} \\
P &= \begin{pmatrix} P_\lambda & 0 \\ b'G_\lambda - G_\lambda b & P_\lambda \end{pmatrix} & G &= \begin{pmatrix} G_\lambda & 0 \\ G_\lambda^2 b - b'G_\lambda^2 & G_\lambda \end{pmatrix}
\end{aligned}
\tag{4}
$$

The last pair have been obtained using $1 - P = \tilde{b}Gd + Gd\tilde{b}$ and $G = G^2 d\tilde{b} + \tilde{b}G^2 d$.

Notice that the identities $\tilde{b}^2 = 0$ and $\tilde{b}\tilde{B} + \tilde{B}\tilde{b} = 0$ on $\bar{\Omega}\tilde{\mathcal{A}}$ are equivalent to the Connes-Tsygan identities

$$
\begin{aligned}
b^2 &= b'^2 = 0 \\
b(1-\lambda) &= (1-\lambda)b' \\
b'N_\lambda &= N_\lambda b
\end{aligned}
\tag{5}
$$

for the operators on C.

As a check on the above calculations, notice that the matrices for $\tilde{b}Gd$ and $Gd\tilde{b}$ clearly give idempotent operators such that $(\tilde{b}Gd)(Gd\tilde{b}) = 0$. For $(Gd\tilde{b})(\tilde{b}Gd) = 0$, we need $G_\lambda b P_\lambda^\perp = P_\lambda^\perp b' G_\lambda$. But the former is $G_\lambda b(1-\lambda)G_\lambda$, the latter is $G_\lambda(1-\lambda)b'G_\lambda$, and these agree by (5). Thus the matrices for $\tilde{b}Gd$ and $Gd\tilde{b}$ give mutually annihilating idempotents, so the matrix for $P = 1 - \tilde{b}Gd - Gd\tilde{b}$ is idempotent and annihilates these two idempotents.

Having described our operators on $\bar{\Omega}\tilde{\mathcal{A}}$ in terms of operators on C, we turn next to the homology associated to the mixed complex $(\bar{\Omega}\tilde{\mathcal{A}}, \tilde{b}, \tilde{B})$.

Recall that the cyclic homology $HC_n(\mathcal{A})$ of $\mathcal{A}$ is by definition ([LQ]) the homology of the total complex T of the cyclic bicomplex:

$$\begin{array}{ccccccc}
\downarrow{\scriptstyle b} & & \downarrow{\scriptstyle -b'} & & \downarrow{\scriptstyle b} & & \\
C_2 & \xleftarrow{1-\lambda} & C_2 & \xleftarrow{N_\lambda} & C_2 & \xleftarrow{1-\lambda} & \\
\downarrow{\scriptstyle b} & & \downarrow{\scriptstyle -b'} & & \downarrow{\scriptstyle b} & & \\
C_1 & \xleftarrow{1-\lambda} & C_1 & \xleftarrow{N_\lambda} & C_1 & \xleftarrow{1-\lambda} & \\
\downarrow{\scriptstyle b} & & \downarrow{\scriptstyle -b'} & & \downarrow{\scriptstyle b} & & \\
C_0 & \xleftarrow{1-\lambda} & C_0 & \xleftarrow{N_\lambda} & C_0 & \xleftarrow{1-\lambda} &
\end{array} \tag{6}$$

where the lower left corner has bidegree $(0,0)$.

If we let S on T be the obvious translation two steps to the left, then T becomes an S-module, which can be identified with $\mathcal{B}(\bar{\Omega}\tilde{A})$, (cf. [LQ], 4.2). Under this identification $u^m\bar{\Omega}\tilde{A}$ corresponds to the subspace of T consisting of elements supported in columns $p = 2m, 2m+1$, and the two terms $\tilde{b}$, $S\tilde{B}$ of the differential of $\mathcal{B}(\bar{\Omega}\tilde{A})$ are made up of the b, $-b'$, $1-\lambda$ arrows and the N_λ arrows respectively.

Thus we have

$$HC_n(\mathcal{A}) \;=\; HC_n(\bar{\Omega}\tilde{A}) \;=\; \bar{H}C_n(\tilde{A}) \tag{7}$$

and similarly for Hochschild homology.

We know as a consequence of the harmonic decomposition that $\bar{\Omega}\tilde{A}$ has Connes' property, but this fact follows directly from the exactness of the rows in the cyclic bicomplex.

Indeed, the complex $(\bar{\Omega}\tilde{A}, \tilde{b})$ may be identified with the mapping cone for the map of complexes $1-\lambda : (C, b') \to (C, b)$. Similarly $(\operatorname{Ker}\tilde{B}/\operatorname{Im}\tilde{B}, \tilde{b})$ may be viewed as the mapping cone for the map $(C/\operatorname{Im}N_\lambda, b') \to (\operatorname{Ker}N_\lambda, b)$ induced by $1-\lambda$. But the latter map is an isomorphism by the exactness of the rows in the cyclic bicomplex, so the complex $(\operatorname{Ker}\tilde{B}/\operatorname{Im}\tilde{B}, \tilde{b})$ is acyclic, in fact, it is contractible in a canonical way.

Having understood Connes' property for $\bar{\Omega}\tilde{A}$ in terms of the cyclic bicomplex, let us now consider the homology arising from this mixed complex.

The reduced cyclic complex $\bar{C}^\lambda(\tilde{A})$ is by definition $\bar{\Omega}\tilde{A}/\operatorname{Ker}\tilde{B}$ with differential induced by $\tilde{b}$. By the matrix formulas for $\tilde{b}, \tilde{B}$ we have the identification

$$\bar{C}^\lambda(\tilde{A}) \;=\; C^\lambda(\mathcal{A}) \tag{8}$$

where the *cyclic complex* $C^\lambda(\mathcal{A})$ of $\mathcal{A}$ is defined to be $C/\operatorname{Im}(1-\lambda)$ with differential induced by b. The results §4 (3), (4) in the case of $\tilde{A}$ yield the formulas

$$HC_n(\mathcal{A}) \;=\; H_n(C^\lambda(\mathcal{A})) \qquad\qquad H_\nu(\bar{\Omega}\tilde{A}, b+B) \;=\; 0 \tag{9}$$

But the former follows immediately by diagram chasing from the exactness of the rows in the cyclic bicomplex, and the same is true for the latter using the periodic cyclic bicomplex.

The moral then to be learned from this discussion is that Connes' property is the appropriate notion for general mixed complexes, which corresponds to the exactness of the rows in the cyclic bicomplex in the case of mixed complexes of the form $\bar{\Omega}\tilde{\mathcal{A}}$.

Finally, 4.2 in the case of $\tilde{\mathcal{A}}$ becomes the unreduced form of the Connes-Karoubi theorem

$$H_n(\bar{\Omega}\tilde{\mathcal{A}}_\natural, d) = \mathrm{Ker}\{B : HC_n(\mathcal{A}) \to HH_{n+1}(\mathcal{A})\} \qquad (10)$$

$P\bar{\Omega}\tilde{\mathcal{A}}$ and the S operator on $C^\lambda(\mathcal{A})$

We next examine the harmonic space $P\bar{\Omega}\tilde{\mathcal{A}}$.

Combining the criterion 2.2 with the matrix expressions for $\tilde{b}, d, \kappa$ we find that $P\bar{\Omega}\tilde{\mathcal{A}}$ consists of $\binom{x}{y}$ such that both x and $bx + (1-\lambda)y$ are λ-invariant.

Since $bx + (1-\lambda)(-G_\lambda bx) = P_\lambda bx$ is λ-invariant, it follows that we have a decomposition

$$P\bar{\Omega}\tilde{\mathcal{A}} = \left\{ \begin{pmatrix} x \\ -G_\lambda bx \end{pmatrix} \middle| P_\lambda x = x \right\} \oplus \left\{ \begin{pmatrix} 0 \\ y \end{pmatrix} \middle| P_\lambda y = y \right\} \qquad (11)$$

Let C^λ be the cyclic complex (8), let $\pi : C \to C^\lambda$ be the canonical surjection, and let $\bar{N}_\lambda : C^\lambda \to C$ be the map induced by N_λ, i.e. $\bar{N}_\lambda \pi = N_\lambda$. Clearly $b\pi = \pi b$, $b'\bar{N}_\lambda = \bar{N}_\lambda b$, and $\bar{N}_\lambda$ is injective.

The exact sequence §4 (2) in the case $A = \tilde{\mathcal{A}}$ is then easily identified with

$$0 \to C^\lambda[1] \xrightarrow{i} P\bar{\Omega}\tilde{\mathcal{A}} \xrightarrow{j} C^\lambda \to 0 \qquad (12)$$

where

$$i = \begin{pmatrix} 0 \\ \bar{N}_\lambda \end{pmatrix} \qquad j = \begin{pmatrix} \pi & 0 \end{pmatrix} \qquad (13)$$

Thus one has $Pi = i$, $\tilde{b}i = i(-b)$, $j\tilde{b} = bj$, $ji = 0$, $ij = \tilde{B}$, which may be checked directly. The exactness of the sequence is also clear from (11).

Moreover, (11) implies the sequence has an obvious splitting when the differentials are ignored. Specifically

$$l\pi = P\begin{pmatrix} P_\lambda \\ 0 \end{pmatrix} = \begin{pmatrix} P_\lambda \\ -G_\lambda b P_\lambda \end{pmatrix}$$

defines a lifting l for j.

According to the discussion preceding 3.3, the lifting l gives rise to an operator S of degree -2 on C^λ defined by $-iS = \tilde{b}l - lb$, which commutes with b and induces the Connes S operator on cyclic homology. Using $\tilde{b}P = P\tilde{b}$ we have

$$(\tilde{b}l - lb)\pi = P\left\{\tilde{b}\begin{pmatrix} P_\lambda \\ 0 \end{pmatrix} - \begin{pmatrix} P_\lambda \\ 0 \end{pmatrix} b\right\} = \begin{pmatrix} P_\lambda & 0 \\ b'G_\lambda - G_\lambda b & P_\lambda \end{pmatrix}\begin{pmatrix} [b, P_\lambda] \\ 0 \end{pmatrix}$$

Thus we have

$$\begin{pmatrix} 0 \\ \bar{N}_\lambda S\pi \end{pmatrix} = \begin{pmatrix} P_\lambda[b, P_\lambda] \\ (-b'G_\lambda + G_\lambda b)[b, P_\lambda] \end{pmatrix}$$

This proves

Proposition 5.1: *The formula*

$$\bar{N}_\lambda S\pi = (-b'G_\lambda + G_\lambda b)[b, P_\lambda] \tag{14}$$

defines an operator S on the cyclic complex C^λ of degree -2, which commutes with b and induces the Connes S operator on cyclic homology.

Applying P_λ to (14) yields

$$\bar{N}_\lambda S\pi = P_\lambda(-b')G_\lambda bP_\lambda \tag{15}$$

Then applying π and using the fact that $\pi\bar{N}_\lambda z = (n+1)z$ for $z \in C_n^\lambda$, we obtain

$$S(\pi x) = \frac{1}{n-1}\pi(-b')G_\lambda bP_\lambda x \qquad x \in C_n \tag{16}$$

This formula identifies S with the periodicity operator on the cyclic complex constructed by Kassel ([K3], 4.2).

We next show that S corresponds by duality to the periodicity operator on the complex of cyclic cochains constructed by Connes ([C1], p. 322, 11). His construction is based on the degree -2 operator on C corresponding to the map

$$\begin{aligned} a_0da_1\cdots da_n \mapsto\ & a_0a_1a_2da_3\cdots da_n \\ & + \sum_{j=2}^{n-1} a_0da_1\cdots da_{j-1}a_ja_{j+1}da_{j+2}\cdots da_n \end{aligned} \tag{17}$$

under the identification $a_0da_1\cdots da_n \leftrightarrow (a_0,\ldots,a_n)$. Using

$$\begin{aligned} a_0da_1\cdots da_{j-1}a_ja_{j+1} = {} & (-1)^{j-1}a_0a_1\cdots d(a_ja_{j+1}) \\ & + \sum_{i=1}^{j-2}(-1)^{j-1-i}a_0\cdots d(a_ia_{i+1})\cdots d(a_ja_{j+1}) \\ & + a_0\cdots d(a_{j-1}a_ja_{j+1}) \end{aligned}$$

where $\cdots$ on the right stands for the appropriate string $da_k \cdots da_l$, one sees easily that (17) corresponds to the operator $-b'^{[2]}$ on C such that

$$b'^{[2]} = \sum_{0 \le i < j < n} (-1)^{i+j} d_i d_j$$

on elements of C_n, where d_i is the standard face operator

$$d_i(a_0, .., a_n) = \begin{cases} (a_0, .., a_i a_{i+1}, .., a_n) & 0 \le i < n \\ (a_n a_0, a_1, .., a_{n-1}) & i = n \end{cases} \tag{18}$$

(The notation $b'^{[2]}$ comes from Loday (*[L]*, E.2.2.6), who also defines an operator $b^{[2]}$ (*[L]*, 2.2.5.2) by the same formula with the summation over $0 \le i < j \le n$.)

The next result identifies the transpose of our S on C^λ with Connes' periodicity operator on cyclic cochains. This provides another proof of the fact that his operator commutes with b.

Proposition 5.2: *One has*

$$S(\pi x) = \frac{1}{(n-1)n} \pi b'^{[2]} P_\lambda x \qquad x \in C_n \tag{19}$$

Proof: By (16) it suffices to show

$$P_\lambda(-b')G_\lambda b P_\lambda = \frac{1}{n} P_\lambda b'^{[2]} P_\lambda \tag{20}$$

on (elements of) C_n.

We need the relations

$$\begin{aligned} P_\lambda b G_\lambda &= G_\lambda b' P_\lambda = 0 \\ P_\lambda c N_\lambda c P_\lambda &= 0 \end{aligned} \tag{21}$$

Since P_λ and N_λ are proportional in each degree, it suffices to prove these with N_λ replacing P_λ. Now $N_\lambda b G_\lambda = b' N_\lambda G_\lambda = 0$ and similarly $G_\lambda b' N_\lambda = 0$. Finally, using $N_\lambda c N_\lambda = N_\lambda b$, which is half of Tsygan's proof ([T]) of the identity $N_\lambda b = b' N_\lambda$, we have $N_\lambda c N_\lambda c N_\lambda = N_\lambda c N_\lambda b = N_\lambda b^2 = 0$.

Using these relations we have on C_n

$$\begin{aligned} P_\lambda(-b')G_\lambda b P_\lambda &= P_\lambda(-b+c)G_\lambda(b'+c)P_\lambda \\ &= P_\lambda c G_\lambda c P_\lambda \\ &= \frac{1}{n}\sum_{k=0}^{n-1}\left(\frac{n-1}{2} - k\right) P_\lambda c \lambda^k c P_\lambda \\ &= \frac{1}{n}\sum_{k=0}^{n-1}(-k) P_\lambda c \lambda^k c P_\lambda \end{aligned}$$

On the other hand, by Tsygan's formula $\lambda^i c\lambda^{-i-1} = (-1)^i d_i$ for the face operators we have

$$b'^{[2]} = \sum_{0\le i<j<n} \lambda^i c\lambda^{-i-1}\lambda^j c\lambda^{-j-1}$$

on C_n, hence

$$\begin{aligned} P_\lambda b'^{[2]} P_\lambda &= \sum_{0\le i<j<n} P_\lambda c\lambda^{j-i-1} c P_\lambda \\ &= \sum_{k=0}^{n-1} (n-1-k) P_\lambda c\lambda^k c P_\lambda \\ &= \sum_{k=0}^{n-1} (-k) P_\lambda c\lambda^k c P_\lambda \end{aligned}$$

completing the proof. (As $P_\lambda b^{[2]} P_\lambda - P_\lambda b'^{[2]} P_\lambda = P_\lambda c N_\lambda c P_\lambda = 0$, we also get the formula [L], 2.2.7 except for sign.)

To conclude this section, we recall that the periodicity operator S on C_λ arises from the canonical splitting l of (12). By the discussion preceding 3.3 this means that one has the following description of the harmonic space $P\bar{\Omega}\tilde{A}$.

Proposition 5.3: *The mixed complex $P\bar{\Omega}\tilde{A}$ is canonically isomorphic to the mixed complex $\Lambda \tilde{\otimes} C^\lambda$ associated to the S-module given by the cyclic complex C^λ equipped with the explicit periodicity operator S discussed above.*

6 The Mixed Complex $C(A)$

In this section we discuss the chain picture dual to the cochain formalism developed by Connes in *[1]*, II.

We begin with a general splitting result for mixed complexes with Connes' property.

Let M be a mixed complex, and let K and I denote the kernel and image of B. Recall that Connes' property for M says that K/I is b-acyclic. Examples of mixed complexes with this property are those which are B-acyclic, those which have $B = 0$ and are b-acyclic, and also a direct sum of these two types, such as $\bar{\Omega}A = P\bar{\Omega}A \oplus P^\perp\bar{\Omega}A$ as in §4. The next proposition shows this exhausts the possibilities.

Proposition 6.1: *M has Connes' property iff $M = P \oplus Q$ where P, Q are mixed subcomplexes such that P is B-acyclic, $B(Q) = 0$, and Q is b-acyclic.*

Before proving this, we note that if M has such a decomposition, then Q is a contractible mixed complex, and so M is homotopy equivalent to the B-acyclic mixed complex P. On the other hand, if M is homotopy equivalent to a mixed complex with Connes' property, then M also has this property, because homotopic maps of mixed complexes induce homotopic maps on the B-homology $\operatorname{Ker} B/\operatorname{Im} B$ equipped with differential b. Thus we deduce the first statement of

Corollary 6.2: *A mixed complex has Connes' property iff it is homotopy equivalent to a B-acyclic mixed complex. A map between mixed complexes with Connes' property is a quasi-isomorphism iff it is a homotopy equivalence.*

For the second statement it suffices to show that a quasi-isomorphism between B-acyclic mixed complexes is a homotopy equivalence, and this is *[13]*, 1.8.

We now prove 6.1. Since the direction $\Leftarrow$ is clear from the preceding discussion, we only have to construct the required splitting of M assuming Connes' property.

Consider the exact sequences of complexes

$$0 \to K/I \to M/I \to M/K \to 0 \tag{1}$$

$$0 \to I \to K \to K/I \to 0 \tag{2}$$

where the differential is b. Since K/I is b-acyclic by hypothesis, there exist splittings of these sequences compatible with b. Indeed, because we work over a field there exist splittings not necessarily respecting b. The obstruction to modifying these into b-compatible splittings are respectively homotopy classes of maps $M/K \to (K/I)[1]$ and $K/I \to I[1]$, which vanish as K/I is contractible.

Now a b-compatible splitting of (1) is given by a b-subcomplex P of M, such that $P \supset I$ and $P/I \oplus K/I = M/I$, or equivalently, $P + K = M$ and $P \cap K = I$. Also, a b-splitting of (2) is given by a b-subcomplex Q of K such that $I \oplus Q = K$. One then has a diagram

$$\begin{array}{ccc} P & \to & M \\ \uparrow & & \uparrow \\ I & \to & K \\ \uparrow & & \uparrow \\ 0 & \to & Q \end{array} \tag{3}$$

where the arrows are inclusions and the squares are bicartesian.

We claim that P, Q have the desired properties. They are mixed subcomplexes of M since $Q \subset K$, hence $B(Q) = 0$, and $B(P) \subset B(M) = I \subset P$. Q is b-acyclic since $Q \simeq K/I$. Furthermore, the kernel of B on P is $P \cap K = I$, and $I = B(P + Q) = B(P)$, so P is B-acyclic. Finally $M = P \oplus Q$ is clear from the above diagram.

The next proposition gives a variant of the preceding construction where Q is partially specified.

Proposition 6.3: *Assume M has Connes' property, and let Q' be a mixed subcomplex contained in K which is b-acyclic and such that $I \cap Q' = 0$. Then $M =$*

$P \oplus Q' \oplus Q''$, where P, Q'' are mixed subcomplexes such that P is B-acyclic, $B(Q'') = 0$, and Q'' is b-acyclic.

In particular, $M/Q' \simeq P \oplus Q''$ has Connes' property and Q' is a summand of the mixed complex M.

Proof: As $I \cap Q' = 0$ we have the exact sequence

$$0 \to Q' \to K/I \to K/(I \oplus Q') \to 0$$

Since Q' and K/I are b-acyclic by hypothesis, $K/(I \oplus Q')$ is b-acyclic. Consequently the exact sequence

$$0 \to I \oplus Q' \to K \to K/(I \oplus Q') \to 0$$

splits compatibly with b, yielding a b-subcomplex Q'' such that $K = I \oplus Q' \oplus Q''$. We can then take $Q = Q' \oplus Q''$ in (3) and obtain the desired splitting $M = P \oplus Q' \oplus Q''$. Notice that $B(Q'') = 0$ and that Q'' is b-acyclic since it is a summand of K/I.

The mixed complex $C(A)$ when A is H-unital

We next recall a construction of Kassel (*[9]*, §5) in a suitable form for our purposes.

Let $C = C(A)$ be the graded vector space with $C(A)_n = A^{\otimes n+1}$ for $n \geq 0$ and $C(A)_n = 0$ for $n < 0$. Let $C' = C'(A)$ the same vector space equipped with the shifted grading $C'(A)_n = C(A)_{n-1}$. In §5 we identified the mixed complex $\bar{\Omega}\tilde{A}$ with $C \oplus C'$ equipped with the operators

$$\tilde{b} = \begin{pmatrix} b & 1-\lambda \\ 0 & -b' \end{pmatrix} \qquad \tilde{B} = \begin{pmatrix} 0 & 0 \\ N_\lambda & 0 \end{pmatrix}$$

So far A can be any nonunital algebra. We now suppose given an operator h of degree 1 on C satisfying $[b', h] = 1$. Such an h exists iff A is H-unital in the sense of Wodzicki.

We introduce the linear maps

$$i = \begin{pmatrix} -B_0 \\ 1 \end{pmatrix} : C' \to \bar{\Omega}\tilde{A}$$

$$j = \begin{pmatrix} 1 & B_0 \end{pmatrix} : \bar{\Omega}\tilde{A} \to C$$

where $B_0 = (1-\lambda)h$. Using the identity

$$1-\lambda = (1-\lambda)(b'h + hb') = bB_0 + B_0 b' \tag{4}$$

one finds $\tilde{b}i = -ib'$ and $\tilde{B}i = 0$, whence i is an injection of mixed complexes when C' is equipped with the differentials $-b'$, 0 induced by $\tilde{b}$, $\tilde{B}$ respectively. One also has $j\tilde{b} = bj$ and $j\tilde{B} = Bj$, where B on C is defined by $B = (1-\lambda)hN_\lambda = B_0 N_\lambda$. Thus j is a surjection of mixed complexes when C is equipped with the differentials b, B.

With this understood, it is clear that one has an exact sequence of mixed complexes

$$0 \to C' \xrightarrow{i} \bar{\Omega}\tilde{A} \xrightarrow{j} C \to 0 \tag{5}$$

Furthermore C' is a contractible mixed complex, $-h$ being a contraction. As a consequence j is a quasi-isomorphism of mixed complexes, so it induces isomorphisms on cyclic type homology and cohomology.

This quasi-isomorphism property of j has been refined by Kassel as follows.

Proposition 6.4: *The exact sequence of S-modules*

$$0 \to \mathcal{B}(C') \longrightarrow \mathcal{B}(\bar{\Omega}\tilde{A}) \longrightarrow \mathcal{B}(C) \to 0 \tag{6}$$

associated to (5) *has a canonical splitting. Consequently, one has an explicit deformation retraction of $\mathcal{B}(\bar{\Omega}\tilde{A})$ onto $\mathcal{B}(C)$.*

Proof: Consider the maps

$$l = \begin{pmatrix} 1 - B_0 S h N_\lambda \\ S h N_\lambda \end{pmatrix} : \mathcal{B}(C) \to \mathcal{B}(\bar{\Omega}\tilde{A})$$

$$r = \begin{pmatrix} -ShN_\lambda & 1 \end{pmatrix} : \mathcal{B}(\bar{\Omega}\tilde{A}) \to \mathcal{B}(C')$$

One computes that $ri = 1$, $jl = 1$, $ir + lj = 1$, and $(\tilde{b} + S\tilde{B})l = l(b + SB)$, $r(\tilde{b} + S\tilde{B}) = -b'r$, which yields the desired splitting. On the other hand $\mathcal{B}(C')$ is a contractible S-module with contraction given by $-h$. The splitting and this contraction then give the required deformation retraction.

By specializing S to 1 in the formulas for l, r we obtain

Corollary 6.5: *There is a canonical splitting of* (5) *which is compatible with the $\mathbb{Z}/2$-grading and the total differential $\tilde{b} + \tilde{B}$.*

We do not know whether in general the sequence (5) has a splitting compatible with the mixed complex structure. However, this turns out to be true in Connes' situation to which we now turn.

Connes' mixed complex $C(A)$ for A unital

We suppose now that A is unital. Then there are two standard choices for h, namely

$$\begin{aligned} s(a_0, \ldots, a_n) &= (1, a_0, \ldots, a_n) \\ (-\lambda^{-1}s)(a_0, \ldots, a_n) &= (-1)^n (a_0, \ldots, a_n, 1) \end{aligned} \tag{7}$$

The latter leads to the Connes B operator on $C(A)$, namely $B = B_0 N_\lambda$ where $B_0 = (1-\lambda^{-1})s$:

$$B_0(a_0, .., a_n) = (1, a_0, .., a_n) + (-1)^n (a_0, .., a_n, 1) \tag{8}$$

From now on we assume $h = -\lambda^{-1}s$. However, all the results are easily seen to hold also in the case $h = s$, requiring only minor changes in the arguments.

Lemma 6.6: *One has* $i(C') \cap B\bar{\Omega}\tilde{A} = 0$.

Proof: Let

$$\xi = \begin{pmatrix} -B_0 x \\ x \end{pmatrix} = \begin{pmatrix} 0 \\ N_\lambda y \end{pmatrix}$$

be an element in the intersection. One has the identities

$$cs = -\lambda \qquad\qquad c\lambda^{-1}s = 1 \tag{9}$$

where $c(a_0, .., a_n) = (-1)^n (a_n a_0, .., a_{n-1})$ is the crossover term. Since $B_0 x = 0$ we have $0 = cB_0 x = c(1-\lambda^{-1})sx = -(\lambda+1)x$. But $x = N_\lambda y$ implies $\lambda x = x$, so $x = 0$ and $\xi = 0$.

Proposition 6.7: *Assuming h chosen as above, one has:*
1) *The mixed complex* $C(A)$ *has Connes' property.*
2) *The mixed complex* $\bar{\Omega}\tilde{A}$ *splits*

$$\bar{\Omega}\tilde{A} = P\bar{\Omega}\tilde{A} \oplus i(C') \oplus Q'' \tag{10}$$

where Q'' *is a b-acyclic mixed subcomplex killed by B. In particular, the exact sequence* (5) *splits in a way respecting the mixed complex structure.*

Proof: This is a consequence of the lemma and 6.3 applied to $M = \bar{\Omega}\tilde{A}$ and $Q' = i(C')$. By the proof of 6.3 the mixed subcomplex $P \subset M$ can be arbitrary such that the top square in the diagram (3) is bicartesian, hence we can take it to be $P\bar{\Omega}\tilde{A}$.

Remarks

1. Assertion 1) is Connes' fundamental lemma *[1]*, p. 337, 36. By 3.1 it implies that the homology of $C(A)$ considered as supercomplex with differential $b+B$ vanishes, whence by duality the fact (*[2]*) that cochains of arbitrary support yield a trivial cohomology with respect to $b+B$.
2. In theory it is possible to construct an explicit decomposition as in (10) by first carrying out the proof of Lemma 6.6 with maps rather than elements.

7 The Operator κ on $C(A)$ and Simplicial Normalization

Let $C = C(A)$, $\Omega = \Omega A$, and let $p : C \to \Omega$ be the canonical surjection such that $p(a_0, .., a_n) = a_0 da_1 \cdots da_n$. One has $pb = bp$ and $ps = dp$. We are going to prove that $p : C \to \Omega$ is a homotopy equivalence of complexes with differential b, by introducing a suitable Karoubi operator on C.

Define the operator κ on C by $1 - \kappa = [b, s]$. This operator is a lifting of the Karoubi operator on Ω, i.e. $p\kappa = \kappa p$. As $b^2 = 0$, κ commutes with b, but it need not commute with s.

One has $[b, s] = [b', s] + cs + sc = 1 - \lambda + sc$, hence $\kappa = \lambda - sc$, which means that κ on C is also given by the formula §1 (16).

We now examine κ on C_n. The kernel of p on C_n is the degenerate subspace $D_n = W_1 + \cdots + W_n$, where W_i is the subspace spanned by $(a_0, \ldots, a_n)$ with $a_i = 1$. One has $\kappa W_i \subset W_{i+1}$ for $1 \leq i \leq n-1$ and $\kappa W_n = 0$. Thus $\kappa^n D_n = 0$.

On the other hand we know by §1 (21) that $(\kappa^n - 1)(\kappa^{n+1} - 1)$ kills $C_n/D_n = \Omega^n$. Thus κ satisfies the polynomial relation

$$\kappa^n(\kappa^n - 1)(\kappa^{n+1} - 1) = 0 \tag{1}$$

on C_n.

Put $P_0 = (\kappa^n - 1)(\kappa^{n+1} - 1)$. Then $P_0 C_n \subset D_n$, and $P_0 = 1$ on D_n, since the polynomial in κ defining P_0 is congruent to 1 modulo κ^n. It follows that P_0 is a projection operator with $P_0 C_n = D_n$; it is the spectral projection for κ and the eigenvalue 0. The subspace $P_0^\perp C_n$ is the unique κ-invariant complement to D_n.

Combining these operators for different n we get the spectral projection P_0 on C for κ and the zero eigenvalue. The projection P_0 commutes with any operator commuting with κ, in particular with b. Moreover its image is the degenerate subspace $D = \bigoplus D_n$, so we obtain the splitting

$$C = D \oplus P_0^\perp C \tag{2}$$

compatible with the differential b.

Now D is closed under s, and the operator $[b, s] = 1 - \kappa$ is invertible on D as κ is locally nilpotent. Thus $[b, (1 - \kappa)^{-1}s] = 1$ showing that the complex D is contractible.

We then deduce that $p : C \to \Omega$ is a homotopy equivalence of complexes.

This result is usually derived by applying the simplicial normalization theorem to the standard simplicial structure on C. According to this theorem one has a splitting $C = D \oplus E$ into subcomplexes, where $E_n \subset C_n$ is the intersection of the kernels of the face operators $d_i : C_n \to C_{n-1}$ for $0 \leq i < n$, with d_i as in §5 (18). Now on elements of C_n one has $d_0\kappa = 0$ and $d_i\kappa = -\kappa d_{i-1}$ for $1 \leq i < n$. Consequently E_n is closed under κ, so it coincides with $P_0^\perp C_n$, because this is the unique κ-invariant subspace of C_n complementary to D_n. Thus the splitting (2) is the same as the one given by the simplicial normalization theorem.

8 Supertraces on QA

In this section we discuss the algebraic formalism introduced by Connes ([2], §2) for infinite dimensional cycles.

Let $(\Omega A)^*$ be the dual space of linear functionals on ΩA. An element $f \in (\Omega A)^*$ is the same as a sequence (f_n), where f_n is a linear functional on

$\Omega^n A$. Furthermore, f_n can be identified with a multilinear functional of degree $n+1$ on A

$$(a_0, \ldots, a_n) \mapsto f_n(a_0 da_1 \cdots da_n) \tag{1}$$

which is (simplicially) normalized in the sense that it vanishes when $a_i = 1$ for some $i \geq 1$. In this way f can be identified with an inhomogeneous normalized cochain on A. For brevity we will refer to elements of $(\Omega A)^*$ simply as cochains in this section.

Any linear operator T on ΩA gives rise to a transpose operator $f \mapsto fT$ on cochains.

Let QA be the superalgebra given by ΩA equipped with the even-odd grading and the Fedosov product

$$\omega \circ \eta = \omega\eta - (-1)^{|\omega|} d\omega d\eta \tag{2}$$

There are two homomorphisms $a \mapsto a \pm da$ from A to QA, which give rise to a canonical isomorphism $A * A \xrightarrow{\sim} QA$, cf. *[3]*, §1.

Proposition 8.1: *The following conditions are equivalent for* $\tau \in (\Omega A)^*$:
a) τ *is a supertrace on* QA, *i.e. it vanishes on all Fedosov product supercommutators*

$$[\omega, \eta]_\circ = \omega\circ\eta - (-1)^{|\omega||\eta|}\eta\circ\omega$$

b) *Connes' infinite cycle identity holds*:

$$\tau([\omega, \eta]) = (-1)^{|\omega|} 2\tau(d\omega d\eta) \tag{3}$$

where $[\omega, \eta]$ *is the supercommutator for the ordinary product on* ΩA.
c) *One has* $\tau\kappa = \tau$ *and* $\tau_{n-1} b = 2\tau_{n+1} d$ *for* $n > 0$.

Proof: If τ satisfies (3), then $\tau([d\omega, d\eta]) = 0$, hence using the identity

$$[\omega, \eta]_\circ = [\omega, \eta] - (-1)^{|\omega|} 2 d\omega d\eta + (-1)^{|\omega|}[d\omega, d\eta]$$

we see τ is a supertrace on QA. Conversely, if τ is a supertrace on QA, then τ kills $[d\omega, d\eta]_\circ = [d\omega, d\eta]$ and this identity yields (3). Thus a) $\Leftrightarrow$ b).

Since QA is generated by the elements a, da for $a \in A$, τ is a supertrace on QA iff it vanishes on the Fedosov product supercommutators

$$\begin{aligned}
\omega\circ a - a\circ\omega &= [\omega, a] - (-1)^{|\omega|} d\omega da + da d\omega \\
&= (-1)^{|\omega|}\{b(\omega da) - d\omega da - \kappa(d\omega da)\} \\
&= (-1)^{|\omega|}(b - (1+\kappa)d)(\omega da) \\
\omega\circ da - (-1)^{|\omega|} da\circ\omega &= \omega da - (-1)^{|\omega|} da\omega \\
&= (1-\kappa)(\omega da)
\end{aligned}$$

for all $a \in A$, $\omega \in \Omega A$. Thus τ is a supertrace on QA iff $\tau b - \tau(1+\kappa)d$ and $\tau(1-\kappa)$ vanish on forms of degree > 0, i.e. $\tau_{n-1} b = \tau_{n+1}(1+\kappa)d$ and $\tau_n \kappa = \tau_n$

for $n > 0$. These conditions are equivalent to c) since $\kappa = 1$ in degree zero. Thus a) $\Leftrightarrow$ c).

Remark: In c) the condition that τ be κ-invariant is redundant. Indeed, assuming $\tau_{n-1}b = 2\tau_{n+1}d$ for $n > 0$, then $\tau_{n-1}bd = 0$ and $\tau_{n+1}db = 0$ for $n > 0$. Thus $\tau_n(1-\kappa) = 0$ for $n \geq 2$ since $1 - \kappa = bd + db$. For $n = 1$, besides $\tau_1 bd = 0$ one also has

$$\tau_1 db \;=\; \tau_1 Bb \;=\; -\tau_1 bB \;=\; -2\tau_3 dB \;=\; 0$$

so τ_1 is also κ-invariant. Finally this holds for τ_0 trivially.

Let r be the scaling operator on ΩA which multiplies by r_n on $\Omega^n A$, where

$$r_{2m} = \frac{(-1)^m}{m!} \qquad\qquad r_{2m+1} = \frac{(-1)^m 2^m}{1\cdot 3\cdots(2m+1)} \tag{4}$$

Setting $\tau^r = \tau r$ and using the fact that $\tau_n B = n\tau_n d$ when τ_n is κ-invariant, one sees easily that c) above is equivalent to $\tau^r\kappa = \tau^r$ and $\tau^r_{n-1}b + \tau^r_{n+1}B = 0$ for $n > 0$. Thus we obtain

Corollary 8.2: *τ is a supertrace on QA iff the rescaled cochain τ^r is κ-invariant and such that $\tau^r(b+B) = 0$ except in degree zero.*

Notice that the degree zero component of $\tau^r(b+B)$ is $\tau_1^r B = \tau_1 d$, which is the linear functional $a \mapsto \tau(da)$ on A. When τ is a supertrace on QA, one has $\tau_1^r Bb = -\tau_1^r bB = \tau_3^r B^2 = 0$, so $\tau_1 d$ is a trace on A which is reduced, i.e. vanishes on 1. Let $\bar{H}C^0(A)$ be the vector space of reduced traces on A.

Proposition 8.3: *One has an exact sequence*

$$0 \to \left\{ \begin{matrix} \kappa - \text{invariant} \\ b + B \text{ cocycles} \end{matrix} \right\} \xrightarrow{r^{-1}} \left\{ \begin{matrix} \text{supertraces} \\ \text{on } QA \end{matrix} \right\} \xrightarrow{j} \bar{H}C^0(A) \to 0 \tag{5}$$

where r^{-1} denotes the map $f \mapsto fr^{-1}$ and $j(\tau) = \tau_1 d$.

By splitting this into even and odd parts we obtain

Corollary 8.4: *One has a bijection*

$$r^{-1} : \left\{ \begin{matrix} \text{even } \kappa - \text{invariant} \\ b + B \text{ cocycles} \end{matrix} \right\} \xrightarrow{\sim} \left\{ \begin{matrix} \text{even supertraces} \\ \text{on } QA \end{matrix} \right\}$$

and an exact sequence

$$0 \to \left\{ \begin{matrix} \text{odd } \kappa - \text{invariant} \\ b + B \text{ cocycles} \end{matrix} \right\} \xrightarrow{r^{-1}} \left\{ \begin{matrix} \text{odd traces} \\ \text{on } QA \end{matrix} \right\} \xrightarrow{j} \bar{H}C^0(A) \to 0$$

Proof: The sequence (5) is clearly left exact, so it suffices to prove j is surjective.

From §4 (4) we know $P\bar{\Omega}A$ has trivial homology as supercomplex with differential $b + B$, hence by duality $(P\bar{\Omega}A)^*$ has trivial cohomology with respect to $b+B$. We can identify $(P\bar{\Omega}A)^*$ with the subspace of cochains $f \in (\Omega A)^*$ which are reduced: $f(1)=0$ and harmonic in the sense that $f = fP$.

Let T be a reduced trace on A. Then T is a reduced cochain homogeneous of degree zero such that $Tb=0$, $TB=0$, and $TP=T$, hence T is an even reduced harmonic $b+B$ cocycle. Thus there exists an odd reduced harmonic cochain f such that $f(b+B) = T$. The surjectivity of j will follow if we show that f is κ-invariant, for then 8.2 shows $f = \tau^r$, where τ is a supertrace such that $j(\tau) = T$.

Now $fbB = f(b+B)B = TB = 0$, so f is κ-invariant by condition (b) of the following characterization of κ-invariant cochains.

Proposition 8.5: *For $f \in (\Omega A)^*$ the following are equivalent:*
(a) $f(1-\kappa)=0$.
(b) $fbB = 0$ *and* $fP = f$.
(c) $fbd = 0$ *and* $fd(1-\kappa) = 0$.

Proof: We can assume f homogeneous of degree n, where $n > 0$.

From §1 (25) and §2 (4) we have $f(\kappa^{n(n+1)} - 1) = fbB$ and $f(1-\kappa)G = f(1-P)$, hence (a) implies (b).

Assuming (b) and using $B = NPd$ from §2 (11), we have $0 = fbB = fbNPd = (n+1)fbd$ and $0 = fB(1-\kappa) = fNPd(1-\kappa) = nfd(1-\kappa)$, so (b) implies (c).

Finally assuming (c) we have

$$nfdb = \sum_{i=0}^{n-1} fd\kappa^i b = fBb = -fbB = -fbd\sum_{i=0}^{n}\kappa^i = 0$$

hence fdb and fbd both vanish, yielding (a) since $1-\kappa = bd+db$.

We next discuss Connes' version (*[2]*, §2) of the preceding ideas. Instead of our ΩA and QA he uses the universal nonunital DG algebra generated by A and the free product of A with itself in the category of nonunital algebras, even though A is unital.

To begin with we consider a nonunital algebra $\mathcal{A}$. As remarked earlier the augmentation ideal $\bar{\Omega}\tilde{\mathcal{A}}$ of $\Omega\tilde{\mathcal{A}}$ is the universal nonunital DG algebra generated by $\mathcal{A}$.

Let $\bar{Q}\tilde{\mathcal{A}}$ be the augmentation ideal of $Q\tilde{\mathcal{A}}$, i.e. $\bar{\Omega}\tilde{\mathcal{A}}$ equipped with Fedosov product. From the universal property of $Q\tilde{\mathcal{A}} = \tilde{\mathcal{A}} * \tilde{\mathcal{A}}$ it follows readily that $\bar{Q}\tilde{\mathcal{A}}$ is the free product of $\mathcal{A}$ with itself in the category of nonunital algebras.

We may identify $f_n \in (\bar{\Omega}^n\tilde{\mathcal{A}})^*$ with a pair (ψ_n, ϕ_{n-1}), where $\psi_n \in C(\mathcal{A})^*_n$ and $\phi_{n-1} \in C(\mathcal{A})^*_{n-1}$. The next proposition describes those linear functionals $f = (\psi, \phi)$ on $\bar{\Omega}\tilde{\mathcal{A}}$ which are respectively κ-invariant and supertraces on $\bar{Q}\tilde{\mathcal{A}}$.

Proposition 8.6: *For $(\psi, \phi) \in (\bar{\Omega}\tilde{\mathcal{A}})^*$ one has:*
1) *(ψ, ϕ) is κ-invariant iff $\psi(1-\lambda) = \phi b'$ and $\phi(1-\lambda) = 0$. In this case $\phi b = 0$.*
2) *(ψ, ϕ) is a supertrace on $\bar{Q}\tilde{\mathcal{A}}$ iff $\psi(1-\lambda) = \phi b'$ and $\psi_n b = 2\phi_{n+1}$ for $n \geq 0$. In this case $\phi(1-\lambda) = 0$.*

Proof: One has

$$\begin{aligned}(\psi,\phi)\tilde{b}d &= (\psi(1-\lambda)-\phi b',0)\\ (\psi,\phi)d(1-\kappa) &= (\phi(1-\lambda),0)\\ (\psi,\phi)d\tilde{b} &= (\phi b,\phi(1-\lambda))\end{aligned}$$

From 8.5 we know that (ψ,ϕ) is κ-invariant iff the first two expressions vanish, in which case the third also vanishes since $1-\kappa = d\tilde{b}+\tilde{b}d$. This proves 1).

By 8.1 and the following remark, $f = (\psi,\phi)$ is a supertrace on $\bar{Q}\tilde{A}$ iff for $n \geq 0$ one has $f_n\tilde{b} = 2f_{n+2}d$, in components:

$$\psi_n b = 2\phi_{n+1} \qquad \psi_n(1-\lambda) = \phi_{n-1}b'$$

This proves the first assertion of 2), and the second is clear because a supertrace on $\bar{Q}\tilde{A}$ is κ-invariant.

Finally we consider Connes' situation *[2]*, where A is a unital algebra A and we have the quotient mixed complex $C(A)$ of $\bar{\Omega}\tilde{A}$ with B_0 as in §6 (8). We can then describe the κ-invariant linear functionals $(\psi,\phi) \in (\bar{\Omega}\tilde{A})^*$ in terms of their first components.

Proposition 8.7: *(ψ,ϕ) is κ-invariant iff $\phi = \psi B_0$, $\psi b B_0 = 0$, and $\psi B_0(1-\lambda) = 0$.*

Proof: If $\phi = \psi B_0$, then by the identity §6 (4) we have

$$\psi(1-\lambda)-\phi b' = \psi b B_0 \qquad \phi(1-\lambda) = \psi B_0(1-\lambda)$$

Using 8.5 we get the implication $\Leftarrow$.

For the converse it suffices to show that if (ψ,ϕ) is κ-invariant, then $\phi = \psi B_0$. But this is clear from

$$(1-\kappa)\begin{pmatrix}\lambda^{-1}s\\ 0\end{pmatrix} = \begin{pmatrix}1-\lambda & 0\\ c & 1-\lambda\end{pmatrix}\begin{pmatrix}\lambda^{-1}s\\ 0\end{pmatrix} = \begin{pmatrix}-B_0\\ 1\end{pmatrix}$$

In other words, referring to the exact sequence §6 (6), we have $(1-\kappa)C(A) \supset i(C')$, so a κ-invariant linear functional on $\bar{\Omega}\tilde{A}$ descends to $C(A)$.

The next result is Connes' description (*[2]*, §2) of supertraces on $\bar{Q}\tilde{A}$ in terms of a single cochain $\psi \in C(A)^*$ rather than a pair (ψ,ϕ).

Proposition 8.8: *Supertraces on $\bar{Q}\tilde{A}$ are of the form $(\psi,\psi B_0)$ with $\psi \in C(A)^*$ satisfying $\psi B_0(1-\lambda) = 0$ and $\psi_n b = 2\psi_{n+2}B_0$ for $n \geq 0$.*

Proof: By 8.6 supertraces on $\bar{Q}\tilde{A}$ are of the form (ψ,ϕ), where $\psi(1-\lambda) = \phi b'$, $\phi(1-\lambda) = 0$, and $\psi_n b = 2\phi_{n+1}$ for $n \geq 0$. The first two conditions mean that (ψ,ϕ) is κ-invariant. Applying 8.7 we see that supertraces are of the form $(\psi,\psi B_0)$, where $\psi b B_0 = 0$, $\psi B_0(1-\lambda) = 0$, and $\psi_n b = 2\psi_{n+2}B_0$ for $n \geq 0$. But the condition $\psi b B_0 = 0$ is redundant, since $(\psi_n b)B_0 = 2(\psi_{n+2}B_0)(1-\lambda)(-\lambda^{-1}s)$ and $\psi B_0(1-\lambda) = 0$.

References

[1] A. Connes, Non–commutative differential geometry, Publ. Math. I.H.E.S. **62** (1985), 257–360.

[2] A. Connes, Entire cyclic cohomology of Banach algebras and characters of θ-summable Fredholm modules, K-theory **1** (1988), 519–548.

[3] J. Cuntz and D. Quillen, Algebra extensions and nonsingularity, J. American Math. Soc. (to appear)

[4] J. Cuntz and D. Quillen, Cyclic homology and nonsingularity, J. American Math. Soc. (to appear)

[5] J.D.S. Jones and C. Kassel, Bivariant cyclic theory, K-theory **3** (1989), 339–365.

[6] M. Karoubi, Homologie cyclique et K–théorie, Astérisque **149** (1987).

[7] C. Kassel, Cyclic homology, comodules, and mixed complexes, J. Alg. **107** (1987), 195–216.

[8] C. Kassel, Caractère de Chern bivariant, K-Theory **3** (1989), 367–400.

[9] C. Kassel, Homologie cyclique, caractère de Chern, et lemme de perturbation, J. Reine Angew. Math. **408** (1990), 159–180.

[10] D. Kastler, Cyclic cohomology within the differential envelope, Travaux en cours 30, Hermann (1988).

[11] J.-L. Loday, Cyclic Homology, Springer-Verlag (1992).

[12] J.-L. Loday and D. Quillen, Cyclic homology and the Lie algebra homology of matrices, Comment. Math. Helvetici **59** (1984), 565–591.

[13] D. Quillen, Bivariant cyclic cohomology and models for cyclic homology types, J. Pure Appl. Algebra, (to appear).

[14] B. L. Tsygan, Homology of matrix algebras over rings and Hochschild homology, Uspehi Mat. Nauk **38, 2** (1983), 217–218; Russian Math. Surveys **38** (1983), 198–199.

[15] D. Kastler, $\mathbb{Z}/2$-graded cyclic cohomology within the differential envelope, Contemp. Math. **62**, 221–235.

A Discrete Morse Theory for Cell Complexes

Robin Forman[1]

Department of Mathematics
Rice University
Houston, TX 77251

In this paper we will present a very simple discrete Morse theory for cell complexes. This Morse theory first appeared in *[6]*. In addition to proving analogues of the main theorems of Morse theory, we also present discrete analogues of such (seemingly) intrinsically smooth notions as the gradient vector field and the corresponding gradient flow associated to a Morse function. Using these, we define a Morse complex, a differential complex built out of the critical points of our discrete Morse function, which has the same homology as the underlying manifold. We conclude by presenting discrete analogues of Witten's Hodge theoretic approach to Morse theory *[21]*. Most of this paper will be an informal exposition of the contents of *[6]* and *[7]*.

This discrete Morse theory takes on particular significance in the context of PL manifolds. To clarify this statement, we take a small historical digression. In 1961, Smale *[16]* proved the h-cobordism theorem for smooth manifolds (and hence its corollary, the Poincaré conjecture in dimension ≥ 5) using a combination of handlebody theory and Morse theory. In *[12]*, Milnor presented a completely Morse theoretic proof of the h-cobordism theorem. In (*[9]*, *[1]*, *[17]*) Mazur, Barden and Stallings generalized Smale's theorem by replacing Smale's hypothesis that the cobordism be simply-connected by a weaker simple-homotopy condition. Along the way, this more general theorem (the s-cobordism theorem) was extended to other categories of manifolds. In particular, a PL s-cobordism theorem was established. In this case, it was necessary to prove "by hand" the sort of geometric decomposition theorem that follows from Morse theory in the smooth category. The Morse theory presented in this paper can be used to give a completely Morse theoretic proof of the PL s-cobordism theorem, along the lines of the proof in *[12]*.

Although this discrete Morse theory can be defined for any CW complex, to avoid minor complications we will restrict attention, in this paper, to simplicial complexes. Let M be any finite simplicial complex. We emphasize that M need not be a triangulated manifold, nor have any other special property. Denote by K the set of simplices of M, and by K_p the simplices of dimension p. A discrete Morse function on M will actually be a function on K. That is, we assign a single real number to each simplex in M. A simple example is shown in Figure 1.

[1] Partially supported by the National Science Foundation.

Geometry, Topology, and Physics for Raoul Bott

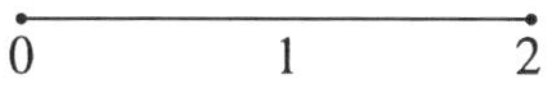

Figure 1.

Here, M consists of 3 simplices, and the value of the Morse function on a simplex is written below the corresponding simplex. This function has a minimum and a maximum, occurring at the 2 vertices. However, since the interval is homotopy equivalent to a point, standard Morse theory indicates that only the minimum is topologically significant. Loosely speaking (everything will be defined precisely in a moment), we achieve this by declaring that a critical simplex is topologically significant only if its index (as a critical point) is equal to its dimension. To guarantee that we are not missing any significant critical points, we must restrict the functions we consider. Again loosely speaking, we require that higher dimensional simplices be assigned higher values, with at most one exception locally at each simplex.

We now work more precisely. Write $\sigma^{(p)}$ if σ has dimension p, and $\tau > \sigma$ if σ lies in the boundary of τ.

Definition: We say a function

$$f : K \longrightarrow \mathbf{R}$$

is a discrete Morse function if for every $\sigma^{(p)} \in K_p$

(1) $\#\{\tau^{(p+1)} > \sigma \mid f(\tau) \leq f(\sigma)\} \leq 1$.

(2) $\#\{v^{(p-1)} < \sigma \mid f(v) \geq f(\sigma)\} \leq 1$.

For example, in Figure 2, the function (i) is not a discrete Morse function as the edge $f^{-1}(0)$ violates rule (2) and the vertex $f^{-1}(3)$ violates rule (1). The function (ii) is a Morse function.

Definition: We say $\sigma^{(p)}$ is critical (with index p) if

(1) $\#\{\tau^{(p+1)} > \sigma \mid f(\tau) \leq f(\sigma)\} = 0$.

(2) $\#\{v^{(p-1)} < \sigma \mid f(v) \geq f(\sigma)\} = 0$.

For example, in Figure 1, $f^{-1}(0)$ is a critical point of index 0 and there are no other critical simplices. In Figure 2 (ii), $f^{-1}(0)$ is a critical point of index 0, $f^{-1}(3)$ is a critical simplex of index 1 and there are no other critical simplices. Note that if $\sigma^{(p)}$ is critical, it is necessarily critical of index p.

Figure 2.

The above definition provides a discrete analogue of the smooth notion of a critical point of index p. For example, suppose x is a critical point of index 1 of a smooth Morse function F on a smooth manifold of dimension n. Then the Morse Lemma (see *[11]*, Lemma 2.2) states that there are coordinates $(t_1, \ldots, t_n)$, with x corresponding to $(0, \ldots, 0)$, such that in these coordinates

$$F(t_1, \ldots, t_n) = F(x) - t_1^2 + \sum_{i=2}^{n} t_i^2.$$

That is, beginning at the point x, F decreases to both sides in the t_1 direction, and increases in the transverse directions. Now suppose σ is a critical edge of a discrete Morse function f. Then $f(\sigma)$ is greater than f at either boundary vertex, and less than f at any 2-simplex with σ in its boundary (see, for example, Figure 3). That is, f decreases as one moves from the edge to either boundary component, and increases in every transverse direction. We can see that this is, in fact, a discrete analogue of the smooth situation. Moreover we see that, heuristically, if $\sigma^{(p)}$ is critical then the simplex σ can be thought of as representing the p-dimensional "unstable" directions at a smooth critical point of index p.

Before stating the main theorems, we need to present a definition. Suppose f is a discrete Morse function on a simplicial complex M. For any $c \in \mathbf{R}$ define the level subcomplex $M(c)$ by

$$M(c) = \bigcup_{f(\tau) \leq c} \bigcup_{\sigma \leq \tau} \sigma.$$

That is, $M(c)$ is the subcomplex of M consisting of all simplices τ with $f(\tau) \leq c$, as well as all of their faces.

Theorem: *Suppose the interval $[a, b]$ contains no critical values of f. Then $M(a)$ is a deformation retract of $M(b)$. Moreover, $M(b)$ simplicially collapses onto $M(a)$.*

To define simplicial collapse, suppose M is a simplicial complex, $\sigma^{(p)}$ is a p-simplex of M which is not the face of any other simplex, and $v^{(p-1)} < \sigma$ is a $(p-1)$-face of σ which is not the face of any other simplex. Then we say that M simplicially collapses onto $M - (\sigma \cup v)$. More generally, a simplicial collapse

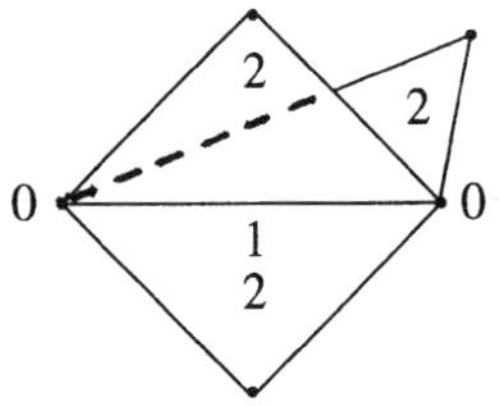

Figure 3.

Figure 4.

is any sequence of such operations (see Figure 4 for the simplicial collapse of a 2-simplex onto a vertex).

We note that the equivalence relation generated by simplicial collapse is called *simple homotopy equivalence.* This indicates that our Morse theory is particularly well suited to handling questions in this category.

Theorem: *Suppose $\sigma^{(p)}$ is a critical simplex with $f(\sigma) \in [a, b]$, and there are no other critical simplices with values in $[a, b]$. Then $M(b)$ is homotopy equivalent to*

$$M(a) \bigcup_{\dot{e}^{(p)}} e^{(p)}$$

where $e^{(p)}$ is a p-cell, and it is glued to $M(a)$ along its boundary $\dot{e}^{(p)}$.

Corollary 1.1: *Suppose M is a simplicial complex with a discrete Morse function. Then M is homotopy equivalent to a CW complex with exactly one cell of dimension p for each critical simplex of dimension p.*

The Morse inequalities follow immediately from this corollary (see *[11]*).

As another corollary, we have a PL sphere theorem.

Corollary: *Suppose M is a PL n-manifold without boundary with a Morse function with exactly 2 critical points. Then M is a PL n-sphere.*

The proof of this corollary rests heavily on Whitehead's theorem on the uniqueness of regular neighborhoods *[20]* which implies, as a special case, that a PL n-manifold with boundary which simplicially collapses onto a vertex is a PL n-ball. Suppose M is a PL n-manifold with a discrete Morse function f with exactly 2 critical points. Then the maximum of f must occur at an n-simplex $\sigma^{(n)}$ which is a critical point of index n. Thus, f restricts on $M - \sigma$ to a Morse function with exactly 1 critical point, which must occur at a vertex. By the first main theorem, M simplicially collapses to that vertex. Now we apply Whitehead's theorem to conclude that $M - \sigma$ is a PL n-ball. It follows that $M = (M - \sigma) \cup \sigma$ is a PL n-sphere.

Rather than prove the main theorems here (rigorous proofs appear in *[6]*), we will illustrate the main ideas with an example (see Figure 5). Here $f^{-1}(0)$ is a critical simplex of index 0, $f^{-1}(8)$ is a critical simplex of index 1, and there

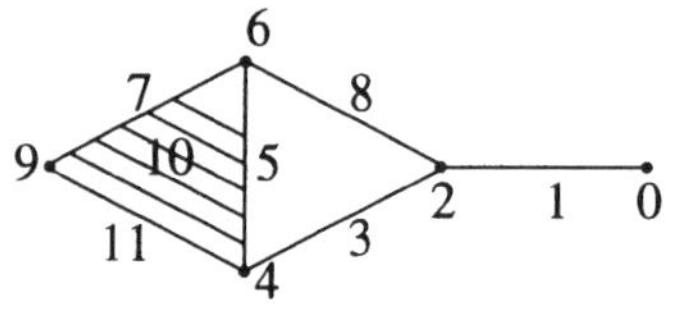

Figure 5.

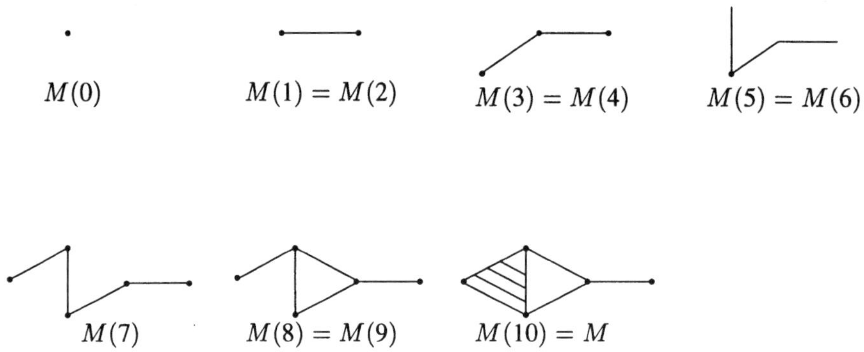

Figure 6.

are no other critical simplices. Thus, it follows from the above corollary that M is homotopy equivalent to a circle, as is evident from the picture. Consider the sequence of level subcomplexes (Figure 6).

We can see why the theorems are true. If σ is not critical, then one of two cases must occur. The first possibility is that σ has a codimension one face v with $f(v) \geq f(\sigma)$ (for example $\sigma = f^{-1}(1)$). In this case, when σ is added to M, v is added at the same time. The new subcomplex collapses onto the previous complex by pushing in σ along the free face v. (There are still a few things to check. For example, that σ is not a face of any simplex τ with $f(\tau) \leq f(\sigma)$, but this is true.) The other possibility is that there is a simplex τ of dimension 1 greater with σ as a face and with $f(\tau) \leq f(\sigma)$ (for example $\sigma = f^{-1}(2)$). In this case, σ is added to M when τ is added to M, and again the new complex collapses onto the previous complex by pushing in τ along the free face σ.

If σ is critical (for example $f^{-1}(8)$), then for every simplex τ with $\tau > \sigma$ we have $f(\tau) > f(\sigma)$ (this requires a bit of proof). Thus σ appears first in $M(f(\sigma))$. On the other hand, for every face v of σ, $f(v) < f(\sigma)$, and is added earlier. Therefore, when σ is added, it is simply glued to $M(f(\alpha) - \epsilon)$ along its boundary. This completes our "proof" of the main theorems.

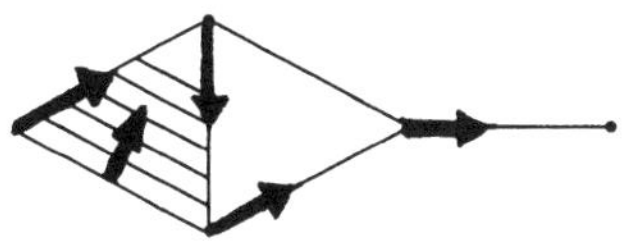

Figure 7.

Returning to the example in Figure 3, it is useful to indicate pictorially the simplicial collapse referred to in the theorem. Suppose $\sigma^{(p)}$ is a non-critical simplex with $\tau^{(p+1)} > \sigma$ satisfying $f(\tau) \leq f(\sigma)$. We then draw an arrow from σ to τ. The resulting diagram can be seen in Figure 7. A simplex is critical if and only if it is neither the tail nor the head of an arrow. These arrows can be viewed as the discrete analogue of the gradient vector field of the Morse function.

It is better to think of the gradient vector field, which we now call V, as a map of *oriented* simplices. That is, if v is a boundary vertex of an edge e with $f(e) \leq f(v)$, we want to think of $V(v)$ as a discrete tangent vector *leaving* v, i.e. with e given the orientation indicated by the arrow in Figure 8.

More generally, if $\tau^{(p+1)} > \sigma^{(p)}$ satisfies $f(\tau) \leq f(\sigma)$ then we set $V(\sigma) = \pm\tau$ with the sign chosen so that

$$< \sigma, \partial V(\sigma) > = \; -1$$

where $<,>$ is the canonical inner product on oriented chains with respect to which the simplices are orthonormal. Now V can be extended linearly to a map

$$V : C_p(M, \mathbf{Z}) \longrightarrow C_{p+1}(M, \mathbf{Z})$$

where, for each p, $C_p(M, \mathbf{Z})$ is the space of integer p-chains on M.

The next step is to define the discrete-time gradient flow. We define a map

$$\Phi : C_p(M, \mathbf{Z}) \longrightarrow C_p(M, \mathbf{Z}),$$

the discrete-time flow, by

$$\Phi \;=\; 1 + \partial V + V\partial.$$

We illustrate by an example. Consider the following complex (Figure 9), with the indicated gradient vector field V. Let e be the top edge oriented from left to right.

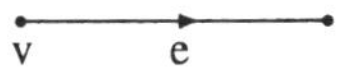

Figure 8.

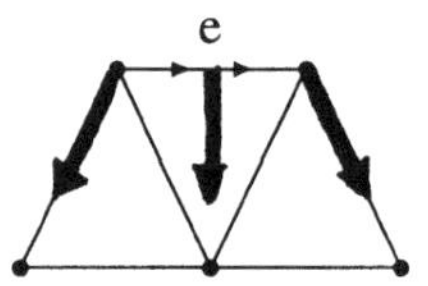

Figure 9.

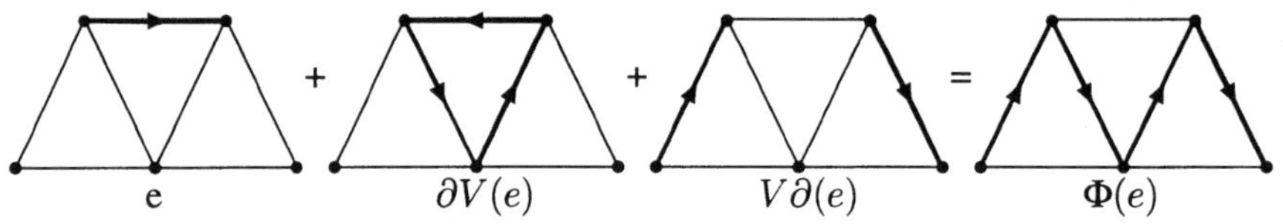

Figure 10.

We calculate

$$\Phi(e) \; = \; e + \partial V(e) + V\partial(e)$$

in Figure 10.

The map Φ has many of the properties that one expects from the gradient flow. For example, if σ is not a critical simplex, then $\Phi(\sigma)$ is the sum of edges on which f is less than $f(\sigma)$, so, in particular, Φ decreases f. The discrete gradient flow differs from the gradient flow in the smooth case primarily in that it stabilizes in finite time, i.e. there is an N such that

$$\Phi^N = \Phi^{N+1} = \ldots = \Phi^\infty .$$

This observation can be quite useful (for example, see the proof of the following theorem).

Let $C_p^\Phi \subseteq C_p(M, \mathbf{Z})$ denote the Φ-invariant p-chains, i.e., those p-chains c such that $\Phi(c) = c$. Since

$$\partial\Phi \; = \; \Phi\partial,$$

the Φ-invariant chains form a differential complex

$$\mathcal{C}^\Phi : 0 \longrightarrow C_n^\Phi \xrightarrow{\partial} C_{n-1}^\Phi \xrightarrow{\partial} C_{n-2}^\Phi \xrightarrow{\partial} \cdots .$$

Moreover, we have

Theorem: $H_*(\mathcal{C}^\Phi) = H_*(M, \mathbf{Z})$.

That is, this complex, the discrete Morse complex, has the same homology as the underlying manifold. The proof of this theorem is not difficult. To prove that the stabilization map $\Phi^\infty : C_*(M, \mathbf{Z}) \to C_*^\Phi$ induces an isomorphism on homology (with the inverse isomorphism induced by the natural injection), it is sufficient to find an algebraic homotopy operator, i.e. an operator $L : C_*(M, \mathbf{Z}) \to C_{*+1}(M, \mathbf{Z})$ such that $\Phi^\infty - 1 = \partial L + L\partial$. We now use the fact that $\Phi^\infty = \Phi^N$ for some N. If $N = 1$, then $L = V$ works. The general case is not much harder.

The Morse complex can also be defined in terms of critical simplices. For each p, let $\mathcal{M}_p \subseteq C_p(M, \mathbf{Z})$ denote the span of the critical p-simplices. We then have

Theorem: *For each p, the map*

$$\Phi^\infty : \mathcal{M}_p \to C_p^\Phi.$$

is an isomorphism.

Thus, the Morse complex can be defined equivalently as the complex

$$\mathcal{M} : 0 \longrightarrow \mathcal{M}_n \xrightarrow{\tilde{\partial}} \mathcal{M}_{n-1} \xrightarrow{\tilde{\partial}} \mathcal{M}_{n-2} \xrightarrow{\tilde{\partial}} \cdots$$

where $\tilde{\partial}$ is the differential induced by the above isomorphism. This differential is described explicitly in the following theorem.

Theorem: *For any oriented critical simplex* $\tau^{(p)}$

$$\tilde{\partial}\tau = \sum_{\tilde{\sigma}^{(p-1)} < \tau} < \partial\tau, \tilde{\sigma} > \sum_{\substack{\text{oriented} \\ \text{critical } \sigma^{(p-1)}}} \left[\sum_{\gamma \in \Gamma(\tilde{\sigma}, \sigma)} m(\gamma) \right] \sigma$$

where $\Gamma(\tilde{\sigma}, \sigma)$ *is the set of gradient paths from* $\tilde{\sigma}$ *to* σ.

A *gradient path* from $\tilde{\sigma}$ to σ is a sequence of simplices

$$\tilde{\sigma} = \sigma_0^{(p)}, \tau_0^{(p+1)}, \sigma_1^{(p)}, \tau_1^{(p+1)}, \sigma_2^{(p)}, \ldots, \tau_r^{(p+1)}, \sigma_{r+1}^{(p)} = \sigma$$

such that for each $i = 1, \ldots r$, $f(\sigma_i) \geq f(\tau_i) > f(\sigma)$. In particular, we must have $V(\sigma) = \pm\tau$. In Figure 11 we show a single gradient path from the boundary of a critical 2-simplex τ to a critical edge σ, where the arrows indicate the gradient vector field.

The coefficient $m(\gamma)$ is equal to ± 1 and is determined by whether the orientation on σ induces the chosen orientation on τ, or the opposite orientation.

It is often desirable, for example when proving the s-cobordism theorem, to deform one Morse function into another, simpler, Morse function. It is convenient to work instead with the gradient vector field. When deforming the gradient vector

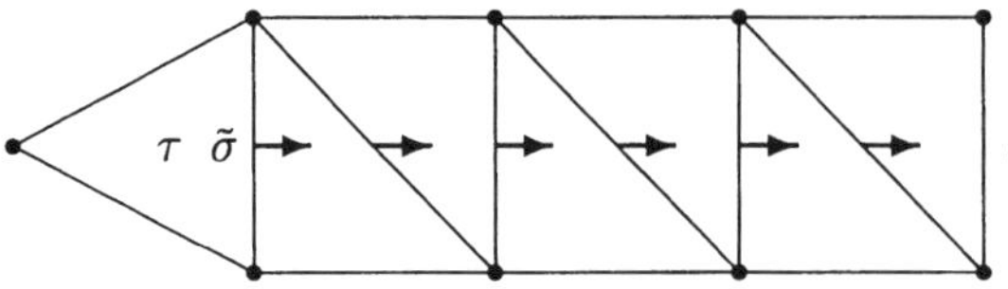

Figure 11.

field, it is necessary to know that the resulting vector field is, in fact, the gradient vector field of a Morse function. Thus, it is important to characterize those vector fields which are gradient vector fields of Morse functions. In the smooth case, Smale provided such a characterization in *[15]*. We can provide an analogous characterization of discrete gradient vector fields. First, we define an object we call a discrete vector field. Such objects have been previously studied in *[5]* and *[17]* (under a different name) and the references therein.

Definition: We define a discrete vector field to be any collection (one for each p) of one-to-one maps

$$U_p : K_p \to K_{p+1} \cup \{0\}$$

satisfying for each $\sigma^{(p)}$

(1) $U_p(\sigma) = 0$ or $\sigma < U_p(\sigma)$

(2) If $\sigma \in \text{Image}(U_{p-1})$ then $U_p(\sigma) = 0$.

A discrete gradient vector field is not quite such an object, since it was defined for oriented simplices, but for now we will ignore such subtleties. We define a U−path to be any path as in Figure 11 (where now the arrows indicate the map U).

Theorem: *A discrete vector field U is the gradient vector field of a discrete Morse function if and only if there are no non-trivial closed U-paths.*

As an application, we easily prove a "cancellation" theorem, a discrete analogue of Theorem 5.4 in *[12]*. That is, if $\sigma^{(p)}$ and $\tau^{(p+1)}$ are 2 critical simplices, and if there is exactly 1 gradient path from $\partial\tau$ to σ, then σ and τ can be cancelled. More precisely,

Theorem: *Suppose f is a discrete Morse function on M such that $\tau^{(p+1)}$ and $\sigma^{(p)}$ are critical, and there is exactly one gradient path from $\partial\tau$ to σ. Then there is another Morse function g on M with the same critical points except that σ and τ are no longer critical. Moreover, the gradient vector field associated to g is equal to the gradient vector field associated to f except along the unique gradient path from $\partial\tau$ to σ.*

In the smooth case, the proof, either as presented originally by Morse in *[13]* or as presented in *[12]*, is rather technical. In our discrete case the proof is simple.

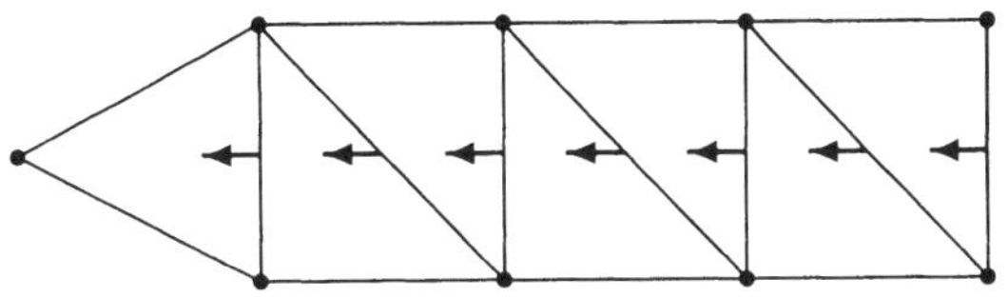

Figure 12.

If, in Figure 11, the indicated gradient path is the only gradient path from $\partial\tau$ to σ, then we can reverse the gradient vector field along this path, replacing the figure by the vector field in Figure 12. Our characterization of gradient vector fields immediately implies that this new vector field is the gradient vector field associated to some Morse function, and σ and τ are no longer critical.

In *[21]*, Witten provided a Hodge-theoretic proof of the main theorems of smooth Morse theory, as well as an analytic derivation of the Morse complex. In *[7]*, we provide a discrete Hodge-theoretic derivation of the main theorems of discrete Morse theory and the Morse complex. To explain this, we must first restrict our class of Morse functions. Say a discrete Morse function is *flat* if for all $\sigma^{(p)} \in K_p$

$$\begin{aligned} \#\{\tau^{(p+1)} > \sigma \mid f(\tau) < f(\sigma)\} &= 0 \\ \#\{u^{(p-1)} < \sigma \mid f(u) > f(\sigma)\} &= 0. \end{aligned}$$

For example, the Morse function in Figure 5 is not flat, but the function in Figure 13 is flat. We prove in *[7]* that for any discrete Morse function, there is an equivalent (i.e., one that has the same critical points and associated gradient vector field) flat Morse function. From now on, we assume f is flat.

Beginning with the real chain complex

$$0 \longrightarrow C_n(M,\mathbf{R}) \stackrel{\partial}{\longrightarrow} C_{n-1}(M,\mathbf{R}) \longrightarrow \cdots \longrightarrow C_0(M,\mathbf{R}) \longrightarrow 0$$

we replace the boundary operator ∂ by

$$\partial_t = e^{tf}\partial e^{-tf}$$

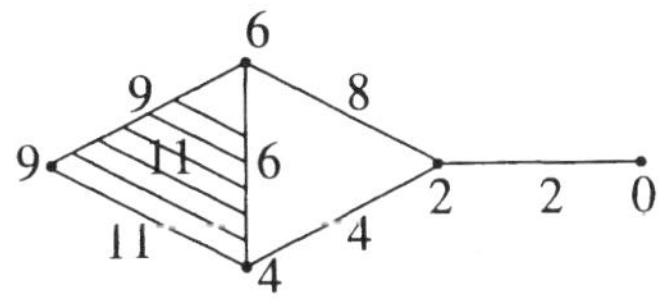

Figure 13.

where e^{tf} multiplies each simplex σ by $e^{tf(\sigma)}$ and is extended linearly to all chains. The resulting differential complex has the same homology for all $t \in \mathbf{R}$. We will now investigate this homology using ideas from Hodge theory. Let ∂_t^* denote the adjoint of ∂_t with respect to the canonical inner product $<,>$ on chains, and $\Delta_p(t)$ the induced Laplacian on p-chains

$$\Delta_p(t) = \partial_t \partial_t^* + \partial_t^* \partial_t : C_p(M, \mathbf{R}) \longrightarrow C_p(M, \mathbf{R}).$$

More explicitly, choose an orientation for each simplex. Then

$$\partial_t \sigma^{(p)} = \sum_{u^{(p-1)} < \sigma} \pm e^{t(f(u) - f(\sigma))} u$$

(where the $\pm$ sign is determined by whether the orientations on u and σ agree or not)

$$\partial_t^* \sigma^{(p)} = \sum_{\tau^{(p+1)} > \sigma} \pm e^{t(f(\sigma) - f(\tau))} \tau$$

$$\Delta_p(t)\sigma^{(p)} = \sum_{\tilde{\sigma}^{(p)}} \left[\sum_{\substack{u^{(p-1)} < \sigma \\ u < \tilde{\sigma}}} \pm e^{t(2f(u) - f(\sigma) - f(\tilde{\sigma}))} + \sum_{\substack{\tau^{(p+1)} > \sigma \\ \tau > \tilde{\sigma}}} \pm e^{t(f(\sigma) + f(\tilde{\sigma}) - 2f(\tau))} \right] \tilde{\sigma}.$$

It follows from our restrictions on f that if $\tilde{\sigma} \neq \sigma$, then the coefficient of $\tilde{\sigma}$ in the above expression is $O(e^{-tc})$ as $t \to \infty$, for some $c > 0$. Thus, up to exponentially small errors as $t \to \infty$, the Laplacian $\Delta_p(t)$ is diagonalized by the p-simplices. A closer look reveals

$$\left\langle \Delta_p(t)\sigma^{(p)}, \sigma^{(p)} \right\rangle = \begin{array}{ll} O(e^{-tc}) & \text{if } \sigma \text{ is critical} \\ 1 + O(e^{-tc}) & \text{if } \sigma \text{ is not critical.} \end{array} \tag{14}$$

In particular, we learn

Theorem: *As $t \to \infty$, $\#\{$eigenvalues of $\Delta_p(t)$ which are $O(e^{-tc})\} = \#\{$critical p simplices$\}$. All other eigenvalues of $\Delta_p(t)$ are $1 + O(e^{-tc})$.*

This theorem is sufficient to imply the Morse inequalities (over $\mathbf{R}$). See, for example, *[2]* (where the proof is discussed in the smooth category) or *[7]* for the completion of the argument.

We now take a closer look at the small eigenspaces. Let $W_p(t) \subseteq C_p(M, \mathbf{R})$ denote the span of the eigenvectors of Δ_t^p corresponding to the eigenvalues which are $O(e^{-tc})$. Then

$$\partial_t(W_p(t)) \subseteq W_{p-1}(t)$$

so for all t we have a differential complex (the Witten complex) with the same homology as M

$$\mathcal{W}(t) : 0 \longrightarrow W_n(t) \xrightarrow{\partial_t} W_{n-1}(t) \longrightarrow \cdots \longrightarrow W_0(t) \longrightarrow 0.$$

It follows from (14) that

$$\lim_{t\to\infty} W_p(t) = \mathcal{M}_p.$$

We can say more. In fact, after a simple change of coordinates, the differential ∂_t on $\mathcal{W}(t)$ converges to the differential $\widetilde{\partial}$ on $\mathcal{M}$. To describe this change of coordinates, let π denote the orthogonal projection onto $\mathcal{M}_p$. For large t

$$\pi : W_p(t) \longrightarrow \mathcal{M}_p$$

is an isomorphism. Define a function θ on $W_p(t)$ by setting, for any critical simplex $\sigma^{(p)}$

$$\theta(\pi^{-1}(\sigma)) = e^{tf(\sigma)}$$

and extending linearly. Let $\widetilde{\partial}_t = \theta^{-1}\partial_t\theta$.

Theorem: $\lim_{t\to\infty} (\mathcal{W}(t), \widetilde{\partial}_t) = (\mathcal{M}, \widetilde{\partial})$.

In *[7]* we use this theorem to give a second proof, along the lines suggested by Witten in *[21]* that the discrete Morse complex $\mathcal{M}$ has the same homology as the underlying manifold.

As a final remark, we emphasize that all eigenvalues which do not converge to 0 converge to 1. This has an interesting application. Given a representation

$$\rho : \pi_1(M) \longrightarrow O(n, \mathbf{R})$$

we can twist the chain complex by ρ to define a complex

$$\mathcal{C}(\rho) : 0 \longrightarrow C_n(M,\rho) \xrightarrow{\partial} C_{n-1}(M,\rho) \xrightarrow{\partial} \cdots \longrightarrow C_0(M,\rho) \longrightarrow 0.$$

There is a canonical inner product on each $C_p(M,\rho)$ and hence we can define an adjoint operator ∂^* and a Laplacian

$$\Delta_p(\rho) = \partial\partial^* + \partial^*\partial : C_p(M,\rho) \longrightarrow C_p(M,\rho).$$

If each Laplacian $\Delta_p(\rho)$ is invertible, we can define the Reidemeister torsion of M with respect to the representation ρ by

$$T(M,\rho) = \prod_{p=0}^{n} (Det\, \Delta_p(\rho))^{(-1)^p \frac{p}{2}}$$

(this formula is due to Ray and Singer *[14]*). Now replace the differential ∂ by ∂_t as before, form the induced Laplacians, and consider

$$T(M,\rho,t) = \prod_{p=0}^{n} (Det\, \Delta_p(\rho,t))^{(-1)^p \frac{p}{2}}.$$

As with the "untwisted" Laplacian, all eigenvalues of $\Delta_p(\rho, t)$ which do not converge to 0 converge to 1. Thus, as $t \to \infty$, the right hand side "collapses" onto the complex $\mathcal{W}(\rho, t)$ of small eigenspaces. By letting $t \to \infty$, and keeping track of how T varies with t, and the effect of conjugating ∂_t by θ, we prove

Theorem: $T(M,\rho) = \mathit{Torsion}(\mathcal{M}(\rho), \widetilde{\partial})$

where $\mathcal{M}(\rho)$ denotes the canonical twisting of the Morse complex $\mathcal{M}$ by ρ, and the torsion of $\mathcal{M}(\rho)$ is

$$\prod_{p=0}^{n} (Det\, \widetilde{\Delta}_p(\rho))^{(-1)^p \frac{p}{2}}$$

where

$$\widetilde{\Delta}_p(\rho) \;=\; \widetilde{\partial}\widetilde{\partial}^* + \widetilde{\partial}^*\widetilde{\partial} : \mathcal{M}_p(\rho) \longrightarrow \mathcal{M}_p(\rho).$$

The smooth analogue of this theorem, proved by techniques in *[8]*, plays a key role in recent proofs of the equality of the Reidemeister torsion and Ray and Singer's analytic torsion (*[19]*, *[3]*, *[4]*).

References

[1] D. Barden, The structure of manifolds, Thesis, Cambridge University, 1963.

[2] R. Bott, Morse theory indomitable, Publ. Math. I.H.E.S. **68** (1988), 99–117.

[3] J.M. Bismut and W. Zhang, An extension of a theorem by Cheeger and Müller, Astérisque No. 205, Société Mathématique de France, 1992.

[4] D. Burghelea, L. Friedlander and T. Kappeler, Analytic torsion equals Reidemeister torsion, a new proof, preprint.

[5] A. Duval, A combinatorial decomposition of simplicial complexes, Israel J. of Math. **87**, 77–87.

[6] R. Forman, Morse theory for cell complexes, preprint.

[7] ——, Witten-Morse theory for cell complexes, preprint.

[8] B. Helffer, J. Sjöstrand, Puits multiples en mécanique semi-classique, IV etude du complexe de Witten, Comm. P.D.E. **10** (1985), 245–340.

[9] B. Mazur, Differential topology from the viewpoint of simple homotopy theory, Publ. Math. I.H.E.S. **15** (1963), 5–93.

[10] ——, Morse theory, in: *Differential and Combinatorial Topology (A Symposium in Honor of M. Morse)*, Princeton University Press (1965), 145–166.

[11] J. Milnor, Morse theory, Annals of Mathematics Study No. 51, Princeton University Press, 1962.

[12] ——, Lectures on the h-Cobordism Theorem, Princeton Mathematical Notes, Princeton University Press, 1965.

[13] M. Morse, Bowls of a non-degenerate function on a compact differentiable manifold, in: *Differential and Combinatorial Topology (A Symposium in Honor of M. Morse)* Princeton University Press (1965), 81–104.

[14] D.B. Ray and I.M. Singer, R-torsion and the Laplacian on Riemannian manifolds, Adv. in Math. **17** (1971), 145–210.

[15] S. Smale, On gradient dynamical systems, Annals of Math. **74** (1961), 199–206.

[16] ———, The generalized Poincaré conjecture in dimensions greater than four, Annals of Math. **74** (1961), 391–406.

[17] J.R. Stallings, Lectures on polyhedral topology, Tata Institute of Fundamental Research, Bombay, 1968.

[18] R. Stanley, A combinatorial decomposition of acyclic simplicial complexes, Discrete Math. **118** (1993).

[19] F. Tangerman, Reidemeister torsion and analytic torsion, preprint, 1988.

[20] J.H.C. Whitehead, Simplicial spaces, nuclei, and m-groups, Proc. London Math. Soc.

[21] E. Witten, Supersymmetry and Morse theory, J. Diff. Geom. **17** (1982), 661–692.

Characteristic Numbers and Generalized Path Integrals

Daniel S. Freed [1]

Department of Mathematics
University of Texas
Austin, TX 78712

I hope in this exposition to touch on two themes which are among Professor Bott's many mathematical interests. The first—characteristic numbers—appears throughout his work. My second topic, path integrals, have been around in physics since they were introduced by Feynman in the late 1940s, but have only recently been applied to problems of purely geometric interest. Edward Witten has led the way in this topological quantum field theory, which has attracted the enthusiasm of many mathematicians. The ideas of quantum field theory, though far from mathematically understood, seem to provide a unified framework for many recent invariants in low dimensional topology (à la Donaldson, Jones, Casson, Floer, ...) and introduce some new invariants as well. Our first goal is to explain that ideas of (classical) field theory inspire new insight into characteristic numbers. The second goal is to explain how summing over fields—the path integral—produces diffeomorphism invariants which satisfy *gluing laws*, and how a generalization of this idea explains some of the algebraic structure behind these invariants in three dimensions. We can only do this rigorously in a "toy model", where the path integral reduces to a finite sum, but we hope that the ideas here will shed light on more interesting examples as well. The toy model is based on a characteristic number for finite principal covering spaces. As I have explained most of these ideas elsewhere, my goal here is to give an elementary account of the basic concepts and the simplest cases.

The oldest, and in some sense most intriguing, characteristic number is the Euler number of a manifold. More generally, an oriented real vector bundle of rank n over a closed oriented manifold of dimension n has an Euler number, which takes values in the integers. Real and complex vector bundles over closed oriented manifolds have other characteristic numbers due to Chern, Pontrjagin, and Stiefel-Whitney. These integers are *primary* characteristic numbers in *cohomology*; they are topological invariants. There are similar *primary* characteristic numbers in K-*theory* for real and complex bundles over compact oriented (or spin) manifolds.

[1] The author is supported by NSF grant DMS-9307446, a Presidential Young Investigators award DMS-9057144, and by the O'Donnell Foundation. He thanks the University of Warwick and the Geometry Center at the University of Minneapolis for their hospitality while this note was written.

Geometry, Topology, and Physics for Raoul Bott

Cambridge, MA 02138 USA

The topological definition *[1]* of these characteristic numbers uses Bott Periodicity in an essential way. They include the signature of a manifold, the $\hat{A}$-genus, etc. What I term *secondary characteristic numbers* are *geometric* invariants which depend on a metric or connection and take values in $\mathbb{R}/\mathbb{Z}$, or, after exponentiating, in the circle group $\mathbb{T}$. In cohomology they are the Chern-Simons numbers and in K-theory they are the η-invariants. We summarize in Table 1. Our observation is that these classical invariants extend to invariants of manifolds "below the top dimension" and of manifolds with boundary. Thus, for example, the characteristic number c_4 is an integer associated to a complex vector bundle over a closed oriented 8-manifold. There are invariants of complex vector bundles over compact oriented manifolds (possibly with boundary) in all dimensions ≤ 8, but now the invariants are not simply numbers. For manifolds of dimension 7 they are sets, for manifolds of dimension 6 they are categories, and so on. These sets and categories have more structure, of course. These extended invariants obey gluing laws, so are in some sense *local* invariants. That is, in some respect they behave in some respects like the integral over the manifold of a locally computed density. In §1 we explain the simplest case of these ideas: the Euler number of a complex line bundle over a surface. The structure of the invariants in other cases is similar, though the actual constructions are quite different. I do not know all of the constructions for the K-theory invariants.

	cohomology	**K-theory**
primary	Chern numbers, Pontrjagin numbers	signature, $\hat{A}$-genus
secondary	Chern-Simons invariants	η-invariants

Table 1: Classical characteristic numbers

Because the invariants in Table 1 obey gluing laws, they are appropriate classical actions for a field theory. Whereas the classical action is an "integral" over a finite dimensional manifold, the quantum path integral is an "integral" over a space of fields on that manifold, which typically is infinite dimensional. Although mathematical physicists have made great progress in understanding the technology of these integrals, the cases of topological significance discussed here remain elusive. Our main idea in §2 is to extend this (ill-defined) notion of integration over fields to the *extended* classical action, i.e., to fields on manifolds below the top dimension. For example, in a three dimensional topological field theory the path integral for a closed oriented 3-manifold produces a complex number, whereas canonical quantization assigns a complex Hilbert space to a closed oriented 2-manifold. These are usually considered as different processes. We reinterpret this Hilbert space as the result of an integration process over the space of fields on the 2-manifold. The extended classical action has values which are hermitian lines, and the integral (direct sum) of hermitian lines is a Hilbert space. This is very strange—integrals are usually numbers, not Hilbert spaces! Nonetheless, this strange notion of integration leads immediately to a gluing law for the Hilbert spaces which, for example, is closely related to the "Verlinde formula" in conformal field theory. This idea is a pure formality in general; we emphasize that so far this whole scheme only works rigorously for our toy model. After all, the usual path integral is also not rigorously defined in topological theories. Our extended notion of path integral applies to 1-manifolds (in a three dimensional topological theory) and leads to a braided monoidal category, and so ultimately to a quantum group. We view this as the *solution* to the theory. That is, we start with the classical action to define the theory, then compute the quantum group (actually its category of representations) from the theory, and finally derive formulas for invariants from the gluing laws of the path integral (e.g. in terms of a presentation as surgery on a link).

	cohomology	**K-theory**
primary	Donaldson invariants, Floer homology, Gromov invariants	???
secondary	Reshetikhin-Turaev-Witten invariants, Jones invariants	???

Table 2: Quantum characteristic numbers

I find it useful to organize many of the topological quantum field theories floating around into Table 2. I like to think of the invariants in these theories as *quantum characteristic numbers*. Often people distinguish between two types of topological field theories, which here is the distinction between primary and secondary invariants. Notice that on the quantum level the geometric data used to define the secondary characteristic number has been integrated out, so we are left with topological invariants in both types of theory. The main example here is Witten's definition *[19]* of an invariant of 3-manifolds as a path integral of Chern-Simons. This is most closely analogous to the usual path integral in physics. Less clear is the integration process which leads from primary characteristic numbers (which are already topological invariants) to Donaldson invariants, etc. Witten introduced *supersymmetric* path integrals *[20, 21]* to explain these invariants, and Baulieu-Singer reinterpreted this in terms of primary characteristic numbers and gauge fixing *[4]*. We should also mention that these invariants are in some sense an infinite dimensional Euler number, and that the path integral representation is an infinite dimensional version of the Mathai-Quillen formula *[16, 2]*. From this point of view the primary characteristic numbers in K-theory which I placed in Table 1 could be pigeon-holed in the upper left hand corner of Table 2, thanks to supersymmetric quantum mechanics. Just as with the classical invariants in Table 1, the geometric and topological applications of *primary* quantum invariants (Donaldson) have been more striking than the applications of *secondary* invariants (Chern-Simons).

The idea that in a three dimensional topological quantum field theory one should attach certain types of categories to 1-manifolds has been discussed by Kazhdan, Segal, Lawrence, Yetter, Crane, and many others. The construction by generalized path integrals is new. I refer the reader to *[9]* for more details about generalized path integrals and for the detailed computations for finite group gauge theory. The appendix to *[14]* defines the characteristic numbers used in this model, and *[14]* contains many more details about the basics of the theory. There is another expository account of some of this material in *[10]*. In particular, the derivation of the quantum group is explained in more detail there.

It is a great pleasure and honor to dedicate this paper to Raoul Bott. His birthday is a wonderful occasion to celebrate his youthful exuberance for life and for mathematics. *L'chaim*, Raoul!

1 Characteristic Numbers

Consider a complex line bundle $L \to X^2$ over a closed oriented 2-manifold X. Its Euler number $e_X(L) \in \mathbb{Z}$ can be computed in several ways. For example, choose any section $s : X \to L$ which is transverse to the zero section. Then s has isolated zeros, and the orientations determine a sign $\operatorname{sgn}(y)$ for each zero y. A standard argument shows that

$$e_X(L) = \sum_{y \in \mathrm{Zero}(s)} \operatorname{sgn}(y) \tag{1.1}$$

is independent of s, and so defines a topological invariant $e_X(L) \in \mathbb{Z}$. As we deform s two zeros of opposite sign can simultaneously die or simultaneously appear. But the total signed number of zeros is constant.

Consider now the same situation over a compact oriented 2-manifold X with nonempty boundary. Assume that the section s does not vanish on ∂X. It is still true that (1.1) does not change if we modify s in the interior of X, but now (1.1) depends on the restriction of s to the boundary. If s is deformed allowing zeros on the boundary, then (1.1) changes according to the number of zeros (counted with sign, of course) which flow through the boundary during the deformation. More formally, define $e_X(L, t)$ for any nonvanishing section $t : \partial X \to L$ by extending to a section $s : X \to L$ and counting zeros as in (1.1). Then for two such section t, t' we have

$$e_X(L, t') - e_X(L, t) = \deg_{\partial X}(t'/t). \tag{1.2}$$

Here the ratio t'/t is a map $\partial X \to \mathbb{C}^\times$ which has a degree, or winding number, computed using the induced orientation of ∂X. So the relative Euler number depends on the section t (up to homotopy). This is the traditional point of view on relative characteristic numbers—they depend on a trivialization on the boundary.

Now the new twist: We can define a relative Euler number which is independent of t if we abandon the idea that it should be a number. Namely, let

Sect* denote the set of nonzero sections, and set

$$T_{\partial X}(\partial L) = \{e : \text{Sect}^*(\partial L) \longrightarrow \mathbb{Z} \text{ which obey } e(t') - e(t) = \deg_{\partial X}(t'/t)\}. \tag{1.3}$$

Notice that if $e, e' \in T_{\partial X}(\partial L)$ then $e' - e$ is a constant integer. Likewise, if $e \in T_{\partial X}(\partial L)$ and $n \in \mathbb{Z}$ then $e + n \in T_{\partial X}(\partial L)$. So $T_{\partial X}(\partial L)$ is a principal homogeneous space for the integers, a so-called $\mathbb{Z}$-*torsor*. It is an "affine" copy of $\mathbb{Z}$—a copy of $\mathbb{Z}$ without a preferred origin. In other terms it is a principal $\mathbb{Z}$ bundle over a point. It is *not* a group. The function $e_X(L, \cdot)$ lies in this torsor, by (1.2), and so we obtain a relative Euler number

$$e_X(L) \in T_{\partial X}(\partial L) \tag{1.4}$$

which only depends on the bundle L.

We now explain the sense in which (1.4) is a topological invariant. First, notice that (1.3) defines a $\mathbb{Z}$-torsor $T_Y(K)$ for any line bundle $K \to Y$ over any closed oriented 1-manifold Y. It is a topological invariant of K. Usually topological invariants are *numbers*, and equivalent objects have *equal* invariants. Here the topological invariant is a *set*, and equivalent objects have *isomorphic* invariants. Precisely, if $K' \xrightarrow{\psi} K$ is an isomorphism of line bundles over Y, it induces an isomorphism of $\mathbb{Z}$-torsors $T_Y(K') \xrightarrow{\psi_*} T_Y(K)$. This is familiar from algebraic topology, where invariants like the fundamental group or homology groups of a space are sets and homeomorphisms on spaces induce isomorphisms of these sets. The relative Euler number (1.4) is a topological invariant in this sense: If $L' \xrightarrow{\varphi} L$ is an isomorphism of circle bundles over X, then $(\partial\varphi_*)\big(e(L')\big) = e(L)$. Most importantly, the Euler number obeys a *gluing law*. To see this, first note that if $-Y$ denotes Y with the opposite orientation, then for any $K \to Y$ there is a natural pairing

$$+ : T_Y(K) \times T_{-Y}(K) \longrightarrow \mathbb{Z} \tag{1.5}$$

by addition. Now suppose X_1, X_2 are oriented surfaces with boundaries $\partial X_1 = Y$ and $\partial X_2 = -Y$. Let $X = X_1 \cup_Y X_2$ denote the surface formed by gluing along the boundary. Suppose we are also given circle bundles $L_i \to X_i$ and an isomorphism $\partial L_1 \to \partial L_2$. Let $L \to X$ denote the glued bundle. Then the gluing law asserts

$$e_X(L) = e_{X_1}(L_1) + e_{X_2}(L_2). \tag{1.6}$$

The proof is direct from the definitions.

There are alternative constructions of these invariants. For example, fix a metric on L and let $\mathcal{A}_L$ denote the space of unitary connections on L. Let $F(\theta) \in \Omega^2_X$ denote $i/2\pi$ times the curvature of a connection $\theta \in \mathcal{A}_L$. Then

$$e_X(L, \theta) = \int_X F(\theta) \tag{1.7}$$

is a real-valued function on $\mathcal{A}_L$. If $\partial X = \emptyset$ it is a constant integer equal to $e_X(L)$. If $\partial X \neq \emptyset$ then

$$\exp\Big(2\pi i \int_X F(\theta)\Big) = \text{hol}_{\partial X}(\theta),$$

where $\mathrm{hol}_{\partial X}(\theta)$ is the holonomy of θ around ∂X. In this situation we define the $\mathbb{Z}$-torsor

$$T'_{\partial X}(\partial L) = \{f : \mathcal{A}_{\partial L} \to \mathbb{R} : e^{2\pi i f(\theta)} = \mathrm{hol}_{\partial X}(\theta)\}. \tag{1.8}$$

Then (1.7) defines the relative Euler number. With these definitions the gluing law (1.6) follows from the additivity of integration:

$$\int_X = \int_{X_1} + \int_{X_2} .$$

The Euler number is an example of a primary characteristic number (see Table 1). More generally, if G is a Lie group and $\lambda \in H^n(BG;\mathbb{Z})$ a universal characteristic class, then λ determines[2] a characteristic number $\lambda_X(P)$ for a G bundle $P \to X$ over a closed oriented n-manifold X. One can define a $\mathbb{Z}$-torsor $T_Y(Q)$ for a G bundle $Q \to Y$ over a closed oriented $(n-1)$-manifold Y, and a relative invariant for n-manifolds with boundary. The story continues to higher codimensions. For example, the invariant of a bundle over a closed oriented $(n-2)$-manifold is a certain type of category, called a *gerbe*.

We illustrate the codimension 2 invariant in the simplest case of the Euler number. Recall the definition (1.3) of the $\mathbb{Z}$-torsor associated to a line bundle $K \to Y$ over a closed oriented surface. Now suppose Y is a compact oriented 1-manifold with boundary, i.e., a finite union of circles and closed intervals. Let $u : \partial Y \to \partial K$ be a nonzero section over the boundary and let $\mathrm{Sect}^*(K,u)$ be the set of nonzero sections $t : Y \to K$ with $\partial t = u$. Then define the $\mathbb{Z}$-torsor

$$T_Y(K,u) = \{e : \mathrm{Sect}^*(K,u) \to Z : e(t') - e(t) = \deg_{\partial X}(t'/t)\}.$$

The degree is well-defined since $\partial t' = \partial t$. We need to determine the dependence of $T_Y(K,u)$ on u. Now if u' is any other trivialization of $\partial K \to \partial Y$, then the set of nonzero paths joining u to u', up to homotopy, is a $\mathbb{Z}$-torsor $T_{\partial Y}(\partial K, u, u')$. Such a path gives a 1:1 correspondence $\mathrm{Sect}^*(K,u) \to \mathrm{Sect}^*(K,u')$, and so there is an isomorphism

$$T_Y(K,u) \times T_{\partial Y}(\partial K, u, u') \longrightarrow T_Y(K,u').$$

This leads us to define

$$\begin{aligned}\mathcal{G}_{\partial Y}(\partial K) = \big\{&T : \mathrm{Sect}^*(\partial K) \to \{\mathbb{Z}\text{-torsors}\} \text{ with given isomorphisms} \\ &T(u) \times T_{\partial Y}(\partial K, u, u') \to T(u') \text{ which satisfy various conditions}\big\}.\end{aligned} \tag{1.9}$$

In this definition T is a set-valued function, so $\mathcal{G}_{\partial Y}(\partial K)$ is a collection of set-valued functions, in particular, a category. The "various conditions" are related to the category structure and describe an associativity constraint for the isomorphism associated to three nonzero sections u, u', u''. Think of the *gerbe* (1.9) as an

[2] In fact, we must choose a particular representative of the cohomology class $\lambda \in H^n(BG;\mathbb{Z})$ to carry out the constructions which follow *[14, Appendix B], [9, §2]*.

affine copy of the category of $\mathbb{Z}$-torsors. The relative invariant $T_Y(K)$ lies in the gerbe $\mathcal{G}_{\partial Y}(\partial K)$ by definition.

This invariant admittedly contains little information. However, if we start with a 4 dimensional characteristic class, then the gerbe attached to a surface leads to a central extension of the diffeomorphism group of the surface which arises in quantum Chern-Simons theory *[19]*. (It is better here to start with the *signature* of a 4-manifold *[10]*.) The central extension is often realized in terms of "2-framings".

Our interest in §2 is in 3 dimensional invariants of principal bundles with *finite* gauge group G. But if G is finite then $H^n(BG;\mathbb{Z})$ consists entirely of torsion elements for $n > 0$, and so the primary integral characteristic numbers vanish. Rather, we start with

$$\lambda \in H^3(BG;\mathbb{R}/\mathbb{Z}) \cong H^4(BG;\mathbb{Z}) \tag{1.10}$$

and obtain a characteristic number $\lambda_X(P) \in \mathbb{R}/\mathbb{Z}$ for G bundles $P \to X^3$ over a closed oriented 3-manifold. Exponentiating, the invariant

$$e^{2\pi i \lambda_X(P)} \in \mathbb{T} \tag{1.11}$$

lies in the circle group $\mathbb{T}$ of unit norm complex numbers. The relative picture is similar: To a G bundle $Q \to Y$ over a closed oriented surface Y we attach a $\mathbb{T}$-torsor, and to a bundle $P \to X$ over a 3-manifold with boundary the relative invariant lives in the $\mathbb{T}$-torsor of the boundary. These invariants obey a gluing law (1.6) which we now write multiplicatively. Notice that any $\mathbb{T}$-torsor is the set of unit norm elements in a hermitian line, which here we denote $L_Y(Q)$. So the relative invariant of $P \to X$ is

$$e^{2\pi i \lambda_X(P)} \in L_{\partial X}(\partial P). \tag{1.12}$$

Similarly, the invariant of a G bundle $R \to S$ over a closed oriented 1-manifold S (finite union of circles) is a $\mathbb{T}$-gerbe $\mathcal{G}(R)$, which we think of as an affine copy of the category of hermitian lines. (Imagine definition (1.9) with hermitian line-valued functions.) Then the relative invariant of a G bundle $Q \to Y$ over an oriented surface with boundary is an element

$$L_Y(Q) \in \mathcal{G}_{\partial Y}(\partial Q). \tag{1.13}$$

Although (1.11) is a topological invariant, and in that sense is primary (Table 1), because of the transgression in (1.10) we can think of it as a Chern-Simons invariant for bundles with finite structure group. More generally, for any compact group G and class $\lambda \in H^n(BG;\mathbb{Z})$ there are secondary *geometric* invariants for connections on bundles over compact oriented manifolds of dimension at most $n-1$. For a closed oriented $(n-1)$-manifold the invariant lies in $\mathbb{T}$, for a closed oriented $(n-2)$-manifold the invariant *is* a $\mathbb{T}$-torsor (or hermitian line), etc. For a family of connections on an $(n-2)$-manifold there is a connection on the line bundle over the parameter space. We develop these ideas in *[11]* using a variant of the *Čech-de Rham complex* (more fondly "tic-tac-toe" *[5]*) which is related to Deligne cohomology *[6]*. This amounts to a generalization of the theory of differential forms, including both differentiation and integration. The case $n - 1 = 3$ is of special interest *[12]* because of the relation with special geometric structures in low dimensional gauge theory and with the corresponding quantum invariants.

Let us briefly consider the K-theory side of Table 1. From the point of view of Dirac operators the primary characteristic number—in $\mathbb{Z}$—is the index of a Dirac operator on an even dimensional closed spin manifold, and the secondary characteristic number—in $\mathbb{T}$—is the exponentiated ξ-invariant of a Dirac operator on an odd dimensional closed spin manifold. (Recall that the ξ-invariant is half the η-invariant plus half the dimension of the kernel.) Notice that the former is a topological invariant whereas the latter depends on geometric data, i.e., a metric and possibly a connection. One can define relative invariants in this context. First, a reinterpretation of the work of Atiyah-Patodi-Singer *[3]* constructs the index of a Dirac operator on a manifold with boundary as a topological invariant *[13]*. It lives in a $\mathbb{Z}$-torsor (which is an invariant of the Dirac operator on the boundary) and satisfies a gluing law. This is completely analogous to (1.4) and (1.6). In recent work with Xianzhe Dai *[7]* we construct a relative exponentiated ξ-invariant (analogous to (1.12)) which lives in the determinant line of the Dirac operator on the boundary and satisfies a gluing law. However, I do not know how to extend these constructions to manifolds of lower dimension (except for the next step on the "primary side"). Also, it would be interesting to give topological constructions of the relative primary invariants which generalize the direct image *[1]*.

2 Generalized Path Integrals

Consider a finite group G and a class $\lambda \in H^3(BG; \mathbb{R}/\mathbb{Z})$. In the last section we indicated that there are invariants $\lambda_X(P) \in \mathbb{T}$ for each principal G bundle $P \to X^3$ over a closed oriented 3-manifold. If we are interested in constructing topological invariants of X, then we need to eliminate the dependence on P. One way—the physicists' way—is to integrate over P. In quantum field theory P is called a *field* and this integral over the space of fields is called the *path integral*. It is in this sense that our discussion here is a field theory. More specifically, it is the quantum Chern-Simons theory *[19]* for finite gauge group, first introduced by Dijkgraaf and Witten *[8]*. Fortunately, the integral reduces to a finite sum in this theory and there are no analytic problems to worry about.

Let $\mathcal{C}_X$ denote the collection of G bundles over X. It is a category—morphisms are bundle maps which cover the identity. Let $\overline{\mathcal{C}_X}$ denote the finite set of equivalence classes of bundles, and define a measure on $\overline{\mathcal{C}_X}$ by

$$\mu_X(P) = \frac{1}{\# \mathit{Aut}\, P} \tag{2.1}$$

for $P \in \mathcal{C}_X$. The topological invariant of a closed oriented 3-manifold X we wish

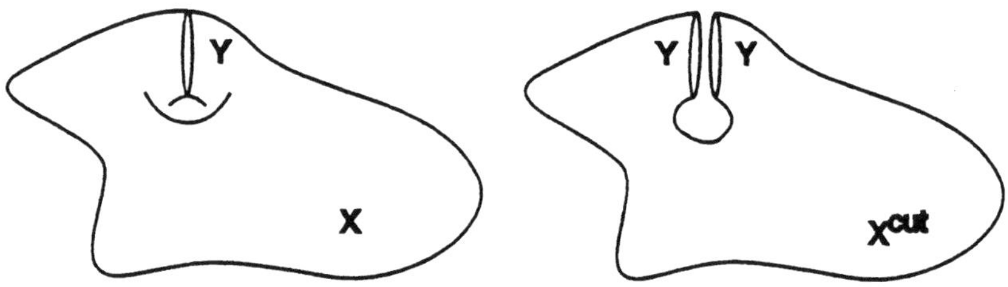

Figure 2.1:

to study is

$$Z_X = \int_{\mathcal{C}_X} e^{2\pi i \lambda_X(P)}\, d\mu_X(P). \tag{2.2}$$

We view the integrand (1.11) as a complex number and perform the integral by summing complex numbers. So $Z_X \in \mathbb{C}$. Although we use integral notation, (2.2) is a finite sum. The fact that $e^{2\pi i \lambda_X(P)}$ is a topological invariant of P quickly leads to a proof that Z_X is a topological invariant of X.

Just as with the Euler number in §1, we would like to define an invariant of a 3-manifold with boundary which satisfies a gluing law analogous to (1.6). Let X be a closed oriented 3-manifold and $Y \hookrightarrow X$ an embedded closed oriented surface. Denote by X^{cut} the manifold obtained by cutting X along Y. (See Figure 2.1.) Notice that $\partial X^{\text{cut}} = Y \sqcup -Y$ is a disjoint union of two copies of Y. The fields (bundles) fit into the following diagram:

$$\begin{array}{ccc} \mathcal{C}_X & \xrightarrow{\ c\ } & \mathcal{C}_{X^{\text{cut}}} \\ \Big\downarrow{\scriptstyle r_1} & & \Big\downarrow{\scriptstyle r_2} \\ \mathcal{C}_Y & \xrightarrow{\ \Delta\ } & \mathcal{C}_Y \times \mathcal{C}_{-Y} \end{array}$$

The vertical arrow r_1 is restriction to Y, the arrow r_2 is restriction to ∂X^{cut}, the arrow Δ is the diagonal inclusion, and c is the pullback under the gluing map. For the moment we ignore symmetries and pretend that (2.2) is an integral over $\mathcal{C}_X$. Then we propose to do the integral over $\mathcal{C}_X$ in two stages using Fubini's theorem: First integrate over the fibers of r_1 and then over $\mathcal{C}_Y$. Now the gluing law (1.9) for λ says that if $P \in \mathcal{C}_X$ and $P^{\text{cut}} = c(P) \in \mathcal{C}_{X^{\text{cut}}}$, then

$$e^{2\pi i \lambda_X(P)} = e^{2\pi i \lambda_{X^{\text{cut}}}(P^{\text{cut}})}. \tag{2.3}$$

The right hand side of (2.3) lives in $L_Y(Q) \otimes L_{-Y}(Q)$ (cf. (1.12)) which is identified with $\mathbb{C}$ via a pairing analogous to (1.5). Our hope, then, is to make the following computation:

$$\begin{aligned} Z_X &= \int_{\mathcal{C}_X} e^{2\pi i\lambda_X(P)}\, d\mu_X(P) \\ &= \int_{\mathcal{C}_Y}\int_{r_1^{-1}(Q)} e^{2\pi i\lambda_X(P)}\, d\mu_{r_1^{-1}(Q)}(P)\, d\mu_Y(Q) \\ &= \int_{\mathcal{C}_Y}\int_{r_2^{-1}(Q,Q)} e^{2\pi i\lambda_{X^{\text{cut}}}(P^{\text{cut}})}\, d\mu_{r_2^{-1}(Q,Q)}(P^{\text{cut}})\, d\mu_Y(Q). \end{aligned} \tag{2.4}$$

For this to be a valid computation we need the measures to work out properly. Also, we must include the symmetries. Both are easily handled *[14, §2]*.

Let's reinterpret the last line of (2.4). For a moment suppose that X' is any compact oriented 3-manifold with boundary. Then for $Q \in \mathcal{C}_{\partial X'}$ define

$$\mathcal{C}_{X'}(Q) = \{P \to X' \text{ such that } \partial P = Q\}.$$

There are symmetries here as well, and we are deliberately vague in order to keep the ideas as simple as possible. Note that $\mathcal{C}_{X'}(Q)$ is the space of fields with a given fixed boundary value. We generalize the path integral (2.2) to manifolds with boundary by defining

$$Z_{X'}(Q) = \int_{\overline{\mathcal{C}_{X'}(Q)}} e^{2\pi i\lambda_{X'}(P)}\, d\mu_{X'}(P). \tag{2.5}$$

The right hand side takes values in the hermitian line $L_{\partial X'}(Q)$ (cf. (1.12)). So $Z_{X'}$ is a section of the hermitian line bundle $L_{\partial X'} \to \mathcal{C}_{\partial X'}$. In an appropriate sense it is invariant under symmetries. For any closed surface Y set

$$E(Y) = \text{invariant sections of } L_Y \to \mathcal{C}_Y. \tag{2.6}$$

Then (2.5) determines a relative invariant

$$Z_{X'} \in E(\partial X').$$

We impose an L^2 inner product on (2.6) using the measure (2.1) on *equivalence classes* of bundles.

With these definitions the gluing law (2.4) (with symmetries restored) takes the form

$$\begin{aligned} Z_X &= \int_{\overline{\mathcal{C}_{\partial X}}} Z_{X^{\text{cut}}}(Q,Q)\, d\mu_{\partial X}(Q) \\ &= \text{Trace}_{\partial X}(Z_{X^{\text{cut}}}), \end{aligned} \tag{2.7}$$

where $\text{Trace}_{\partial X} : E(\partial X) \otimes \underline{E(-\partial X)} \to \mathbb{C}$ is formed using the L^2 inner product. (Notice that $L_{-\partial X}(Q) \cong \overline{L_{\partial X}(Q)}$ and so $E(-\partial X) \cong \overline{E(\partial X)}$.)

What we have recounted so far is the standard argument that path integrals give numerical invariants which satisfy gluing laws. The relative invariants in quantum theories do not live in "one dimensional" torsors, as in the classical case (1.3), (1.8), but rather in Hilbert spaces (2.6). Now we want to go further and derive gluing laws for the Hilbert spaces.

Here is the main idea: Re-express the Hilbert space $E(Y)$ as an integral over the space of fields $\overline{\mathcal{C}_Y}$. Then repeat the argument which leads from (2.2) to the gluing law (2.7). In the process we will define an invariant $\mathcal{E}(S)$ of a closed oriented 1-manifold S, by analogy with (2.6). If the latter can be re-expressed as an integral we can again iterate the process.

To find the integral which computes $E(Y)$ recall that the characteristic class λ associates to each G bundle $Q \to Y$ a hermitian line $L_Y(Q)$. Furthermore, an isomorphism $Q' \to Q$ of bundles induces an isomorphism $L_Y(Q') \to L_Y(Q)$ of hermitian lines. So there is a quotient hermitian line bundle

$$L_Y \longrightarrow \overline{\mathcal{C}_Y}$$

(This bundle degenerates where automorphisms of $Q \in \mathcal{C}_Y$ act nontrivially on $L_Y(Q)$.) Our formula for the Hilbert space is:

$$\boxed{E(Y) = \int_{\overline{\mathcal{C}_Y}} L_Y(Q)\, d\mu_Y(Q)} \tag{2.8}$$

This is a boldly presented formula and it requires some explanation. The right hand side has the form

$$\sum_{\text{finite}} (\text{positive number}) \cdot (\text{hermitian line}). \tag{2.9}$$

If $\mu > 0$ and L is a hermitian line, define $\mu \cdot L$ to be the hermitian line with the same underlying vector space as L but with an inner product which is μ times the inner product on L. Interpret the finite sum as the direct sum of hermitian lines. This is the sense in which the right hand side of (2.8) defines a Hilbert space—as a sum of hermitian lines—and it is easy to see that (2.6) and (2.8) define the same Hilbert space. As explained in §1 the hermitian line $L_Y(Q)$ is the two dimensional counterpart of the three dimensional invariant $e^{2\pi i \lambda_X(P)}$. So the integrals (2.2) and (2.8) have the same form: they are the integral of the exponentiated classical action over the space of fields.

By now it should be clear how to continue. If Y is an oriented surface with boundary, then by analogy with (2.5) we define

$$E(Y)(R) = \int_{\overline{\mathcal{C}_Y(R)}} L_Y(Q)\, d\mu_Y(Q), \qquad R \in \mathcal{C}_{\partial Y}.$$

Recall from (1.13) that $L_Y(Q)$ is an element of the gerbe $\mathcal{G}_{\partial Y}(R)$. Think of the latter as an affine copy of the category of hermitian lines. Then just as we can add hermitian lines to obtain a Hilbert space (2.9), we can add elements of $\mathcal{G}_{\partial Y}(R)$. What do we get? If $\mathcal{G}_{\partial Y}(R)$ were trivial we would get a Hilbert space. Then $E(Y)$ would be an invariant function from $\mathcal{C}_{\partial Y}$ to the category of Hilbert spaces. The nontriviality of $\mathcal{G}_{\partial Y}(R)$ means that $E(Y)$ is an invariant function from $\mathcal{C}_{\partial Y}$ to an affine version of the category of Hilbert spaces, a so-called *2-Hilbert space [15]*. The space $\mathcal{E}(\partial Y)$ of all such functions is again a 2-Hilbert space and $E(Y) \in \mathcal{E}(\partial Y)$. If we cut a surface along a circle, then we can formulate a gluing law analogous to (2.7). Finally, for any closed oriented 1-manifold S we write the 2-Hilbert space $\mathcal{E}(S)$ as an integral:

$$\mathcal{E}(S) = \int_{\overline{\mathcal{C}_S}} \mathcal{G}_S(R)\, d\mu_S(R). \tag{2.10}$$

In principle, we can continue to even lower dimensions.

We stop here, but refer the reader to *[10]* for an expository account which begins with the notion of a 2-Hilbert space. There we give a general argument to show how in a 3 dimensional topological quantum field theory a "quantum group" arises from the 2-Hilbert space. In *[9]* we carry out detailed computations for gauge theory with finite gauge group.

We end with some remarks about Chern-Simons theory with *continuous* gauge group. Then (2.2), (2.8), and (2.10) are replaced by integrals over the space of connections mod equivalence. Equation (2.2) is then the path integral heuristic given by Witten *[19]*. Reshetikhin and Turaev *[17]* subsequently gave an explicit computable formula in terms of quantum group data and proved that it is a topological invariant. Note that the formal integral (2.2) is over an infinite dimensional space. By contrast, the integral (2.8) reduces, after symplectic reduction, to an integral over the finite dimensional moduli space $\mathcal{M}_Y$ of flat connections on Y. The integrand is a line bundle with connection whose curvature is the symplectic form on $\mathcal{M}_Y$. In this situation we may imagine that (2.8) is a formal expression for geometric quantization. The use of an integral is justified by the existence of gluing laws—Verlinde's formula *[18]*. By now there are various proofs of at least special cases of these gluing laws.

What about (2.10)? If S is a circle then up to equivalence a connection is determined by its holonomy, which is a conjugacy class in the gauge group G. So (2.10) is an integral over a compact Lie group G with some invariance under the adjoint action. The integrand is a geometric object over G, "invariant" under the adjoint action, with a "connection" whose curvature is the canonical biinvariant 3-form on G constructed from the starting data $\lambda \in H^4(BG)$. (See *[6]* for one description.) I think it is reasonable to imagine that there is a geometric process—analogous to geometric quantization—which constructs the 2-Hilbert space $\mathcal{E}(S)$ from this "gerbe bundle with connection", but for now this remains a mystery.

References

[1] M. F. Atiyah and F. Hirzebruch , Riemann-Roch theorems for differentiable manifolds, Bull. Amer. Math. Soc. **65** (1959), 276–281.

[2] M. F. Atiyah and L. Jeffrey, Topological lagrangians and cohomology, J. Geom. Phys. **7** (190), 119–136.

[3] M. F. Atiyah, V. K. Patodi, and I. M. Singer, Spectral asymmetry and Riemannian geometry I, Math. Proc. Cambridge Philos. Soc. **77** (1975), 43–69.

[4] L. Baulieu and I. M. Singer, Topological Yang-Mills symmetry, Nuclear Phys. B Proc. Suppl. **5b** (1988), 12–19.

[5] R. Bott and L. Tu, *Differential Forms in Algebraic Topology*, Springer-Verlag, New York, (1982).

[6] J.-L. Brylinski, *Loop Spaces, Characteristic Classes and Geometric Quantization*, Birkhäuser, Boston (1993).

[7] X. Dai and D. S. Freed, η-invariants and determinant lines, J. Math. Phys. **35** (1994), 5155–5194.

[8] R. Dijkgraaf and E. Witten, Topological gauge theories and group cohomology, Commun. Math. Phys. **129** (1990), 393–429.

[9] D. S. Freed, Higher algebraic structures and quantization, Commun. Math. Phys. **159**, 343–398.

[10] ——, Extended structures in topological quantum field theory, in: *Quantum Topology*, L.H. Kauffman, R.A. Baadhiv, eds., World Scientific, 1993, 162–173.

[11] ——, Higher line bundles, in preparation.

[12] ——, Classical Chern-Simons theory, Part 1, Adv. Math., to appear.

[13] ——, A gluing law for the index of Dirac operators, in: *Global Analysis in Modern Mathematics*, Karen Uhlenbeck, ed., Publish or Perish, 1993, 5–14.

[14] D. S. Freed and F. Quinn, Chern-Simons theory with finite gauge group, Commun. Math. Phys., **151** (1993), 435–472.

[15] M. M. Kapranov and V. A. Voevodsky, 2-Categories and Zamolodchikov tetrahedra equations, preprint, 1992.

[16] V. Mathai and D. Quillen, Superconnections, Thom classes, and equivariant differential forms, Topology **25** (1986), 85–110.

[17] N. Reshetikhin and V. G. Turaev, Invariants of 3-manifolds via link polynomials and quantum groups, Invent. math. **103** (1991), 547–597.

[18] E. Verlinde, Fusion rules and modular transformations in 2d conformal field theory, Nucl. Phys. **B300** (1988), 360–376.

[19] E. Witten, Quantum field theory and the Jones polynomial, Commun. Math. Phys. **121** (1989), 351–399.

[20] ——, Topological quantum field theory, Commun. Math. Phys. **117** (1988), 353–386 .

[21] ——, Topological sigma models, Commun. Math. Phys. **118** (1988), 411.

Valentin Poenaru's Program For The Poincare Conjecture

David Gabai[1]
California Institute of Technology
Pasadena, CA 91125 USA

-1 Disclaimer

My goal is to expose some of the central ideas and landmarks of Poénaru's monumental program, despite the fact that I have yet to go through many of the details.

0 Introduction

The following is known as the

Poincaré conjecture: *If Σ is a closed simply connected 3-manifold, then Σ is homeomorphic to S^3.*

Definition 0.1: A manifold is *geometrically simply connected* if it has a handlebody decomposition without 1-handles. Thus the Poincaré conjecture is equivalent to saying that every simply connected closed 3-manifold is geometrically simply connected.

Valentin Poénaru (henceforth known as Po) proposes to establish the Poincaré Conjecture via the following two steps.

Step 1. Stable Poincaré Conjecture: If Σ is a closed simply connected 3-manifold, then $\Sigma \times I$ is geometrically simply connected.

Step 2: The Stable Poincaré conjecture implies the Poincaré conjecture.

Remarks: 0) The arguments in support of both steps are carried out in the smooth category. According to Corollary 1.3 *[2]*, $\Sigma \times \mathbb{R}$ is homeomorphic to $S^3 \times \mathbb{R}$.

i) Compare Steps 1-2 with these two outstanding open problems in smooth 4-manifold topology. Are all smooth homotopy 4-spheres geometrically simply connected? Does the geometrical simple connectivity of the homotopy 4-sphere Σ imply that Σ is S^4?

[1] *Note added in Proof (1 April 1995): Poenaru has withdrawn the bulk of [13]. The "only if" direction of Theorem 2.14 is now conjectural, thereby nullifying lines 6-7, 16-17 of p. 141. Theorem 2.10 and Corollary 2.11 remain intact. Since [13] is independent of the other works, the Poenaru program would now reduce the Poincare Conjecture to the above conjecture which we call the "Po V conjecture".

[2] Partially supported by NSF Grant DMS-9200584

Geometry, Topology, and Physics for Raoul Bott

Cambridge, MA 02138 USA

ii) If $\Sigma^3 \times I$ is geometrically simply connected, then it has a handle decomposition consisting of one 0-handle, p 2-handles and $p+1$ 3-handles. For homological reasons the 2-handles are attached with 0-framings to a regular neighborhood of a p component link L in S^3 all of whose linking numbers are trivial.

iii) Taking boundaries, we see that zero framed surgery on L yields $\Sigma\#(-\Sigma)\#$ (p $S^2 \times S^1$'s). One approach to proving Step 2 is by establishing the generalized Poénaru conjecture (1967) "If $M\#(p$ $S^2 \times S^1$'s) is obtained by zero framed surgery on a link L of p components in S^3, then $M = S^3$". Here # denotes ordinary connected sum. See *[3]* for a proof in the case $p = 1$. The case $p > 1$ is open. Po follows a different route for Step 2.

iv) See *[15]* for early efforts along these lines.

How is Poénaru's program different? The typical abortive proof of the Poincaré Conjecture proceeds as follows. First one notes that a homotopy 3-sphere Σ can be described in terms of certain data. Second, one develops a calculus for the data, i.e. operations which change data into new data which also describes the same 3-manifold. Third, one obtains a characterization of S^3 in terms of that data, i.e. a proposition which asserts that if the data is of a certain type then the manifold in question is S^3. Fourthly and finally one shows how to transform via the calculus the initial data describing Σ into data describing S^3. The first three steps often appear especially seductive and clever, and are usually true. The fourth step usually involves long and difficult geometric operations requiring a huge stabilization of the data. The procedure is thus susceptible to a myopic flaw, e.g. a wrong sign or inequality embedded in a great quantity of such. In gross terms, one is proving that blood is water, by mixing a drop into a swimming pool and then observing that it is devoid of blood.

Po's program for the Stable Poincaré conjecture starts off in a similar manner. First *[8]* gives a geometric characterization (called a *collapsible pseudospine representation*) for a homotopy 3-sphere. (This will be discussed in §1.) Then *[9]* shows that a particularly good (i.e. *coherent*) pseudospine decomposition of Σ implies that $\Sigma \times I$ is geometrically simply connected (See §2). *[9]–[10]* shows that a collapsable pseudospine representation can be transformed into a particularly nice one called a *honeycomb representation* and develops a calculus for transforming a honeycomb for Σ into a new honeycomb for Σ. This operation is analogous to a huge stabilization (See §3). Here is the big innovation. Po recognizes that a honeycomb that fails to yield the geometric simple connectivity of $\Sigma \times I$ will, under his moves, create new honeycombs which also fail. He knows exactly how the difficulty is passed from one honeycomb to the next under the calculus. Via the operations, Po inductively constructs an infinite sequence of honeycombs in such a way that the difficulty is pushed off to ∞. A coherent honeycomb representation of Σ implies that $(\Delta^3 - (n \mathring{B}^3\text{'s})) \times I = B^4 \cup (n$ 2-handles), where Δ^3 denotes the homotopy 3-cell $\Sigma - \mathring{B}^3$. Said another way $\Delta^3 \times I\#(n\ D^2 \times S^2\text{'s}) = B^4 \cup (n$ 2-handles), where # denotes boundary connected sum. The object of *[11]* is to show that pushing the difficulty off to ∞ implies that $int(\Delta^3 \times I\#(\cup_1^\infty D^2 \times S^2\text{'s})) = int(W^4 \cup_1^\infty (\text{2-handles}))$, where $\mathring{W}^4 = \mathbb{R}^4_{standard}$

and ∂W^4 is a union of countably many $\mathring{D}^2 \times S^1$'s. (See §4.) Furthermore if W_N is $\mathring{W}$ together with the first N boundary components, then W_N smoothly compactifies to B^4, i.e. W_N is diffeomorphic to B^4–(closed subset of ∂B^4). This implies that $\Delta^3 \times I - \mathring{N}(\partial(\Delta^3 \times I)) \subset X \subset \Delta^3 \times I$ where X is geometrically simply connected, which implies, by *[13]*, that $\Delta^3 \times I$ is geometrically simply connected and hence so is $\Sigma \times I$.

In 1966 Stallings [25] said "we always run against the problem of trying to simplify, by some geometric trick the situation. But any little homotopy that would simplify the picture always, in fact, greatly complicates it." These words ring true for Po. However, through his infinite honeycomb calculus *[9]–[12]* he succeeds in corralling the complications in the 'end', i.e. he obtains the proposition "$\Delta^3 \times I - \mathring{N}(\partial(\Delta^3 \times I)) \subset X \subset \Delta^3 \times I$ where X is geometrically simply connected." Now, dropping a complicated data set full of infinite numbers of holes and handles, he starts afresh with the conclusion of the proposition as hypothesis and establishes in *[13]* the geometrical simple connectivity of $\Delta^3 \times I$, via a (long) more or less self contained argument. Except for some *[9]* basics it does not rely on the previous works.

Remark: The titles of Po's papers *Infinite processes and the Poincaré Conjecture 1-6* are references to the process of pushing difficulties out to ∞. There is one such process in each of *[11]*, *[12]*, and *[14]*.

One more difference: Po's technology has produced interesting results of independent mathematical interest, see *[16]– [19]*, *[23, 24]*.

Other surveys: The difference between this paper and Po's own very brief expositions *[20, 21]* is more a matter of exposition and organization than it is content. This paper also owes a lot to the hours Po spent patiently explaining some of the elements of the program. For a more detailed outline of the whole program see *[22]*.

Notation: $N(E)$ will denote a regular neighborhood of E, $\mathring{E}$ will denote interior of E and unless obviously otherwise # will denote boundary connect sum.

1 Homotopy 3-Spheres have Collapsable Pseudospine Representations

One level of Po's program is centered around the study of generic PL maps of finite 2-complexes into homotopy 3-spheres. Away from a finite number of isolated points the mappings will be locally injective (i.e. are *immersions*) and the images of the maps will always be quite tame, appearing locally as in Figures 1.1.

Remarks 1.1: Let $f : K \to \Sigma$, where K is a finite 2-complex with boundary, i.e. has free edges.

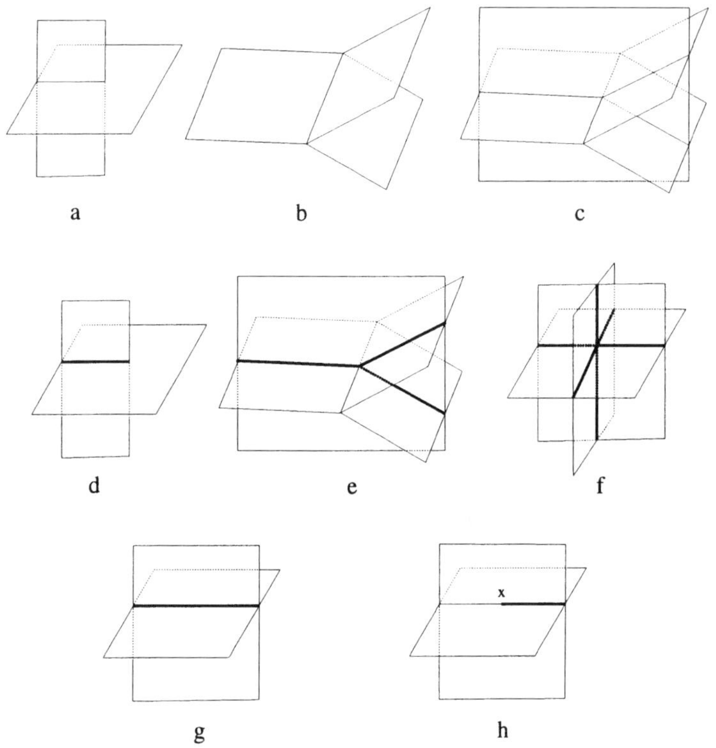

Figure 1.1

i) The heavy lines indicate the *double curves*, i.e. the points whose preimage consists of either two points or three points. The latter, called *triple points*, appear exactly as in Figure 1.1 f). Note that one can reconstruct K from $f(K)$ and the labelling of the double curve locus. Thus one can talk about points of the domain and range simultaneously.

ii) For simplicity Figures 1.1 b), c), e) are drawn with three 2-cells branched from a 1-cell. However, these mappings can have $n > 0$ cells branched from that 1-cell. In fact the cases for $n = 1, 2$ for Figure 1.1 e) are respectively drawn in 1.1 d), g).

iii) The map f is not an immersion at the point x in Figure 1.1 h). Such a point x will be called a *singularity* or an *undrawable singularity*. An open neighborhood P of x is homeomorphic to the complex obtained by gluing together two planes along properly embedded rays. However, $f(P)$ is the union of two "orthogonal" planes.

iv) A mapping whose local images appear as in Figures 1.1 a)–h) will be called a *singular immersion*.

Figure 1.2 shows several natural ways, called $O(i)$ moves, that the singular immersion $f : K \to \Sigma$ can be transformed into a new singular immersion $f_1 : K_1 \to \Sigma$.

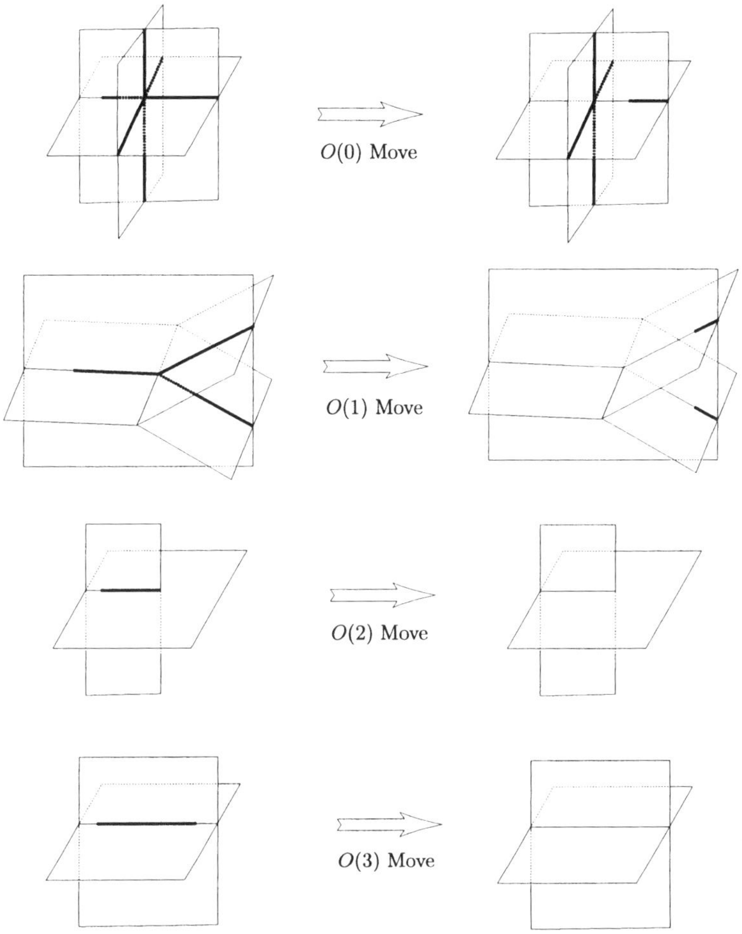

Figure 1.2

Remarks 1.2: i) $f(K) = f_1(K_1)$.

ii) After an $O(0)$ move $K = K_1$, but the configuration of the double curve locus has changed.

iii) After either an $O(1)$ or $O(2)$ move K and K_1 are different but homotopy equivalent.

iv) An $O(3)$ move changes the homotopy type of K; it corresponds to adding a 2-handle.

v) $O(i)$ moves are also called *zippings*. Since the combinatorial length of the double curve locus is finite, a finite number of $O(i)$ moves transforms a singular immersion into an immersion.

Theorem 1.3: *[8] If Σ is a homotopy 3-sphere, then there exists a sequence of singular immersions $f_i : K_i \to \Sigma, i = 0, 1, \cdots, n$ such that f_{i+1} is obtained from f_i by* *an $O(i)$ move, K_0 is a finite collapsible 2-complex, f_n is an embedding and $\Sigma - f_n(K_n)$ is a union of $\mathring{B}^3$'s.*

Remarks 1.4: i) A *spine* (resp. *pseudospine*) is a 2-complex whose complement consists of a single (resp. more than one) $\mathring{B}^3$ (resp. $\mathring{B}^3$'s). A sequence of maps as in the conclusion of the theorem is called a *collapsable pseudospine representation*.

ii) The collapsability of K_0 and Remark 1.2 imply that $f_n(K_n)$ is simply connected, hence a closed 3-manifold with a collapsable pseudospine representation is a homotopy 3-sphere.

2 Four Dimensional Thickenings and the Stable PC

In this section we outline why the geometric simple connectivity of $\Sigma \times I$ is equivalent to Σ possessing a particularly nice collapsable pseudospine representation. Consider

$$K \xrightarrow{f} \Sigma \longrightarrow \Sigma \times \tfrac{1}{2} \longrightarrow \Sigma \times I, \qquad F : K \to \Sigma \times I$$

where f is a singular immersion and F is the induced map into $\Sigma \times I$. Emanating from each singularity is a double curve, which locally is the image of distinct curves in K called *branches*. For each singularity choose one branch and push a small neighborhood (disjoint from the other branch) into $\Sigma \times (1/2, 1]$. This changes F into an immersion $F_\phi : K \to \Sigma \times I$. See Figure 2.1.

Definition 2.1: Define $\Theta^4(K, \phi)$ to be the regular neighborhood of the immersion F_ϕ, more precisely a compact regular neighborhood of the zero section of the normal bundle of that map. $\Theta^4(K, \phi)$, which is called a *4-dimensional thickening of K*, depends only on which branch was pushed into $\Sigma \times (1/2, 1]$. Hence the 4-dimensional thickenings are parametrized (possibly with multiplicity) by the functions $\phi : \mathcal{B} \to \{S, N\}$, where $\mathcal{B}$ denotes the set of local branches associated to the singularities and if b, b' are the branches emanating from the same singularity, then $\phi(b) \neq \phi(b')$. The function ϕ is called a *desingularization*, the S denotes the *specified branch* and $\phi(b) = S$ if and only if a neighborhood of b was pushed into $\Sigma \times (1/2, 1]$. Associated to (K, ϕ) is the *desingularized* 2-complex $\hat{K}$, the 2-complex obtained by cutting out from K a little open 2-disc around the specified branch. Therefore F_ϕ could have been chosen so that $F \mid \hat{K} = F_\phi \mid \hat{K}$ and $F_\phi(\text{little disc}) \subset \Sigma \times (1/2, 1]$. Among other things $\hat{K}$ is a good pictorial notation for keeping track of which branch has been specified. Figures 2.2 a), b) respectively show the $\hat{K}$'s associated to the F_ϕ's of Figure 2.1.

Remark 2.2: There is another description in *[9]* of $\Theta^4(K, \phi)$ given purely in

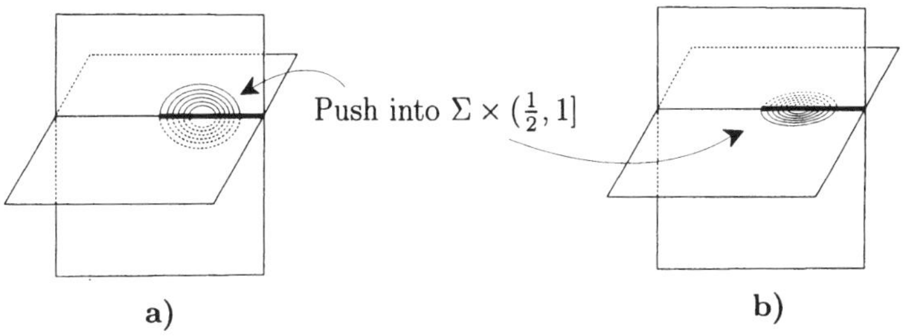

Figure 2.1

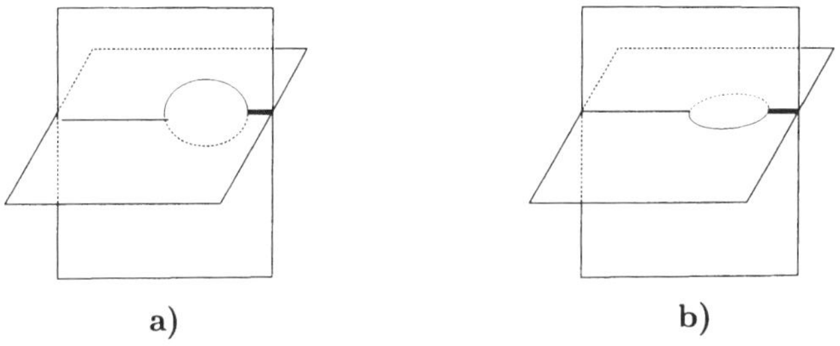

Figure 2.2

terms of ϕ and the singular immersion f and without reference to $\Sigma \times I$. Associated to this description is a canonical embedding $\hat{K} \to \partial\Theta^4(K, \phi)$.

Definition 2.3: A desingularization ϕ_i of K_i and an $O(i)$ move $(f_i, K_i) \to (f_{i+1}, K_{i+1})$ induce a desingularization ϕ_{i+1} of K_{i+1}. The desingularization is unchanged away from the support of the $O(i)$ move, the induced desingularization near the locus of zipping is described in Figure 2.3.

Definition 2.4: An $O(3)$ move involves zipping away a short embedded double curve with endpoints on singularities. The preimage of the curve consists of two curves with the same endpoints. The $O(3)$ move is *coherent* if the two specified branches associated to the singularities involved in the $O(3)$ move lie on the same curve. It is *noncoherent* otherwise. See Figure 2.4. A collapsable pseudospine representation is coherent if there exists a desingularization of (f_0, K_0) such that, with respect to the induced desingularizations of the K_i's, all the $O(3)$ movements are coherent.

Lemma 2.5 *[9] : i) $\Theta^4(K, \phi) = B^4$, if K is collapsable.*

ii) $\Theta^4(K_{i+1}, \phi_{i+1}) = \Theta^4(K_i, \phi_i)$, if K_{i+1} is obtained from K_i by an $O(i)$ move, $i \leq 2$.

iii) $\Theta^4(K_{i+1}, \phi_{i+1}) = \Theta^4(K_i, \phi_i) \cup$ 2-handle, if K_{i+1} is obtained from K_i by a coherent $O(3)$ move.

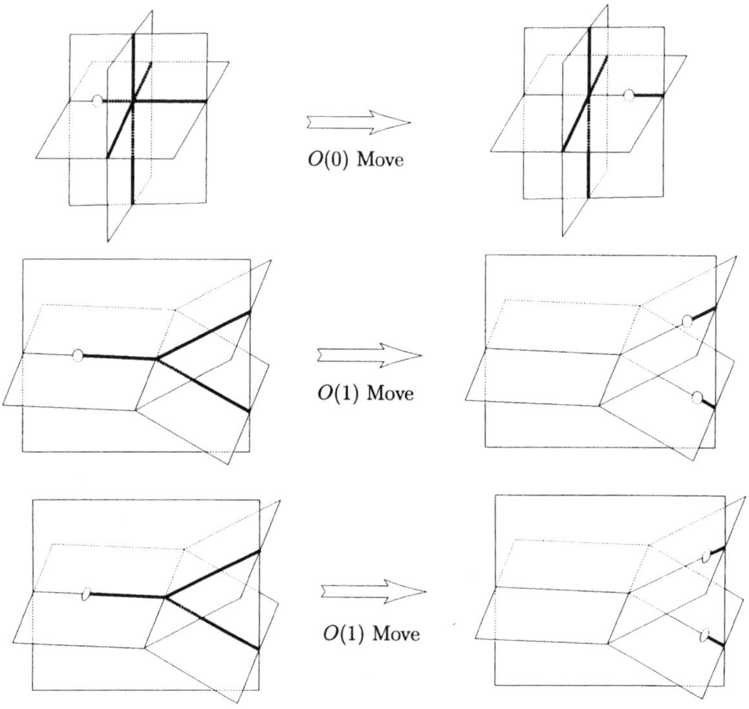

Figure 2.3

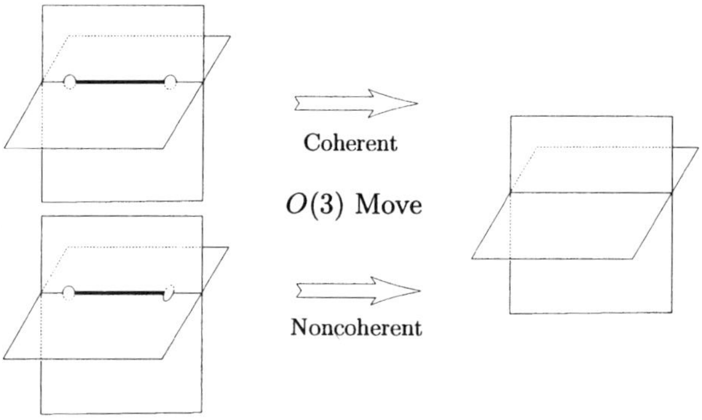

Figure 2.4

iv) $\Theta^4(K_n) = N(K_n) \times I$ *if* $f_n : K_n \to \Sigma$ *is an embedding and* $N(K_n)$ *is the regular neighborhood of* $K_n \subset \Sigma$.

Remark 2.6: Figure 2.5 a), b) respectively show local 2-dimensional cross sections of $\Sigma \times I$ before and after coherent and noncoherent $O(3)$ moves. The

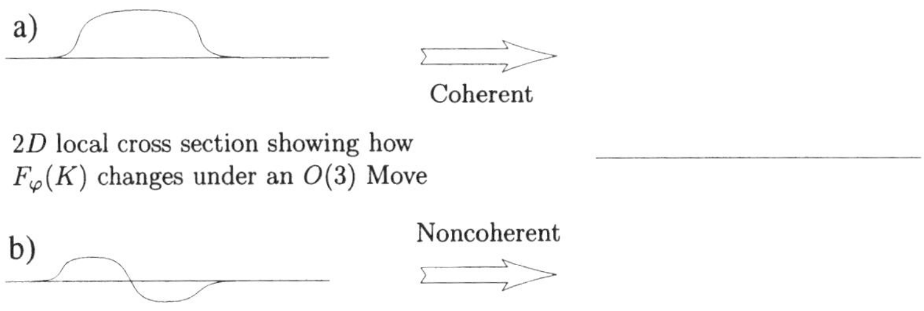

Figure 2.5

proof of 2.5 iii) becomes transparent after staring at Figure 2.5 a). The local map of F_ϕ in the noncoherent case is not an embedding and the relation between $\Theta^4(K_{i+1}, \phi_{i+1})$ and $\Theta^4(K_i, \phi_i)$ is nonobvious. Nevertheless there is one situation, of great importance, when a noncoherent $O(3)$ move corresponds, at the 4-dimensional level, to the attachment of a 2-handle.

Lemma 2.7 *[9]:* $\Theta^4(K_{i+1}, \phi_{i+1}) = \Theta^4(K_i, \phi_i) \cup$ *2-handle if* K_{i+1} *is obtained from* K_i *by a* $O(3)$ *move and there exists an embedded arc* $\eta : [0,1] \to K_i$ *so that* $\eta(1) \in \partial K_i$ *(i.e. in a free edge) and* $\eta(0) \in$ *(zipping locus of the* $O(3)$ *move). See Figure 2.6 a).*

Proof: Change the singular immersion (f_i, K_i) to $(f_{i+1/2}, K_{i+1/2})$ as in Figure 2.6 b). Perform two $O(2)$ moves to obtain $(f_{i+3/4}, K_{i+3/4})$ and then attach a 2-cell to obtain $(f_{i+7/8}, K_{i+7/8})$ as in Figure 2.6 c). Using Lemma 2.5 its clear that

$$\Theta^4(K_i, \phi_i) = \Theta^4(K_{i+1/2}, \phi_{i+1/2}) = \Theta^4(K_{i+3/4}, \phi_{i+3/4})$$

and

$$\Theta^4(K_{i+3/4}, \phi_{i+3/4}) \cup \text{2-handle} = \Theta^4(K_{i+7/8}, \phi_{i+7/8}) = \Theta^4(K_{i+1}, \phi_{i+1}).$$

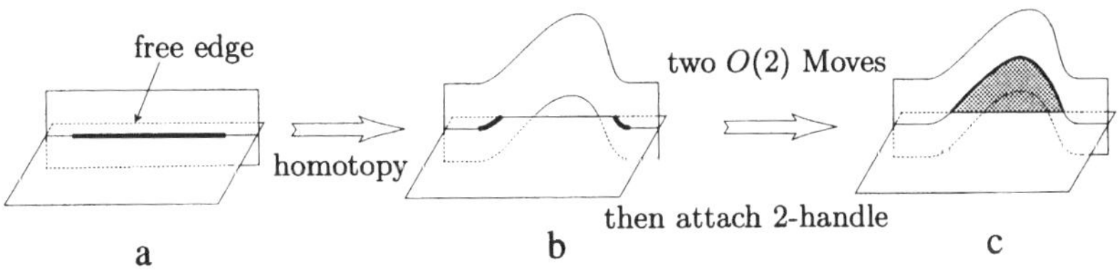

Figure 2.6

Definition 2.8: Call such a disc η an $O(3)$ *boundary compressing* disc and the $O(3)$ *boundary compressible.*

The following result is an immediate corollary of Lemma 2.5.

Theorem 2.9 *[9]: If Σ has a coherent collapsible pseudospine representation, then for some m, $(\Delta - m\ \mathring{B}^3\text{'s}) \times I = B^4 \cup m$ 2-handles.*

Proof: For Euler characteristic reasons the number of 3-cells deleted is the number of 2-handles attached. If $\Sigma - K_n = m + 1$ open 3-balls, then $N(K_n) = \Delta - m\ \mathring{B}^3$'s. Thus Theorem 2.9 follows from Lemma 2.5 and the definition of coherent collapsible pseudospine representation.

Theorem 2.10 *[13]: If $\Sigma \times I$ is geometrically simply connected, then Σ has a coherent collapsible pseudospine decomposition.*

Idea of Proof: The geometric simple connectivity of $\Sigma \times I$ implies that $\Sigma \times I$ has a cell decomposition with almost collapsible 2-skeleton (i.e. after removing a finite number of discs from 2-cells the resulting complex is collapsible.) On the other hand $\Sigma \times I$ has a product cell structure Ψ (i.e. there exists $0 = a_1 < a_2 < \cdots < a_n = 1$ and a cell structure ψ of Σ such that the 2-skeleton of Ψ restricts to $\psi^2 \times a_i$ at the a_i-levels and is of the form $\psi^1 \times [a_i, a_{i+1}]$ on $\Sigma \times [a_i, a_{i+1}]$, where ψ^i denotes the i-skeleton of ψ). Combining the smooth Hauptvermuétung of Whitehead with other PL manipulations one shows that there exists a cell structure L on $\Sigma \times I$ with both properties, i.e. has an almost collapsible 2-skeleton and is a product cell structure on $\Sigma \times I$. Let K be the 2-skeleton (with tiny holes so as to be collapsible) of Ψ. By projecting K into Σ and then perturbing slightly both the mapping and K, one obtains a singular immersion of a collapsible 2-complex K_0 into Σ. Provided the 2-complex K was sufficiently dense in $\Sigma \times I$, then by "walking up the I factor" one can zip up the entire double curve locus of K_0 to obtain an embedding K_n. The height function on I induces a natural desingularization K_0 as follows. A double curve emanating from a singularity of K_0 is associated to two curves in K, one of which is higher than the other. Decree that the higher branch be specified. With some care this property will persist under zipping and hence all $O(3)$ moves will be coherent.

Corollary 2.11 *[9, 13]: The following are equivalent*

i) $\Sigma \times I$ is geometrically simply connected.

ii) $(\Sigma - (m+1)\ \mathring{B}^3\text{'s}) \times I = B^4 \cup m$ 2-handles for some $m \geq 0$.

iii) Σ has a coherent collapsible pseudospine decomposition.

Definition 2.12: A manifold M is *geometrically simply connected at long distance* if each compact set in $\mathring{M}$ is contained in a handlebody $X \subset M$ without 1-handles.

Remarks 2.13: i) The words *undrawable singularity*, *geometrically simply connected*, *geometrically simply connected at long distance*, and *sort of link* are due to Barry Mazur.

ii) A compact manifold M is geometrically simply connected at long distance if and only if $M - \mathring{N}(\partial M) \subset X \subset M$, where X is a handlebody without 1-handles.

iii) The following result is of crucial importance to Po's program.

Theorem 2.14 *[13]* **:** *$\Delta \times I$ is geometrically simply connected at long distance if and only if $\Delta \times I$ is geometrically simply connected.*

Remark 2.15: The proof of this pure 4-dimensional topology theorem exploits the product structure of $\Sigma \times I$. One wonders for what other compact 4-manifolds does geometric simple connectivity at long distance imply geometric simple connectivity. For example Andrew Casson has observed that extending Theorem 2.14 to all smooth homotopy 4-balls would have the following relevance to the smooth 4-dimensional Schoenflies conjecture. Let X be a smooth compact submanifold of S^4 bounded by a manifold diffeomorphic to S^3. X is geometrically simply connected at long distance for by Mazur *[5]* $\mathring{X}$ is diffeomorphic to $\mathbb{R}^4_{standard}$. Therefore if Theorem 2.14 held for such X, then X would have a handle decomposition without 1-handles.

3 Three Dimensional Thickening and Honeycomb Representations

This chapter first describes Po's theory of singular 3-manifolds in Σ, a concept which directly generalizes his theory of singular 2-complexes presented in §1-2. Second, it states the honeycomb representation theorem, a much refined version of Theorem 1.3. Third, it defines how a "nice" honeycomb gives rise to a directed graph and outlines why the nonexistence of cycles implies the geometric simple connectivity of $\Sigma \times I$.

Definition 3.1: Associate to a singular 2-complex $K \subset \Sigma$ a singular 3-manifold $\Theta^3(K)$ called the *3-dimensional thickening of* K. Away from singular points, $\Theta^3(K)$ is obtained by taking a (closed) regular neighborhood of K. Near singular points $\Theta^3(K)$ appears as in Figure 3.1, where one sees that the singular point becomes a singular square. So the singular squares are both the points where $\Theta^3(K)$ fails to be a 3-manifold and the points where the mapping into Σ fails to be an immersion. After thickening, local arcs of double curve locus are replaced by square$\times I$ and the isolated triple points are replaced by cubes of triple points.

The zipping operations $O(0) - O(3)$ of §1 naturally extend to zipping operations $O(0) - O(3)$ of $\Theta^3(K)$. The $O(1)$ move now being the operation of

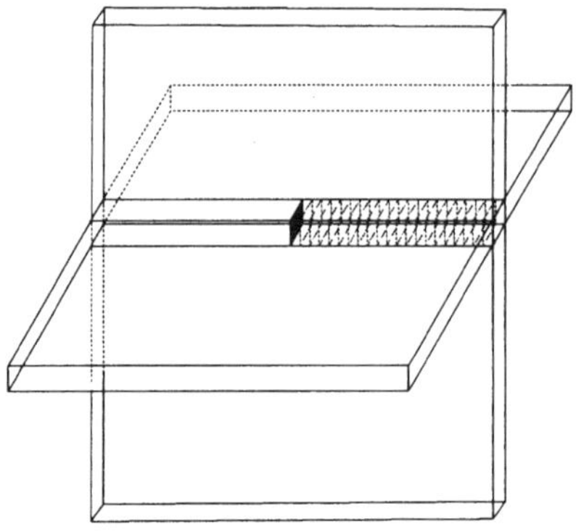

Figure 3.1

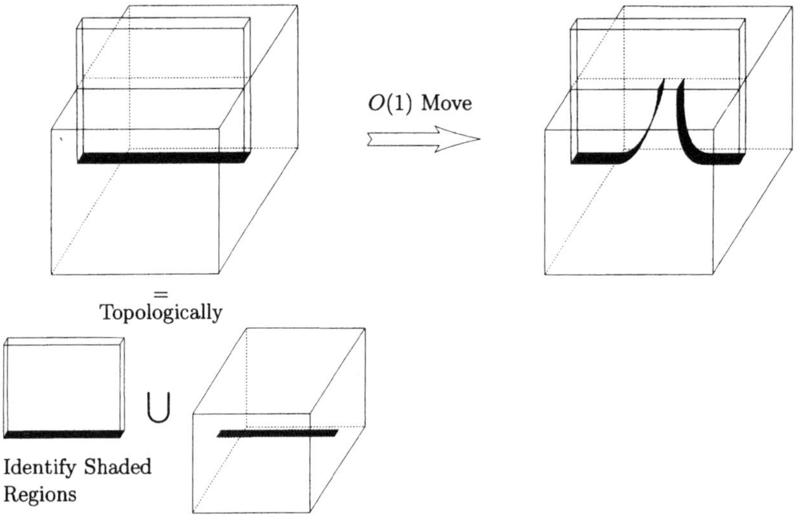

Figure 3.2

Figure 3.2. Thus a collapsible pseudospine representation of Σ gives rise to a *3-dimensional collapsible pseudospine representation*, i.e. in the statement of Theorem 1.3 K_i's are now singular 3-complexes.

Convention 3.2: The singular 3-manifolds that appear in Po's works are not always thickenings of singular 2-complexes, rather they are more generally 3-manifolds with only square and annulus singularities. Neighborhoods of the squares will always be conjugate to the standard model of Figure 3.1, i.e. M near squares is a thickening of the standard 2-complex singularity. If A is an annulus singularity, then the singular immersion restricted to a neighborhood $U \subset M$ of A is conjugate to the singular immersion of a neighborhood of the singular annulus A of Figure 3.3. (Because the mapping is an immersion there may be triple points

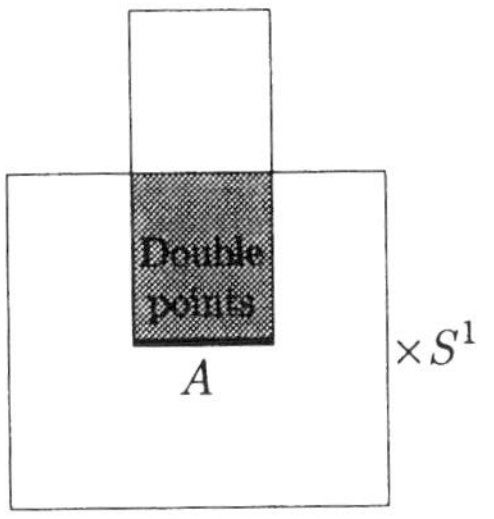

Figure 3.3

near A.) As we shall soon see, annulus singularities will only appear momentarily, for they will get quickly zipped to square singularities via the $O(1)$ move.

Let $\sigma(M)$ denote the singularities of M.

Definition 3.3: Figure 3.4 describes the $O(4)$ move, the one new zipping operation needed in the singular 3-manifold setting. A move involving a "crossing change" of the singular locus. All singular 3-manifolds can be transformed into immersed 3-manifolds via a finite number of $O(i)$ movements, $i \leq 4$.

Definition 3.4: *(4-dimensional thickening, desingularization)* Given a singular 3-manifold M, a small perturbation of specified branches of the local double curve locus near singularities into $\Sigma \times (1/2, 1]$, as in §2, yields an immersion $F_\phi : M \to \Sigma \times [0,1]$. Again by taking regular neighborhoods, one obtains the *4-dimensional thickening* $\Theta^4(M, \phi)$. As in the 2-complex case the various local branches of the double curve locus are labeled $\{S, N\}$ depending on whether or not they are specified and such a function is called a *desingularization*. $\Theta^4(M, \phi)$ is uniquely determined by ϕ, for different choices of F_ϕ differ by small regular

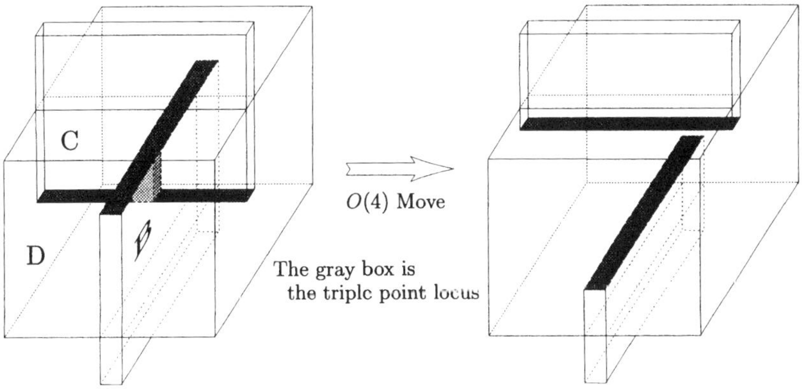

Figure 3.4

homotopies. If M has a desingularization ϕ_M and N is obtained by an $O(i)$ move, then N naturally inherits from M a desingularization ϕ_N. A desingularization ϕ on a 2-complex K induces a desingularization ϕ on $\Theta^3(K)$.

Definition 3.5: If M has a desingularization ϕ, then the $O(4)$ move is *coherent* if $\phi(C) \neq \phi(B)$ and is *noncoherent* otherwise, where B, C are the branches denoted in Figure 3.4.

Lemmas 3.6, 3.8 are the analogues of Lemmas 2.5, 2.7 for thickenings of singular 3-manifolds.

Lemma 3.6 *[9]: 0) The immersion $F_\phi : K \to \Sigma \times I$ extends to $F_\phi : \Theta^3(K) \to \Sigma \times I$ and $\Theta^4(\Theta^3(K), \phi) = \Theta^4(K, \phi)$.*

i) $\Theta^4(M, \phi) = B^4$, if M is a collapsible singular 3-manifold.

ii) $\Theta^4(M_{i+1}, \phi_{i+1}) = \Theta^4(M_i, \phi_i)$, if M_{i+1} is obtained from M_i by an $O(i)$ move, $i \leq 2$, or a coherent $O(4)$ move.

iii) $\Theta^4(M_{i+1}, \phi_{i+1}) = \Theta^4(M_i, \phi_i) \cup$ 2-handle, if M_{i+1} is obtained from M_i by a coherent $O(3)$ move.

iv) $\Theta^4(M_n) = M_n \times I$ if $f_n : M_n \to \Sigma$ is an embedding.

Remark 3.7. Comments on ii) **for the case of the** $O(4)$ **move:** Let B, C, D be subsets of the singular 3-manifold M as shown in Figure 3.4. So D is the cube and B (resp. C) is a subset of M lying below (resp. above) a singularity. After a regular homotopy $F_\phi \mid D = id$ and $F_\phi(C - D) \subset \Sigma \times (1/2, 1]$ (resp. $\Sigma \times [0, 1/2)$) if and only if C is a specified branch (resp. non specified branch). A similar statement holds with C replaced by B. Figure 3.5 shows the 3-dimensional cross sectional plane cutting through $C \times I \subset \Sigma \times I$, before and after coherent and noncoherent $O(4)$ moves. In the case of a coherent $O(4)$, F_ϕ is an embedding near the support of the $O(4)$ and the effect on $\Theta^4(M)$ is an isotopy. In the noncoherent $O(4)$ case F_ϕ is only an immersion near the support of the $O(4)$ and the effect on $\Theta^4(M)$ is nonobvious.

The following is the analogue of Lemma 2.7 for singular 3-manifolds.

Lemma 3.8 *[9]: $\Theta^4(M_{i+1}, \phi_{i+1}) = \Theta^4(M_i, \phi_i) \cup$ 2-handle if M_{i+1} is obtained from M_i by an $O(3)$ and there exists an embedded disc $\eta : I \times I \to M_i$ such that $\eta(I \times 1 \cup I \times 0 \cup 1 \times I) \subset \partial M_i$, $\eta(I \times I) \cap \sigma(M_i) = \emptyset$ and η^{-1}(zipping locus of $O(3)$ move)$= 0 \times I$.*

Let T_n denote the 3-dimensional handlebody of genus n parametrized as a 3-ball with handles $V_1, \cdots, V_n$ attached as in Figure 3.6. Each V_i is endowed with a standard core curve γ_i, a standard annulus C_i whose boundary consists of γ_i and another curve in ∂V_i, and a standard compressing disc Δ_i which intersects C_i in a single arc.

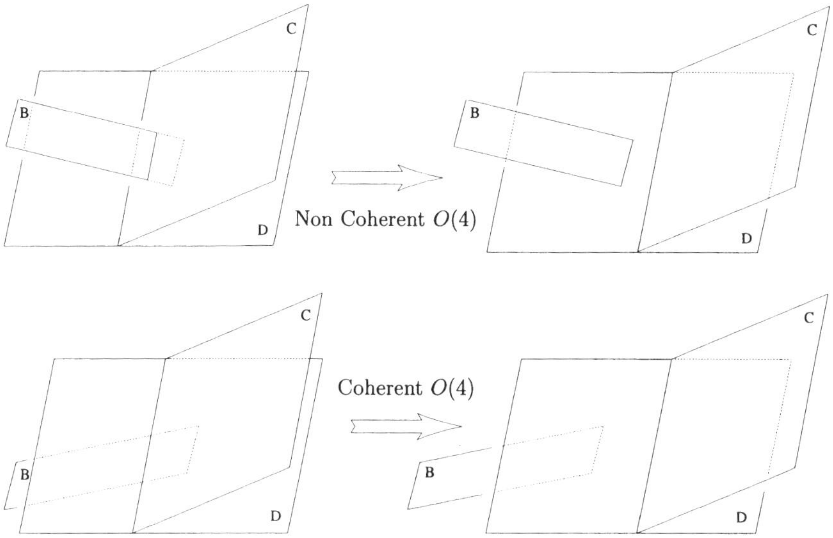

Figure 3.5

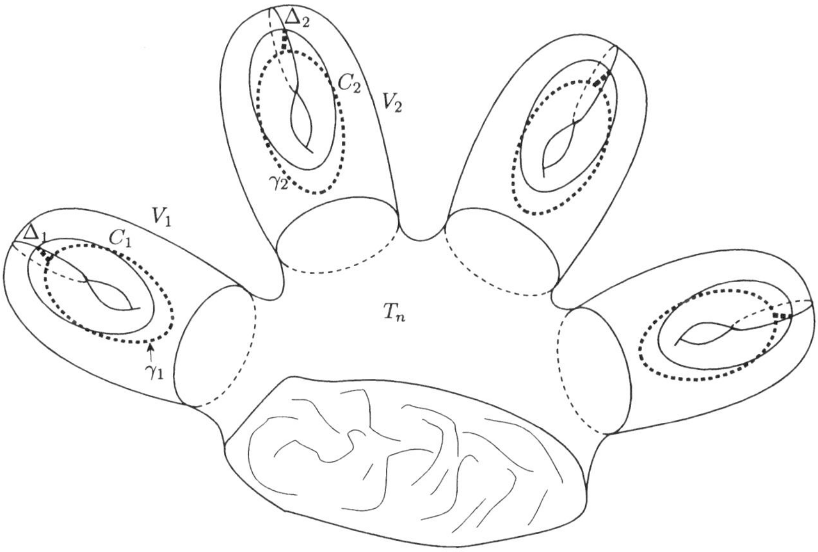

Figure 3.6

Definition 3.9: A *honeycomb representation* for Σ consists of a parametrized handlebody T_n embedded in Σ together with a collection $\{F_i\}$ of annuli embedded in T_n such that for every i, $\partial F_i = \gamma_i \cup c_i$, where $c_i \subset \partial T_n$ and

i) *Relations between* F_i, C_i*:* There exist $\rho \leq p \leq n/2$ with the following properties. Letting $\hat{F}_i$ denote the component of $F_i \cap V_i$ which contains γ_i, then

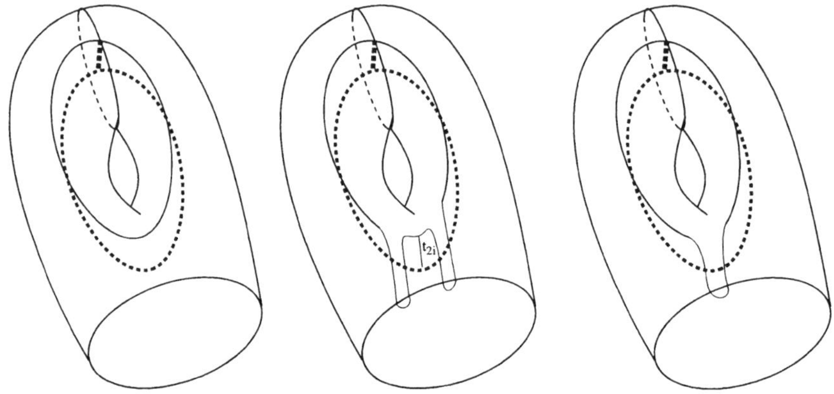

Figure 3.7

$\hat{F}_i$ appears as in Figure 3.7. In particular $F_i \cap \Delta_i = C_i \cap \Delta_i$ and the germs of F_i, C_i agree near both γ_i and Δ_i. Also for $i \leq \rho$, there exists a vertical arc $t_{2i} \subset C_{2i} \cap F_{2i} - \Delta_{2i}$ such that the germs of F_{2i}, C_{2i} agree near t_{2i}. The arcs $t_{2i}, \Delta_{2i} \cap F_{2i}$ divide F_{2i} into two halves.

ii) *Control of c_i, c_j intersections:* If $x < y$, then $c_x \cap c_y = \emptyset$ unless $x = 2i - 1, y = 2i$ and $i \leq p$, in which case $c_x \cap c_y = 2$ points of opposite sign. The components of $F_r \cap V_s$ which hit γ_s where $r, s \in \{2i-1, 2i\}, r \neq s, i \leq p$ appear as in Figure 3.8.

iii) *control of F_i, F_j intersections:* a) There exists $r > 2p$ such that if $x < y$ and $F_x \cap F_y \neq \emptyset$, then these intersections either occur as in Figure 3.8 or $x \leq r$ and $y > r$.

b) There are no triple points of intersection among the F_i's.

c) Each component of $F_i \cap F_j$ has both endpoints in $\gamma_i \cup \gamma_j$ (i.e. is either a clasp or ribbon), unless $F_i \cap F_j$ already has appeared in Figure 3.8.

d) For each $x \neq y, F_x \cap (F_y \cap \Delta_y) = \emptyset$. Also for $i \leq \rho, F_x \cap t_{2i} = \emptyset$ if $x \neq 2i$.

e) Each F_i intersects at most one F_j in a ribbon.

iv) *How to obtain Σ:* Each c_i bounds a properly embedded disc d_i in $\Sigma - \mathring{T}_n$. These discs are pairwise disjoint, except for d_{2i-1}, d_{2i} which intersect along a diameter. Finally $T_n \cup_1^n N(d_i) = \Sigma - (\mathring{B}^3\text{'s})$.

Theorem 3.10 *[9, 10]*: *Every homotopy 3-sphere has a honeycomb representation.*

Much of *[9]* is spent establishing a "weak" honeycomb theorem, the full honeycomb theorem being established in *[10]*.

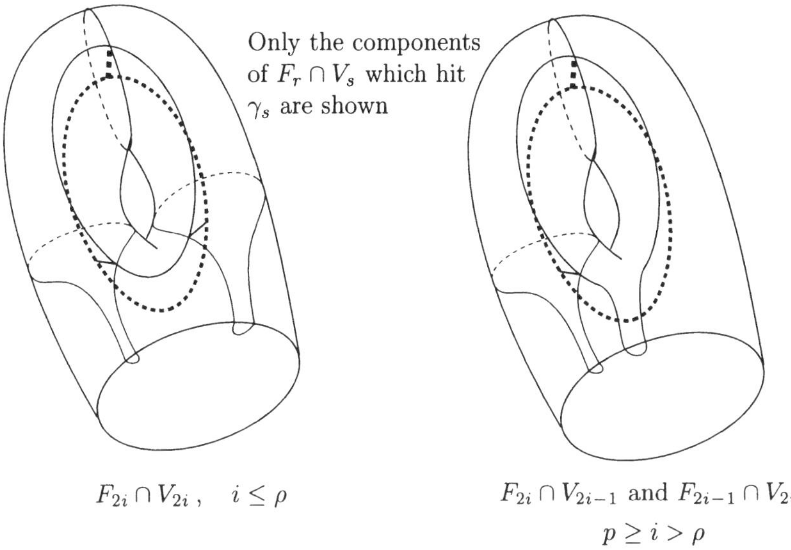

Figure 3.8

Proposition 3.11 *[9]: Associated to each honeycomb representation is a 3-dimensional collapsible pseudospine representation of* Σ.

Proof: Let $D_i = F_i \cup d_i$. Let M_{-1} be the quotient space of $(T_n \amalg_1^n D_i \times I)$ where D_i is equated with $D_i \times 1/2$, $\amalg$ denotes disjoint union and $(\partial D_i) \times I$ is identified with its associated annulus in T_n. Define $f_{-1} : M_{-1} \to \Sigma$ to be the mapping which restricts to the natural embeddings on T_n and $D_i \times I$'s respectively.

Let M_0 be the singular 3-complex obtained by zipping up M_{-1} along the whole image of $N(F_i) = N(C_i) = (D_i \times I) \cap T_n, i = 1, 3, \cdots, 2\rho - 1$, and via $O(1)$ moves (as in Figure 3.2) near each of $F_i \cap t_i$ for $i = 2, 4, \cdots, 2\rho$, and near each of $F_i \cap \Delta_i$ for all i. By Definition 3.9 iii) such regions are clean of other F_i's. M_0 is still collapsible and has n singularities, one associated to each "half" of $F_{2i}, i = 1, \cdots, \rho$ and one associated to each of $F_i, i > 2\rho$. Let $f_0 : M_0 \to \Sigma$ be the induced map.

We now describe the canonical (up to obviously commuting $O(i)$ movements) zipping of the double curve locus. First do $O(4)$ moves to undo each clasp (from 3.9 iii c)), second do more zipping (i.e. $O(0)$ moves) which amounts to sliding off the ribbons, third zip up, with $O(3)$ moves, the (diameter) intersections between D_{2i-1} and D_{2i}, $1 \leq i \leq p$, and finally do $O(2)$ moves ($n - 2p$ of them) to obtain an embedding.

Definition 3.12: Give M_0 a desingularization. By Lemma 3.6 $\Sigma \times I$ is geometrically simply connected provided each of the $O(4)$ and $O(3)$ movements are coherent. Any such noncoherence is called an *accident*.

Definition 3.13: A *sufficient family* is a collection of pairs, $\{(F_{2i-1}, F_{2i})\}, 1 \leq i \leq s \leq \rho$ (after possibly reindexing $1, \cdots, \rho$), in a honeycomb satisfying

i) There exists a desingularization ϕ of M_0 such that every accident corresponds to an $O(3)$ movement along the diameter of some member of the sufficient family.

ii) Let N be the singular 3-manifold obtained by zipping up M_0 as above except along the (diameter) intersections between D_{2i-1} and D_{2i}, $s < i \leq p$. (This possibly involves noncoherent $O(3)$'s.) Then $\Theta^4(N)\#(D^2 \times S^2)$'s $= (\Sigma - \mathring{B}^3$'s$) \times I$.

Remark 3.14: By 3.13 i) N with the induced desingularization can be zipped up using only coherent movements, hence by Lemma 3.6 $\Theta^4(N) \cup$ (2-handles) = $(\Sigma - \mathring{B}^3$'s$) \times I$, thus ii) appears to be a much stronger statement than i).

Remarks 3.15: i) The bulk of *[10]* discusses $\mathcal{C}$ operations which transform a honeycomb H into new one H'. These $\mathcal{C}$ operations have a number of remarkable features, the least of which is that the sufficient family for H' is associated 1-1 with that of H and the correspondence restricts to a bijection between accidents, so in particular preserves the quality of being coherent. Also, among other things, $\mathcal{C}^2$ (the operation iterated twice) induces an isomorphism between the directed graphs constructed in Definition 3.17.

ii) We have suppressed the fact that associated to a honeycomb is a 2-dimensional collapsible pseudospine representation. The $\mathcal{C}$ operations are carried out at this 2-dimensional level.

iii) I added condition 3.9 iii) e) to simplify the proof of Proposition 3.11 which is true without 3.9 iii) e).

Lemma 3.16: *After a $\mathcal{C}$ operation the following three conditions hold.*

i) For $i \leq s$, each V_{2i-1} is flanked by $V_x, V_y; x, y > r$, where r is as in Definition 3.9 iii). See Figure 3.9 showing the normal form for $\cup_1^s F_{2j} \cap (V_{2i-1} \cup V_x \cup V_y)$ together with $F_{2i-1} \cup \hat{F}_x \cup \hat{F}_y$. Furthermore for $i \leq s$ each feeler emanating from $\hat{F}_{2i}$ lies very close to ∂T^n until it eventually enters the region of Figure 3.9 and grabs F_{2i-1} and one of F_x, F_y. Figure 3.10 shows $F_{2i} \cap ((\cup_1^s \Delta_{2j-1}) \cup F_{2i-1} \cup F_x \cup F_y)$.

ii) For each $j \leq s, \Delta_{2j-1} \cap (\cup_1^s F_{2i})$ appears as in Figure 3.11, i.e. is a union of pairwise disjoint boundary parallel arcs, no one of which separates the others. More generally, for $i \leq s, F_{2i}$ intersects only F_{2i-1}, F_x and F_y. Furthermore $(\cup_1^s F_{2i} \cap (V_{2i-1} \cup V_x \cup V_y))$ consists of the two subsets of F_{2i} (described in i) above) together with a disjoint collection of adjacent pairwise disjoint boundary parallel bands following the thick line of Figure 3.9.

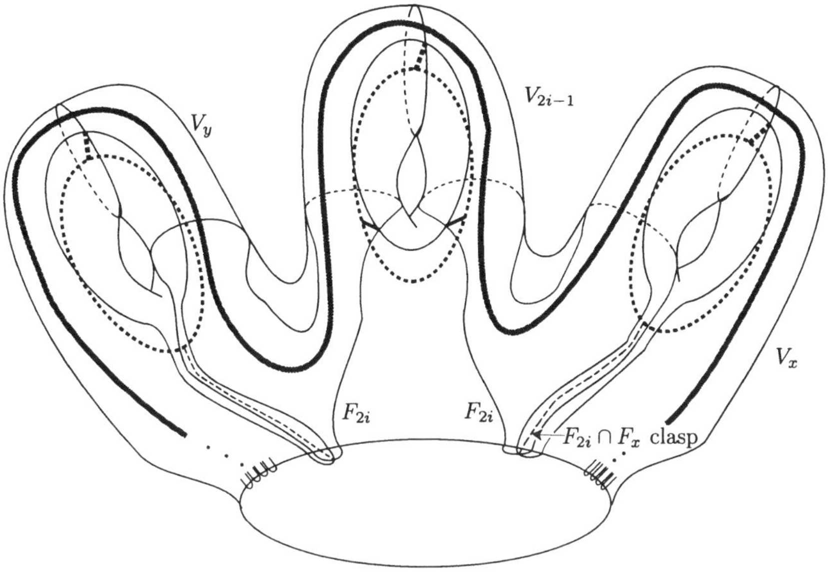

Figure 3.9

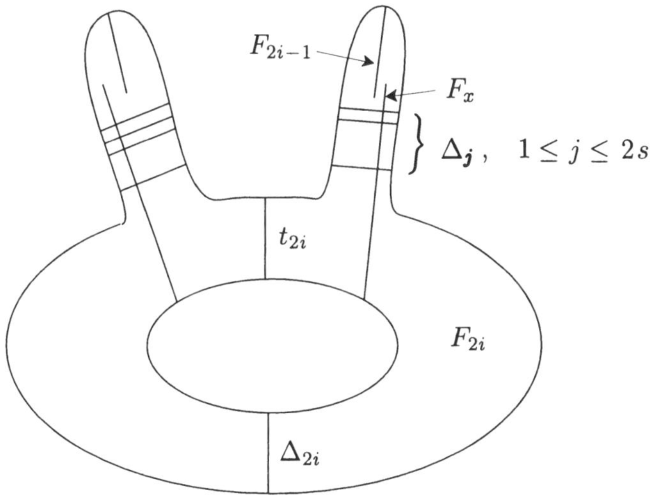

Figure 3.10

iii) Each F_i intersects at most one F_j in a ribbon.

Definition 3.17: Associated to a sufficient family there exists a directed graph $\mathcal{M}$ with one vertex for each member $s_i = (F_{2i-1}, F_{2i})$ of the sufficient family and k directed edges from s_j to s_i if $\Delta_{2i-1} \cap F_{2j} = k$ components. If $\mathcal{M}$ has no directed cycles, then it has a natural partial ordering, by declaring $s_i > s_j$ if there exists a directed path from s_i to s_j.

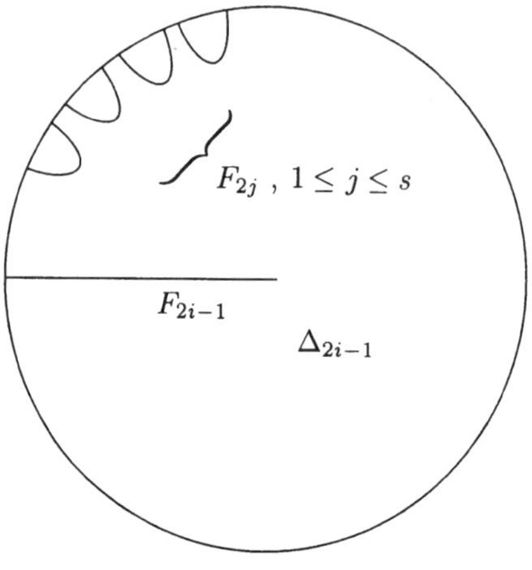

Figure 3.11

Theorem 3.18: Baby Smooth Tameness Theorem *[11].If the graph $\mathcal{M}$ associated to a honeycomb representation of Σ has no (directed) cycles, then $\Sigma \times I$ is geometrically simply connected.*

Proof: It suffices to show that $\Theta^4(N) = B^4 \cup$ 2-handles. For that result, coupled with 3.13 ii), yields $(\Delta^3 \times I)\#(S^2 \times D^2\text{'s}) = B^4 \cup$ 2-handles and hence $\Sigma^3 \times I = B^4 \cup$ 2,3-handles.

The idea of the proof is to show that M_0 can be zipped up to N in such a way that each $O(3)$ move is boundary compressible. We assume that the honeycomb satisfies the additional properties of Lemma 3.16.

Figure 3.12 shows how a neighborhood (in Σ^3) of a D_{2i} disc $i \leq s$ (i.e. belongs to a sufficient family) gets zipped up. We now show that associated to the $O(3)$ move corresponding to a diagonal intersection between D_{2i} and D_{2i-1} there exists an $O(3)$-boundary compressing disc η_i which has the following property. If $s_i > s_j$, then $f_0(\eta_i) \cap f_0(D_{2j}) = \emptyset$ where $f_0 : M_0 \to \Sigma^3$ is the map defined in the proof of Proposition 3.11.

We show how to construct η_1, η_2, η_3 in the case that $\mathcal{M}$ has 3 vertices, two edges and $s_1 > s_2 > s_3$. The general case is similar. Figure 3.13 a) shows how to construct η_1, i.e. η_1 is Δ_1 together with a band in D_1 connecting it to the $O(3)$ zipping locus. Figure 3.13 b shows how to begin to construct η_2. Instead of fully using Δ_3 and crashing through F_2, η_2 follows F_2 until it approaches the vicinity of V_1 whereupon it finishes as in Figure 3.14 a (which should be compared with Figure 3.9). Notice that $\eta_1 \subset \eta_2$. To construct η_3, mimic the construction of η_2, i.e. start out as in Figure 3.13 b, where all the indices are incremented by 2. Now η_3 follows F_4 until it reaches the vicinity of V_3 as in Figure 3.14 b where upon it encounters F_2 two times. Starting at each encounter η_3 follows F_2 and finishes up in Figure 3.14 a. Notice that η_3 has two subsets parallel to η_2. These somewhat complicated η_i's are used to more closely emulate the *[11]* proof of the baby theorem which provides a hint for the proof of Theorem 4.2 i).

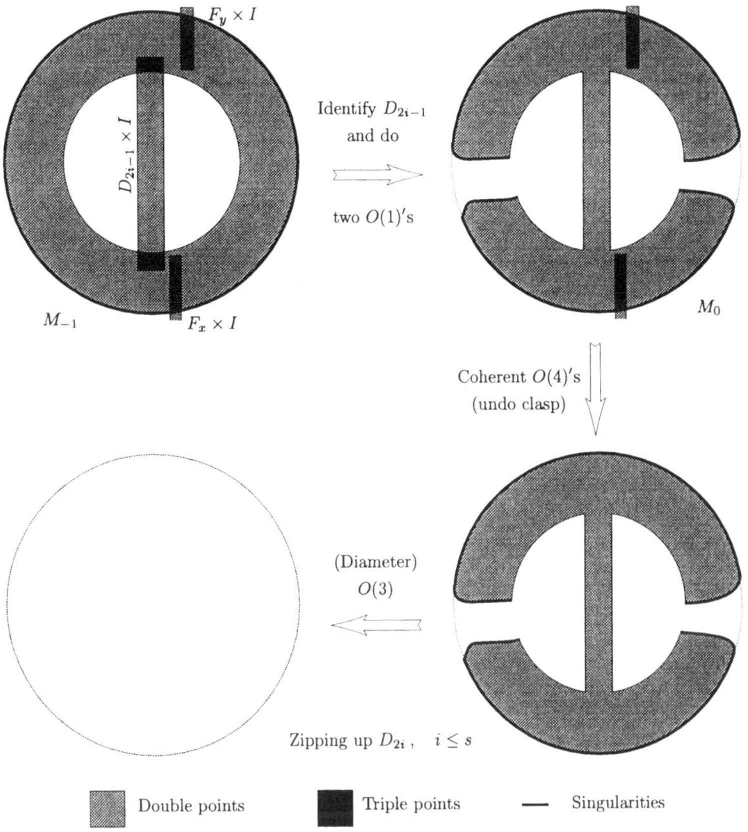

Figure 3.12

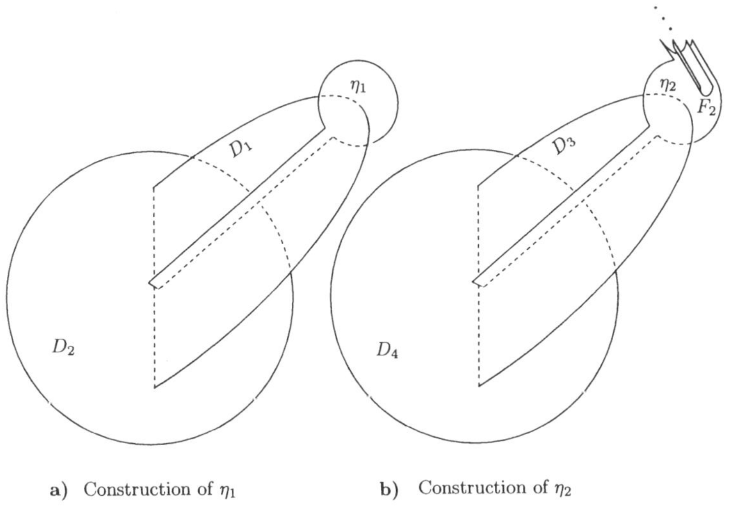

Figure 3.13

To obtain N from M_0, first consider a minimal (with respect to the partial order on $\mathcal{M}$) vertex $t_1 = (F_{2j-1}, F_{2j})$ and zip up D_{2j} as in Figure 3.12 to obtain the singular manifold N_1. Thanks to the existence of $\eta_j, \Theta^4(N_1) = \Theta^4(M_0) \cup$ 2-handle $= B^4 \cup$ 2-handle, by Lemma 3.8. By construction each $\eta_i, i \neq j$, is a boundary compressing disc in N_1 for its associated $O(3)$ movement. Inductively construct $N_2, \cdots, N_s$ where N_k is obtained from N_{k-1} by zipping up a D_{2i}, where $i = t_k$ is minimal in $\{1, 2, \cdots, s\} - \{t_1, \cdots, t_{k-1}\}$. Since the diameter $O(3)$ associated to D_{2i} is boundary compressible in N_{k-1} via η_i, it follows that $\Theta^4(N_k) = B^4 \cup$ 2-handles. Since N is obtained from N_s by $O(0), O(2)$ and coherent $O(4)$ moves, $\Theta^4(N) = B^4 \cup$ 2-handles.

Definition 3.19: There is another graph $\mathcal{M}^*$ associated to a sufficient family, defined as follows. The vertices of $\mathcal{M}^*$ are the elements of the sufficient family. There is a 1-1 correspondence between directed edges of $\mathcal{M}^*$ from s_i to s_j and components of intersection between F_{2i} and Δ_{2j}; however there is no edge corresponding to the canonical intersection of F_{2i} and Δ_{2i}. Note that $\mathcal{M}$ (respectively $\mathcal{M}^*$) can be viewed as an $s \times s$ matrix with entry s_{ij} equal to the number of directed edges from s_i to s_j. Thus $\mathcal{M}$ is nilpotent as a matrix if and only if the associated graph has no directed cycles. The sufficient family is said to be *binilpotent* if $\mathcal{M} + \mathcal{M}^*$ is nilpotent.

Remarks 3.20: i) The proof of the baby theorem also works if $\mathcal{M}^*$ is nilpotent, the only change being in the choice of η_i's. However this uses the fact that the description of how the V_{2i}'s and Δ_{2i}'s relate to the various F_{2j}'s $1 \leq i, j \leq s$ is similar to the one described in Lemma 3.16.

ii) As mentioned in Remark 3.15 the $\mathcal{C}$ operation associates to a sufficient

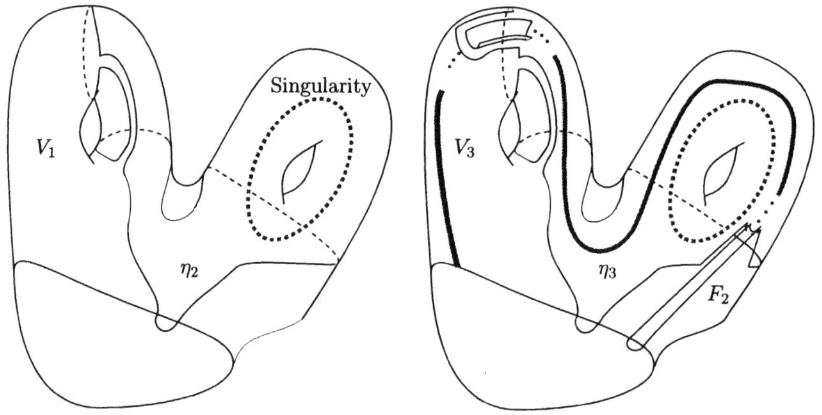

a) Construction of η_2 b) Construction of η_3

Figure 3.14

family of H a sufficient family of H'. The effect of $\mathcal{C}$ on these matrices is as follows $\mathcal{M} \to \mathcal{M}'^*$ and $\mathcal{M}^* \to \mathcal{M}'$.

4 The Smooth Tameness Theorem

Definition 4.1: A *sort of link* is a smooth 4-manifold W^4 such that $\overset{\circ}{W}{}^4 = \mathbb{R}^4_{standard}$ and $\partial W^4 = \overset{\circ}{D}{}^2 \times S^1$'s (possibly infinitely many). Thus a Casson handle is a *sort of knot.*

Theorem 4.2: Smooth Tameness Theorem *[11, 12]. There exists a sort of link W^4 such that*

i) $W^4 \cup$ 2-handles $= int(\Delta^3 \times I\#(\infty\ S^2 \times D^2$'s))

ii) If $W_N = W^4 \cup$ (first N boundary components), then W_N smoothly compactifies to B^4, i.e. there exists a smooth embedding of W_N into B^4 with $\overset{\circ}{W}{}^4 = \overset{\circ}{B}{}^4$.

Remarks 4.3: i) More precisely $int(\Delta^3 \times I\#(\infty\ S^2 \times D^2$'s)) is W^4 with a 2-handle attached to each $\overset{\circ}{D}{}^2 \times S^1$ component of ∂W^4, attached via the zero framing.

ii) Conclusion i) by itself is called the *Sort of Links Theorem* and is the content of *[11]*.

iii) The compactifications are not shown to be compatible. It is possible that $\partial W_N \subset \partial B^4_N$ represents a different link than $\partial W_n \subset \partial B^4_{N+1}$ where B^4_M denotes the compactification of W_M.

iv) To obtain the second conclusion, Po argues that W_N smoothly embeds in a 4-manifold A^4 such that $\overset{\circ}{A}{}^4 = \mathbb{R}^4_{\text{standard}}$ and $\partial A^4 = \mathbb{R}^3$ and uses the standard fact that A^4 compactifies to B^4.

v) There exists a Casson handle that is an example of a sort of knot which topologically but not smoothly compactifies to B^4. See *[4]*.

Corollary 4.4: *[12] $\Delta^3 \times I$ is geometrically simply connected at long distance (See 2.12).*

Proof: The standard compactness argument applied to the conclusion of the Smooth Tameness Theorem gives rise to embeddings $\Delta^3 \times I \to B^4 \cup$ (n 2-handles)$\to \Delta^3 \times I\#(m\ S^2 \times D^2$'s)$= (\Delta^3 - m\overset{\circ}{B}{}^3$'s)$\times I \subset \Delta^3 \times I$, so that the latter $\Delta^3 \times I$ is a regular neighborhood of the former $\Delta^3 \times I$.

4.5: A Hint of how to Find the Sort of Link W^4. The idea is to construct a sequence of honeycombs H_i with sufficient families S_i in which the directed cycles of the graphs $\mathcal{M}_i$ are pushed off to infinity. Carrying out a variant of the proof of the Baby Theorem in that context yields W^4.

Here is a hint of what "pushing out to infinity" means. For each i there is a honeycomb H_i with associated $Z_i \subset S_i$, where S_i is a sufficient family and the subset Z_i is binilpotent. Furthermore $(Z_1, S_1) \subset (Z_2, S_2) \subset \cdots$ and if $Z_\infty = \cup Z_i, S_\infty = \cup S_i$, then $Z_\infty = S_\infty$, so each state eventually becomes part of the binilpotent states. Also (among other things) at the level of matrices $\mathcal{Z}_i$ is a submatrix of $\mathcal{Z}_{i+1}$ and $\mathcal{Z}_i^*$ is a submatrix of $\mathcal{Z}_{i+1}^*$, where $\mathcal{Z}_i$ is the square submatrix of $\mathcal{M}_i$ associated to $Z_i \subset S_i$ and $\mathcal{Z}^*$ is defined similarly. So Z_i binilpotent means $\mathcal{Z}_i + \mathcal{Z}_i^*$ is nilpotent. Furthermore it is completely understood how one passes from H_i to H_{i+1}. (Thus one knows exactly the relation between the associated 4-dimensional thickenings, before and after various geometric operations, making the whole infinite process amenable to various compact exhaustion type arguments).

There are many inductive methods to transform a finite directed graph to an infinite directed graph without cycles. Much of the argument is involved with showing that some methods can be realized at the level of honeycombs, in ways compatible with various other geometric operations.

Remarks 4.6: i) Binilpotency rather than just nilpotency is needed to circumvent local finiteness difficulties in the above infinite process.

ii) There is some flexibility in creating the sort of link W^4. However, the construction must proceed with great care to create the one said to be amenable to the B^4 compactification.

Theorem 4.7: *If Σ is a homotopy 3-sphere, then $\Sigma \times I$ is geometrically simply connected.*

Proof: Combine the conclusions of Corollary 4.4 and Theorem 2.14.

Corollary 4.8: $(\Delta^3 - m\ \mathring{B}^3\text{'s}) \times I = B^4 \cup m$ *2-handles for some* $m \geq 0$.

By now applying Corollary 2.11, we have finally obtained the characterization of homotopy 3-spheres, which to Poénaru's mind represents the great and all important monument demarcating the two halves of the program to establish the Poincaré Conjecture.

Coherence Theorem 4.9: *Every homotopy 3-sphere has a coherent collapsible pseudospine repsentation.*

Remark 4.10: A consequence of the *[9, 10]* technology is that this representation can be chosen to be a honeycomb satisfying all the properties described in §3, e.g. has a sufficient family, satisfies Lemma 3.16, $\cdots$.

5 From Coherence To The Poincare Conjecture

Definition 5.1: A *stable sort of link* is a smooth noncompact 4-manifold Z^4 such that $\mathring{Z}^4 = int(B^4\#(\infty\ S^2 \times D^2\text{'s}))$ and $\partial Z^4 = p\ \mathring{D}^2 \times S^1\text{'s}$, $p < \infty$. In the obvious way, associated to any link (B^4, V), where V is a disjoint union of finitely many $D^2 \times S^1$'s in ∂B^4, there is a stable sort of link Z^4 with $\partial Z^4 = \mathring{V}$.

By Corollary 4.8 there exists a finite collection of pairwise disjoint solid tori $V \subset \partial B^4$ satisfying

(5.1) $(\Delta^3 \times I)\#(p\ S^2 \times D^2\text{'s}) = B^4 \cup (p$ 2-handles attached to $V)$

Theorem 5.2: *For each Σ^3 there exists a link V satisfying (5.1) with the following additional property. There exists a smooth embedding $Q : (Z^4, \mathring{V}) \to (\mathring{Z}^4 \cup \mathbb{R}^3, \mathbb{R}^3)$ such that $Q(\mathring{V}) = \mathring{U}$ where U is a regular neighborhood of the unlink of p components in $\mathbb{R}^3$. Furthermore $Q \mid \mathring{Z}^4$ =id and $\mathring{Z}^4 \cup \mathbb{R}^3$ is a smooth 4-manifold with boundary $\mathbb{R}^3$.*

Proof that Theorem 5.2 implies the Poincaré Conjecture: Via boundary connect sum attach to B^4, in an organized way, an infinite sequence of $S^2 \times D^2$'s, the attaching sites being away from some $\mathbb{R}^3 \subset \partial B^4$ which contains V. Taking the union of the interior of this space T and $\mathring{V}$ one obtains Z^4. Up to diffeomorphism there is only one way to attach an $\mathbb{R}^3$ boundary component to $\mathring{Z}^4$; thus Theorem 5.2 implies that Z^4 is also diffeomorphic to the interior of T and $\mathring{U}$, where U is a regular neighborhood of the unlink of p components in this same $\mathbb{R}^3$. Thus Z^4 has two exhaustions by compact sets $A_0 \subset A_1 \subset \cdots$ and $B_0 \subset B_1 \subset \cdots$. Here $A_0 = B^4 - \mathring{N}(\partial B^4 - \mathring{V})$ and $A_{i+1} - \mathring{A}_i$ is diffeomorphic to $(\delta A_i \times I)\#(S^2 \times D^2)$, where δ denotes relative boundary. Also B_0 is diffeomorphic to $B^4 - \mathring{N}(\partial B^4 - \mathring{U})$ and $B_{i+1} - \mathring{B}_i$ is diffeomorphic to $(\delta B_i \times I)\#(S^2 \times D^2)$.

Choose i, j such that $A_0 \subset B_i \subset A_j$. Now $\pi_1(Z^4 - A_r) \cong \pi_1(\partial B^4 - V) * F(r)$ and $\pi_1(Z^4 - B_i) \cong F(p) * F(i)$ where $F(s)$ denotes the free group on s letters and $*$ denotes free product. The inclusions $\alpha : Z^4 - A_j \to Z^4 - B_i$, $\beta : Z^4 - B_i \to Z^4 - A_0$ induce π_1 maps with same names such that $\beta \circ \alpha \mid \pi_1(\partial B^4 - \mathring{V})$ =id. This implies that $\alpha \mid \pi_1(\partial B^4 - \mathring{V})$ is an injective map into a free group and hence (since subgroups of free groups are free) $\pi_1(\partial B^4 - \mathring{V})$ is a free group. The standard Dehn's lemma argument *[7]* implies that V is the unlink of p components. Taking the boundary of (5.1) we see that $\Sigma\#(-\Sigma)\#(p\ S^2 \times S^1\text{'s}) = \#(p\ S^2 \times S^1)$'s so that $\Sigma = S^3$ by the uniqueness theorem *[6]* for (ordinary) connected sums. Po alternately notes that $\#(p\ S^2 \times S^1)$ contains no fake 3-cells for its universal covering embeds in $\mathbb{R}^3$ and the latter is irreducible by *[1]*.

Theorem 5.2 is a corollary of the main technical result of *[14]*

Theorem 5.3. Strange Compactification: *For every homotopy 3-sphere Σ there exists a link V satisfying (5.1) with the following additional property. There exists a smooth embedding $P : (Z^4, \mathring{V}) \to (Y^4, Y^3)$ where $P \mid \mathring{Z}^4$ is a diffeomorphism*

onto $\mathring{Y}^4, Y^3 = \partial Y^4$ *and* $P(\mathring{V})$ *is an open regular neighborhood of the trivial link of p components in* Y^3.

Remarks 5.4: i) L being the trivial link in Y^3 means that the components of L bound pairwise disjoint embedded discs in Y^3. Thus one obtains an $\mathbb{R}^3$ by taking the union of $P(\mathring{V})$ and regular neighborhoods of these discs together with 1-handles connecting them up. Theorem 5.2 thus follows from Theorem 5.3.

ii) According to Po, measured by both volume and difficulty, the argument for Theorem 5.3 approximates that of the coherence theorem.

iii) A long argument, involving no infinite processes, applied to a coherent honeycomb representation produces a collection of pairs $((F_1, F_2), \cdots (F_{2s-1}, F_{2s}))$ called a *strongly sufficient family* whose definition is somewhat akin to that of a sufficient family. Associated to this family is a link V satisfying (5.1) (compare 2.11) and it is this link which reputedly satisfies the conclusion of the strange compactification theorem.

iv) Somehow this space Y^4, constructed by an infinite process, has the following three features.

a) A simply connected boundary Y^3 with infinitely generated π_2 and possibly $\pi_1^\infty(Y^3) \neq 1$.

b) If L is the link consisting of the cores of $P(V)$, then each component L_i of L bounds a properly embedded (in Y^3) planar surface E_i which is disjoint from $L - L_i$.

c) A 2-dimensional singular lamination $\lambda \subset Y^3$, by planes, which is disjoint from each L_j.

A deeper understanding of c) shows that each component of $Y^3 - \lambda$ is simply connected at ∞, hence by Dehn's lemma each E_i can be truncated to a compact disc and so L is the unlink.

v) Here is a description of how the pair (Y^3, λ) may arise. Let $T = \#(\infty\ S^1 \times D^2\text{'s})\#(\infty\ S^2 \times I\text{'s})$. (The complexity of the end structure of this manifold with boundary depends on how the boundary connected sums are performed.) Associated with T are

a) An infinite collection of pairwise disjoint simple closed curves $\{a_i\}$ in ∂T, which is conjugate to a system of generators of $\pi_1(T)$.

b) An infinite collection of pairwise disjoint embedded discs $\{D_i\}$ in T such that for each i,

$$D_i \cap a_j = \begin{cases} 1 \text{ point if } i = j \\ \emptyset \text{ otherwise.} \end{cases}$$

c) The accumulation points of $\cup \partial D_i$ which are disjoint from $\cup_1^\infty a_i$.

Define $Y^3 = int\{T \cup 2\text{-handles attached along } \{a_i\}\}$ and let λ be the restriction of the accumulation points of $\cup D_i$ to Y^3. The set $\cup a_i$ may not be closed in T and it is this property which suggests that Y^3 is wild. Nevertheless λ has "tame", i.e. $\pi_1^\infty = 0$, complementary regions. These complementary regions when completed with respect to the path metric (so the boundary components correspond with leaves of λ) may not be "standard". The "wicks" of the boundary components may be Artin-Fox curves.

vi) To ensure that for each $i, L_i \cap \lambda = \emptyset$ the following difficulty must be overcome. How to construct Y^4 so that for each j, L_j nontrivially intersects only finitely many D_i's?

References

[1] J. W. Alexander, On the subdivision of the 3-sphere by a polyhedron, Proc. Nat. Acad. Sci. **10** (1924), 6–8.

[2] M. Freedman, The topology of 4-dimensional manifolds, J. Diff. Geom. **17** (1982), 357–453.

[3] D. Gabai, Foliations and the topology of 3-manifolds III, J. Diff. Geom. **26** (1987), 479–536.

[4] R. Gompf, Three exotic $\mathbb{R}^4$'s and other anomalies, J. Diff. Geom. **18** (1983), 313–328.

[5] B. Mazur, On embeddings of spheres, Acta Math. **105** (1961), 1–17.

[6] J. Milnor, A unique factorization theorem for 3-manifolds, Amer. J. Math. **84** (1962), 1–7.

[7] C. D. Papakyriakopoulos, On Dehn's Lemma and the asphericity of knots, Ann. of Math. **66** (1957).

[8] V. Poénaru, Infinite processes and the Poincaré Conjecture, I; The collapsible pseudospine representation theorem, Topology **31** (1992), 625–636.

[9] ———, Infinite processes and the Poincaré Conjecture, II; The honeycomb representation theorem, 2nd version, Orsay preprint 93-14, 1993.

[10] ———, Infinite processes and the Poincaré Conjecture, III; The algorithm, Orsay preprint 92-10, 1992.

[11] ———, Infinite processes and the Poincaré Conjecture, IV; Theoreme de non-sauvagerie lisse (The smooth tameness theorem) Part A, Orsay preprint 92-83, 1992.

[12] ———, Infinite processes and the Poincaré Conjecture, IV; Theoreme de non-sauvagerie lisse (The smooth tameness theorem) Part B, preprint expected Dec. 1994.

[13] ———, Infinite processes and the Poincaré Conjecture, V; Differential topology in dimension 3+1, Orsay preprint, 94–25 (1994).

[14] ———, Infinite processes and the Poincaré Conjecture, VI; The strange compactification theorem, manuscript.

[15] ———, Produits cartesiens de varietes differentielles par un disque, *Proc. Inter. Cong. Math. Stockholm 1962*, 481–489.

[16] V. Poénaru, Killing handles of index one stably and π_1^∞, Duke Math. J. **63** (1991), 431–447.

[17] ———, Almost convex groups, Lipschitz combings and π_1^∞ for universal coverings of closed 3-manifolds, J. Diff. Geom. **35** (1992), 103–130.

[18] ———, Geometry" á la Gromov" for the fundamental group of a closed 3-manifold M^3 and the simple connectivity at infinity of $\tilde{M}^3$, Topology **33** (1994), 181–196.

[19] ——, The handles of index one of the product of an open simply connected 3-manifold with a high dimensional ball, manuscript.

[20] ——, Infinite processes and the 3-dimensional Poincaré Conjecture: An outline of the outline of the proof, Orsay preprint 89-06, 1989.

[21] ——, 1992 letter to Barry Mazur.

[22] ——, Infinite processes and the 3-dimensional Poincaré Conjecture (An outline of the proof) Second Version, 1991.

[23] V. Poénaru and C. Tanasi, Representations of the Whitehead manifold Wh^3 and Julia sets, Orsay preprint 92-57, 1992.

[24] ——, k-weakly almost convex groups and $\pi_1^\infty(\tilde{M}^3)$, Geom. Ded. **48** (1993), 57–81.

[25] J. Stallings, How not to prove the Poincaré Conjecture, Ann. of Math. Stud. **60** (1966), 83–88.

Cyclic Operads and Cyclic Homology

E. Getzler

Department of Mathematics
Massachusetts Institute of Technology
Cambridge, Massachusetts 02139

M.M. Kapranov

Department of Mathematics
Northwestern University
Evanston, Illinois 60208

The cyclic homology of associative algebras was introduced by Connes [**4**] and Tsygan [**22**] in order to extend the classical theory of the Chern character to the non-commutative setting. Recently, there has been increased interest in more general algebraic structures than associative algebras, characterized by the presence of several algebraic operations. Such structures appear, for example, in homotopy theory [**18**], [**3**] and topological field theory [**9**]. In this paper, we extend the formalism of cyclic homology to this more general framework.

This extension is only possible under certain conditions which are best explained using the concept of an operad. In this approach to universal algebra, an algebraic structure is described by giving, for each $n \geq 0$, the space $\mathcal{P}(n)$ of all n-ary expressions which can be formed from the operations in the given algebraic structure, modulo the universally valid identities. Permuting the arguments of the expressions gives an action of the symmetric group $\mathbb{S}_n$ on $\mathcal{P}(n)$. The sequence $\mathcal{P} = \{\mathcal{P}(n)\}$ of these $\mathbb{S}_n$-modules, together with the natural composition structure on them, is the operad describing our class of algebras. In order to define cyclic homology for algebras over an operad $\mathcal{P}$, it is necessary that $\mathcal{P}$ is what we call a *cyclic operad*: this means that the action of $\mathbb{S}_n$ on $\mathcal{P}(n)$ extends to an action of $\mathbb{S}_{n+1}$ in a way compatible with compositions (see Section 2). Cyclic operads are a natural generalization of associative algebras with involution (see (2.2)).

For associative algebras over a field of characteristic 0, it is a result of Feigin and Tsygan [**6**] that cyclic homology is the non-abelian derived functor of the functor $A \mapsto A/[A, A]$, the target of the universal trace on A. The notion of a trace does not make sense for more general algebras. However, for an associative algebra A *with unit,* there is a bijection between traces T and invariant bilinear forms B on A, obtained by setting $T(x) = B(1, x)$. It turns out that the structure of a cyclic operad on $\mathcal{P}$ is precisely the data needed to speak about invariant bilinear forms on algebras over an operad $\mathcal{P}$. The cyclic homology of an algebra over a

Geometry, Topology, and Physics for Raoul Bott

Cambridge, MA 02138 USA

cyclic operad $\mathcal{P}$ is now defined as the non-abelian derived functor of the target of the universal invariant bilinear form.

In making this definition, we were strongly influenced by ideas of M. Kontsevich: he emphasized that invariant bilinear forms are more fundamental than traces [**15**]. Furthermore, he constructed an action of $\mathbb{S}_{n+1}$ on the space $\mathcal{L}ie(n)$ of n-ary operations on Lie algebras.

We construct an explicit complex $\mathrm{CA}(\mathcal{P}, A)$ calculating the cyclic homology $\mathrm{HA}_\bullet(\mathcal{P}, A)$. There is a long exact sequence which involves not only $\mathrm{HA}_\bullet(\mathcal{P}, A)$ but two other functors $\mathrm{HB}_\bullet(\mathcal{P}, A)$ and $\mathrm{HC}_\bullet(\mathcal{P}, A)$:

$$\dots \to \mathrm{HC}_{n+1}(\mathcal{P}, A) \to \mathrm{HA}_n(\mathcal{P}, A) \to \mathrm{HB}_n(\mathcal{P}, A) \to \mathrm{HC}_n(\mathcal{P}, A) \to \dots$$

In the general case, $\mathrm{HC}(\mathcal{P}, A)$ may be identified with the cyclic homology of the dual algebra over the dual operad (in the sense of [**12**] and [**11**]). For unital associative algebras, this sequence may be identified with the sequence of Connes and Tsygan

$$\dots \to H^\lambda_{n+2}(A) \xrightarrow{S} H^\lambda_n(A) \xrightarrow{B} H_{n+1}(A, A) \xrightarrow{I} H^\lambda_{n+1}(A) \to \dots,$$

by an isomorphism $\mathrm{HA}_n(\mathcal{U}\mathcal{A}ss, A) \cong \mathrm{HC}_{n+1}(\mathcal{U}\mathcal{A}ss, A)$ which has no analogue for a general operad. For Lie algebras, we obtain the long exact sequence

$$\dots \to H_{n+2}(\mathfrak{g}) \to \mathrm{HA}_n(\mathcal{L}ie, \mathfrak{g}) \to H_n(\mathfrak{g}, \mathfrak{g}) \to H_{n+1}(\mathfrak{g}) \to \dots,$$

where $\mathrm{HA}_\bullet(\mathcal{L}ie, \mathfrak{g})$ is the Lie cyclic homology of the Lie algebra $\mathfrak{g}$.

Using the long exact sequence for the commutative operad $\mathcal{C}om$, we are able to prove an isomorphism of $\mathbb{S}_{n+1}$-modules

$$\mathcal{L}ie(n+1) \cong_{\mathbb{S}_{n+1}} \mathcal{L}ie(n) \otimes V_{n,1},$$

where $V_{n,1}$ is the n-dimensional irreducible representation of $\mathbb{S}_{n+1}$ on the hyperplane $x_0 + \dots + x_n = 0$.

Another example of a cyclic operad is the operad whose algebras are A_∞-algebras: cyclic homology of A_∞-algebras was studied in [**10**].

The research of both authors is partially supported by NSF grants and A.P. Sloan Research Foundation. This paper was written while the authors were visitors of the Mathematical Sciences Research Institute, Berkeley, whose hospitality and financial support are gratefully acknowledged.

1 Operads and trees

Here, we review background material, referring the reader to May [**18**] and to the introductory chapters of [**11**], [**12**] for more details.

(1.1) Notation: Throughout this paper, we work over a fixed field $\mathbf{k}$ of characteristic 0. We denote the symmetric group $\mathrm{Aut}\{1, \dots, n\}$ by $\mathbb{S}_n$. We will also need the group $\mathrm{Aut}\{0, 1, \dots, n\}$ which we will denote $\mathbb{S}_{n+1}$. To avoid confusion, we will always explicitly write "+1" in the subscript when we deal with this latter group. For example, $\mathbb{S}_{2+1}$ will denote the group $\mathrm{Aut}\{0, 1, 2\}$.

For $n \geq 0$, we denote by $\tau_n \in \mathbb{S}_{n+1}$ the cycle $(0\,1 \ldots n)$; we will often write it τ if n is clear from the context. The element τ_n generates a cyclic group $\mathbb{Z}_{n+1} \subset \mathbb{S}_{n+1}$ which, together with $\mathbb{S}_n$, generates $\mathbb{S}_{n+1}$.

The category of *graded* $\mathbf{k}$-*vector spaces* is a symmetric monoidal category, with tensor product

$$(V \otimes W)_k = \bigoplus_{i+j=k} V_i \otimes W_j$$

and symmetry maps $S : V \otimes W \to W \otimes V$ given by the formula

$$v \otimes w \mapsto (-1)^{|v|\,|w|} w \otimes v;$$

here, we write $|v| = i$ if $v \in V_i$.

A *chain complex* (dg-vector space) is a graded vector space $V_\bullet$ together with a differential $\delta : V_i \to V_{i-1}$, such that $\delta^2 = 0$. The category of chain complexes is again a symmetric monoidal category with tensor product and symmetry defined as for graded vector spaces and the differential given by the standard Leibniz formula (see, for example, Dold [**5**]). Given two chain complexes V and W, there is a chain complex $\mathrm{hom}(V, W)$, such that $\mathrm{hom}_n(V, W)$ is the space of all linear maps homogeneous of degree n from V to W [**5**].

The *suspension* $\Sigma V_\bullet$ of a chain complex $V_\bullet$ has components $(\Sigma V)_n = V_{n-1}$, and differential equal to minus that of $V_\bullet$. By $\Sigma^n V_\bullet$, $n \in \mathbb{Z}$, we denote the n-fold iterated suspension of $V_\bullet$.

The *determinant* $\det(V)$ of a finite dimensional vector space V is its highest exterior power $\Lambda^{\dim(V)} V$.

(1.2) $\mathbb{S}$-modules: An $\mathbb{S}$-*module* is a sequence of vector spaces

$$\mathcal{V} = \{\mathcal{V}(n) \mid n \geq 0\},$$

together with an action of $\mathbb{S}_n$ on $\mathcal{V}(n)$ for each n. Denote the category of $\mathbb{S}$-modules, with the evident notion of morphisms, by $\mathbb{S}$-mod.

A *cyclic* $\mathbb{S}$-*module* is a sequence of vector spaces $\mathcal{V} = \{\mathcal{V}(n) \mid n \geq 0\}$, together with an action of $\mathbb{S}_{n+1}$ on $\mathcal{V}(n)$. The category of cyclic $\mathbb{S}$-modules will be denoted $\mathbb{S}_+$-mod. Since $\mathbb{S}_n$ is a subgroup of $\mathbb{S}_{n+1}$, any cyclic $\mathbb{S}$-module may be regarded as an $\mathbb{S}$-module.

An $\mathbb{S}$-module $\mathcal{V}$ gives rise to a functor, which we also denote by $\mathcal{V}$, from the category of finite sets and their bijections to the category of vector spaces. Explicitly, if I is a set with n elements, we define

$$\mathcal{V}(I) = \Big(\bigoplus_{f:\{1,\ldots,n\} \to I} \mathcal{V} \Big)_{\mathbb{S}_n}, \tag{1.3}$$

where the sum is over all bijections f and the space of coinvariants is taken with respect to the total action. For example, if $V = W^{\otimes n}$ is the n-th tensor power of another vector space W, with the natural action of $\mathbb{S}_n$, then the formula (1.3) gives the space which we will denote $W^{\otimes I}$. In a similar way one defines the tensor product $\bigotimes_{i\in I} W_i$ of a family of vector spaces labelled by a set I.

A similar construction to (1.3) extends a cyclic $\mathbb{S}$-module to a functor on the category of non-empty finite sets and bijections: if I is a set with $n+1$ elements, we define

$$\mathcal{V}((I)) = \Big(\bigoplus_{f:\{0,\dots,n\}\to I} \mathcal{V} \Big)_{\mathbb{S}_{n+1}}. \tag{1.4}$$

(1.5) Operads: A $\mathbf{k}$-linear *operad* is an $\mathbb{S}$-module $\mathcal{P}$ together with a map $\eta : \mathbf{k} \to \mathcal{P}(1)$ (called the unit map) and maps

$$\mu_{i_1\dots i_k} = \mu^{\mathcal{P}}_{i_1,\dots,i_k} : \mathcal{P}(k) \otimes \mathcal{P}(i_1) \otimes \dots \otimes \mathcal{P}(i_k) \to \mathcal{P}(i_1 + \dots + i_k)$$

(called composition maps) given for any $k \geq 1$ and $i_1, \dots, i_k \geq 0$. These data are required to satisfy the axioms of equivariance, associativity and unit (see May [**18**], Hinich-Schechtman [**13**], and Getzler-Jones [**11**]).

We denote the product $\mu_{i_1\dots i_k}(p \otimes q_1 \otimes \dots q_k)$ by $p(q_1, \dots, q_k)$.

A *morphism of operads* $\Phi : \mathcal{P} \to \mathcal{Q}$ is a morphism of $\mathbb{S}$-modules preserving the unit and composition maps.

The *endomorphism operad* $\mathcal{E}_V$ of a vector space V consists of the spaces $\mathcal{E}_V(n) = \mathrm{Hom}(V^{\otimes n}, V)$ with the obvious action of $\mathbb{S}_n$, unit and composition maps.

An *algebra* over an operad $\mathcal{P}$ (or $\mathcal{P}$-algebra) is a vector space A together with a morphism of operads $\mathcal{P} \to \mathcal{E}_A$.

An important example of a $\mathcal{P}$-algebra is the *free $\mathcal{P}$-algebra* generated by a vector space V: this is the direct sum

$$\mathsf{F}(\mathcal{P}, \mathcal{V}) = \bigoplus_{n=0}^{\infty} \mathcal{P}(n) \otimes_{\mathbb{S}_n} V^{\otimes n},$$

where $\mathbb{S}_n$ acts on $V^{\otimes n}$ by permuting the factors.

Classes of algebras such as associative, commutative, Lie and Poisson, can be regarded as algebras over suitable operads, as we explain in Section 2.

(1.6) Operads in monoidal categories: The notion of an operad makes sense in any symmetric monoidal category $(\mathcal{C}, \otimes)$ in the sense of [**17**]. The cases of interest in this paper are as follows:

(a) $\mathbf{k}$-*linear operads:* $\mathcal{C}$ is the category of $\mathbf{k}$-vector spaces, with $\otimes$ the tensor product, the case described above.

(b) *topological operads:* $\mathcal{C}$ is the category of topological spaces, with $\otimes$ the Cartesian product.

(c) *graded operads:* $\mathcal{C}$ is the category of graded vector spaces. If $\mathcal{P}$ is a topological operad, the homology spaces $H_\bullet(\mathcal{P}(n), \mathbf{k})$ form a graded operad, called the *homology operad* of $\mathcal{P}$ and denoted $H_\bullet(\mathcal{P}, \mathbf{k})$.

(d) *dg-operads:* $\mathcal{C}$ is the category of chain complexes $(V_\bullet, \delta)$ over $\mathbf{k}$.

(1.7) Trees: A *tree* T is a contractible graph with finitely many vertices and edges. We denote the set of vertices of a tree by $\mathrm{Vert}(T)$. There are two sorts of edges: those terminated at both ends by a vertex are called internal, and the set of these is denoted Edge(T), while those terminated at either one end or none by a vertex are called legs, and the set of these is denoted Leg(T). (Legs are called leaves in [**11**].) For any vertex $v \in T$, the set of internal edges and legs incident with v is denoted Leg(v), and the *valence* of the vertex v is the cardinality of Leg(v).

If T is a tree, we denote by $\det(T)$ the determinant of the vector space

$$\bigoplus_{e \in \mathrm{Edge}(T)} \mathrm{Or}(e),$$

where $\mathrm{Or}(e)$ is the orientation line of the edge e. Since the chain complex of the tree T may be written

$$0 \to \bigoplus_{e \in \mathrm{Edge}(T)} \mathrm{Or}(e) \to \mathbf{k}^{\mathrm{Vert}(T)} \to 0$$

has $H_0 \cong \mathbf{k}$ and $H_1 \cong 0$, we see that $\det(T)$ may be canonically identified with the determinant of the vector space $\mathbf{k}^{\mathrm{Vert}(T)}$.

The tree T with no vertices (and hence no internal edges) is slightly exceptional, in that its edge is not canonically oriented. We define $\det(T)$ in this case to be the orientation line of this edge.

A *rooted tree* is a tree together with a choice of one element $\mathrm{Out}(T) \in \mathrm{Leg}(T)$ (the root). In a rooted tree, there is a canonical orientation of the edges, in such a way that the unique path from any edge to the root is oriented positively. This orientation has the following properties:

(1) for every vertex $v \in \mathrm{Vert}(T)$, there is exactly one edge $\mathrm{Out}(v) \in \mathrm{Leg}(v)$ emerging from v;

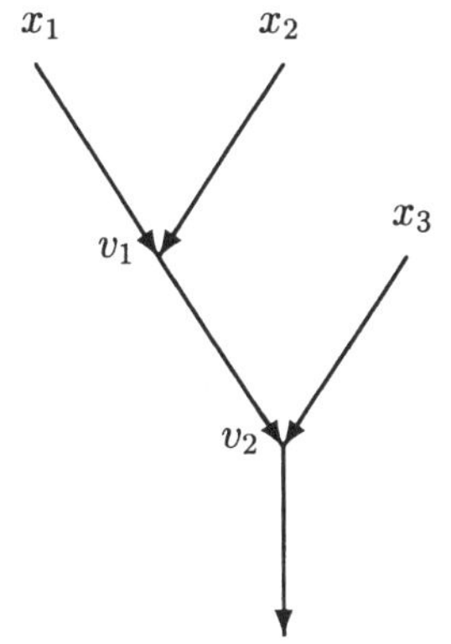

Figure 1.1: A rooted tree

(2) $\mathrm{Out}(T)$ is the unique leg which emerges from T.
This orientation is illustrated in Figure 1.1. Denote by $\mathrm{In}(T) = \mathrm{Leg}(T) \backslash \{\mathrm{Out}(T)\}$ the set of incoming legs of the rooted tree T, and by $\mathrm{In}(v) = \mathrm{Leg}(v) \backslash \{\mathrm{Out}(v)\}$ the set of legs at a vertex v.

If I is a finite set, an unrooted (respectively rooted) I-tree is a tree T together with a bijection $I \to \mathrm{Leg}(T)$ (resp. tree T together with a bijection $I \to \mathrm{In}(T)$). Clearly, a rooted I-tree is the same thing as an unrooted I_+-tree, where I_+ is the union of I and an additional base point. We will refer to rooted $\{1, \dots, n\}$-trees as rooted n-trees, and to unrooted $\{0, \dots, n\}$-trees as unrooted $n+1$-trees, following the same convention on subscripts as in (1.1).

A tree T is *binary* if all of its vertices have valence three, that is, $|\,\mathrm{Leg}(v)| = 3$ for all vertices $v \in \mathrm{Vert}(T)$.

(1.8) Triples associated to trees: Let $\mathcal{V}$ be an $\mathbb{S}$-module, and let T be a rooted tree. Define

$$\mathcal{V}(T) = \bigotimes_{v \in \mathrm{Vert}(T)} \mathcal{V}(\mathrm{In}(v)),$$

where $\mathcal{V}(\mathrm{In}(v))$ is defined by (1.3). We now define a functor $\mathbb{T}$ from $\mathbb{S}$-mod to itself, by summing over isomorphism classes of rooted trees:

$$\mathbb{T}\mathcal{V}(n) = \bigoplus_{\text{rooted } n\text{-trees T}} \mathcal{V}(T). \tag{1.9}$$

Similarly, let $\mathcal{W}$ be a cyclic $\mathbb{S}$-module, and let T be an unrooted tree. We define $\mathcal{W}(\!(T)\!)$ to be the tensor product over the vertices $v \in \mathrm{Vert}(T)$ of $\mathcal{W}(\!(\mathrm{Leg}(v))\!)$. We now define a functor $\mathbb{T}_+$ from $\mathbb{S}_+$-mod to itself, by summing over isomorphism classes of unrooted trees:

$$\mathbb{T}_+\mathcal{W}(n) = \bigoplus_{\text{unrooted } n+1\text{-trees T}} \mathcal{W}(\!(T)\!). \tag{1.10}$$

Under the forgetful functor $\mathbb{S}_+$-mod $\to$ $\mathbb{S}$-mod, the functor $\mathbb{T}_+$ reduces to the functor $\mathbb{T}$.

Grafting of trees defines, for any $\mathbb{S}$-module $\mathcal{V}$, a natural map $\mathbb{T}\mathbb{T}\mathcal{V} \to \mathbb{T}\mathcal{V}$. We also have embeddings $\mathcal{V} \to \mathbb{T}\mathcal{V}$, which identify $\mathcal{V}(n)$ with the summand in (1.9) corresponding to the unique rooted n-tree with one vertex and $n+1$ legs.

The following result is proved in Section 1.4 of [**11**].

(1.11) Theorem: *The maps $\mathbb{T}\mathbb{T}\mathcal{V} \to \mathbb{T}\mathcal{V}$ and $\mathcal{V} \to \mathbb{T}\mathcal{V}$ make the functor $\mathbb{T}$: $\mathbb{S}$-mod $\to$ $\mathbb{S}$-mod into a triple (or monad). An algebra over the triple $\mathbb{T}$ in the category of $\mathbb{S}$-modules is the same as an operad.*

This theorem amounts to the statement that if $\mathcal{P}$ is an operad and T is a rooted n-tree, there is induced a map $\mu_T : \mathcal{P}(T) \to \mathcal{P}(n)$, and these maps determine the operad structure of $\mathcal{P}$.

(1.12) Free operads: If $\mathcal{V}$ is an $\mathbb{S}$-module, the $\mathbb{S}$-module $\mathbb{T}\mathcal{V}$ has a natural structure of an operad, by Theorem (1.11): it is called the *free operad* generated by $\mathcal{V}$. It is convenient to represent elements of $\mathbb{T}\mathcal{V}(n)$ in a more algebraic fashion as follows. An element $p \in \mathcal{V}(n) \subset \mathbb{T}\mathcal{V}(n)$ is thought of as a formal "operation" $p(x_1, ..., x_n)$ on n indeterminates. If T is a rooted n-tree and for each vertex $v \in T$ we have assigned a vector $p_v \in \mathcal{V}(\mathrm{In}(v))$, then the element $\bigotimes p_v \in \mathcal{V}(T) \subset \mathbb{T}\mathcal{V}(n)$ is represented as an operation obtained by iterated application of the p_v, taking the tree T as a scheme of calculation. As an example, with the 3-tree T of Fig. 1.1 and $p_{v_1}, p_{v_2} \in E(2)$, the product $p_{v_1} \otimes p_{v_2} \in \mathbb{T}E(3)$ is represented by the expression $p_{v_2}(p_{v_1}(x_1, x_2), x_3)$.

2 Cyclic operads

In the same way as we defined in (1.8) a structure of a triple on the functor $\mathbb{T}$, we may define a structure of a triple on $\mathbb{T}_+$, acting on the category of cyclic $\mathbb{S}$-modules: grafting of unrooted trees defines, for any cyclic $\mathbb{S}$-module $\mathcal{V}$, a natural map $\mathbb{T}_+\mathbb{T}_+\mathcal{V} \to \mathbb{T}_+\mathcal{V}$, and we also have embeddings $\mathcal{V} \to \mathbb{T}_+\mathcal{V}$.

(2.1) Definition: A $\mathbf{k}$-linear cyclic operad is an algebra over the triple $\mathbb{T}_+$ in the category of cyclic $\mathbb{S}$-modules.

Similarly, we may define a cyclic operad in an arbitrary symmetric monoidal category $(\mathcal{C}, \otimes)$.

A cyclic operad $\mathcal{A}$ may be considered to be an ordinary operad by forgetting part of its structure. A cyclic operad structure on an operad $\mathcal{P}$ is determined by an extension of the $\mathbb{S}$-module structure of $\mathcal{P}$ to a cyclic $\mathbb{S}$-module structure, such that certain compatibility conditions are satisfied. For simplicity, we restrict attention to ungraded operads. As in (1.1), τ_n is the cycle $(0\,1 \dots n)$ in $\mathbb{S}_{n+1}$.

(2.2) Theorem: *A cyclic operad $\mathcal{P}$ is an operad $\mathcal{P}$ together with an extension of the $\mathbb{S}$-module structure of $\mathcal{P}$ to a cyclic $\mathbb{S}$-module structure, such that:*

(1) $\tau_1(1) = 1$, *where* $1 \in \mathcal{P}(1)$ *is the unit of the operad* $\mathcal{P}$*;*

(2) If $p \in \mathcal{P}(m)$ *and* $q \in \mathcal{P}(n)$*, then*

$$\tau_{m+n-1}(p(1, \dots, 1, q)) = (\tau_n q)(\tau_m(p), 1, \dots, 1).$$

(If $\mathcal{P}$ is a graded operad, there is an additional factor of $(-1)^{|p|\,|q|}$.*)*

If $\mathcal{P}$ is an operad, the space $\mathcal{P}(1)$ is an algebra, with product $\mu_1 : \mathcal{P}(1) \otimes \mathcal{P}(1) \to \mathcal{P}(1)$. The above theorem implies that if $\mathcal{P}$ is a cyclic operad, $\mathcal{P}(1)$ is an algebra with involution $\tau = \tau_1$; that is, $\tau^2 = 1$ and $\tau(pq) = \tau(q)\tau(p)$ for all $p, q \in \mathcal{P}(1)$.

We give the proof of Theorem (2.2) later in this section, but first we present some examples of cyclic operads.

(2.3) Examples:

(a) *Endomorphism operads:* If V is a finite-dimensional $\mathbf{k}$-vector space with a non-degenerate inner product $B(x, y)$, the endomorphism operad $\mathcal{E}_V$ of V is a cyclic operad. The cyclic $\mathbb{S}$-module structure of $\mathcal{E}_V$ is defined as follows: using

B, we identify $\mathcal{E}_V(n) = \mathrm{Hom}(V^{\otimes(n)}, V)$ with $\mathrm{Hom}(V^{\otimes(n+1)}, \mathbf{k})$, and give this space the obvious action of $\mathbb{S}_{n+1}$.

(b) *The associative operad:* Associative **k**-algebras (possibly without unit) are algebras over an operad $\mathcal{A}ss$ defined as follows: $\mathcal{A}ss(n)$, $n \geq 1$, is the subspace of the space of non-commutative polynomials $\mathbf{k}\langle x_1, \ldots, x_n\rangle$ spanned by monomials of the form $x_{\sigma(1)} \ldots x_{\sigma(n)}$, where $\sigma \in \mathbb{S}_n$, in other words, monomials linear in each letter x_i, while $\mathcal{A}ss(0) = 0$. Thus, $\mathcal{A}ss(n)$ has dimension $n!$ and is isomorphic to the regular representation of $\mathbb{S}_n$. There is also an operad $\mathcal{UA}ss$ which describes the category of unital associative algebras: $\mathcal{UA}ss$ differs from $\mathcal{A}ss$ only in that $\mathcal{UA}ss(0) = \mathbf{k}$, the space of non-commutative polynomials in zero variables.

(2.4) Proposition: *The operads $\mathcal{A}ss$ and $\mathcal{UA}ss$ are cyclic.*

Proof: We give an explicit formula for the action of $\mathbb{S}_{n+1}$ on $\mathcal{UA}ss(n)$. The action of the subgroup $\mathbb{S}_n$ being known, it suffices to define the action of the cycle $\tau = (0\,1 \ldots\, n)$, which is

$$\tau(x_{\sigma(1)} \ldots x_{\sigma(n)}) = x_{\sigma(\sigma^{-1}(n)+1)+1} \ldots x_{\sigma(n)+1} x_1 x_{\sigma(1)+1} \ldots x_{\sigma(\sigma^{-1}(n)-1)+1}. \tag{2.5}$$

Since $x_1 \ldots x_n$ is invariant under τ, we see that

$$\mathcal{UA}ss(n) \cong \mathrm{Ind}_{\mathbb{Z}_{n+1}}^{\mathbb{S}_{n+1}} \mathbb{1}. \tag{2.6}$$

It is not hard to see that the conditions of Theorem (2.2) are verified, and the proposition follows.

(c) *The A_∞-operad:* From the operad $\mathcal{UA}ss$ we may construct an operad $\mathcal{M}$ in the category of sets, whose n-th space $\mathcal{M}(n)$ may be identified with the symmetric group $\mathbb{S}_n$ itself: the product on $\mathcal{M}$ is defined by identifying $\mathcal{M}(n)$ with the subset

$$\{x_{\sigma(1)} \ldots x_{\sigma(n)} \mid \sigma \in \mathbb{S}_n\} \subset \mathcal{UA}ss(n).$$

It is clear that $\mathcal{M}$ is a cyclic operad in the category of sets, since the composition maps and $\mathbb{S}_{n+1}$-actions preserve the basis vectors. An $\mathcal{M}$-algebra in the category of sets is a monoid with unit.

Stasheff [**21**] introduced an operad A_∞ in the category of regular cellular complexes with the homotopy type of $\mathcal{M}$; an algebra over A_∞ in the category of topological spaces is an H-space associative up to an explicit collection of higher homotopies.

The topological operad A_∞ is defined as follows: let K_n be the nth *Stasheff polytope*, whose vertices correspond to complete parenthesizations of the product $x_1 \ldots x_n$ (Stasheff [**21**]). Such parenthesizations are in bijective correspondence with triangulations of an $(n+1)$-gon P_{n+1} into triangles whose vertices are among vertices of P_{n+1}, while faces of K_n correspond to decomposition of P_{n+1} into polygons (see, for example, Kapranov [**14**]). Taking P_{n+1} to be a regular polygon, we obtain an action of $\mathbb{Z}_{n+1}$ on P_{n+1} by rotation through multiples of the angle $2\pi/(n+1)$, inducing an action on K_n. We now define $A_\infty(n)$ to be the induced $\mathbb{S}_{n+1}$-space $\mathrm{Ind}_{\mathbb{Z}_{n+1}}^{\mathbb{S}_{n+1}} K_n$, analogous to (2.6). The operad structure on A_∞ is induced by the embeddings

$$K_n \times K_{a_1} \times \cdots \times K_{a_n} \hookrightarrow K_{a_1+\ldots+a_n} \tag{2.7}$$

described in [**21**].

(2.8) Proposition: *The topological operad A_∞ and the dg-operad $C_\bullet(A_\infty, \mathbf{k})$ are cyclic.*

Proof: The cyclicity of A_∞ follows from the fact that the embeddings (2.7) satisfy the conditions of Theorem (2.2), as may be explicitly checked. Since the action of $\mathbb{S}_{n+1}$ on $A_\infty(n)$ and all of the compositions of the operad A_∞ are cellular maps, it follows immediately that $C_\bullet(A_\infty, \mathbf{k})$ is a cyclic dg-operad.

Another proof that $C_\bullet(A_\infty, \mathbf{k})$ is cyclic will be given in (5.4), using the cobar-construction from dg-cooperads to dg-operads.

(d) *Moduli spaces of stable curves:* Let $\overline{\mathcal{M}}_{g,n+1}$ be the Grothendieck-Knudsen moduli space (or rather, stack) of stable $(n+1)$-pointed curves of genus g (for genus $g = 0$, this is described in [**12**], and the higher genus case is similar). By definition, a point of $\overline{\mathcal{M}}_{g,n+1}$ is a system $(C, x_0, \ldots, x_n)$, where C is a projective curve of arithmetic genus g with at most nodal singularities, and x_i are distinct smooth points; it is required that C has no infinitesimal automorphisms preserving the points x_i. The cyclic $\mathbb{S}$-space $\overline{\mathcal{M}}$ has

$$\overline{\mathcal{M}}(n) = * \cup \coprod_{2(g-1)+n>0} \overline{\mathcal{M}}_{g,n+1}$$

with $\mathbb{S}_{n+1}$ acting by permuting the points $\{x_i\}$. Then $\overline{\mathcal{M}}$ is a cyclic operad, with unit the point $* \in \overline{\mathcal{M}}(1)$, and composition induced by gluing curves together along marked points (see [**12**], 1.4.3). We may also define a cyclic suboperad $\overline{\mathcal{M}}_0$ of $\overline{\mathcal{M}}$ by restricting attention to curves of genus 0.

(e) *Moduli space of curves with holes:* We will only define the genus 0 version of this cyclic operad, although there is an analogue with unrestricted genus.

Let $\widehat{\mathcal{M}}_0(n)$ be the moduli space of complex analytic embeddings

$$\phi : \coprod_{i=0}^{n} \bar{D} \to S^2,$$

where $\bar{D}$ is the closed unit disk in $\mathbb{C}$; we let $\widehat{\mathcal{M}}_0(1)$ be a single point and $\widehat{\mathcal{M}}_0(0)$ be the empty set. Then the spaces $\widehat{\mathcal{M}}_0(n)$ assemble to form a cyclic $\mathbb{S}$-space. Although $\widehat{\mathcal{M}}_0(n)$ is infinite-dimensional, it has the homotopy type of a finite-dimensional CW complex: in fact, for $n > 1$, we have a fibration in the homotopy category

$$\widehat{\mathcal{M}}_0(n) \longrightarrow \{(z_1, \ldots, z_n) \in \mathbb{C}^n \mid z_i \neq z_j \text{ for } i \neq j\},$$

with homotopy fiber the torus $(S^1)^{n+1}$.

The cyclic $\mathbb{S}$-module $\widehat{\mathcal{M}}_0$ is a cyclic operad: the composition is given by gluing along the boundaries of embedded disks (see [**9**]).

(f) *The Batalin-Vilkovisky operad:* This is the operad $\mathcal{BV} = H_\bullet(\widehat{\mathcal{M}}_0, \mathbf{k})$ in the category of graded vector spaces over $\mathbf{k}$, defined by taking the homology of the topological operad $\widehat{\mathcal{M}}_0$. Since $\widehat{\mathcal{M}}_0$ is cyclic, it is automatic that $\mathcal{BV}$ is as well.

More explicitly, a $\mathcal{BV}$-algebra (Batalin-Vilkovisky algebra) is a graded commutative algebra A together with a linear operator $\Delta : A \to A$ of degree 1 such that $\Delta^2 = 0$ and

$$\begin{aligned}\Delta(abc) = \Delta(ab)c + (-1)^{|a|}a\Delta(bc) + (-1)^{(|a|-1)|b|}b\Delta(ac) \\ - (\Delta a)bc - (-1)^{|a|}a(\Delta b)c - (-1)^{|a|+|b|}ab(\Delta c).\end{aligned}$$

The graded vector space $\mathcal{BV}(1)$ may be identified with the homology of the circle $H_\bullet(S^1, \mathbf{k})$: the cyclic group $\mathbb{S}_{1+1}$ acts by the antipode map on $\widehat{\mathcal{M}}_0(1) \simeq S^1$, and hence acts trivially on $\mathcal{BV}(1)$. See [**9**] for more information on Batalin-Vilkovisky algebras and some applications to mathematical physics.

(2.9) Proof of Theorem (2.2): Let $\mathcal{P}$ be an operad, and suppose that we are given an extension of the $\mathbb{S}$-module structure of $\mathcal{P}$ to a cyclic $\mathbb{S}$-module structure. In order for $\mathcal{P}$ to be a cyclic operad, we must define, for each unrooted $(n + 1)$-tree T, a morphism of $\mathbb{S}_{n+1}$-modules $\mu_T : \mathcal{P}(T) \to \mathcal{P}(n)$ such that the maps

$$\mu_n : \mathbb{T}(\mathcal{P})(n) = \bigoplus_{\text{unrooted } (n+1)\text{-trees } T} \mathcal{P}(T) \xrightarrow{\oplus\mu_T} \mathcal{P}(n)$$

make $\mathcal{P}$ into a $\mathbb{T}$-algebra in the category $\mathbb{S}_+$-mod. Any unrooted $(n + 1)$-tree may be considered as a rooted n-tree. For any vertex $v \in \operatorname{Vert}(T)$, we have $\mathcal{P}((\operatorname{Leg}(v))) = \mathcal{P}(\operatorname{In}(v))$, where on the left-hand side, we use the cyclic $\mathbb{S}$-module structure, while on the right-hand side we use the $\mathbb{S}$-module structure. This implies that the structure of an operad on $\mathcal{P}$ determines the maps μ_T. Moreover, $\mathcal{P}$ is a cyclic operad if and only if each μ_n thus constructed is equivariant with respect to $\mathbb{S}_{n+1}$.

In terms of the individual μ_T, equivariance means that for any $\sigma \in \mathbb{S}_{n+1}$ and $x \in \mathcal{P}(T) \cong \mathcal{P}(\sigma(T))$, we have

$$\mu_{\sigma(T)}(x) = \sigma(\mu_T(x)), \tag{2.10}$$

where $\sigma(T)$ is the unrooted $(n+1)$-tree obtained from T by renumbering the legs according to σ. Condition (2) of the theorem is just (2.10) for the unrooted $(m+l)$-tree T depicting the composition $p(1, \ldots, 1, q)$. Thus it is certainly necessary: we will now prove that it is sufficient.

By construction, (2.10) already holds for any $\sigma \in \mathbb{S}_n$. We have the factorization $\mathbb{S}_{n+1} = \mathbb{S}_n \cdot \mathbb{Z}_{n+1}$. By means of the action of $\mathbb{S}_n$, we may transform T to a planar tree, that is, an unrooted tree embeddable in the plane in such a way that the labels $\{0,\ldots,n\}$ on the legs occur in clockwise order. Thus, it suffices to prove (2.10) for the case where T is planar and $\sigma=\tau$ is the cycle generating $\mathbb{Z}_{n+1}$, and from now on we make this assumption. Let $\Gamma(T)$ be the unique path between the legs labelled 0 and n, and let $\gamma(T)$ be the number of vertices in $\Gamma(T)$. We will prove invariance of μ_T by induction on $\gamma(T)$.

The cases $\gamma(T)=0$ and $\gamma(T)=1$ are hypotheses (1) and (2) of the theorem. If $\gamma(T)\geq 2$, let e be the internal edge of $\Gamma(T)$ adjacent to the leg labelled by 0. Cut e in two, and denote by T' and T'' the components of the resulting graph containing the legs labelled 0 and n respectively. If

$$x=\bigotimes_{v\in T} x_v,\quad x_v\in\mathcal{P}(\mathrm{Leg}(v)),$$

is a decomposable element of $\mathcal{P}(T)$, let $x'=\bigotimes_{v\in\mathrm{Vert}(T')} x_v\in\mathcal{P}(T')$ and $x''=\bigotimes_{v\in\mathrm{Vert}(T'')} x_v\in\mathcal{P}(T'')$. Let S be the tree with two vertices, obtained by contracting all of the vertices of T' into one vertex, and all of the vertices of T'' into another. Then we can write

$$\begin{aligned}\tau(\mu_T(x)) &= \tau(\mu_S(\mu_{T'}(x')\otimes\mu_{T''}(x'')))\\ &= \mu_{\tau(S)}(\mu_{T'}(x')\otimes\mu_{T''}(x''))) = \mu_{\tau(T)}(x).\end{aligned}$$

Here, the first equality follows from the associativity of composition in $\mathcal{P}$, the second from hypothesis (2) of the theorem, and the third from the inductive assumption that (2.10) holds for T' and T'', and associativity. This proves Theorem (2.2).

(2.11) Anticyclic operads and suspension: The *suspension* of a dg-operad $\mathcal{P}$ is the dg-operad $\Lambda\mathcal{P}$ whose n-th space is

$$\Lambda\mathcal{P}(n)=\Sigma^{1-n}\,\mathrm{sgn}_n\otimes\mathcal{P}(n),$$

Here, sgn_n is the sign representation of $\mathbb{S}_n$. The compositions in $\Lambda\mathcal{P}$ are defined as follows: if $q_0\in\mathcal{P}(n)$ and $q_j\in\mathcal{P}(i_j)$, $1\leq j\leq n$, and we denote by ε_n the standard basis vector of sgn_n, then

$$\begin{aligned}&\mu^{\Lambda\mathcal{P}}_{i_1\ldots i_n}((\varepsilon_n\otimes q_0)\otimes(\varepsilon_{i_1}\otimes q_1)\otimes\ldots\otimes(\varepsilon_{i_n}\otimes q_n))\\ &\qquad=(-1)^{\sigma}\varepsilon_{i_1+\ldots+i_n}\otimes\mu^{\mathcal{P}}_{i_1\ldots i_n}(q_0\otimes q_1\otimes\ldots\otimes q_n),\end{aligned}$$

where $\sigma=\sum_{j=0}^{n}(i_{j+1}+\cdots+i_n+n-j)|q_j|$. If A is a dg-algebra over a dg-operad $\mathcal{P}$, then ΣA is naturally an algebra over $\Lambda\mathcal{P}$. We may summarize the above discussion by saying that there is a canonical isomorphism of triples $\Lambda^{-1}\mathbb{T}\Lambda\cong\mathbb{T}$.

There is an evident extension of the *suspension* functor Λ to the category of cyclic $\mathbb{S}$-modules:

$$\Lambda\mathcal{P}(n)=\Sigma^{1-n}\,\mathrm{sgn}_{n+1}\otimes\mathcal{P}(n).$$

However, the triple $\mathbb{T}_-=\Lambda^{-1}\mathbb{T}_+\Lambda$ is *not* isomorphic to the triple $\mathbb{T}_+$, since

$$\mathbb{T}_-\mathcal{W}(n)\cong\bigoplus_{\text{unrooted } n+1\text{-trees } T}\det(T)\otimes\bigotimes_{v\in\mathrm{Vert}(T)}\mathcal{W}((\mathrm{Leg}(v))).$$

An *anticyclic* operad is an algebra over the triple $\mathbb{T}_-$. The analogue of Theorem (2.2) for anticyclic operads is obtained by multiplying the right-hand side of conditions (1) and (2) by -1.

3 Cyclic quadratic operads

(3.1) Quadratic operads and quadratic duality: Let E be a graded vector space with action of $\mathbb{S}_2$. We regard it as an $\mathbb{S}$-module with $E(n) = E$ for $n = 2$ and $E(n) = 0$ otherwise. Let $\mathbb{T}E$ be the free graded operad generated by E and let $R \subset \mathbb{T}E(3)$ be an $\mathbb{S}_3$-invariant subspace. The *quadratic operad* $\mathcal{P}(E, R)$ is formed by factoring $\mathbb{T}E$ by the ideal generated by the relations R.

This definition differs from the one in **[12]** in that we allow vector spaces to be graded, but restrict ourselves to the case of operads with $\mathcal{P}(1) = \mathbf{k}$. The case where the grading on E and R is trivial, so that $\mathcal{P}(E, R)$ is a $\mathbf{k}$-linear operad, is the most important for us.

If $\mathcal{P}$ is a quadratic operad, the graded vector spaces E and R may be recovered from $\mathcal{P}$ by

$$E = \mathcal{P}(2), \quad R = \ker(\mathbb{T}E(3) \to \mathcal{P}(3)).$$

Note that $\mathbb{T}E(3) \cong \operatorname{Ind}_{\mathbb{S}_2}^{\mathbb{S}_3} E \otimes E$ is isomorphic to the sum of three copies of $E \otimes E$, corresponding to the three binary trees with 3 leaves.

If $\mathcal{P} = \mathcal{P}(E, R)$ is a quadratic operad, its *quadratic dual* is the operad $\mathcal{P}^! = \mathcal{P}(E^* \otimes \operatorname{sgn}, R^\perp)$, where sgn is the sign representation of $\mathbb{S}_2$ and $R^\perp$ is the orthogonal complement of $R \subset \mathbb{T}E(3)$ in $\mathbb{T}(E^* \otimes \operatorname{sgn})(3)$.

If $\mathcal{P}$ is a quadratic operad (regarded as a dg-operad with trivial differential) then the suspension $\Lambda\mathcal{P}$ is also quadratic.

(3.2) Cyclic quadratic operads: Let $\mathcal{P} = \mathcal{P}(E, R)$ be a quadratic operad. The action of $\mathbb{S}_2$ on E extends naturally to an action of $\mathbb{S}_{2+1}$ via the sign representation:

$$\operatorname{sgn} : \mathbb{S}_{2+1} \to \{\pm 1\} = \mathbb{S}_2. \tag{3.3}$$

This makes the $\mathbb{S}$-module E into a cyclic $\mathbb{S}$-module, so that the free operad $\mathbb{T}E$ is a cyclic operad. We say that $\mathcal{P}$ is a *cyclic quadratic* operad if R is an $\mathbb{S}_{3+1}$-invariant subspace of $\mathbb{T}E(3)$. It follows immediately that $\mathcal{P}$ is a cyclic operad. The reason for this definition will be clearer when we discuss invariant bilinear forms in Section 4.

Observe that $\mathcal{P}^!$ is cyclic quadratic if and only if $\mathcal{P}$ is.

(3.4) Cyclic quadratic operads with one generator: We now turn to the construction of examples of cyclic quadratic operads. We use the parametrization of representations of $\mathbb{S}_n$ by Young diagrams **[8]**: if $\lambda_1 \geq \ldots \lambda_m \geq 1$ and $\lambda_1 + \cdots + \lambda_m = n$, we denote by $V_{\lambda_1,\ldots,\lambda_m}$ the corresponding irreducible representation of $\mathbb{S}_n$. For example, $V_n = \mathbb{1}$ is the trivial 1-dimensional representation, $V_{1^n} = \operatorname{sgn}$ is the sign representation, and $V_{n-1,1}$ is the $(n-1)$-dimensional representation on the hyperplane

$$\{(x_1, \ldots, x_n) \mid x_1 + \cdots + x_n = 0\} \tag{3.5}$$

We will make use the character table of $\mathbb{S}_4$ (or, in our notation, $\mathbb{S}_{3+1}$); for the convenience of the reader, we display it here.

	I	$(0\ 1)$	$(0\ 1\ 2)$	(0123)	$(01)(23)$
$1\!\!1$	1	1	1	1	1
sgn	1	-1	1	-1	1
$V_{2,2}$	2	0	-1	0	2
$V_{3,1}$	3	1	0	-1	-1
$V_{2,1,1}$	3	-1	0	1	-1

Note that the first three representations in this table can be obtained from the representations $1\!\!1$, sgn, $V_{2,1}$ of $\mathbb{S}_3$ via the natural surjection $\mathbb{S}_{3+1} \to \mathbb{S}_3$.

(3.6) Proposition: *Every ungraded quadratic operad* $\mathcal{P} = \mathcal{P}(E, R)$ *such that* $\dim(E) = 1$ *is cyclic quadratic.*

Proof: The representation E of $\mathbb{S}_2$ is either the trivial representation or the sign representation: the operad $\mathcal{P}$ thus has either a commutative or an anticommutative product. We consider these two cases separately.

We start with $E = 1\!\!1$, the trivial representation. We use the shorthand notation for elements of $\mathbb{T}E(3)$ explained in (1.12). Thus, we represent a basis vector of E by a formal product (x_1, x_2); the fact that $\mathbb{S}_2$ acts trivially means that $(x_1, x_2) = (x_2, x_1)$; elements of $\mathbb{T}E(3)$ are then represented by expressions in x_1, x_2, x_3 obtained by iterating this product. We see that the space $\mathbb{T}E(3)$ is a three-dimensional representation of $\mathbb{S}_{3+1}$, spanned by formal expressions $((x_1, x_2), x_3)$, $((x_2, x_3), x_1)$ and $((x_3, x_1), x_2)$. We claim that as a representation of $\mathbb{S}_{3+1}$,

$$\mathbb{T}E(3) \cong 1\!\!1 \oplus V_{2,2}. \tag{3.7}$$

To see this, observe that the trace of a transposition (for example, $(1\ 2)$) equals 1, the trace of a cycle of length 3 (for example $(1\ 2\ 3)$) vanishes, and the trace of the cycle $\tau = (0\ 1\ 2\ 3)$ equals 1. We leave it to the reader to verify that

$$\tau((x_1, x_2), x_3) = ((x_2, x_3), x_1), \quad \tau((x_2, x_3), x_1) = ((x_1, x_2), x_3),$$
$$\tau((x_3, x_1), x_2) = ((x_3, x_1), x_2),$$

from which this last fact follows.

On restriction to the subgroup $\mathbb{S}_3$, the isomorphism (3.7) becomes

$$\mathbb{T}E(3) \cong 1\!\!1 \oplus V_{2,1}.$$

Thus, the $\mathbb{S}_{3+1}$-invariant and $\mathbb{S}_3$-invariant subspaces of $\mathbb{T}E(3)$ are the same, so that any space of relations R automatically defines a cyclic quadratic operad.

The case $E = \text{sgn}$, the sign representation, is similar. Here, we simply give the answer:

$$\mathbb{T}E(3) \cong_{\mathbb{S}_{3+1}} \text{sgn} \oplus V_{2,2} \cong_{\mathbb{S}_3} \text{sgn} \oplus V_{2,1}, \tag{3.8}$$

where the first isomorphism is that of $\mathbb{S}_{3+1}$-modules and the second one is an isomorphism of $\mathbb{S}_3$-modules.

(3.9) List of cyclic quadratic operads with one generator:

(a) *The commutative operad* $\mathcal{C}om$ has one commutative product (x_1, x_2), and relations $R = V_{2,2}$ in (3.7), implementing the associativity relations

$$((x_1, x_2), x_3) = ((x_2, x_3), x_1) \text{ and } ((x_2, x_3), x_1) = ((x_3, x_1), x_2).$$

We have $\mathcal{C}om(n) = \mathbb{1}$ for all $n \geq 1$, and all of the composition maps are isomorphisms of the trivial representation. $\mathcal{C}om$-algebras are nonunital commutative (associative) algebras in the usual sense.

By contrast, unital commutative algebras are described by the operad $\mathcal{UC}om$, which has $\mathcal{UC}om(n) = \mathbb{1}$ for all $n \geq 0$, together with the evident structure maps; however, this operad is not quadratic.

(b) *The Lie operad* $\mathcal{L}ie$ has one anticommutative product $[x_1, x_2]$ and relations $R = \text{sgn}$ in (3.8), implementing the Jacobi rule

$$[[x_1, x_2], x_3] + [[x_2, x_3], x_1] + [[x_3, x_1], x_2] = 0.$$

We have $\mathcal{L}ie \cong \mathcal{C}om^!$. The $\mathbb{S}_n$-module $\mathcal{L}ie(n)$ may be realized as the subspace in the free Lie algebra on generators $x_1, \ldots, x_n$ spanned by Lie monomials which contain each x_i exactly once; it has dimension $(n-1)!$. Since $\mathcal{L}ie$ is a cyclic operad, we see that $\mathcal{L}ie(n)$ has a natural $\mathbb{S}_{n+1}$-action, as was first observed by Kontsevich [**15**].

(c) *The mock-Lie operad* $\widetilde{\mathcal{L}ie}$, with one commutative product and relations $R = \mathbb{1}$ in (3.7): this implements the relation

$$((x_1, x_2), x_3) + ((x_2, x_3), x_1) + ((x_3, x_1), x_2) = 0.$$

(d) *The mock commutative operad* $\widetilde{\mathcal{C}om}$ has one anticommutative product and relations $R = V_{2,2}$ in (3.8), implementing the relations

$$[[x_1, x_2], x_3] = -[x_1, [x_2, x_3]] \text{ and } [[x_2, x_3], x_1] = -[x_2, [x_3, x_1]].$$

These relations easily imply that $\widetilde{\mathcal{C}om}(3) \cong \text{sgn}$, while $\widetilde{\mathcal{C}om}(n) = 0$ for $n > 3$. We have $\widetilde{\mathcal{L}ie} \cong \widetilde{\mathcal{C}om}^!$. The operads $\widetilde{\mathcal{L}ie}$ and $\widetilde{\mathcal{C}om}$ are quite pathological[1].

[1] For instance, these operads are not Koszul in the sense of [**12**]. This is because the exponential generating function $g(t) = \sum_{n=1}^{\infty} \dim \widetilde{\mathcal{C}om}(n) t^n/n!$ is equal to $g(t) = t + t^2/2 + t^3/6$; if $\widetilde{\mathcal{C}om}$ was Koszul, the generating function of $\widetilde{\mathcal{L}ie}$ would coincide with the solution $h(t)$ of the functional equation $h(-g(-t)) = t$ (by Theorem 3.3.2 of [**12**]). However

$$h(t) = t + t^2/2 + t^3/3 + 5t^4/24 + t^5/12 - 7t^6/144 - 13t^7/72 + \ldots$$

has some negative coefficients, giving a contradiction.

(d) The trivial and free operads are obtained by imposing as relations respectively $R = \mathbb{T}E(3)$ and $R = 0$.

(3.10) The associative and Poisson operads: The operad $\mathcal{A}ss$ describing nonunital associative algebras (2.3) (b) is quadratic, since it is generated by the binary operations x_1x_2 and x_2x_1, while the space R of relations is the image of the relation $x_1(x_2x_3) - (x_1x_2)x_3 = 0$ under the action of $\mathbb{S}_3$. The operad $\mathcal{UA}ss$ describing unital associative algebras is not quadratic. We will omit the adjective nonunital in referring to nonunital associative algebras.

A *Poisson algebra* is a commutative algebra P with a Lie bracket $\{-,-\}$ such that each operator $\{x,-\}$ is a derivation of the product. The operad $\mathcal{P}ois$ associated to Poisson algebras is defined in a similar way to the associative operad: $\mathcal{P}ois(n)$, $n \geq 1$, is a subspace in the free Poisson algebra generated by letters $\{x_1, \dots, x_n\}$ spanned by expressions which contain each letter just once, while $\mathcal{P}ois(0) = 0$. It is easily seen that $\mathcal{P}ois$ is a quadratic operad.

(3.11) Proposition: *The operads $\mathcal{A}ss$ and $\mathcal{P}ois$ are cyclic quadratic.*

Proof: In (2.5), we defined an explicit action of $\mathbb{S}_{n+1}$ on $\mathcal{A}ss(n)$. It is straightforward to show that the natural map $\mathbb{T}(\mathcal{A}ss(2))(3) \to \mathcal{A}ss(3)$ is $\mathbb{S}_{3+1}$-equivariant: thus, its kernel R, the space of relations in $\mathcal{A}ss$, is $\mathbb{S}_{3+1}$-invariant, and $\mathcal{A}ss$ is cyclic quadratic.

We now turn to $\mathcal{P}ois$. Consider the basis of $\mathcal{A}ss(2)$ given by the two operations

$$(x_1, x_2) = x_1x_2 + x_2x_1 \quad , \quad [x_1, x_2] = x_1x_2 - x_2x_1 .$$

Any elements in $\mathcal{A}ss(n)$ may be expressed as a sum of monomials in $x_1, \dots, x_n$ using the two operations $(-,-)$ and $[-,-]$. Filter $\mathcal{A}ss(n)$ by subspaces $F_r\mathcal{A}ss(n)$ spanned by words involving the product $(-,-)$ not more than r times (Getzler-Jones [11], Section 5.3): $F_0\mathcal{A}ss(n) \cong \mathcal{L}ie(n)$, while $F_{n-1}\mathcal{A}ss(n) \cong \mathcal{A}ss(n)$. An alternative description of F is obtained by identifying $\mathbf{k}\langle x_1, \dots, x_n\rangle$ with the universal enveloping algebra of the free Lie algebra generated by the x_i, and taking the intersection of $\mathcal{A}ss(n) \subset \mathbf{k}\langle x_1, \dots, x_n\rangle$ with the standard filtration of the enveloping algebra.

The filtration F is compatible with the operad structure of $\mathcal{A}ss$: $F_r\mathcal{A}ss(n)$ is invariant under the action of $\mathbb{S}_n$, and

$$\mu_{i_1 \dots i_k}(F_r\mathcal{A}ss(k) \otimes F_{r_1}\mathcal{A}ss(i_1) \otimes \dots \otimes F_{r_k}\mathcal{A}ss(i_k)) \subset F_{r+r_1+\dots r_k}\mathcal{A}ss(i_1 + \dots i_k).$$

Using the characterization of F in terms of the filtration on the universal enveloping algebra, we see that the operad formed by the spaces[2] $\mathrm{gr}^F\mathcal{A}ss(n)$ is naturally isomorphic to the Poisson operad $\mathcal{P}ois$.

It suffices to show that this filtration is preserved under the action of $\mathbb{S}_{n+1}$. Let $g \in \mathcal{A}ss(n)$ be a monomial in $x_1, \dots, x_n$ obtained by iterated application

[2] taken with trivial grading, not the grading induced by F

of the product $(-,-)$ and $[-,-]$. As in (1.12), we represent g as a binary rooted n-tree T in which each vertex is labelled by one of the product $(-,-)$ or $[-,-]$. If $\sigma \in \mathbb{S}_{n+1}$ is a permutation, the element σg is represented, up to sign, by the same tree with legs renumbered by σ, and as its new root the leg whose label is now 0; in particular, the assignment of products to the vertices remains the same. In particular, the number of occurrences of the product $(-,-)$ in σg is the same as in g.

(3.12) Cyclic quadratic operads generated by the regular representation: We will now classify all ungraded cyclic quadratic operads with $E = 1\!\!1 \oplus \mathrm{sgn}$ the regular representation of $\mathbb{S}_2$. The associative operad $\mathcal{A}ss$ and the Poisson operad $\mathcal{P}ois$ will emerge naturally from this classification.

(3.13) Proposition: *If* $E = 1\!\!1 \oplus \mathrm{sgn}$, *then we have an isomorphism of* $\mathbb{S}_{3+1}$-*modules*

$$\mathbb{T}E(3) \cong 1\!\!1 \oplus \mathrm{sgn} \oplus 2V_{2,2} \oplus V_{3,1} \oplus V_{2,1,1}.$$

Proof: We denote the commutative product in E by (a,b), and the anticommutative product by $[a,b]$. The torus $\mathrm{Aut}_{\mathbb{S}_2}(E) = \mathbf{k}^\times \oplus \mathbf{k}^\times$ acts on $\mathbb{T}E$, by rescaling the two products: we say that a monomial has weight k if there are k occurrences of the anticommutative bracket $[-,-]$ in it.

The space $\mathbb{T}E(3)$ has dimension 12, and is spanned by monomials

weight 0 $((1,2),3)$, $((2,3),1)$ and $((3,1),2)$;

weight 1 $([1,2],3)$, $([2,3],1)$, $([3,1],2)$, $[(1,2),3]$, $[(2,3),1]$ and $[(3,1),2]$;

weight 2 $[[1,2],3]$, $[[2,3],1]$ and $[[3,1],2]$.

We have already seen that the monomials of weight 0 and 2 contribute the irreducible representations $1\!\!1 \oplus V_{2,2}$ and $\mathrm{sgn} \oplus V_{2,2}$ respectively. On the monomials of weight 1, the permutations (1 2) and (1 2 3) both have trace 0, permitting us to conclude (by looking at the character table) that it is isomorphic to a direct sum of either $V_{3,1}$ or $1\!\!1 \oplus V_{2,2}$, and either $V_{2,1,1}$ or $\mathrm{sgn} \oplus V_{2,2}$. To distinguish between these different possibilities, we calculate the action of the involution $\sigma = (0\ 1)(2\ 3)$ and find it to be:

$$([1,2],3) \longleftrightarrow -[(1,2),3] \quad , \quad ([3,1],2) \longleftrightarrow [(3,1),2],$$
$$([2,3],1) \to -([2,3],1) \quad , \quad [(2,3),1] \to -[(2,3),1].$$

It follows that σ has trace -2 on the monomials of weight 1, and hence that they are isomorphic to $V_{3,1} \oplus V_{2,1,1}$ as a representation of $\mathbb{S}_{3+1}$.

(3.14) Theorem: *There are exactly* 80 *cyclic quadratic operads with* $\mathcal{P}(2) = 1\!\!1 \oplus \mathrm{sgn}$*:*

(1) 32 *operads corresponding to subsets of* $\{1\!\!1, \mathrm{sgn}, V_{2,2}, V_{3,1}, V_{2,1,1}\}$.

(2) 48 *operads corresponding to a choice of a subset of* $\{1\!\!1, \mathrm{sgn}, V_{3,1}, V_{2,1,1}\}$ *and a choice of one of the three symbols* $0, 1, \infty$.

Proof: By the previous proposition, the $\mathbb{S}_{3+1}$-invariant subspaces of $\mathbb{T}E(3)$ may be classified as follows:

(1) 32 subspaces corresponding to subsets of the set of irreducible representations $\{\mathbb{1}, \mathrm{sgn}, V_{2,2}, V_{3,1}, V_{2,1,1}\}$ of $\mathbb{S}_{3+1}$.

(2) 16 families of subspaces corresponding to subsets of $\{\mathbb{1}, \mathrm{sgn}, V_{3,1}, V_{2,1,1}\}$; each family is parametrized by a projective line P^1, which specifies a proper $\mathbb{S}_{3+1}$-invariant subspace of $V_{2,2} \oplus V_{2,2}$.

The torus $\mathrm{Aut}_{\mathbb{S}_2}(E) = \mathbf{k}^\times \oplus \mathbf{k}^\times$ fixes any subspace of the first type, and acts in a non-trivial way on P^1. This action has three orbits, two points denoted by symbols 0 and ∞, and one isomorphic to $\mathbf{k}^\times$, denoted by the symbol 1.

Thus, the symbol 0 means that we are imposing the associativity relations on the commutative product $(-,-)$, the symbol ∞ means that we are imposing the mock associativity relations on the anticommutative product, while the symbol 1 means that we impose some mixture of the two.

Let us show how the associative operad $\mathcal{A}ss$ and Poisson operad $\mathcal{P}ois$ fit into this classification. As an $\mathbb{S}_{3+1}$-module,

$$\mathcal{A}ss(3) \cong \mathrm{Ind}_{\mathbb{Z}_{3+1}}^{\mathbb{S}_{3+1}} \mathbb{1} \cong \mathbb{1} \oplus V_{2,2} \oplus V_{2,1,1}.$$

Thus, the associative operad corresponds in our classification to the subset

$$\{\mathbb{1}, V_{2,1,1}\} \subset \{\mathbb{1}, \mathrm{sgn}, V_{3,1}, V_{2,1,1}\}.$$

Furthermore, the orbit of the torus $\mathbf{k}^\times \times \mathbf{k}^\times$ is the one labelled by $1 \in \{0, 1, \infty\}$.

The Poisson operad lies in the same family as the associative operad, since $\mathcal{P}ois(3) \cong \mathcal{A}ss(3)$ as a representation of $\mathbb{S}_{3+1}$. However, now the set of relations forming a representation $V_{2,2}$ correspond to the orbit labelled by $0 \in \{0, 1, \infty\}$.

The third operad lying in this family, labelled by $\infty \in \{0, 1, \infty\}$, is rather pathological. It is generated by two operations: a commutative but non-associative product $x \circ y$ and a non-commutative operation $[x, y]$ such that $[[x, y], z] = 0$. These operations satisfy the Leibniz rule

$$[x, y \circ z] = [x, y] \circ z + y \circ [x, z].$$

(3.15) The Leibniz operad is not cyclic quadratic: A Leibniz algebra (Loday [**16**]) is an algebra with a product xy which satisfies the identity

$$x_1(x_2x_3) - (x_1x_2)x_3 + (x_1x_3)x_2 = 0. \tag{3.16}$$

Note that an anticommutative Leibniz algebra is the same thing as a Lie algebra. There is a quadratic operad $\mathcal{L}eib$, with $\mathcal{L}eib(2) \cong \mathbb{1} \oplus \mathrm{sgn}$, describing Leibniz algebras. However, this operad is not cyclic, since the cycle $\tau \in \mathbb{S}_{3+1}$ may be checked to send the relation (3.16) to

$$(x_1x_2)x_3 - x_1(x_2x_3) + (x_3x_1)x_2 = 0,$$

which does not hold in a general Leibniz algebra.

(3.17) The braid operad is not cyclic quadratic: The *braid operad* $\mathcal{B}r$ is the suboperad of the Batalin-Vilkovisky operad generated by the operation x_1x_2, of degree 0, and the operation

$$\{x_1, x_2\} = \Delta(x_1x_2) - (\Delta x_1)x_2 - (-1)^{|x_1|}x_1(\Delta x_2),$$

of degree 1; both of these operations are invariant under the action of $\mathbb{S}_2$ on $\mathcal{B}r(2)$ (Getzler [9], Getzler-Jones [11]). In fact, $\mathcal{B}r$ is a quadratic operad, with relations $(x_1x_2)x_3 = x_1(x_2x_1)$, $\{x_1, x_2x_3\} = \{x_1, x_2\}x_3 + (-1)^{(|x_1|-1)|x_2|}x_2\{x_1, x_3\}$, and

$$\{\{x_1, x_2\}, x_3\} + (-1)^{|x_1|(|x_2|+|x_3|)}\{\{x_2, x_3\}, x_1\} + (-1)^{|x_3|(|x_1|+|x_2|)}\{\{x_3, x_1\}, x_2\} = 0.$$

A braid algebra is similar to a Poisson algebra, except that the bracket $\{-,-\}$ has degree 1. The reason for calling $\mathcal{B}r$ the braid operad is that if V is a graded vector space, the free braid algebra generated by V may be expressed in the form

$$\mathsf{F}(\mathcal{B}r, V) = \bigoplus_{n=1}^{\infty} H_\bullet(\mathbb{B}_n, V^{\otimes n}),$$

where $\mathbb{B}_n$ is the braid group on n strands, acting on $V^{\otimes n}$ through the homomorphism $\mathbb{B}_n \to \mathbb{S}_n$. This isomorphism is related to the fact that the homology of a double loop space $H_\bullet(\Omega^2 M)$ is a braid algebra. More precisely, as shown in Cohen [3], $\mathcal{B}r$ is isomorphic to the homology operad $H_\bullet(\mathcal{C}_2, \mathbf{k})$ of the little square operad $\mathcal{C}_2$, for which double loop spaces are algebras.

(3.18) Proposition: *The braid operad is not cyclic quadratic.*

Proof: We show that the Poisson identity is not preserved by the action of $\mathbb{S}_{3+1}$ on $\mathbb{T}E(3)$, where $E = \mathcal{B}r(2) \cong \mathbb{1} \oplus (\Sigma\mathbb{1})$. The space $\mathbb{T}E(3)_1$ is spanned by the monomials $\{x_1, x_2\}x_3$, $\{x_2, x_3\}x_1$, $\{x_3, x_1\}x_2$, $\{x_1x_2, x_3\}$, $\{x_2x_3, x_1\}$ and $\{x_3x_1, x_2\}$. We calculate that $\tau = (0\ 1\ 2\ 3)$ acts by

$$\{x_1, x_2\}x_3 \to (-1)^{|x_1|}x_1\{x_2, x_3\} \to \{x_1x_2, x_3\} \to \{x_1, x_2x_3\} \to \{x_1, x_2\}x_3,$$
$$\{x_3, x_1\}x_2 \longleftrightarrow \{x_3x_1, x_2\}.$$

It follows that the Poisson relation

$$\{x_1, x_2x_3\} - \{x_1, x_2\}x_3 - (-1)^{(|x_1|-1)|x_2|}x_2\{x_1, x_3\}$$

is mapped to

$$\{x_1, x_2\}x_3 - (-1)^{|x_1|}x_1\{x_2, x_3\} - (-1)^{|x_1||x_2|}\{x_2, x_1x_3\}$$
$$= (-1)^{|x_1||x_2|}(-\{x_2, x_1x_3\} + \{x_2, x_1\}x_3 - (-1)^{(|x_2|-1)|x_1|}x_1\{x_2, x_3\}),$$

which is not one of the defining relations of the braid operad $\mathcal{B}r$.

4 Invariant bilinear forms

(4.1) Definition: Let $\mathcal{P}$ be a cyclic (or anticyclic) dg-operad and let A be a $\mathcal{P}$-algebra. A bilinear form $B : A \otimes A \to V$ with values in a graded vector space V is *invariant* (with respect to the cyclic structure on $\mathcal{P}$) if for all $n \geq 0$, the map $B_n : \mathcal{P}(n) \otimes A^{\otimes(n+1)} \to V$ defined by the formula

$$B_n(p \otimes x_0 \otimes x_1 \otimes \ldots \otimes x_n) = (-1)^{|x_0|\,|p|} B(x_0, p(x_1, \ldots, x_n))$$

is invariant under the action of $\mathbb{S}_{n+1}$ on $\mathcal{P}(n) \otimes A^{\otimes(n+1)}$.

(4.2) Remarks: (a) The special case of ungraded operads and algebras, in which the signs disappear from the formulas, arises frequently in applications.

(b) Since the map B_n of the above definition is automatically invariant under $\mathbb{S}_n$, it suffices to check that it is invariant under the action of the cycle $\tau \in \mathbb{S}_{n+1}$.

(c) Taking $p = 1 \in \mathcal{P}(1)$, we learn that an invariant bilinear form is (anti)symmetric if $\mathcal{P}$ is (anti)cyclic, that is,

$$B(x, y) = \pm(-1)^{|x|\,|y|} B(y, x),$$

with $+$ in the cyclic case and $-$ in the anticyclic case. This reflects the fact that symmetric bilinear forms B on a graded vector space V correspond to antisymmetric bilinear forms $\tilde{B}$ on its suspension ΣV, by means of the formula

$$\tilde{B}(\Sigma v, \Sigma w) = (-1)^{|v|} B(v, w).$$

(4.3) Proposition: *Let $\mathcal{P}$ be a cyclic quadratic operad. A bilinear form B on a $\mathcal{P}$-algebra is invariant if and only if it is symmetric and for each $p \in \mathcal{P}(2)$,*

$$B(p(x_0, x_1), x_2) = (-1)^{|p|\,|x_0|} B(x_0, p(x_1, x_2)). \tag{4.4}$$

Proof: We will simplify the discussion by supposing that $\mathcal{P}$ and A are concentrated in degree 0. Invariance of $B_2 : \mathcal{P}(2) \otimes A^{\otimes 3} \to V$ under the cycle $\tau = (0\ 1\ 2)$ implies that

$$B(x_0, (\tau p)(x_1, x_2)) = B(x_2, p(x_0, x_1)).$$

By the definition (3.3) of the $\mathbb{S}_{2+1}$ action on $\mathcal{P}(2)$, we know that $\tau p = p$. Thus, (4.4) is necessary and sufficient for invariance of B_2 under τ. Invariance of $B_n : \mathcal{P}(n) \otimes A^{\otimes(n+1)} \to V$ for $n > 2$ follows from invariance of B_2 since $\mathcal{P}(2)$ generates $\mathcal{P}(n)$ for $n > 2$.

(4.5) Examples: (a) *Associative algebras:* Let $\mathcal{P} = \mathcal{A}ss$ be the associative operad. A bilinear form on an associative algebra A is invariant if and only if it is symmetric and $B(x_0, x_1 x_2) = B(x_0 x_1, x_2)$: this implies that $B(x_0, x_1 \ldots x_n)$ is invariant under the action of τ on $A^{\otimes(n+1)}$. This explains the formula (2.5) for the action of the cycle $\tau \in \mathbb{S}_{n+1}$ on the space $\mathcal{A}ss(n)$.

If A has a unit, there is an identification between the invariant bilinear forms on A with values in V and traces, that is, linear maps $T : A \to V$ such that $T(xy) = (-1)^{|x|\,|y|}T(yx)$. The correspondence associates to an invariant bilinear form B the linear map $T(x) = B(1,x)$; the bilinear form may be recovered by setting $B(x,y) = T(xy)$.

(b) *Commutative algebras:* Let $\mathcal{P} = \mathcal{C}om$ be the commutative operad. If A is a commutative algebra, an invariant bilinear form on A is a symmetric map $B : A \otimes A \to V$ such that $B(x_0 x_1, x_2) = B(x_0, x_1 x_2)$. In this case the expression $B(x_0, x_1 \dots x_n)$ is totally symmetric.

(c) *Lie algebras:* Let $\mathcal{P} = \mathcal{L}ie$ be the Lie operad. If $\mathfrak{g}$ is a Lie algebra, an invariant bilinear form on $\mathfrak{g}$ is a symmetric map $B : \mathfrak{g} \otimes \mathfrak{g} \to V$ such that $B([x,y],z) = B(x,[y,z])$. If $\mathfrak{g}$ is finite dimensional, an example is given by the Killing form.

(d) *Poisson algebras:* If P is a Poisson algebra, an invariant bilinear form on P is a symmetric map $B : P \otimes P \to V$ such that $B(xy,z) = B(x,yz)$ and $B(\{x,y\},z) = B(x,\{y,z\})$. To give a geometric example, let $\mathbf{k} = \mathbb{R}$ and (M,ω) be a connected $2n$-dimensional symplectic manifold. Let $P = C_c^\infty(M)$ be the Poisson algebra of smooth real functions on M of compact support. We equip this algebra with the standard C^∞-topology.

(4.6) Proposition: *The space of continuous invariant bilinear forms on $C_c^\infty(M)$ with values in $\mathbb{R}$ is spanned by the Liouville bilinear form*

$$B(f,g) = \int_M fg\,\omega^n.$$

Proof: We first check that the Liouville bilinear form is invariant: it is obvious that it is symmetric, and that $B(fg,h) = B(f,gh)$, while

$$\begin{aligned} B(f,\{g,h\}) &= \int_M f\{g,h\}\omega^n = \int_M fH_g(h)\omega^n = -\int_M H_g(f)h\omega^n \\ &= -B(\{g,f\},h) = B(\{f,g\},h). \end{aligned}$$

Here, we denote by H_g the Hamiltonian vector field associated to the function g, and we have used Liouville's theorem, which says that the Lie derivative $\mathcal{L}_{H_g}(\omega^n) = n\omega^{n-1}\mathcal{L}_{H_g}(\omega)$ vanishes.

It remains to check that all continuous invariant bilinear forms are proportional to the Liouville form. A continuous bilinear form on P is the same thing as a current on $M \times M$, while the condition $B(fg,h) = B(f,gh)$ says that this current is the direct image $\Delta_*\gamma$ of a current γ on M by the inclusion of the diagonal $\Delta : M \to M \times M$. The condition $B(f,\{g,h\}) = B(\{f,g\},h)$ implies that the Lie derivative of γ with respect to any Hamiltonian vector field vanishes. Any such current is proportional to ω^n, as follows by calculation in Darboux coordinates.

(e) *Batalin-Vilkovisky algebras:* Let $\mathcal{P} = \mathcal{BV}$ be the Batalin-Vilkovisky operad ((2.3)) (f). If A is a $\mathcal{BV}$-algebra, an invariant bilinear form on A is a symmetric map $B : A \otimes A \to V$ such that $B(xy,z) = B(x,yz)$ and $B(\Delta x, y) =$

$(-1)^{|x|}B(x,y)$. One important source of Batalin-Vilkovisky algebras is the space of functions on a *Batalin-Vilkovisky supermanifold* (M,ω,ν): this is an odd symplectic supermanifold (M,ω) together with a nowhere-vanishing section ν of the Berezinian bundle such that the operator $\Delta f = \operatorname{div}_\nu H_f$ satisfies the Batalin-Vilkovisky equation $\Delta^2 = 0$ (see **[9]**). By an analysis not too different from that which we gave above for the Poisson algebra of a symplectic manifold, it may be shown that any continuous invariant bilinear form on the Batalin-Vilkovisky algebra $C_c^\infty(M)$ associated to a Batalin-Vilkovisky manifold with values in $\mathbb{R}$ is proportional to $B(f,g) = \int_M f g \nu$.

(4.7) The universal invariant bilinear form: Let $\mathcal{P}$ be a cyclic graded operad and A be a graded $\mathcal{P}$-algebra. We define the (graded) vector space $\lambda(\mathcal{P},A)$ to be the quotient of $A\otimes A$ by the subspace spanned by the images of all the maps

$$\mathcal{P}(n)\otimes A^{\otimes(n+1)} \xrightarrow{1-\sigma} \mathcal{P}(n)\otimes A^{\otimes(n+1)} \xrightarrow{B_n} A\otimes A, \quad n\geq 1, \sigma\in\mathbb{S}_{n+1}$$

where B_n is as in (4.1) and σ acts diagonally. There is a tautological bilinear form $A\otimes A \to \lambda(\mathcal{P},A)$, which is clearly universal among all the invariant bilinear forms on A.

For the cases $\mathcal{P} = \mathcal{A}ss$, $\mathcal{C}om$, and $\mathcal{L}ie$, the space $\lambda(\mathcal{P},A)$ was introduced by Kontsevich **[15]**, who denoted it $\Omega^0(A)$.

Thus $\lambda(\mathcal{P},-)$ is a functor from the category of $\mathcal{P}$-algebras to the category of graded vector spaces. If $\mathcal{P}$ is a dg-operad and A is a dg-algebra over $\mathcal{P}$ then $\lambda(\mathcal{P},A)$ acquires a natural differential, becoming a chain complex. Of course, if $\mathcal{P}$ and A are ungraded, then so is $\lambda(\mathcal{P},A)$.

(4.8) Example: If $\mathcal{P} = \mathcal{A}ss$ and A is a unital associative algebra, then $\lambda(\mathcal{A}ss,A) = A/[A,A]$ is the quotient of A by its commutant. If $\mathcal{P} = \mathcal{C}om$ and A is a unital commutative algebra, then $\lambda(\mathcal{C}om,A) = A$. If $\mathcal{P} = \mathcal{P}ois$ and P is a unital Poisson algebra (this means that 1 is a unit for the product and $\{1,x\}=0$ for all $x\in P$), then $\lambda(\mathcal{P}ois,P) = P/\{P,P\}$ is the quotient of P by the span of all Poisson brackets.

In the next proposition, we calculate the universal invariant bilinear form of a free algebra $\mathsf{F}(\mathcal{P},V)$.

(4.9) Proposition: *Let $\mathcal{P}$ be a dg-operad, and let A be a dg-algebra over $\mathcal{P}$. There is a natural isomorphism of chain complexes*

$$\lambda(\mathcal{P},\mathsf{F}(\mathcal{P},V)) \cong \bigoplus_{n=0}^{\infty} \mathcal{P}(n)\otimes_{\mathbb{S}_{n+1}} V^{\otimes(n+1)}.$$

Proof: If $p\in\mathcal{P}(n)$ and $q\in\mathcal{P}(m)$, and $x_i, y_i\in V$, the image of the element

$$p(x_1,\dots,x_n)\otimes q(y_1,\dots,y_n) \in \lambda(\mathsf{F}(\mathcal{P},V))$$

is equal, up to a sign, to the image of

$$y_1\otimes(\tau p)(y_2,\dots,y_m,q(x_1,\dots,x_n)).$$

We see that $\lambda(\mathcal{P}, \mathsf{F}(\mathcal{P}, V))$ is a quotient of

$$V \otimes \mathsf{F}(\mathcal{P}, V) \cong \bigoplus_{n=0}^{\infty} \mathcal{P}(n) \otimes_{\mathbb{S}_n} V^{\otimes(n+1)}.$$

Using once more the coinvariance under the action of the cyclic group $\mathbb{Z}_{n+1}$, we get a surjection

$$\bigoplus_{n=0}^{\infty} \mathcal{P}(n) \otimes_{\mathbb{S}_{n+1}} V^{\otimes(n+1)} \to \lambda(\mathcal{P}, \mathsf{F}(\mathcal{P}, V)).$$

It is easily seen that this is in fact an isomorphism.

For the cases $\mathcal{P} \in \{\mathcal{A}ss, \mathcal{C}om, \mathcal{L}ie\}$, this result was proved by Kontsevich [**15**].

5 Cyclic homology

(5.1) Non-abelian derived functors: All operads and algebras over them in the remainder of this paper will lie in the symmetric monoidal category of positively graded chain complexes ($V_n = 0$ for $n < 0$), unless otherwise mentioned; of course, ungraded vector spaces form a subcategory of this category. The reason for this is that we wish to use the methods of Quillen's homotopical algebra [**19**], even if we are primarily interested in ungraded operads and algebras.

(5.2) Definition: Let $\mathcal{P}$ be a cyclic operad. The cyclic homology $\mathrm{HA}_\bullet(\mathcal{P}, A)$ us the non-abelian (left) derived functor of the functor $\lambda(\mathcal{P}, A)$.

This definition is motivated by the result of Feigin and Tsygan [**6**] who show that cyclic homology of a unital associative algebras (over a field of characteristic zero) are the derived functors of the functor $A \mapsto A/[A, A]$.

Let us briefly explain the construction of non-abelian derived functors. If V is a chain complex, denote by $V^\#$ the graded vector space obtained from V by forgetting the differential. The same notation will be used for dg-operads and dg-algebras. We say that a map $A \to B$ of $\mathcal{P}$-algebras (or chain complexes) is a *weak equivalence* if it induces an isomorphism on homology.

If A is an $\mathcal{P}$-algebra, an *almost free resolution* of A is a weak equivalence of $\mathcal{P}$-algebras $\tilde{A} \to A$ such that the $\mathcal{P}^\#$-algebra $\tilde{A}^\#$ is free. In Chapter 2 of [**11**], it is shown that every $\mathcal{P}$-algebra A has a natural almost free resolution.

Let F be a functor from $\mathcal{P}$-algebras A to chain complexes. The non-abelian derived functor $\mathsf{L}_\bullet F(A)$ is $H_\bullet(F(\tilde{A}))$, where $\tilde{A} \to A$ is an almost free resolution of A.

For this definition to be sensible, we must of course check that the result is independent of the choice of almost free resolution. A sufficient condition for this is that whenever $A \to B$ is a weak equivalence of almost free $\mathcal{P}$-algebras, the induced map $F(A) \to F(B)$ is a weak equivalence of chain complexes. For the functor $\lambda(\mathcal{P},-)$ this condition will be implied by the following theorem.

(5.3) Theorem: *There is a functor* $\mathrm{CA}(\mathcal{P},A)$ *from* $\mathcal{P}$*-algebras to chain complexes, together with a natural transformation* $\mathrm{CA}(\mathcal{P},A) \to \lambda(\mathcal{P},A)$*, having the following two properties:*

(1) If $A \to B$ *is a weak equivalence of* $\mathcal{P}$*-algebras, the induced map* $\mathrm{CA}(\mathcal{P},A) \to \mathrm{CA}(\mathcal{P},B)$ *is a weak equivalence of chain complexes.*

(2) If A *is an almost free algebra, the induced map* $\mathrm{CA}(\mathcal{P},A) \to \lambda(\mathcal{P},A)$ *is a weak equivalence.*

The proof of Theorem (5.3) occupies the remainder of this section.

A standard way of constructing non-abelian derived functors is the method of simplicial resolutions (Quillen [**19**]). Indeed, if the field **k** over which we were working had positive characteristic, this would be the only method available. (This is because the category of algebras is not in general a closed model category unless the characteristic is 0, whereas the category of simplicial algebras is always a simplicial category.) Another setting in which it might be of interest to investigate simplicial resolutions is that of cyclic homology for algebras over cyclic operads in categories of spectra — this would allow one to relate the theory of topological Hochschild and cyclic homology for "rings up to homotopy" developed by Bökstedt, Hsiang and Madsen [**2**] to our work.

(5.4) Cooperads: A (cyclic) *cooperad* in a symmetric monoidal category $\mathcal{C}$ is a (cyclic) operad in the opposite category $\mathcal{C}^{\mathrm{op}}$ (see Getzler-Jones [**11**]): thus, a cooperad $\mathcal{Z}$ has cocomposition maps

$$\nu_{i_1\dots i_k} : \mathcal{Z}(i_1+\dots+i_k) \to \mathcal{Z}(k)\otimes\mathcal{Z}(i_1)\otimes\dots\mathcal{Z}(i_k).$$

If $\mathcal{Z}$ is a **k**-linear cooperad (with grading concentrated in degree 0), its dual $\mathbb{S}$-module $\mathcal{Z}^* = \{\mathcal{Z}(n)^*\}$ is an operad; the converse is true provided $\mathcal{Z}(n)$ is finite-dimensional for all $n \geq 0$. The definition of cyclic and anticyclic cooperads is similar.

An augmented dg-operad is a dg-operad $\mathcal{P}$ together with a map $\varepsilon : \mathcal{P}(1) \to \mathbf{k}$. Denote the kernel of the augmentation ε by $\bar{\mathcal{P}}$.

An important example of a dg-cooperad is the *bar cooperad* $\mathcal{BP}$ of an augmented dg-operad $\mathcal{P}$, whose underlying graded cooperad is the cofree cooperad (in the sense dual to that of (1.12)) generated by the $\mathbb{S}$-module $\Sigma\bar{\mathcal{P}} = \{\Sigma\bar{\mathcal{P}}(n)\}$:

$$\mathcal{BP}(n) = \bigoplus_{\text{rooted } n\text{-trees } T} (\Sigma\bar{\mathcal{P}})(T).$$

The differential δ in $\mathcal{BP}(n)$ is a sum $\delta_1+\delta_2$, where δ_1 is induced by the differential of $\mathcal{P}$ and δ_2 is defined as follows (see Chapter 2 of [**11**] or Chapter 3 of [**12**]).

If T is a rooted tree and $e \in \mathrm{Edge}(T)$ is an internal edge of T, let T/e be the tree obtained from T by removing e and merging its two ends to a single vertex. The composition map $\mu_{T,e} : \mathcal{P}(T) \to \mathcal{P}(T/e)$ induces a map

$$\partial_{T,e} : (\Sigma\bar{\mathcal{P}})(T) \to (\Sigma\bar{\mathcal{P}})(T/e),$$

and δ_2 is the sum over all rooted trees T and internal edges $e \in \mathrm{Edge}(T)$ of the operators $\partial_{T,e}$.

The tensor product of the suspensions of complexes $\{V(i) \mid i \in I\}$ is related to the tensor product of the complexes themselves by the canonical isomorphism

$$\bigotimes_{i\in I}(\Sigma V(i)) \cong \det(\mathbf{k}^I) \otimes \Sigma^{|I|}\left(\bigotimes_{i\in I} V(i)\right).$$

Therefore we can write

$$\mathcal{BP}(n) = \bigoplus_{\text{rooted } n\text{-trees } T} \det(T) \otimes \Sigma^{|\operatorname{Vert}(T)|}\bar{\mathcal{P}}(T), \tag{5.5}$$

where $\det(T)$ is as in (1.7). This shows that if $\mathcal{P}$ is a cyclic (resp. anticyclic) dg-operad, then $\mathcal{BP}$ is an anticyclic (resp. cyclic) dg-cooperad.

There is a functor $\mathcal{B}^*$ (called the *cobar-construction*) from dg-cooperads to dg-operads, adjoint to $\mathcal{B}$ [**11**]. Similarly to $\mathcal{B}$, it sends cyclic cooperads to anti-cyclic operads and vice versa. It is proved in [**11**] (and, in dual language, in [**12**]) that $\mathcal{B}$ and $\mathcal{B}^*$ are inverse to each other, up to weak equivalence. The dg-operad $C_\bullet(A_\infty, \mathbf{k})$ of (2.3) (c) is isomorphic to $\mathcal{B}^*(\Lambda\mathcal{A}ss)^*$, and we obtain another proof that it is cyclic; furthermore, the two cyclic structures agree.

(5.6) Twisting cochains: If $\mathcal{Q}$ and $\mathcal{P}$ are operads, their tensor product $\mathcal{Q} \otimes \mathcal{P} = \{\mathcal{Q}(n) \otimes \mathcal{P}(n)\}$ is again an operad in an obvious way. Similarly, if $\mathcal{Z}$ is a dg-cooperad and $\mathcal{P}$ is a dg-operad, the $\mathbb{S}$-module $\hom(\mathcal{Z}, \mathcal{P}) = \{\hom(\mathcal{Z}(n), \mathcal{P}(n))\}$ is a dg-operad; however, this operad is not necessarily concentrated in positive degrees, even if $\mathcal{Z}$ and $\mathcal{P}$ are.

A *twisting cochain* from a dg-cooperad $\mathcal{Z}$ to a dg-operad $\mathcal{P}$ is a sequence of maps $\Phi_n \in \hom_{\mathbb{S}_n}(\mathcal{Z}(n), \mathcal{P}(n))$ of degree -1 such that

$$\delta\Phi_n + \sum_{i+j+k=n} \Phi_{i+k+1}(\underbrace{1, \dots, 1}_{i \text{ times}}, \Phi_j, \underbrace{1, \dots, 1}_{k \text{ times}}) = 0,$$

where the composition is with respect to the structure of an operad on $\hom(\mathcal{Z}, \mathcal{P})$. (See Section 2.3 of [**11**].)

If $\mathcal{Z}$ is an anticyclic cooperad and $\mathcal{P}$ is a cyclic operad, a *cyclic twisting cochain* Φ from $\mathcal{Z}$ to $\mathcal{P}$ is a twisting cochain from $\mathcal{Z}$ to $\mathcal{P}$ such that each Φ_n is equivariant with respect to $\mathbb{S}_{n+1}$.

(5.7) Examples: (a) *The universal twisting cochain:* Let $\mathcal{P}$ be an augmented dg-operad and $\mathcal{BP}$ its bar cooperad. There is a natural twisting cochain $\Phi^{\mathcal{P}}$ from $\mathcal{BP}$ to $\mathcal{P}$ defined as follows: it vanishes on summands of $\mathcal{BP}$ associated to trees with more than one vertex, while on trees with one vertex, it is the inclusion $\Sigma\bar{\mathcal{P}}(n) \hookrightarrow \mathcal{P}(n)$. As explained in [**11**], $\Phi^{\mathcal{P}}$ is universal, in a natural sense, among twisting cochains to $\mathcal{P}$.

If the dg-operad $\mathcal{P}$ is cyclic, then $\Phi^{\mathcal{P}}$ is a cyclic twisting cochain, as may be easily verified.

(b) *The quadratic twisting cochain:* Let $\mathcal{P} = \mathcal{P}(E, R)$ be a quadratic operad, such that E is finite dimensional, and let $\mathcal{P}^!$ be its quadratic dual. Then $\mathcal{P}^\perp = \Lambda^{-1}(\mathcal{P}^!)^*$ is a cooperad, and there is a twisting cochain $\Psi^{\mathcal{P}}$ from $\mathcal{P}^\perp$ to $\mathcal{P}$, such that $\Psi^{\mathcal{P}}_n = 0$ for $n \neq 2$, and

$$\Psi^{\mathcal{P}}_2 : \mathcal{P}^\perp(2) \cong \Sigma E \to E = \mathcal{P}(2)$$

is the identity. If $\mathcal{P}$ is cyclic quadratic, then $\mathcal{P}^\perp$ is an anticyclic cooperad, and $\Psi^{\mathcal{P}}$ is a cyclic twisting cochain.

(c) *The unital-associative twisting cochain:* There is a modification of the quadratic twisting cochain of (b) for the operad $\mathcal{U}\mathcal{A}ss$ describing unital associative algebras (3.10). Let $\mathcal{UC}oass = \Lambda^{-1}\mathcal{U}\mathcal{A}ss^*$; the only difference between the cooperads $\mathcal{UC}oass$ and $\mathcal{A}ss^\perp$ is that $\mathcal{UC}oass(0) = \mathbf{k}$ whereas $\mathcal{A}ss^\perp(0) = 0$. We define the twisting cochain $\Psi^{\mathcal{U}\mathcal{A}ss}$ from $\mathcal{UC}oass$ to $\mathcal{U}\mathcal{A}ss$ by setting $\Psi^{\mathcal{U}\mathcal{A}ss}(2)$ be the identity map as in (b) and the other components of $\Psi^{\mathcal{U}\mathcal{A}ss}$ to be zero.

(5.8) Three complexes: Let $\mathcal{P}$ be a cyclic dg-operad, A be a $\mathcal{P}$-algebra, $\mathcal{Z}$ be a cyclic dg-cooperad, and Φ be a cyclic twisting cochain from $\mathcal{Z}$ to $\mathcal{P}$. We now define a natural short exact sequence of chain complexes

$$0 \to \mathrm{CA}(\Phi, A) \to \mathrm{CB}(\Phi, A) \to \mathrm{CC}(\Phi, A) \to 0. \tag{5.9}$$

The underlying graded vector spaces of these complexes are as follows:

$$\mathrm{CA}(\Phi, A) = \bigoplus_{n=1}^{\infty} \mathcal{Z}(n) \otimes_{\mathbb{S}_{n+1}} (V_{n,1} \otimes A^{\otimes(n+1)}),$$
$$\mathrm{CB}(\Phi, A) = \bigoplus_{n=0}^{\infty} \mathcal{Z}(n) \otimes_{\mathbb{S}_{n+1}} (\mathbf{k}^{n+1} \otimes A^{\otimes(n+1)}),$$
$$\mathrm{CC}(\Phi, A) = \bigoplus_{n=0}^{\infty} \mathcal{Z}(n) \otimes_{\mathbb{S}_{n+1}} A^{\otimes(n+1)},$$

where $\mathbf{k}^{n+1}$ is the standard permutation representation of $\mathbb{S}_{n+1}$, and $V_{n,1} \subset \mathbf{k}^{n+1}$ is the irreducible representation defined in (3.5). The short exact sequence (5.9) is induced by the short exact sequence of $\mathbb{S}_{n+1}$-modules

$$0 \to V_{n,1} \to \mathbf{k}^{n+1} \to \mathbb{1} \to 0.$$

In the special case of the universal twisting cochain $\Phi^{\mathcal{P}}$ from $\mathcal{BP}$ to $\mathcal{P}$, we denote the corresponding complexes by $\mathrm{CA}(\mathcal{P}, A) = \mathrm{CA}(\Phi^{\mathcal{P}}, A)$, etc.

The differential δ on $\mathrm{CB}(\Phi, A)$ is the sum $\delta_{\mathcal{Z}} + \delta_A + \delta_\Phi$, where $\delta_{\mathcal{Z}}$ and δ_A are induced by the differentials of $\mathcal{Z}$ and A, and δ_Φ is defined as follows.

If $L \coprod M$ is a partition of $\{0, \dots, n\}$ into disjoint subsets, let $L^+ = L \coprod \{M\}$ and let $M^+ = M \coprod \{L\}$. Using the cocomposition of the anticyclic cooperad $\mathcal{Z}$, we obtain a map

$$\mathcal{Z}(n) \to \mathcal{Z}((L^+)) \otimes \mathcal{Z}((M^+)),$$

which induces a map

$$\mathcal{Z}(n) \otimes A^{\otimes(n+1)} \otimes \mathbf{k}^{n+1} \to (\mathcal{Z}((L^+)) \otimes A^{\otimes L}) \otimes (\mathcal{Z}((M^+)) \otimes A^{\otimes M}) \otimes \mathbf{k}^{n+1}.$$

The twisting cochain $\Phi : \mathcal{Z}((L^+)) \to \mathcal{P}((L^+))$ is a map of degree -1 which, when composed with the structure map

$$\mathcal{P}((L^+)) \otimes A^{\otimes L} \cong \mathcal{P}(L) \otimes A^{\otimes L} \to A$$

of the $\mathcal{P}$-algebra A induces a map

$$\mathcal{Z}(n) \otimes A^{\otimes(n+1)} \otimes \mathbf{k}^{n+1} \to A \otimes \mathcal{Z}(M) \otimes A^{\otimes M} \otimes \mathbf{k}^{n+1}.$$

There is a map $\mathbf{k}^{n+1} \to \mathbf{k}^{M^+}$ defined by the composition

$$\mathbf{k}^{n+1} \cong \mathbf{k}^L \oplus \mathbf{k}^M \xrightarrow{\text{sum}\oplus\text{id}} \mathbf{k} \oplus \mathbf{k}^M \cong \mathbf{k}^{M^+}, \tag{5.10}$$

where sum : $\mathbf{k}^L \to \mathbf{k}$ is the summation map. Composing these maps, we obtain a map

$$\mathcal{Z}(n) \otimes A^{\otimes(n+1)} \otimes \mathbf{k}^{n+1} \xrightarrow{\delta_M} \mathcal{Z}(M) \otimes A^{\otimes M^+} \otimes \mathbf{k}^{M^+}.$$

Summing over all subsets $M \subset \{0, \ldots, n\}$ of cardinality m, we obtain a map

$$\mathcal{Z}(n) \otimes A^{\otimes(n+1)} \otimes \mathbf{k}^{n+1} \xrightarrow{\delta_m} \bigoplus_{|M|=m} \mathcal{Z}(M) \otimes A^{\otimes M^+} \otimes \mathbf{k}^{M^+}.$$

The group $\mathbb{S}_{n+1}$ acts on both sides, and the map δ_m is equivariant — here, we use the fact that the twisting cochain is cyclic. Thus, it descends to a map from $\mathcal{Z}(n) \otimes_{\mathbb{S}_{n+1}} (A^{\otimes(n+1)} \otimes \mathbf{k}^{n+1})$ to the space of $\mathbb{S}_{n+1}$-coinvariants of the right-hand side, which we compose with the projection

$$\left(\bigoplus_{|M|=m} \mathcal{Z}(M) \otimes A^{\otimes M^+} \otimes \mathbf{k}^{M^+} \right)_{\mathbb{S}_{n+1}} \to \mathcal{Z}(m) \otimes_{\mathbb{S}_{m+1}} (A^{\otimes(m+1)} \otimes \mathbf{k}^{m+1}).$$

The sum of these maps over all n and $m \leq n$ gives an endomorphism δ_Φ of degree -1 of $\mathrm{CB}(\Phi, A)$.

(5.11) Lemma: (1) *The endomorphism* $\delta_{\mathcal{Z}} + \delta_A + \delta_\Phi$ *of* $\mathrm{CB}(\Phi, A)$ *is a differential.*

(2) *The differential* $\delta_{\mathcal{Z}} + \delta_A + \delta_\Phi$ *of* $\mathrm{CA}(\Phi, A)$ *preserves* $\mathrm{CA}(\Phi, A)$, *and projects to a differential on* $\mathrm{CC}(\Phi, A)$.

Proof: Part (1) is a straightforward consequence of the fact that Φ is a twisting cochain, and is left to the reader.

Part (2) is a consequence of the fact that the map of (5.10) takes the subspace $V_{n,1}$ onto $V_{|L|,1}$.

For any cyclic twisting cochain Φ from an anticyclic cooperad $\mathcal{Z}$ to a cyclic operad $\mathcal{P}$, there is a natural transformation

$$\mathrm{CA}(\Phi, A) \to \lambda(\mathcal{P}, A),$$

which sends $\mathcal{Z}(n) \otimes_{\mathbb{S}_{n+1}} (V_{n,1} \otimes A^{\otimes(n+1)})$ to zero for $n \neq 1$, and for $n = 1$, is the composition

$$\mathcal{Z}(1) \otimes_{\mathbb{S}_2} (V_{1,1} \otimes A \otimes A) \xrightarrow{\varepsilon} (A \otimes A)_{\mathbb{S}_2} \to \lambda(\mathcal{P}, A),$$

induced by the counit $\varepsilon : \mathcal{Z}(1) \to \mathrm{sgn} \cong V_{1,1}$ of the anticyclic cooperad $\mathcal{Z}$. It is obvious that this natural transformation intertwines the differential $\delta_{\mathcal{Z}} + \delta_A + \delta_\Phi$ which we have constructed with the natural differential on $\lambda(\mathcal{P}, A)$.

(5.12) Proof of Theorem (5.3): Part (1) follows by a simple spectral sequence argument; convergence is ensured by the assumption that all complexes in question are situated in non-negative degrees.

To prove part (2), we consider the spectral sequence relating the homology of the complex $\mathrm{CA}(\mathcal{P}, A)$ with that of the complex $\mathrm{CA}(\mathcal{P}^\#, A^\#)$ with differential δ_Φ. Comparing this with a similar spectral sequence for the homology of $\lambda(\mathcal{P}, A)$, we find that all we need to prove is the following lemma.

(5.13) Lemma: *Let $\mathcal{P}$ be a graded operad (with trivial differential, situated in degrees ≥ 0), and let V be a graded vector space (also situated in degrees ≥ 0). Let $A = \mathsf{F}(\mathcal{P}, V)$ be the free graded $\mathcal{P}$-algebra generated by V. Then the map*

$$\mathrm{CA}(\mathcal{P}, A) \to \lambda(\mathcal{P}, A) \cong \bigoplus_{n=0}^{\infty} \mathcal{P}(n) \otimes_{\mathbb{S}_{n+1}} A^{\otimes(n+1)}$$

is a weak equivalence.

Proof: We have

$$\mathrm{CA}(\mathcal{P}, A) = \bigoplus_{n=1}^{\infty} \left(V_{n,1} \otimes \mathcal{BP}(n) \otimes A^{\otimes(n+1)}\right)_{\mathbb{S}_{n+1}}. \tag{5.14}$$

For a finite set I, denote by $\mathbf{k}_0^I$ the vector space of functions $I \to \mathbf{k}$ whose sum is zero; thus, $\mathbf{k}_0^n \cong V_{n-1,1}$. By (5.5), and bearing in mind that rooted n-trees are the same as unrooted $(n+1)$-trees, we may rewrite the n-th summand of (5.14) as

$$\bigoplus_{\substack{\text{unlabeled, unrooted} \\ (n+1)\text{-trees } T}} \left(\mathbf{k}_0^{\mathrm{Leg}(T)} \otimes (\Sigma\bar{\mathcal{P}})(T) \otimes A^{\otimes \mathrm{Leg}(T)}\right)_{\mathrm{Aut}(T)}.$$

Since the free algebra A has the form

$$A = \bigoplus_{n \geq 0} \mathcal{P}(n) \otimes_{\mathbb{S}_n} V^{\otimes n},$$

we can express $\mathrm{CA}(\mathcal{P}, A)$ directly in terms of trees. First, we need some terminology.

A vertex v of a tree T is called *extremal,* if the number of internal edges incident to v is one or zero. By an *extended tree* (T, S), we mean a tree T together with a subset S of its set of extremal vertices; we denote by $T\backslash S$ the tree obtained from T by removing all vertices in S together with all of the legs attached to them. For an extended tree (T, S), let

$$\tilde{\mathcal{P}}(T,S) = \bigotimes_{v\in\mathrm{Vert}(T\backslash S)} (\Sigma\bar{\mathcal{P}})(\!(\mathrm{Leg}(v))\!) \otimes \bigotimes_{v\in S} \mathcal{P}(\!(\mathrm{Leg}(v))\!).$$

Using this notation, we may write

$$\mathrm{CA}(\mathcal{P}, A) = \bigoplus_{(T,S)} \Big(\mathbf{k}_0^{\mathrm{Leg}(T\backslash S)} \otimes \tilde{\mathcal{P}}(T,S) \otimes V^{\otimes \mathrm{Leg}(T)}\Big)_{\mathrm{Aut}(T,S)}, \tag{5.15}$$

where the summation is over isomorphism classes of extended trees.

Introduce an increasing filtration G on $\mathrm{CA}(\mathcal{P}, A)$ by setting

$$G_l\mathrm{CA}(\mathcal{P}, A) = \bigoplus_{\{(T,S)\mid |\mathrm{Vert}(T)|\le l\}} \Big(\mathbf{k}_0^{\mathrm{Leg}(T\backslash S)} \otimes \tilde{\mathcal{P}}(T,S) \otimes V^{\otimes \mathrm{Leg}(T)}\Big)_{\mathrm{Aut}(T,S)}$$

This filtration is preserved by the differential $\delta = \delta_{\mathcal{BP}} + \delta_\Phi$ of $\mathrm{CA}(\mathcal{P}, A)$. The associated graded complex $\mathrm{gr}^G \mathrm{CA}(\mathcal{P}, A)$ may be identified, as a graded vector space, with $\mathrm{CA}(\mathcal{P}, A)$, while its differential $\mathrm{gr}(\delta)$ is the sum of those pieces of δ which preserve the number of vertices in an extended tree (T, S). Thus $\delta_{\mathcal{BP}}$ does not contribute at all to $\mathrm{gr}(\delta)$, while δ_Φ contributes a sum, over extremal vertices of T not in S, of the operation which adjoins v to S and in (5.15) attaches the sum of the elements of $\mathbf{k}$ labelling edges in $\mathrm{Leg}(v)$ to the leg of the tree $T \setminus (S \cup \{v\})$ corresponding to v.

On the summand of $\mathrm{gr}^G \mathrm{CA}(\mathcal{P}, A)$ corresponding to the extended tree (T, S), define an operator H of degree $+1$ as a sum over vertices $v \in S$ of the operation which removes v from S and divides the element of $\mathbf{k}$ attached to the leg of $T \setminus S$ corresponding to v equally among the edges in $\mathrm{Leg}(v)$. It is easily seen that $I - (\mathrm{gr}(\delta)H + H\,\mathrm{gr}(\delta))$ is the projection onto summands in (5.15) corresponding to extended trees (T, S) such that T has one vertex v and $S = \{v\}$. The spectral sequence associated to the filtration on $G_l\mathrm{CA}(\Phi, A)$ collapses, and the homology of $\mathrm{CA}(\Phi, A)$ equals the homology of $\mathrm{gr}^G \mathrm{CA}(\Phi, A)$, which is

$$\bigoplus_{n\ge 0} \mathcal{P}(n) \otimes_{\mathbb{S}_{n+1}} V^{\otimes(n+1)}.$$

By Proposition (4.9), this is the same as $\lambda(\mathcal{P}, A)$, completing the proof of Theorem (5.3).

6 Examples

(6.1) Associative algebras: We first consider the theory of associative algebras with unit. Let $\mathcal{P} = \mathcal{U}\mathcal{A}ss$ be the unital associative operad (see (3.10)), and let $\mathcal{U}\mathcal{C}oass$ be the cooperad introduced in (5.7) (c); thus,

$$\mathcal{U}\mathcal{C}oass(n) \cong \Sigma^{n-1} \operatorname{Ind}_{\mathbb{Z}_{n+1}}^{\mathbb{S}_{n+1}} \operatorname{sgn}$$

for $n \geq 0$. Let $\Psi = \Psi^{\mathcal{U}\mathcal{A}ss}$ be the unital-associative twisting cochain.

Let A be a unital associative algebra, which for simplicity we will suppose to have trivial grading. We abbreviate the complex $\mathrm{CA}(\Psi, A)$ by $\mathrm{CA}(A)$, and similarly for $\mathrm{CB}(\Psi, A)$ and $\mathrm{CC}(\Psi, A)$. We will now identify these complexes, relating them to the usual Connes-Tsygan complexes for cyclic homology of associative algebras, whose definition we now recall (see Loday [**16**]).

The *Hochschild complex* $C_\bullet(A, A)$ of an associative algebra A has $C_n(A, A) = A^{\otimes(n+1)}$ with differential b given by the formula

$$b(x_0 \otimes \ldots \otimes x_n) = \sum_{i=0}^{n-1} (-1)^i x_0 \otimes \ldots \otimes x_i x_{i+1} \ldots \otimes x_n + (-1)^n x_n x_0 \otimes \ldots \otimes x_{n-1}.$$

The homology of the complex $C_\bullet(A, A)$ may be identified with $\operatorname{Tor}_\bullet^{A \otimes A^{\mathrm{op}}}(A, A)$. We also have the complex $C_\bullet(A, b')$, with $C_n(A, b') = A^{\otimes(n+1)}$ but with the modified differential

$$b'(x_0 \otimes \ldots \otimes x_n) = \sum_{i=0}^{n-1} (-1)^i x_0 \otimes \ldots \otimes x_i x_{i+1} \ldots \otimes x_n.$$

The complex $C_\bullet(A, b')$ is contractible if A is unital.

Let $t = (-1)^n \tau \in \mathbf{k}[\mathbb{S}_{n+1}]$, and let $N = \sum_{i=0}^n t^i$ The *cyclic double complex* of A, denoted $C_{\bullet\bullet}(A)$, is the double complex

$$0 \leftarrow C_\bullet(A, A) \xleftarrow{1-t} C_\bullet(A, b') \xleftarrow{N} C_\bullet(A, A) \xleftarrow{1-t} C_\bullet(A, b') \xleftarrow{N} \ldots$$

The homology of the total complex of $C_{\bullet\bullet}(A)$ is the classical *cyclic homology* of the algebra A, denoted $H_\bullet^\lambda(A)$.

We will also need the cyclic complex $C_\bullet^\lambda(A)$, which is the cokernel of the map of complexes $C_\bullet(A, A) \xleftarrow{1-t} C_\bullet(A, b')$. Since $C_\bullet(A, b')$ is acyclic, $C_\bullet^\lambda(A)$ is weakly equivalent to the total complex of the double complex $C_{\bullet\bullet}(A)$.

We now identify the complex $\mathrm{CB}(A)$. Note that

$$\mathcal{U}\mathcal{C}oass(n) \otimes \mathbf{k}^{n+1} \cong \Sigma^{n-1} \operatorname{Ind}_{\mathbb{Z}_{n+1}}^{\mathbb{S}_{n+1}} \mathbb{1} \otimes \mathbf{k}^{n+1} \cong \Sigma^{n-1} \mathbf{k}[\mathbb{S}_{n+1}]$$

is the $(n-1)$-fold suspension of the regular representation of $\mathbb{S}_{n+1}$. Thus

$$\mathrm{CB}(A) \cong \bigoplus_{n=0}^{\infty} \Sigma^{n-1} A^{\otimes(n+1)}.$$

The differential δ_Ψ of this complex may be identified with the differential b of the Hochschild complex $C_\bullet(A,A)$ of A. Thus, $\mathrm{CB}(A) \cong \Sigma^{-1}C_\bullet(A,A)$ is the desuspension of the Hochschild complex of A.

The complex $\mathrm{CC}(A)$ is identified with the cokernel of the operator $1-t$ acting on $\mathrm{CB}(A)$:

$$\mathrm{CC}(A) \cong \bigoplus_{n=1}^{\infty} A^{\otimes(n+1)} / \operatorname{Im}(1-t) = \Sigma^{-1}C_\bullet(A,A)/\operatorname{Im}(1-t).$$

It follows that $\mathrm{CC}(A) \cong \Sigma^{-1}C^\lambda_\bullet(A)$, and that $\mathrm{HC}_n(A) \cong H^\lambda_{n+1}(A)$ is the cyclic homology of A.

Finally, we identify $\mathrm{CA}(A)$. Note that $V_{n,1}$, regarded as a module over the group algebra $\mathbf{k}[\mathbb{S}_{n+1}]$, is generated by one element u, invariant under $\mathbb{S}_n \subset \mathbb{S}_{n+1}$, and with the relation

$$(1+\tau+\cdots+\tau^{n-1}+\tau^n)u = 0.$$

This implies that

$$\mathrm{CA}(A) \cong \bigoplus_{n=1}^{\infty} \Sigma^{n-1} A^{\otimes(n+1)} / \operatorname{Im}(N).$$

Unraveling the definition of the differential δ_Ψ, we see that we may identify it with b'; thus, $\Sigma\mathrm{CA}(A)$ is the cokernel of the map of complexes $C_\bullet(A,A) \xrightarrow{N} C_\bullet(A,b')$, or equivalently, of the map of complexes $C^\lambda_\bullet(A) \xrightarrow{N} C_\bullet(A,b')$. It turns out that this complex again calculates the cyclic homology.

(6.2) Proposition: *If A is unital, the complex* $\mathrm{CA}(A)$ *is weakly equivalent to the total complex of the double complex* $C_{\bullet\bullet}(A)$.

Proof: By unitality of A, the complex $C_\bullet(A,b')$ is contractible. The result follows from the short exact sequence of complexes

$$0 \to C^\lambda_\bullet(A) \xrightarrow{N} C_\bullet(A,b') \to \Sigma\mathrm{CA}_\bullet(A) \to 0.$$

To summarize our results, if A is a unital associative algebra, $\mathrm{HA}_\bullet(A)$ and $\Sigma\mathrm{HC}_\bullet(A)$ are both naturally isomorphic to the cyclic homology $H^\lambda_\bullet(A)$ of A, and the long exact sequence of homology associated to our sequence of complexes (5.9), is nothing other than the Connes-Tsygan *periodicity exact sequence:*

$$\begin{array}{ccccccccc}
\longrightarrow & \mathrm{HC}_{n+1}(A) & \longrightarrow & \mathrm{HA}_n(A) & \longrightarrow & \mathrm{HB}_n(A) & \longrightarrow & \mathrm{HC}_n(A) & \longrightarrow \\
& \| & & \| & & \| & & \| & \\
\longrightarrow & H^\lambda_{n+2}(A) & \xrightarrow{S} & H^\lambda_n(A) & \xrightarrow{B} & H_{n+1}(A,A) & \xrightarrow{I} & H^\lambda_{n+1}(A) & \longrightarrow
\end{array}$$

In particular, the map $S : H^{\lambda}_{n+2}(A) \to H^{\lambda}_{n}(A)$ (the additive analog of the inverse of the Bott map in K-theory) is identified with the boundary map $\mathrm{HC}_{n+1}(A) \to \mathrm{HA}_n(A)$ of our long exact sequence, after we take into account the "accidental" isomorphism $\mathrm{HC}_n(A) \cong \mathrm{HA}_{n+1}(A)$.

The relation between HA and HC for more general cyclic operads will be studied in (6.10). We will see that these theories are in a certain sense dual.

Let us now briefly consider the operad $\mathcal{A}ss$; there is a surjective map of operads $\mathcal{UA}ss \to \mathcal{A}ss$ which is an isomorphism $\mathcal{UA}ss(n) \cong \mathcal{A}ss(n)$ except for $n = 0$, where $\mathcal{A}ss(0) = 0$. The operad $\mathcal{A}ss$ is cyclic quadratic, the cooperad $\mathcal{A}ss^{\perp}$ (see (5.7) (b)) embeds in $\mathcal{UC}oass$, and $\mathcal{UC}oass(n) \cong \mathcal{C}oass(n)$ except for $n = 0$, where $\mathcal{C}oass(0) = 0$. The quadratic twisting cochain $\Psi = \Psi^{\mathcal{A}ss}$ from the cooperad $\mathcal{A}ss^{\perp}$ to the operad $\mathcal{A}ss$ is induced by the cyclic twisting cochain from $\mathcal{UC}oass$ to $\mathcal{UA}ss$ by means of these maps. If A is an associative algebra (not necessarily with unit), the three complexes are closely related to those which we described above:

(1) $\mathrm{CA}(\Psi, A)$ is identical to $\mathrm{CA}(A)$;

(2) $\mathrm{CB}_n(\Psi, A)$ and $\mathrm{CC}_n(\Psi, A)$ are identical to $\mathrm{CB}_n(A)$ and $\mathrm{CC}_n(A)$, respectively, if $n \geq 0$, but the bottom of the complex, the space A in degree -1, is discarded in each case.

Thus, the long exact sequence again agrees with that of Connes and Tsygan, except at the last few terms.

In a similar way, we can put the theory of cyclic homology of A_∞-algebras developed in [**10**] into the framework of the present paper by using the cyclic structure of the operad A_∞ described in (2.3) (c).

(6.3) Koszul operads: Let $\mathcal{P} = \mathcal{P}(E, R)$ be a cyclic quadratic operad with quadratic dual $\mathcal{P}^!$. There is a natural projection of operads $\Lambda^{-1}(\mathcal{BP})^* \to \mathcal{P}^!$ defined as follows (see [**12**], 4.1.1):

$$\Lambda^{-1}(\mathcal{BP})^*(n) \cong (\mathcal{B}\Lambda\mathcal{P})^*(n) = \bigoplus_{\text{rooted } n\text{-trees } T} (\Sigma\Lambda\bar{\mathcal{P}})(T)^*$$

$$\longrightarrow \bigoplus_{\text{binary } n\text{-trees } T} (\Sigma\Lambda\bar{\mathcal{P}})(T)^* = \mathbb{T}(E^*)(n) \longrightarrow \mathcal{P}^!(n).$$

The operad $\mathcal{P}$ is called *Koszul,* if all of these maps are weak equivalences. The operads $\mathcal{A}ss$, $\mathcal{C}om$ and $\mathcal{L}ie$ are Koszul [**12**], as is $\mathcal{P}ois$ [**11**].

Let $\mathcal{P}^{\perp}$ be the cooperad $\Lambda^{-1}(\mathcal{P}^!)^*$ (5.7) (b). The dual of the above projection of operads is an embedding of cooperads $j : \mathcal{P}^{\perp} \hookrightarrow \mathcal{BP}$.

(6.4) Proposition: *Let $\mathcal{P}$ be a Koszul cyclic quadratic operad, and let $\Psi = \Psi^{\mathcal{P}}$ be the quadratic twisting cochain from $\mathcal{P}^{\perp}$ to $\mathcal{P}$. Then the functor $H_i(\mathrm{CA}(\Psi, A))$ is naturally equivalent to the cyclic homology $\mathrm{HA}_i(\mathcal{P}, A)$.*

Proof: The morphism of cooperads $j : \mathcal{P}^{\perp} \to \mathcal{BP}$ being a weak equivalence, we easily infer that the induced morphism of complexes $\mathrm{CA}(\Psi^{\mathcal{P}}, A) \to \mathrm{CA}(\Phi^{\mathcal{P}}, A)$, where $\Phi^{\mathcal{P}}$ is the universal twisting cochain, is as well. By Theorem (5.3), the complex $\mathrm{CA}(\Phi^{\mathcal{P}}, A)$ calculates the cyclic homology of A, and the corollary follows.

In the remainder of this section, we illustrate this corollary in the cases of commutative and Lie algebras; we take up Poisson and Batalin-Vilkovisky algebras elsewhere.

(6.5) Commutative algebras: The dual cooperad $\mathcal{C}om^{\perp}$ of the commutative operad $\mathcal{C}om$ is isomorphic to $\Lambda^{-1}\mathcal{L}ie^{*}$. We now identify the complexes $\mathrm{CA}(\Psi, A)$ and $\mathrm{CC}(\Psi, A)$, where $\Psi = \Psi^{\mathcal{C}om}$ is the canonical cyclic twisting cochain from $\mathcal{C}om^{\perp}$ to $\mathcal{C}om$.

Recall the definition of the *Harrison complex* $\mathrm{CHarr}_{\bullet}(A, M)$ of a commutative algebra A with coefficients in a module M [**1**], [**11**]: as a graded vector space,

$$\mathrm{CHarr}_{\bullet}(A, M) \cong \bigoplus_{n=1}^{\infty} \mathcal{L}ie(n)^{*} \otimes_{\mathbb{S}_n} (\Sigma A)^{\otimes n} \otimes M,$$

while its differential is induced from that of the Hochschild complex $C_{\bullet}(A, M)$, of which it is naturally a subcomplex. Denote the cohomology of the Harrison complex by $\mathrm{Harr}_{\bullet}(A, M)$. We will be especially concerned with the cases where $M = A$, and where $M = \mathbf{k}$ with trivial action of A. In particular, we will need the theorem of Barr [**1**] and Quillen [**20**], that $H_{\bullet}(A, \mathbf{k})$ are the left non-abelian derived functors of the functor $A \mapsto A/A^2$ from commutative algebras to vector spaces (see also [**11**]).

The complex $\mathrm{CC}_{\bullet}(\Phi, A)$ is isomorphic to $\Sigma\,\mathrm{CHarr}_{\bullet}(A, A)$; the identification of the differential proceeds as in the case of associative algebras. In particular, if A is concentrated in degree 0, we see that $\mathrm{CC}_n(\Phi, A) \cong \mathrm{CHarr}_{n+1}(A, A)$.

We now identify the complex $\mathrm{CA}(\Phi, A)$. If A is an almost free commutative algebra, the natural map from $\lambda(\mathcal{C}om, A) \to A^2$ induced by sending $x \otimes y$ to xy is an isomorphism. It follows that if A is almost free, there is a short exact sequence

$$0 \to \lambda(\mathcal{C}om, A) \to A \to A/A^2 \to 0.$$

The homology long exact sequence of this short exact sequence may be identified with

$$\ldots \to \mathrm{Harr}_{n+1}(A, \mathbf{k}) \to \mathrm{HA}_n(\mathcal{C}om, A) \to H_n(A) \to \mathrm{Harr}_n(A, \mathbf{k}) \to \ldots, \tag{6.6}$$

since the homology of $\lambda(\mathcal{C}om, A)$ is by definition the cyclic homology of A, while the homology of A/A^2 is isomorphic to $\mathrm{Harr}_{\bullet}(A, \mathbf{k})$ by the theorem of Barr and Quillen. Since the long exact sequence (6.6) is natural, it actually holds for all A. So we get the following proposition.

(6.7) Proposition: *If A is an ungraded commutative algebra and $n > 0$,*

$$\mathrm{HA}_n(\mathcal{C}om, A) \cong \mathrm{Harr}_{n+1}(A, \mathbf{k}).$$

This proposition sheds some light on the structure of the $\mathbb{S}_{n+1}$-module $\mathcal{L}ie(n)$.

(6.8) Corollary: *There is a natural isomorphism of* $\mathbb{S}_{n+1}$*-modules*

$$\mathcal{L}ie(n) \otimes V_{n,1} \cong \mathcal{L}ie(n+1).$$

Proof: Let A be the commutative algebra spanned by vectors $\{x_0, \ldots, x_n\}$, with vanishing product. Then the complexes $\mathrm{CA}_\bullet(\Phi, A)$ and $\mathrm{CHarr}_\bullet(A, \mathbf{k})$ have vanishing differentials, and

$$\mathrm{CA}_n(\Phi, A) = \mathrm{HA}_n(\mathcal{C}om, A) \cong \mathrm{Harr}_{n+1}(A, \mathbf{k}) = \mathrm{CHarr}_{n+1}(A, \mathbf{k}).$$

Both complexes decompose into a direct sum of multi-homogeneous components according to the degree in each generator x_i. The summand of $\mathrm{CA}_\bullet(\Phi, A)$ and $\mathrm{CHarr}_\bullet(A, \mathbf{k})$ of multi-degree $(1, \ldots, 1)$ are isomorphic to

$$\mathcal{C}om^\perp(n) \cong \Sigma^{n-1}\mathcal{L}ie(n)^* \otimes \mathrm{sgn} \otimes V_{n,1}$$

and $\Sigma^n \mathcal{L}ie(n+1)^* \otimes \mathrm{sgn}$ respectively, and the result follows.

We do not identify the complex $\mathrm{CB}(\Phi, A)$ explicitly, but simply state the long exact sequence for the three homology theories:

$$\begin{array}{ccccccc}
\longrightarrow & \mathrm{HC}_{n+1}(\mathcal{C}om, A) & \longrightarrow & \mathrm{HA}_n(\mathcal{C}om, A) & \longrightarrow & \mathrm{HB}_n(\mathcal{C}om, A) & \longrightarrow \\
& \| & & \| & & & \\
& \mathrm{Harr}_n(A, A) & \xrightarrow{S} & \mathrm{Harr}_{n+1}(A, \mathbf{k}) & & &
\end{array}$$

Note that the map S, analogous to the periodicity map of cyclic homology, now maps between two distinct homology theories.

(6.9) Lie algebras: The dual cooperad $\mathcal{L}ie^\perp$ of the Lie operad $\mathcal{L}ie$ is isomorphic to $\Lambda^{-1}\mathcal{C}om^*$. Let us identify the complexes $\mathrm{CA}(\Psi, \mathfrak{g})$ and $\mathrm{CC}(\Psi, \mathfrak{g})$, where Ψ is the canonical cyclic twisting cochain from $\mathcal{L}ie^\perp$ to $\mathcal{L}ie$. Explicitly, $\Psi_2 : \mathcal{L}ie^\perp \cong \Sigma\mathcal{C}om(2)^* \otimes \mathrm{sgn} \to \mathcal{L}ie(2)$ is the natural isomorphism, while $\Psi_n = 0$ for $n \neq 2$.

Since $\mathcal{L}ie^\perp(n) \cong \Sigma^{n-1}\,\mathrm{sgn}$, we see that

$$\mathrm{CB}_n(\Psi, \mathfrak{g}) \cong \mathfrak{g} \otimes \Sigma^n \Lambda^n \mathfrak{g} \cong C_n(\mathfrak{g}, \mathfrak{g})$$

is the Chevalley-Eilenberg chain complex of $\mathfrak{g}$ with coefficients in the adjoint module, while $\mathrm{CC}_n(\Psi, \mathfrak{g}) \cong C_{n+1}(\mathfrak{g}, \mathbf{k})$ is the truncated Chevalley-Eilenberg complex of $\mathfrak{g}$ with trivial coefficients. As a consequence, $\mathrm{CA}_n(\Psi, \mathfrak{g})$ is the kernel of the map of complexes $C_n(\mathfrak{g}, \mathfrak{g}) \to C_{n+1}(\mathfrak{g}, \mathbf{k})$ given by the formula

$$g_0 \otimes (g_1 \wedge \ldots \wedge g_n) \mapsto g_0 \wedge g_1 \wedge \ldots \wedge g_n.$$

(6.10) Duality between HA **and** HC**:** By Theorem (5.3), the complex $\mathrm{CA}(\mathcal{P}, A)$ calculates the cyclic homology of an algebra A over a cyclic operad $\mathcal{P}$. We will now give a similar interpretation of the homology of $\mathrm{CC}(\mathcal{P}, A)$.

Let $\mathcal{Z}$ be a dg-cooperad. A *dg-coalgebra* over $\mathcal{Z}$ is a chain complex C with coaction maps $C \to \mathcal{Z}(n) \otimes C^{\otimes n}$ satisfying conditions dual to those for an algebra over an operad ([**11**], Section 1.7). If V is a chain complex then the *cofree $\mathcal{Z}$-coalgebra* generated by V is defined as

$$\mathsf{G}(\mathcal{Z}, V) = \bigoplus_{n=0}^{\infty} (\mathcal{Z}(n) \otimes V^{\otimes n})^{\mathbb{S}_n}.$$

If Φ is a twisting cochain from a cooperad $\mathcal{Z}$ to an operad $\mathcal{P}$, there is a functor $\mathbb{B}(\Phi, A)$ from $\mathcal{P}$-algebras to $\mathcal{Z}$-coalgebras called the *bar-construction.* As a coalgebra, $\mathbb{B}(\Phi, A) = \mathsf{G}(\mathcal{Z}, A)$ is the cofree $\mathcal{Z}$-coalgebra generated by A. The differential in $\mathbb{B}(\mathcal{Z}, A)$ is the sum of the differential of $\mathsf{G}(\Phi, A)$ and a differential involving the twisting cochain ([**11**], Section 2.3).

In the special case that Φ is the universal twisting cochain from $\mathcal{BP}$ to $\mathcal{P}$, we denote the bar construction by $\mathbb{B}(\mathcal{P}, A)$. It is proved in [**11**] that the functor $A \mapsto \mathbb{B}(\mathcal{P}, A)$ induced an equivalence of categories

$$\mathrm{Ho}\{\mathcal{P}\text{-algebras}\} \equiv \mathrm{Ho}\{\mathcal{BP}\text{-coalgebras}\}$$

where Ho is the homotopy category, obtained by inverting weak equivalences.

The cyclic cohomology of a dg-coalgebra C over a cyclic dg-cooperad $\mathcal{Z}$ has a definition dual to that of the cyclic homology of an algebra over a cyclic dg-operad. We define the chain complex $\lambda(\mathcal{Z}, C)$ to be the kernel of the map

$$C \otimes C \to \mathcal{Z}(n) \otimes C^{\otimes(n+1)} \to \mathcal{Z}(n) \otimes_{\mathbb{S}_n^+} C^{\otimes(n+1)},$$

where the first map is induced by the coaction $C \to \mathcal{Z}(n) \otimes C^{\otimes n}$ of C, and the second is projection to coinvariants; the differential is induced by the differential on $C \otimes C$. For a cofree coalgebra $\mathsf{G}(\mathcal{Z}, V)$, have

$$\lambda(\mathcal{Z}, \mathsf{G}(\mathcal{Z}, V)) \cong \bigoplus_{n=0}^{\infty} \mathcal{Z}(n) \otimes_{\mathbb{S}_{n+1}} V^{\otimes(n+1)}, \tag{6.11}$$

in a way analogous to Proposition (4.9). The cyclic cohomology groups of a coalgebra C are the right non-abelian derived functors $\mathsf{R}_i\lambda(\mathcal{Z}, C))$; they may be calculated by applying the functor $\lambda(\mathcal{Z}, -)$ to an *almost cofree resolution* $C \to \tilde{C}$ of the coalgebra C; that is, a weak equivalence of $\mathcal{Z}$-coalgebras with $\tilde{C}^{\#}$ a cofree $\mathcal{Z}^{\#}$-coalgebra.

Since the bar coalgebra $\mathbb{B}(\mathcal{P}, A)$ is almost cofree, the following proposition follows immediately from (6.11).

(6.12) Proposition: *Let $\mathcal{P}$ be an augmented cyclic dg-operad, and let A be a $\mathcal{P}$-algebra. Then the homology of the complex* $\mathrm{HC}(\mathcal{P}, A)$ *coincides with the cyclic cohomology of the coalgebra* $\mathbb{B}(\mathcal{P}, A)$ *over the cooperad* $\mathbb{B}\mathcal{P}$.

Note that the result of Feigin and Tsygan relating the cyclic homology of a Koszul associative algebra with the cyclic homology of its dual ([**7**], Theorem 2.4.1 (b)) is a special case of this proposition, corresponding to $\mathcal{P} = \mathcal{A}ss$.

References

[1] M. Barr, Harrison homology, Hochschild homology and triples, J. Algebra **8** (1968), 314–323.

[2] M. Bökstedt, W.S. Hsiang, I. Madsen, The cyclotomic trace and algebraic K-theory of spaces, Inv. Math. **111** (1993), 465-540.

[3] F.R. Cohen, The homology of $\mathcal{C}_{n+1}$-spaces, $n \geq 0$, Lecture Notes in Math. **533** (1976), 207–351.

[4] A. Connes, Non-commutative differential geometry, Inst. Hautes Etudes Sci. Publ. Math. **62** (1985), 257–360.

[5] A. Dold, Lectures on algebraic topology, Grundlehren Math. Wiss. **200**, 1981.

[6] B.L. Feigin and B.L. Tsygan, Additive K-theory, Lecture Notes in Math. **1289** (1987), 97–209.

[7] ———, Cyclic homology of algebras with quadratic relations, Lecture Notes in Math. **1289** (1987), 210–239.

[8] W. Fulton and J. Harris, Representation theory, Graduate Texts in Math. **129**, 1991.

[9] E. Getzler, Batalin-Vilkovisky algebras and two-dimensional topological field theories, Commun. Math. Phys. **159** (1994), 265–285.

[10] E. Getzler and J.D.S. Jones, A_∞-algebras and the cyclic bar complex, Ill. Jour. Math. **34** (1989), pp. 256–283.

[11] ———, Operads, homotopy algebra, and iterated integrals for double loop spaces, preprint, 1994.

[12] V.A. Ginzburg and M.M. Kapranov, Koszul duality for operads, Duke Math. J. **76** (1994), 203–272.

[13] V. Hinich and V. Schechtman, Homotopy Lie algebras, in "I.M. Gelfand Seminar," Adv. Soviet Math. **16**, part 2 (1993), 1–28.

[14] M.M. Kapranov, Permuto-associahedron, MacLane's coherence theorem and asymptotic zones for the KZ equation, J. Pure Appl. Algebra **85** (1993), 119–142.

[15] M. Kontsevich, Formal (non)-commutative symplectic geometry, in "The Gelfand mathematics seminars, 1990–1992," L. Corwin, I. Gelfand, J. Lepowsky (eds.), Birkhäuser, Boston-Basel-Berlin, 1993.

[16] J.L. Loday, Cyclic homology, Grundlehren Math. Wiss. **301**, 1992.

[17] S. Maclane, Categories for the working mathematician, Graduate Texts in Math. **5**, 1971.

[18] J.P. May, The geometry of iterated loop spaces, Lecture Notes in Math. **271**, 1972.

[19] D. Quillen, Homotopical algebra, Lecture Notes in Math. **43**, 1967.

[20] D. Quillen, On the (co-) homology of commutative rings, Proc. Sympos. Pure Math., Vol. XVII, 65–87.

[21] J.D. Stasheff, Homotopy associativity of H-spaces I, Trans. AMS **108** (1963), 275–292.

[22] B.L. Tsygan, The homology of matrix algebras over rings and Hochschild homology, Russian Math. Surveys **38** no. 2 (1983), 198–199.

The Coefficients of the Duistermaat-Heckman Polynomial and the Cohomology Ring of Reduced Spaces

Victor Guillemin
Department of Mathematics
Massachusetts Institute of Technology
Cambridge, MA 02139

Shlomo Sternberg
Department of Mathematics
Harvard University
Cambridge, MA 02138

1 Introduction

Let (M, ω) be a symplectic manifold of dimension $2d$, G a compact connected Lie group, $\tau : G \to \text{Diff}(M, \omega)$ a Hamiltonian action of G on M with moment map $\Phi : M \to g^*$ and $Z = \Phi^{-1}(0)$ the zero level set of the moment map. If 0 is a regular value of Φ, then the action of G on Z is locally free, and the reduced space

$$X = Z/G$$

is a symplectic orbifold of dimension $2(d - n)$ where $n = \dim G$. Let us assume that Φ is proper. Then X is compact, so its cohomology ring, $H^*(X, \mathbf{C})$ is finite dimensional and satisfies Poincaré duality. In the early 1980's, Kirwan, *[23]*, showed how to compute the Betti numbers of X, using Morse theoretic techniques, and recently, quite a bit of progress has been made on the much more difficult problem of understanding the cohomology ring structure of X. Some relevant papers are *[25]*, *[22]*, and *[20]*. (See also the survey paper, *[7]*.) In particular, Jeffrey and Kirwan have found a general formula for pairings of cohomology classes on X which enables them, in principle, to determine the ring structure of X when the ambient space, M is compact. (In practice, the problem of decoding the ring structure from their formula is non-trivial.)

The purpose of this note is to point out that a good deal of information about the cohomology ring structure of X can be extracted from the Duistermaat-Heckman

Geometry, Topology, and Physics for Raoul Bott

Cambridge, MA 02138 USA

theorem. The version of this theorem that we will use is the following: Let $G = T^n$ be the standard n-dimensional torus, and for $\xi \in g^*$ close to zero, set

$$Z_\xi = \Phi^{-1}(\xi) \quad \text{and} \quad X_\xi = Z_\xi / G. \tag{1}$$

Theorem 1.1: *As a differentiable manifold $X = X_\xi$ and*

$$[\mu_\xi] = [\mu] + \sum \xi_i c_i \tag{2}$$

where $[\mu_\xi]$ is the cohomology class of the symplectic form on X_ξ, $[\mu] = [\mu_0]$ is the cohomology class of symplectic form on X, and $c = (c_1, \ldots, c_n)$ is the Chern class of the fibration $Z \to X$.

Let

$$v(\xi) = \int_{X_\xi} \exp([\mu_\xi]) = \int_X \exp([\mu] + \sum \xi_i c_i) \tag{3}$$

be the symplectic volume of X_ξ. From (3) it follows that $v(\xi)$ is a polynomial of degree $d - n$ and that

$$\left(\frac{\partial^\alpha}{\partial \xi^\alpha} v\right)(0) = \frac{1}{k!} \int_X [\mu]^k \cdot c_1^{\alpha_1} \cdots c_n^{\alpha_n},$$
$$|\alpha| = \alpha_1 + \cdots + \alpha_n = d - n - k, \quad 0 \le k \le d - n. \tag{4}$$

In particular, if $|\alpha| = d - n$ then

$$\left(\frac{\partial^\alpha}{\partial \xi^\alpha} v\right)(0) = \int_X c_1^{\alpha_1} \cdots c_n^{\alpha_n} \tag{5}$$

Thus the coefficients of degree $d - n - k$ of the polynomial $v(\xi)$ determine the cohomology pairings (4) and, in particular, the leading coefficients determine the cohomology pairings (5).

The identities (4) and (5) are more or less well known, but they haven't been used very much as a tool for computing the ring structure of $H^*(X, \mathbf{C})$. Possibly this is because they give no information if the fibration $Z \to X$ is trivial, which is frequently the case. On the other hand, if the c_i generate the cohomology ring, one can read off from (5) all multiplicative relations of the form

$$\sum_{|\beta|+|\gamma|=d-n} a_\beta c^{\beta+\gamma} = 0, \quad |\beta| + |\gamma| = d - n,$$

and hence, by Poincaré duality, all relations of the form

$$\sum a_\beta c^\beta = 0, \quad 0 \le |\beta| \le d - n.$$

Writing the identity above as

$$Q\left(\frac{\partial}{\partial \xi}\right) \frac{\partial^\gamma}{\partial \xi^\gamma} v_0(\xi) = 0$$

where $v_0(\xi)$ is the homogeneous component of degree $d-n$ of v and $Q(x) = \sum a_\beta x^\beta$ one obtains:

Theorem 1.2: *If* $c_1, \ldots, c_n$ *generate* $H^*(X, \mathbf{C})$, *then* $H^*(X, \mathbf{C})$ *is isomorphic as an abstract ring to*

$$\mathbf{C}[x_1, \ldots, x_n]/\mathrm{ann}(v_0) \tag{6}$$

where $Q(x_1, \ldots, x_n) \in \mathrm{ann}(v_0)$ *if and only if* $Q(\frac{\partial}{\partial \xi_1}, \ldots, \frac{\partial}{\partial \xi_n})v_0(\xi) = 0$.

In this note we discuss two examples of applications of this theorem in the first of which X is the generic flag variety and in the second X is the toric variety associated with a simplicial fan. In neither of these case are our results new. The cohomology rings of the flag manifolds were determined by Borel in his classic paper, *[2]*, and for toric varieties by Danilov in *[8]*. Nevertheless it is interesting to note that the structure of $H^*(X, \mathbf{C})$ in both these examples is given by the general recipe (6).

Let us return to the case of an arbitrary (M, ω). Let p be a fixed point of the $G = T^n$ action on M, and let $\alpha_1, \ldots, \alpha_d$ be the weights of the isotropy representation of G on the tangent space at p. We will call p an *extremal* fixed point if there exists a $v \in g$ such that

$$\alpha_i(v) > 0, \ \forall i.$$

(See *[14]*, Section 3.) If p has this property, we will prove in Section 4 that if ξ is a regular value of Φ close to $\Phi(p)$ then X_ξ is a toric variety, and we will show how to compute its moment polytope in terms of the α_i. This has the following useful corollary. Suppose that M is compact and the fixed point set of G is finite. By the convexity theorem (*[1]* or *[16]*) the image, $\Delta = \Phi(M)$, is a convex polytope. Let $\Delta^0 \subset \Delta$ denote the set of regular values of Φ. The connected components, $\Delta_1, \ldots, \Delta_N$ of Δ^0 are themselves open convex polytopes. By the Duistermaat-Heckman theorem, the diffeotype of the reduced space, X_ξ is constant as ξ varies over each Δ_i. In particular, the cohomology ring of this reduced space depends only on Δ_i.

Theorem 1.3: *Suppose the closure of* Δ_i *contains a vertex of* Δ. *Then its associated reduced space is a toric variety.*

Thus for a Δ_i whose closure contains a vertex, we can compute the cohomology ring of its reduced space by theorem 1.2. In general, there will be many connected components of Δ^0 which *don't* have this property. This brings up the following interesting question: How does the cohomology ring of the reduced space, X_ξ, change, as ξ passes through a common $(n-1)$-dimensional face of two adjacent Δ_i's? Quite a lot is known about how the additive structure changes, cf. *[5]*, *[17]*, *[15]*, or *[19]*.

2 Flag Manifolds

Let K be a compact semi-simple Lie group and T its Cartan subgroup. The adjoint action of T on the lie algebra k has a T equivariant splitting

$$k = t \oplus t^{\perp}. \tag{7}$$

Here t is the Lie algebra of T and $t^{\perp}$ is its orthocomplement with respect to the Killing form. Using the right action of T on K we get the fibration

$$K \to K/T \tag{8}$$

mapping an element of K into its right T coset. This gives us a fibration of K over the flag variety K/T. The splitting (7) gives us a splitting of the tangent space to K at e into vertical and horizontal components, which can be extended via the left action of K to the whole of K, giving an intrinsic $K \times T$ invariant connection, θ, on the bundle (8). By minimal coupling, cf. *[24]*, one gets a presymplectic form

$$d\langle \theta, \mathrm{pr}_2 \rangle \tag{9}$$

on $K \times t^*$ where

$$\mathrm{pr}_2 : K \times t^* \to t^* \tag{10}$$

denotes projection onto the second factor. The presymplectic form given by (9) is symplectic on the set

$$K \times (t^*_+)^0, \tag{11}$$

where t^*_+ denotes the positive Weyl chamber and $(t^*_+)^0$ its interior. Letting $K \times T$ act trivially on the second factor in (11) gives a Hamiltonian action of $K \times T$ on $K \times (t^*_+)^0$. The T moment map for this action is (the restriction of) pr_2.

For $\xi \in (t^*_+)^0$ let O_ξ denote the coadjoint orbit of K through ξ. The stabilizer of ξ in K is T, so

$$O_\xi \sim K/T$$

as a K-homogeneous space. The reduction of $K \times (t^*_+)^0$ at ξ (with respect to T) is also K/T as a homogeneous K space. We claim

Proposition 2.1: *The reduced space of $K \times (t^*_+)^0$ with respect to T at ξ is isomorphic as a Hamiltonian K space to O_ξ.*

Proof: At the point $(e, \xi) \in K \times (t^*_+)^0$ the symplectic form defined by minimal coupling satisfies

$$\omega(x, y) = -\langle \xi, [x, y] \rangle, \quad x, y \in k/t. \tag{12}$$

But the expression on the right is precisely the Kirillov-Kostant form evaluated on $x, y \in k/t$ at $\xi \in O_\xi$.

Let c be the Chern class of the T fibration $K \to K/T$. By definition, this is a t-valued cohomology class of degree two which we can write as

$$c = \sum H_i \otimes c_i, \quad c_i \in H^2(K/T, \mathbf{R}),$$

where H_i is the standard Weyl basis of t.

Proposition 2.2: *The $c_1, \ldots, c_n$ generate the cohomology ring $H^*(K/T, \mathbf{C})$.*

Proof: See Borel *[2]* or Bott-Samelson *[6]*.

As a corollary, we see that the identities (5) completely determine the cohomology ring structure of K/T. To make these identities more explicit, we will use the following result, cf. *[4]* p. 232, giving a formula for the symplectic volume of a coadjoint orbit, and hence an explicit formula for the left hand sides of (4) and (5):

Proposition 2.3: *The symplectic volume, $v(\xi)$ of the coadjoint orbit, O_ξ is given by the formula*

$$v(\xi) = \frac{1}{\gamma} \prod_{\alpha \in \Pi +} (\xi, \alpha) \tag{13}$$

where α ranges over the positive roots, where (ξ, α) is the inner product of ξ and α with respect to the form induced on t^* by the negative of the Killing form and where the constant γ is given by

$$\gamma = \prod_{\alpha \in \Pi +} (\rho, \alpha), \quad \rho = \frac{1}{2} \sum_{\alpha \in \Pi +} \alpha . \tag{14}$$

Notice that the right hand side of (13) is a polynomial in ξ of degree $d - n$ as required, as $2d = \dim K + n$, $\dim K = 2p + n$ where p denotes the number of positive roots. Plugging (13) into (5) gives

$$\int c_1^{\beta_1} \cdots c_n^{\beta_n} = \frac{1}{\gamma} \frac{\partial^{|\beta|}}{\partial \xi^\beta} \prod_{\alpha \in \Pi +} (\xi, \alpha). \tag{15}$$

We will say a few words about how these results are related to the theorem of Borel which we mentioned above. Borel's theorem says that the cohomology ring of K/T is isomorphic to the ring

$$\mathbf{C}[x_1, \ldots, x_n] / \mathbf{C}[x_1, \ldots, x_n]^W , \tag{16}$$

$\mathbf{C}[x_1, \ldots, x_n]^W$ being the ideal generated by the Weyl group invariant polynomials of degree greater than zero. To deduce this from theorem 1.2, one has to show that if $Q(x_1, \ldots, x_n)$ is a homogeneous Weyl group invariant polynomial of degree greater than zero and $v(\xi)$ is the function (12) then

$$Q\left(\frac{\partial}{\partial \xi}\right) v(\xi) = 0 .$$

Notice, however, that if α is a simple root and $\sigma_\alpha : t \to t$ is reflection through the hyperplane $\alpha(\xi) = 0$, then σ_α maps all of the positive roots except α into themselves and maps α onto $-\alpha$. Hence

$$\sigma_\alpha^* v(\xi) = -v(\xi) ,$$

and since Q is Weyl group invariant,

$$\sigma_\alpha^* Q\left(\frac{\partial}{\partial\xi}\right) v(\xi) = -v(\xi) .$$

In particular, $Q\left(\frac{\partial}{\partial\xi}\right) v(\xi)$ vanishes on the hyperplane $\alpha(\xi) = 0$, so the monomial $\alpha(\xi)$ divides $Q\left(\frac{\partial}{\partial\xi}\right) v(\xi)$. Since every positive root is Weyl group conjugate to a simple root, it follows that, for every root α, $\alpha(\xi)$ divides $Q\left(\frac{\partial}{\partial\xi}\right) v(\xi)$. Hence $v(\xi)$ itself divides $Q\left(\frac{\partial}{\partial\xi}\right) v(\xi)$, and this is impossible unless $Q\left(\frac{\partial}{\partial\xi}\right) v(\xi) = 0$.

Remark: For this argument we are indebted to David Vogan.

3 Delzant Spaces

We will refer to the toric varieties in this section as Delzant spaces since we will be thinking of them as symplectic manifolds (or orbifolds) rather than as complex projective varieties. Let us briefly review the definition of these spaces as given by Delzant *[9]*. Let V be a real n-dimensional vector space and $L \subset V$ and n-dimensional lattice, with dual space V^* and dual lattice L^*. Let T be the torus $T = V/L$. Given a convex polytope, $\Delta \subset V^*$ satisfying certain axioms listed below, we will associate a symplectic orbifold, X_Δ of dimension $2n$ to Δ. We will equip X_Δ with a Hamiltonian T action so that Δ is the image of its moment map. By the uniqueness theorem of Delzant this characterizes X_Δ up to isomorphism.

The assumptions about Δ are the following:

1. Exactly n edges of Δ meet at every vertex.

2. The edges meeting at any vertex p point in rational directions (relative to L^*), i.e. lie along half rays of the form

$$p + t\beta_i, \quad t \geq 0, \quad \beta_i \in L^*, \quad i = 1, \ldots, n.$$

3. The β_i, $i = 1, \ldots, n$, form a basis of V^*.

If one wants X_Δ to be a manifold rather than merely an orbifold, condition **3** should be replaced by the stronger

3*. The β_i, $i = 1, \ldots, n$, form a basis of L^*.

Let d denote the number of $(n-1)$-dimensional faces of Δ. Condition **2** implies that these faces can be defined by equations of the form

$$\langle \xi, u_i \rangle = \lambda_i \quad i = 1, \ldots, d,$$

where the vector u_i belongs to L. The vector u_i can be normalized be requiring it to be a primitive element of L, which then determines it up to sign. The sign can be fixed by requiring that Δ be contained in the half space

$$\langle \xi, u_i \rangle \geq \lambda_i, \quad i = 1, \ldots, d. \tag{17}$$

These normalizations determine the u_i and λ_i uniquely. Conversely, Δ is given as the intersection of the half spaces (17).

Define the map

$$\pi : \mathbf{R}^d \to V, \quad \pi : e_i \mapsto u_i \tag{18}$$

where $\{e_1, \ldots, e_d\}$ is the standard basis of $\mathbf{R}^d$. Since

$$\pi : \mathbf{Z}^d \to L,$$

we get an induced map

$$\Pi : T^d \to T, \quad T^d = \mathbf{R}^d/\mathbf{Z}^\mathbf{d}. \tag{19}$$

We have the standard linear action, ρ, of the torus t^d on $\mathbf{C}^d$ given by

$$\rho(x)z = (e^{2\pi i x_1} z_1, \ldots, e^{2\pi i x_d} z_d), \tag{20}$$

where

$$x = (x_1, \ldots, x_d) \in T^d \quad \text{and} \quad z = (z_1, \ldots, z_d) \in \mathbf{C}^d.$$

This action preserves the symplectic form

$$-i \sum_j dz_j \wedge d\overline{z}_j$$

and is a Hamiltonian action with moment map

$$J(z) = \frac{1}{2} \sum |z_j|^2 e_j^*. \tag{21}$$

Let G be the kernel of the map Π (cf. (19). Restricting ρ to G gives a Hamiltonian action of G on $\mathbf{C}^d$ with moment map

$$\Phi(z) = \frac{1}{2} \sum |z_i|^2 \alpha_i \tag{22}$$

where each α_i is the weight of G on the one-dimensional subspace of $\mathbf{C}^d$ spanned by e_i, $i = 1, \ldots, d$. Set

$$\lambda = \sum \lambda_i \alpha_i$$

where the λ_i are the same as those entering into the description (17) of Δ. Let Z_λ be the λ level set of the map Φ

$$Z_\lambda = \{z : \sum |z_i|^2 \alpha_i = \lambda\}.$$

We refer to *[9]* or *[12]* for the following:

Proposition 3.1: *If Δ satisfies conditions* **1-3** *the action of G on Z_λ is locally free. If, in addition,* **3*** *holds, then this action is free.*

We now define

$$X_\Delta = Z_\lambda / G. \tag{23}$$

By the proposition, this is an orbifold, and is a manifold if **3*** holds. Since G is the kernel of Π, the group T acts on X_Δ in Hamiltonian fashion. We refer again to *[9]* or *[12]* for a proof of

Proposition 3.2: *The image of the moment map for the T action on X_Δ is the polytope Δ.*

The aim of this section is to give a description of the cohomology ring of X_Δ along the lines indicated in the introduction. To this end we need to determine the Chern class of the fibration

$$Z_\lambda \to X_\Delta.$$

Since Z_λ is a principal G bundle, a one-dimensional representation of G gives rise to a line bundle over X_Δ. Let L_i be the line bundle associated to the representation of G on the one-dimensional subspace of $\mathbf{C}^d$ spanned by e_i (with weight α_i). Let c_i be the corresponding Chern class.

Proposition 3.3: *The $\mathbf{R}^d$-valued cohomology class*

$$c = (c_1, \ldots, c_d)$$

takes values in $\mathfrak{g}$, the Lie algebra of G, and is the Chern class of the fibration $Z_\lambda \to X_\Delta$.

Proof: We content ourselves with proving the first of these two assertions. We must show that if $\kappa = (\kappa_1, \ldots, \kappa_d) \in (\mathbf{R}^d)^*$ lies in the annihilator space, $\mathfrak{g}^0$ of $\mathfrak{g}$ then $\sum \kappa_i c_i = 0$. It is enough to prove this when the $\kappa_1, \ldots, \kappa_d$ are integers. But then $\sum \kappa_i c_i$ is the Chern class of the line bundle

$$L_1^{\kappa_1} \otimes \cdots \otimes L_d^{\kappa_d}$$

which is the line bundle associated with the weight $\sum \kappa_i c_i$. But

$$\kappa \in \mathfrak{g}^0 \iff \sum \kappa_i \alpha_i = 0,$$

so this line bundle is the trivial bundle and hence its Chern class is zero. QED. For the following see *[8]* Section 11:

Proposition 3.4: *The Chern classes c_i generate the cohomology ring $H^*(X_\Delta, \mathbf{C})$.*

If we now let λ vary slightly about a fixed value, the symplectic volume varies, and we can apply (5). Once again, to apply this formula we need an alternative computation of the symplectic volume: Here we have a strikingly simple description, for whose proof we refer to *[13]*, Section 6:

Theorem 3.1: *The symplectic volume of X_Δ is the Euclidean volume of the polytope Δ.*

So if we let $v(\lambda)$ denote this volume, we can write (5) as

$$\frac{\partial^\beta v(\lambda)}{\partial \lambda^\beta} = \int_{X_\Delta} c_1^{\beta_1} \cdots c_d^{\beta_d} \tag{24}$$

for any multi-index β with $|\beta| = n$. (Note that since $v(\lambda)$ is a polynomial of degree n the left hand side of (24) is a constant.)

We will briefly describe the tie-in between this result and the result of Danilov which we alluded to in the introduction. Let $\mathfrak{g}$ be the Lie algebra of G and ι the inclusion of $\mathfrak{g}$ into $\mathbf{R}^n$. From (16) one gets an exact sequence

$$0 \leftarrow \mathfrak{g}^* \overset{\iota^*}{\leftarrow} (\mathbf{R}^d)^* \overset{\pi^*}{\leftarrow} V^* \leftarrow 0 \,.$$

For λ and λ' in $\mathbf{R}^d$ let Δ and Δ' be the polytopes defined by

$$\langle \xi, u_i \rangle \geq \lambda_i \quad \text{and} \quad \langle \xi, u_i \rangle \geq \lambda_i' \,.$$

We leave as an exercise the following result:

Proposition 3.6: *Δ and Δ' are congruent if and only if $\lambda - \lambda' \in \ker \iota^*$ in which case*

$$\Delta' = \Delta + v$$

where $\lambda - \lambda' = \pi^* v$.

From proposition 3.5 one gets n first order partial differential equations satisfied by $v(\lambda)$ namely

$$D_{w_i} v(\lambda) = 0 \,, \; i = 1, \ldots, n \,, \tag{25}$$

where $w_i = \pi^* v_i$ and v_i, $i = 1, \ldots, n$, is a basis of V; and from these equations one gets n homogeneous generators of degree two of the ideal, $\text{ann}(v)$. However, in addition to these generators, there are also some generators of higher order. Namely by (17) one can associate with each of the standard basis vectors, e_i, of $\mathbf{R}^d$ the $(n-1)$-dimensional face of Δ defined by

$$\langle \xi, u_i \rangle = \lambda_i \,.$$

Denoting this face by F_i it is not hard to show that for a multi-index, $I = (i_1, \ldots, i_k)$, $1 \leq i_1 < i_2 < \ldots < i_k \leq d$,

$$\frac{\partial}{\partial \lambda_I} v(\lambda) = 0 \tag{26}$$

if and only if $F_{i_1} \cap \ldots \cap F_{i_k} = \emptyset$. We claim (but will not attempt to prove here):

Theorem 3.2: *$ann(v)$ is generated by the n generators of degree two associated with the n equations (25) and the generators of degree $2|I|$ associated with the equations (26).*

Combining this with theorem 1.2 one ends up with the theorem of Danilov:

Theorem 3.3: *$H^*(X_\Delta, \mathbf{C})$ is a ring with d generators (each of degree 2) satisfying the relations (25) and (26).*

4 Reduction: The Linear Case

Let V be a complex d-dimensional vector space equipped with a positive definite Hermitian form, H and let ω be the symplectic form $\omega = \operatorname{Im} H$. Let G be an r-dimensional torus and ρ a unitary representation of G on V. We can ignore the linear structure on V and consider ρ as a Hamiltonian action. In this section we discuss what reduced spaces look like in this linear setting. Let $e_1, \ldots, e_d$ be an orthonormal basis of V with the property that the one-dimensional subspaces spanned by the e_i are G-invariant. Each such subspace has an associated weight, call it α_i. This basis gives us a coordinate system $(z_1, \ldots, z_d)$ on V.

Proposition 4.1: *The moment map for the action ρ is*

$$\Phi : z \mapsto \sum |z_i|^2 \alpha_i. \tag{27}$$

We will henceforth assume that the action of G is effective. (This implies that the weights α_i span $\mathfrak{g}^*$.) For the following see *[14]*, Section 3:

Proposition 4.2: *The moment map (27) is proper if and only if there exists a vector $v \in g$ for which $\alpha_i(v) > 0$, $\forall i$.*

The action ρ is said to be *quasifree* if, for all $p \in V$ the stabilizer group, G_p is either of dimension greater than zero or is trivial. It turns out that for linear actions quasifree is equivalent to the condition that G_p be connected for every $p \in G$; cf. *[10]*. We also refer to *[10]* for the following

Proposition 4.3: *ρ is quasi-free if and only if every r-element subset of $\{\alpha_1, \ldots, \alpha_d\}$ which spans $\mathfrak{g}^*$ is also a set of generators for the weight lattice of G.*

It is also clear from (27) that the image of Φ is the cone

$$\{x_1\alpha_1 + \cdots + x_d\alpha_d;\ \ x_i \geq 0\}. \tag{28}$$

For the following more refined result see *[14]*, Prop.3.28,

Proposition 4.4: *If $\alpha_{i_1}, \ldots, \alpha_{i_m}$ span a proper subspace of $\mathfrak{g}^*$ every point on the cone*

$$\{\sum y_q \alpha_{i_q};\ \ y_q \geq 0\}$$

is a critical value of the moment map (27). Conversely every critical value lies on one of these cones.

Now let λ be a regular value of the moment map. If the moment map is proper, the level set

$$Z_\lambda = \Phi^{-1}(\lambda)$$

is a compact submanifold of V. If the action is also quasi-free, the action of G on Z_λ is free, and the reduced space

$$X_\lambda = Z_\lambda / G \tag{29}$$

is a compact symplectic manifold. We will prove that this manifold is a Delzant space (of the type considered in the preceding section): Let $\mathbf{R}^d_+$ denote the positive orthant

$$\{(x_1, \ldots, x_d),\ \ x_i \geq 0\}$$

and let

$$\tau : \mathbf{R}^d \to \mathfrak{g}^*$$

be the linear map sending the ith standard basis element into α_i. Let

$$\Delta_\lambda = \tau^{-1}(\lambda) \cap \mathbf{R}^d_+.$$

Theorem 4.1: *X_λ is a Delzant space and its moment polytope is Δ_λ.*

Proof: In terms of our coordinates, the representation given by (20) is a representation of T^d on V and we can regard G as a subgroup of T^d. The representation given by (20) restricts to the given representation of G, so there is no harm in denoting this extended representation also by ρ. This action of T^d commutes with the action of G and induces a Hamiltonian action of T^d/G on X_λ. Now T^d acts freely near any point $z \in V$ if all its coordinates $z_i \neq 0$, and hence T^d acts freely on an open dense subset of Z_λ. Consequently T^d/G act freely on an open dense subset of X_λ. Since $2n = \dim X_\lambda = 2\dim T^d/G$ where $n = d - r$ the action of T^d/G is a Delzant action, cf. *[12]*. The computation of its moment polytope, which we will omit, involves staring carefully at the Delzant construction which we outlined in the preceding section. QED.

Finally, let (M, ω) be an arbitrary Hamiltonian G-space with moment map $\Phi : M \to g^*$. Let p be an extremal fixed point of G. Near the point p the action of G is isomorphic to the linear isotropy action of G on the tangent space T_p by the equivariant Darboux theorem. Therefore, taking $V = T_p$, the conclusions of the preceding theorem are valid for regular values, λ, of Φ providing that λ is sufficiently close to $\Phi(p)$. This proves the assertion we made in the introduction about reduced spaces associated with such values. In particular, for such values λ we can compute the cohomology ring of X_λ by the algorithm described in Section 3.

References

[1] M. Atiyah, Convexity and commuting Hamiltonians, Bull. Lon. Math. Soc. **14** (1982), 1–15.

[2] A. Borel, Sur la cohomologie des espaces fibrés principaux et des espaces homogènes des groupes de Lie compacts, Ann. of Math. **57** (1953), 115–207.

[3] I.N. Bernstein, I.M. Gelfand, and S.I. Gelfand Schubert, Cells and the cohomology of the spaces G/P, Russian Math. Surveys (1972), 169–174.

[4] N. Berline, E. Getzler, and M. Vergne, *Heat Kernels and Dirac Operators*, Springer-Verlag, Heidelberg (1991).

[5] M. Brion and C. Procesi, Action d'un tore dans une variété projective, in: *Operator Algebras, Unitary Representations Enveloping Algebras, and Invariant Theory*, 509–539, Birkhäuser (1989).

[6] R. Bott and H. Samelson, The cohomology ring of G/T, Proc. NAS **41** (1955) ,490–493.

[7] J.J. Duistermaat, Equivariant cohomology and stationary phase, to appear in: *Proc. of the conference on symplectic geometry and its applications, Keio U.*, June, 1993.

[8] V.I. Danilov, The geometry of toric varieties, Russian Math. Surveys **33** (1978), 97–154.

[9] T. Delzant, Hamiltoniens périodiques et image convexe de l'application moment, Bull. Soc. Math. France **116** (1988), 315–339.

[10] R. da Sousa,V. Guillemin, and E. Prato, Consequences of quasi-free, Ann. Global Anal. Geom. **8** (1990), 77–85.

[11] J.J. Duistermaat and G. Heckman, On the variation in the cohomology of the symplectic form of the reduced phase space, Invent. Math. **69** (1982), 259–269.

[12] V. Guillemin, Woudschoten lectures on geometric multiplicity theory, M.I.T. preprint (1992).

[13] V. Guillemin, Kaehler structures on toric varieties, J. Diff. Geom., **40**, 285–309.

[14] V. Guillemin, E. Lerman, and S.Sternberg, On the Kostant multiplicity formula, Jour. Geom. Phy. **5** (1988), 2–30.

[15] R.M. Goresky and R.D. MacPherson, On the topology of algebraic torus actions, Lecture Notes in Math. **1271** (1986), 73–90.

[16] V. Guillemin and S. Sternberg, Convexity properties of the moment map, Invent. Math. **67** (1982), 491–513.

[17] V. Guillemin and S. Sternberg, Birational equivalence in the symplectic category, Invent. Math. **97** (1989), 485–522.

[18] V. Guillemin and S. Sternberg, *Symplectic Techniques in Physics*, Cambridge University Press, Cambridge, New York (1984).

[19] Y. Hu, The geometry and topology of quotient varieties, Thesis M.I.T. (1991).

[20] L. Jeffrey and F. Kirwan, Localization for non-abelian group actions, Technical report, Balliol College, Oxford (May 1993).

[21] B. Kostant, Lie algebra cohomology and generalized Schubert cells, Ann. of Math. **77** (1963), 72–144.

[22] J. Kalkman, Cohomology rings of symplectic quotients, Technical report 795 Math. Inst. Univ. Utrecht, (April 1993).

[23] F. Kirwan, *Cohomology of quotients in symplectic and algebraic geometry*, Princeton Univ. Press (1984).

[24] S. Sternberg, On minimal coupling and the symplectic mechanics of a classical particle in the presence of a Yang-Mills field, Proc. Nat. Acad. Sci. **74** (1977), 5253–4.

[25] E. Witten, Two-dimensional gauge theories revisited, J. Geom. Phys. **9** (1992), 303–368.

Hyperpolar Actions on Symmetric Spaces

Ernst Heintze
Institut für Mathematik
Universität Augsburg
D-86159 Augsburg, Germany

Richard S. Palais
Department of Mathematics
Brandeis University
Waltham, MA 02254-9110

Chuu-lian Terng[1]
Department of Mathematics
Northeastern University
Boston, MA 02115

Gudlaugur Thorbergsson[2]
Mathematisches Institut
Universität Köln
Weyertal 86-90
D-50931 Köln, Germany

1 Introduction

An isometric action of a compact Lie group G on a Riemannian manifold M is called *polar* if there exists a closed, connected submanifold Σ of M that meets all G-orbits and meets orthogonally. Such a Σ is called a *Section*. A section is automatically totally geodesic in M, and if it is also flat in the induced metric then the action is called *hyperpolar*. In this paper we study hyperpolar actions on compact symmetric spaces, prove some structure and classification theorems for them, and study their relation to polar actions on Hilbert space and to involutions of affine Kac-Moody algebras.

[1] research support in part by NSF grant DMS 9304285
[2] research support in part by NSF grant DMS 9100179.

Geometry, Topology, and Physics for Raoul Bott

Cambridge, MA 02138 USA

The origin of the theory of hyperpolar actions can be traced back at least to the 1950's, with Bott *[4]* and Bott and Samelson's *[5]* introduction of the closely related concept of a "variationally complete" action. We will give a brief review below of what we know of the history of this subject, but first we recall some basic notation and terminology from the theory of symmetric spaces that will be needed throughout the paper.

Let G be a compact, semi-simple, connected Lie group, and $\mathfrak{g} = TG_e$ its Lie algebra. A subgroup K of G is called a *symmetric subgroup* of G (and the pair (G, K) is called a *symmetric pair*) if there is an involution σ of G such that $(K_\sigma)_0 \subset K \subset K_\sigma$, where K_σ is the fixed point set of σ and $(K_\sigma)_0$ is the connected component of K_σ. In this case G/K equipped with a G-invariant metric induced from any Ad_G invariant inner product for $\mathfrak{g}$ is a *symmetric space.* The corresponding orthogonal decomposition $\mathfrak{g} = \mathfrak{K} \oplus \mathfrak{P}$ is called the *Cartan decomposition* associated to σ. It is completely determined by σ (independent of the inner product on $\mathfrak{g}$); namely $\mathfrak{K}$ and $\mathfrak{P}$ are respectively the $+1$ and -1 eigenspaces of $d\sigma_e : \mathfrak{g} \to \mathfrak{g}$. It follows that $\mathfrak{K}$ and $\mathfrak{P}$ satisfy the characteristic bracket relations $[\mathfrak{K}, \mathfrak{K}] \subseteq \mathfrak{K}$, $[\mathfrak{K}, \mathfrak{P}] \subseteq \mathfrak{P}$, and $[\mathfrak{P}, \mathfrak{P}] \subseteq \mathfrak{K}$.

Definition 1.1: If (G, K) is a symmetric pair and $\mathfrak{g} = \mathfrak{K} \oplus \mathfrak{P}$ the corresponding Cartan decomposition, then the Adjoint representation of K on $\mathfrak{P}$ is called an *s-representation.*

Of course, up to equivalence the s-representation can be identified with the isotropy representation of the symmetric space G/K at eK (from which it gets its name).

Remark 1.2: It is known (cf. *[15, 21]*) that if $M = G/K$ is a Riemannian symmetric space corresponding to a symmetric pair (G, K), then G is the connected component $\mathrm{Isom}_0(M)$ of the group of all isometries of M.

Remark 1.3: A compact Lie group G with a bi-invariant metric is a symmetric space, and $(G \times G, \Delta(G))$ is a symmetric pair. For $G \times G$ acts transitively on G by $(g_1, g_2) \cdot x = g_1 x g_2^{-1}$ and the isotropy group at e is $\Delta(G) = \{(g, g) \mid g \in G\}$; and $\sigma(x, y) = (y, x)$ is an involution of $G \times G$ with fixed point set $\Delta(G)$. These compact symmetric spaces are referred to as being of Type II or "group type".

Remark 1.4: There is a well-known criterion for a Riemannian manifold M to be symmetric, namely the existence for each p in M of an isometry ϕ that fixes p and satisfies $d\phi_p(v) = -v$ for all v in TM_p. However it does *not* then necessarily follow that if G is a transitive group of isometries of M and $K = G_p$ then (G, K) must be a symmetric pair. By 1.2 this is only so if $G = \mathrm{Isom}_0(M)$. For example, two possible ways of representing the symmetric space S^{2n-1} are as $SO(2n)/SO(2n-1)$ and as $SU(n)/SU(n-1)$. And while $(SO(2n), SO(2n-1))$ is a symmetric pair, $(SU(n), SU(n-1))$ is not. This distinction between symmetric space and symmetric pair will be important in what follows.

Remark 1.5: We have been slightly sloppy in the above remarks and, to avoid long and awkward statements of theorems, we will continue to allow ourselves the liberty of the same slight imprecision that we now describe. In general we can *not* really regard G as a group of diffeomorphisms of $M = G/K$, because the action of G on G/K may not be effective. The kernel, N, of this action is the intersection of all the conjugates gKg^{-1} of K, i.e., the largest subgroup of K that is normal in G, and this can be non-trivial. But for most purposes we can simply replace G by G/N and K by K/N. For example, if (G, K) is a symmetric pair then so is $(G/N, K/N)$, and what 1.2 really means is that $G/N = \mathrm{Isom}_0(M)$. (A typical example of such behavior is $M = S^n = Spin(n+1)/Spin(n) = SO(n+1)/SO(n)$.) In particular, if $Z(G)$ denotes the center of G, then $\Delta(Z(G))$ is the kernel of the action of $G \times G$ on G, so a connected group of isometries of G should really be regarded as a subgroup of $G \times G/\Delta(Z(G))$. Nevertheless, we will frequently speak as if any group of isometries of G were a subgroup of $G \times G$.

We now turn to our historical remarks on the origins of the theory of polar and hyperpolar actions, which as we said begins with the papers *[4]* and *[5]*. First we recall the definition of variational completeness. Let M be a complete Riemannian G-manifold. A geodesic γ in M is called *G-transversal* if $\gamma(t)$ is perpendicular to the orbit $G \cdot \gamma(t)$ for all t, and a Jacobi field J along a G-transversal geodesic γ is *G-transversal* if it is the deformation of G-transversal geodesics. The G-action on M is called *variationally complete* if every G-transversal Jacobi field J along γ satisfying the condition that there exists $0 < t_0$ such that $J(t)$ is tangent to the G-orbit at $\gamma(t)$ for $t = 0$ and $t = t_0$, is of the form $J(t) = \xi(\gamma(t))$ for some Killing vector field $\xi \in \mathfrak{g}$. Bott and Samelson also gave an informal definition: "Intuitively we like to think of variational completeness as an absence of conjugate points in the orbit space M/G".

Now let X be a closed submanifold of $M \times M$, $P(M, X)$ the Hilbert manifold defined by

$$P(M,X) = \{\gamma \in H^1([0,1], M) \mid (\gamma(0), \gamma(1)) \in X\},$$

and $E : P(M, X) \to R$ the energy functional

$$E(\gamma) = \int_0^1 \| \gamma'(t) \|^2 dt.$$

The first main theorem in *[5]* (Theorem I, page 976) can be stated as follows:

Theorem 1.6: *([5]) Suppose the action of G on M is variationally complete, $x_0 \in M$ and N is a G-orbit. Then the energy functional $E : P(M, \{x_0\} \times N) \to R$ is a perfect Morse function and a homology basis can be constructed explicitly.*

The second main theorem in *[5]* (Theorem II, page 986) states that:

Theorem 1.7: *([5]) If (G, K) is a symmetric pair with Cartan decomposition $\mathfrak{g} = \mathfrak{K} \oplus \mathfrak{P}$, then three natural actions associated to (G, K) are variationally complete;*

namely the action of K on the symmetric space G/K, the action of $K \times K$ on G, and finally the s-representation of K on $\mathfrak{P}$ (or equivalently the isotropy representation of G/K).

Shortly after, in *[16]*, R. Hermann generalized Theorem 1.7 as follows:

Theorem 1.8: *([16]) If K_1 and K_2 are two symmetric subgroups of the same compact Lie group G, then the action of K_1 on G/K_2 is variationally complete and so is the action of $K_1 \times K_2$ on G.*

A couple of years later still, L. Conlon in *[6]* associated to each automorphism σ of G, the subgroup $G(\sigma) := \{(g, \sigma(g)) \mid g \in G\}$ of $G \times G$. As a subgroup of $G \times G = \mathrm{Isom}_0(G)$, $G(\sigma)$ acts isometrically on G. As a consequence of Hermann's theorem, Conlon obtained:

Corollary 1.9: *([6]) Let G be a compact, connected Lie group with a bi-invariant metric and let σ be any automorphism of G. Then $G(\sigma)$ is a symmetric subgroup of $G \times G$, and hence the action of $G(\sigma)$ on G is variationally complete.*

Proof: The map $(x, y) \mapsto (\sigma^{-1}(y), \sigma(x))$ is clearly an involution of $G \times G$, and its fixed point set is $G(\sigma)$. In particular $G(\mathrm{id}) = \Delta(G)$, and since $G \times G/\Delta(G) = G$, we can apply Hermann's theorem to the symmetric subgroups $K_1 = G(\sigma)$ and $K_2 = \Delta(G)$ of $G \times G$.

Moreover, Conlon noted that all the known examples of variationally complete actions admit flat submanifolds that meet every orbit orthogonally. He called these flat submanifolds *K-transversal domains*, and the existence of such is exactly our definition of an action being hyperpolar. He proved:

Theorem 1.10: *([8]) If an isometric action of G on M is hyperpolar then it is variationally complete.*

Theorem 1.10 is generalized in *[27]* where it is shown that if the action of G on M is polar and sections have no conjugate points then the action is variationally complete. This agrees with the intuitive definition of variational completeness mentioned above.

Conlon also made some progress on the classification of hyperpolar actions on Euclidean space (*[9]*), which was finally completed by J. Dadok (*[10]*):

Theorem 1.11: *([10]) Suppose H is a compact, connected Lie group, and $\rho : H \to SO(n)$ is a representation such that the action of H on R^n is hyperpolar. Then there exist a symmetric space G/K and a linear isometry $A : R^n \to \mathfrak{P}$ such that A maps H-orbits isometrically onto K-orbits of the s-representation on $\mathfrak{P}$ (where $\mathfrak{g} = \mathfrak{K} + \mathfrak{P}$ is the Cartan decomposition).*

Remark 1.12: Let N be a K-orbit in $\mathfrak{P}$ for the s-representation of K on $\mathfrak{P}$, and $x_0 \in \mathfrak{P}$. Then N is a generalized (partial) flag manifold. Identifying N as the submanifold of geodesic in $\mathfrak{P}$ joining x_0 to a point in N, then the map defined by $r(\gamma) = \gamma(1)$ is a deformation retract of the path space $P(\mathfrak{P}, \{x_0\} \times N)$ to N and the energy functional E on $P(\mathfrak{P}, \{x_0\} \times N)$ restricted to N is the square of the Euclidean distance function $f_{x_0}(x) = \| x - x_0 \|^2$. It is known that $f_{x_0} : N \to R$ is a perfect Morse function, and the images of the Morse cycles explicitly constructed by Bott and Samelson under the map r are in fact the Schubert cycles.

It is known that (cf. *[12, 25, 27]*) if N is an orbit of the action of $G(\sigma)$ on G for some automorphism σ on G, then the path space $P(G, \{e\} \times N)$ is an Adjoint orbit of the affine Kac-Moody group associated to σ, it can be viewed as an infinite dimensonal complex (partial) flag manifold, the energy function E on the path space corresponds to the distance function in the affine Kac-Moody algebra, and the Morse cycles constructed for E are analogues of the Schubert cycles. In this paper, we prove that if K_1, K_2 are symmetric subgroups of G and N is an orbit of the $K_1 \times K_2$-action on G then the path space $P(G, \{e\} \times N)$ is an infinite dimensional analogue of the finite dimensional s-representations, i.e., there exist an affine algebra $\mathfrak{U}$ and an involution σ on $\mathfrak{U}$ such that $\mathfrak{U}$ is the direct sum of the $+1$-eigenspace $\hat{\mathfrak{K}}$ and -1-eigenspace $\hat{\mathfrak{P}}$ of σ, the Adjoint action of $\hat{K}$ leaves $\hat{\mathfrak{P}}$ invariant, and $P(G, \{e\} \times N)$ equipped with the Sobolev H^1-metric is isometric to an orbit of $\hat{K}$ on $\hat{\mathfrak{P}}$ via the Adjoint action. Moreover, $P(G, \{e\} \times N)$ has the analogous geometric and topological properties as stated in Remark 1.12.

Next we recall the definition of infinite dimensional hyperpolar action (*[28]*). An action of a Hilbert Lie group $\mathfrak{G}$ on a Hilbert space V is called *proper* if the map $\mathfrak{G} \times V \to V \times V$ defined by $(g, v) \mapsto (g \cdot v, v)$ is proper, i.e., the preimage of a compact set is compact. The $\mathfrak{G}$-action on V is called *Fredholm* if given any $v \in V$ the orbit map $\mathfrak{G} \to V$ defined by $g \mapsto g \cdot v$ is a Fredholm map, i.e., the differential at every point g is a linear Fredholm operator. A proper, Fredholm, isometric $\mathfrak{G}$-action on V is called *hyperpolar* if there exists a closed affine subspace Σ of V that meets every $\mathfrak{G}$-orbit orthogonally. It is known that if G is a compact, semi-simple Lie group and H is a closed subgroup such that the action of H on G is hyperpolar then the action of $P(G, H)$ on the Hilbert space $H^0([0.1], \mathfrak{g})$ by gauge transformations is hyperpolar. Moreover, identifying $P(G, \{e\} \times G)$ with $H^0([0, 1], \mathfrak{g})$ via $g \mapsto g^{-1}g'$, the path space $P(G, \{e\} \times N)$ is identified with a $P(G, H)$-orbit, where N is an H-orbit N in G.

The above considerations naturally lead us to study the following closely related problems:

(1) the classification of hyperpolar actions on compact symmetric spaces,

(2) the geometry and topology of orbits of hyperpolar actions on compact symmetric spaces,

(3) the classification of hyperpolar actions on Hilbert space,

(4) the classification of isometric involutions of affine Kac-Moody algebras.

One of the main results of this paper is to reduce (1) to a problem in Lie algebra. First we prove that the action of a closed subgroup H of G on the symmetric space G/K is hyperpolar if and only if the action of $H \times K$ on G is hyperpolar. This reduces (1) to the classification of hyperpolar actions on compact

Lie groups. It is known (*[23]*) that if the action of H on G is hyperpolar then the slice representation of H_e on the normal space $\nu(H \cdot e)_e$ is hyperpolar with an abelian subalgebra as a section. Although the converse is not in general true, we prove that it is true under the assumption that G is a compact, semi-simple Lie group equipped with the bi-invariant metric induced from the Killing form on $\mathfrak{g}$. In fact, we prove that the action of a closed subgroup H of $G \times G$ on G is hyperpolar if the Lie algebra $\mathfrak{H}$ of H satisfies the following condition (A):

(A) There exists $g_0 \in G$ such that the orthogonal complement of $\{g_0^{-1} x g_0 - y \mid (x, y) \in \mathfrak{H}\}$ in $\mathfrak{g}$ is an abelian subalgebra.

In particular, this proves that a cohomogeneity one action on G is automatically hyperpolar. Although this reduces (1) to the problem of finding all subalgebras $\mathfrak{H}$ of $\mathfrak{g} + \mathfrak{g}$ satisfying condition (A), it remains a difficult open problem. To suggest the level of complexity of this problem, we note that finding all subgroups H_1 of G such that the action of $H_1 \times H_1$ on G has cohomogeneity one is equivalent to the problem of finding all the two-point homogeneous spaces.

There has been lot of progress made for problem (2) (for details see *[29]*). We first note that the submanifold geometry of one principal orbit determines the whole orbit foliation of the hyperpolar action, and a principal orbit N of a hyperpolar action on a symmetric space X satisfies the following conditions:

(a) the normal bundle $\nu(N)$ is globally flat and $\exp(\nu(N)_x)$ is contained in some flat of X for all $x \in N$,

(b) if v is a parallel normal field on N then $\Gamma(v(x)) = \Gamma(v(y))$ for all $x, y \in N$, where $\Gamma(v(x))$ is the focal data at x, defined as the set of all pairs (t, k) such that $\exp(tv(x))$ is a multiplicity k focal point of N with respect to x.

A submanifold N of a symmetric space X is called *equifocal* if N satisfies the conditions (a) and (b) above. It is proved in *[29]* that there are equifocal submanifolds of symmetric spaces that are not orbits of any isometric action, and Theorem 1.6 holds for general equifocal submanifold in compact symmetric space. In fact, we proved that if N is an equifocal submanifold of a compact symmetric space X and $x_0 \in X$ is a non-focal point of N, then there exist a point $y \in N$ and an affine Coxeter group $\hat{W}$ acting on the normal space $\nu(N)_y$ such that $x_0 = \exp(v)$ for some $v \in \nu(N)_y$ and the energy functional $E : P(X, \{x_0\} \times N) \to R$ is a perfect Morse function whose critical set is equal to

$$\{\gamma_w(t) = \exp(t(w \cdot 0) + (1-t)v) \mid w \in \hat{W}\}.$$

Moreover, an analogue of the Schubert cycle is explicitly constructed for each critical point of E.

Now we turn to problem (3). We have mentioned above that if H is a closed subgroup of $G \times G$ and the action of H on G is hyperpolar then the action of the Hilbert Lie group $P(G, H)$ on the Hilbert space $H^0([0,1], \mathfrak{g})$ by gauge transformations

$$g * u = gug^{-1} - g'g^{-1}$$

is hyperpolar. In particular, using examples in Theorem 1.7, 1.8 and 1.9 we conclude that the actions of $P(G, K \times K)$, $P(G, K_1 \times K_2)$ and $P(G, G(\sigma))$ on $H^0([0,1], \mathfrak{g})$ are hyperpolar. However, problem (3) remains unsolved.

To explain problem (4), we need to give a rough description of affine Kac-Moody algebras (cf. *[18]*, and a more detailed review in Section 4). Let $\mathfrak{g}$ be a compact, semi-simple Lie algebra, σ an order k automorphism of $\mathfrak{g}$, and $L(\mathfrak{g}, \sigma)$ the space of loops $u : [0, 2\pi] \to \mathfrak{g}$ satisfying the condition $\sigma(u(t)) = u(t + \frac{2\pi}{k})$ for all t. Let $\hat{L}(\mathfrak{g}, \sigma) = L(\mathfrak{g}, \sigma) + Rc + Rd$ denote the two dimensional extension of $L(\mathfrak{g}, \sigma)$ such that c is the center with 2-cocyle defined by

$$\omega(u, v) = \int_0^{2\pi} (u'(t), v(t))dt, \tag{1.1}$$

and $[d, u] = u'$. Then $\hat{L}(\mathfrak{g}, \sigma)$ is an affine (Kac-Moody) algebra of type ℓ, where ℓ is the order of $[\sigma] \in \mathrm{Aut}(\mathfrak{g})/\mathrm{Int}(\mathfrak{g})$. Moreover, all affine algebras can be obtained this way. Let $\langle\ ,\ \rangle$ denote the non-degenerate, ad-invariant bi-linear form on $\hat{L}(\mathfrak{g}, \sigma)$ defined by

$$\langle u_1 + r_1 c + s_1 d, u_2 + r_2 c + s_2 d\rangle = r_1 s_2 + r_2 s_1 + \int_0^{2\pi} (u(t), v(t))dt, \tag{1.2}$$

where $(\ ,\ )$ is the negative of the Killing form on $\mathfrak{g}$. Since the bi-linear form $\langle\ ,\ \rangle$ has index one, the horosphere

$$V_0 = \{u + d - \tfrac{\langle u,u\rangle + 1}{2} c \mid u \in L(\mathfrak{g})\}$$

is isometric to $L(\mathfrak{g}, \sigma)$ (equipped with the L^2-inner product) via the map

$$u + d - \tfrac{\langle u,u\rangle + 1}{2} c \quad \mapsto u.$$

Involutions on complex affine algebras have been classified (cf. *[20, 1]*). In this paper we are interested in finding all isometric involutions of the real affine algebra $\hat{L}(\mathfrak{g}, \sigma)$ that leave $Rc + Rd$ invariant. Since the classification results for complex affine algebras are rather complicated, it is not easy to translate them into the real case, and we give a direct and independent classification of such involutions. We first note that if ϕ is such an involution on $\hat{L}(\mathfrak{g}, \sigma)$ then the restriction of ϕ to $Rc + Rd$ is either equal to id or $-$id. Such ϕ will be called *involutions of the first and second kind* respectively. Moreover, the following statements are true:

(i) $\hat{L}(\mathfrak{g}, \sigma)$ can be written as a direct sum of $+1$ and -1 eigenspaces $\hat{\mathfrak{K}}$ and $\hat{\mathfrak{P}}$ of ϕ, which will be called the *Cartan decomposition* of $\hat{L}(\mathfrak{g}, \sigma)$ defined by ϕ,

(ii) If ϕ is of the first kind then $\hat{\mathfrak{K}}$ is a subalgebra of $\hat{L}(\mathfrak{g}, \sigma)$, and is isomorphic to some $\hat{L}(\mathfrak{g}_0, \tau)$ for some subalgebra $\mathfrak{g}_0$ of $\mathfrak{g}$ and an automorphism τ on $\mathfrak{g}_0$. In particular, this says that the $\mathrm{Ad}(\hat{K})$-orbit in $\hat{\mathfrak{P}}$ is equivalent to the Adjoint action of some affine algebra.

(iii) If ϕ is of the second kind then there exist an automorphism σ and involutions ρ_1, ρ_2 on $\mathfrak{g}$ such that $\sigma\rho_1 = \rho_2$ and the Adjoint action of $\hat{K}$ on $\hat{\mathfrak{P}}$ is equivalent to the $\hat{K}_1$-action on $\hat{\mathfrak{P}}_1$, where $\hat{L}(\mathfrak{g},\sigma) = \hat{\mathfrak{K}}_1 + \hat{\mathfrak{P}}_1$ is the Cartan decomposition of the involution ρ_1^* of the second kind defined by $\rho_1^*(u)(t) = \rho_1(u(-t))$.

(iv) The Adjoint action leaves the Hilbert space $\hat{L}(\mathfrak{g},\sigma) \cap V_0$ invariant. If ϕ is an involution of $\hat{L}(\mathfrak{g},\sigma)$ of the second kind and $\hat{L}(\mathfrak{g},\sigma) = \hat{\mathfrak{K}} + \hat{\mathfrak{P}}$ is the corresponding Cartan decomposition, then the Adjoint action of $\hat{\mathfrak{K}}$ leaves the Hilbert space $\hat{\mathfrak{P}} \cap V_0$ invariant.

These actions on Hilbert space obtained from involutions of the first and second kind will be called *affine s-representations*. Next we want to describe all affine s-representations, and state results analogous to those in Remark 1.12 for orbits of affine s-representations. It is known that (cf. *[28]*) an affine s-representation arising from the Adjoint action of $\hat{L}(\mathfrak{g},\sigma)$ is equivalent to the $P(G, G(\sigma))$-action on $H^0([0,1],\mathfrak{g})$. Moreover, every Adjoint orbit in $V_0 \cap \hat{L}(\mathfrak{g},\sigma)$ can be viewed as an infinite dimensional analogue of a complex (partial) flag manifold, is isometric to $P(G, \{e\} \times N)$ for some $G(\sigma)$-orbit N of G, and the restriction of the height function

$$u + d - \tfrac{\langle u,u\rangle + 1}{2}\, c \quad \mapsto \quad \langle u + d - \tfrac{\langle u,u\rangle+1}{2}\, c,\ -d + \tfrac{1}{2}\, c\rangle$$

to the Adjoint orbit corresponds to the perfect energy functional E on $P(G, \{e\} \times N)$. In this paper, we prove that an affine s-representation arising from an involution of $\hat{L}(\mathfrak{g},\sigma)$ of the second kind is equivalent to the $P(G, K_1 \times K_2)$-action on $H^0([0,1],\mathfrak{g})$ for some symmetric subgroups K_1, K_2 of G. Moreover, the orbit of $\hat{K}$ on $\hat{\mathfrak{P}} \cap V_0$ can be viewed as an infinite dimensional analogue of a real (partial) flag manifold, is isometric to $P(G, \{e\} \times N)$ for some $K_1 \times K_2$-orbit N of G, and the restriction of the above height function to the Adjoint orbit corresponds to the perfect energy functional E on $P(G, \{e\} \times N)$.

We also prove that given any symmetric subgroups K_1, K_2 of G, the action of $P(G, K_1 \times K_2)$ on $H^0([0,1],\mathfrak{g})$ is equivalent to some affine s-representations of the second type. Since there exists a cohomogeneity one action on G that are not examples given by Theorem 1.8, the above discussion implies that Dadok's theoerm 1.11 is not true for Hilbert space, i.e., there exist hyperpolar actions on Hilbert space that are not affine s-representations.

This paper is organized as follows: In Section 2 we prove that, given a closed subgroup H of $G \times G$, the condition (A) above is necessary and sufficient for the H-action on G to be hyperpolar. In Section 3 we list the known examples of hyperpolar actions on G and give characterizations of some of these examples. In Section 4 we construct all isometric involutions of affine algebras that leave $Rc + Rd$ invariant, and discuss their relations to hyperpolar actions on compact Lie groups and on Hilbert space. Finally, in Section 5 we state some open problems.

The authors would like to thank the Max-Planck Institute in Bonn and Academia Sinica in Taipei for their hospitality while this paper was being prepared.

2 Algebraic Criteria and Rational Tori

The main purpose of this section is to give a simple algebraic criterion for an isometric action on a symmetric space to be hyperpolar.

We will use the convention that for $g \in G$ and $y \in \mathfrak{g}$, $gy := (L_g)_*(y)$, $yg := (R_g)_*(y)$, and $e^y := \exp(y)$, where L_g and R_g are left and right translation by g respectively. Recall that the *cohomogeneity* of an action of G on M is the dimension of the orbit space M/G, or equivalently the codimension of a principal orbit.

Theorem 2.1: *Let G be a compact, semi-simple Lie group with the bi-invariant metric induced from the negative of the Killing form on $\mathfrak{g}$. Let H be a closed subgroup of $G \times G$, acting isometrically on G by $(g_1, g_2) \cdot g = g_1 g g_2^{-1}$. Then the following statements are equivalent:*

(i) the H-action on G is hyperpolar,

(ii) there exists $g_0 \in G$ such that

$$g_0^{-1}\nu(H \cdot g_0)_{g_0} = \{g_0^{-1} x g_0 - y \,|\, (x, y) \in \mathfrak{H}\}^{\perp}$$

is abelian,

(iii) $\nu(H \cdot e)_e$ contains a k-dimensional abelian subalgebra, where k is the cohomogeneity of the H-action on G.

To prove this theorem, we first need a lemma:

Lemma 2.2: *Let G and H be as in Theorem 2.1. Suppose $\nu(H \cdot e)_e = \mathfrak{A}$ is abelian. Then*

(i) $T(H \cdot e^a) \perp e^a\mathfrak{A}$ for all $a \in \mathfrak{A}$,

(ii) if $\exp(\mathfrak{A})$ is closed, then the H-action on G is hyperpolar.

Proof: It is obvious that

$$T(H \cdot e)_e = \{x - y \,|\, (x, y) \in \mathfrak{H}\},$$
$$e^{-a}T(H \cdot e^a) = \{e^{-a} x e^a - y \,|\, (x, y) \in \mathfrak{H}\}.$$

Since $\mathfrak{A}$ is abelian, $\langle\, ,\, \rangle$ is Ad-invariant, and $\nu(H \cdot e)_e = \mathfrak{A}$, we get

$$\langle e^{-a} x e^a - y, b\rangle = \langle e^{-a} x e^a, b\rangle - \langle y, b\rangle = \langle x, b\rangle - \langle y, b\rangle = \langle x - y, b\rangle = 0$$

for all $(x, y) \in \mathfrak{H}$ and $b \in \mathfrak{A}$. This proves (i). Since H is compact, every orbit meets $\exp(\nu(H \cdot e)_e)$, and (ii) follows.

It follows from the fact that $\exp(\nu(H \cdot g)_g)$ meets every H-orbit and Lemma 2.2 that statements (ii) and (iii) in Theorem 2.1 are equivalent. Moreover, if $g_0 \in G$ and $H \cdot g_0$ is a principal orbit, then e is a regular point of the action of

$$\{(g_0^{-1} g_1 g_0, g_2) \,|\, (g_1, g_2) \in H\}$$

on G. So we may assume that e is a regular point, and then Theorem 2.1 follows from:

Theorem 2.3: *Let G, H be as in Theorem 2.1. If $\nu(H \cdot e)_e = \mathfrak{A}$ is abelian, then the H-action on G is hyperpolar.*

Thus, to prove Theorem 2.3, it suffices to show that $\exp(\mathfrak{A})$ is a torus. To do this, we introduce the notion of a rational lattice. A *full lattice* Λ of an n-dimensional real vector space V is a discrete subgroup of V isomorphic to Z^n. Let T be a torus (i.e., a compact, connected, abelian Lie group), $\mathfrak{T}$ its Lie algebra, and $\exp : \mathfrak{T} \to T$ the exponential map. Then $\Lambda(T) = \ker(\exp)$ is a full lattice of $\mathfrak{T}$, called the *lattice of* T. A basis for $\Lambda(T)$ over Z is called a Z-basis for Λ. The following Proposition is easy and well-known:

Proposition 2.4: *Let T be a torus, $\Lambda(T)$ the lattice of T, and $\mathfrak{A}$ a linear subspace of $\mathfrak{T}$. Then $\exp(\mathfrak{A})$ is a subtorus of T if and only if $\mathfrak{A} \cap \Lambda(T)$ is a full lattice of $\mathfrak{A}$.*

Definition 2.5: A full lattice Λ of a (real) Euclidean vector space V will be called a *rational lattice* if the ratio of non-zero inner products of any two elements of Λ is rational. (It clearly suffices to check the rationality of ratios of non-zero inner products of elements from some Z basis for Λ).

Proposition 2.6: *Let Λ be a full lattice of an Euclidean vector space V, and Λ' a full, sublattice of Λ. Then Λ is rational if and only if Λ' is rational.*

Proof: Of course, any sublattice of a rational lattice is rational. If Λ' has the same rank as Λ, then the integer matrix expressing the elements of a Z-basis for Λ' in terms of a Z-basis of Λ can be inverted, and the inverse is a matrix of rational numbers. So if Λ' is rational then Λ is also rational.

We call a torus T a *Riemannian torus* if it has a specified translation invariant Riemannian metric. Such a metric is of course uniquely determined by the induced Euclidean structure $\langle\ ,\ \rangle$ on $\mathfrak{T}$, and we will usually not distinguish between $\langle\ ,\ \rangle$ and the Riemannian structure. A Riemannian torus is called *rational* if its lattice $\Lambda(T)$ is rational in $\mathfrak{T}$.

Proposition 2.7: *Any subtorus A of a rational torus T is rational.*

Proof: For, as noted above, the lattice of A is a sublattice of the lattice of T.

Theorem 2.8: *If $\mathfrak{A}$ is a linear subspace of the Lie algebra $\mathfrak{T}$ of a rational torus T, then $\exp(\mathfrak{A})$ is a subtorus of T if and only if $\exp(\mathfrak{A}^\perp)$ is.*

Proof: Since rescaling an inner product does not change orthogonal complements, we can assume that the non-zero inner product of any two elements in $\mathfrak{T}$ is rational.

Assume $A = \exp(\mathfrak{A})$ is a subtorus of T. Then $\Lambda(A) = \Lambda(T) \cap \mathfrak{A}$ is a full lattice of $\mathfrak{A}$. Choose a basis $a_1, \ldots, a_k$ for $\mathfrak{A}$ with $a_i \in \Lambda(A)$, and extend it to a basis $a_1, \ldots, a_n$ for $\mathfrak{T}$ with $a_{k+i} \in \Lambda(T)$. Let $v_1, \ldots, v_n$ denote the orthogonal basis for $\mathfrak{T}$ obtained by applying the Gram-Schmidt process to $a_1, \ldots, a_n$, omitting the renormalization steps. That is, $v_1 = a_1$, and recursively,

$$v_j = a_j - \sum_{i=1}^{j-1} \left\langle a_j, \frac{v_i}{\langle v_i, v_i \rangle} \right\rangle v_i.$$

Then clearly $v_1, \ldots, v_k$ is an orthogonal basis for $\mathfrak{A}$, and $v_{k+1}, \ldots, v_n$ is an orthogonal basis for $\mathfrak{A}^\perp$. Moreover, since all $\langle a_i, a_j \rangle$ are rational, it follows by an obvious induction on j that v_j is a *rational* linear combination $\sum_{i \le j} r_{ij} a_i$ of a_i with $i \le j$. If m is the product of the denominators of the non-zero r_{ij}, then $mv_1, \ldots, mv_n$ is an *integral* linear combinations of the a_i, hence in $\Lambda(T)$. But of course $mv_{k+1}, \ldots, mv_n$ is still a basis for $\mathfrak{A}^\perp$, so $\mathfrak{A}^\perp \cap \Lambda(T)$ has rank $n - k$, i.e., it is a full lattice in $\mathfrak{A}^\perp$.

Next we want to show that any torus subgroup of G is rational in the induced metric. Since a subtorus of a rational torus is rational, and since every torus of G is included in a maximal torus, it will suffice to prove that a maximal torus T of G is rational. By Proposition 2.6, it will suffice to show that there exists a full sublattice Λ_0 of $\Lambda(T)$, that is rational. We will show below that Λ_0 can be chosen to be the co-root lattice.

Let $\mathfrak{T}$ be a maximal abelian subalgebra of $\mathfrak{g}$, $\mathfrak{g}_C = \mathfrak{g} \otimes_C C$, $R \subset \mathfrak{T}^*$ the root space of $\mathfrak{g}_C$, and let the root space decomposition of $\mathfrak{g}_C$ be:

$$\mathfrak{g}_C = \mathfrak{T}_C \oplus \bigoplus_{\alpha \in R} \mathfrak{g}_\alpha,$$

where $[a, z_\alpha] = i\alpha(a) z_\alpha$ for all $\alpha \in R$ and $z_\alpha \in \mathfrak{g}_\alpha$. Dualizing α with respect to the Killing form gives elements $\vec{\alpha}$ of $\mathfrak{T}$ satisfying $\alpha(X) = \langle X, \vec{\alpha} \rangle$, and the co-roots $\alpha^* \in \mathfrak{T}$ are defined by $\alpha^* = 2\vec{\alpha}/\langle \vec{\alpha}, \vec{\alpha} \rangle$. The co-root lattice Λ_0 is then the lattice generated by the co-roots $\alpha^*, \alpha \in R$. It is part of the elementary theory of simple Lie algebras (cf. *[15]*) that Λ_0 is a full sublattice of $\Lambda(T)$, and also that if α and β are roots, then:

(1) $\alpha(\alpha^*) = 2$

(2) $\beta(\alpha^*) \in Z$

(3) $\beta - \beta(\alpha^*)\alpha$ is also a root.

By (2) we have $2\langle \vec{\beta}, \vec{\alpha} \rangle / \langle \vec{\alpha}, \vec{\alpha} \rangle = \langle \vec{\beta}, \alpha^* \rangle = \beta(\alpha^*)$ is an integer, and $\| \vec{\alpha} \|^2 / \| \vec{\beta} \|^2 = \alpha(\beta^*)/\beta(\alpha^*)$ is rational for any $\alpha, \beta \in R$. So given any α, β, γ in R we have

$$\langle \alpha^*, \beta^* \rangle / \langle \gamma^*, \gamma^* \rangle = \frac{\langle \vec{\alpha}, \vec{\beta} \rangle}{\| \vec{\alpha} \|^2} \frac{\| \vec{\gamma} \|^2}{\| \vec{\beta} \|^2}$$

is rational. This proves that Λ_0 is rational, and we have proved:

Theorem 2.9: *If G is a compact, semi-simple Lie group equipped with the bi-invariant metric given by the Killing form on $\mathfrak{g}$, then any torus subgroup of G is rational.*

2.10 Proof of Theorem 2.3: It suffices to prove that $A = \exp(\mathfrak{A})$ is closed, i.e., a torus. In any case $B = \overline{A}$ is a torus, so its Lie algebra $\mathfrak{B}$ is abelian. Let $\mathfrak{A}_1$ denote the orthogonal complement of $\mathfrak{A}$ in $\mathfrak{B}$, and $A_1 = \exp(\mathfrak{A}_1)$. First we claim that B is transversal to the orbit $H \cdot e$. To see this, we note that by Lemma 2.2, $T(H \cdot a)_a \perp a\mathfrak{A}$, i.e.,

$$(a^{-1}xa - y, \mathfrak{A}) = 0, \quad \forall\, a \in A,\ (x, y) \in \mathfrak{H}.$$

Since B is the closure of A, it follows by continuity that $T(H \cdot b)_b \perp b\mathfrak{A}$ for all $b \in \mathfrak{B}$. If $b \in (H \cdot e) \cap B$, then $TB_b = b\mathfrak{B} \supset b\mathfrak{A} = \nu(H \cdot b)_b$, which proves that B is transversal to $H \cdot e$. By transversality, $B \cap (H \cdot e)$ is a compact submanifold of G. Next we show that $B \cap (H \cdot e) = A_1$. In fact, since $T(B \cap (H \cdot e))_b \subset b\mathfrak{B} \cap T(H \cdot e)_b$ and $T(B \cap (H \cdot e))_b \perp b\mathfrak{A}$, we have that $T(B \cap (H \cdot e))_b = b\mathfrak{A}_1$. Hence $B \cap (H \cdot e) = A_1$. Since B and A_1 are tori, Theorem 2.8 implies that A is a torus.

In the remainder of this section, we will derive an algebraic criterion for an isometric action on a symmetric space to be hyperpolar. First, we reduce the classification of hyperpolar actions on G/K to that of G:

Proposition 2.11: *Let (G, K) be a symmetric pair, $M = G/K$ the corresponding symmetric space, and H a closed connected subgroup of G. Then the H-action on M is hyperpolar if and only if the action of $H \times K$ is hyperpolar on G.*

Proof: Let $\pi : G \to G/K$ be the natural Riemannian submersion, $\bar{g} = \pi(g) = gK$, and $\mathfrak{g} = \mathfrak{K} + \mathfrak{P}$ the Cartan decomposition. By conjugation, we may assume that $\bar{e}$ is a regular point with respect to the H-action. Identifying $T(G/K)_{\bar{e}}$ with $\mathfrak{P}$, we have $\nu(H \cdot \bar{e})_{\bar{e}} = \nu((H \times K) \cdot e)_e$. If the H-action on G/K is hyperpolar, then $\nu(H \cdot \bar{e})_{\bar{e}}$ is abelian, which implies that $\nu((H \times K) \cdot e)_e$ is abelian, so by Theorem 2.3, the $H \times K$-action on G is hyperpolar. Conversely, if the action of $H \times K$ on G is hyperpolar and e is a regular point, then $A = \exp(\nu((H \times K) \cdot e)_e)$ is a torus. So $\pi(A)$ is a flat, closed submanifold of G/K that is a section of the H-action.

As a consequence of Theorem 2.1 and the above Proposition, we have:

Corollary 2.12: *Let G be as in Theorem 2.1, (G, K) a symmetric pair, $M = G/K$ the corresponding symmetric space equipped with the quotient metric, and $\mathfrak{g} = \mathfrak{K} + \mathfrak{P}$ the Cartan decomposition. Let $\pi : G \to G/K$ be the natural Riemannian submersion, and $\bar{g} = \pi(g)$. Let H be a closed connected subgroup of G, and k the cohomogeneity of the H-action on G/K. Then the following statements are equivalent:*

(i) the H-action on G/K is hyperpolar,

(ii) *$\nu(H \cdot \bar{e})_{\bar{e}}$ contains a k-dimensional abelian subalgebra,*

(iii) *there exists $g \in G$ such that $g_*^{-1}(\nu(H \cdot \bar{g})_{\bar{g}})$ is a k-dimensional abelian subalgebra.*

Corollary 2.13: *Let G/K be as in Corollary 2.12, and H a closed subgroup of G. If the H-action on G/K is of cohomogeneity one, then it is hyperpolar.*

Remark 2.14: It follows from the proof of Theorem 2.1 that this theorem still holds if G is equipped with a bi-invariant metric such that maximal tori of G are rational.

3 Examples and Characterizations

In this section, we give some known examples of hyperpolar actions on compact Lie groups and symmetric spaces, and give characterizations and conjugacy relations for some of these examples. We also discuss the maximality of H in G when the action of H on the symmetric space G/K is hyperpolar.

Examples 3.1: Example (2) is given in (*[6]*), which is a special case of the example (4) of Hermann (*[16]*).

(1) **The Adjoint Action**. Let $H = \{(g, g) \mid g \in G\}$. Then the H-action on G is the Adjoint action of G, which is hyperpolar with maximal tori as sections.

(2) **The σ-action**. Let σ be an automorphism of G, and

$$H = G(\sigma) = \{(g, \sigma(g)) \mid g \in G\}.$$

The action of $G(\sigma)$ on G is called the σ-action, and is hyperpolar.

(3) **The $K \times K$-action**. If G/K is a symmetric space, then the action of $K \times K$ on G is hyperpolar.

(4) **Hermann's examples**. Let ϕ and ρ be involutions on G, and K_ρ and K_ϕ the fixed point set of ρ and ϕ respectively. Then the action of $H = K_\rho \times K_\phi$ on G is hyperpolar. To see this, we note that the isotropy subgroup is $H_e = \Delta(K_\phi \cap K_\rho)$ the diagonal group of $K_\phi \times K_\rho$. Furthermore, $T(H \cdot e)_e = \{X - Y \mid X \in \mathfrak{K}_\rho \text{ and } Y \in \mathfrak{K}_\phi\} = \mathfrak{K}_\rho + \mathfrak{K}_\phi$ and $\nu_e = \mathfrak{P}_\rho \cap \mathfrak{P}_\phi$, where $\mathfrak{P}_\rho$ and $\mathfrak{P}_\phi$ are the orthogonal complements of $\mathfrak{K}_\rho$ and $\mathfrak{P}_\phi$ respectively. Notice that on the Lie group $G^{\phi\rho} = \{g \in G \mid \phi\rho g = g\}$ we have the involution $\rho = \phi$ with fixed point group $K_\rho \cap K_\phi$ and Cartan decomposition $\mathfrak{G}^{\phi\rho} = (\mathfrak{K}_\rho \cap \mathfrak{K}_\phi) \oplus (\mathfrak{P}_\rho \cap \mathfrak{P}_\phi)$. The cohomogeneity of the slice representation H_e on ν_e is therefore equal to the dimension of a maximal abelian subalgebra of $\mathfrak{P}_\rho \cap \mathfrak{P}_\phi$. Then it follows from Theorem 2.1 that the H-action on G is hyperpolar.

(4*) Let $\tau : G \times G \to G \times G$ be an involution, and H the fixed point set of τ. Then the H-action on G is hyperpolar. For the action of $H \times \Delta(G)$ is hyperpolar on $G \times G$ by (4) and hence the action of H on $G = G \times G/\Delta(G)$ is hyperpolar by Proposition 2.12.

(5) **Cohomogeneity one actions.** As noted in Corollary 2.13, any cohomogeneity one action on a symmetric space is hyperpolar. For example, if $\rho : K \to SO(n)$ is the isotropy representation of a rank 2 symmetric space, then the K-action on $S^{n-1} = SO(n)/SO(n-1)$ has cohomogeneity one. Hence the action of $\rho(K) \times SO(n-1)$ on $SO(n)$ is hyperpolar by Proposition 2.11. Notice that this example does not in general arise as a special case of Hermann's example (4). (For example, the isotropy representation of the Grassmannian $G_{2,m}(R)$ of two planes in R^m for $m > 2$: $\rho : SO(2) \times SO(m) \to SO(2m)$.)

Remark 3.2: The group H in Example 3.1 (3) is of the form $K \times K$, where K is a symmetric subgroup of G. This example can be characterized as follows (*[14]*): If L is a closed subgroup of G such that the action of $H = L \times L$ on G is hyperpolar, then G/L is a symmetric space of rank r, where r is the cohomogeneity of the H-action. But (G, L) is not necessarily a symmetric pair. For example, $(Spin(7), G_2)$ or $(G_2, SU(3))$.

Next we give a characterization of Example 3.1 (2):

Theorem 3.3: *Let G be a compact, simply connected Lie group, and H a compact, connected, simple subgroup of $G \times G$. If the H-action on G is non-transitive and hyperpolar, then G is simple and there is an automorphism σ of G such that $H = \{(g, \sigma(g)) \mid g \in G\}$.*

First we need a Lemma.

Lemma 3.4: *Let K be a proper subgroup of a compact, connected, simple Lie group G of rank r. Then* $\dim(G/K) > \sqrt{2}\, r$.

Proof: Let G be endowed with a bi-invariant metric and G/K with the quotient metric. Then G acts by isometries on G/K. The set N of elements of G that act trivially is a normal subgroup of G. So N is finite. It follows that $\dim G \leq \dim \mathrm{Iso}(G/K)$. Set $m = \dim(G/K)$. The dimension of the isometry group of a compact m-dimensional manifold is at most $m(m+1)/2$. It follows from the classification of compact simple Lie groups of rank r that their dimension is at least $r^2 + 2r$. Hence

$$r^2 + 2r \leq \frac{m(m+1)}{2},$$

which implies $m > \sqrt{2}\, r$.

3.5 Proof of Theorem 3.3: Let $p_i : H \to G$ be the projection onto the i-th factor for $i = 1, 2$. Since H is simple and $\ker p_i$ is a normal subgroup of H, $\ker p_i$ is either H or a finite subgroup of H. We have the following cases:
(i) $\ker p_2 = H$. Then $H = H_1 \times \{e\}$ for some subgroup H_1 of G. This means that H acts on G by left translations. All orbits are principal orbits, in particular the

one through e. The section is a torus T and the Lie algebra G splits orthogonally into $\mathfrak{G} = \mathfrak{H}_1 + \mathfrak{T}$, i.e., $\mathfrak{T} = \mathfrak{H}_1^{\perp}$. Since $\langle\ ,\ \rangle$ is ad-invariant, $\langle [\mathfrak{H}_1, \mathfrak{T}], \mathfrak{T}\rangle = 0$ and $\langle [\mathfrak{H}_1, \mathfrak{T}], \mathfrak{H}_1\rangle = 0$. So we have $[\mathfrak{H}_1, \mathfrak{T}] = 0$ and thus $\mathfrak{T} = 0$ by the semisimplicity of G. Hence H acts transitively on G, a contradiction.

(ii) $\ker p_1 = H$. Then it follows similarly that H is transitive.

(iii) Both $\ker p_1$ and $\ker p_2$ are finite. Let $H' = p_1(H)$. Let $\mathfrak{G} = \mathfrak{G}_1 + \cdots + \mathfrak{G}_k$ be the decomposition of $\mathfrak{G}$ into simple ideals, and let G_i be the subgroup of G corresponding to $\mathfrak{G}_i$. Let $H_i' \subset G_i$ be the projection of H' into G_i. Using that the codimension of the orbits of H cannot exceed the rank $r(G)$ of G, we get $\dim H + r(G) \geq \dim G$. Thus

$$\sum_{i=1}^{k} \dim H_i' + \sum_{i=1}^{k} r(G_i) \geq \dim H + \sum_{i=1}^{k} r(G_i) \geq \sum_{i=1}^{k} \dim G_i. \qquad (*)$$

If $H_i' \subset G_i$ is a proper subgroup of G_i, then $\dim H_i' + \sqrt{2}r(G_i) < \dim G_i$ by Proposition 3.4. By $(*)$ we thus have $H_i' = G_i$ for some i, and without loss of generality we can assume $i = 1$. Then $H' \to G_1$ is a covering map, and so in fact an isomorphism, since G and therefore all the G_i are simply-connected. It also follows that $p_i : H \to p_i(H)$ are isomorphisms and hence $H = \{(g, \sigma(g)) \mid g \in H'\}$, where $\sigma : H' \to G$ is the injective homomorphism $\sigma = p_2 p_1^{-1}$. It is therefore left to prove that $H' = G$. Using (*),

$$\dim(H) + r(G) \geq \dim(G), \qquad \dim(G_i) \geq r(G_i)^2 + 2r(G_i)$$

and the fact that $H \simeq H' \simeq G_1$, we get

$$\sum_{i=1}^{k} r(G_i) \geq \sum_{i=2}^{k} \dim G_i \geq \sum_{i=2}^{k} (r(G_i)^2 + 2r(G_i)).$$

Hence

$$r(G_1) \geq \sum_{i=2}^{k} (r(G_i)^2 + r(G_i)).$$

In particular $r(H') = r(G_1) > r(G_i)$ for all $i \geq 2$. Therefore the projection π_i of H' to G_i, $i \geq 2$, cannot have finite kernel and thus π_i must be trivial. This implies $H' = G_1$.

Similarly we get $p_2(H) = G_i$ for some i. Since $p_2(H) \simeq H \simeq G_1$ and $r(G_i) < r(G_1)$ for $i \geq 2$ we must have $p_2(H) = G_1$ as well. Thus H acts only on G_1 and leaves the G_i, $i \geq 2$, fixed. Since the action of H is hyperpolar, G_i must be flat for $i \geq 2$. Hence $k = 1$ and $H' = G_1 = G$.

Next we recall the notion of conjugacy for isometric actions.

Definition 3.6: Let X_1 be a Riemannian G_1-manifold, and X_2 a Riemannian G_2-manifold. Then X_1 and X_2 are called *conjugate* if there is an isomorphism

$\phi : G_1 \to G_2$ and an isometry $f : X_1 \to X_2$ such that $\phi(g) \cdot f(x) = f(g \cdot x)$ for all $g \in G_1$ and all $x \in X_1$.

We now discuss the conjugacy of Examples 3.1 (4). Let G be a simple, compact Lie group. Recall that there is a standard list of (finitely many) involutions on G such that any involution on G can be conjugated to one of the standard list by an inner automorphism (cf. *[21]*, p.101). For example, the standard list of involutions for the classical groups is:

$$\begin{aligned} A_n &= SU(n+1): \ \tau, \ \mathrm{Ad}(I_{p,q}), \ \tau\,\mathrm{Ad}(I_{p,q}) \\ B_n &= SO(2n+1): \ \mathrm{Ad}(I_{p,q}) \\ C_n &= Sp(n): \ \tau, \ \mathrm{Ad}(I_{p,q}) \\ D_n &= SO(2n): \ \mathrm{Ad}(I_{p,q}), \ \mathrm{Ad}(J_n), \end{aligned}$$

where $I_{p,q} = \mathrm{diag}(1, \ldots, 1, -1, \ldots, -1)$ is the diagonal matrix with p entries equal to 1 and q entries equal to -1, J_n is the $2n \times 2n$ matrix corresponding to the standard symplectic form on R^{2n}, and τ denotes complex conjugation.

In particular, if τ_1, τ_2 are two involutions on G, then there exist σ_1, σ_2 from the standard list of involutions on G and $b_i \in G$ such that $\tau_i = \mathrm{Ad}(b_i)\sigma_i \,\mathrm{Ad}(b_i)^{-1}$. Note that $K_\sigma = \rho(K_\tau)$ if $\sigma = \rho\tau\rho^{-1}$ for some $\rho \in \mathrm{Aut}(G)$, where K_σ and K_τ are the fixed point sets of σ and τ respectively. So to classify the examples in 3.1(iv) up to conjugacy leads us naturally to the following result:

Proposition 3.7: *Suppose H_1 and H_2 are two subgroups of $G \times G$ and there exists a fixed $b \in G$ such that*

$$H_2 = \{(bh_1b^{-1}, h_2) \mid (h_1, h_2) \in H_1\}.$$

Then the H_1-actions and H_2-actions on G are conjugate via the isometry L_b given by the left multiplication $g \mapsto bg$. Similarly, if

$$H_2 = \{(h_1, bh_2b^{-1}) \mid (h_1, h_2) \in H_1\},$$

then the H_1 and H_2 actions on G are conjugate via the isometry $R_{b^{-1}}$, the right multiplication by b^{-1}.

Proof: In the first case note that $bh_1b^{-1}bxh_2^{-1} = L_b(h_1xh_2^{-1})$ implies that the actions are conjugate. The other case follows similarly.

Corollary 3.8: *Let G be a simple, simply connected, compact Lie group, and τ_1, τ_2 involutions on G. Then there exist σ_1, σ_2 from the standard list of involutions on G such that the actions of $K_{\tau_1} \times K_{\tau_2}$ and $K_{\sigma_1} \times K_{\sigma_2}$ on G are conjugate.*

Next we discuss the conjugacy of Examples 3.1 (2). Note that the σ-action is hyperpolar even when σ is not of finite order. We will prove that every σ-action on G is conjugate to a τ-action for some automorphism τ of order 1, 2 or 3.

In order to proceed, we need to review the diagram automorphism group of a simply connected, compact, semisimple Lie group G (cf. *[21]* p. 44 and *[15]* Chap 9, 10). It is known that each automorphism of the Dynkin digram of G gives rises to an automorphism of G, which will be called a *diagram automorphism.* Let $E(G)$ denote the group of diagram automorphisms. Then $E(G)$ is isomorphic to $\mathrm{Aut}(R)/W$, where R is the root system of $\mathfrak{g}$ and W the corresponding Weyl group. Let $\mathrm{Aut}(G)$ and $\mathrm{Int}(G)$ denote the group of automorphisms and inner automorphisms of G respectively. Then $\mathrm{Int}(G)$ is the identity component of $\mathrm{Aut}(G)$ and the quotient group is isomorphic to the group $E(G)$. Furthermore, $\mathrm{Aut}(G)$ is the semidirect product of $\mathrm{Int}(G)$ and $E(G)$. It is known that $E(G)$ is as follows (see *[21]*, p. 99):

$$E(G) = 1, \quad \text{if } G = A_1, B_r, C_r, E_7, E_8, F_4, G_2$$
$$E(G) = Z_2, \quad \text{if } G = A_r\ (r > 1), D_r (r \neq 4), E_6$$
$$E(G) = S_3, \quad \text{if } G = D_4$$

Note that every element in Z_2 or S_3 has order $1, 2$ or 3. For $\sigma \in \mathrm{Aut}(G)$, we let $[\sigma]$ denote the left coset $\mathrm{Int}(G)\sigma$ in $\mathrm{Aut}(G)/\mathrm{Int}(G)$. Using that $\mathrm{Aut}(G)$ is the semidirect product of $\mathrm{Int}(G)$ and $E(G)$, one sees easily that there exists an automorphism $\mu \in [\sigma]$ of the same order as $[\sigma]$ in $E(G)$ (i.e, of order 1, 2 or 3). So there is an element $b \in G$ such that $\sigma = \mathrm{Ad}(b)\mu$. Hence

$$G(\sigma) = \{(g, \sigma(g)) \mid g \in G\} = \{(g, b\mu(g)b^{-1}) \mid g \in G\}.$$

Then it follows from Proposition 3.7 that the σ-action and the μ-action on G are conjugate. This proves the following theorem:

Theorem 3.9: *Let G be a simply connected, compact Lie group, $\sigma \in \mathrm{Aut}(G)$. Then there exists a diagram automorphism μ of G that represents the same element as σ in $\mathrm{Aut}(G)/\mathrm{Int}(G)$ and has order* 1, 2, *or* 3. *Moreover, the σ-action on G is conjugate to the μ-action.*

As a consequence, Examples 3.1 (2) on a simply connected, semi-simple Lie group G are classified up to conjugacy by the diagram automorphisms of G.

Remark 3.10: Let G be a non-connected, compact Lie group, G_0 the identity component, and G_1 some other component. Let $h_1 \in G_1, \sigma : G_0 \to G_0;\ g \to h_1 g h_1^{-1}$, and $f : G_1 \to G_0;\ h \to h h_1^{-1}$. Then $f(ghg^{-1}) = gf(h)\sigma(g)^{-1}$. This proves that the action of G_0 on G_1 by conjugation is hyperpolar, and is conjugate to the σ-action on G_0. This also gives the generalization of the maximal torus theorem to nonconnected groups given in *[26]*.

Motivated by Dadok's classification of polar representations *[10]*, we define the notion of ω-equivalence—a weaker equivalence relation than conjugacy.

Definition 3.11: *Let X_1 be a Riemannian G_1-manifold, and X_2 a Riemannian G_2-manifold. Then X_1 and X_2 are called ω-equivalent (or orbital equivalent), if there exists an isometry $f : X_1 \to X_2$ such that $f(G_1 \cdot x) = G_2 \cdot f(x)$.*

We will show below that a group whose action is hyperpolar is often maximal in the sense that the action of a larger group is either ω-equivalent or transitive.

Theorem 3.12: *Let $H \subset L \subset SO(n)$ be connected, closed subgroups. If the action of H on R^n is irreducible and polar, then the L-action is either transitive on S^{n-1} or is ω-equivalent to the H-action.*

Proof: We will use some basic results from the theory of isoparametric submanifolds that can be found in *[24]*. The proof will be by induction on $k = \dim(R^n/H)$. The theorem is obviously true if $k = 1$. Now assume the theorem is proved for $k-1$, and assume $\dim(R^n/H) = k$. Choose $p \in S^{n-1}$ such that the slice representation of H_p has rank $k-1$ and its associated Dynkin diagram is connected. This can be done by chosing p to be a non-zero point on the intersection of the hyperplanes $\ell_2, \ldots, \ell_k$, where $\ell_1, \ldots, \ell_k$ are hyperplanes corresponding to simple roots and ℓ_1 corresponds to the left-end vertex of the Dynkin diagram. Then it follows from the slice theorem that the slice representation of H_p is irreducible on $\nu_0 = \nu(Hp)_p \cap (\mathbf{R}p)^\perp$. Since L_p leaves $\nu_1 = \nu(Lp)_p \cap (\mathbf{R}p)^\perp$ invariant, $\nu_1 \subset \nu_0$ and $H_p \subset L_p$, ν_1 is an invariant subspace of ν_0 under H_p. But H_p on ν_0 is irreducible. So one of the following cases is true:

(1) $\nu_1 = 0$. This implies that $\nu(Lp)_p = \mathbf{R}p$ is one-dimensional. Then the L-action is transitive on S^{n-1}.

(2) $\nu_1 = \nu_0$. Then $\nu = \nu(Hp)_p = \nu(Lp)_p$ and Hp and Lp have the same dimension. But $Hp \subset Lp$, so $Hp = Lp$. By the slice theorem we have that H_p is irreducible and polar with cohomogeneity $k-1$ on ν_0. Also $H_p \subset L_p \subset SO(\nu_0)$. So, by the induction hypothesis, the L_p-action on ν_0 is polar and either ω-equivalent to the H_p-action or transitive on the unit sphere in ν_0. In the first case it follows that the dimension of the orbits of H and L are equal and hence the orbits coincide since $H \subset L$. In the second case the orbits of L are hypersurfaces in spheres and it follows that L is polar. The orbits of both the H- and the L-action give rise to isoparametric foliations with the orbit $Hp = Lp$ in common. It is proved in *[13]* that a full leaf in an isoparametric foliation determines the foliation. This proves that the orbits of H and L coincide, i.e., H and L are ω-equivalent.

Theorem 3.13: *Let $X = G/K$ be a compact, connected and irreducible symmetric space, and $H \subset L \subset G$ closed, connected subgroups. Suppose the H-action on X is hyperpolar and there exists a $p \in X$ such that $Hp = Lp$. Then the L-action on X is ω-equivalent to the H-action.*

Proof: If A is a section of H passing through p, then A meets one orbit of L, $Lp = Hp$, orthogonally. Corollary 2.12 implies that the action of L on X is hyperpolar with A as a section. The principal orbits of H and L are therefore equal since their dimensions are equal and $H \subset L$. The principal orbits of a hyperpolar action determine the other orbits. It follows that the H- and L-actions are ω-equivalent.

It is known that (cf. *[8, 23]*) if the action of H on the compact symmetric space G/K is hyperpolar and A is a section through eK then $W(A) = N(A)/Z(A)$ is a finite group, where $N(A)$ and $Z(A)$ are the normalizer and centralizer of A in H respectively. Let $\mathfrak{A}$ denote the Lie algebra of A, and $\Lambda = \exp^{-1}(e) \cap \mathfrak{A}$ the unit lattice in $\mathfrak{A}$. Then $W(A)$ acts on Λ and the corresponding semi-direct product $\hat{W}(A) = W(A) \ltimes \Lambda$ is an affine Weyl group. The group $W(A)$ and $\hat{W}(A)$ will be called the *generalized Weyl group* and the *affine Coxeter group associated to the H-action*.

Corollary 3.14: *Let $X = G/K$ be a compact, connected and irreducible symmetric space, and $H \subset L \subset G$ closed, connected subgroups. Suppose the H-action on X is hyperpolar and the affine Coxeter group associated to the H-action is irreducible. Then the L-action on X is either transitive or is ω-equivalent to the H-action.*

Proof: It is known that the slice representation of a polar action is polar (*[23]*). Since the associated affine Coxeter group is irreducible, there exists $p \in X$ such that the slice representation of H_p on $\nu = \nu(Hp)_p$ is irreducible. Since $H \subset L$, $\nu_1 = \nu(Lp)_p \subset \nu$. Because ν is an irreducible H_p-space, we have the following two cases:

(1) $\nu_1 = 0$, this implies that the L-action is transitive on X.

(2) $\nu_1 = \nu$, then $Hp = Lp$. So by Theorem 3.13, the H and L actions are ω-equivalent.

4 Involutions of Kac-Moody Algebras and Polar Actions on Hilbert Space

In this section, we classify isometric, involutive automorphisms of affine algebras, and prove that up to ω-equivalence the affine s-representations associated to these involutions are either the $P(G, G(\sigma))$-actions or $P(G, K_1 \times K_2)$-actions on $H^0([0,1], \mathfrak{g})$, where σ is some automorphism and K_1, K_2 are some symmetric subgroups of G.

First we review some basic facts about affine algebras (cf. *[18]*). Let $\mathfrak{g}$ be the Lie algebra of a compact, semi-simple Lie group, $L(\mathfrak{g}) = H^0(S^1, \mathfrak{g})$, $t \in [0, 2\pi]$ the angle variable in S^1, and let u' denote du/dt (for those u for which du/dt is defined, i.e., $u \in H^1(S^1, \mathfrak{g})$). Let $[u, v](t) = [u(t), v(t)]$. Then $L(\mathfrak{g})$ is a Lie algebra. There are three types of affine algebras, which are two-dimensional extensions of $L(\mathfrak{g})$ (cf. *[18]*, Ch 7 and 8). We begin with the algebras of type one, which are extensions of $L(\mathfrak{g})$:

$$\hat{L}(\mathfrak{g}) = L(\mathfrak{g}) + Rc + Rd$$

with the bracket operation defined on a dense subspace of $\hat{L}(\mathfrak{g})$ by

$$\begin{aligned} [u,v]^\wedge &= [u,v] + \omega(u,v)c, \\ [d,u]^\wedge &= u' \\ [c,u]^\wedge &= [c,d]^\wedge = 0, \end{aligned}$$

where ω is the 2-cocyle defined on $L(\mathfrak{g})$ by formula (1.1). Let $\langle\,,\,\rangle$ be the bi-linear form (1.2) defined on $\hat{L}(\mathfrak{g})$. Then $\langle\,,\,\rangle$ is non-degenerate and ad-invariant, i.e., for all $\xi, \eta, \zeta \in \hat{L}(\mathfrak{g})$ we have

$$\langle [\xi, \eta]^\wedge, \zeta\rangle = \langle \xi, [\eta, \zeta]^\wedge\rangle.$$

It is known that (cf. *[25]*) there exists a group $\hat{L}(G)$ and a torus bundle $\pi : \hat{L}(G) \to L(G)$ such that $d\pi_e$ is the natural projection of $\hat{L}(\mathfrak{g})$ to $L(\mathfrak{g})$. Moreover, the Adjoint action of $\hat{L}(G)$ on its Lie algebra is given as follows (cf. *[18, 25]*):
(1) For $\hat{g} \in \pi^{-1}(g)$ and $u \in L(\mathfrak{g})$ then

$$\mathrm{Ad}(\hat{g})(d + u - \frac{\| u \|^2 + 1}{2} c) = d + g * u - \frac{\| g * u \|^2 + 1}{2} c,$$

where $g * u = gug^{-1} - g'g^{-1}$ is the gauge transformation and $\| u \|$ is the L^2-norm of u.
(2) Let $\{\exp(sd) \mid s \in [0, 2\pi]\}$ denote the one parameter circle subgroup in $\hat{L}(G)$ generated by the element d. Then

$$\mathrm{Ad}(\exp(sd))(u)(t) = u(t + s).$$

Although the bracket $[\,,\,]^\wedge$ is defined on a dense subset of $\hat{L}(\mathfrak{g})$ and $\hat{L}(G)$ is not a Lie group, the Adjoint action is well-defined on $\{d + u - \frac{\| u \|^2+1}{2} c \mid u \in L^2(\mathfrak{g})\}$.

Next we recall some explicit constructions of automorphisms on affine algebras in terms of automorphisms on $\mathfrak{g}$.

Proposition 4.1: *Let σ be an automorphism of $\mathfrak{g}$ of order k, m respectively. Then $\hat{\sigma} : \hat{L}(\mathfrak{g}) \to \hat{L}(\mathfrak{g})$ defined by*

$$\hat{\sigma}(u)(t) = \sigma(u(-\tfrac{2\pi}{k} + t)), \quad \hat{\sigma}(c) = c, \quad \hat{\sigma}(d) = d$$

is an automorphism of order k, is an isometry with respect to the ad-invariant form $\langle\,,\,\rangle$, and the fixed point set of $\hat{\sigma}$ is

$$\hat{L}(\mathfrak{g}, \sigma) = L(\mathfrak{g}, \sigma) + Rc + Rd,$$

where

$$L(\mathfrak{g}, \sigma) = \{u \in L(\mathfrak{g}) \mid u(t) = \sigma(u(-\tfrac{2\pi}{k} + t))\}.$$

Proposition 4.2: *Let ρ be an involution on $\mathfrak{g}$. Then $\rho^* : \hat{L}(\mathfrak{g}) \to \hat{L}(\mathfrak{g})$ defined by*

$$\rho^*(u)(t) = \rho(u(-t)), \quad \rho^*(c) = -c, \quad \rho^*(d) = -d,$$

is an involution, and is an isometry with respect to the ad-invariant form $\langle\,,\,\rangle$.

If the order of $[\sigma]$ in $\mathrm{Aut}(\mathfrak{g})/\mathrm{Int}(\mathfrak{g})$ is equal to ℓ, then $\hat{L}(\mathfrak{g}, \sigma)$ is called an *affine algebra of type ℓ*. The following two propositions give constructions of automorphisms of $\hat{L}(\mathfrak{g}, \sigma)$:

Proposition 4.3: *Let σ be an order k automorphism and ρ an involution on $\mathfrak{g}$. Then $\rho^* : \hat{L}(\mathfrak{g}) \to \hat{L}(\mathfrak{g})$ leaves $\hat{L}(\mathfrak{g}, \sigma)$ invariant if and only if $\sigma\rho$ is an involution, or equivalently if and only if $\sigma\rho = \rho\sigma^{-1}$.*

Proof: It suffices to prove that $\rho^*(L(\mathfrak{g}, \sigma)) \subset L(\mathfrak{g}, \sigma)$. To see this, we assume that $u \in L(\mathfrak{g})$ satisfies the condition that $u(t) = \sigma(u(-2\pi/k + t))$, and we want to show that

$$\rho^*(u)(t) = \sigma(\rho^*(u)(-2\pi/k + t)).$$

By the definition of $\rho^*(u)$, we see that the left hand side (LHS) and the right hand side (RHS) of the above equation are equal to

$$LHS = \rho(u(-t)),$$
$$RHS = \sigma(\rho(u(2\pi/k - t))) = \sigma(\rho(\sigma(u(-t)))).$$

Hence $RHS = LHS$ if and only if $\sigma\rho\sigma = \rho$.

Proposition 4.4: *Let σ and ρ be finite order automorphisms on $\mathfrak{g}$ of order k and m respectively, with $k = 2$ or 3. Then the map $\hat{\rho} : \hat{L}(\mathfrak{g}) \to \hat{L}(\mathfrak{g})$ leaves $\hat{L}(\mathfrak{g}, \sigma)$ invariant if and only if $\rho\sigma = \sigma\rho$. Moreover, if $\rho\sigma = \sigma\rho$ and $\hat{\mathfrak{K}}$ denotes the set of fixed points of $\hat{\rho}$ on $\hat{L}(\mathfrak{g}, \sigma)$, then*

(1) if $m = kn$ for some n, then $\hat{\mathfrak{K}} = \hat{L}(\mathfrak{g}_0, \theta)$, where $\mathfrak{g}_0$ is the fixed point set of $\sigma^{-1}\rho^n$ and $\theta = \rho \,|\, \mathfrak{g}_0$, or

(2) if k is not a factor of m, then there exist integers a, b such that $ak + bm = 1$ and $\hat{\mathfrak{K}} = \hat{L}(\mathfrak{g}, \rho^a\sigma^b)$.

Proof: The first part of the Proposition is obvious. To prove the rest, let $u \in L(\mathfrak{g}) \cap \hat{\mathfrak{K}}$. Then we have

$$\sigma(u(t)) = u(\tfrac{2\pi}{k} + t), \qquad \rho(u(t)) = u(\tfrac{2\pi}{m} + t).$$

(1) If $m = kn$, then

$$\rho^n(u(t)) = u(2n\pi/m + t) = u(2n\pi/kn + t) = u(2\pi/k + t) = \sigma(u(t)).$$

This implies that $\hat{K} = \hat{L}(K_\tau, \theta)$, where $\tau = \sigma^{-1}\rho^n$, and $\theta = \rho \,|\, K_\tau$. (2) If k does not divide m, then since k is a prime, k, m are relatively prime. Hence there exist integers a, b such that $ak + bm = 1$, or equivalently $a/m + b/k = 1/km$. Then we have

$$u(2\pi/km + t) = u(2\pi(a/m + b/k) + t) = \rho^a\sigma^b(u(t)).$$

This proves that $\hat{K} = \hat{L}(\mathfrak{g}, \rho^a\sigma^b)$.

The following theorem gives a characterization of the involutions on the affine algebras constructed in Proposition 4.3 and 4.4:

Theorem 4.5: *Let $\mathfrak{g}$ be the Lie algebra of a compact Lie group such that $\mathfrak{g}_C = \mathfrak{g} \otimes C$ is simple, σ an order k diagram automorphism of $\mathfrak{g}$, and ϕ an involution of the affine algebra $\hat{L}(\mathfrak{g}, \sigma)$. Suppose ϕ is an isometry with respect to the ad-invariant form $\langle , \rangle$ and ϕ leaves $Rc + Rd$ invariant. Then ϕ must be one of the following two types:*

(1) there exist $s \in [0, 2\pi]$ and an involution ρ on $\mathfrak{g}$ such that $\sigma\rho = \rho\sigma$ and ϕ is the restriction of $\mathrm{Ad}(\exp(sd)) \circ \hat{\rho}$ *to $\hat{L}(\mathfrak{g}, \sigma)$,*

(2) there exist $s \in [0, 2\pi]$ and an involution ρ of $\mathfrak{g}$ such that $\sigma\rho\sigma = \rho$ and ϕ is the restriction of $\mathrm{Ad}(\exp(sd)) \circ \rho^*$ *to $\hat{L}(\mathfrak{g}, \sigma)$.*

Proof: It follows easily from the fact that Rc is the center and ϕ is an isometry that the restriction of ϕ to $Rc + Rd$ is either equal to id or $-$ id. Now we assume that $\phi(c) = -c, \phi(d) = -d$, and we will prove below that statement (2) is true. A similar proof will also show that (1) is true if $\phi(c) = c$ and $\phi(d) = d$.

Let $\mathfrak{G}$ the complex Lie algebra $\mathfrak{g} \otimes C$, and $\tau : \mathfrak{G} \to \mathfrak{G}$ the conjugation map. Then σ can be extended uniquely to an order k automorphism of $\mathfrak{G}$, which will still be denoted by σ. Let $\epsilon = e^{\frac{2\pi}{k}}$, and $\mathfrak{G}_j$ the ϵ^j-eigenspace of σ. Then

$$\mathfrak{G} = \mathfrak{G}_0 + \mathfrak{G}_1 + \ldots + \mathfrak{G}_{k-1}.$$

For $j \in Z$, we define $\mathfrak{G}_j = \mathfrak{G}_i$ if $0 \leq i \leq k-1$ and $i \equiv j \pmod k$.

Given $u(t) = \sum u_n e^{int} \in L(\mathfrak{G})$, then $u \in L(\mathfrak{g}, \sigma)$ if and only if

$$\tau(u_n) = u_{-n}, \qquad u_n \in \mathfrak{G}_n \qquad \forall n.$$

Since

$$\phi([d, [d, u]^\wedge]^\wedge) = [d, [d, \phi(u)]^\wedge]^\wedge,$$

ϕ leaves the eigenspaces of d^2 invariant. So there exist conjugate linear maps $F_n : \mathfrak{G}_n \to \mathfrak{G}_{-n}$ such that

$$\phi(xe^{int} + \tau(x)e^{-int}) = F_n(x)e^{-int} + \tau(F_n(x_n))e^{int}.$$

where $x \in \mathfrak{G}_n$. Since ϕ is an automorphism of $\hat{L}(\mathfrak{g}, \sigma)$, we have

$$[F_n(X), F_m(Y)] = F_{n+m}([X, Y]), \qquad X \in \mathfrak{G}_m, Y \in \mathfrak{G}_n, \qquad (*_{m,n})$$

$$F_{-n} \circ F_n = \mathrm{id}\,.$$

We divide the proof of (2) into the following steps:

(i) There exist a non-trivial simple ideal $\mathfrak{h}$ in $\mathfrak{G}_0$ and $a \in C$ such that $F_{nk}(x) = a^n F_0(x)$ for all $x \in \mathfrak{h}$.

To see this, we set $A_n = F_0^{-1} F_{nk}$. Then A_n is complex linear map on $\mathfrak{G}_0$. Applying F_0^{-1} to equation $(*_{0,nk})$, we get

$$A_n([x, y]) = [x, A_n(y)]. \qquad (**)$$

Since σ is a diagram automorphism, there is a non-trivial simple ideal $\mathfrak{h}$ of $\mathfrak{g}_0$ (cf. *[15]*). Then equation (**) implies that $A_n(\mathfrak{h}) \subset \mathfrak{h}$. By Schur's lemma, there exists $a_n \in C$ such that $A_n \,|\, \mathfrak{h} = a_n \,\mathrm{id}$. But equation $(*_{m,n})$ implies that $a_n a_m = a_{n+m}$. So we have $A_n \,|\, \mathfrak{h} = a^n \,\mathrm{id}$ with $a = a_1$. This implies that $F_n(x) = a^n F_0(x)$ for all $x \in \mathfrak{h}$.

(ii) $F_{j+nk} = a^n F_j$.

For $0 \leq j < k$, we let

$$\mathfrak{G}_{j,a} = \{X \in \mathfrak{G}_j \,|\, F_{j+nk}(X) = a^n F_j(X) \ \forall n \in Z\}$$

$$\mathfrak{G}_a = \sum_{j=0}^{k-1} \mathfrak{G}_{j,a}.$$

It is easy to check that $\mathfrak{G}_a$ is an ideal of $\mathfrak{G}$. But we have proved that $\mathfrak{G}_{0,a} \supset \mathfrak{h}$ in (i). So $\mathfrak{G}_a$ is a non-zero ideal of the simple algebra $\mathfrak{G}$, which implies that $\mathfrak{G}_a = \mathfrak{G}$. Hence we have proved that

$$F_{j+nk}(X) = a^n F_j(X), \qquad \forall\ X \in \mathfrak{g}_C, 0 \leq j < k.$$

(iii) Let $\beta = a^{-\frac{1}{k}}$, and $\rho : \mathfrak{G} \to \mathfrak{G}$ be the map defined by

$$\rho(X) = \beta^j F_j(X), \quad \text{if } X \in \mathfrak{G}_j.$$

Then ρ is an automorphism of $\mathfrak{G}$. To see this, we let $X \in \mathfrak{G}_i$ and $Y \in \mathfrak{G}_j$:

(a) if $i + j < k$, then

$$\begin{aligned}[\rho(X), \rho(Y)] &= [\beta^i F_i(X), \beta^j F_j(Y)] = \beta^{i+j}[F_i(X), F_j(Y)] \\ &= \beta^{i+j} F_{i+j}([X,Y]) = \rho([X,Y]),\end{aligned}$$

(b) if $i + j > k - 1$, then

$$\begin{aligned}[\rho(X), \rho(Y)] &= [\beta^i F_i(X), \beta^j F_j(Y)] = \beta^{i+j} F_{i+j}([X,Y]) \\ &= \beta^{i+j} a F_{i+j-k}([X,Y]) = \rho([X,Y]).\end{aligned}$$

(iv) ρ is an involution and $\sigma\rho\sigma = \rho$.

It follows from the construction of ρ that $\sigma\rho\sigma = \rho$. So it remains to prove that ρ is an involution. To see this, we compute directly for $X \in \mathfrak{G}_i$:

$$\begin{aligned}\rho(\rho(X)) &= \rho(\beta^i F_i(X)) = \beta^i \rho(F_i(X)) \\ &= \beta^i \beta^{k-i} F_{k-i}(F_i(X)) = \beta^k a F_{-i}(F_i(X)) = X.\end{aligned}$$

(v) Let $X \in \mathfrak{G}_n$, and $u(t) = Xe^{int} + \tau(X)e^{-int}$. Then

$$\phi(u) = \beta^{-n}\rho(X)e^{-int} + \beta^n \rho(\tau(X))e^{int}. \qquad (***)$$

In particular, this implies that $\beta^{-n} = \bar{\beta}^n$, i.e., $|\,\beta\,| = 1$. So there exists $s \in [0, 2\pi]$ such that $\beta = e^{is}$.

(vi) $\phi = \mathrm{Ad}(\exp(sd))\rho^*$.

This follows from (v) and formula (***).

Next we recall a theorem proved in *[28]*:

Theorem 4.6: *([28]) Let G be a compact, semi-simple Lie group equipped with a bi-invariant metric, and H a closed subgroup of $G \times G$. Let $P(G, H)$ denote the space of H^1-paths $g : [0, 1] \to G$ such that $(g(0), g(1)) \in H$, and let $P(G, H)$ act on $H^0([0, 1], \mathfrak{g})$ by gauge transformations:*

$$g * u = gug^{-1} - g'g^{-1}.$$

Let $\phi : H^0([0, 1], \mathfrak{g}) \to G$ be the holonomy map defined by $\phi(u) = g(1)$, where $g : [0, 1] \to G$ is the solution to the initial value problem

$$g^{-1}g' = u, \qquad g(0) = e.$$

Then

(i) the orbit $P(G, H) * u = \phi^{-1}(H \cdot \phi(u))$,

(ii) if the action of H on G is hyperpolar and A is a section through e, then the action of $P(G, H)$ on $H^0([0, 1], \mathfrak{g})$ is polar with the space $\mathfrak{A}$ of constant paths in the Lie algebra $\mathfrak{A}$ of A as a section.

Applying this theorem to Examples 3.1, we obtain

Corollary 4.7: *The actions of the following groups on the Hilbert space $H^0([0, 1], \mathfrak{g})$ via gauge transformations are hyperpolar:*

(1) $P(G, G(\sigma))$, where σ is an automorphism on G and $G(\sigma) = \{(g, \sigma(g)) \mid g \in G\}$,

(2) $P(G, K \times K)$, where K is the fixed point set of some involution σ on G,

(3) $P(G, K_1 \times K_2)$, where K_i is the fixed point set of the involution σ_i for $i = 1, 2$,

(4) $P(G, H)$, if the cohomogeneity of the H-action on G is one.

As a consequence of Theorem 4.6 (i), we see that if the actions of H_1 and H_2 on G are ω-equivalent then the actions of $P(G, H_1)$ and $P(G, H_2)$ are ω-equivalent. The following Proposition shows that if the actions of H_i on G are conjugate, then so are the actions of $P(G, H_i)$ on $H^0([0, 1], \mathfrak{g})$.

Proposition 4.8: *Let H_1 and H_2 be two closed subgroups of $G \times G$. If there exists an element $(a, b) \in G \times G$ such that*

$$H_2 = \{(ah_1a^{-1}, bh_2b^{-1}) \mid (h_1, h_2) \in H_1\},$$

then the $P(G, H_1)$-action and $P(G, H_2)$-action on $V = H^0([0, 1], \mathfrak{g})$ are conjugate.

Proof: It suffices to prove the proposition for $b = e$. Since G is connected and compact, there exists $c \in \mathfrak{g}$ such that $a = \exp(c)$. Given $g \in P(G, H_1)$, i.e., $g : [0,1] \to G$ such that $(g(0), g(1)) \in H_1$, let

$$\tilde{g}(t) = e^{(1-t)c} g(t) e^{-(1-t)c}.$$

It is easy to see that $\tilde{g} \in P(G, H_2)$ because

$$\tilde{g}(0) = \exp(c) g(0) \exp(-c) = a g(0) a^{-1}, \quad \tilde{g}(1) = g(1).$$

For $u \in V$, a direct computation implies that

$$\tilde{g} * u = e^{(1-t)c} \{ g(e^{-(1-t)c} u e^{(1-t)c} - c) g^{-1} - g' g^{-1} \} e^{-(1-t)c} + c.$$

Let $V = H^0([0,1], \mathfrak{g})$, and $F : V \to V$ the map defined by

$$F(u) = \exp((1-t)c) u \exp(-(1-t)c) + c.$$

Then F is an isometry of V and

$$\tilde{g} * F(u) = F(g * u),$$

i.e., the $P(G, H_1)$-action and $P(G, H_2)$-action on V are conjugate via the isometry F.

As an immediate consequence of the above Proposition, Corollary 3.8 and Theorem 3.9, we obtain

Corollary 4.9: *The following statements are true:*

(i) Given two involutions τ_1, τ_2 on G, the $P(G, K_{\tau_1} \times K_{\tau_2})$-action on V is conjugate to $P(G, K_{\sigma_1} \times K_{\sigma_2})$-action on V for some involutions σ_1, σ_2 from the standard list of involutions on G, where K_{τ_i} and K_{σ_i} denote the fixed point sets of τ_i and σ_i respectively.

(ii) Given $\tau \in \mathrm{Aut}(G)$, the $P(G, G(\tau))$-action on V is conjugate to the action of $P(G, G(\mu))$ for some diagram automorphism μ. Moreover, the order of μ is 1, 2 or 3, and $[\tau]$ and $[\mu]$ are the same element in $\mathrm{Aut}(G)/\mathrm{Int}(G)$.

Proposition 4.10: *([25]) Let $V_0 = \{u + rc + d \mid u \in L(\mathfrak{g}), r = \frac{\|u\|^2+1}{2}\}$. Then*

(1) the map $u + rc + d \mapsto u$ defines an isometry from V_0 to $H^0([0, 2\pi], \mathfrak{g})$,

(2) the Adjoint action of $\hat{L}(G)$ on $\hat{L}(\mathfrak{g})$ leaves V_0 invariant, and induces an action of $L(G)$ on V_0,

(3) the induced action of $L(G)$ on V_0 is conjugate to the $P(G, G(\mathrm{id}))$ action on $H^0([0, 2\pi], \mathfrak{g})$, where $G(\mathrm{id}) = \{(g, g) \mid g \in G\}$.

Given an order k automorphism σ and involution ρ of G, we can define in the same way as for $L(\mathfrak{g})$ an automorphism $\hat{\sigma}$ and an involution ρ^* on $L(G)$. Moreover, (cf. *[27]*):

(1) both $\hat{\sigma}$ and ρ^* can be lifted to the affine group $\hat{L}(G)$,

(2) the Lie algebra of the fixed point set $\hat{L}(G,\sigma)$ of $\hat{\sigma}$ on $\hat{L}(G)$ is $\hat{L}(\mathfrak{g},\sigma)$ (here we still use σ to denote the automorphism $d\sigma_e$ on $\mathfrak{g}$),

(3) the loop part of $\hat{L}(G,\sigma)$, i.e., $\pi\hat{L}(G,\sigma)$, is

$$L(G,\sigma) = \{g \in H^1(S^1,G) \mid g(t) = \sigma(g(-2\pi/k + t))\}.$$

Proposition 4.11: *([28]) Let σ be an order k automorphism of G, $\hat{\sigma}$ the automorphism on $\hat{L}(G)$ associated to σ, and $\hat{L}(G,\sigma)$ the fixed point set of $\hat{\sigma}$. Let V_0 be as in Proposition 4.10, and $V(\sigma) = V_0 \cap \hat{L}(\mathfrak{g},\sigma)$. Then*

(1) the Adjoint action of $\hat{L}(G,\sigma)$ on its Lie algebra $\hat{L}(\mathfrak{g},\sigma)$ leaves $V(\sigma)$ invariant, and hence induces an action of $L(G,\sigma)$ on $V(\sigma)$,

(2) let $H^0([0,\frac{2\pi}{k}],\mathfrak{g})$ denote the Hilbert space with the inner product defined by

$$\langle u,v\rangle = k\int_0^{\frac{2\pi}{k}} (u(t),v(t))dt,$$

then the map

$$V(\sigma) \to H^0([0,\tfrac{2\pi}{k}],\mathfrak{g}), \qquad u + rc + d \mapsto u \mid [0,\tfrac{2\pi}{k}]$$

is an isometry,

(3) the induced action of $L(G,\sigma)$ on $V(\sigma)$ is conjugate to the $P(G,G(\sigma))$-action on $H^0([0,\frac{2\pi}{k}],\mathfrak{g})$.

Remark 4.12: We saw in Corollary 4.9 that the action of $P(G,G(\tau))$ on $H^0([0,1],\mathfrak{g})$ is conjugate to that of $P(G,G(\mu))$ if μ is an automorphism of G such that $[\tau]$ and $[\mu]$ are equal in $\mathrm{Aut}(G)/\mathrm{Int}(G)$. This is related to Theorem 8.5 on p.133 in *[18]* where it is proved that if σ is an automorphism of finite order, then the the affine algebra $\hat{L}(\mathfrak{g}_C,\sigma)$ is isomorphic to the affine algebra $\hat{L}(\mathfrak{g}_C,\mu)$ where μ is the diagram automorphism of G such that $[\sigma]$ and $[\mu]$ are equal in $\mathrm{Aut}(\mathfrak{g}_C)/\mathrm{Int}(\mathfrak{g}_C)$. The order k of the diagram automorphism μ coincides with the order of $[\tau] = [\sigma]$ in $\mathrm{Aut}(\mathfrak{g}_C)/\mathrm{Int}(\mathfrak{g}_C)$ and the type of the corresponding Kac-Moody algebra $\hat{L}(\mathfrak{g}_C,\sigma)$ is k. Note that τ in Corollary 4.9 (ii) does not have to be of finite order.

Proposition 4.13: *([28]) Let ρ be an involution of $\mathfrak{g}$, ρ^* the induced involution on $\hat{L}(\mathfrak{g})$, and $\hat{L}(\mathfrak{g}) = \hat{\mathfrak{K}} + \hat{\mathfrak{P}}$ the decomposition of $\hat{L}(\mathfrak{g})$ into ± 1-eigenspaces of ρ^*, i.e.,*

$$\hat{\mathfrak{K}} = \{u \in L(\mathfrak{g}) \mid \rho(u(-t)) = u(t)\},$$
$$\hat{\mathfrak{P}} = \{u \in L(\mathfrak{g}) \mid \rho(u(-t)) = -u(t)\} + Rc + Rd.$$

Then

(1) the Adjoint action of $\hat{K}$ leaves $V_0 \cap \hat{\mathfrak{P}}$ invariant, where V_0 is as in Proposition 4.10,

(2) *the map* $u + rc + d \mapsto u \mid [0, \pi]$ *is an isometry from* $V_0 \cap \hat{\mathfrak{P}}$ *to* $H^0([0, \pi], \mathfrak{g})$, *where* $H^0([0, \pi], \mathfrak{g})$ *is equipped with the inner product*

$$\langle u, v \rangle = 2 \int_0^\pi (u(t), v(t))\, dt,$$

(3) *the Adjoint action of* $\hat{K}$ *on* $V_0 \cap \hat{\mathfrak{P}}$ *is conjugate to the action of* $P(G, K \times K)$ *on* $H^0([0, \pi], \mathfrak{g})$.

Next we want to show that the third family of examples in Corollary 4.7 also arises from some involutions of affine algebras.

Proposition 4.14: *Let* σ *be an automorphism of order* k *and* ρ *an involution on* G. *Suppose the automorphism* $\varphi = \sigma\rho$ *is an involution. Then the restriction of* ρ^* *to* $\hat{L}(G, \sigma)$ *gives an involution on* $\hat{L}(G, \sigma)$. *Moreover,*

(i) *the fixed point set* $\tilde{K}$ *of* ρ^* *on* $\hat{L}(G, \sigma)$ *is*

$$\tilde{K} = \{g \in H^1(S^1, G) \mid g(-t) = \rho(g(t)),\ \sigma(g(-\tfrac{2\pi}{k} + t)) = g(t)\},$$

which is isomorphic to the Hilbert Lie group $P(G, K_\rho \times K_\varphi)$,

(ii) *the Adjoint action of* $\tilde{K}$ *leaves* $V(\sigma) = V \cap \hat{L}(\mathfrak{g}, \sigma)$ *invariant, where* V *is as in Proposition 4.10,*

(iii) *the* $\mathrm{Ad}(\tilde{K})$ *action on* $V(\sigma)$ *is conjugate to the action of* $P(G, K_\rho \times K_\varphi)$ *on* $H^0([0, \frac{\pi}{k}], \mathfrak{g})$.

Proof: Let $g \in \tilde{K}$. Then $g(0) = \rho(g(0))$ implies that $g(0) \in K_\rho$; and

$$g(\pi/k) = \sigma(g(-\pi/k)) = \sigma(\rho(g(\pi/k))) = \varphi(g(\pi/k)),$$

where the last equality is true because $\varphi = \sigma\rho$. So $g(\pi/k) \in K_\varphi$. Next we claim that for $g \in \tilde{K}$, g is determined by $g \mid [0, \pi/k]$. To see this, for $t \in [0, \pi/k]$ we have

$$g(\tfrac{\pi}{k} + t) = \sigma(g(-\tfrac{\pi}{k} + t)) = \sigma(\rho(g(\tfrac{\pi}{k} - t)).$$

This proves (i). Statements (ii) and (iii) follow.

Remark 4.15: Let φ, ρ be two involutions on G such that $\sigma = \varphi\rho$ has finite order. Then Proposition 4.14 implies that the $P(G, K_\rho \times K_\varphi)$-action on $H^0([0, 1], \mathfrak{g})$ is essentially the Adjoint action of $\tilde{K}$ on $\hat{\mathfrak{P}}$, where $\tilde{K}$ is the fixed point set of the involution ρ^* on $\hat{L}(G, \sigma)$ and $\hat{\mathfrak{P}}$ is the (-1)-eigenspace of ρ^* on $\hat{L}(\mathfrak{g}, \sigma)$. In other words, the $P(G, K_\rho \times K_\varphi)$-action is the isotropy representation of the "symmetric space" $\hat{L}(G, \sigma)/\tilde{K}$.

Proposition 4.16: *Let* G *be a compact, simple Lie group,* σ_1, σ_2 *two involutions on* G, *and* K_1, K_2 *the corresponding fixed point sets. If the action of* $K_1 \times K_2$ *on* G *is* *not transitive, then there exists $x \in G$ such that the composition of the involutions $\mathrm{Ad}(x)\sigma_1 \mathrm{Ad}(x)^{-1}$ and σ_2 has finite order.*

Proof: It is proved in *[7]* that there exists $x \in G$ such that the two involutions $\sigma_1^x = \mathrm{Ad}(x)\sigma_1 \mathrm{Ad}(x)^{-1}$ and σ_2 commute except when $(\mathfrak{g}, \mathfrak{K}_1, \mathfrak{K}_2)$ is one of the following three triples:

$$\begin{cases} (su(2n), sp(n), su(2q-1) \times su(2(n-q)+1)) \\ (so(2n), su(n), so(2q+1) \times so(2(n-q)-1) \\ (so(8), so(3) \times so(5), \omega(so(3) \times so(5))), \end{cases}$$

where ω is the triality automorphism of $so(8)$. In particular, this says that the order of $\sigma_1^x \circ \sigma_2$ is equal to 2 if $(\mathfrak{g}, \mathfrak{K}_1, \mathfrak{K}_2)$ is not one the above three triples. But for the above three triples, it is easy to check that the order of $\sigma_1 \circ \sigma_2$ is equal to $4, 4$ and 3 respectively.

Recall that in the Introduction, we call the action of $L(G, \sigma)$ on $V(\sigma)$ given in Proposition 4.11 an affine s-representation of the first kind, and the action of $\hat{K}$ on $V(\sigma)$ in Proposition 4.14 an affine s-representation of the second kind. Since $\sigma_1^x = \mathrm{Ad}(x)\sigma_1 \mathrm{Ad}(x)^{-1} = \mathrm{Ad}(x\sigma_1(x)^{-1}) \circ \sigma_1$, $[\sigma_1^x] = [\sigma_1]$ in $\mathrm{Aut}(\mathfrak{g})/\mathrm{Int}(\mathfrak{g})$. As consequence of Corollary 4.9, Proposition 4.14 and 4.16, we have

Corollary 4.17: *Let G be a simply connected, compact, simple Lie group, and K_1, K_2 symmetric subgroups of G. If the action of $K_1 \times K_2$ is not transitive on G, then the action of $P(G, K_1 \times K_2)$ on $H^0([0,1], \mathfrak{g})$ is w-equivalent to an affine s-representation of the second kind.*

5 Open Problems

The classification of hyperpolar actions on symmetric spaces and on Hilbert space remains unsolved. In this section, we will give some open problems, which we think are needed for achieving such classifications. In order to explain our approach, we first give an outline of a proof different from the one given by Dadok for Theorem 1.11. It is not difficult to see that if $\rho : G \to SO(n)$ is a decomposable polar representation then there exist polar representations $\rho_1 : G_1 \to SO(V_1)$ and $\rho_2 : G_2 \to SO(V_2)$ such that the G-action on R^n is ω-equivalent to the product action of $G_1 \times G_2$ on $V_1 \times V_2$. So the classification of polar representations reduces to that of irreducible ones. By Theorem 3.12 the classification of irreducible polar representations can be carried out by checking whether the action of H on S^{n-1} is polar for a closed subgroup H of $SO(n)$, where H is one of the following cases:

(i) H is a maximal subgroup of $SO(n)$,

(ii) H is a maximal subgroup of a closed subgroup G on $SO(n)$, where G acts on S^{n-1} transtively.

This can be done because the H's in (i) and (ii) are known (cf. *[2, 3, 22, 11, 19]*).

Since we have a maximality result similar to Theorem 3.12 for hyperpolar actions on symmetric spaces, it is a natural to study the problem of decomposability of a hyperpolar action. To explain this notion, we make the following definition:

Definition 5.1: *An isometric action of G on M is called decomposable if there exist a Riemannian G_1-manifold M_1 and G_2-manifold M_2 such that the action of G on M is ω-equivalent to the product action of $G_1 \times G_2$ on $M_1 \times M_2$; and say that the action on M decomposes as the product of two isometric actions.*

Conjecture 5.2: Let G be a compact, simply connected Lie group, and H a closed subgroup of $G \times G$. Suppose the action of H on G is hyperpolar, A is a section through e, and $W(A) = N(A)/Z(A)$ is the generalized Weyl group associated to the H-action, where $N(A)$ and $Z(A)$ are the normalizer and centralizer of A in H respectively. Then the following statements are equivalent:

(a) the H-action on G decomposes as product of two isometric actions,

(b) the H-action on G decomposes as the product of two hyperpolar actions,

(c) the action of $W(A)$ on A is decomposable.

The above conjecture is true for the following two families of hyperpolar actions:

(i) $H = G(\sigma)$ for some automorphism σ of G. By Theorem 3.9, we may assume that s is a diagram automorphism of G. Then this conjecture is true because of the decomposability of affine Kac-Moody algebras (cf. *[18]*).

(ii) $H = K \times K$, where K is a closed subgroup of G. This is proved in *[14]*.

Conjecture 5.3: If the action of H on G is hyperpolar and indecomposible, then G is simple.

If the above two conjectures are true, then the classification of hyperpolar actions on a compact, simply connected, simple Lie group G is reduced to the classification of indecomposable ones. A closed subgroup H of the isometry group of a symmetric space X is called *transitive* if the action of H on X is transitive. It follows from Corollary 3.14 that every indecomposable hyperpolar action on the symmetric space G/K is ω-equivalent to one of the following:

(i) H_1-action on G/K, where H_1 is a maximal subgroup of some transitive, maximal subgroup H of G and the H_1-action on G is hyperpolar,

(ii) H-action on G/K, where H is maximal in G and the H-action on G/K is not transtive but is hyperpolar.

Since the classification of maximal subgroups of G is known (cf. *[11, 19]*), the classification of indecomposable hyperpolar actions on the symmetric space G/K can be achieved following the above two steps.

Similarly, every indecomposable hyperpolar action on G is ω-equivalent to one of the following:

(i) H_1-action on G, where H_1 is a maximal subgroup of some transitive, maximal subgroup H of $G \times G$ and the H_1-action on G is hyperpolar,

(ii) H-action on G, where H is maximal in $G \times G$ and the H-action on G is not transtive but is hyperpolar.

So in order to classify hyperpolar actions on G, we pose the following two problems:

(a) The classification of all maximal subgroups of $G \times G$.

(b) The classification of all maximal subgroups of a transitive sugroup of $G \times G$.

Since the geometry of the orbit foliation of a variationally complete actions is very similar to that of a hyperpolar action on a symmetric space, the following is a natural question:

Question 5.4: Is a variationally complete action on a compact symmetric space hyperpolar?

We suspect there is also a decomposition theorem for polar actions on a Hilbert space V. In fact, we think a more general conjecture should be true. To explain this, we recall that a submanifold M of V is *isoparametric* if the normal bundle is globally flat, the restriction of η to any finite disk normal bundle is proper and Fredholm, and the shape operators along any parallel normal field are conjugate, where $\eta : \nu(M) \to V$ is the end point map defined by $\eta(v) = x + v$ if $v \in \nu(M)_x$. It is known that the principal orbits of polar action on Hilbert space are isoparametric (*[28]*). We have also associated to each isoparametric submanifold M an affine Coxeter group W such that W acts on each normal plane $x_0 + \nu(M)_{x_0}$ isometrically and the action is equivariant under the normal parallel translation. An isoparametric submanifold M of V is called *decomposable* if there exist two closed subspaces V_1, V_2 of V and isoparametric submanifolds M_i in V_i for $i = 1, 2$ such that $V = V_1 + V_2$ and M is equal to the product of $M_1 \times M_2$. Otherwise, M is called *indecomposable*.

Conjecture 5.5: If M is an isoparametric submanifold of a Hilbert space V, then M is indecomposable if and only if the affine Coxeter group associated to M is irreducible.

In particular, this will give the complete decomposition theorem, i.e., every isoparametric submanifold in V can be written as direct sum of irreducible ones.

Note that if the action of H on G is hyperpolar then the action of the path group $P(G, H)$ on the Hilbert space $H^0([0,1], \mathfrak{g})$ via gauge transformation is polar. Moreover, a principal orbit of the $P(G, H)$-action is decomposable as an isoparametric submanifold if and only if the $P(G, H)$-action is decomposable, which is also equivalent to the condition that the H-action on G is decomposable. So Conjecture 5.5 implies Conjecture 5.2.

References

[1] J. Bausch and G. Rousseau, Involutions de prèmiere espèce des algèbres affines, Revue de l'Institut Elie Cartan, no. 11, Nancy, 1989.

[2] A. Borel, Some remarks about Lie groups transitive on spheres and tori, Bull. Amer. Math. Soc. **55** (1949), 580–587.

[3] A. Borel, Le plan projectif des octaves et les sphères comme espaces homogènes, C.R. Acad. Sc. Paris **230** (1950), 1378–1380.

[4] R. Bott, An application of the Morse theory to compact Lie groups, Bulletin de la Société Mathématique de France **84** (1956), 251–282.

[5] R. Bott, and H. Samelson, Applications of the theory of Morse to symmetric spaces, Amer. J. Math. **80** (1958), 964–1029.

[6] L. Conlon, The topology of certain spaces of paths on a compact symmetric space, Trans. Amer. Math. Soc. **112** (1964), 228–248.

[7] ——, Remarks on commuting involutions, Bull. Amer. Math. Soc. **22** (1969), 255–257.

[8] ——, Variational completeness and K-transversal domains, J. Differential Geometry **5** (1971), 135–147.

[9] ——, A class of variationally complete representations, J. Differential Geometry **7** (1972), 149–160.

[10] J. Dadok, Polar coordinates induced by actions of compact Lie groups, Trans. Amer. Math. Soc. **288** (1985), 125–137.

[11] E.B. Dynkin, The maximal subgroups of the classical groups, AMS Translation from Trudy Moskov. Mat. **6** (1957), 245–378.

[12] H. Garland and M.S. Raghunathan, A Bruhat decomposition for the loop space of a compact group: A new approach to results of Bott, Proc. Nat. Acad. Sci. USA **72** (1975), 4716–4717.

[13] E. Heintze, C. Olmos, and G. Thorbergsson, Submanifolds with constant principal curvatures and normal holonomy groups, International J. Math. **2** (1991), 167–175.

[14] E. Heintze, R.S. Palais, C.L. Terng, and G. Thorbergsson, Hyperpolar Actions and k-flat Homogeneous Spaces, J. reine angew. Math. **454** (1994), 163–179 .

[15] S. Helgason, *Differential Geometry, Lie Groups, and Symmetric Spaces*, Academic Press, NY, 1978.

[16] R. Hermann, Variational completeness for compact symmetric spaces, Proc. Amer. Math. Soc. **11** (1960), 544–546.

[17] ——, Totally geodesic orbits of groups of isometries, Nederl. Akad. Wetensch. **65** (196), 291–298.

[18] V.G. Kac, *Infinite Dimensional Lie Algebras*, Third Edition, Cambridge University Press, Cambridge, 1990.

[19] Krämer, Eine Klassifikation bestimmter Untergruppen kompakter zusammenhängender Liegruppen, Comm. in Algebra **3** (1975), 691–737.

[20] F. Levstein,A classification of involutive automorphisms of an affine Kac-Moody Lie algebra, J. Algebra **114** (1988), 489–518.

[21] O. Loos, *Symmetric Spaces*, I,II, Benjamin Inc., New York, 1969.

[22] D. Montgomery and H. Samelson,Transformation groups on spheres, Annals of Math. **44** (1943), 454–470.

[23] R.S. Palais and C.L. Terng,A general theory of canonical forms, Trans. Amer. Math. Soc. **300** (1987), 771–789.

[24] R.S. Palais and C.L. Terng, *Critical Point Theory and Submanifold Geometry*, Lecture Notes in Math., vol. 1353, Springer-Verlag, Berlin and New York, 1988.

[25] A. Pressley and G. Segal, *Loop Groups*, Clarendon Press, Oxford, 1986.

[26] G. Segal, The representation ring of a compact Lie group, Publ. Math. IHES **34** (1968), 13–128.

[27] C.L. Terng, Variational completeness and infinite dimensional goemtry, Proceedings of Leeds Conference, Geometry and Topology of submanifolds **3** (1990), 279–293.

[28] ———, Polar Actions on Hilbert Space, to appear in Journal of Geometric Analysis.

[29] C.L. Terng and G. Thorbergsson, Submanifold Geometry in Symmetric spaces, preprint.

The Smooth Classification of Elliptic Surfaces

John W. Morgan
Department of Mathematics
Columbia University
New York, NY 10024

Tomasz S. Mrowka
Department of Mathematics
California Institute of Technology
Pasadena, CA 91125

1 Introduction

The purpose of this paper is to extend the classification results for elliptic surfaces with $p_g = 1$ established in *[15]* and *[1]* to elliptic surfaces with $p_g \geq 1$. Our main result is the following:

Theorem 1.1: *Let S and S' be possibly blown-up elliptic surfaces whose relatively minimal models have positive Euler characteristic. Suppose that $p_g(S) > 0$. Then S and S' are diffeomorphic if and only if S and S' are deformation equivalent in the sense of Definition 1.1 of Chapter I of [7]. That is to say S and S' are diffeomorphic if and only if they have the same Euler characteristic and first Betti number, have the same number of multiple fibers with the same multiplicities, and have been blown up from their relative minimal model the same number of times.*

This result completes the diffeomorphism classification of elliptic surfaces with the exception of those of geometric genus 0 which have exactly two multiple fibers with multiplicities being relatively prime.

According to the results of Chapter VI of *[7]* to prove Theorem 1.1 one needs only establish the following result.

Theorem 1.2: *Let S and S' be relatively minimal elliptic surfaces over $\mathbf{C}P^1$ with at most two multiple fibers and with $p_g(S) = p_g(S') \geq 1$. Suppose that there is an orientation-preserving homotopy equivalence $f\colon S \to S'$ such that the pullback of the Donaldson polynomial invariants of S under f agree with the Donaldson polynomial invariants of S' (e.g., if f is a diffeomorphism form S to S'). Then S and S' have the same number of multiple fibers and the multiplicities of the multiple fibers of S agree with those of S'.*

Geometry, Topology, and Physics for Raoul Bott

Cambridge, MA 02138 USA

Theorem 1.2 is established by extending the partial computations of Donaldson polynomials of elliptic surfaces with $p_g = 1$ done in *[15]* and *[1]* to elliptic surfaces with $p_g \geq 1$. Recall from Corollary 2.14 of Chapter VI of *[7]* and *[17]* that the $SU(2)$-Donaldson polynomials $\gamma_k(S)$ of an elliptic surface S with two multiple fibers of multiplicities m_1, m_2 have the form

$$\gamma_k(S) = \sum_{i=0}^{n} \alpha(k,i) Q_S^{(i)} \kappa_S^{d-2i} \tag{1}$$

where Q_S is the quadratic intersection form of S, where $Q_S^{(i)}$ is an appropriate symmetrized version of the ith power of Q_S (see below for the precise definition), where κ_S is $[F]/m_1m_2$ where F is a generic fiber, where $d = 4k - 3(1 + p_g(S))$, and where $n = (d - p_g(S) + 1)/2$. Furthermore, according to Theorem 2.1 of Chapter VIII of *[7]* we have

$$\alpha(k,n) = \frac{d!}{(2n)!}(m_1m_2)^{p_g(S)}. \tag{2}$$

It follows immediately that the product m_1m_2 is a diffeomorphism invariant.

The classification of elliptic surfaces with $p_g = 1$ was accomplished by the computation of $\alpha(3,2)$ in *[15]* and by a similar computation of a related $SO(3)$-invariant in *[1]*. In *[15]* it was shown that

$$\alpha(3,2) = \frac{m_1m_2}{6}\left(2m_1^2m_2^2 - (m_1^2 + m_2^2)\right).$$

This paper is devoted to giving the following generalization of this result.

Theorem 1.3: *Let S be a relative minimal elliptic surface over $\mathbf{C}P^1$ with positive Euler characteristic. Suppose that S has at most two multiple fibers of multiplicity*[1] *$m_1, m_2 \geq 1$. Let Q_S denote the quadratic form of S and for any $r > 0$ set*

$$Q_S^{(r)} = \frac{(2r)!}{r!2^r} Q_S^r.$$

Set $d = 4k - 3(1 + p_g(S))$ and $n = (d - p_g(S) + 1)/2$. Finally, let κ_S be the rational cohomology class Poincaré dual to the generic fiber divided by m_1m_2, Then for any $k > 0$ the $SU(2)$-Donaldson polynomial invariant $\gamma_k(S)$ for S takes the following form:

$$\gamma_k(S) = \sum_{i=0}^{n} \alpha(k,i) Q_S^{(i)} \kappa_S^{d-2i}$$

where

$$\alpha(k,n) = \frac{d!}{(2n)!}(m_1m_2)^{p_g}$$

[1] If S has one multiple fiber, then we take $m_1 = 1$ and m_2 equal to this multiplicity; if S has none, then we set $m_1 = m_2 = 1$.

and

$$\alpha(k, n-1) = \frac{d!}{(2n-2)!}\frac{(m_1 m_2)^{p_g}}{6}\left((p_g+1)(m_1 m_2)^2 - (m_1^2 + m_2^2)\right).$$

Clearly, from this result if follows that m_1, m_2 are determined by the Donaldson polynomial invariants of S, and hence Theorem 1.2 follows immediately.

Method of proof

Our proof is a generalization of the proof of the connected sum theorem in the sense that we pull the elliptic surface apart and relate the situation for a finite stage of the pulling-apart process with that for the geometric limit which is a manifold with cylindrical ends. There are two crucial differences here from the connected sum case. The one that has occurred be.ore, for example in the work of *[14]* and *[9]*, is that the manifold that we are pulling apart along is not simply connected, so that one has to invoke the analysis of of *[14]* and gluing theorems given in *[13]*. The second difference is a phenomenon which is just beginning to get attention, cf, *[2]*. This is a study of divisors associated to surfaces embedded in the four-manifold which cut across the splitting three-manifold. This study is quite intricate and, in our view, quite interesting. There are many new phenomena which are not encountered in other contexts. The explication of these phenomena accounts for most of the work of this paper.

Let us consider the case when p_g is odd, say $p_g = 2t-1$. In this case we decompose the surface into a fiber-connected sum of t elliptic surfaces with $p_g = 1$, all but one of which is a $K3$-surface. This decomposition is along a disjoint union Y of three-tori. We consider a family of Riemannian metrics on S, constant outside a product neighborhood $Y \times I$ of Y, but with the length of the interval in this product structure going to infinity. Then using the calculus for comparing invariants of the entire surface with those for the constituent pieces we are able to deduce our main theorem in this case from the previously known results for the case $p_g = 1$.

To study the case of even geometric genus we first consider the case when $p_g = 2$. We handle this case backwards. The fiber-connected sum of two elliptic surfaces with $p_g = 2$ is an elliptic surface with $p_g = 5$. Hence, from the case that we have already treated, we know the result for the fiber-connected sum. We can turn our computations around to deduce the result for the case $p_g = 2$ in the case when there are no mulitple fibers. Next, we treat the case of an elliptic surface with $p_g = 2t > 2$ by decomposing the surface into a fiber-connected sum of $t-1$ elliptic surfaces with $p_g = 1$ and one with $p_g = 2$. The result is established by using the method by which we treated surfaces with p_g odd, and invoking the computations for the surface with $p_g = 2$ as well as the previously known results in the case $p_g = 1$. Finally, we work backwards again to complete the case when $p_g = 2$.

Comparison with other methods

Since we announced the results contained here, there have been several other proofs of these and related results. There is an argument similar in spirit, but different in the details, due to Stipsicz and Szabo *[16]*. They decompose in much the same way, but use the boundary of the neighborhood of a cusp fiber instead of a smooth fiber. There is a more classical argument using only Floer-Donaldson theory of decomposing along homology spheres due to Lisca, *[11]*. Friedman *[6]* has recently announced an algebro-geometric computation of closely related $SO(3)$-invariants, obtaining very similar formulas and the same classification result. In fact, Friedman's results also cover the case omitted here when $p_g = 0$. In the case when there are no multiple fibers Kronheimer-Mrowka *[10]* computed the entire Donaldson polynomial invariants. They also gave a conjectured formula in general. More recently, Fintushel-Stern *[5]* have established this conjecture using the technique of cutting along rational homology three-spheres.

Outline of the paper

In Section 2 we decompose our elliptic surface S as a fiber-connected sum of elliptic surfaces S_i of $p_g = 1$ and $p_g = 2$. This means that we consider a decomposition along a disjoint union of three-tori, called *splitting tori*. Given such a decomposition we discuss collections of embedded Riemann surfaces some of which are disjoint from the splitting three-tori and some of which are multi-sections of the elliptic fibration and hence cut across all the splitting three-tori. Under appropriate numerical restrictions on the number of Riemann surfaces of the various types, we consider weak limits of ASD connections in the intersection of the divisors associated to Riemann surfaces that do not cross the splitting three-tori. Our first result is that with our numerical restrictions any sequence of such connections converges strongly to a connection on the cylindrical-end manifold Z obtained as a geometric limit of the pulling-apart process. Furthermore, the energy of the limiting connection on each component of Z is determined by the number of surfaces contained in that component. Then we discuss the conditions satisfied by the limiting connection if every connection in the sequence is contained in a divisor associated to a surface τ meeting the splitting three-tori. It turns out that either the resulting connection must lie in an analogous divisor for one of the components of Z or one of the boundary values of the connection along a splitting three-torus must be exceptional for the intersection of τ with that three-torus. Given all of this we can then describe entirely the set of weak limits of sequences of ASD connections in the intersection of all the divisors.

In Section 3 we consider the relationship between the set of weak limits studied in Section 2 and the ASD connections on S with respect to a metric which has stretched-out the collar-neighborhoods of the splitting three-tori. We show that for sufficiently stretched-out metrics there is exactly one ASD connection (counted with sign) for each point of intersection described in Section 2. That is to say the intersection number described in Section 2 is equal to the value of the Donaldson invariant on classes represented by the given set of Riemann surfaces.

Then in Section 4 we begin the analysis of the terms described in Section 2.

From the definition, the intersection number is a sum of products of invariants associated with the components Z_i of the limiting cylindrical-end manifold Z. In order to prove Theorem 1.3 we must identify these invariants with known invariants of the elliptic surfaces S_i obtained from the Z_i by filling in solid tori. Section 4 considers the general relationship of invariants for a cylindrical-end manifold with boundary components tori to the invariants for the closed manifold obtained by gluing in solid tori along the boundary components. We consider questions of orientation of the moduli space; questions of the preimage of points and two-dimensional subsets of the character variety of the boundary and what they represent in terms of the closed manifold, and we give some of the simplest cases where the invariants of the cylindrical-end manifold can be identified with those of the related closed four-manifold.

In Section 5 we continue the analysis begun in Section 4. We introduce a new two-dimensional class ϵ which is the first Chern class of a natural reduction of the based $SO(3)$-bundle over the moduli space for any Z_i to an S^1-bundle. This class plays a fundamental role in our analysis, for it turns out that this class measures the disconitunity of our divisors as we cross exceptional levels. Once we have introduced this class and studied its basic properties, we can relate all the invariants that we will need of the cylindrical-end manifolds Z_i to the invariants of the closed elliptice surfaces S_i. These results are enumerated in Section 5.

Then in Section 6 we apply all of this machinery to prove Theorem 1.3. First, we consider the case when p_g is odd. In this case we decompose S into a fiber-connected sum of elliptic surfaces with $p_g = 1$. Given the known results for these surfaces and the calculus for relating the invariants of S to a sum of products of invariants of the constituent pieces, it is a simple combinatorial exercise to complete the proof in this case. Next, we consider the special case when $p_g = 2$ and there are no muliple fibers. We consider an elliptic surface S with $p_g = 5$ and decompose it into two ellitpic surfaces S_1, S_2 with $p_g = 2$. In this case we know the value of the invariant for S and our calculus tells us how to express this answer as a sum of products of invariants associated with S_1 and S_2. Using the symmetry between S_1 and S_2 it turns out that we can then deduce the result for S_1 from that for S. Once we know the result for $p_g = 2$ and no multiple fibers, we can establish it for p_g even and $p_g > 2$ with multiple fibers. To do this we decompose the surface as a fiber-connected sum $S_1 \#_f \cdots \#_f S_t$ where $p_g(S_i) = 1$ for $i < t$ and $p_g(S_t) = 2$ and where all the multiple fibers are contained in S_1. The results in the case $p_g = 1$ and the just established result for $p_g = 2$ and no mulitple fibers can then be used in the calculus to deduce the answer in this case. Lastly, we consider the case when $p_g = 2$ and there are multiple fibers. This case is done by decomposing a surface S with $p_g = 4$ into a fiber connected sum $S_1 \#_f S_2$ where $p_g(S_1) = 2$ and S_1 contains the multiple fibers. We have already established the result for S and S_2. Using the calculus we can then deduce the result for S_1, hence completing the proof of Theorem 1.3.

Acknowledgements

In the early stages of this work we benefitted from several conversations with Bob Friedman. Also, comparison of our results with those of Andras Stipsicz and Zoltan Szabo proved to be an invaluable psychological boost. It is a pleasure to thank these people for their help.

2 Weak Limits

The main result of this section (Proposition 2.9) concerns weak limits of ASD connections contained in the divisors associated with certain sets of embedded riemann surfaces in S.

2.1 The basic setup

In this subsection S is a regular elliptic surface over $\mathbf{C}P^1$. We write S as a fiber-connected sum

$$S = S_1 \#_f \cdots \#_f S_t$$

where each S_i is a regular elliptic surface over $\mathbf{C}P^1$ with at most two multiplic fibers, and where $p_g(S_i) \leq 2$ and $p_g(S_i) = 1$ if $i \neq 1, t$. We let X_i be the complement in S_i of an open tubular neighborhood of a smooth fiber for $i = 1, t$ and of two smooth fibers for $1 < i < t$. We do this in such a way that the X_i are also embedded in S with disjoint interiors so that their union covers S. This gives us a presentation of S as being obtained by attaching the X_i together along their boundaries. Each X_i is a compact manifold with boundary consisting or either one or two three-tori. By convention we define $p_g(X_i) = p_g(S_i)$, so that $p_g(X_i) = 1$ for $1 < i < t$ and $p_g(X_1)$ and $p_g(X_t)$ are each equal to either one or two. For each $1 < i < t$ we denote by $\partial^{\pm}(X_i)$ the component of X_i which is glued to $X_{i\pm 1}$. Analogously, we have $\partial^+(X_1)$ and $\partial^-(X_t)$. We denote by $\lambda_i^{\pm} \subset \partial^{\pm}(X_i)$ the loop which bounds a disk in S_i in a regular neighborhood of the smooth fiber which was removed in forming X_i, a disk which meets this fiber transversely in a single point. In the gluing process we glue λ_i^+ to λ_{i+1}^-.

We let $\tilde{Y} = \coprod_i \partial X_i$. Then the gluing identification of the X_i to form S is given by an involution $\iota\colon \tilde{Y} \to \tilde{Y}$ interchanging components. The quotient $Y = \tilde{Y}/\iota$ is the image of $\tilde{Y}$ in S. We denote by $\chi(\tilde{Y})$ and $\chi(Y)$ the character varieties of representations of the fundamental groups into $SU(2)$. The involution on $\tilde{Y}$ defines an involution, also called ι, on $\chi(\tilde{Y})$ whose quotient is identified with $\chi(Y)$.

Let $\{g_m\}_{m=1}^{\infty}$ be a sequence of generic Riemannian metrics on S, constant outside a regular neighborhood ν of Y, and isometric to products $Y \times (-m, m)$ on ν. Let Z be the Riemannian manifold which is the geometric limit of these Riemannian manifolds. Then Z is a complete Riemannian manifold with cylindrical ends. It is diffeomorphic to $\coprod_i \operatorname{int} X_i$. We denote by Z_i the component of Z corresponding to X_i We denote by $\tilde{\nu}$ the subset of Z corresponding to $\nu \subset S$. The subset $\tilde{\nu}$ is isometric to the product cylinder $[0, \infty) \times \tilde{Y}$. We also suppose that the metric on Z is a generic cylindrical-end metric.

2.2 Weak limits of divisors associated to surfaces $\Sigma \subset S$ that do not cut across Y

Suppose that $\Sigma_1, \ldots, \Sigma_r$ are closed, oriented surfaces smoothly embedded in $S - Y = \coprod_{i=1}^{t} \operatorname{int} X_i$. The first step is to limit, under appropriate hypotheses on the number of Σ_a contained in each X_i, for all sufficiently large m, the nature of any point of intersection of the divisors $D_{\Sigma_1}(m), \ldots, D_{\Sigma_r}(m)$ in $\mathcal{M}_k(S, g_m)$ representing the image of the $[\Sigma_i] \in H_2(S)$ under the μ-map.

First, let us fix as usual the way in which we choose these divisors. We begin with surfaces Σ_a in $S - Y$ which meet transversely and have the property that the image of the restriction map from $\mathcal{M}(Z) \to \mathcal{B}(\Sigma_a)$ lies in $\mathcal{B}^*(\Sigma_a)$ (see Chapter 7 of *[4]*). We then choose generic divisors in the spaces $\mathcal{B}^*(\Sigma_a)$ representing the images under the μ-map of the fundamental classes $[\Sigma_a]$. These divisors will be transverse to the restriction mappings from the $\mathcal{M}(S, g_m)$ and from $\mathcal{M}(Z, g)$. Pulling back via the restriction maps gives rise to divisors $D_{\Sigma_a}(m)$ in $\overline{\mathcal{M}}(S, g_m)$ and D_{Σ_a} in $\overline{\mathcal{M}}(Z, g)$ (see Section 3.1 of *[7]* or Chapter 4 of *[4]*) Then the usual arguments show that the $D_{\Sigma_a}(m)$ in the compactified moduli space $\overline{\mathcal{M}}(S, g_m)$ converge as $m \mapsto \infty$ to D_{Σ_a}. By choosing the original divisors generically we can also assume that the D_{Σ_a} meet transversely in compactified moduli spaces $\overline{\mathcal{M}}(Z)$ and hence in $\overline{\mathcal{M}}(S, g_m)$ for all $m >> 0$. From now on we assume that we have chosen the divisors in $\mathcal{B}^*(\Sigma_a)$ so that this holds.

Lemma 2.1: *Let Z_i be the Riemannian manifold with cylindrical-end which is the component of Z corresponding to X_i. Then the dimension of the moduli spaces $\mathcal{M}_\ell(Z_i, g)$ are given as follows:*

- $2(4\ell - 5)$ *if* $\ell > 0$ *and* $1 < i < t$,
- ≤ 4 *if* $\ell = 0$ *and* $1 < i < t$,
- $2\left(4\ell - 3\left(1 + p_g(X_i)\right) - \frac{1}{2}\right)$ *if* $\ell > 0$ *and* $i = 1, t$, *and*
- ≤ 1 *if* $\ell = 0$ *and* $i = 1, t$.

Proof: This is immediate from the assumptions on the X_i, the dimension formula given in Theorem 9.0.5 of *[14]* and the fact that g is a generic metric.

Definition 2.2: We say that an ASD connection ω on a four-manifold X is *ϵ-concentrated* if there exist a disjoint union of balls $B_1, \ldots, B_c$ radius ϵ in X and positive integers $n_1, \ldots, n_c$ such that the energy of $\omega|_{B_i}$ is within ϵ of $8\pi^2 n_i$ and such that the energy of ω on $X - \cup_i B_i$ is less than ϵ.

Proposition 2.3: *Let $\Sigma_1, \ldots, \Sigma_{4r}$ be closed oriented surfaces in $\coprod_{i=1}^{t} X_i$. Let $4r_i$ be the number of the Σ_j contained in X_i. Suppose that:*

- *for each $i \leq t$ we have $r_i \geq 3$ and*
- $4r + 4p_g(S) + 4 \geq 4k$

Then for any $\epsilon > 0$ the following holds for all $n >> 0$. Let the intersection of the divisors $D_{\Sigma_1}(m), \ldots, D_{\Sigma_{4r}}(m)$ in the compactified moduli space $\overline{\mathcal{M}}_k(S, g_m)$ be denoted by $\Delta_\Sigma(m)$. Then every point of $\Delta_\Sigma(m)$ is represented by a connection *which is not ϵ-concentrated on any component X_i of X and which has energy less that ϵ on ν.*

Proof: We prove this result by showing under the given conditions on the r_i that any connection on Z which is the weak limit of a sequence of ideal connections $\omega(m)$ whose classes lie in $\Delta_\Sigma(m)$ in $\overline{\mathcal{M}}_k(S, g_m)$ has the property that its energy is $8\pi^2 k$ and that it does not have a flat background connection on any component of Z. From this it is a standard argument to establish the proposition.

Let $\omega(m)$ be a sequence of ideal connections as described in the previous paragraph. Suppose that ω is an ideal connection on Z which is the weak limit of the $\omega(m)$. Let ω_i be the restriction of ω to the component of Z corresponding to X_i. Then this connection lies in the intersection of r_i divisors in $\mathcal{M}(Z_i, g)$. From the previous lemma we see that this implies that the energy k_i of ω_i satisfies:

- $4k_i \geq \min(4r_i + 5, 2r_i - 4)$ if $1 < i < t$.
- $4k_i \geq \min(4r_i + 3(1 + p_g(X_i)), 2r_i)$ if $i = 1, t$.

Thus, since each $r_i \geq 3$, we see that in all cases

$$4k_i \geq 4r_i + 4(1 + p_g(X_i)).$$

Of course, the total energy of ω is $8\pi^2 \sum_{i=1}^t k_i$. In light of the above estimates and the fact that $p_g(S) = \sum_{i=1}^t p_g(X_i) + (t-1)$ we have

$$\sum_{i=1}^t 4k_i \geq \sum_{i=1}^t 4r_i + 4(p_g(S) + 1) \geq 4k.$$

Since we know that the energy of ω is at most $8\pi^2 k$, it follows that $\sum_{i=1}^t k_i = k$; i.e., that ω has the same energy as the ω_t.

This means that $4k_i = 4r_i + 4(1 + p_g(X_i))$ for all $1 \leq i \leq t$, and hence that k_i is greater than $(r_i - \dim(\chi(X_i)/2)/2$. This implies that for all i the background connection of ω_i is not flat. This completes the proof.

For any limit ω of a sequence $\omega(m)$ on (S, g_m), satisfying the conclusion of Proposition 2.3, the boundary values of ω are invariant under ι. Thus, we can speak of the boundary values of ω along a component Y_j of Y.

2.3 Weak limits of divisors that intersect Y

In this subsection we continue the study of the elliptic surface S decomposed along Y, which is a union of $(t-1)$ three-tori. We keep all the general assumptions of the previous section. Our purpose here is to also consider the nature of divisors in the moduli space $\mathcal{M}(S, g_n)$ associated to certain Riemann surfaces τ in S that cut across the splitting three-manifold Y. We assume that we have already restricted to the intersection $\Delta_\Sigma(n)$ of divisors associated to surfaces $\Sigma_1, \ldots \Sigma_r$ in S that do not intersect Y. We assume that these surfaces satisfy the numerical conditions of Proposition 2.3, so that its conclusion also holds for $\Delta_\Sigma(m)$ for all $m >> 0$. We shall now consider the nature of divisors associated to surfaces embedded in S in such a way that they intersect Y. We shall concentrate attention on a neighborhood of $\Delta_\Sigma(m)$ in the compactified moduli space $\overline{\mathcal{M}}_k(S, g_m)$. First though, let us list the assumptions that we shall make on these surfaces.

We shall assume that each surface τ_n intersects each component Y_j of Y in a single simply closed curve which is homotopic to a positive multiple $c_j(\tau_n)$ of λ_j. In the case of an elliptic surface it follows that for all n, j, j' we have $c_j(\tau_n) = c_{j'}(\tau_n)$. In this case we denote the common value by $c(n)$. It is the intersection number of τ_n with the generic fiber of the elliptic fibration. As in the case of the Σ_a we choose the τ_n to meet each other and the Σ_a transversely and to be such that the image of the restriction map $\mathcal{M}(Z) \to \mathcal{B}(\tau_n \cap Z)$ lies in $\mathcal{B}^*(\tau_n \cap Z)$. We choose a generic representative divisor in $\mathcal{B}^*(\tau_n)$ for $\mu([\tau_n])$. Then for each m we have the divisor $D_{\tau_n}(m) \subset \overline{\mathcal{M}}(S, g_m)$ induced by pullback. This divisor represents $\mu([\tau_n])$.

2.3.1 Exceptional levels

Let T be a three-torus. Then the $SU(2)$-character variety $\chi(T)$ is identified with the quotient of the S^1-character variety of T by the involution induced by complex conjugation on S^1. Thus $\chi(T)$ is identified with $T^3/\{x \cong -x\}$. This space has eight singular points, the eight ad-trivial flat connections, and other than that is a three-manifold. A neighborhood of each singular point is isomorphic to the cone on $\mathbf{R}P^2$, and hence the manifold of smooth points is non-orientable. Now suppose that $\lambda \subset T$ is an oriented, homotopically non-trivial simple closed curve, whose homology class is indivisible. We have the function

$$\mathrm{trace}_\lambda \colon \chi(T) \to [-2, 2]$$

given by assigning to each $[\rho] \in \chi(T)$ the trace of the holonomy of ρ around λ. The level sets of trace_λ give a "foliation" of $\chi(T)$. All leaves of this foliation except the extreme ones where $\mathrm{trace}_\lambda = \pm 2$ are two-tori. Each of the extremal leaves is a two-sphere containing four singular points. Let T^2 be a two-torus in T dual to λ. Restricting gives us a mapping $\chi(T) \to \chi(T^2)$. This map is an isomorphism on each of the extremal levels and is a double branched covering, branched over the four singular points on each of the other levels. It is easy to see that the complement of the extremal levels of trace_λ forms a manifold diffeomorphic to $(-2, 2) \times T^2$, and an orientation for the dual two-torus T^2 determines an orientation for this manifold.

Definition 2.4: Suppose that $\gamma \subset T$ is an oriented, simple closed curve whose homology class is non-trivial, but not necessarily indivisible. Then an *exceptional level associated to* γ is a component of the subset

$$\{\rho \in \chi(T) \,\big|\, \mathrm{trace}\, \rho(\gamma) = -2\}.$$

Suppose that λ is the primitive element in $H_1(T; \mathbf{Z})$ some positive multiple of which is γ. Then each exceptional level for γ is a leaf of the foliation determined by λ. If γ is homotopic to $2n$ times λ then there are n exceptional levels for γ, each of which is a two-torus leaf of the foliation. In this case, the normal bundle of each exceptional level is trivial. We trivialize it so that the oriented unit normal points toward the subset of classes of representations which have trace $+2$ on λ.

More generally, suppose that $Y = \coprod_{j=1}^{r} Y_j$ is a disjoint union of three-tori, and $\gamma = (\gamma_1, \ldots, \gamma_r)$ is a family of simple closed curves with $\gamma_j \subset Y_j$ homotopically non-trivial. We define an *exceptional level for* γ in $\chi(Y) = \prod_{j=1}^{r} \chi(Y_j)$ to be the preimage of an exceptional level for some γ_j in $\chi(Y_j)$ under the natural projection.

Remark 2.5: We shall always arrange that each component of γ is an even multiple of a primitive class. This means that each τ_n is of multiplicity $c(n) \equiv 0 \pmod 2$.

2.3.2 The divisor in the complement of the exceptional levels

Lemma 2.6: *Let (A, g) be a complete, cylindrical-end Riemannian four-manifold with boundary a disjoint union $\coprod_j Y_j$ of three-tori. Let*

$$\partial: \mathcal{M}_k(A, g) \to \chi(Y) = \prod_j \chi(Y_j)$$

be the boundary map which associates to each ASD connection the aymptotic flat connections at infinity. Let $\tau \subset Z$ be a generic smoothly embedded complete surface with cylindrical-end whose boundary at infinity meets each Y_j in a single simple closed curve γ_j which is homotopically non-trivial. Let $\gamma = \{\gamma_j\}_j$ and let $E_{\partial\tau} \subset \chi(Y)$ be the union of all the exceptional levels for γ. Then there is a determinant line bundle associated to the twisted Dirac operator over $\mathcal{M}(A, g) - \partial^{-1}(E_{\partial\tau})$. Its first Chern class is a two dimensional cohomology class defined on this subspace which we call $\mu([\tau])$.

Proof: Let Σ be an oriented Riemann surface with cylindrical ends. The twisted Dirac operator

$$\not{D}_\omega: L^2(\Sigma; S_+ \otimes E) \to L^2(\Sigma; S_- \otimes E) \tag{3}$$

is Fredholm of index zero provided that the Dirac operator

$$\not{D}_{\partial(\omega)}: L^2(\partial\Sigma; S \otimes E) \to L^2(\partial\Sigma; S \otimes E)$$

is invertible. Since the bounding spin structure on S^1 is non-trivial, the twisted Dirac operator of Equation 3 is Fredholm of index zero for connections in $\mathcal{M}(A, g) - \partial^{-1}(E_{\partial\tau})$. From this, the standard arguments applied to the determinant bundle of this operator give the result.

Remark 2.7: *In the next section we shall show that the class $\mu([\tau])$ does not extend across the exceptional levels, and in fact we shall explicitly identify the obstruction to extending it.*

Proposition 2.8: *Suppose that $\tau \subset S$ is a surface satisfying the properties listed at the beginning of this section, with the induced divisors $D_\tau(m)$ in the moduli space $\mathcal{M}(S, g_m)$. Suppose that for each component Y_j of Y the intersection $\gamma_j = \tau \cap Y_j$ is a non-zero class. Suppose that $[\omega(m)] \in D_\tau(m)$ for all $m \geq 1$, and that the energy of each $\omega(m)|_\nu$ is less than ϵ. Let $\hat{\tau} \subset Z$ be the geometric limit of τ as we pull the metric apart. Suppose that the $\omega(m)$ converge weakly to an ideal connection ω on Z. Then one of the following two things must be true:*

- *$\partial[\omega]$ is contained in an exceptional level for $\partial(\hat{\tau})$.*
- *The first case does not hold and the ideal connection $[\omega]$ is contained in $\mu(\hat{\tau})$.*

Proof: Suppose that $\omega(m)$ converges on compact sets to ω on Z. Let α_m be elements of norm one in the kernel of the Dirac operator twisted by $\omega(m)$. After passing to a subsequence, we can suppose that the $\alpha(m)$ converge on compact sets to an element α in the kernel of the Dirac operator for $\hat{\tau}$ twisted by ω. Then only thing to show is that $\alpha \neq 0$. If on a cylinder of the form $[-m, m] \times Y_j$ the connection $\omega(m)$ is sufficiently close to a flat connection Γ_j then a standard argument shows that for some constant $C > 0$ we have

$$|\alpha(t)| \leq C(|\alpha(m)|e^{\mu(t-m)} + |\alpha(-m)|e^{\mu(m-t)})$$

where μ is the absolute value of the eigenvalue of the Dirac operator on S^1 twisted by $\Gamma_j|_{\gamma_j}$ closest to zero with γ_j being $\tau \cap Y_j$. Of course, μ is non-zero provided that the trace of Γ_j on γ_j is not -2, and hence, if none of the traces is -2, then the limit α is non-zero. That is to say the only way that the limit α can be zero is for the resulting limiting flat connection on Y_j to be in an exceptional level for γ_j.

2.4 The first step in the proof of Theorem 1.3

In addition to our running assumptions on the X_i we make the following assumptions about our family of surfaces. Suppose that we have $\Sigma_1, \ldots, \Sigma_{4r}$ divisors in S that do not meet Y and divisors $\tau_1, \ldots, \tau_s$ in S which do meet Y. We assume that these surfaces are in general position and meet Y transversally. Furthermore, we assume that:

- $4r + s = d(k)$.
- If $4r_i$ denotes the number of the Σ_j contained in X_i then r and the r_i satisfy the hypotheses of Proposition 2.3.
- The intersection of τ_n with each component Y_j of Y is a simple closed curve which is a multiple $c(n)$ of λ_j.

- For each n the number $c(n)$ is even and positive.
- If $n \neq n'$ then the maximal power of 2 that divides $c(n)$ is different from the maximal power of 2 that divides $c(n')$.

Proposition 2.9: *Under the above assumptions, if $[\omega(m)] \in \mathcal{M}_k(S, g_m)$ is a sequence of connections contained in the intersection*

$$\Delta(m) = D_\Sigma(m) \cap D_{\tau_1}(m) \cap \cdots \cap D_{\tau_s}(m)$$

with a weak limit ω, an ASD connection on Z, then:

- *The $\omega(m)$ converge strongly to ω, and ω has compatible boundary values.*
- *For each component Y_j of Y there is a unique $\tau_{n(j)}$ such that the boundary value of ω along Y_j is in an exceptional level for $\tau_{n(j)}$.*
- *For $j \neq j'$ we have $n(j) \neq n(j')$.*

Proof: It follows from Proposition 2.3 that the limit ω is a strong limit in the tubes ν and hence that its boundary values are compatible, i.e., invariant under ι. Because we have chosen the $c(n)$ to be divisible by different powers of 2, it follows that the boundary value of ω along Y_j can lie in an exceptional level of at most one of the $\partial\tau_n$. Next, we show that it does indeed lie in one. Suppose that of the $t-1$ flat connections at infinity, exactly a of them lie in an exceptional level for one of the τ_n, for some $0 \leq a \leq t-1$. An easy computation from the formula in Theorem 9.0.5 of *[14]* shows that the dimension of the subspace of $\mathcal{M}_k(Z, g)$ with compatible boundary values is $2d(k) - t + 1$. Since we can assume that for generic metrics the boundary map from the subspace of ASD connections with compatible boundary values to $\chi(Y) = \prod_j \chi(Y_j)$ is transverse to all products of exceptional levels, imposing the condition that the boundary values lie in a exceptional levels gives a subspace of ASD connections on Z of dimension $2d(k) - t + 1 - a$. Our limiting connection ω lies in this subspace and also lies in the intersection of at least $d(k) - a$ divisors (those divisors associated with $\Sigma_1, \ldots, \Sigma_{4r}$ as well as the divisors, of which there are at least $s - a$, associated with the τ_n for which ω is not exceptional on any component of infinity). By general position, we see that $2(d(k) - a) \leq 2d(k) - t + 1 - a$ or $t \leq a + 1$. Since $a \leq t - 1$, it follows that $a = t - 1$, and that ω lies in exactly $d(k) - t + 1$ divisors. This implies the last two items of the proposition. It also implies that the set of Uhlenbeck limits of sequences of connections in $\Delta(m)$ is zero dimensional. The usual compactness arguments show that this set is in fact finite and that a limit point ω can have no points on concentrated curvature, i.e., ω is the strong limit of the $\omega(m)$.

The following is simply a reformulation of the above result.

Corollary 2.10: *Let us keep the assumptions as above. Given $\epsilon > 0$ then for all m sufficiently large each ω with $[\omega] \in \Delta(m)$ in the compactified moduli space satisfies the following:*

- *ω has no points of highly concentrated curvature, i.e. the pointwise norm of F_ω is uniformly bounded independent of ω and m.*

- *The energy of ω on ν is less than ϵ.*
- *There is an injective function $\varphi_\omega\colon\{1,\ldots,t-1\}\to\{1,\ldots,s\}$ denoted $j\mapsto n(j)$ such that for each $1\le j\le t-1$ the restriction of ω to the component ν_j of ν corresponding to Y_j is close to product connection Γ_j which representsa point in an exceptional level for $\tau_{n(j)}\cap Y_j$.*
- *$[\omega]$ is close to a point of intersection of the divisors associated to the Σ_ℓ, $1\le\ell\le 4r$, and the τ_n for n not in the image of φ_ω.*

Furthermore, the function φ_ω and the point of intersection of the other divisors are uniquely determined by $[\omega]$. The estimate on the distance between $[\omega]$ and this point of intersection, measured in the L^2_2-topology of the restrictions to the compact subset $\overline{S-\nu}$, goes to zero as $m\mapsto\infty$. Lastly, we can choose Γ_j in the exceptional level for $\tau_{n(j)}$ such that the estimate on difference between the restriction of ω to ν_j and the flat connection Γ_j, measured in the $L^2_{2,\delta}$-topology on ν, also goes to zero as $m\mapsto\infty$.

3 A "Product Formula" Evaluating the Donaldson Polynomial of S in Terms of Invariants of the X_i

We carry over the notation and assumptions from the last section. In this section we use the result on weak limits from the last section, together with some gluing results to evaluate

$$\gamma_k(S)([\Sigma_1],\ldots,[\Sigma_{4r}],[\tau_1],\ldots,[\tau_s]).$$

At this point the answer will be in terms of invariants of the individual components Z_i of Z, i.e., cylindrical-end manifolds diffeomorphic to the interiors of the X_i. In Section 6 we shall identify the invariants that arise in this section with values of Donaldson polynomials of the surfaces S_i obtained from the Z_i by filling in solid tori.

The character variety $\chi(Y)$ is identified with the ι-invariant characters in $\chi(\tilde{Y})$. We denote by $\mathcal{M}_k(Z,\iota)\subset\mathcal{M}_k(Z)$ the preimage $\partial^{-1}(\chi(Y))$. We have a commutative diagram

$$\begin{array}{ccc}\mathcal{M}_k(Z) & \xrightarrow{\partial} & \chi(\tilde{Y})\\ \uparrow & & \uparrow\\ \mathcal{M}_k(Z,\iota) & \xrightarrow{\partial} & \chi(Y).\end{array}$$

3.1 The first main result

The character variety $\chi(Y)=\prod_{i=1}^{t-1}\chi(Y_j)$ is of dimension $3(t-1)$. For any injective function $\varphi\colon\{1,\ldots,t-1\}\to\{1,\ldots,s\}$ let us consider the subset E_φ consisting of the intersection of the exceptional levels for $\gamma_j=\tau_{n(j)}\cap Y_j$, for $i=1,\ldots t-1$. It is a smooth submanifold of codimension $(t-1)$ contained in the smooth point of $\chi(Y)$, and it has a trivial normal bundle. In fact its normal bundle is naturally a product of interval bundles, each of which is oriented according to our convention that the positive normal vector for an exceptional level in $\chi(Y_j)$ point toward the level $\{\rho|\text{trace}\,\rho(\lambda_j) = +2\}$. Since the metric on Z is a generic cylindrical-end metric,

$$\partial\colon \mathcal{M}_k(Z,\iota) \to \chi(Y)$$

is transverse to this submanifold. We set

$$\mathcal{N}_k(Z,\iota,\varphi) \subset \mathcal{M}_k(Z,\iota)$$

equal to the preimage. It is of dimension $2(d(k)-t+1)$. The choice of orientation of $H^2_+(Z)$ and of the fibers of the elliptic fibration determine an orientation of $H^2_+(S)$ and $H_1(Z)$, and hence give an orientation of $\mathcal{N}_k(Z,\iota,\varphi)$.

We also denote by Δ_φ the intersection of the divisors D_{Σ_a} and the D_{τ_n} for those n not in the image of φ. Since we have chosen the D_{Σ_a} and D_{τ_n} generically, and since the image of φ consists of $t-1$ elements, this means that $\Delta_\varphi \subset \mathcal{M}_k(Z,\iota)$ is of codimension $2(d(k)-t+1)$, has an oriented normal bundle, and meets $\mathcal{N}_k(Z,\varphi)$ transversally in a finite set of points. Each point has a sign coming from the orientation of $\mathcal{N}_k(Z,\iota,\varphi)$ and the normal orientations of the divisors associated to the surfaces. Here is the main technical result of this section.

Proposition 3.1: *For each injective function φ as above, for each point $p \in \Delta_\varphi \cap \mathcal{N}_k(Z,\varphi)$, and for all m sufficiently large, there is exactly one point of $\Delta(m)$, counted with sign, whose assoicated function is φ and whose associated point of intersection is p. The sign of this point in $\Delta(m)$ is the same as the sign of p as a point of intersection of Δ_φ and $\mathcal{N}_k(Z,\iota,\varphi)$.*

Combining this with Corollary 2.10 gives us our product formula.

Theorem 3.2: *We have the following expression*

$$\gamma_k(S)([\Sigma_1],\ldots,[\Sigma_{4r}],[\tau_1],\ldots,[\tau_s]) = \sum_\varphi \Delta_\varphi \cap \mathcal{N}_k(Z,\iota,\varphi)$$

where φ ranges over the injective functions $\{1,\ldots,t-1\} \to \{1,\ldots,s\}$.

Now let us turn to the proof of Proposition 3.1.

Proof: By a standard product argument we can reduce the proof of Proposition 3.1 to the following special case.

Lemma 3.3: *Suppose that $t=2$. Suppose that $E \subset \chi(T)$ is an exceptional level for τ_s. Then the intersection of*

$$D_{\Sigma_1} \cap \cdots \cap D_{\Sigma_r} \cap D_{\tau_1} \cap \cdots \cap D_{\tau_{s-1}}$$

inside the fibered product

$$\mathcal{M}_{k_1}(Z_1) \times_{\chi(T)} \mathcal{M}_{k_2}(Z_2)$$

is an oriented one-manifold A which is transverse to E. (Here, $k_1 + k_2 = k$.) Consider a small interval in this one-manifold which cuts E transversely. The preimage of A in the fibered product

$$\mathcal{M}^0_{k_1}(Z_1) \times_{\mathcal{R}(T)} \mathcal{M}^0_{k_2}(Z_2)/SO(3)$$

is a cylinder. For each $m >> 0$, this cylinder deforms slightly to give an oriented cylinder $C(m)$ in the intersection of the divisors

$$\Delta_\Sigma(m) \cap D_{\tau_1}(m) \cap \cdots \cap D_{\tau_{s-1}}(m)$$

in $\mathcal{M}_k(S, g_m)$. There is exactly one point of $C(m)$ which lies in the divisor D_{τ_n}. The sign of this point is the same as the sign of p as a point of intersection of A with E.

Proof: The fact that the intersection $D_{\Sigma_1} \cap \cdots \cap D_{\Sigma_r} \cap D_{\tau_1} \cap \cdots \cap D_{\tau_{s-1}}$inside the fibered product

$$\mathcal{M}_{k_1}(Z_1) \times_{\chi(T)} \mathcal{M}_{k_2}(Z_2)$$

is an oriented one-manifold A which is transverse to E is a straightforward dimension count. Let $\gamma = \tau \cap Y$. By deforming the one-manifold in the fibered product by a homotopy not changing any of the intersections with the exceptional levels we can assume that the arc A has both end points in levels where the holonomy around γ is the identity and crosses exactly one exceptional level for γ, let us say in the positive direction. As we have noted, A in the fibered product corresponds to a cylinder C in the space of connections on τ. We split τ along γ into $\tau^+ \cup_\gamma \tau^-$. The arc A is represented by an interval of connections $\omega^\pm(t), 0 \leq t \leq 1$, on bundles $P_\pm$ over $\tau^\pm$. The restriction of $P_\pm$ to $\partial\tau^\pm$ is split as a sum $L_\pm \oplus L_\pm^{-1}$ of complex line bundles. The connections $\omega^\pm(t)$ preserve these decompositions over $\partial\tau^\pm$. The holonomy of $\omega^\pm|_{L_\pm}$ around the oriented loop $\partial\tau^\pm$ is $\exp(2\pi i t)$. The double covering $\tilde{C}$ of the cylinder $C = A \times S^1$ parameterizing all a pairs (t, φ) where φ is an isomorphism of bundles with connections

$$(L_+, \omega^+(t)|_{L_+}) \cong (L_-^{-1}, \omega^-(t)|_{L_-^{-1}}).$$

The universal $SO(3)$-bundle on $\tau \times C$ lifts to a universal $SU(2)$-bundle on $\tau \times \tilde{C}$. We shall complete this universal bundle to an $SU(2)$-bundle over $\tau \times S^2$ by filling in a bundle over $\tau \times D^2$ at each end of $\tau \times \tilde{C}$. We do this so that the universal connection extends. Let us consider one end. The circle of bundles over τ is obtained by gluing together P_+ and P_- by a circle of bundle isomorphism preserving the splittings and connections over the boundaries. But since we are over an end of $\tilde{C}$, the connections restricted to the boundary are trivial. Hence, the circle of bundle isomorphism described by the boundary component of $\tilde{C}$ in fact lies in a space isomorphic to $SU(2)$ of bundle isomorphisms preserving the connections on the boundary. Hence, we can extend the S^1 of bundle isomorphism to a two-disk inside this $SU(2)$-family. Since these bundle isomorphisms preserve the connections over the boundary, there is a natural extension over this family of the universal connection over $\tau \times \tilde{C}$.

Since this extended family is described as a family of glued-up connections, and since the traces of the holonomy over the disks that we have added are all $+2$, and hence not exceptional, it follows that the intersection of the divisor given by the jumping locus of the kernel of the twisted Dirac operator with the cylinder $\tilde{C}$ is equal to the second Chern class of the universal bundle over $\tau \times S^2$. Since $\tilde{C}$ double covers C, the intersection number of the divisor with C is one-half of this number.

It is clear that the second Chern form of the universal connection is zero outside the product of a neighborhood of γ in τ with $\tilde{C}$. Taking coordinates (t, θ) on $\tilde{C}$ and (s, ψ) with $0 \leq s \leq 1$ on the subset of τ one sees that over this subset the connection splits as a sum of line bundles and is given in one of the line bundle factors by

$$i\theta ds - itd\psi.$$

A simple computation shows that the integral of the second Chern form over this subset is 2, giving the final result of $+1$ for the intersection number with C. This complete the proof of Theorem 3.2

4 Preliminary Results Needed to Identify the Terms in Theorem 3.2

We shall deduce Theorem 1.3 from Theorem 3.2. The idea is to relate the individual terms in the product formula given in Theorem 3.2 to invariants for the closed four-manifolds S_i obtained by filling in each boundary component with a solid torus. In our case, the filled in manifolds are elliptic surfaces of geometric genus 1 for $1 < i < t$ and with $1 \leq p_g \leq 2$ for $i = 1, t$. Hence, the invariants that we need to make our computation are computed in *[15]* and *[1]*. We also use a periodicity relation due to Kronheimer (see Section 4.7).

In this section we discuss some of the gluing results that are necessary in order to carry out this argument.

4.1 Orientations of the Moduli Spaces of cylindrical-end manifolds

Let Z be a complete oriented Riemannian manifold with a cylindrical end isometric to $[0, \infty) \times T$ where T is the three-torus. We take the orientation on T so that this isometry is orientation-preserving.

One fact will be used in several places.

Claim 4.1: *$H^2(T; \mathbf{R})$ is canonically oriented.*

Proof: This is immediate from thc cquality

$$H^2(T; \mathbf{R}) = \Lambda^2 H^1(T; \mathbf{R}).$$

Lemma 4.2: *Suppose that $b_1(Z) = 0$. Let $[\rho]$ be a smooth point of $\chi(T)$ and let U be an open subset of the smooth points of $\chi(T)$. Then, for a generic metric with cylindrical end on X the following hold for the boundary map $\partial\colon \mathcal{M}_k(Z) \to \chi(T)$:*

1. *the intersection of the fiber $\partial^{-1}([\rho])$ with the subspace of irreducible connections is a smooth orientable manifold, and*
2. *the intersection of the preimage $\mathcal{M}_k(Z,U) = \partial^{-1}(U)$ with the subspace of irreducible connections is a smooth manifold.*

Furthermore, for a metric for which the previous two conditions hold, a choice of orientation for $H^2_+(Z;\mathbf{R})$ determines:

1. *an orientation for $\partial^{-1}([\rho])$ away from the reducible connections, and*
2. *an isomorphism between the orientation bundle of $\mathcal{M}_k(Z,U)$ away from the reducible connections and pullback via ∂ of the orientation bundle of U.*

Proof: In Section 8.8 of *[14]* it was proved that the first two items of the lemma hold for a generic metric on Z with cylindrical end. It was also proved that the intersection of the fiber over a smooth point of the character variety with the subspace of irreducible connections is orientable. We consider the question of orienting the tangent bundle along the fibers. By what we just asserted this bundle is orientable along each fiber. Thus, to show that this bundle is orientable, it suffices to show that it is orientable over a generating set of loops in the open subset of smooth points of $\chi(T)$. Let X be the compact manifold with boundary whose interior is diffeomorphic to Z. We can choose a generating set such that for each element α in this set there is an identification $i_\alpha\colon \partial D^2 \times T^2 \cong T$ with α homotopic to the image under i_α of a loop in the subset of smooth points of $\chi(T^2)$. Let $\hat{X}_\alpha$ be the closed manifold obtained from X by gluing in $D^2 \times T^2$ by the map i_α. The gluing theorem of *[13]* gives an identification of the preimage P_α under ∂ of the image under i_α of the smooth points of $\chi(T^2)$ with an open subset of $\mathcal{M}_k(\hat{X}_\alpha)$. The moduli space $\mathcal{M}_k(\hat{X}_\alpha)$ is orientable and is oriented by a choice of orientation for $H^2_+(\hat{X}_\alpha)$. Thus, P_α is orientable. Since the subset of smooth points of $\chi(T^2)$ is also orientable, this implies that the tangent bundle along the fibers is orientable over the image under i_α of the smooth points of $\chi(T^2)$, and hence is orientable over the loop α.

To orient the tangent bundle along the fiber it suffices to orient it at one point. We consider a point in the preimage under ∂ of the image under i_α of a smooth point of $\chi(T^2)$. As we have seen, an orientation of $H^2_+(\hat{X}_\alpha)$ determines an orientation of P_α. Clearly, an orientation T^2 determines an orientation of $\chi(T^2)$. Together, these give an orientation for the fiber.

On the other hand, these two orientations induce an orientation for $H^2_+(X) = H^2_+(Z)$. Changing the orientation of T^2 changes both the induced orientation on $H^2_+(Z)$ and on $\chi(T^2)$ but leaves the orientation of P_α invariant. It follows that the fiber orientation is determined by the induced orientation of $H^2_+(Z)$.

We must show that this fiber orientation is independent of the way we glue in the $D^2 \times T^2$. To do this we will compare the orientation defined above with an orientation of the fiber complex over the trivial connection. The cohomology groups of the fiber complex $F_\delta(\omega)$ for ω a trivial connection are $H^0 = H^1 = 0$ and $H^2 = H^2_{\geq 0}(Z)$. Thus to orient the fiber complex we must orient $H^2_{\geq 0}(Z)$. From the long exact sequence of the pair, we see that this is the same thing as orienting $H^2_+(Z)$ since, by Claim 4.1, the space $H^2(T)$ is canonically oriented. This of course is a formal orientation of the deformation complex, but anytime the fiber of $\mathcal{M}_k(Z)$ over the trivial connection is smooth, it is oriented. Now we assume that the fiber over the trivial connection is smooth, and we wish to consider the fiber over a nearby non ad-trivial representation. A local desription of the thickened moduli space (see Theorem 9.0.5 of *[14]*) near a smooth point of the fiber over the trivial connection is $H^1(F_\delta(\omega)) \times (H^1(T) \otimes su(2))$. The representation variety sits inside the second factor as the decomposable elements of the tensor product. The fiber over a nearby non ad-trivial connection is the preimage of the stable set of a two-sphere orbit of the natural action of $SO(3)$ on the tensor product. Thus, given an orientation of the fiber over the trivial connection, to orient this nearby fiber we must orient the stable set of a two-sphere orbit. The normal space to the stable set is identified with the tangent space to the character variety times the unstable set of an non ad-trivial flat connection. This character variety is locally oriented by the choice of orientation of T and the stablizer. The unstable set is oriented by the choice of orientation of the stablilizer. Thus the product is oriented once an orientation of T is chosen. But as we noted at the beginning of this discussion, we orient T as the boundary at infinity of Z. This gives a second way in which an orientation of $H^2_+(Z)$ leads to an orientation of a fiber over an non ad-trivial connection.

Now, fixing an orientation of $H^2_+(Z)$, let us compare the two induced orientations of the tangent bundle along the fibers over the smooth points of $\chi(T)$. The gluing theorem of *[13]* at an irreducible connection in the fiber over the trivial connection tells us that the local model for $\mathcal{M}_k(\hat{X}_\alpha)$ is the product

$$H^1(F_\delta(\omega)) \times (H^1(T^2) \otimes su(2)).$$

The first term in this product is oriented as described above from an orientation of $H^2_+(Z)$. The second term is oriented by a choice of orientation for T^2. Using the cohomology sequence relating X and $\hat{X}_\alpha$ it is easy to check that the induced orientation on $\mathcal{M}_k(\hat{X}_\alpha)$ obtained in this way agrees with the one determined in the usual way form the orientation of $H^2_+(\hat{X}_\alpha)$ determined by the chosen orientations of $H^2_+(Z)$ and T^2.

4.2 Orientations and the gluing theorem

We turn now to the gluing theorem for manifolds decomposed along three-tori. We examine this result first in the simplest case of a single three-torus splitting the manifold into two pieces.

Suppose that M is a closed oriented 4-manifold with $b_1(M) = 0$. Let $T \subset M$ be a smooth embedding with $M - T$ orientation-preserving diffeomorphic to $Z \coprod Z'$ where Z and Z' are oriented complete Riemannian manifolds with cylindrical ends. Let X and X' be the compact manifolds with boundary whose interiors are diffeomorphic respectively to Z and Z',

Lemma 4.3: *Orientations for any two of $H^2_+(Z)$, $H^2_+(Z')$, and $H^2_+(M)$ determine an orientation for the third.*

Proof: There is an exact sequence

$$H^2(Z, \partial X) \oplus H^2(X', \partial X') \to H^2(M) \to H^2(T) \to 0.$$

Any maximal positive definite subspace of $H^2(M)$ projects onto $H^2(T)$ under the last map in this sequence. For an appropriate choice of such a subspace the kernel of this projection is the injective image of the direct sum of maximal positive definite subspaces in $H^2(X, \partial X)$ and $H^2(X', \partial X')$. The result is now immediate from Claim 4.1.

Lemma 4.4: *Suppose that we have points $a \in \mathcal{M}_k(Z)$ and $b \in \mathcal{M}_\ell(Z')$ with the following properties:*

- *$\partial(a) = \partial(b)$ is a smooth point of $\chi(T)$,*
- *a and b are smooth points of the moduli spaces and are irreducible, and*
- *the differential of ∂ is surjective at a and b, i.e., a and b are smooth points of the fibers of ∂.*

Then orientations for the fibers of ∂ at a and b determine an orientation for the fiber product of the based moduli spaces

$$\mathcal{M}^0_k(Z) \times_{\mathcal{R}(T)} \mathcal{M}^0_\ell(Z')$$

at any point $[a^0, b^0]$ where a^0 and b^0 project to a and b, respectively.

Proof: We pick arbitrarily one of the two cylindrical manifolds, say Z. This choice determines an orientation of T (as the boundary of X) and hence an orientation for $\mathcal{R}(T)$. Choose an orientation for the stabilizer of $\partial(a)$. The fiber of $\mathcal{M}^0_k(Z) \to \mathcal{R}(T)$ at $\partial(a)$ has a free action of Stab$(\partial(a))$ with quotient identified with the fiber of $\mathcal{M}_k(Z) \to \chi(T)$. Thus, we orient the fiber of $\mathcal{M}^0_k(Z) \to \mathcal{R}(T)$ by using the sum of the given orientations on the fiber of $\mathcal{M}_k(Z) \to \chi(T)$ and the orientation on the stabilizer of $\partial(a)$. (Notice that since the fiber of $\mathcal{M}_k(Z) \to \chi(T)$ is even dimensional, the order of the two factors is unimportant for the final orientation.) We orient $\mathcal{M}^0_\ell(Z')$ using the given orientation on T and using the orientation of the stabilizer of $\partial(b)$ induced from the chosen orientation of $\partial(a)$ and the isomorphism between these stabilizers.

We have now oriented all the pieces of the fibered product, and hence of the fibered product itself. If we reverse the orientation on Stab$(\partial(a))$ then we reverse the orientation of each of the based fibers but not of $\mathcal{R}(T)$. Hence, the orientation of the fibered product is unchanged. If we use Z' instead of Z, but the same orientation on the stabilizers, then we reverse the orientation of

T and hence of $\mathcal{R}(T)$ but leave invariant the orientations of the based fibers. On the other hand we are changing the order of the factors, replacing the order (based fiber for $Z, \mathcal{R}(T)$, based fiber for Z') with (based fiber for $Z', \mathcal{R}(T)$, based fiber for Z). Since each of the pieces is odd dimensional, this reverses the orientation. Hence, the final orientation is unchanged.

Lemma 4.5: *Suppose that we have chosen orientations for $H^2_+(Z)$ and $H^2_+(Z')$. According to the above two lemmas this determines orientations for $H^2_+(M)$ and for the fibered product*

$$\mathcal{M}^0_k(Z) \times_{\mathcal{R}(T)} \mathcal{M}^0_\ell(Z').$$

The orientation for $H^2_+(M)$ determines an orientation for $\mathcal{M}^0_{k+\ell}(M)$. The gluing construction gives an embedding of an open subset of $\mathcal{M}^0_k(Z) \times_{\mathcal{R}(T)} \mathcal{M}^0_\ell(Z')$ around the point $[a^0, b^0]$ onto an open subset of $\mathcal{M}^0_{k+\ell}(M)$. This map is an orientation-preserving with respect to the orientations we have specified here. There is the similar result for the unbased gluing map.

Proof: Since the orientation of the fiber over a smooth point of $\chi(T)$ determined by an orientation of $H^2_+(Z)$ is compatible with the orientation of the fiber over the trivial connection, it suffices to consider formal orientations when gluing together at points in the fibers over the trivial connection. The model space for the space of patching parameters is

$$H^1(F_\delta(a^0)) \times (H^1(T) \otimes su(2)) \times H^1(F_\delta(b^0)).$$

As before, a straightforward cohomological argument shows that the induced orientation agrees with the orientation of the moduli space inherited from the orientation on $H^2_+(M)$.

We have the following straightforward generalization of Lemma 4.5.

Lemma 4.6: *With notation as above, suppose that $b_1(M) = 0$. Choose orientations for $H^1(Z)$ and $H^2_+(Z)$. This determines an orientation for $\mathcal{M}_k(Z, \iota)$. The orientations for $H^1(Z)$ and $H^2_+(Z)$ also determine an orientation for $H^+_2(M)$ and hence for $\mathcal{M}_k(M)$. With these induced orientations on the moduli spaces, the gluing map is an orientation-preserving diffeomorphism onto an open subset.*

4.3 Comparison of invariants for closed manifolds and cylindrical-end manifolds

Now let us apply this general discussion to a situation relevant to the computations that we wish to do. Throughout this subsection we suppose that we have a closed, oriented four-manifold S with $b_1(S) = 0$ and an embedded oriented two-torus $F \subset S$ with trivial self-intersection representing a homology class which is of infinite order. Let Z be the cylindrical manifold determined by $S - F$. With our hypotheses, it is the case that $b_1(Z) = 0$. We have the boundary map

$$\partial \colon \mathcal{M}_k(Z) \to \chi(T).$$

Let $\lambda \subset T$ be the loop bounding a disk in the neighborhood of F, a disk which intersects F transversely in one point, a point of positive intersection. Let $\chi_2(T) \subset \chi(T)$ denote the level set $\{[\rho] \in \chi(T) \big| \text{trace}\,\rho(\lambda) = 2\}$.

By the gluing theorem we identify the preimage $\partial^{-1}(\chi_2(T))$ in $\mathcal{M}_k(Z)$ with an open subset of $\mathcal{M}_k(S)$. The given orientation for F determines an orientation for this level set. Thus, a choice of orientation for $H^2_+(Z)$ detemines an orientation of the preimage of this level set in $\mathcal{M}_k(Z)$. Furthermore, the orientation of F and $H^2_+(Z)$ determine an orientation of $H^2(S)$ and hence of the moduli space $\mathcal{M}_k(S)$. With these orientations, the gluing diffeomorphism is orientation-preserving. This leads immediately to the next result.

Proposition 4.7: *Suppose the dimension of $\mathcal{M}_k(Z)$ is $2d(k)+1$ and that $\Sigma_1, \ldots, \Sigma_{d(k)}$ are oriented surfaces embedded in Z. Then the Σ_i define divisors D_{Σ_i} in $\mathcal{M}_k(Z)$, obtained by restricting universal divisors in $\mathcal{B}^*(\Sigma_i)$. Choosing these universal divisors generically, the intersection of the induced divisors is a one-manifold C in $\mathcal{M}_k(Z)$ with an oriented normal bundle. The intersection of the image of C in $\chi(T)$ with the level $\chi_2(T)$, oriented by the choice of orientation of F, is equal to the value of the Donaldson invariant*

$$\gamma_k(S)([\Sigma_1], \ldots, [\Sigma_{d(k)}]).$$

Notice that the choice of orientation for F in fact orients any level

$$\chi_a(T) = \{[\rho] \in \chi(T) \big| \text{trace}\,\rho(\lambda) = a\}.$$

The intersection number of the image of C with any level $\chi_a(T)$ with $-2 < a < 2$ is equal to

$$2\gamma_k(S)([\Sigma_1], \ldots, [\Sigma_{d(k)}]).$$

Proof: The first part is immediate from Lema 4.5. The second statement is a reflection of the fact that the divisors D_{Σ_i} are defined on all of $\mathcal{M}_k(Z)$ and that as $a \mapsto 2$ the level $\chi_a(T)$ converges to an oriented double covering of the level $\chi_2(T)$.

4.4 The $SO(3)$-case

Lemma 4.8: *There is a projectively flat connection on a $U(2)$-bundle over $D^2 \times T^2$ whose restriction to the boundary reduces to a flat $SU(2)$-connection whose holomony is trivial on $\pi_1(T^2)$ and is $-\mathrm{Id}$ on the boundary of D^2.*

Proof: Let ω be a connection on the trivial line bundle L over D^2 whose curvature has integral $i\pi$ over the disk and vanishes near the boundary. Consider the connection $\omega \oplus \omega$ on $L \oplus L$. Clearly, this is a projectively flat connection whose restriction to ∂D^2 is flat with holomony $-\mathrm{Id}$. Tensoring this connection with the trivial connection on the trivial line bundle over F gives the required connection.

Notice that the connection in this lemma yields a trivial $SO(3)$-connection The existence of such a connection leads to the following gluing theorem.

Proposition 4.9: *Let F, S, Z, λ be as in the previous subsection. Then there is an identification between the preimage in $\mathcal{M}_k(Z)$ of the smooth points of $\chi_{-2}(T)$ with an open subset of $SO(3)$-moduli space $\mathcal{M}_{-4k,[F]}(S)$, as defined in [12].*

Now we wish to consider the questions on orientations for the $SO(3)$-moduli space. Fix an orientation on $H^2_+(Z)$ and an orientation of F. As before this determines an orientation of $H^2_+(S)$ and hence an orientation for $\mathcal{M}_{-4k,[F]}(S)$. On the other hand, as we have seen, an orientation of $H^2_+(X)$ determines an orientation of the fibers of $\mathcal{M}_k(F) \to \chi(T)$ above any smooth point. Using the orientation of F to orient $\chi_{-2}(T)$, then yields an orientation of the preimage of the smooth points of $\chi_{-2}(T)$.

Lemma 4.10: *With respect to the orientations given in the above paragraph, the gluing isomorphism of Proposition 4.9 is orientation-preserving.*

Proof: As before, there is also a gluing theorem for gluing near smooth points in the fibers over an ad-trivial connection in $\chi(T)$. An open subset of the moduli space $\mathcal{M}_{-4k,[F]}(\hat{X})$ is identified with the product of an open subset of $H^1(F) \otimes su(2)$ with the fiber over the ad-trivial connection. The convention in the $SO(3)$-case is to write the cohomology as $H^1(F) \otimes (\mathbf{R} \oplus \mathbf{C})$, to orient the first factor via the orientation of $H^1(F)$, and to orient the second factor as a complex vector space. Since the element $1 \in \mathbf{R}$ acts via bracket as multiplication by i on $\mathbf{C}$, these orientations together give the standard orientation of $su(2)$. From this, the argument follows as in the $SU(2)$ case.

Corollary 4.11: *With notation as above, suppose that the dimension of $\mathcal{M}_k(Z)$ is $2d(k)+1$ and that $\Sigma_1, \ldots, \Sigma_{d(k)}$ are oriented surfaces embedded in X. Let D_{Σ_i} be the associated divisors which we choose generically. Then their intersection is a one-manifold C whose normal bundle is oriented. The orientation of F determines an orientation of the level $\chi_{-2}(T)$ and the intersection number of the image of C with this level is equal to*

$$\gamma_{-4k,[T]}(S)([\Sigma_1], \ldots, [\Sigma_{d(k)}]).$$

We have the following corollary of this result and Proposition 4.7.

Corollary 4.12: *With S as above, $\Sigma_1, \ldots, \Sigma_{d(k)}$ are embedded surfaces which are disjoint from an embedded torus F, then*

$$\gamma_k(S)([\Sigma_1], \ldots, [\Sigma_{d(k)}]) = \gamma_{-4k,[T]}(S)([\Sigma_1], \ldots, [\Sigma_{d(k)}]).$$

In particular, if S is an elliptic surface over $\mathbf{C}P^1$ with geometric genus 1 and with multiple fibers of multiplicity m_1, m_2, and if $F \subset S$ is a generic fiber of the elliptic *fibration, then the invariant* $\gamma_{-4k,[F]}(S)$ *has an expression of the form*

$$\gamma_{-4k,[F]}(S) = \sum_{i=0}^{d/2} \alpha^{SO(3)}(k,i) Q_S^{(i)} \kappa_S^{d-2i}.$$

In this expression the coefficient of $Q_S^{(d/2)}$ *is equal to the coefficient of* $Q_S^{(d/2)}$ *in Equation 1, namely it is equal to* $m_1 m_2$.

Remark 4.13: One can establish that the leading coefficient of both these polynomials is $m_1 m_2$ directly using the machinery developed here, (cf, Theorem 3.3 of *[8]*)

4.5 The class represented by the preimage of a point

In this brief subsection we keep the assumptions on Z and we consider the class represented by the codimension-three subset

$$D_{[\rho]} = \partial^{-1}([\rho]) \subset \mathcal{M}_k(Z)$$

where $[\rho] \in \chi_2(T)$. The subset $D_{[\rho]}$ is codimension-three in $\mathcal{M}_k(Z)$ but it is codimension-two in the preimage of the level set $\chi_2(T)$. As we have seen, the preimage of this level set is identified with an open subset of $\mathcal{M}_k(S)$. Under this identification $D_{[\rho]}$ becomes the class of D_F, the divisor associated with the fiber F. The reason is that $\mu(F)$ restricted to $\chi(F)$ is represented by a smooth point in $\chi(F)$.

This leads us to the following corollary.

Corollary 4.14: *Suppose the dimension of* $\mathcal{M}_k(Z)$ *is* $2d(k)+1$ *and that* $\Sigma_1, \ldots, \Sigma_{d(k)-1}$ *are oriented surfaces embedded in* Z*, with corresponding divisors* D_{Σ_i}*. Choosing these universal divisors generically, the intersection of the induced divisors is a three-manifold* C *in* $\mathcal{M}_k(Z)$ *with an oriented normal bundle. The degree of the boundary map* $C \to \chi(T)$ *over any generic point is equal to the value of the Donaldson invariant*

$$\gamma_k(S)([\Sigma_1], \ldots, [\Sigma_{d(k)-1}], [F]).$$

Remark 4.15: This result holds even if $b_2^+(Z) = 0$ when Z is simply connected since there are only finitely many reducibles in $\mathcal{M}_k(Z)$, see Chapter 9 of *[14]*.

4.6 Another example where the $SO(3)$- and $SU(2)$-invariants agree

Lemma 4.16: *Let* S *be an elliptic surface with at most two multiple fibers over* $\mathbf{C}P^1$ *and with* $p_g(S) = 2$*. Let* $F \subset S$ *be a generic fiber of the elliptic fibration. Then for any classes* $x_1, \ldots, x_{4r+3}$ *with the property that* x_i *is orthogonal to* $[F]$ *for* $1 \leq i \leq 4r+2$ *we have*

$$\gamma_{r+3}(S)(x_1, \ldots, x_{4r+3}) = \gamma_{-4r-12,[F]}(S)(x_1 \ldots, x_{4r+3}).$$

Proof: We split S along a three-torus into two pieces $Z \cup Z'$ where Z and Z' complete (by adding in a solid tori) to elliptic surfaces $\hat{Z}$ with $p_g = 1$ and $\hat{Z}'$. We arrange things so that $\hat{Z}'$ is without multiple fibers and has $p_g = 0$. Let g be the induced metric on $\hat{Z}'$. We take $x_1, \ldots, x_{4r-2}$ to be represented by surfaces lying in Z and $x_{4r-1}, \ldots, x_{4r+2}$ to be represented by surfaces lying in Z'. We also represent x_{4r+3} by a surface τ which intersects the splitting three-torus in a single simple closed curve γ. At the expense of doubling x_{4r+3}, we can assume that this intersection is an even multiple c of an indivisible class in H_1 of the three-torus. It is easy to see that as we pull S apart along this three-torus, the weak limit of any sequence of connections in the intersection of the divisors associated to the first $4r+2$ classes must be a strong limit with energy $r+1$ on the Z-side and energy 2 on the Z'-side. When we cut down by these divisors the moduli space on the Z-side is one-dimensional manifold A and that on the Z'-side is three-dimensional space C. According to Theorem 3.2 the value of $\gamma_{r+3}(S)$ on these classes is then identified with the sum over the exceptional levels of γ of the product of the intersection of the image of A with the exceptional level with the degree over a point of this level of the map from C to $\chi(T)$. According to Proposition 4.7 the intersection number of A with each exceptional level of γ is $2\gamma_{r+1}(\hat{Z})(\Sigma_1, \ldots, \Sigma_{4r-2})$. According to remark following Corollary 4.14 the degree of the boundary map $C \to \chi(T)$ is given by $\gamma_2(\hat{Z}', g)(x_{4r-1}, \ldots, x_{4r+2}, [F'])$ where F' is the core of the solid torus added in forming $\hat{Z}'$.

The result then is

$$c\gamma_{r+1}(\hat{Z})(x_1, \ldots, x_{4r-2}) \cdot \gamma_2(\hat{Z}', g)(x_{4r-1}, \ldots, x_{4r+2}, [F']). \tag{4}$$

(From the description that we have given here it seems that, since $b_2^+(\hat{Z}') = 1$, this invariant may well depend on the metric g. In fact, because one of the classes that we are evaluating on is the fiber and since the energy is two, it turns out that any metric that arises from this construction gives the same value for the invariant. Since we make no use of this fact we do not establish it.)

The important fact here is that the terms of which we take the product, that is to say the value of the invariants for $\hat{Z}$ and $\hat{Z}'$, are independent of the level. We can compute the value of the $SO(3)$-invariant $\gamma_{-4r-12,[F]}(S)$ by using as a model for the relevant moduli space the fibered product of the based moduli spaces for Z and Z' over the maps to the representation variety of T where the map on the Z-side is the usual boundary map whereas the one on the Z'-side is the composition of the usual boundary map with the inversion on the representation variety defined by multiplying any representation by the representation whose image is contained in the center of $SU(2)$, is $+1$ on the loops contained in a torus parallel to F' and is -1 on the loop dual to this parallel torus. Using this description we see that the value of the $SO(3)$-invariant on the given set on classes is given by exactly the same sum over exceptional levels of γ and hence is also given by Equation 4.

4.7 Kronheimer's relations for the four-dimensional class

We shall also need some results due to Kronheimer (unpublished) about the four-dimensional class for elliptic surfaces. Proofs of both of these results can be found in *[17]*.

Lemma 4.17: *Let S be a closed, oriented 4-manifold with $b_2^+(S) > 1$. Suppose that S contains the E_8-plumbing diagram of two-spheres (for example suppose that S is an elliptic surface over $\mathbf{C}P^1$ with $p_g(S) \geq 1$). Then suppose that $\alpha \in H_2(S)$ is an integral class of self-intersection -2 represented by a smoothly embedded two-sphere. Then for any classes $x_1, \ldots, x_r \in (H_0(S) \oplus H_2(S))$ with trivial intersection with α we have*

$$-\gamma_k(x_1, \ldots, x_r, \nu) = \gamma_k(x_1, \ldots, x_r, \alpha, \alpha)$$

where $\nu = \mu(pt)$ is the generating four-dimensional class. The same result holds for the $SO(3)$-invariants.

Lemma 4.18: *With notation as in the previous lemma we have*

$$\gamma_{k+1}(x_1, \ldots, x_r, \nu, \nu) = 4\gamma_k(x_1, \ldots, x_r).$$

The same result holds for the $SO(3)$-invariants.

Lemma 4.17 and the formula for the leading coefficient of the Donaldson polynomial for the $K3$ surfaces lead immediately to the following corollary.

Corollary 4.19: *Let S be a $K3$-surface and let $p_1 \in H^4(\mathcal{M}_k(S))$ be the first Pontrjagin class of the base point bundle. Then for any classes $x_1, \ldots, x_{4r} \in H_2(S)$ we have*

$$\gamma_{r+2}(x_1, \ldots, x_{4r}, p_1) = -8Q_S^{(2r)}(x_1, \ldots, x_{4r}).$$

The same result holds for the $SO(3)$-invariants.

Putting Lemma 4.17 and 4.18 together and using the form of the invariants for elliptic surfaces we have the following corollary.

Corollary 4.20: *Let S be an elliptic surface over $\mathbf{C}P^1$ with $p_g \geq 1$ and with at most two multiple fibers. Then the coefficients $\alpha(k,n)$ given in Formula 2 satisfy the periodicity relation*

$$\frac{\alpha(k,n)}{d(k)!} = \frac{\alpha(k',n)}{d(k')!}.$$

The same periodicity relation holds for the coefficients $\alpha^{SO(3)}(k,n)$ in the $SO(3)$ Donaldson polynomials.

5 Further Results Needed to Identify the Terms in Theorem 3.2

The idea in this section is to relate the individual terms on the right-hand-side of the equation in Theorem 3.2 to a product of Donaldson polynomials of the ellitpic surfaces S_i. The problem is that these terms are written in terms of intersection of divisors in $\mathcal{M}_k(Z,\iota)$ (which is fine for what we want to do) and the preimage under the boundary map of products of exceptional levels in the $\chi(Y_j)$. If these exceptional levels were all replaced by the levels $\{\mathrm{trace}\,\rho(\lambda_i^{\pm}) = \pm 2\}$, then it would be an easy matter to identify this intersection with the product of the values of Donaldson polynomials for the surfaces S_i. Unfortunately, it is not possible to make such a change of level and leave the intersection number invariant. Thus, what we must do here is to understand how these intersection numbers vary as we change the exceptional levels. It turns out that there is a single, crucial class which measures all the changes. Let us begin by describing this class.

5.1 The class ϵ

Let $\lambda \subset T$ be an oriented simple closed curve representing a primitive homology class in the three-torus T. Let $\mathcal{R}(T)$ denote the $SU(2)$-representation variety based at point $p \in T$. Let $\chi(T,\lambda) \subset \chi(T)$ be the open subset of classes of representations into $SU(2)$ whose traces on λ are not ± 2. Now suppose that Z is a complete, cylindrical-end Riemannian four-manifold and that one of this components at infinity is identified with T. We denote by $\hat{\mathcal{M}}_k(Z,\lambda)$ the preimage $\partial^{-1}(\chi(T,\lambda))$ inside $\mathcal{M}_k(Z)$. Suppose that an ASD connection ω on an $SU(2)$-bundle P over Z represents a point of $\hat{\mathcal{M}}_k(Z,\lambda)$. Then the restriction to infinity of the fundamental complex two-plane bundle η associated to P splits uniquely as a sum of complex line bundles $L \oplus L^{-1}$ invariant under ω in such a way that the holonomy of the restriction of ω to L around the loop λ has positive imaginary part. Let the base point p lie at infinity. Associating to ω the subset of P_p consisting of all base frames over p that have the first basis vector lying in $L|_p$ determines a reduction of the $SO(3)$-bundle $\hat{\mathcal{M}}_k^0(Z,\lambda) \to \hat{\mathcal{M}}_k(Z,\lambda)$ to an S^1-bundle. We let $\epsilon(\lambda) \in H^2(\hat{\mathcal{M}}_k(Z,\lambda);\mathbf{Z})$ be the first Chern class of this reduction. Of course, by definition, the class $\epsilon(\lambda)$ satisfies

$$\epsilon(\lambda)^2 = p_1|_{\hat{\mathcal{M}}_k(Z,\lambda)}$$

where p_1 is the first Pontrjagin class of the base bundle $\mathcal{M}_k^0(Z) \to \mathcal{M}_k(Z)$. If we reverse the direction of λ then we change the sign of $\epsilon(\lambda)$.

Proposition 5.1: *Let Z be a complete Riemannian four-manifold with a cylindrical end and with boundary at infinity a three-torus T. Suppose that $\tau \subset Z$ is a complete surface with cylindrical end whose boundary circle is homotopic to c times λ, for some integer $c > 0$, $c \equiv 0 \pmod 2$. Let $E \subset \chi(T)$ be an exceptional level for $\partial\tau$. Since $\partial\tau$ is an even multiple of λ, in fact $E \subset \chi(T,\lambda)$. Suppose that the boundary map ∂ is transverse to E. Then $\partial^{-1}(E) \subset \mathcal{M}_k(Z)$ is a codimension-1 submanifold. Let $\mathcal{M}^+$ and $\mathcal{M}^-$ be neighborhoods of the two sides* *with the convention that the trace of the holonomy about λ is larger on $\mathcal{M}^+$ than it is on $\mathcal{M}^-$. Then there is a class*

$$\alpha \in H^2(\mathcal{M}^+ \cup \partial^{-1}(E) \cup \mathcal{M}^-; \mathbf{Z})$$

which agrees with $\mu([\tau])$ on $\mathcal{M}^-$ and with $\mu([\tau]) + \epsilon(\lambda)$ on $\mathcal{M}^+$.

Proof: Let $\mathcal{B}^*(\tau)$ denote the space of gauge equivalence classes of irreducible $SU(2)$-connections over τ which are exponentially approaching a flat connection at infinity. For $I \subset [-2,2]$ let $\mathcal{B}_I^*(\tau)$ denote the subset where the flat connection at infinity has trace of holonomy around $\partial\tau$ contained in I. Then there is a universal class $\mu(\tau) \in H^2(\mathcal{B}^*_{[2,-2)}(\tau); \mathbf{Z})$. We also have the class $\epsilon_\tau \in H^2(\mathcal{B}^*_{(2,-2)}(\tau); \mathbf{Z})$ defined in the same way as ϵ and using the same convention for the choice of the sign.

Let N be a of a regular neighborhood of $\mathcal{B}^*_{-2}(\tau)$ in $\mathcal{B}^*(\tau)$. Then ∂N is the two-sphere bundle over $\mathcal{B}^*_{-2}(\tau)$ associated to the base point bundle $\mathcal{B}^0_{-2}(\tau) \to \mathcal{B}^*_{-2}(\tau)$. Let $p_1 \in H^4(\mathcal{B}^*_{-2}; \mathbf{Z})$ denote the Pontryagin class of base point bundle. The restriction of ϵ_τ to this boundary is the Euler class of the tangents along the fibers of this sphere bundle. (We do not actually need to know the sign of the class for our calculation of the Donaldson invariants so we will leave to the interested reader to check that with our convention the sign works as we have claimed.) Since ∂N is an oriented two-sphere bundle its rational cohomology ring is isomorphic to

$$H^*(\mathcal{B}^*_{-2}(\tau); \mathbf{Q})[\epsilon_\tau]/(\epsilon_\tau^2 - p_1)$$

There is a fiber preserving diffeomorphism $a: N \to N$ which restricts to the two-sphere fibers of ∂N as the antipodal map. Thus a acts on the cohomology ring $H^*(\partial N; \mathbf{Q})$ by fixing the classes pulled up from the base and $a(\epsilon_\tau) = -\epsilon_\tau$.

Lemma 5.2: $a^*(\mu(\tau)|_{\partial N}) = \big(\mu(\tau) - \epsilon\big)|_{\partial N}$.

Proof: By the discussion of the cohomology ring we can write $\mu(\tau)|_{\partial N} = x + r\epsilon|_{\partial N}$ where x is a class pulled up from the base and r is a rational number. Since $\epsilon|_{\partial N}$ is the Euler class of the tangents along the fibers, $2r$ is the evaluation of $\mu(\tau)$ on a fiber S^2. We also see that $a^*(\mu(\tau)|_{\partial N}) = x - r\epsilon|_{\partial N} = \big(\mu(\tau) - 2r\epsilon\big)|_{\partial N}$. Thus to prove the lemma it suffices to show the following claim.

Claim 5.3: *For any fiber S^2 of $\partial N \to \mathcal{B}^*_{-2}(\tau)$ we have $\int_{S^2} \mu(\tau) = 1$.*

Remark 5.4: For a similar result in a slightly different context see *[3]*.

Proof: Fix a connection ω on the trivial $SU(2)$-bundle over τ representing a point of $\mathcal{B}^*_{-2}(\tau)$ which is constant on the cylinderical end of τ. We construct a two-sphere of connections homotopic to a fiber of the projection in the following way. Take the two-sphere S_t of radius $0 < t < 2\pi$ in the Lie algebra of su_2. For each $\xi \in S_t$, we add to ω the one form $\beta\xi d\theta$ where β is a cutoff function which is identically one outside a sufficiently large compact set in τ and zero off the cylinder and θ is an angular coordinate for the cylinder. This produces a family $\mathcal{F}_t$ of connections on the trivial $SU(2)$ bundle over τ parameterized by S_t. For $0 < t < 2\pi$ the families $\mathcal{F}_t$ are all homotopic so the value of $\mu(\tau)$ on each of these families is the same. Consider the family $\mathcal{F}_\pi$. The family of connections at infinity for this family is trivial and hence $\mathcal{F}_\pi$ extends over the one point compactification $\hat{\tau}$ of τ. Thus we have a bundle over $\hat{\tau} \times S^2$ with a partial connection along the $\hat{\tau}$ factor. From the construction we see that the corresponding universal bundle is isomorphic to the bundle obtained in the following way. Let $P_1 \to \hat{\tau}$ be a non-trivial $SO(3)$ bundle. Let $P_2 \to S^2 \times S_\pi$ be the $SO(3)$ bundle obtained from by clutching by the map from $S^1 \times S_\pi \to SO(3)$ by $(t, \xi) \mapsto \exp(t\xi)$. Here $S^1 = [0,1]/(0 \cong 1)$. Now identify the bundles $P_1 \times S_\pi|_{\infty \times S_\pi}$ with the bundle $P_2|_{pt \times S_\pi}$ over $\infty \times S_\pi$ obtaining a bundle over $(\hat{\tau} \wedge S^2) \times S_\pi$. The pull back of this bundle to $\hat{\tau} \times S_\pi$ by the map which collapses a circle around the point at infinity in $\hat{\tau}$ to a point is isomorphic to the universal bundle. From this description it is clear that the $\int_{S_\pi} \mu(\tau)$ is equal to $-\frac{1}{4}p_1(P_2)$ or half the degree of the clutching map. This gives the desired result.

This finishes the proof of Lemma 5.2.

We can now finish the proof of Proposition 5.1. Fix K a relatively compact, open subset of $\partial^{-1}(E)$. Since we are assuming that the moduli space $\mathcal{M}_k(Z)$ is transverse to the exceptional level E there is a neighborhood U of K in $\mathcal{M}_k(Z)$ diffeomorphic to $(-1,1) \times K$. Let $U_\pm = \mathcal{M}_\pm \cap U$. The restriction of α to U is simply the pullback of $\mu(\tau)|_{-1 \times K}$ by the product structure to U. It remains to check that α has the claimed restriction to U^+. This follows from the following claim and Lemma 5.2.

Claim 5.5: *Identify $(-1,0) \times K$ with $(0,1) \times K$ by the map $(t,k) \mapsto (-t,k)$. The restriction map r_τ on $(-1,0) \times K$ is homotopic in $\mathcal{B}^*_{(-2,2)}$ to the composition $a \circ r_\tau$ on $(0,1) \times K$*

Proof: The transversality of ∂ to the exceptional level translates into the fact that the restriction $r_\tau : U \to N \subset \mathcal{B}^*(\tau)$ is diffemorphic to a fiber-wise injective bundle map from from the I bundle $U \to K$ into the 3-disk bundle $N \to \mathcal{B}^*_{-2}$.

Given the claim, the proof of Proposition 5.1 is immediate for we have

$$\alpha|_{U_+} = r_\tau^* \circ a^*(\mu(\tau)) = r_\tau^*(\mu(\tau)) - r_\tau^*(\epsilon_\tau),$$

and by our conventions on U_- we have $r_\tau^*(\epsilon_\tau) = -\epsilon(\lambda)$,

Corollary 5.6: *Let Z, τ, λ be as above. Suppose that $\partial\tau = c\lambda$ for some integer $c > 0$, $c \equiv 0 \pmod 2$. Let $\hat{\mathcal{M}}_k(Z,\lambda) \subset \mathcal{M}_k(Z)$ be the preimage of $\chi(T,\lambda)$. Then there is a class $\alpha \in H^2(\hat{\mathcal{M}}_k(Z,\lambda))$ which agrees with the restriction of $\mu([\tau])$ near the preimage of the level $\chi_{-2}(T)$ and agrees with the class $\mu([\tau]) + \frac{c}{2}\epsilon(\lambda)$ on the preimage of*

$$\{[\rho] | \, 2 - \delta < \operatorname{trace} \rho(\lambda) < 2\}$$

for some $\delta > 0$ sufficiently small.

5.2 The case of two fiber classes

Now we are ready to start identifying the invariants of Z_i with those of the closed manifold S_i obtained by filling in the solid tori. Let us introduce some notation which will simplify the expressions that we deal with. We shall consider surfaces $\Sigma_1, \ldots, \Sigma_{4r_i} \subset X_i$ which meet transversely. As always we assume that these surfaces are sufficiently generic so that any irreducible ASD connection on Z_i restricts to an irreducible connection on each of the Σ_i. Because of this we can pull back divisors from $\mathcal{B}^*(\Sigma_i)$ to induce divisors D_{Σ_i} on $\mathcal{M}(Z_i)$. We let $A = \{1, \ldots, 4r_i\}$ be the index set for this set of surfaces. Then we set

$$D_{\Sigma(A)} = D_{\Sigma_1} \cap \cdots \cap D_{\Sigma_{4r_i}}.$$

Similarly, we set

$$Q_{S_i}^{(2r_i)}([\Sigma(A)]) = Q_{S_i}^{(2r_i)}([\Sigma_1], \ldots, [\Sigma_{4r_i}])$$

and for any set of cohomology classes $x_1, \ldots, x_b \in H_*(S_i)$ we set

$$\gamma_k([\Sigma(A)], x_1, \ldots, x_b) = \gamma_k([\Sigma_1], \ldots, [\Sigma_{4r_i}], x_1, \ldots, x_b).$$

Lemma 5.7: *Let X_i, $1 < i < t$, be as in Section 2.1. Let Z_i be the cylindrical manifold modeled on X_i. Suppose that we have classes $\Sigma_1, \ldots, \Sigma_{4r_i}$ in X_i as described above. Let $A = \{1, \ldots, 4r_i\}$ Let $\partial \colon \mathcal{M}_{r+2}(Z_i) \to \chi(\partial Z_i)$ be the boundary map. Then for any generic $([\rho^-], [\rho^+]) \in \chi(\partial Z_i^-) \times \chi(\partial Z_i^+)$ the intersection*

$$\mathcal{M}_{r+2}(Z_i) \cap D_{\Sigma(A)} \cap \partial^{-1}([\rho^-], [\rho^+])$$

is a finite set of points whose algebraic sum is

$$\gamma_{r_i+2}([\Sigma(A)], [F^+], [F^-]) = 0.$$

Proof: First of all note that by Theorem 9.0.5 of *[14]* the dimension of $\mathcal{M}_{r_i+2}(Z_i)$ is $8r_i + 6$ and the dimension of $\chi(\partial Z_i)$ is six. Hence, for generic $([\rho^-], [\rho^+]) \in \chi(\partial Z_i^-) \times \chi(\partial Z_i^+)$ the dimension of the fiber $\partial^{-1}([\rho])$ is of dimension $8r_i$. Thus, the intersection in the statement of the lemma is 0-dimensional. By the usual compactness arguments it is a finite set of points.

Since all the divisors are defined globally on $\mathcal{M}_{r_i+2}(Z_i)$, the intersection number of the given intersection is independent of $([\rho^-], [\rho^+]) \in \chi(\partial Z_i^-) \times \chi(\partial Z_i^+)$. According to Corollary 4.14 (or rather the obvious generalization of it to cover the case of more than one boundary component) this intersection number is identified with the intersection

$$D_{\Sigma(A)} \cap D_{[F^-]} \cap D_{[F^+]}$$

in $\mathcal{M}_{r_i+2}(S_i)$ where F^+, F^- which are the cores of the solid tori which are filled in to form S_i from X_i. Thus, we see that the intersection number we are interested in is exactly

$$\gamma_{r_i+2}(S_i)([\Sigma(A)], [F^+], [F^-]).$$

Since S_i is an elliptic surface with $p_g = 1$, and since the Σ_j are disjoint from the $F^\pm$, it follows easily from Formula 1 that this number is zero.

5.3 The case of one fiber class and one multi-section

Lemma 5.8: *With notation as in the previous subsection, suppose that we also have a surface τ with boundary in X_i, $1 < i < t$. Let M be the product of the mulitplicities of the multiple fibers of X_i. Suppose that $\partial^{\pm}(\tau) = c\lambda_i^{\pm}$ for some positive integer $c \equiv 0 \pmod 2$. Let $\ell^- \subset \chi(\partial^-(Z_i), \lambda_i^-)$ be a level which is not an exceptional level for $\partial^-(\tau)$. Let $[\rho^+] \in \chi(\partial^+(Z_i))$ be a generic point. Then the intersection of*

$$D_{\Sigma(A)} \cap D_\tau \cap \partial^{-1}(\ell^- \times \{[\rho^+]\})$$

inside $\mathcal{M}_{r_i+2}(Z_i)$ is a finite set of points whose algebraic sum is equal to

$$2\gamma_{r_i+2}(S_i)([\Sigma(A)], [\hat{\tau}], [F^+]) = 2cMQ_{S_i}^{(2r_i)}([\Sigma(A)]).$$

Proof: As in the previous lemma the intersection of $D_{\Sigma(A)}$ inside $\mathcal{M}_{r_i+2}(Z_i)$ is a compact four-dimensional space B. We suppose that $[\rho^+]$ is in a level ℓ^+ separated by e^+ exceptional levels from $\chi_2(\partial^+ Z_i)$. Let e^- be the number of exceptional levels for $\partial^-(\tau)$ which separate ℓ^- from $\chi_2(\partial^- Z_i)$. According to Proposition 5.1, there is a two-dimensional cohomology class α in $\hat{\mathcal{M}}_{r_i+2}(Z_i, \lambda)$ which agrees with $\mu(\tau)$ on the preimage of the level $\ell^- \times \ell^+$, and agrees with $\mu(\tau) + e^-\epsilon^- + e^+\epsilon^+$ on the preimage of the level $\chi_{2-\delta}(\partial^- Z_i) \times \chi_{2-\delta}(\partial^+ Z_i)$, for any $\delta > 0$ sufficiently small. (Here, we use the notation $\epsilon^{\pm} = \epsilon(\lambda_i^{\pm})$.)

Thus, the number that we are trying to compute is equal to

$$\langle \mu(\tau) + e^-\epsilon^- + e^+\epsilon^+, D_{\Sigma(A)} \cap \partial^{-1}\left(\chi_{2-\delta}(\partial^-(Z)) \times \{[\psi^+]\}\right)\rangle$$

where $[\psi^+] \in \chi_{2-\delta}(\partial^+ Z_i)$ and $\delta > 0$ is sufficiently small. We write this number as the sum of

$$\langle \mu(\tau), D_{\Sigma(A)} \cap \partial^{-1}\left(\chi_{2-\delta}(\partial^-(Z)) \times \{[\psi^+]\}\right)\rangle$$

plus two other terms:

$$\langle e^{\pm}\epsilon^{\pm}, D_{\Sigma(A)} \cap \partial^{-1}\left(\chi_{2-\delta}(\partial^-(Z)) \times \{[\psi]\}\right)\rangle.$$

By Proposition 4.7 and Corollary 4.14, the first term is identified with

$$2\gamma_{r_i+2}(S_i)([\Sigma(A)], [F^+], [\hat{\tau}])$$

where $\hat{\tau}$ is the completion in S_i of τ. By Equation 1 this term is equal to

$$2cMQ_{S_i}^{(2r)}([\Sigma(A)]).$$

On the other hand, according to Corollary 5.6 we have

$$\langle(\mu([\hat{\tau}]) + \frac{c}{2}\epsilon^-, D_{\Sigma(A)} \cap \partial^{-1}\left(\chi_{2-\delta}(\partial^-(Z)) \times \{[\psi^+]\}\right)\rangle$$

is equal to

$$\langle \mu([\hat{\tau}]), D_{\Sigma(A)} \cap \partial^{-1}\left(\chi_{-2+\delta}(\partial^-(Z)) \times \{[\psi^+]\}\right)\rangle.$$

Invoking Corollary 4.14, Proposition 4.7, and Corollary 4.11 it follows that

$$\frac{c}{2}\langle \epsilon^-, D_{\Sigma(A)} \cap \partial^{-1}\left(\chi_{2-\delta}(\partial^-(Z)) \times \{[\psi^+]\}\right)\rangle$$

is equal to

$$2\gamma_{-4r-8,[F^-]}([\Sigma(A)],[F^+],[\hat{\tau}]) - 2\gamma_{r+2}(S_i)([\Sigma(A)],[F^+],[\hat{\tau}]).$$

But by Corollary 4.12 these latter two invariants are equal. Thus,

$$\frac{c}{2}\langle \epsilon^-, D_{\Sigma(A)} \cap \partial^{-1}\left(\chi_{2-\delta}(\partial^-(Z)) \times \{[\psi^+]\}\right)\rangle = 0.$$

A similar argument shows

$$\frac{c}{2}\langle \epsilon^+, D_{\Sigma(A)} \cap \partial^{-1}\left(\chi_{2-\delta}(\partial^-(Z)) \times \{[\psi^+]\}\right)\rangle = 0.$$

Putting all this together proves the lemma.

A similar argument shows the following analogous result for X_1 and X_t.

Lemma 5.9: *With notation as above, suppose that $1 \leq p_g(X_1) \leq 2$ and that we have a surfaces $\tau_1, \ldots, \tau_{p_g(X_1)}$ with boundary in X_1. Suppose that $\partial^+(\tau_n) = c(n)\lambda_1^+$ for integers $c(n) > 0$, $c(n) \equiv 0 \pmod 2$. Let c be the product of the $c(n)$ for $1 \leq n \leq p_g(X_1)$. Let $\ell^+ \subset \chi(\partial^+(Z_1), \lambda_1^+)$ be a level which is not an exceptional level for $\partial^+(\tau_n)$ for any $n \leq p_g(X_1)$. Let $[\rho^+] \in \chi(\partial^+(Z_1))$ be a generic point in ℓ^+. Then the intersection of*

$$D_{\Sigma(A)} \cap D_{\tau_1} \cap \cdots \cap D_{\tau_{p_g}} \cap \partial^{-1}(\{[\rho^+]\})$$

inside $\mathcal{M}_{r_1+1+p_g(X_1)}(Z_1)$ is a finite set of points whose algebraic sum is equal to

$$2\gamma_{r_1+2}(S_1)([\Sigma(A)],[\hat{\tau}_1],\ldots,[\hat{\tau}_{p_g}],[F^+]) = 2cM^{p_g(X_1)}Q_{S_1}^{(2r_1)}([\Sigma(A)]).$$

Exactly, the same statement holds for subscript index t replacing the subscript index 1 and the superscript $-$ replacing the superscript $+$.

Proof: We consider the case of X_1, the other is symmetric. When $p_g(X_1) = 1$, the proof follows exactly along the lines of the other case. When $p_g(X_1) = 2$ and we have two sections τ_1 and τ_2, we get an expression with constant, linear, and quadratic terms in $\epsilon(\lambda_1^+)$. As before, the constant term gives the result claimed in the statement. The new ingredient is the quadratic term. But since $\epsilon(\lambda_1^+)^2 = p_1$, the Pontrjagin class of the based bundle, we show that these terms vanish by showing the following claim:

$$\gamma_{r_1+3}([\Sigma(A)],[F^+],p_1) = 0$$

This term is equal to zero because of an invariant theoretic argument – we know from invariant theory (see Chapter VII in *[7]*) that the Donaldson invariant must be a polynomial in p_1^*, the class dual to p_1, the quadratic form Q_{S_1} and the canonical class κ_{S_1}. Since κ_{S_1} vanishes on $[F^+]$ and since all the Σ_i have zero intersection with $[F^+]$, the claim follows easily.

Once we have shown that the quadratic term in $\epsilon(\lambda_1^+)$ vanishes, we identify the linear terms with a difference of the $SU(2)$-invariant and an $SO(3)$-invariant. The proof is completed by showing the following claim which establishes that these invariants have the same value on the given set of classes.

Claim 5.10: *Suppose that $p_g(S_1) = 2$. Let $\hat{\tau}_1$ and $\hat{\tau}_2$ be multi-sections of the elliptic fibration structure for S_1. Then*

$$\gamma_{r_1+3}(S_1)([\Sigma(A)], [\hat{\tau}_1], [\hat{\tau}_2], [F^+]) = \gamma_{-4r-12,[F^+]}(S_1)([\Sigma(A)], [\hat{\tau}_1], [\hat{\tau}_2], [F^+]).$$

Proof: This statement is equivalent to the statement that the first non-zero coefficients of the $SU(2)$-invariant and the $SO(3)$-invariant of S_1 agree. This is immediate from Lemma 4.16.

This completes the proof of Lemma 5.9.

5.4 The case of two multi-sections

Now we need to consider the case of two multi-sections in the X_i.

Lemma 5.11: *Suppose that $p_g(X_1) = 1$ and that X_1 has no multiple fibers. Suppose that we have surfaces τ_1, τ_2 in X_1. Suppose that $\partial(\tau_n) = c(n)\lambda$ for some even integers $c(n) > 0$. Let $c = c(1) \cdot c(2)$. Let $\ell \subset \chi(\partial(Z_1), \lambda_1^+)$ be a level which is not an exceptional level for $\partial(\tau_n)$ for $n = 1, 2$. Let e_n be the number of exceptional levels for $\partial\tau_n$ separating ℓ from $\chi_2(\partial Z_1)$. Suppose that the τ_n are disjoint near infinity and that they are disjoint from the Σ_a. Then the intersection of*

$$D_{\Sigma(A)} \cap D_{\tau_1} \cap D_{\tau_2} \cap \partial^{-1}(\ell)$$

inside $\mathcal{M}_{r_1+2}(Z_1)$ is a finite set of points whose algebraic sum is equal to

$$\left(2(\tau_1 \cdot \tau_2) + 4(e_1 c(2) + e_2 c(1)) - 16 e_1 e_2\right) Q_{S_1}^{(2r_1)}([\Sigma(A)]).$$

Proof: By Corollary 5.6 for any $\delta > 0$ sufficiently small, we have that

$$\langle (\mu(\tau_1) + \frac{c(1)}{2}\epsilon) \cup (\mu(\tau_2) + \frac{c(2)}{2}\epsilon), D_{\Sigma(A)} \cap \partial^{-1}\left(\chi_{2-\delta}(\partial Z_1)\right)\rangle$$

is equal to

$$\langle \mu(\tau_1) \cap \mu(\tau_2), D_{\Sigma(A)} \cap \partial^{-1}\left(\chi_{-2+\delta}(\partial Z_1)\right)\rangle.$$

This tells us that

$$\frac{c(1)}{2}\langle \epsilon \cup \mu(\tau_2), D_{\Sigma(A)} \cap \partial^{-1}\left(\chi_{2-\delta}(\partial Z_1)\right)\rangle$$

$$
\begin{aligned}
&+\frac{c(2)}{2}\langle \epsilon \cup \mu(\tau_1),\ D_{\Sigma(A)} \cap \partial^{-1}\left(\chi_{2-\delta}(\partial Z_1)\right)\rangle \\
&+\frac{c(1)c(2)}{4}\langle \epsilon^2, D_{\Sigma(A)} \cap \partial^{-1}\left(\chi_{2-\delta}(\partial Z_1)\right)\rangle \\
&= 2\gamma_{-4r_1-8,[F]}(S_1)(\Sigma(A), \hat{\tau}_1, \hat{\tau}_2) \\
&-2\gamma_{4r_1+2}(S_1)([\Sigma(A)], [\hat{\tau}_1], [\hat{\tau}_2]) = 0
\end{aligned}
$$

where $\hat{\tau}_n$ is the completion of τ_n in S_1 and where we have used the fact that since S_1 is a $K3$ surface these Donaldson invariants are equal.

By symmetry, this yields

$$
\begin{aligned}
&\frac{c(1)}{2}\langle \epsilon \cup \mu(\tau_2), D_{\Sigma(A)} \cap \partial^{-1}\left(\chi_{2-\delta}(\partial Z_1)\right)\rangle \\
&\quad = -\frac{c(1)c(2)}{8}\langle \epsilon^2, D_{\Sigma(A)} \cap \partial^{-1}\left(\chi_{2-\delta}(\partial Z_1)\right)\rangle.
\end{aligned}
$$

We have the symmetric statement with the indices 1 and 2 reversed.

Now let us consider

$$\langle \mu(\tau_1) \cap \mu(\tau_2), D_{\Sigma(A)} \cap \partial^{-1}(\ell)\rangle.$$

This is equal to

$$\langle \left(\mu(\tau_1) + e_1\epsilon\right) \cup \left(\mu(\tau_2) + e_2\epsilon\right), D_{\Sigma(A)} \cap \partial^{-1}\left(\chi_{2-\delta}(\partial Z_1)\right)\rangle.$$

Since

$$\langle \mu(\tau_1) \cup \mu(\tau_2), D_{\Sigma(A)} \cap \partial^{-1}(\chi_{2-\delta}(\partial Z_1))\rangle = 2\gamma_{r_1+2}(S_1)([\Sigma(A)], [\hat{\tau}_1], [\hat{\tau}_2])$$

plugging in the relations given in the previous paragraph we see that this is equal to

$$
\begin{aligned}
&2\gamma_{r_1+2}(S_1)([\Sigma(A)], [\hat{\tau}_1], [\hat{\tau}_2]) \\
&\quad + \left(2e_1e_2 - \frac{c(1)e_2 + e_1c(2)}{2}\right)\gamma_{4r_1+2}(S_1)([\Sigma(A)], p_1).
\end{aligned}
$$

It follows from Lemma 4.19 that this expression is equal to the expression claimed in the statement of the lemma.

Now let us consider the case when X_1 has multiple fibers.

Lemma 5.12: *Let us keep the notation and assumptions as in the previous lemma, except that X_1 has at most two multiple fibers of multiplicities m_1, m_2. Then*

$$\langle \mu(\tau_1) \cup \mu(\tau_2), D_{\Sigma(A)} \cap \partial^{-1}(\ell)\rangle$$

is equal to

$$
\begin{aligned}
Q_{S_1}^{(2r_1)}([\Sigma(A)])\,&(2m_1m_2\tau_1 \cdot \tau_2 + \frac{1}{3}(2m_1^2m_2^2 - m_1^2 - m_2^2) + (4(e_1c(2) + e_2c(1)) \\
&- 16e_1e_2)m_1m_2). \qquad (5)
\end{aligned}
$$

Proof: The argument goes exactly as in the previous case and uses the computation given in *[15]* for the value of the Donaldson polynomial of S_1 on the given classes. We also need to know that the value of the $SO(3)$-polynomial $\gamma_{-4r_1-8,[F]}(S_1)$ on these classes is equal to the $SU(2)$-polynomial $\gamma_{r_1+2}(S_1)$ on these classes. This result is the subject of the next claim.

Claim 5.13: *Let S_1 be a regular elliptic surface over $\mathbf{C}P^1$ with at most two multiple fibers. Let $\Sigma_1, \ldots, \Sigma_{4r_1}$ be classes orthogonal to the generic fiber and let $A = \{1, \ldots, 4r_1\}$. Then for any classes τ_1, τ_2 we have*

$$\gamma_{r_1+2}(S_1)([\Sigma(A)], [\tau_1], [\tau_2]) = \gamma_{-4r_1-8,[F]}(S_1)([\Sigma(A)], [\tau_1], [\tau_2]).$$

Proof: By Kronheimer's periodicity (Corollary 4.20) it suffices to prove this result for $r_1 = 2$. We split S_1 as the fiber connected sum of two elliptic surfaces A and B with $p_g = 0$ along a three-torus T. We suppose that there are exactly 4 of the Σ_i in say with index set a in A and four with index set b in B. The limit of any sequence of ASD connections in the intersection of the Σ_i must have energy $16\pi^2$ on each of A and B. There are two possilities for the limiting connection vis-a-vis τ_1. They are:

- The limiting boundary value is in an exceptional level for one of the τ_n and the connection on one of the two sides lies in the divisor associated to τ_{3-n}.
- The limiting connection on one of the sides is reducible and lies in the closure of the divisors associated to both the τ_n.

Case 1 is exactly like the one that we have already encountered. Suppose that the boundary value is exceptional for τ_1 and the limiting connection $\omega|_B$ lies in D_{τ_2}. The contribution of such terms is the product of the degree of the map

$$\mathcal{M}_2(A) \cap D_{\Sigma_a} \to \chi(T)$$

and an intersection number from the moduli space of B.

Let us consider a term of the second type, say reducible on the B-side. The local model for $\mathcal{M}_4(S)$ around such connections is

$$\tilde{\mathcal{M}}_2(A) \times_{S^1} \mathbf{C}^6$$

where $\tilde{\mathcal{M}}_2(A)$ is the total space of the principal S^1-bundle whose first Chern class is $\epsilon(\lambda)$. Each divisor from the B-side intersects this neighborhood in a multiple of the divisor $\tilde{\mathcal{M}}_2(A) \times_{S^1} \mathbf{C}^5$. Thus, the contribution of such a term is the product of multiplicities of the divisors from the B-side near the given reducible connection times the degree of the map

$$\mathcal{M}_2(A) \cap D_{\Sigma_a} \to \chi(T).$$

The point is that terms of each of these types contribute in exactly the same way if we glue the moduli spaces together over the involution on $\chi(T)$ which reverses the trace of the loop λ. The reason is that every term is a product of invariants of A and B at least one of which is independent of the level at which we are computing. Ths completes the proof of the claim.

From the claim, the lemma is proved in the same way as the previous case.

Of course, both of these results hold for X_t replacing X_1.

Lemma 5.14: *With notation as above, suppose that we also have surfaces τ_1, τ_2 with boundary in X_i, $1 < i < t$. Suppose that X_i has no mulitple fibers. Suppose that $\partial^{\pm}(\tau_n) = c(n)\lambda_i^{\pm}$ for some even integers $c(1), c(2) > 0$. Let $c = c(1)c(2)$. Let $\ell^{\pm} \subset \chi(\partial^{\pm}(Z_i), \lambda_i^{\pm})$ be levels which are not exceptional levels for $\partial^{\pm}(\tau_n)$. Suppose that the τ_n are disjoint near infinity and that they are disjoint from the Σ_i. Then*

$$\langle \mu(\tau_1) \cap \mu(\tau_2), D_{\Sigma(A)} \cap \partial^{-1}(\ell^- \times \ell^+) \rangle$$

is equal to

$$4Q_{S_i}^{(2r_i)}([\Sigma(A)])(\tau_1 \cdot \tau_2)$$
$$+ \left(4(e_1^+ + e_1^-)c(2) + 4(e_2^+ + e_2^-)c(1) - 16e_1^+e_2^+ - 16e_1^-e_2^-\right) Q_{S_i}^{(2r_i)}([\Sigma(A)])$$

where $e_n^{\pm}$ is the number of exceptional levels of $\partial^{\pm}\tau_n$ separating $\ell^{\pm}$ from $\chi_2(\partial^{\pm}Z_i)$.

Proof: This result is proved exactly as Lemma 5.11. Here, we can move the level in either $\chi(\partial^{\pm}Z_i)$ and we have two classes $\epsilon(\lambda_i^{\pm})$ measuring the differences when we cross an exceptional level. The preimage of the product of the $+2$ levels for both boundary components gives the $SU(2)$-invariant, the preimage of the $+2$-level for one boundary component cross the -2 level for the other gives the $SO(3)$-invariant, and the preimage of the product of the -2-levels gives the $SU(2)$-invariant. Using these facts, and the fact that on the given set of classes the $SO(3)$ and the $SU(2)$-invariant agree, it follows that $\epsilon(\lambda_i^+) \cdot \epsilon(\lambda_i^-) = 0$. Of course, $(\epsilon(\lambda_i^+))^2 = (\epsilon(\lambda_i^-))^2 = p_1$. From these relations it is easy to deduce the formula in the lemma.

6 Proof of Theorem 1.3

We have now done all the preliminary work and we are ready to attack Theorem 1.3. The first case, when $p_g(S)$ is odd, we do directly. The second case when $p_g(S) = 2t > 0$ is done in a slightly more round-about manner.

6.1 The case when $p_g = 2t - 1$

In this case we divide S into a fiber connected sum

$$S = S_1 \#_f \ldots \#_f S_t$$

where each S_i is an elliptic surface over $\mathbf{C}P^1$ with $p_g = 1$ and S_i has no multiple fibers for $i > 1$. We denote by m_1, m_2 the multiplicity of the multiple fibers of

S, and we let $M = m_1 m_2$. We fix integers $r_i \geq 3$. We set $r = \sum_{i=1}^{t} r_i$, and we set $s = 2t$. We define k so that the dimension of $\mathcal{M}_k(S)$ is equal to $8r + 4t$. That is to say so that $4k - 3(p_g(S) + 1) = 4r + 2t$. This means that $k = r + 2t$ so that $r + 4p_g(S) + 4 = 4k$. We choose $4r_i$ closed, oriented surfaces $\Sigma(A(i))$ in X_i and closed, oriented surfaces τ_n, $1 \leq n \leq 2t$, meeting each splitting torus in a simple closed curve of even multiplicity $c(n) > 0$. As indicated before, we arrange that the maximal power of two that divides $c(n)$ is distinct from that which divides $c(n')$ for all $n \neq n'$. In this way we construct closed, oriented surfaces $\Sigma_1, \ldots, \Sigma_{4r}, \tau_1, \ldots, \tau_{2t}$ satisfying the conditions given at the beginning of Subsection 2.4. We can also suppose that the τ_n are disjoint near the union Y of the splitting three-tori and that the τ_n are disjoint from the Σ_m. We set $k_i = r_i + 2$ so that $\sum_{i=1}^{t} k_i = k$, and so that $\mathcal{M}_{k_i}(Z_i) \cap D_{\Sigma(A(i))}$ is of dimension 6 when $1 < i < t$ and is of dimension 5 which $i = 1, t$. All connections in these intersections have uniformly bounded curvature on fixed neighborhoods of infinity.

According to Theorem 3.2 what we must do is to compute a sum over the set of injective functions $\varphi\colon \{1, \ldots, t-1\} \to N = \{1, \ldots, 2t\}$, written $n = \varphi(j)$. The term associated to an injective function φ is the intersection number of Δ_φ with the preimage $\mathcal{N}_k(Z, \iota, \varphi)$ under the boundary map

$$\mathcal{M}_k(Z,\iota) = \mathcal{M}_{k_1}(Z_1) \times_{\chi(Y_1)} \mathcal{M}_{k_2}(Z_2) \times \cdot \times_{\chi(Y_{t-1})} \mathcal{M}_{k_t}(Z_t) \to \prod_{j=1}^{t} \chi(Y_j)$$

of the union of the exceptional levels $E_\varphi \subset \prod_{j=1}^{t-1} \chi(Y_j)$.

Let us fix φ and examine the nature of the terms associated with it. Let us fix a component $\ell_1 \times \cdots \times \ell_{t-1}$ of E (i.e., ℓ_j is an exceptional level for $\tau_{\varphi(j)} \cap Y_j$). A point in the intersection of

$$\Delta_\varphi = \Delta_\Sigma \cap D_{\tau_{N - \mathrm{Image}\,\varphi}}$$

with the preimage in $\mathcal{M}_k(Z, \iota)$ of this component is given as follows. Fix a function $n \mapsto i(n)$ from the subset N – Image φ to the set $\{1, \ldots, t\}$ indexing the components of Z. Furthermore, for each boundary component of each Z_i we take either the preimage of a point in the given level or the preimage of the entire level. We do this in such a way that of the two components of ∂Z associated to any Y_j in one we take the preimage of a point and in the other we take the preimage of the entire level. Then we sum up over all possible ways of making these choices. The number associated to a given function $n \mapsto i(n)$ and choice of boundary data is the product from $i = 1, \ldots, t$ of the intersection number in $\mathcal{M}_{k_i}(Z_i)$ of the divisors associated to the Σ_m contained in Z_i, the divisors associated to the τ_n such that $i(n) = i$, and the preimages of the points and/or exceptional levels associated to the boundary components of Z_i. Because of dimension considerations we see that for each Z_i, $1 < i < t$, we have three possible situations:

- There are no $n \in (N$ – Image $\varphi)$ such that $i(n) = i$ and for each boundary component of Z_i we take the preimage of a point in the given level.

- There is exactly one $n \in (N - \text{Image}\,\varphi)$ such that $i(n) = i$ and for one boundary component of Z_i we take the preimage of a point and for the other we take the preimage of the entire level.
- There are exactly two $n \in (N - \text{Image}\,\varphi)$ such that $i(n) = i$ and for both boundary components of Z_i we take the preimage of the given level.

As for $i = 1, t$ we have two possibilities:

- There is exactly one $n \in (N - \text{Image}\,\varphi)$ such that $i(n) = i$ and for the boundary component of Z_i we take the preimage of a point in the given level.
- There are exactly two $n \in (N - \text{Image}\,\varphi)$ such that $i(n) = i$ and for the boundary component of Z_i we take the preimage of the given level.

These are the only possibilities which lead to zero dimensional intersections in each $\mathcal{M}_{k_i}(Z_i)$.

By Lemma 5.7 the intersection number in Z_i, $1 < i < t$, of the subvarieties given by the first possibility is zero. Hence, those associations contribute zero to the final answer and thus can be ignored. From now on we assume that we are considering associations in which this possibility does not arise. Because of this we have:

Corollary 6.1: *The function from* $N - \text{Image}\,(\varphi)$ *to* $\{1, \ldots, t\}$ *given by* $n \mapsto i(n)$ *is onto.*

Invoking Lemmas 5.8 and 5.9, we see that the intersection number $\Delta_\varphi \cap \mathcal{N}_k(Z, \iota, \varphi)$ is given as:

$$\sum_{\ell_1,\ldots,\ell_{t-1}} \sum_{t_0=1}^{t} \sum_{\psi} \left(2^{t-2} \prod_{i \neq t_0} c(\psi(i)) M(S_i) Q^{(2r_i)}([\Sigma(A(i))]) \right) \cdot D_{\Sigma(A(t_0))} \cap D_{\tau_p} \cap D_{\tau_q} \cap \partial_{t_0}^{-1}(\ell_{t_0-1} \times \ell_{t_0}). \tag{6}$$

Here, $(\ell_1, \ldots, \ell_{t-1})$ varies in such a manner that $\ell_1 \times \cdots \times \ell_{t-1}$ ranges over the components of E_φ, the functions ψ vary over injective maps from $\{1, \ldots, t_0 - 1, t_0 + 1, \ldots t\}$ to $(N - \text{Image}\,\varphi)$ and $\{p, q\}$ is the complement of the union of the image of φ and the image of ψ in N. Also, ℓ_{t_0-1} and ℓ_{t_0} are identified with levels in $\chi(\partial^-(Z_{t_0}))$ and $\chi(\partial^+(Z_{t_0}))$, respectively, in the obvious way, and they range over the exceptional levels for $\tau_{\varphi(t_0-1)}$ and $\tau_{\varphi(t_0)}$, respectively. (We regard (ℓ_0, ℓ_1) as simply $\ell_1 \subset \partial^+ Z_1$ and similarly for (ℓ_{t-1}, ℓ_t).) Lastly, $M(S_i)$ is the product of the multiplicities of S_i, i.e., $M(S_1) = M$ and $M(S_i) = 1$ for $i > 1$.

Since there are $c(\tau_{\varphi(j)})/2$ exceptional levels for $\tau_{\varphi(j)}) \cap Y_j$, we can rewrite Formula 6 as

$$\sum_{t_0=1}^{t} \sum_{(\ell_{t_0-1}, \ell_{t_0})} \sum_{\psi} \prod_{j \neq t_0-1, t_0} c(\varphi(j)) \left(\prod_{i \neq t_0} c(\psi(i)) M(S_i) Q^{(2r_i)}([\Sigma(A(i))]) \right) \cdot D_{\Sigma(A(t_0))} \cap D_{\tau_p} \cap D_{\tau_q} \cap \partial_{t_0}^{-1}(\ell_{t_0-1} \times \ell_{t_0}) \tag{7}$$

where ℓ_{t_0-1} and ℓ_{t_0} range over the exceptional levels for $\tau_{\varphi(t_0-1)} \cap Y_{t_0-1}$ and $\tau_{\varphi(t_0)} \cap Y_{t_0}$, respectively.

According to Lemma 5.14, if $1 < t_0 < t$, then

$$D_{\Sigma_{a(t_0)}} \cap D_{\tau_p} \cap D_{\tau_q} \cap \partial_{t_0}^{-1}(\ell_{t_0-1} \times \ell_{t_0}) = \tag{8}$$
$$4Q^{(2r_{t_0})}([\Sigma(A(t_0))]) \Big((\tau_p|_{Z_{t_0}}) \cdot (\tau_q|_{Z_{t_0}}) + (e_p^+ + e_p^-)c(q)$$
$$+(e_q^+ + e_q^-)c(p) - 4e_p^+e_q^+ - 4e_p^-e_q^-\Big)$$

where e_p^- is the number of exceptional levels separating ℓ_{t_0-1} from $\chi_2(\partial^- Z_{t_0})$, and e_p^+ is the number of exceptional levels separating ℓ_{t_0} from $\chi_2(\partial^+ Z_{t_0})$, and similarly for $e_q^{\pm}$.

Let us consider the case when $t_0 = 1, t$. First, notice that we have

$$\gamma_{r_{t_0}+2}(S_{t_0})([\Sigma(A(t_0))],[\hat{\tau}_p],[\hat{\tau}_q]) = M(S_{t_0})Q_{S_{t_0}}^{(r_{t_0})}([\Sigma(A(t_0))])$$
$$\left(\hat{\tau}_p \cdot \hat{\tau}_q + \left(\frac{M(S_{t_0})^2 - m_1(S_{t_0})^2 - m_2(S_{t_0})^2}{3}\right)\right.$$
$$\left.\left(\frac{c(p)c(q)}{M(S_{t_0})^2}\right)\right).$$

Thus, we have by Lemma 5.12

$$D_{\tau_p} \cap D_{\tau_q} \cap D_{\Sigma(A(t_0))} \cap \partial^{-1}(\ell)$$

is equal to

$$\left(Q^{(2r_{t_0})}([\Sigma(A(t_0))])\right) \cdot \Big(2M(S_{t_0})(\tau_p|_{Z_{t_0}}) \cdot (\tau_q|_{Z_{t_0}})$$
$$+ \frac{1}{3M}(2M^2(S_{t_0}) - m_1^2(S_{t_0}) - m_2^2(S_{t_0}))c(p)c(q)$$
$$+ \left(4(e_pc(q) + e_qc(p)) - 16e_pe_q\right) M(S_{t_0})\Big).$$

where e_p, e_q are the number of exceptional levels separating ℓ from $\chi_2(Y_1)$ or $\chi_2(Y_{t-1})$.

Plugging in these values we see that the number that we are computing by summing up the expressions in Equations 7 and 8 is given as the sum of three terms. Let us examine the first term, that is the term involving the $(\tau_p|_{Z_i}) \cdot (\tau_q|_{Z_i})$. Given φ, for each subset $\{p,q\} \in (N - \text{Image}\,\varphi)$ there are $(t-1)!$ injective functions ψ with the complement of the union of image of φ and the image of ψ being $\{p,q\}$. Thus, we see that one of the first term is

$$(t-1)!MQ^{(2r)}([\Sigma_1],\ldots,[\Sigma_{4r}]) \sum_{(p,q)} \left(\sum_{t_0=1}^{t} (\tau_p|_{Z_{t_0}}) \cap (\tau_q|_{Z_{t_0}})\right) \prod_{n \neq p,q} c(n). \tag{9}$$

Of course,

$$\sum_{t_0=1}^{t} (\tau_p|_{Z_{t_0}}) \cap (\tau_q|_{Z_{t_0}}) = [\tau_p] \cap [\tau_q].$$

Thus, the expression in Equation 9 is equal to

$$(t-1)!MQ^{(2r)}([\Sigma_1],\ldots,[\Sigma_{4r}])\left(\sum_{(p,q)}[\tau_p]\cap[\tau_q]\right)\prod_{n\neq p,q}c(n).$$

We also have

$$\langle\kappa_S,[\tau_n]\rangle=\frac{c(n)}{M(S)}.$$

Thus, since the τ_j are disjoint from the Σ_m, we can rewrite this expression as

$$(t-1)!M^{2t-1}\sum_{p,q}Q^{(2r+1)}([\Sigma_1],\ldots,[\Sigma_{4r}],[\tau_p],[\tau_q])\prod_{n\neq p,q}\langle\kappa_S,[\tau_n]\rangle.$$

Given (p,q) there are $(2t-2)!/(t-1)!$ injective functions φ which do not have (p,q) in their image. Hence, summing these expressions up over the injective functions φ gives

$$(2t-2)!M^{2t-1}\sum_{p,q}Q^{(2r+1)}([\Sigma_1],\ldots,[\Sigma_{4r}],[\tau_p],[\tau_q])\prod_{n\neq p,q}\langle\kappa_S,[\tau_n]\rangle.$$

This in turn is exactly the value of the leading term in the Donaldson polynomial

$$\frac{(4r+2t)!}{(4r+2)!}M^{2t-1}Q_S^{(2r+1)}\kappa_S^{2t-2}([\Sigma_1],\ldots,[\Sigma_{4r}],[\tau_1],\ldots,[\tau_{2t}]).$$

Now let us examine the other terms that arise in summing up the terms in Equations 7 and 8. In order to do this we need the following elementary combinatorial lemma whose proof we leave to the reader.

Lemma 6.2: *Let T be a three-torus and λ an indivisible element in $H_1(T)$. Let c_0, c_1 and c_2 be even positive integers divisible by different powers of two. (Here the indices are considered modulo three.) For any exceptional level ℓ for c_i we denote by $b(\ell, c_{i+1})$ and $b(\ell, c_{i+2})$ be the number of exceptional levels for $\lambda^{c_{i+1}}$ and $\lambda^{c_{i+2}}$ spearating ℓ from the level $\chi_2(T)$. Then*

- $$\sum_{\ell}b(\ell,c_{i+1})=\frac{c_ic_{i+1}}{8}$$
- $$\sum_{i=0}^{2}\sum_{\ell}b(\ell,c_{i+1})b(\ell,c_{i+2})=\frac{c_ic_{i+1}c_{i+2}}{8}$$

where in case the sums are performed over all exceptional levels ℓ of λ^{c_i}.

From this lemma, it is easy to see that the other terms in Equation 7 for $1<t_0<t$ sum up to

$$\sum_{\psi}M(S)Q^{(2r)}([\Sigma_1],\ldots,[\Sigma_{4r}])\frac{2t-4}{3}\prod_{n=1}^{2t}c(n).$$

Since there are $(t+1)!/2$ functions ψ for a given function φ and there are $(2t)!/(t+1)!$ injective functions φ, adding these terms up over all φ and ψ gives

$$(2t)!M(S)\prod_{n=1}^{2t} c(n)Q^{(2r)}([\Sigma_1],\ldots,[\Sigma_{4r}])\frac{2t-4}{6}.$$

Using the fact that $c(n)/M = \langle \kappa_S, [\tau_n]\rangle$, we can rewrite this as

$$(2t)!M(S)^{2t+1}Q^{(2r)}([\Sigma_1],\ldots,[\Sigma_{4r}])\frac{2t-4}{6}\prod_{n=1}^{2t}\langle \kappa_S,[\tau_n]\rangle. \tag{10}$$

In a completely similar manner we evaluate the terms associated to $t_0 = t$ and $t_0 = 1$ respectively as

$$\frac{1}{6}(2t)!M(S)^{2t+1}Q^{(2r)}([\Sigma_1],\ldots,[\Sigma_{4r}])\prod_{n=1}^{2t}\langle \kappa_S,[\tau_n]\rangle \tag{11}$$

and

$$(2t)!M(S)^{2t}Q^{(2r)}([\Sigma_1],\ldots,[\Sigma_{4r})]\left(\frac{2(M(S)^2-m_1^2-m_2^2)}{6M}+\frac{M}{6}\right)\prod_{n=1}^{2t}\langle \kappa_S,[\tau_n]\rangle. \tag{12}$$

Adding up the expressions in Formulas 10, 11, and 12 gives

$$(2t)!M(S)^{2t-1}Q^{(2r)}([\Sigma_1],\ldots,[\Sigma_{4r}])\frac{(2tM(S)^2-m_1^2-m_2^2)}{6}\prod_{n=1}^{2t}\langle \kappa_S,[\tau_n]\rangle.$$

On the other hand, this is exactly the value of

$$\frac{(4r+2t)!}{(4r)!}M(S)^{2t-1}\frac{(2tM(S)^2-m_1^2-m_2^2)}{6}Q^{(2r)}\kappa_S^{2t}$$

on the classes $[\Sigma_1],\ldots,[\Sigma_{4r}],[\tau_1],\ldots,[\tau_{2t}]$.

This proves Theorem 1.3 in the case when $p_g = 2t-1$.

6.2 Facts about the case $p_g = 2$

In order to complete the argument in the case when p_g is odd, we first must establish some facts about the case when $p_g = 2$.

We need the following extension of Lemma 4.16.

Lemma 6.3: *Let S be an elliptic surface with at most two multiple fibers over $\mathbf{C}P^1$ and with $p_g(S) = 2$. Let $F \subset S$ be a generic fiber of the elliptic fibration. Then for any classes $x_1,\ldots,x_{4r+3}$ with the property that x_i is orthogonal to $[F]$ for $1 \le i \le 4r$ we have*

$$\gamma_{r+3}(S)(x_1,\ldots,x_{4r+3}) = \gamma_{-4r-12,[F]}(S)(x_1\ldots,x_{4r+3}).$$

Proof: We decompose S into a fiber connected sum $S_1\#_f S_2$ of an elliptic surface S_1 with $p_g = 1$ and one S_2 without multiple fibers with $p_g = 0$. What we must show is that if $\Sigma_1, \ldots, \Sigma_{4r-4}$ are surfaces in X_1, and if $\Sigma_{4r-3} \ldots, \Sigma_{4r}$ are surfaces in X_2, and τ_1, τ_2, τ_3 are multisections, then the value of the $SU(2)$-invariant and the $SO(3)$-invariant on these classes agree. To simplify the notation we set $a(1) = \{1, \ldots, 4r-4\}$ and $a(2) = \{4r-3, \ldots, 4r\}$.

Using the calculus developed in the previous sections, we see that as we pull S apart along the splitting torus T, it must be the case that any sequence of ASD connections ω_m in the intersection of the divisors associated to the Σ_i must converge strongly to a connection ω on Z with energy $8(r+1)\pi^2$ on S_1 and energy $16\pi^2$ on S_2. We have the following possibilities vis-a-vis the τ_n.

- The connection $\omega|_{Z_1}$ lies in exactly two of the divisors associated to $\tau_n|_{Z_1}$ and the boundary value is in an exceptional level for the third τ_n.
- The connections $\omega|_{Z_1}$ and $\omega|_{Z_2}$ each lie in the divisor associated to one of the τ_n and the boundary value is in an exceptional level for the third.
- The connection $\omega|_{Z_2}$ is reducible and lies in the closure of two or more of the divisors associated with the τ_n.

Each term of the first type is given as a product of an invariant of Z_1 with the degree of the map

$$\partial \colon \mathcal{M}_2(Z_2) \cap D_{\Sigma(A(2))} \to \chi(T).$$

Each term of the second type is given as a product of an invariant of Z_2 with the degree of the map

$$\partial \colon \mathcal{M}_{r+1}(Z_1) \cap D_{\Sigma(A(1))} \cap D_{\tau_n} \to \chi(T).$$

Terms of the last type are the special terms, similar to the ones that we encountered in Claim 5.13. The reason that they can arise here is that $b_2^+(X_2) = 0$, and hence we cannot necessarily avoid reducible connections on the Z_2-side.

The argument now follows that of Lemma 4.16 and Claim 5.13. At least one of the two multiplicative factors in each of the terms of the first two types is independent of the level. Hence, when we compute the $SO(3)$-invariant by gluing the moduli spaces together over the involution of the character variety instead of the identity corresponding terms of these types evaluate the same in the two cases.

Let us examine the nature of the moduli space for S near a reducible connection on Z_2. A neighborhood of the reducible connection in $\mathcal{M}_2(Z_2)$ is $\mathbf{C}^6/S^1$ where the S^1-action is the standard one. An open subset of $\mathcal{M}_{r+3}(S)$ of connections close to the given reducible on Z_2 then is of the form

$$\tilde{\mathcal{M}}_{r+1}(Z_1) \times_{S^1} \mathbf{C}^6$$

where $\tilde{\mathcal{M}}_{r+1}(Z_1)$ is the total space of the principal S^1-bundle which reduces the based $SO(3)$-bundlle and whose first Chern class is ϵ. Each divisor associated to a surface in Z_2 restricted to this neighborhood is some integral multiple of the divisor $\tilde{\mathcal{M}}_{r+1}(Z_1) \times_{S^1} \mathbf{C}^5$. We call this multiple the *multiple of the divisor at the reducible connection*.

Let us examine a term of the third type when the reducible connection on Z_2 lies in the closure of exactly two of the divisors associated to the τ_n. In this case we see that the result is the product of the degree of

$$\partial\colon \mathcal{M}_{r+1}(Z_1) \cap D_{\Sigma(A(1))} \cap D_{\tau_n} \to \chi(Y) \tag{13}$$

over the boundary value of the reducible connection with the product of the multiples of the divisors on the Z_2-side at the reducible. As before, this product is the same whether we glue the moduli spaces together over the involution of $\chi(T)$ or over the identity of $\chi(T)$. The reason is that the degree of the map in Expression 13 is independent of the point at which we compute the degree.

Lastly, we must consider reducibles in the closure of all three of the divisors D_{τ_n}. If we take the intersection of the divisors $D_{\Sigma(A(1))}$ on the Z_1-side and then the intersection with the fiber over the boundary value of the reducible connection, we obtain a compact two-dimensional submanifold A of $\mathcal{M}_{r+1}(Z_1)$. Let $\tilde{A} \to A$ be the restriction of $\tilde{\mathcal{M}}_{r+1}(Z_1)$ to A. Inside $\tilde{A} \times_{S^1} \mathbf{C}^6$ we are intersecting seven divisors, each of which is a multiple of the divisor $\tilde{A} \times_{S^1} \mathbf{C}^5$. The result is then equal to the product of the multiplicities times the value of the first Chern class $c_1(\tilde{A})$ on A. The value of this Chern class is equal to

$$\langle \epsilon, \mathcal{M}_{r+1}(Z_1) \cap D_{\Sigma(A(1))} \cap \partial^{-1}([\rho])\rangle.$$

But this expression is equal to the difference

$$\frac{1}{c(\tau)}\gamma_{r+1}(S_1)([\Sigma(A(1))],[F],[\tau]) - \gamma_{-4r-4,[F]}(\Sigma(A(1))],[F],[\tau])$$

which by Corollary 4.12 is equal to zero. This shows that terms of the last type with three divisors in the closure of the reducible connection on the Z_2-side contribute zero to the value of the Donaldson invariant. This is fortunate since when we compute the $SO(3)$-invariant the class ϵ in the above expression is replaced by its negative.

Thus, in all cases the contributions of each term to the $SU(2)$-invariant and to the $SO(3)$-invariant are equal.

The following lemma is a consequence of the computation of the leading coefficients given in *[7]* and Lemma 4.17.

Lemma 6.4: *Let S be an elliptic surface with at most two multiple fibers over $\mathbf{C}P^1$ and with $p_g(S) = 2$. Let the multiplicities of the multiple fibers be m_1, m_2, and let M be their product. Then for any classes $x_1, \ldots, x_{4r+1}$ with the property that x_i is orthogonal to $[F]$ for $1 \le i \le 4r$, letting ν denote the class represented by a point of S, we have*

$$\gamma_{r+3}(S)(x_1,\ldots,x_{4r+1},\nu) = 2M^2Q^{(2r)}(x_1,\ldots,x_{4r})\langle\kappa_S, x_{4r+1}\rangle.$$

6.3 The case when $p_g = 2$ and there are no multiple fibers

Now we are ready to prove the first case of Theorem 1.3 when p_g is even.

Proposition 6.5: *Let S be an elliptic surface with no multiple fibers over $\mathbf{C}P^1$ and with $p_g(S) = 2$. Fix $k > 0$ and set $d = 4k - 9$ and $n = (d-1)/2$. Then the Donaldson polynomial of S is given by*

$$\gamma_k(S) = \sum_{i=0}^{n} \alpha(k,i) Q_S^{(i)} \kappa_S^{d-2i}$$

where

$$\alpha(k,n) = \frac{d!}{(2n)!}$$

and

$$\alpha(k,n-1) = \frac{d!}{(2n-2)!}\frac{1}{6}.$$

Proof: We already know that the formula for the leading coefficient is correct; we concentrate on the second coefficient. By Kronheimer's periodicity (Corollary 4.20) it suffices to prove this result for all $r >> 0$. Let S be an elliptic surface over $\mathbf{C}P^1$ with no mulitple fibers and with $p_g(S) = 5$. We divide S into a fiber connected sum of S_1 and S_2, each of which is an elliptic surface with $p_g = 2$. We take surfaces $\Sigma(A(1)) = \{\Sigma_1, \ldots, \Sigma_{4r_1}\}$ embedded in X_1, $\Sigma(A(2)) = \{\Sigma_{4r_1+1}, \ldots, \Sigma_{4r}\}$ embedded in X_2 and multi-sections $\tau_1, \ldots, \tau_6$ with $\tau_n \cdot F = c(n)$. We set $r = r_1 + r_2$. For simplicity we assume that the Σ_i are disjoint from the τ_j and that the τ_j are disjoint among themselves.

By Theorem 1.3 in the case of p_g odd we know that

$$\gamma_{r+3}(S)([\Sigma_1], \ldots, [\Sigma_{4r}], [\tau_1], \ldots, [\tau_6]) = \frac{4}{6} 6! Q_S^{(2r)}([\Sigma(A(1))]) \prod_{n=1}^{6} c(n).$$

On the other hand, according to the calculus developed at the beginning of this section, this invariant is given as the sum of two terms. One is

$$\sum_{n=1}^{6} \sum_{\ell_n} \sum_{I_1, I_2} \left[\mathcal{M}_{r_1+3}(Z_1) \cap D_{\Sigma(A(1))} \cap \left(\cap_{n \in I_1} D_{\tau_n}\right) \cap \partial^{-1}([\rho])\right] \tag{14}$$

$$\cdot \left[\mathcal{M}_{r_2+3}(Z_2) \cap D_{\Sigma(A(2))} \cap \left(\cap_{n \in I_2} D_{\tau_n}\right) \cap \partial^{-1}(\ell_n)\right]$$

where ℓ_n ranges over the exceptional levels for $\tau_n \cap Y$, where I_1 and I_2 form a partition of $(\{1, \ldots, 6\} - n)$ with I_1 being of cardinality two, and where $[\rho] \in \ell_n$. The second term is the symmetric one where I_1 has cardinality three. Clearly, by symmetry the value of the invariant is equal to twice the sum given in Expression 14.

By Lemma 5.9 we have

$$\mathcal{M}_{r_1+3}(Z_1) \cap D_{\Sigma(A(1))} \cap \left(\cap_{n \in I_1} D_{\tau_n}\right) \cap \partial^{-1}([\rho])$$

$$= \gamma_{r_1+3}(S_1)([\Sigma(A(1))],[\tau_{I_1}],[F])$$
$$= 2Q_{S_1}^{(2r_1)}([\Sigma(A(1))]) \prod_{n\in I_1} c(n)$$

Plugging this in and using the symmetry we see that

$$\frac{\prod_{n=1}^{6} c(n)}{6} 6! Q_{Z_2}^{(2r_2)}([\Sigma(A(2))]) = \sum_{n=1}^{6}\sum_{\ell_n}\sum_{I_1,I_2}\left(\prod_{i\in I_1} c(i)\right)$$
$$\mathcal{M}_{r_2+3}(Z_2)\cap D_{\Sigma(A(2))]}\cap\left(\cap_{n\in I_2} D_{\tau_n}\right)\cap\partial^{-1}(\ell_n).$$

Arguing as in the case when p_g is odd we can rewrite the right-hand-side of this expression as:

$$\sum_{n=1}^{6}\sum_{\ell_n}\sum_{I_1,I_2}\left(\prod_{i\in I_1} c(i)\right)\Big[2\gamma_{r_2+3}(S_2)([\Sigma(A(2))],[\tau_{I_2}]) \tag{15}$$
$$+\sum_{q\in I_3}\frac{c(q)}{2}\langle\epsilon,\mathcal{M}_{r_2+3}(Z_2)\cap D_{\Sigma(A(2))}\cap\tau_{I_2-\{q\}}\cap\partial^{-1}(\ell_n)\rangle$$
$$+\sum_{p\neq q\in I_2}\frac{c(p)c(q)}{4}\langle\epsilon^2,\mathcal{M}_{r_2+3}(Z_2)\cap D_{\Sigma(A(2))}\cap D_{\tau_{I_2-\{p,q\}}}\cap\partial^{-1}(\ell_n)\rangle$$
$$+\frac{\prod_{p\in I_2} c(p)}{8}\langle\epsilon^3,\mathcal{M}_{r_2+3}(Z_2)\cap D_{\Sigma(A(2))}\cap\partial^{-1}(\ell_n)\rangle\Big].$$

Now let us examine the individual terms in the previous expresssion

Claim 6.6: *Given $a > 0$ there is a neighborhood of infinity in Z_2 such that the energy of any connection in $\mathcal{M}_{r_2+3}(Z_2)\cap D_{\Sigma(A(2))}$ on this neighborhood is less than a. Furthermore,*

$$\langle\epsilon^3,\mathcal{M}_{r_2+3}(Z_2)\cap D_{\Sigma(A(2))}\cap\partial^{-1}(\ell_n)\rangle = 0.$$

Proof: The first statement is an easy consequence of the fact that the intersection in question is of dimension 7. The second expression is equal to

$$(-4)\langle\epsilon\cup\mu(pt),\mathcal{M}_{r_2+3}(Z_2)\cap D_{\Sigma(A(2))}\cap\partial^{-1}(\ell_n)\rangle.$$

From this expression it is clear that the result is independent of the level and is given by

$$(-4)\left(\gamma_{r_2+3}([\Sigma(A(2))],\nu,[\sigma]) - \gamma_{-4r_2-12,[F]}([\Sigma(A(2))],\nu,[\sigma])\right)$$

where σ is a section of the elliptic fibration. But by Lemma 4.17 and Lemma 4.16 this difference is zero.

Claim 6.7:

$$\langle\epsilon^2,\mathcal{M}_{r_2+3}(Z_2)\cap D_{\Sigma(A(2))}\cap D_{\tau_{I_2-\{p,q\}}}\cap\partial^{-1}(\ell_n)\rangle = -16c(s)Q_{Z_2}^{(2r_2)}([\Sigma(A(2))])$$

where $\{p, q, s\} = I_2$.

Proof: This follows from the fact that $\epsilon^2 = \mu(pt)$ and from Lemma 4.17.

Claim 6.8: *Let* σ *be a section of* S_2*, i.e.,* $\sigma \cdot F = 1$*. Then*

$$\langle \epsilon, \mathcal{M}_{r_2+3}(Z_2) \cap D_{\Sigma(A(2))} \cap D_\sigma \cap D_\sigma \cap \partial^{-1}(\ell_n) \rangle = 8Q^{(2r_2)}([\Sigma(A(2))]).$$

Proof: This is established using the previous two claims and the fact that the $SO(3)$- and $SU(2)$-invariants evaluate the same on $([\Sigma(A(2))], [\tau_{I_2}])$, cf, the argument in the proof of Lemma 5.11.

Now using these three claims we go back to Expression 15 and see that it is equal to

$$\sum_{n=1}^{6}\sum_{\ell_n}\sum_{I_1,I_2}\left(\prod_{i\in I_1} c(i)\right)\Big[2\gamma_{r_2+3}(S_2)([\Sigma(A(2))],[\tau_{I_2}])$$
$$+\sum_{q\in I_3}\frac{c(q)}{2}8\left(\prod_{p\in(I_2-\{q\})} c(p)\right)Q^{(2r_3)}_{Z_2}([\Sigma(A(2))])$$
$$+\sum_{p\neq q\in I_2}\frac{c(p)c(q)}{4}(-16)c(s)Q^{(2r_3)}_{Z_2}([\Sigma(A(2))])\Big].$$

Simplifying using Lemma 6.2 we see that Expression 15 is equal to

$$\sum_{n=1}^{6}\sum_{\ell_n}\sum_{I_1,I_2}\left(\prod_{i\in I_1} c(i)\right)\Big[2\gamma_{r_2+3}(S_2)([\Sigma(A(2))],[\tau_{I_2}])$$
$$+\frac{3}{2}\prod_{p\in I_2} c(p)Q^{(2r_2)}_{Z_2}([\Sigma(A(2))]) - \frac{1}{2}\prod_{p\in I_2} c(p)Q^{(2r_2)}_{Z_2}([\Sigma(A(2))])\Big]$$
$$=\frac{6!}{12}\prod_{n=1}^{6} c(n)\left(\gamma_{r_2+3}(S_2)([\Sigma(A(2))],[\sigma_1],[\sigma_2],[\sigma_3]) + Q^{(2r_2)}_{Z_2}([\Sigma(A(2))])\right)$$

where σ_n are sections of S disjoint from all the Σ_i and from each other.

This gives us the equation

$$\frac{1}{6}Q^{(2r_2)}_{S_2}([\Sigma(A(2))])$$
$$=\frac{1}{12}\left(\gamma_{r_2+3}(S_2)([\Sigma(A(2))],[\sigma_1],[\sigma_2],[\sigma_3]) + Q^{(2r_2)}_{S_2}([\Sigma(A(2))])\right).$$

Thus, we have

$$Q^{(2r_2)}_{S_2}([\Sigma(A(2))]) = \gamma_{r_2+3}(Z_2)([\Sigma(A(2))],[\sigma_1],[\sigma_2],[\sigma_3]).$$

This establishes the given formula for $\alpha(k, n-1)$ in the case of an elliptic surface S without multiple fibers and with $p_g(S) = 2$. This then completes the proof of Proposition 6.5.

6.4 The case when $p_g \equiv 0 \pmod 2$

First, let us consider the case when $p_g > 2$.

Proposition 6.9: *Let S be an elliptic surface over $\mathbf{C}P^1$ with $p_g = p_g(S) > 2$ and $p_g \equiv 0 \pmod 2$. Suppose that S has at most two multiple fibers and that m_1, m_2 are their mutliplicities. Let $M = m_1 m_2$. Then*

$$\alpha(k, n-1) = \frac{d!}{(2n-2)!} M^{p_g} \frac{\left((p_g+1)M^2 - m_1^2 - m_2^2\right)}{6}.$$

Proof: Set $t = p_g/2$. We write $S = S_1 \#_f \cdots \#_f S_t$ where each $S_i, i < t$ has $p_g = 1$, and $p_g(S_t) = 2$, and where S_1 has two multiple fibers and the S_i for $i > 1$ have no multiple fibers. Then we pull S apart along the union Y of the splitting tori. Using the results from the previous section in order to evaluate the terms associated with S_t, the analysis proceeds exactly as in the case when all the S_i have $p_g = 1$.

The last case of Theorem 1.3 is covered by the next result.

Proposition 6.10: *Let S be an elliptic surface over $\mathbf{C}P^1$ with $p_g = p_g(S) = 2$. Suppose that S has at most two multiple fibers and that m_1, m_2 are their mutliplicities. Let $M = m_1 m_2$. Then*

$$\alpha(k, n-1) = \frac{d!}{(2n-2)!} M^{p_g} \frac{\left(3M^2 - m_1^2 - m_2^2\right)}{6}.$$

Proof: Consider an ellitpic surface S over $\mathbf{C}P^1$ with $p_g = 4$. By Proposition 6.10 we know the result for this surface. We decompose $S = S_1 \#_f S_2$ where $p_g(S_1) = 2$, S_1 has two multiple fibers, and S_2 is a $K3$-surface. Again we apply the analysis as in the case when all the S_i have $p_g = 1$ to evaluate this invariant in terms of invariants of S_1 and S_2. The argument is similar to than one given in Section 6.3.

References

[1] S. Bauer, Diffeomorphism types of ellitpic surfaces with $p_g = 1$, J. Reine Angew. Math. **451** (1993), 89–148.

[2] P. Braam and S. Donaldson, Fukaya-Floer homology and gluing formulae for polynomial invariants, in *Andreas Floer Memorial Volume*, C.H. Taubes, A. Weinstein and E. Zehnder (eds.), Springer Verlag, Berlin, 1993, to appear.

[3] S. Donaldson, Gluing techniques in the cohomology of moduli spaces, in: *Topological Methods in Modern Mathematics*, Publish or Perish, 1993, 137–170.

[4] S. Donaldson and P. Kronheimer, *The Geometry of Four-Manifolds*, Clarendon Press, Oxford, 1990.

[5] R. Fintushel and R. Stern, 4-manifolds and their rational blow downs, in preperation.

[6] R. Friedman, Vector bundles and SO(3)-invariants for elliptic surfaces, J. Amer. Math. Soc. **8** (1993), 29–139.

[7] R. Friedman and J. Morgan, Smooth four-manifolds and complex structures, Ergebnisse Mathematik und ihrer Grenzgebiete, (1994), 3Folge Band 27, Springer-Verlag, Heidelberg (1993).

[8] R. Gompf and T. Mrowka, Irreducible four manifolds need not be complex, Annals of Math. (1993), 61–111.

[9] P. Kronheimer and T. Mrowka, Gauge theory for embedded surfaces, I, Topology **32** (1993), 773–826; Part II, Topology **34** (1993), 37–97.

[10] ——, Recurrence relations and asymptotics for Donaldson's polynomial invariants, Bull. A.M.S. **30** (1994), 215–221.

[11] P. Lisca, Computations of instanton invariants using Donaldson-Floer theory, preprint.

[12] J. Morgan and T. Mrowka, A note on Donaldson's polynomial invariants, Duke J. Math. **68** (1992), 223–230.

[13] ——, On the diffeomorphism classification of regular elliptic surfaces, IMRN Duke J. (1993), 183–184.

[14] J. Morgan, T. Mrowka, and D. Ruberman, L^2 moduli spaces for manifolds with cylindrical ends and a vanishing theorem for Donaldson invariants, Monographs in Geometry and Topology, (1993), International Press Inc., Boston.

[15] J. Morgan and K. O'Grady, Differential topology of complex surfaces; elliptic surfaces with $p_g = 1$: Smooth classification, Lecture Notes in Math. **1545** (1993).

[16] A. Stipsicz and Z. Szabo, The smooth classification of elliptic surfaces with $b^2 + > 0$, Duke Math. J. **75** (1993), 1–50.

[17] B. Yu, Some computation of Donaldson's invariants via flat connections, PhD thesis, U.C. Berkeley, 1994.

Quasiclassical Asymptotics of Solutions to the KZ Equations

Nicolai Reshetikhin[1]
Department of Mathematics
University of California
Berkeley, California 94720

Alexander Varchenko[2]
Department of Mathematics
University of North Carolina
Chapel Hill, North Carolina 27599

Abstract: The quasiclassical asymptotics of the Knizhnik-Zamolodchikov system is studied. Solutions to this system in this limit are related naturally to Bethe vectors in the Gaudin model of spin chains.

Introduction

The Knizhnik-Zamolodchikov differential equation

$$\frac{\partial F}{\partial z_i}(z) = \frac{1}{\kappa} H_i(z) F(z) , \qquad i = 1, \ldots, n ,$$

appeared first as a holonomic system of differential equations on conformal blocks in a WZW model of conformal field theory *[10]*. Here $F(z_1, \ldots, z_n)$ is a function with values in the tensor product $V_1 \otimes \ldots \otimes V_n$ of representations of a simple Lie algebra $\mathfrak{g}$, $\kappa = k + g$, where k is the central charge of the model, and g is the dual Coxeter number of the simple Lie algebra $\mathfrak{g}$.

One of the remarkable properties of the KZ system is that the coefficient functions $H_i(z)$ commute and that the form $w = \sum_i H_i(z) dz_i$ is closed:

$$\frac{\partial H_i}{\partial z_j} = \frac{\partial H_j}{\partial z_i} , \qquad [H_i, H_k] = 0 .$$

In this work we study asymptotics of solutions to the KZ equation when κ tends to zero. In this limit, solutions to the KZ equation turn into normalized eigenvectors of commuting linear operators $H_1(z), \ldots, H_n(z)$.

[1] The first author was supported by Alfred P. Sloan fellowship and by NSF grant DMS-9296120.
[2] The second author was supported by NSF Grant DMS-9203929.

Geometry, Topology, and Physics for Raoul Bott

Cambridge, MA 02138 USA

There are integral representations for solutions to the KZ equation,

$$F(z) = \int_C \exp(S(t,z)/\kappa)A(t,z)dt\ ,$$

where $S(t,z)$ is some multivalued scalar function of z and t, $A(z,t)$ is a rational vector valued function, and C is a cycle on which $S(z,t)$ is a single-valued function of t, $t=(t_1,\dots,t_k)$ *[16]*. If κ tends to zero, the integral is localized at critical points of S with respect to t. It turns out that every nondegenerate critical point $t(z)$ gives an eigenvector $A(t(z),z)$ of $H_1(z),\dots,H_n(z)$.

The algebraic Bethe Ansatz is a certain construction of eigenvectors for a system of commuting operators. The idea of this construction is to find a vector-valued function of a special form and determine its arguments in such a way that the value of this function will be an eigenvector. The equations which determine these special values of arguments are called the Bethe equations. For more details about the algebraic Bethe ansatz see *[8]*.

One of the systems of commuting operators which can be diagonalized by the ABA is the Gaudin model *[7]* of an inhomogeneous magnetic chain.

It turns out that the functions $A(z,t)$ are exactly those special functions which appear in the ABA for the Gaudin model and that the Bethe equations for the Gaudin model coincide with the equations on critical points of the function $S(z,t)$:

$$\frac{\partial S(z,t)}{\partial t_i} = 0, \qquad i = 1,\dots,k.$$

We show that for every nondegenerate critical point $t(z)$:

$$B(A(t(z),z),A(t(z),z)) = \text{const}\cdot \text{Hess}_t(S(t(z),z))$$

where B is the Shapovalov form on $V_1\otimes\dots\otimes V_n$ and Hess is the Hessian of the function $S(t,z)$ as a function of t at $t=t(z)$. This is an analog of formulae for norms of Bethe vectors in *[9, 11]*.

As an example we consider the case of $\mathfrak{g}=sl_2$. We show that the constructed eigenvectors are pairwise orthogonal with respect to the Shapovalov form and form a basis in the space of singular vectors for generic z. We describe asymptotics of this basis when $|z_1| << |z_2| << \dots << |z_n|$.

The first part of the paper contains some general facts about asymptotically flat sections for families of flat holomorphic bundles. In the second part we recall some basic facts about Knizhnik-Zamolodchikov equations. In section 3 we recall integral representations for solutions to the KZ equations from *[16, 17]*. Section 4 contains the description of asymptotically flat sections of the KZ bundles. Section 5 contains a conjecture about the structure of critical points in the case $n=2$. In section 6 we analyse the asymptotic behaviour of critical points. Sections 7, 8, 9 contain detailed analysis of the sl_2 case. In section 10 we study the Bethe basis for generic configuration of points $z_1,\dots,z_n$. Section 11 contains some remarks about the relation between asymptotically flat sections of the KZ bundle and the

Lamé functions. In the last section 12 we discuss branching properties of the Bethe vectors in the Gaudin model.

The main result of this work is Theorem (4.10) and its Corollary (4.16), it was found in 1990 and has been the subject of several talks given in 1992-1993.

When this manuscript was completed we learned about the papers *[3, 4]* where the relation between asymptotic solutions to the KZ system and the Bethe vectors has been found.

Similarly, quasiclassical assymptotics of q-KZ equations *[6]* reproduces Bethe vectors for corresponding spin chains *[12, 15]*.

The structure of the Bethe vectors for the Gaudin model is naturally related to the representation theory of affine Lie algebras at the critical level. This subject has been studied in *[5]*.

One of us (N.R.) is grateful to B. Feigin , E. Frenkel, I. Frenkel, V. Ginzburg, and to E. Sklyanin for many interesting discussions.

1 Asymptotically Flat Sections

Let $\mathcal{D}$ be a ball in $\mathbb{C}^n$, $\pi : \mathbb{C}^N \times \mathcal{D} \to \mathcal{D}$ the projection. Let

$$\nabla_\kappa = \kappa d - \omega$$

be a *family of flat holomorphic connections* in π, $\kappa \in \mathbb{C}$. Here

$$\omega = H_1 dz_1 + \ldots + H_n dz_n$$

where $\{H_i\}$ are matrix valued functions.

The flatness means that

$$\kappa d\omega + \omega \wedge \omega = 0$$

for all $\kappa \in \mathbb{C}$, or

$$d\omega = 0\,, \qquad \omega \wedge \omega = 0,$$

which implies

$$\frac{\partial H_i}{\partial z_j} = \frac{\partial H_j}{\partial z_i}\,, \qquad [H_i, H_j] = 0\,, \tag{1.1}$$

for all i and j.

We are interested in *asymptotically flat sections* F for our family of flat bundles which have the form

$$F = \exp(S/\kappa)(f_0 + \kappa f_1 + \ldots) \quad \text{for } \kappa \to 0\,. \tag{1.2}$$

Here $S(z_1, \ldots, z_n)$ is a function, $\{f_j(z_1, \ldots, z_n)\}$ are sections of π, and F must be a formal solution to the system of equations

$$\nabla_\kappa F = 0\,. \tag{1.3}$$

We will call $\exp(S/\kappa) f_0$ *an asymptotically flat section of the first order* if there exists a power series (1.2) which provides an asymptotic solution to (1.3) modulo terms of order κ^2.

We describe asymptotically flat sections in terms of fundamental solutions to the system (1.3). First we arrange them into an *asymptotic fundamental solution to* (1.3) which is a matrix with columns being asymptotically flat sections (1.2) such that its first terms form a basis of sections of π. Assume that

- linear operators $\{H_i(z)\}_{i=1}$ are simultaneously diagonalizable for each $z \in \mathcal{D}$,
- the spectrum of operators $\{H_i(z)\}_{i=1}^n$ is simple for each $z \in \mathcal{D}$ ($\lambda_i^\alpha \neq \lambda_i^\beta$ for some $i = 1, \ldots, n$ if $\alpha \neq \beta$).

Let $\phi(z)$ be the matrix with columns being eigenvectors of $H_i(z)$, $i = 1, \ldots n$, and $\Lambda_i(z)$ the diagonal matrix with eigenvalues of $H_i(z)$ on the corresponding places in the diagonal. We have

$$H_i(z)\phi(z) = \Lambda_i(z)\phi(z) \tag{1.4}$$

for each $z \in \mathcal{D}$.

Proposition 1.5: *We have an identity*

$$\partial_i \Lambda_j - \partial_j \Lambda_i = 0 \tag{1.6}$$

Proof: We have the following equalities:

$$\begin{aligned}\partial_i \Lambda_j - \partial_j \Lambda_i &= \partial_j(\phi^{-1} H_i \phi) - \partial_i(\phi^{-1} H_j \phi) \\ &= -\phi^{-1}\partial_j \phi \Lambda_i + \Lambda_i \phi^{-1} \partial_j \phi + \phi^{-1} \partial_i \phi \Lambda_j + \Lambda_j \phi^{-1} \partial_i \phi \\ &= -[\phi^{-1}\partial_j \phi, \Lambda_i] + [\phi^{-1} \partial_i \phi, \Lambda_j] \,.\end{aligned}$$

Clearly the diagonal part of the right-hand side is zero, which implies equality (1.6).

Corollary 1.7: *There exists a function $S(z)$ on $\mathcal{D}$ with values in the diagonal $N \times N$ matrices, such that*

$$\Lambda_i(z) = \partial_i S(z). \tag{1.8}$$

Corollary 1.9: *Let $\{S_\alpha(z)\}$ be diagonal elements of $S(z)$. Then we have*

$$(\phi^{-1}\partial_i \phi)_{\alpha\beta} = N_{\alpha\beta}(\partial_i S_\alpha - \partial_i S_\beta)\,, \qquad \alpha \neq \beta \tag{1.10}$$

for some matrix $N(z)$ which does not depend on i.

This follows from identity (1.6).

The following theorem gives a description of asymptotic fundamental solutions to system (1.3).

Theorem 1.11: *Let $\Phi(z)$ be the following asymptotic series expansion:*

$$\Phi(z) = \phi(z) \sum_{K \geq 0} \kappa^K \psi^{(K)}(z) e^{\frac{S(z)}{\kappa}} C \, . \tag{1.12}$$

Then:
(1) $\Phi(z)$ is an asymptotic fundamental solution to the system

$$\kappa \frac{\partial \Phi}{\partial z_i} = H_i \Phi \, , \qquad i = 1, \dots N \tag{1.13}$$

if
(i) C *is a constant*
(ii) $\phi(z)$ *is as in (1.4)*
(iii) functions $\psi^{(K)}(z)$ *satisfies the following recursive relations:*

$$\psi^{(K)}_{\alpha\beta} = \sum_\gamma N_{\alpha\gamma} \partial_i S_{\alpha\gamma} \psi^{(K-1)}_{\gamma\beta} + \frac{\partial_i \psi^{(K-1)}_{\alpha\beta}}{\partial_i S_\alpha - \partial_i S_\beta} \, , \qquad \alpha \neq \beta \tag{1.14}$$

$$\partial_i \psi^{(K)}_{\alpha\alpha} + \sum_\gamma N_{\alpha\gamma} \psi^{(K)}_{\gamma\alpha} (\partial_i S_\alpha - \partial_i S_\gamma) = 0$$

$$\psi^{(0)}_{\alpha\beta} = \delta_{\alpha\beta} \tag{1.15}$$

where $N_{\alpha\beta}$ is the same as in (1.10).

(iv)

$$(\phi^{-1} \partial_i \phi)_{\alpha\alpha} = 0 \ \ \textit{for each}\ \alpha = 1, \dots, N \, . \tag{1.16}$$

(2) There exists a solution to (1.14) and (1.15) which is unique up to

$$\psi^{(K-1)}_{\gamma\beta} \longrightarrow \text{const} \sum_{S+T=K} \psi^{(S)}_{\alpha\beta} C^{(T)}_\beta \, ,$$

and the matrix $\phi(z)$ has property (iv). Therefore, there exists an asymptotic fundamental solution to system (1.13).

Proof: (1) We just have to substitute (1.12) into the system (1.13). Comparing coefficients of asymptotic expansion with respect to κ we obtain the recursive relations for $\psi^{(K)}$:

$$(\phi^{-1} \partial_i \phi) \psi^{(K-1)} + \partial_i \psi^{(K-1)} = [\partial_i S, \psi^{(K)}]$$

and the condition

$$H_i \phi = \phi \partial_i S \, .$$

Equations (1.14) and (1.15) are the off-diagonal and diagonal matrix elements of (1.14), respectively. Let us prove that the system consisting of (1.14) and (1.15) has solutions. For this we must show:

$$(\alpha) \qquad \frac{\partial_i \psi^{(K)}_{\alpha\beta} + \sum_\gamma \partial_i S_{\alpha\gamma} N_{\alpha\gamma} \psi^{(K)}_{\alpha\beta}}{\partial_i S_{\alpha\beta}} = \frac{\partial_j \psi^{(K)}_{\alpha\beta} + \sum_\gamma \partial_j S_{\alpha\gamma} N_{\alpha\gamma} \psi^{(K)}_{\alpha\beta}}{\partial_j S_{\alpha\beta}}$$

$$(\beta) \qquad \partial_j \sum_\gamma N_{\alpha\gamma} \psi^{(K)}_{\gamma\alpha} \partial_i S_{\alpha\gamma} = \partial_i \sum_\gamma N_{\alpha\gamma} \psi^{(K)}_{\gamma\alpha} \partial_j S_{\alpha\gamma}$$

where $S_{\alpha\beta} := S_\alpha - S_\beta$. Let us first prove (α).

Lemma 1.17:

$$\partial_i N_{\alpha\beta} = L_{\alpha\beta} \partial_i S_{\alpha\beta} \tag{1.18}$$

for some $L_{\alpha\beta}$.

Proof: Let us recall the definition of $N_{\alpha\beta}$ and compute $\partial_j N_{\alpha\beta}$:

$$\partial_j N_{\alpha\beta} = \frac{\partial_j (\phi^{-1} \partial_i \phi)_{\alpha\beta}}{\partial_i S_{\alpha\beta}} - \frac{(\phi^{-1}\partial_i \phi)_{\alpha\beta} \partial_i \partial_j S_{\alpha\beta}}{(\partial_i S_{\alpha\beta})^2}$$

where $S_{\alpha\beta} := S_\alpha - S_\beta$. Then we have:

$$\begin{aligned}
\partial_j (\phi^{-1}\partial_i \phi)_{\alpha\beta} &= -(\phi^{-1}\partial_j \phi \phi^{-1} \partial_i \phi)_{\alpha\beta} + (\phi^{-1}\partial_i \partial_j \phi)_{\alpha\beta} \\
&= -\sum_\gamma N_{\alpha\gamma} N_{\gamma\beta} \partial_j S_{\alpha\gamma} \partial_i S_{\gamma\beta} + (\phi^{-1}\partial_i(\phi[\partial_j S, N]))_{\alpha\beta} \\
&= -\sum_\gamma N_{\alpha\gamma} N_{\gamma\beta} \partial_j S_{\alpha\gamma} \partial_i S_{\gamma\beta} + ([\partial_i S, N][\partial_j S, N])_{\alpha\beta} \\
&\quad + [\partial_i \partial_j S, N]_{\alpha\beta} + [\partial_j S, \partial_i N]_{\alpha\beta} \,.
\end{aligned}$$

Therefore:

$$\partial_j N_{\alpha\beta} = \frac{\partial_j S_{\alpha\beta} \partial_i N_{\alpha\beta}}{\partial_i S_{\alpha\beta}}$$

which implies (1.6).

Now we can prove (α) by induction. Lemma 1.17 provides a base for induction:

$$\psi^{(1)}_{\alpha\beta} = N_{\alpha\beta} \,.$$

Relation (1.18) means that

$$\partial_i \psi^{(1)} = L_{\alpha\beta} \partial_i S_{\alpha\beta}.$$

Assume (1.14) holds for K. Let us prove that this implies (α) and therefore consistency of (1.14) for $K+1$.

$$\partial_i S_{\alpha\beta}\left(\partial_j \psi_{\alpha\beta}^{(K)} + \sum_\gamma \partial_j S_{\alpha\gamma} N_{\alpha\gamma}\psi_{\gamma\beta}^{(K)}\right) - (i \leftrightarrow j) =$$

$$= \partial_i S_{\alpha\beta}\left(\partial_j\left(\frac{\sum_\gamma (\phi^{-1}\partial_i\phi)_{\alpha\gamma}\psi_{\gamma\beta}^{(K-1)} + \partial_i\psi_{\alpha\beta}^{(K-1)}}{\partial_i S_{\alpha\beta}}\right)\right.$$

$$\left. + \sum_\gamma \partial_j S_{\alpha\gamma} N_{\alpha\gamma}\psi_{\gamma\beta}^{(K)}\right) - (i \leftrightarrow j)$$

$$= \left(-\sum_{\gamma,\tau}(\phi^{-1}\partial_i\phi)_{\alpha\gamma}(\phi^{-1}\partial_j\phi)_{\gamma\tau}\psi_{\tau\beta}^{(K-1)} + \sum_\gamma(\phi^{-1}\partial_j\partial_i\phi)_{\alpha\gamma}\psi_{\gamma\beta}^{(K-1)}\right.$$

$$+ \sum_\gamma (\phi^{-1}\partial_i\phi)_{\alpha\gamma}\partial_j\psi_{\gamma\beta}^{(K-1)} + \partial_i\partial_j\psi_{\alpha\beta}^{(K-1)} - \psi_{\alpha\beta}^{(K)}\partial_i\partial_j S_{\alpha\beta}$$

$$\left. + \sum_\gamma \partial_i S_{\alpha\beta}(\phi^{-1}\partial_j\phi)_{\alpha\gamma}\psi_{\gamma\beta}^{(K)}\right) - (i \leftrightarrow j)$$

$$= -\sum_\gamma (\phi^{-1}\partial_j\phi)_{\alpha\gamma}(\phi^{-1}\partial_i\phi)_{\gamma\tau}\psi_{\tau\beta}^{(K-1)}$$

$$+ \sum_\gamma (\phi^{-1}\partial_i\phi)_{\alpha\gamma}\partial_j\psi_{\gamma\beta}^{(K-1)} + \partial_i S_{\alpha\beta}(\phi^{-1}\partial_j\phi)_{\alpha\gamma}\psi_{\gamma\beta}^{(K)} - (i \leftrightarrow j)$$

$$= -\sum_\gamma (\phi^{-1}\partial_j\phi)_{\alpha\gamma}[\sum_\tau (\phi^{-1}\partial_i\phi)_{\gamma\tau}\psi_{\tau\beta}^{(K-1)} + \partial_i\psi_{\gamma\beta}^{(K-1)}]$$

$$+ \sum_\gamma (\phi^{-1}\partial_i\phi)_{\alpha\gamma}[\sum_\tau (\phi^{-1}\partial_j\phi)_{\gamma\tau}\psi_{\tau\beta}^{(K-1)} + \partial_i\psi_{\gamma\beta}^{(K-1)}]$$

$$+ \partial_i S_{\alpha\beta}\sum_\gamma (\phi^{-1}\partial_j\phi)_{\alpha\gamma}\psi_{\gamma\beta}^{(K)} - \partial_j S_{\alpha\beta}\sum_\gamma (\phi^{-1}\partial_i\phi)_{\alpha\gamma}\psi_{\alpha\beta}^{(K)}$$

$$= -\sum_\gamma (\phi^{-1}\partial_j\phi)_{\alpha\gamma}\partial_i S_{\alpha\beta}\psi_{\gamma\beta}^{(K)} + \sum_\gamma (\phi^{-1}\partial_i\phi)_{\alpha\gamma}\partial_j S_{\gamma\beta}\psi_{\gamma\beta}^{(K)}$$

$$+ \sum_\gamma \partial_i S_{\alpha\beta}(\phi^{-1}\partial_j\phi)_{\alpha\gamma}\psi_{\gamma\beta}^{(K)} - \sum_\gamma \partial_j S_{\alpha\beta}(\phi^{-1}\partial_i\phi)_{\alpha\gamma}\psi_{\gamma\beta}^{(K)}$$

$$= \sum_\gamma N_{\alpha\gamma}\psi_{\gamma\beta}^{(K)}$$

$$(-\partial_j S_{\alpha\gamma}\partial_i S_{\gamma\beta} + \partial_i S_{\alpha\gamma}\partial_j S_{\gamma\beta} + \partial_i S_{\alpha\beta}\partial_j S_{\alpha\gamma} - \partial_j S_{\alpha\beta}\partial_i S_{\alpha\gamma}) = 0$$

This proves (α).

Now let us prove (β):

$$\partial_j((\phi^{-1}\partial_i\phi)\psi^{(K-1)}) - \partial_i((\phi^{-1}\partial_j\phi)\psi^{(K-1)}) =$$

$$= -\phi^{-1}\partial_j\phi\phi^{-1}\partial_i\phi\psi^{(K-1)} + \phi^{-1}\partial_i\phi\partial_j\psi^{(K-1)}$$

$$+\phi^{-1}\partial_i\phi\phi^{-1}\partial_j\phi\psi^{(K-1)} - \phi^{-1}\partial_j\phi\partial_i\psi^{(K-1)}$$
$$= -[\phi^{-1}\partial_j\phi, \phi^{-1}\partial_i\phi]\psi^{(K-1)} - \phi^{-1}\partial_i\phi \cdot \phi^{-1}\partial_j\phi\psi^{(K-1)}$$
$$+\phi^{-1}\partial_i\phi[\partial_j S, \psi^{(K)}] + \phi^{-1}\partial_j\phi\phi^{-1}\partial_i\phi\psi^{(K-1)} - \phi^{-1}\partial_j\phi[\partial_i S, \psi^{(K)}]$$
$$= [\partial_i S, N][\partial_j S, \psi^{(K)}] - [\partial_j S, N][\partial_i S, \psi^{(K)}]$$

Diagonal matrix elements of the right-hand side are equal to:

$$\sum_\gamma \partial_i S_{\alpha\beta} N_{\alpha\beta} \partial_j S_{\beta\alpha} \psi^{(K)}_{\beta\alpha} - \sum_\beta \partial_j S_{\alpha\beta} N_{\alpha\beta} \partial_i S_{\beta\alpha} \psi^{(K)}_{\beta\alpha} = 0$$

which proves (β). And finally, it is clear that if each $\psi^{(S)}_{\alpha\beta}$, $S = 0, \ldots, K$, satisfies relations (1.14) and (1.15) then $\psi^{(P)} = \sum_{S=0}^{\rho} \psi^{(S)}_{\alpha\beta} C_\beta^{(P-S)}$ satisfy the same relations. This property reflects the fact that $\psi^{(K)}(z)$ in (1.12) is determined only modulo multiplication from the right by a κ-adic. Theorem 1.11 is proved.

Let V^* be a linear space dual to V and let $H_i^* : V^* \to V^*$ be linear operators dual to H_i. Denote by ∇_i^* the differential operator dual to ∇_i:

$$\nabla_i^* = -\kappa \frac{\partial}{\partial z_i} - H_i^* \tag{1.19}$$

Denote by $\langle \cdot , \cdot \rangle : V^* \otimes V \to \mathbb{C}$ the pairing between V^* and V.
Below we do not assume simplicity of spectrum of H_i.

Proposition 1.20: *(1) If f is a solution to (1.3) and g is a solution to the dual system:*

$$\nabla_i^* g = 0, \qquad i = 1, \ldots, n, \tag{1.21}$$

then

$$\langle g, f \rangle = \text{const.} \tag{1.22}$$

The proof is obvious.
Consider asymptotic solutions f and g to (1.13) and (1.21), respectively:

$$f(z) = \exp\left(\frac{S(z)}{\kappa}\right)(f_0(z) + \kappa\, f_1(z) + \ldots) \tag{1.23}$$

$$g(z) = \exp\left(\frac{T(z)}{\kappa}\right)(g_0(z) + \kappa\, g_1(z) + \ldots) \tag{1.24}$$

Here $S(z)$ and $T(z)$ are two functions which determine eigenvalues of H_i and H_i^* respectively (see (1.8)).

Proposition 1.25: *(1) If $S - T \neq$ const, we have*

$$\langle g_0(z), f_0(z) \rangle = 0$$

(2) If $S - T = const$, we have

$$\langle g_0(z), f_0(z)\rangle = \text{const}.$$

Proof: Indeed, Proposition 1.20 implies implies

$$\langle g(z), f(z)\rangle = \text{const}\ (\kappa).$$

On the other hand,

$$\langle g(z), f(z)\rangle = \exp\left(\frac{S(z) - T(z)}{\kappa}\right)(\langle g_0(z), f_0(z)\rangle + ...)$$

This proves (1) and (2).

Remark 1.26: Assume that V has a nondegenerate bilinear form $(\cdot\,,\,\cdot) : V \otimes V \to \mathbb{C}$ and the linear operators H_i are symmetric with respect to the form. In this case Proposition 1.20 implies

$$(f, g) = \text{const}(\kappa) \tag{1.27}$$

if f is an asymptotic solution to (1.3) and g is an asymptotic solution to (1.3) in which κ is replaced by $-\kappa$.

Proposition 1.25 then implies that for any two asymptotic solutions f and g to (1.3) having forms (1.23) and (1.24), respectively, we have the following:

i $(g_0(z), f_0(z)) = 0$ if $S - T \neq$ const,

ii $(g_0(z), f_0(z)) =$ const if $S - T =$ const.

and, in particular, we have:

$$(f_0(z), f_0(z)) = \text{const}$$

for every asymptotic solution to $\nabla_\kappa f = 0$.

Remark 1.28: Under assumptions of (1.26) one can easily prove the following generalization of (1.26). For any two asymptotically flat sections of the first order f and g we have statements i) and ii) of (1.26).

2 Knizhnik-Zamolodchikov Connections

Let V be a vector space and let $r_{ij} : \mathbb{C}^* \to \text{End}(V)$, $1 \leq i < j \leq n$, be functions satisfying the classical Yang-Baxter equation:

$$[r_{ij}(u), r_{ik}(u+v)] + [r_{ij}(u), r_{jk}(v)] + [r_{ik}(u+v), r_{jk}(v)] = 0 \tag{2.1}$$

for all triples $i < j < k$. Assume the following skew-symmetry condition

$$r_{ij}(u) = -r_{ji}(-u)\,. \tag{2.2}$$

Let $d_i : V \to V,\ i = s, \ldots, n$, be linear maps with the following property:

$$[d_i + d_j, r_{ij}(u)] = 0\,, \qquad [d_i, d_j] = 0\,. \tag{2.3}$$

Consider

$$H_i(z_1, \ldots, z_n) = \sum_{j \neq i} r_{ij}^{V_i V_j}(z_i - z_j) + d_i\,.$$

It is easy to verify that this collection of linear operators provides a family of flat connections on $\mathbb{C}^n \backslash \{\text{diagonals}\}$ which we will call the *KZ family of flat connections* corresponding to $\{r_{ij}(u), d_i\}$.

The *KZ equation* is the equation of flat sections:

$$\nabla_\kappa F = 0.$$

Let $\mathfrak{g}$ be a complex simple Lie algebra, and let $(\rho_1, V_1), \ldots, (\rho_n, V_n)$ be $\mathfrak{g}$-modules, $\rho_i : \mathfrak{g} \to \mathrm{End}(V_i)$. Define $V = V_1 \otimes \ldots \otimes V_n$ and $\Omega \in \mathfrak{g} \otimes \mathfrak{g}$ as the element of $S^2(\mathfrak{g}) \overset{\text{inv}}{\hookrightarrow} \mathfrak{g} \otimes \mathfrak{g}$ corresponding to the Killing form. If $\{I_a\}$ is an orthonormal basis in $\mathfrak{g}$, then $\Omega = \sum_a I_a \otimes I_a$.

Consider the linear operators Ω_{ij} acting in V as

$$\Omega_{ij} = \sum_{a=1}^{\dim \mathfrak{g}} 1 \otimes \ldots \otimes \rho_i(I_a) \otimes \ldots \otimes \rho_j(I_a) \otimes \ldots \otimes 1$$

for $i < j$ and $\Omega_{ij} = \Omega_{ji}$ for $i > j$.

Define

$$r_{ij}(u) = \frac{\Omega_{ij}}{u}\,.$$

Fix any element d of $\mathfrak{g}$ and let $d_i = 1 \otimes \ldots \otimes \underset{i}{\mathrm{d}} \otimes \ldots \otimes 1$.

It is an easy exercise to check that elements $r_{ij}(u)$ and d_i of $\mathrm{End}(V)$ satisfy conditions (2.1)–(2.3) and, therefore, the linear operators

$$H_i = \sum_{j \neq i} \frac{\Omega_{ij}}{z_i - z_j} + d_i$$

determine a family of flat connections in the trivial vector bundle over $\mathbb{C}^n \backslash \{\text{diagonals}\}$ with fiber V.

We consider the case in which all d_i are equal to zero and

$$H_i = \sum_{j \neq i} \frac{\Omega_{ij}}{z_i - z_j}. \tag{2.4}$$

Let $e_1, \ldots, e_r,\ f_1, \ldots, f_r,\ h_1, \ldots, h_r$ be the Chevalley generators of $\mathfrak{g}$ such that

$$[e_i, f_j] = \delta_{ij} h_i\,,$$
$$[h_i, e_j]] = a_{ij} e_j\,,$$
$$[h_i, f_j]] = -a_{ij} f_j\,,$$

where (a_{ij}) is the Cartan matrix.

The map $\tau : \mathfrak{g} \to \mathfrak{g}$, sending $e_1, \ldots, e_r$, $f_1, \ldots, f_r$, $h_1, \ldots, h_r$ to $f_1, \ldots, f_r$, $e_1, \ldots, e_r$, h_1, $\ldots, h_r$, respectively, generate an anti-involution of $\mathfrak{g}$.

Lemma 2.5: *τ preserves the Killing form on $\mathfrak{g}$.*

Corollary:

$$\tau\Omega = \sum \tau(I_a) \otimes \tau(I_a) = \Omega .$$

Let W be a $\mathfrak{g}$ module with highest weight vector w. The *Shapovalov form* B on W is the symmetric bilinear form defined by the conditions:

$$B(w,w) = 1, \qquad B(xu,v) = B(u,\tau(x)v) \tag{2.6}$$

for all $u, v \in W$ and $x \in \mathfrak{g}$.

Lemma 2.7: *Let W_1 and W_2 be modules with highest weight, and let B_1 and B_2 be their Shapovalov forms. Then*

$$B_1 \otimes B_2(\Omega(x \otimes y), u \otimes v) = B_1 \otimes B_2(x \otimes y, \Omega(u \otimes v)) \tag{2.8}$$

for all $x \otimes y,\ u \otimes v \in W_1 \otimes W_2$.

Assume that $V_1, \ldots, V_n$ are $\mathfrak{g}$ modules with highest weight. Let B_i be the Shapovalov form on V_i. Set

$$B = B_i \otimes \ldots \otimes B_n . \tag{2.9}$$

B is a symmetric bilinear form on V.

Lemma 2.10: *The operators $H_1, \ldots, H_n$ in (2.4) are symmetric with respect to B.*

If $V_1, \ldots, V_n$ are irreducible finite-dimensional $\mathfrak{g}$-modules, or Verma modules with generic highest weights, then the form B is nondegenerate.

3 Integral Representation for Solutions of the KZ Equation

Let $V_1, \ldots, V_n$ be $\mathfrak{g}$ modules with highest weight.

Let $\mathfrak{h}$ be the standard Cartan subalgebra of $\mathfrak{g}$, $\alpha_1, \ldots, \alpha_r \in \mathfrak{h}^*$ the simple roots, $(\,\cdot\,,\,\cdot\,)$ the bilinear form on $\mathfrak{h}^*$ induced by the Killing form, $\Lambda_1, \ldots, \Lambda_n$ the highest weights of modules $V_1, \ldots, V_n$, respectively.

Set $\Lambda = \Lambda_1 + \ldots + \Lambda_n$. For $\lambda = (\lambda_1, \ldots, \lambda_r) \in \mathbb{Z}^r_{\geq 0}$, set

$$V_\lambda = \{v \in V \mid h_i v = (\Lambda - \sum_{j=1}^{r} \lambda_j \alpha_j, \alpha_i)v,\ i = 1, \ldots, r\} ,$$

$$\operatorname{Sing} V_\lambda = \{v \in V_\lambda \mid e_i v = 0,\ i = 1, \ldots, r\} . \tag{3.1}$$

Consider the KZ equation

$$\kappa\, dF = \sum_{i<j} \Omega_{ij}\; F\; \frac{d(z_i - z_j)}{(z_i - z_j)}. \tag{3.2}$$

We will recall a construction of solutions to the KZ equation with values in a given
Sing V_λ *[16]*.

An arbitrary solution of the KZ equation is represented as a linear combination of the constructed solutions since the KZ connection commutes with the $\mathfrak{g}$ action on V .

Let $k = \lambda_1 + \ldots + \lambda_n$. Consider the space $\mathbb{C}^k$ with coordinates $t_1(1), t_1(2), \ldots, t_1(\lambda_1), \ldots, t_r(1), t_r(2), \ldots t_r(\lambda_r)$, the space $\mathbb{C}^{n+k}$ with coordinates $t_1(1), t_1(2), \ldots, t_r(\lambda_r),\ z_1, \ldots, z_n$, and the space $\mathbb{C}^n$ with coordinates $z_1, \ldots, z_n$. Let $p : \mathbb{C}^{n+k} \to \mathbb{C}^n$ be the natural projection. Set

$$\Phi(t,z) = \prod_{1\le m<\ell\le n} (z_m - z_\ell)^{(\Lambda_m,\Lambda_\ell)/\kappa}\,. \tag{3.3}$$
$$\cdot \prod_{m,i,j} (z_m - t_i(j))^{-(\Lambda_m,\alpha_i)/\kappa} \prod_{\substack{i<\ell \text{ or} \\ i=\ell \text{ and } j<m}} (t_i(j) - t_\ell(m))^{(\alpha_i,\alpha_\ell)/\kappa}.$$

The function Φ is a multi-valued holomorphic function on $\mathbb{C}^{n+k}$ with singularities at diagonal hyperplanes.

A *monomial of weight* λ is an element $M \in V_\lambda$ of the form

$$M = f_{i_1} \ldots f_{i_{k_i}} v_1 \otimes f_{j_1} \ldots f_{j_{k_2}} v_2 \otimes \ldots \otimes f_{\ell_1} \ldots f_{\ell_{k_n}} v_n\,. \tag{3.4}$$

Here v_i is the highest weight vector of V_i, f's are elements of $\{f_1, \ldots, f_r\}$.

In *[16]*, for any monomial $M \in V_\lambda$, a differential k-form $\eta(M)$ is constructed. The form $\eta(M)$ is a rational form on $\mathbb{C}^{n+k}$ with poles at diagonal hyperplanes.

Consider the V_λ-valued k-form

$$N = \sum_{M\in V_\lambda} \Phi \cdot \eta(M) \otimes M\,. \tag{3.5}$$

In *[16]* it is proved that

3.6 For every i, the form

$$\left(\kappa\frac{\partial}{\partial z_i} - \sum_{j\ne i} \frac{\Omega_{ij}}{z_i - z_j}\right) N$$

is the sum of differential of a suitable $(k-1)$-form and a form which has zero restriction to fibers of the projection p.

3.7 For every j, the form $e_j N = \sum \Psi \cdot \eta(M) \otimes e_j M$ is the sum of differential of a suitable $(k-1)$-form and a form which has zero restriction to fibers of p.

3.8 All forms mentioned in (3.6) and (3.7) have the shape $\sum \Phi\ \omega(M) \otimes M$ where the sum is over all monomials in V, $\{\omega(M)\}$ are suitable rational forms with poles at diagonal hyperplanes.

Now let $z = (z_1, \ldots, z_n)$ be a point in $\mathbb{C}^n$ lying in the complement to the diagonals. Let $\gamma(z)$ be a k-cycle lying in $p^{-1}(z) \subset \mathbb{C}^{n+k}$ outside the diagonals. Assume that $\gamma(z)$ continuously depends on z. Then the function

$$F(z) = \int_{\gamma(z)} N \tag{3.9}$$

takes values in Sing V_λ and satisfies the KZ equation. See a more precise description of $\gamma(z)$ in *[16, 17]*.

4 Asymptotic Solutions to the KZ Equation

Let $\kappa \to 0$. We use the form N defined in (3.5) and the method of steepest descent to construct asymptotic solutions to the KZ equation (3.2). The function Φ can be written in the form

$$\Phi(t, z) = \exp(S(t, z)/\kappa) \tag{4.1}$$

where

$$\begin{aligned} S(t, z) = \sum_{1 \le m < \ell \le n} (\Lambda_m, \Lambda_\ell) \ln(z_m - z_\ell) \quad - \sum_{m,i,j} (\Lambda_m, \alpha_i) \ln(z_m - t_i(j)) \\ + \sum_{\substack{i<\ell \text{ or} \\ i=\ell \text{ and} j<m}} (\alpha_i, \alpha_\ell) \ln(t_i(j) - t_\ell(m)) \end{aligned} \tag{4.2}$$

Its derivatives have the form

$$\begin{aligned} \frac{\partial S}{\partial z_m} &= \sum_{\ell \ne m} (\Lambda_m, \Lambda_\ell) \frac{1}{z_m - z_\ell} - \sum_{i,j} (\Lambda_m, \alpha_i) \frac{1}{z_m - t_i(j)}, \\ \frac{\partial S}{\partial t_i(j)} &= \sum_{(\ell,m) \ne (i,j)} (\alpha_i, \alpha_\ell) \frac{1}{t_i(j) - t_\ell(m)} - \sum_m (\Lambda_m, \alpha_i) \frac{1}{t_i(j) - z_m}, \\ \frac{\partial^2 S}{\partial t_\ell(m) t_i(j)} &= (\alpha_i, \alpha_\ell) \frac{1}{(t_i(j) - t_\ell(m))^2}, \\ \frac{\partial^2 S}{\partial t_i(j)^2} &= - \sum_{(\ell,m) \ne (i,j)} (\alpha_i, \alpha_\ell) \frac{1}{(t_i(j) - t_\ell(m))^2} \\ &\quad + \sum_m (\Lambda_m, \alpha_i) \frac{1}{(t_i(j) - z_m)^2}. \end{aligned} \tag{4.3}$$

Set

$$\text{Hess}_t(-S) = \det\left(- \frac{\partial^2 S}{\partial t_i(j) \partial t_\ell(m)} \right) \tag{4.4}$$

For a fixed $z \in \mathbb{C}^n$, consider the equations of t-critical points for S:

$$\frac{\partial S}{\partial t_i(j)} = 0\,, \qquad i = 1, \ldots, r, \quad j = 1, \ldots, \lambda_r\,. \tag{4.5}$$

Let $t = t(z)$ be a non-degenerate solution to system (4.5) holomorphically depending on z in a neighborhood of a point $z^0 \in \mathbb{C}^n$. Set

$$\overline{S}(z) = S(t(z), z)\,. \tag{4.6}$$

Let $B \subset \mathbb{C}^k$ be a small ball with center at $t(z_0)$. Let $\delta \subset B$ be a k-chain such that

$$\text{Re } S(t(z_0), z_0) > \text{Re } S(t, z_0)|_{\partial\delta} \tag{4.7}$$

where $\partial\delta$ is boundary of δ. This chain is unique in the sense explained in the remark below. For z close to z_0, let $\delta(z) \in p^{-1}(z)$ be the image of the chain δ under the natural isomorphism $p^{-1}(z) \simeq \mathbb{C}^k$. Then for z close to z_0 we have

$$\text{Re } S(t(z), z) > \text{Re } S(t, z)|_{\partial\delta(z)} \tag{4.8}$$

Set

$$F(z) = \kappa^{-\frac{k}{2}} \int_{\delta(z)} N \tag{4.9}$$

where N is given by (3.5). The function F is defined in a neighborhood of z_0.

Theorem 4.10:

1. Let $\kappa \in \mathbb{R}$. As $\kappa \to +0$, the function F has an asymptotic expansion

$$F(z) = \exp(\overline{S}(z)/\kappa) \sum_{m=0}^{\infty} f_m(z)\kappa^m$$

where $\{f_m\}$ are holomorphic functions of z.

2. The function F gives an asymptotic solution to the KZ equation.

3. The functions $\{f_m\}$ take values in Sing V_λ.

Part 1 of the theorem is a direct corollary of the method of steepest descent; see, for example, §11 in *[1]*. Parts 2 and 3 are direct corollaries of (3.6)–(3.8) and (4.7).

Remark: Let

$$B^- = \{t \in B \mid \text{Re } S(t(z_0), z_0) > \text{Re } S(t, z_0)\}\,. \tag{4.11}$$

It is well known that $H_k(B, B^-, \mathbb{Z}) = \mathbb{Z}$ and, moreover, if δ and δ' are two cycles generating the same element in H_k, then $\int_{\delta(z)} N$ and $\int_{\delta'(z)} N$ have the same asymptotic expansions; see §11 in *[1]*. In what follows we assume that δ is a cycle generating H_k.

There exist local coordinates $u_1, \ldots, u_k$ in $\mathbb{C}^k$ centered at $t(z_0)$ such that

$$S(u, z_0) = -u_1^2 - \ldots - u_k^2 + \text{const}\,. \tag{4.12}$$

For such coordinates, δ is a disc

$$\{(u_1, \ldots, u_k) \in \mathbb{R}^k \mid u_1^2 + \ldots + u_k^2 \leq \varepsilon\} \tag{4.13}$$

where ε is a small positive number.

The first term of asymptotics

Let $F(z)$ be given by (4.9) where δ is defined by (4.13). Then the first term of asymptotics can be computed explicitly. Namely, let $M \in V_\lambda$ be a monomial. The form $\eta(M)$ can be written as

$$\eta(M) = A_M(t, z)dt_1(1) \wedge \ldots \wedge dt_r(\lambda_r) \tag{4.14}$$

where A_M is a rational function. Then the standard formula of the method of steepest descent states that

$$f_0(z) = \pm(2\pi)^{k/2}(\text{Hess}_t(-S(t(z), z))^{-1/2} \sum_{M \in V_\lambda} A_M(t(z), z)\, M \tag{4.15}$$

where the sign depends on orientation of δ. According to our previous considerations, we have the following corollary.

Corollary 4.16: *The vector $f_0(z)$ lies in* Sing V_λ. *For any i, $f_0(z)$ is an eigenvector of H_i with eigenvalue $\frac{\partial S}{\partial z_i}(t(z), z)$. The vector $f_0(z)$ has constant length with respect to the Shapovalov form B defined in (2.9).*

The last statement can be reformulated as follows. Set

$$g(t, z) = \sum_{M \in V_\lambda} A_M(t, z)\, M \tag{4.17}$$

We will call $g(t(z), z)$ *the Bethe vector corresponding to the critical point* $(t(z), z)$. Then

$$B(g(t(z), z), g(t(z), z)) = \text{const} \cdot \text{Hess}_t(S(t(z), z))\,. \tag{4.18}$$

This is a rather surprising relation between critical points of S and the Shapovalov form, cf *[9, 11]*. It is shown in *[18]* that const $= 1$ for $\mathfrak{g} = sl_2$.

5 Conjectures

Let us study critical points of the function S in the case $n = 2$. Introduce new variables

$$s_i(\ell) = \frac{t_i(\ell) - z_1}{z_2 - z_1}\,.$$

Obviously,

$$S(t(s,z),z) = T(s) + \text{const}\cdot \ln(z_1 - z_2)$$

where S is the function defined by (4.2) and

$$\begin{aligned} T(s) = & -\sum_{i,j}((\Lambda_1,\alpha_i)\ln(-s_i(j)) + (\Lambda_2,\alpha_i)\ln(1-s_i(j)) \\ & + \sum_{\substack{i<\ell \text{ or} \\ i=\ell \text{ and} j<m}} (\alpha_i,\alpha_\ell)\ln(s_i(j)-s_\ell(m)) \end{aligned}$$

Remarkably, equations (4.5) of $t-$critical points become equations of $s-$critical points for $\{\{s_i(\ell)\}_{\ell=1}^{\lambda_i}\}_{i=1}^{r}$ which do not depend on variables z_1 and z_2 :

$$\frac{(\Lambda_1,\alpha_j)}{s_i(j)} + \frac{(\Lambda_2,\alpha_j)}{s_i(j)-1} = \sum_{(\ell,m)\neq(i,j)} \frac{(\alpha_i,\alpha_\ell)}{s_i(j)-s_\ell(m)}\,. \tag{5.1}$$

Let $g(t,z)$ be the vector defined by (4.17). Set

$$G = g(t,z)/(z_1-z_2)^k.$$

Substitute $t = t(s,z)$. Then G becomes a function of s, z_1, z_2.

Proposition 5.2: *The vector G does not depend on z_1, z_2.*

This proposition follows from the explicit form of functions $A_M(t,z)$ in (4.17) (see *[16]*).

Conjecture 5.3: For generic weights Λ_1 and Λ_2 the following is true:

- Each solution to (5.1) determines a nondegenerate critical point of the function T.
- For each solution s^0 to (5.1), we have
$$B(G(s^0),G(s^0)) = \text{Hess}_s(T(s^0)).$$
- The product of symmetric groups $\mathcal{S} = \mathcal{S}_{\lambda_1}\times\ldots\times\mathcal{S}_{\lambda_r}$ naturally acts on the set of critical points of the function T. We conjecture that for each pair of solutions s^0 and s^1 to system (5.1) which belong to different $\mathcal{S}-$orbits we have
$$B(G(s^0),G(s^1)) = 0.$$
- The set of vectors $\{G(s^0)\}$, where s^0 runs through all $\mathcal{S}-$orbits of critical points of T, forms a basis in $\text{Sing}(V_1\otimes V_2)_\lambda$.

Similarly, for dominant integer weights Λ_1 and Λ_2 we have

Conjecture 5.4: Let $\tilde{V}_1$ and $\tilde{V}_2$ be irreducible highest weight modules with highest weights Λ_1 and Λ_2. Then (1)–(4) from Conjecture 5.3 hold in this case also.

6 Asymptotic Properties of the Critical Points

Let $1 < l < n$ and $v(z)$ be a nondegenerate critical point with the weight $\sum_{i=1}^{l} \Lambda_i - \sum_{i=1}^{r} a_i\alpha_i$ of the function $S(t, z_1, \ldots, z_l)$. Here we assume as above that $z_i \in \mathbb{C}$, $z_i \neq z_j$, and to each point z_i we associate a highest weight Verma module with highest weight Λ_i.

Similarly let $u(x)$ be a nondegenerate critical point with weight $\sum_{i=l+1}^{n} \Lambda_i - \sum_{j=1}^{r} b_j\alpha_j$ of the function $S(t, x_1, \ldots, x_{n-l-1})$. Here we assume that to each point x_i we associate a highest weight Λ_{i+l}.

Let $(\{s_i(j)\}_{j=1}^{c_i})_{i=1}^{r}$ be a solution to (5.1) with Λ_1 replaced by $\Lambda_1 + \ldots + \Lambda_l - \sum_{i=1}^{r} a_i\alpha_i$ and Λ_2 replaced by $\Lambda_{l+1} + \ldots + \Lambda_n - \sum_{j=1}^{r} b_j\alpha_j$.

Theorem 6.1: *(1) There exists a unique nondegenerate critical point of $S(t, z_1, \ldots, z_N)$ with weight $\sum_{i=1}^{n} \Lambda_i - \sum_{j=1}^{r}(a_j + b_j + c_j)\alpha_j$ such that asymptotically, when $z_{l+1}, \ldots, z_n \to \infty$, $z_1, \ldots, z_l, \ldots, z_{l+i} - z_{l+1}, \ldots, z_M - z_{l+1}$ are finite, it has the form*

$$\{t(z)\} = \{v(z_1, \ldots z_l) + \mathcal{O}(\frac{1}{z_{l+1}})\} \cup \{s_i(j)z_{l+1} + \mathcal{O}(1)\}$$
$$\cup\{u(x_1, \ldots, x_{n-l-1}) + z_{l+1} + \mathcal{O}(\frac{1}{z_{l+1}})\}$$

where $x_i = z_{l+i+1} - z_{l+1}$.

We will use the following lemma.

Lemma: *Let $t(z) = \sum_{n\geq 0} t^{(n)} z^{-n}$. Then*

$$\frac{1}{t(z) - w} = \frac{1}{t^{(0)} - w} + \sum_{m\geq 1} z^{-m} \sum_{1\leq n\leq m} \frac{(-1)^n}{(t^{(0)} - w)^{n+1}} \sum_{\substack{p_1+\ldots+p_n=m \\ p_i\geq 1}} t^{(p_1)} \ldots t^{(p_n)} \tag{6.1}$$

Proof: The proof is by straightforward computation.

The proof for an arbitrary simple Lie algebra $\mathfrak{g}$ is absolutely parallel to the one for sl_2. To avoid technicalities we present here the proof for sl_2.

Let $t_\alpha(z)$ be the asymptotic power series of the following type when $z_{l+1}, \ldots, z_n \to \infty$ and $x_i = z_i - z_{l+1}$ is finite, $i = l + 1, \ldots, n$.

$$t_\alpha(z) = v_\alpha(\tilde{z}) + \sum_{p\geq 1} v_\alpha^{(p)}(\tilde{z}, x) z_{l+1}^{-p}, \qquad \alpha = 1, \ldots a$$

$$t_\alpha(z) = z_{K+1}s_{\alpha-a} + \sum_{p\geq 0} s_{\alpha-a}^{(p)}(\tilde{z}, x) z_{l+1}^{-p}, \qquad \alpha = a + 1, \ldots, a + c$$

$$t_\alpha(z) = z_{l+1} + u_{\alpha-a-c}(x) + \sum_{p\geq 1} u_{\alpha-a-c}^{(p)}(\tilde{z}, x) z_{l+1}^{-p}, \qquad \alpha = a + c + 1, \ldots, a + c + b \tag{6.2}$$

where $z = (z_1, \dots, z_n)$, $\tilde{z} = (z_1, \dots z_l)$, $x = (0, x_{l+2}, \dots, x_n)$. Then when $\alpha = 1, \dots, a$ we have the following:

$$\sum_{\substack{\beta \neq \alpha \\ \beta=1}}^{a+b+c} \frac{1}{t_\alpha(z) - t_\beta(z)} = \sum_{\substack{\beta \neq \alpha \\ \beta=1}}^{a} \frac{1}{v_\alpha(\tilde{z}) - v_\beta(\tilde{z})} +$$

$$\sum_{M \geq 1} z_{K+1}^{-M} \left\{ \sum_{1 \leq L \leq M} \frac{(-1)^L}{v_{\alpha\beta}^{L+1}} \sum_{\substack{p_1 + \dots + p_L = M \\ p_i \geq 1}} v_{\alpha\beta}^{(p_1)} \dots v_{\alpha\beta}^{(p_L)} - \right.$$

$$- \sum_{1 \leq L \leq M-1} \frac{(-1)^L}{s_\beta^{L+1}} \sum_{\substack{p_1 + \dots + p_L = M-1 \\ p_i \geq 1}} s_{\alpha\beta}^{(p_1)} \dots s_{\alpha\beta}^{(p_L)} -$$

$$\left. - \sum_{1 \leq L \leq M-1} (-1)^L \sum_{\substack{p_1 + \dots + p_L = M-1 \\ p_i \geq 1}} u_{\alpha\beta}^{(p_1)} \dots u_{\alpha\beta}^{(p_L)} \right\} \tag{6.3}$$

Here $v_{\alpha\beta}^{(p)} = v_\alpha^{(p)} - v_\beta^{(p)}$, $u_{\alpha\beta}^{(p)} = v_\alpha^{(p+1)} - u_\beta^{(p+1)}$ and $s_{\alpha\beta}^{(p)} = v_\alpha^{(p+1)} - s_\beta^{(p+1)}$. We have similar formulas for $\alpha = a+1, \dots, a+c$ and for $\alpha = a+c+1, \dots, a+c+b$, and for expansions of

$$\sum_{i=1}^{n} \frac{\lambda_i}{t_\alpha - z_i}. \tag{6.4}$$

Then one can verify that the equality

$$\sum_{i=1}^{n} \frac{\lambda_i}{t_\alpha - z_i} = \sum_{\beta \neq \alpha}^{a+b+c} \frac{2}{t_\alpha - t_\beta} \tag{6.5}$$

for t_α being an asymptotic power series (6.2) is equivalent to the following recursive system of equations:

$$\sum_{\beta=1}^{a} K_{\alpha\beta} v_\beta^{(m)} = \text{certain function of}$$

$$(v, v^{(1)}, \dots, t^{(m-1)}; s, \dots, s^{(m-1)}; u, \dots, u^{(m-1)})$$

$$\sum_{\beta=c+1}^{a+c} L_{\alpha\beta} s_\beta^{(m)} = \text{certain function of}$$

$$(v, v^{(1)}, \dots, t^{(m-1)}; s, \dots, s^{(m-1)}; u, \dots, u^{(m-1)})$$

$$\sum_{\beta=a+c+1}^{a+b+c} M_{\alpha\beta} u_\beta^{(m)} = \text{certain function of}$$

$$(v, v^{(1)}, \dots, t^{(m-1)}; s, \dots, s^{(m-1)}; u, \dots, u^{(m-1)}) \tag{6.6}$$

Here

$$K_{\alpha\beta} = \left\{ \sum_{i=1}^{l} \frac{\lambda_i}{(v_\alpha - z_i)^2} - \sum_{\beta \neq \alpha} \frac{2}{(v_\alpha - v_\beta)^2} \right\} \delta_{\alpha\beta} - (1 - \delta_{\alpha\beta}) \frac{2}{(v_\alpha - v_\beta)^2}$$

$$L_{\alpha\beta} = \left\{ \frac{\lambda_1 + \ldots + \lambda_l}{s_\alpha^2} + \frac{\lambda_{l+1} + \ldots + \lambda_l}{(1 - s_\alpha)^2} - \sum_{\beta \neq \alpha} \frac{2}{(s_\alpha - s_\beta)^2} \right\}$$

$$\delta_{\alpha\beta} - (1 - \delta_{\alpha\beta}) \frac{2}{(s_\alpha - s_\beta)^2}$$

$$M_{\alpha\beta} = \left\{ \sum_{i=l+1}^{n} \frac{\lambda_i}{(u_\alpha - z_i)^2} - \sum_{\beta \neq \alpha} \frac{2}{(u_\alpha - u_\beta)^2} \right\}$$

$$\delta_{\alpha\beta} - (1 - \delta_{\alpha\beta}) \frac{2}{(u_\alpha - u_\beta)^2} \tag{6.7}$$

These matrices are nondegenerate according to our assumptions and to Conjecture 5.3. Therefore, recurrences determine power series (6.2) uniquely and these series will satisfy equations (6.5).

This finishes the proof of Theorem 6.1 for $\mathfrak{g} = sl_2$.

For other simple Lie algebras the proof is absolutely similar.

7 sl_2-case

Let $\mathfrak{g} = sl_2$. In terms of the standard generators e, f, h we have

$$\Omega = \tfrac{1}{2} h \otimes h + e \otimes f + f \otimes e \,.$$

Let $V_1, \ldots, V_n$ be sl_2-modules with highest weights $\Lambda_1, \ldots, \Lambda_n$ and highest weight vectors $v_1, \ldots, v_n$, respectively. Set $\Lambda = \Lambda_1 + \ldots + \Lambda_n$. For $k \in \mathbb{Z}_{\geq 0}$, set

$$\begin{aligned} V &= V_1 \otimes \ldots \otimes V_n \,, \\ (V)_k &= \{ v \in V \mid hv = (\Lambda - k\alpha, \alpha) v \} \,, \\ \mathrm{Sing}(V)_k &= \{ v \in (V)_k \mid ev = 0 \} \,, \end{aligned}$$

where α is the simple root.

Consider the space $\mathbb{C}^{n+k}$ with coordinates $t_1, \ldots, t_k, \; z_1, \ldots, z_k$. The functions Φ and S, defined in (3.3) and (4.2), now have the form

$$\Phi = \prod_{1 \leq m < \ell \leq n} (z_m - z_\ell)^{(\Lambda_m, \Lambda_\ell)/\kappa} \prod_{m,i} (z_m - t_i)^{-(\Lambda_m, \alpha)/\kappa} \prod_{i<j} (t_i - t_j)^{2/\kappa}$$

$$S = \kappa \ln \Phi. \tag{7.1}$$

A monomial of weight k is an element $M_K \in (V)_k$ of the form

$$M_K = f^{k_1} v_1 \otimes \ldots \otimes f^{k_n} v_n \,, \tag{7.2}$$

$K = (k_1, \ldots, k_n)$, $k_1 + \ldots + k_n = k$. For a monomial M_K, the differential form $\eta(M_K)$, used in (3.5) and (3.9) to construct solutions to the KZ equation, has the form

$$\begin{aligned} \eta(M_K) &= A_{M_K} dt_1 \wedge \ldots \wedge dt_k \,, \\ A_M &= \sum_{\sigma \in S(k, k_1, \ldots, k_n)} \prod_{i=1}^{k} \frac{1}{t_i - z_{\sigma(i)}} \,. \end{aligned} \tag{7.3}$$

The sum is over the set $S(k, k_1, \ldots, k_n)$ of maps σ from $\{1, \ldots, k\}$ to $\{1, \ldots, n\}$ such that for all m the cardinality of $\sigma^{-1}(m)$ is k_m.

Let $t = t(z)$ be a nondegenerate solution to the system of equations

$$\frac{\partial S}{\partial t_i}(t, z) = 0 \,, \qquad i = 1, \ldots, k \,. \tag{7.4}$$

Then formula (4.15) gives a vector $f_0(z) \in \mathrm{Sing}(V)_k$ with properties indicated in (4.16). The function $\exp(S(t(z), z)/\kappa) f_0(z)$ is an asymptotically flat section of first order.

From now on we assume that $V_1, \ldots, V_n$ are Verma modules. In §9 we will show that the eigenvectors constructed by (4.15) form a basis in $\mathrm{Sing}(V)_k$ for generic $\Lambda_1, \ldots, \Lambda_n$, and $z_1, \ldots, z_n$. We will show that these vectors are pairwise orthogonal with respect to the Shapovalov form. §8 contains a preliminary information.

8 Iterated Singular Vectors

Let W_i, $i = 1, 2$, be sl_2 modules. Let $\omega_i \in W_i$ be a singular vector of weight $m_i \in \mathbb{C}$, that is, $e\omega_i = 0$ and $h\omega_i = m_i \omega_i$. For a nonnegative integer ℓ, the vector

$$(\omega_1, \omega_2)_\ell := \left(\prod_{j=1}^{\ell} (m_1 + m_2 + j + 1 - 2\ell) \right) \cdot$$

$$\cdot \sum_{p=0}^{\ell} (-1)^p \, (\ell p) \cdot \prod_{j=0}^{p-1} (m_1 - j)^{-1} \cdot \prod_{j=0}^{\ell-p-1} (m_2 - j)^{-1} f^p \omega_1 \otimes f^{\ell-p} \omega_2 \tag{8.1}$$

is a singular vector in $W_1 \otimes W_2$ of weight $m_1 + m_2 - 2\ell$. We assume that all denominators in this formula are not zero.

Let Ω be the Casimir operator acting on $W_1 \otimes W_2$.

$$\begin{aligned} &(\omega_2, \omega_2)_\ell \text{ is an eigenvector of } \Omega \text{ with eigenvalue } \lambda(m_1, m_2; \ell) \\ &= \tfrac{1}{2} m_1 m_2 - \ell(m_1 + m_2) + \ell(\ell - 1). \end{aligned} \tag{8.2}$$

Now let $V_1, \ldots, V_n$ be sl_2 Verma modules with highest weights $\Lambda_1, \ldots, \Lambda_n$ and highest vectors $v_1, \ldots, v_n$, respectively. Set $m_i = (\Lambda_i, \alpha)$.

For any sequence of non-negative integers $I = (i_2, \ldots, i_n)$, $i_2 + \ldots + i_n = k$, set

$$v_I = (\ldots((v_1, v_2)_{i_2}, v_3)_{i_3}, \ldots, v_n)_{i_n} \tag{8.3}$$

v_I is a singular vector in $V = V_1 \otimes \ldots \otimes V_n$ of weight $m_1 + \ldots + m_n - 2k$. We assume that all denominators in v_I are not zero. The vector v_I will be called the *iterated singular vector.*

For generic $\Lambda_1, \ldots, \Lambda_n$, the iterated vectors form a basis in $\mathrm{Sing}(V)_k$ (8.4)

It is easy to write explicitly the nondegeneracy conditions.

For any $\ell = 2, \ldots, n$, set

$$G_\ell = \sum_{i<\ell} \Omega_{i\ell} \tag{8.5}$$

Lemma 8.6: *For any ℓ and any $I = (i_2, \ldots, i_n)$, the iterated vector v_I is an eigenvector of G_ℓ with eigenvalue*

$$\lambda(m_1 + \ldots + m_{\ell-1} - 2(i_2 + \ldots + i_{\ell-1}), m_\ell; i_\ell).$$

The lemma easily follows from the following property of the Casimir operator:

$$[\Omega, 1 \otimes x + x \otimes 1] = 0$$

for any $x \in sl_2$.

Corollary 8.7: *For generic $\Lambda_1, \ldots, \Lambda_n$ we have the following property:*

Let $I = (i_2, \ldots, i_n)$, $i_2 + \ldots + i_n = k$, and $J = (j_2, \ldots, j_n)$, $j_2 + \ldots + j_n = k$, be different sequences. Then there exists $\ell \in \{2, \ldots, n\}$ such that the eigenvalues of vectors v_I and v_J with respect to G_ℓ are different.

9 Critical Points of Function Φ

Let

$$\Phi_2(t) = \prod_{i=1}^{k} t_i^{-m_1/\kappa} (t_i - 1)^{-m_2/\kappa} \cdot \prod_{1 \le i < j \le k} (t_i - t_j)^{2/\kappa} \tag{9.1}$$

where $m_1, m_2, \kappa \in \mathbb{C}$ are parameters. This function is a special case of the function Φ defined in (7.1). Let $S_2 = \kappa \ln \Phi_2$.

Let

$$\lambda_1 = t_1 + \ldots + t_k, \qquad \lambda_2 = \sum t_i t_j, \qquad \ldots, \qquad \lambda_k = t_1 \cdot \ldots \cdot t_k$$

be the standard symmetric functions. Let

$$\mu_1 = (1-t_1)+\ldots+(1-t_k), \qquad \mu_2 = \sum (1-t_i)(1-t_j), \qquad \ldots$$

$$\mu_k = (1-t_1)\ldots(1-t_k), \qquad \delta = \prod_{1\le i<j\le k} (t_i-t_j)^2.$$

Theorem 9.2 *[18]*, **(cf.** *[14]* **6.7):** *If $t=(t_1,\ldots,t_k)$ is a critical point of Φ_2, then*

$$\lambda_\ell = \binom{k}{\ell}\cdot\prod_{j=1}^{\ell}\frac{(m_1+j-k)}{(m_1+m_2+j+1-2k)},$$

$$\mu_\ell = \binom{k}{\ell}\cdot\prod_{j=1}^{\ell}\frac{(m_2+j-k)}{(m_1+m_2+j+1-2k)},$$

for all ℓ,

$$\delta = \prod_{j=0}^{k-1}(j+1)^{j+1}\frac{(-m_1+j)^j(-m_2+j)^j}{(-m_1-m_2+2k-2-j)^{2k-2-j}},$$

$$\mathrm{Hess}(-S_2(t)) = k!\prod_{j=0}^{k-1}\frac{(-m_1-m_2+2k-2-j)^3}{(-m_1+j)(-m_2+j)}.$$

Let t be a critical point of Φ_2. Consider the vector f_0 given by Theorem (4.10):

$$f_0 = \pm(2\pi)^{k/2}(\mathrm{Hess}(-S_2(t)))^{-\frac{1}{2}}\sum_{\ell=0}^{k} A_{k-\ell,\ell}f^{k-\ell}v_1\otimes f^\ell v_2\,, \tag{9.3}$$

$$A_{k-\ell,\ell} = \sum_{1\le i_1<\ldots<i_\ell\le k}\frac{1}{(t_{i_1}-1)}\cdots\frac{1}{(t_{i_\ell}-1)}\prod_{j\notin(i_1,\ldots,i_\ell)}\frac{1}{t_j}.$$

Here $v_p,\ p=1,2$, is the highest weight vector of the Verma module V_p with highest weight $m_p\in\mathbb{C}$.

Lemma 9.4 *[18]*: *We have*

$$A_{k-\ell,\ell} = (-1)^\ell\,(k\ell)\cdot\prod_{j=1}^{k}(m_1+m_2+j+1-2k)\cdot\prod_{j=0}^{k-\ell-1}(m_1-j)^{-1}\cdot\prod_{j=0}^{\ell-1}(m_2-j)^{-1}$$

for all ℓ.

The vector f_0 is a singular vector of weight m_1+m_2-2k, it is proportional to the vector $(v_1,v_2)_k$ in (8.1).

Now let Φ be the function defined by (7.1). Set $m_\ell = (\Lambda_\ell, \alpha)$ for all ℓ. Consider in $\mathbb{C}^k$ the configuration $\mathcal{C}$ of hyperplanes

$$H_{ij} : t_i = t_j \,, \qquad 1 \le i < j \le k$$
$$H_i^\ell : t_i = z_\ell \,, \qquad i = 1, \dots, k \,, \ell = 1, \dots, n \,. \tag{9.5}$$

Let T be the complement in $\mathbb{C}^k$ to the union of all hyperplanes of $\mathcal{C}$. Set

$$T_{\mathbb{R}} = T \cap \mathbb{R}^k \,.$$

Theorem 9.6 *[18]*: *Assume that $z_1, \dots, z_n$ are real and pairwise different. Assume that $m_1, \dots, m_n$ are negative. Then all critical points of Φ lie in the union of bounded connected components of $T_{\mathbb{R}}$. Each bounded component contains exactly one critical point. This critical point is nondegenerate.*

Assume that $z_1 < z_2 < \dots < z_n$.

We say that a bounded connected component is *admissible* if it lies in the cone $t_1 < t_2 < \dots < t_k$.

Admissible components are enumerated by sequences of nonnegative integers $I = (i_2, \dots, i_n)$, $i_2 + \dots + i_n = k$. The corresponding component has the form

$$\mathcal{D}_I(z) = \{t \in \mathbb{R}^k \,|\, z_1 < t_1 < \dots < t_{i_2} < z_2 < \dots < z_{n-1} < t_{i_2+\dots+i_{n-1}+1} < \dots < t_k < z_n\} \tag{9.7}$$

cf. (8.3). The critical point of Φ in $\mathcal{D}_I(z)$ is denoted by $t_I(z)$.

Consider the Bethe vector

$$g_I(z) = \sum_{k_1+\dots+k_n=k} A_{(k_1,\dots,k_n)}(t_I(z), z) f^{k_1} v_1 \otimes \dots \otimes f^{k_n} v_n \tag{9.8}$$

where $A_k(t,z)$ is given by (7.3). $g_I(z)$ is a vector in $\mathrm{Sing}(V)_k$ where V is the tensor product of Verma modules with highest weights $\Lambda_1, \dots, \Lambda_n$.

It is well known that for generic $\Lambda_1, \dots, \Lambda_n$ the dimension of $\mathrm{Sing}(V)_k$ is equal to the number of iterated singular vectors described in §8 and, therefore, it is equal to the number of admissible domains.

Consider the set of vectors $\{g_I(z)\}_{I \in \mathrm{Adm}}$, where I ranges over the set of sequences of nonnegative integers $(i_2, \dots, i_n)$, $i_2 + \dots + i_n = k$. In *[18]* it is proved that under explicitly written conditions on $\Lambda_1, \dots, \Lambda_n$ the Bethe vectors $\{g_I(z)\}_{I \in \mathrm{Adm}}$, form a basis in $\mathrm{Sing}(V)_k$. Here we'll describe asymptotics of these vectors for $z_1 << z_2 \dots << z_n$.

Theorem 9.9: *Assume that $z_j = s^j$, $j = 1, \dots, n$, and $s \to +\infty$. Then for any $I = (i_2, \dots, i_n)$, $i_2 + \dots + i_n = k$, we have*

$$g_I(z(s)) = s^{d(I)}(v_I + O(s^{-1}))$$

where v_I is given by (8.3) and $d(I)$ is an integer. Moreover, for any $\ell = 2, \dots, n$, the operator $H_\ell(z) = \sum_{j \ne \ell} \Omega_{j\ell}/(z_j - z_\ell)$ has the following asymptotics:

$$H_\ell = s^{-\ell}(\Omega_{1,\ell} + \Omega_{2,\ell} + \dots + \Omega_{\ell-1,\ell} + O(s^{-1})) \,.$$

Corollary 9.10: *The Bethe vectors* $\{g_I(z)\}_{I\in \mathrm{Adm}}$ *form a basis in* $\mathrm{Sing}(V)_k$ *for generic* $z_1, \dots, z_n$ *and generic* $\Lambda_1, \dots, \Lambda_n$.

Corollary 9.11: *Let* $\lambda_I^\ell(z)$ *be the eigenvalue of* $g_I(z)$ *with respect to* $H_\ell(z)$*, then*

$$\gamma_I^\ell(z(s)) = \lambda(m_1 + \ldots + m_{\ell-1} - 2(i_2 + \ldots + i_{\ell-1}), m_\ell; i_\ell) + O(s^{-1})$$

for $\ell > 1$*, see (8.6).*

Corollary 9.12: *For generic* $\Lambda_1, \dots, \Lambda_n$ *we have the following property:*
Let I *and* J *be different admissible sequences. Then there exists* ℓ *such that the eigenvalues of the Bethe vectors* $g_I(z(s))$ *and* $g_J(z(s))$ *with respect to* $H_\ell(z)$ *are different for* $s >> 1$.

Corollary 9.13: Let I and J be different admissible sequences. Then the Bethe vectors $g_I(z(s))$ and $g_J(z(s))$ are orthogonal with respect to the Shapavalov form B.

Remark 9.14: It is shown in *[18]* that

$$B(g_I(z(s)), g_I(z(s))) = \mathrm{Hess}_t(S(t_I(z), z))\,.$$

Proof of the theorem: The statement on H_ℓ is trivial. We prove the statement on g_I. Make a change of variables: $t_j = s^\ell u_j$ if $i_1 + \ldots + i_{\ell-1} < j \le i_2 + \ldots + i_\ell, \ell = 2, ..., n$ (we assume that $i_1 = 0$) . For any $\ell = 2, \dots, n$, introduce a function

$$S_\ell(u) = - \sum_{j=i_{\ell-1}+1}^{i_\ell} (a_\ell \ln u_j + m_\ell \ln(u_j - 1)) + 2 \sum_{i_{\ell-1}+1 \le i < j \le i_\ell} \ln(u_i - u_j)$$

where $a_\ell = m_1 + \ldots + m_{\ell-1} - 2(i_2 + \ldots + i_{\ell-1})$. Then we have

$$S(t(u)) = A \ln s + S_2(u) + \ldots + S_n(u) + O(s^{-1}) \qquad (9.15)$$

for some number A. This formula and the explicit formula for g_I imply the theorem.

Assume that $m_\ell = (\Lambda_\ell, \alpha), \ell = 1, \dots, n$, are positive integers. Let $V_1, \dots, V_n$ be irreducible sl_2 modules with highest weights $\Lambda_1, \dots, \Lambda_n$, respectively.

Apply the construction of Section 4 to this situation. For any $t-$critical point $t = t(z)$ of the function Φ we get a Bethe vector $g(t(z), z) \in \mathrm{Sing}(V)_k$ given by (9.8).

We say that a $t-$critical point $t = t(z)$ is *nontrivial* if $g(t(z), z) \ne 0$, otherwise we will call it *trivial.*

The set of $t-$critical points of Φ is invariant with respect to the group of permutations of coordinates $t_1, \dots, t_k$. Critical points lying in the same orbit give the same eigenvector.

Consider the family $\{g(t(z), z)\}$ of the Bethe vectors where $t(z)$ runs through the set of orbits of critical points.

Theorem 9.16: *For any z every trivial t-critical point is degenerate and all Bethe vectors $\{g(t(z), z)\}$ are pairwise orthogonal with respect to the Shapovalov form. For generic z, all nontrivial critical points are nondegenerate and the corresponding Bethe vectors form a basis in* $\mathrm{Sing}(V)_k$.

Proof: Trivial critical points are degenerate by (9.14). All Bethe vectors are pairwise orthogonal by (9.13). Therefore, it suffices to show that there are at least $\dim \mathrm{Sing}(V)_k$ orbits of nondegenerate critical points for generic z.

Let $I = (i_2, \ldots, i_n)$ be a sequence of nonnegative integers such that $i_1 + \ldots + i_n = k$. We say that I is a *good sequence* if for any $\ell = 2, \ldots, n$ we have

$$i_\ell \leq \min(m_1 + \ldots + m_{\ell-1} - 2(i_2 + \ldots + i_{\ell-1}), m_\ell) .$$

The number N of good sequences is equal to dimension of $\mathrm{Sing}(V)_k$.

We show that there are at least N orbits of nondegenerate critical points if $z_\ell = s^\ell$, $\ell = 1, \ldots, n$, and s tends to $+\infty$.

Let I be a good sequence. Make a change of variables $t_j = s^\ell u_j$ if $i_1 + \ldots + i_{\ell-1} < j \leq i_2 + \ldots + i_\ell$. Then we have formula (9.15).

Critical points of Φ coincide with critical points of $S(t(u))$. The critical point equations for $S(t(u))$ are a deformation of the critical point equations for $S_2(u) + \ldots + S_n(u)$. Critical points of $S_2 + \ldots + S_n$ are described by (9.2).

Let u_I be a critical point of $S_2 + \ldots + S_n$. Let $u_I(s)$ be the critical point of $S(t(u))$ which is the deformation of u_I. Let $t_I(s)$ be the corresponding critical point of Φ and $g_I(t_I(s), z(s))$ the corresponding Bethe vector. Then

$$g(t_I(s), z(s)) = s^{d(I)}(v_I + O(s^{-1}))$$

where $d(I)$ is some integer. This describes asymptotics of $\{g(t(z), z)\}$ and proves the theorem.

10 Trees, Asymptotic Zones, and Bethe Bases

Let $V_1, \ldots, V_n$ be finite-dimensional irreducible sl_2 modules. Consider the family of Bethe vectors $\{g(t(z), z)\} \in \mathrm{Sing}(V)_k$. We proved that for

$$|z_2 - z_1| << |z_3 - z_2| << \ldots << |z_n - z_{n-1}| \qquad (10.1)$$

the Bethe vectors form a basis in $\mathrm{Sing}(V)_k$. Moreover, the Bethe vectors can be enumerated by good sequences $\{I\}$ in such a way that for every I,

$$g(t_I(z), z) \sim \mathrm{const}(z) \cdot v_I \qquad (10.2)$$

where v_I is the corresponding iterated vector, $\mathrm{const}(z)$ is a scalar function, const $\neq 0$.

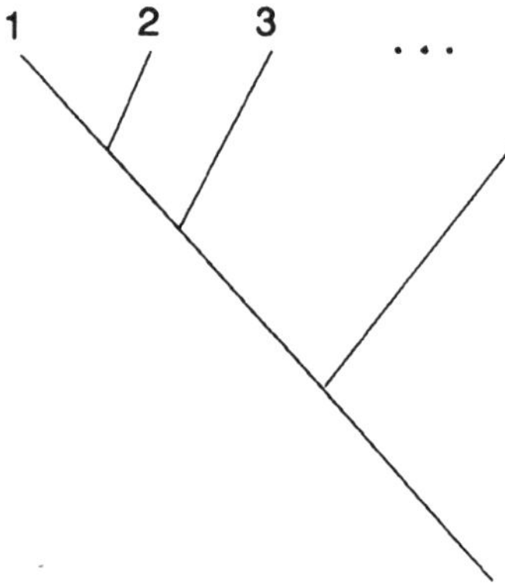

Figure 1:An n-tree T_0.

These facts have the following generalization. *An n-tree* is a planar tree with n tops, one root, and $n-1$ internal triple vertices. See an example in Figure 10.1. The n-tree T_0, shown in the figure, defines an "asymptotic zone" described by (10.1) and a basis $\{v_I\} \in \mathrm{Sing}(V)_k$ where I runs through the good sequences.

Similarly, every n-tree T defines two objects: an asymptotic zone, described by
inequalities similar to (10.1), and a distinguished basis $\mathcal{B}_T$ in $\mathrm{Sing}(V)_k$ consisting of iterated vectors where iterations are defined according to the shape of the tree.

Theorem 10.3: *For any n-tree T, we restrict the Bethe vectors $\{g(t(z), z)\}$ to the asymptotic zone defined by T. Then the Bethe vectors can be enumerated by elements of the basis $\mathcal{B}_T$ in such a way that for every $v \in \mathcal{B}_T$ we have*

$$g_v(t(z), z) \sim \mathrm{const}(z) \cdot v$$

where const(z) *is a scalar function,* const $\neq 0$.

The proof is the same as for (9.16).

Now consider an arbitrary complex simple Lie algebra $\mathfrak{g}$ instead of sl_2. Assume that Conjecture (5.4) is true for $\mathfrak{g}$.

Let $V_1, \ldots, V_n$ be finite-dimensional irreducible $\mathfrak{g}$ modules. Consider the KZ equation with values in Sing $(V)_\lambda$, defined by (3.2). Let $\{g(t(z), z)\} \in \mathrm{Sing}(V_\lambda)$ be the Bethe vectors constructed in Section 4.

For any n-tree T we can construct a distinguished basis $\mathcal{B}_T$ in $\mathrm{Sing}(V)_\lambda$. The basis consists of the iterated vectors where iterations are constructed according to the shape of the tree, and at each step of the iteration we use the eigenvectors discussed in Conjecture (5.4). For example, for the tree shown in the figure and

$g = sl_2$, the iteration procedure is given by (8.3) and the eigenvector used in each step of iteration (8.3) is given by (8.1).

Theorem 10.4: *Assume that Conjecture (5.4) is true for* $\mathfrak{g}$. *For any* n*-tree* T, *we restrict the Bethe vectors* $\{g(t(z), z)\}$ *to the asymptotic zone defined by* T. *Then for every* $v \in \mathcal{B}_T$ *there exists a Bethe vector* $g(t(z), z)$ *such that* $g(t(z), z) \sim \mathrm{const}(z) \cdot v$ *where* $\mathrm{const}(z)$ *is a scalar function,* $\mathrm{const} \neq 0$.

The proof is the same as for (9.16).

11 Bethe Vectors and Lame Functions

Consider the following problem (*[14]*, 6.8).

Problem 11.1: Let $A(t)$ and $B(t)$ be given polynomials of degree n and $n-1$, respectively. To determine a polynomial $C(t)$ of degree $n-2$ such that the differential equation

$$A(t)y''(t) - B(t)y'(t) + C(t)y(t) = 0 \qquad (11.2)$$

has a solution which is a polynomial of preassigned degree k.

A polynomial solution $y(t)$ is called a *Lamé function.* Hein (1878) proved that, in general, there are exactly $\binom{k+n-2}{k}$ determinations of $C(z)$.

Example: Let $A = (1-t^2)$, $B = (\alpha - \beta + (\alpha + \beta + 2)t)$, then $C = k(k+\alpha+\beta+1)$ and the corresponding polynomial solution of degree k, normalized by the condition $y(1) = \binom{k+\alpha}{k}$, is called *the Jacobi polynomial* and is denoted by $P_k^{(\alpha,\beta)}(t)$.

Let $A = (t-z_1)\dots(t-z_n)$,

$$\frac{B(t)}{A(t)} = \frac{m_1}{t-z_1} + \dots + \frac{m_n}{t-z_n} \qquad (11.3)$$

Theorem 11.4 Stieltjes (*[14]*, *§6.8*)**:** *Let* A *and* B *be given polynomials of degree* n *and* $n-1$, *respectively. Then there exists a polynomial* C *of degree* $n-2$ *and a polynomial solution* $y = (t-t_1)\dots(t-t_k)$ *of (11.2) if and only if* $t = (t_1, \dots, t_k)$ *is a critical point of the function*

$$\Psi(t_1, \dots, t_k) = \prod_{j=1}^{k} \prod_{i=1}^{n} (t_j - z_i)^{-m_i/\kappa} \prod_{1 \le i < j \le k} (t_i - t_j)^{2/\kappa}$$

where κ *is an arbitrary nonzero number.*

Therefore, having a Lamé function $y = (t-t_1)\dots(t-t_k)$ we can define a Bethe vector $g(z) \in \mathrm{Sing}(V)_k$ by formulae (9.8) and (7.3). Here $V = V_1 \otimes \dots \otimes V_n$, V_j is an sl_2 highest weight module with highest weight m_j.

It turns out that if V_j, $j = 1, \dots, n$, is the Verma module with highest weight m_j, then, in general, a Bethe vector defines a Lamé function, and, therefore, Problem 11.1 is equivalent to the problem of diagonalization in $\mathrm{Sing}(V)_k$ of the KZ operators

$$H_i(z) = \sum_{j \neq i} \frac{\Omega_{ij}}{z_i - z_j}, \qquad i = 1, \dots, n .$$

Namely, let $g(z)$ be a Bethe vector given by (9.8) and (7.3). Then the vector

$$\prod_{j=1}^{k} \prod_{i=1}^{n} (t_j - z_i) g(z) = \sum_{k_1 + \dots + k_n = k} B_{k_1, \dots, k_n} f^{k_1} v_1 \otimes \dots \otimes f^{k_n} v_n$$

polynomially depends on $t_1, \dots, t_k$, $z_1, \dots, z_n$.

For a sequence $L = (\ell_1 \geq \dots \geq \ell_k)$ of positive integers define a symmetric polynomial p_L by

$$p_L(t_1, \dots, t_k) = \sum t^{\alpha}$$

where the sum is over all permutations α of the sequence $(\ell_1 - 1, \dots, \ell_k - 1)$. We have

$$B_K(z,t) = \sum_L M_{K,L}(z) \cdot p_L(t) .$$

Here $K = (k_1, \dots, k_n)$, $k_1 + \dots + k_n = k$, $k_j \in \mathbb{Z}_{\geq 0}$. The sum is over $L = (\ell_1 \geq \dots \geq \ell_k)$, $n \geq \ell_1$. Denote by $\mathcal{K}_{k,n}$ (respectively $\mathcal{L}_{k,n}$) the set of such K (respectively L).

For every K, L, the coefficient $M_{K,L}$ is a polynomial in z depending only on k, n and independent of $m_1, \dots, m_n$. It is easy to see that $\#\mathcal{K}_{k,n} = \#\mathcal{L}_{k,n}$.

Theorem 11.5: *We have*

$$\det(M_{K,L}(z)) \not\equiv 0.$$

Corollary: *If* $\det(M_{K,L}(z)) \neq 0$ *for some* z*, then the coefficients* $B_K(t,z)$ *of a Bethe vector uniquely determine the symmetric functions* $\{p_L(t)\}$*, the orbit of the critical point* $t = (t_1, \dots, t_k)$*, and the corresponding Lamé function, cf. [13].*

Remark: In several examples, we have

$$\det(M_{K,L}(z)) = \mathrm{const} \prod (z_i - z_j)^a$$

for suitable a.

Proof: Let $z_j = s^j$, $j = 1, \dots, n$. Then $(M_{K,L}(z(s)))$ is a polynomial in s. Let $d_{K,L} = \deg_s M_{K,L}(z(s))$, $d_L = \max_K d_{K,L}$. Let $m_{K,L}$ be the coefficient of s^{d_L} in $M_{K,L}$. We'll show that $\det(m_{K,L}) \neq 0$. This would imply the theorem.

Define the lexicographical order in $\mathcal{L}_{k,n}$ by the rule: $(\ell_1 \geq \dots \geq \ell_k) > (\ell'_1 \geq \dots \geq \ell'_k)$ if $\ell_k = \ell'_k, \dots, \ell_{j+1} = \ell'_{j+1}$, $\ell_j > \ell'_j$ for some $j = 1, \dots, k$.

Define a bijection $\mathcal{L}_{k,n} \to \mathcal{K}_{k,n}$ by the rule $(\ell_1 \geq \ldots \geq \ell_k) \mapsto \sum_{j=1}^{k}(0,\ldots,0,1_{\ell_j},0,\ldots,0)$. This bijection and the lexicographical order on $\mathcal{L}_{k,n}$ induce an order on $\mathcal{K}_{k,n}$. It is easy to see that the matrix $(m_{K,L})$ is a triangular matrix with nonzero diagonal elements and, therefore, $\det(m_{K,L}) \neq 0$.

An explanation of a connection between the Bethe vectors and the Lamé functions is given in the Sklyanin paper *[13]* in which the tensor product V of sl_2 modules is realized in a suitable space of jets of functions of one variable, the sl_2 action is realized by differential operators, and the KZ operators are realized in terms of differential operators (11.2).

12 Branching of the Bethe Vectors, Jordan Blocks of the KZ Operators

The Bethe vectors discussed in this work may have branching outside diagonals $z_i = z_j$. Here is an example. Let V be an sl_2 module with highest weight $m \neq 0$. Consider the Bethe equation corresponding to $(V \otimes V \otimes V)_1$:

$$\frac{1}{t-z_1} + \frac{1}{t-z_2} + \frac{1}{t-z_3} = 0\,.$$

For $(z_1, z_2, z_3) = (0, 1, s)$, this equation gives $3t^2 - 2(1+s) + s = 0$. Its discriminant has two roots s_1 and s_2. Therefore, for $s \neq s_1, s_2$ there are two Bethe vectors and they form a basis in the two-dimensional $\mathrm{Sing}(V \otimes V \otimes V)_1$. When s goes around s_j, $j = 1, 2$, the Bethe vectors interchange positions. For $s = s_j$, the KZ operators $H_\ell(z(s_j))$, $\ell = 1, 2, 3$, have a two-dimensional Jordan block. Their single eigenvector is the Bethe vector corresponding to the single critical point.

References

[1] V. I. Arnold, S. M. Gusein-Zade, and A. M. Varchenko, *Singularities of Differentiable Maps*, volume II, Birkhauser, 1988.

[2] H. Bethe, Z. Phys. **71** (1931), 205.

[3] H.M.Babujian, Off-shell Bethe ansatz equation and N-point correlators in the SU(2) WZNW theory, J. Phys. A **26**, 6981–6990.

[4] H.M.Babujian, R.Flume, Off shell Bethe ansatz equation for Gaudin magnets and solutions of Knizhnik-Zamolodchikov equations, Modern Physics Letters **9**, 2029–2039.

[5] Gaudin model, Bethe ansatz and correlation functions at the critical level, Modern Physics Letters **22** (1994), 2029–2039.

[6] I.Frenkel, N.Reshetikhin, Quantum affine algebras and holonomic difference equations, Comm. Math. Phys. **146** (1992), 1-60.

[7] M. Gaudin, Diagonalizations d'une classe d'hamiltoniens de spin, Jour. de Physique **37**, no. 10 (1976), 1087–1098.

[8] L. Faddeev and L. Takhtajan, Usp. Math. Nauk. **34**, N5 (1979), 13–63 (in Russian).

[9] V. Korepin, Calculation of norms of Bethe wave functions, Comm. Math. Phys. **86** (1982), 391–418.

[10] V. Knizhnik and A. Zamolodchikov, Current algebra and Wess-Zumino models in two dimensions, Nucl. Phys. **B 247** (1984), 83–103.

[11] N. Reshetikhin, Calculation of the norms of Bethe vectors in models with SU(3)-symmetry, Zapiski Nauch. Semin. LOMI **150** (1986), 196–213.

[12] N.Reshetikhin, Jackson type integrals, Bethe vectors, and solutions to a difference analog of the Knizhnik-Zamolodchikov system, Lett. Math. Phys. **26** (1992), 153–165.

[13] E. Skylanin, Separation of variables in the Gaudin model, Jour. of Sov. Math. **47**, 2473–2488.

[14] G.Szegö, Orthogonal Polynomials, AMS, 1939.

[15] V. Tarasov and A. Varchenko, Jackson integral representations for solutions of the KZ equation, St. Petersburg Math. J. **6** (1994), no. 2.

[16] V. Schechtman and A. Varchenko, Arrangements of hyperplanes and Lie algebra homology, Invent. Math. **106** (1991), 139–194.

[17] A. Varchenko, Multidimensional hypergeometric functions and representation theory of Lie algebras and quantum groups, Advanced Series in Mathematical Physics - Vol. 21, 1994, to appear.

[18] A. Varchenko, Critical points of the product of powers of linear functions and families of bases of singular vectors, Compositio Math., to appear.

Descartes and Differential Geometry

Hans Samelson[1]
Department of Mathematics
Stanford University
Stanford, CA 94305

To Raoul Bott –
mathematicus et vir egregius

Let us listen to Descartes and Gauss, discussing the past as they are taking a walk at their present abode (wherever that might be).

Here is Gauss: "Renatus [they talk in Latin of course], at one time of your life you thought about geometry, curves and surfaces and such. Let me tell you an interesting fact I came across once; I considered it quite fundamental and so I called it the Theorema Egregium, the Extraordinary or Outstanding Theorem.

Look at a surface. At any point it has a "curvature", the way it bends in all directions, best understood in terms of the normals to the surface at the point and nearby, the way they tilt on moving away from the point. Well, that curvature is an intrinsic property of the surface; you can find its value by measurements in the surface, you don't have to go out of the surface and look at it. The way I show that is by giving a formula for the curvature in terms of what I call the First Fundamental Form of the surface, which describes measurements on the surface."

And Descartes answers: "Dear Carolus, thanks for telling me this. The fact you describe is very interesting and worth thinking about. This "curvature" that you are considering reminds me that I once used that term in something I worked on. Let me think. Oh, yes. It came up when I found that formula for a convex polyhedron that our Swiss friend Leonhardus found again more than hundred years later; the people down there on earth write it now in the form $v - e + f = 2$, with v, e, f meaning the number of vertices, edges, faces. (I actually used another quantity instead of the number e of edges.)

And now I see that I should not have been surprised by your theorem at all, because I used just that fact in my proof; it was a quite crucial step.

Of course it wasn't exactly your theorem that I used, but the corresponding fact for polyhedra, the piecewise-linear case, which is much easier to prove. I wrote up the results as a short note, but never published it. It was more a memo to myself, with lots of details left out. It got lost later. The people on earth can thank

[1] Support by NSF DMS 91-02078 is acknowledged.

Geometry, Topology, and Physics for Raoul Bott

Cambridge, MA 02138 USA

friend Gottfried Wilhelm for making a copy of it when he saw it, and to Count Foucher de Careil for finding that copy."

Gauss: " Renatus, what you tell me there is fascinating. I certainly didn't know that you had found that formula. As you said, it is definitely a good deal harder to show for surfaces that the curvature is intrinsic then in your case; it took me several years to get it straight. And that reminds me of quiet young Bernhard - he really understood what was going on there. I well remember his first big lecture; I could see that he had opened up a big new subject; I was in deep thought all the way walking back home."

Descartes: "Anyway, it is nice to see that your and my theorem has thrived and blossomed ever since, particularly after Bernhard got his hands on it."

Now for the facts: The story of the survival of the note is unusual enough, so that it bears retelling. Descartes (who lived from 1596 to 1650) spent the last year of his life in Stockholm, at the court of Queen Christina. After his death his belongings were packed up and sent per ship to Paris. On arrival in Paris the ship proceeded to sink in the Seine. The boxes were retrieved, and a friend spread out the manuscripts in his house to dry. Some twenty years later Gottfried Wilhelm Leibniz was in Paris, heard about the existence of Descartes's papers, went to the place, and made a hand written copy of the note. (It is not clear whether he condensed the note, and whether he copied it completely or whether there was more to it.) Leibniz didn't do anything with his copy. He didn't publish it, and he doesn't seem to have worked on the matter.

Both the original and the copy disappeared after a while. Some rumor of the existence of the copy seems to have stayed alive. Then in the 1850's the historian of science Count Foucher de Careil, looking for Leibniziana, made a trip to the Leibniz house in Hannover. What happened there seems to have been like a scene from a bad mystery novel. —"What's in that old cupboard over there ?" —"Oh, nothing much." —"Let's look at it anyway." And there, "under the dust of ages", was Leibniz's copy of Descartes's note. It consists of text, tables and a few simple diagrams, on front and back of one page, and about two thirds of the front of another page.

The note has three distinct parts, one on polyhedra, one on polygonal and polyhedral numbers, and the last one a list of certain data about the regular and some semiregular solids. Relevant for us here are the first six sentences (we number them, for ease of reference):

1. Angulus solidus rectus est qui octavam spherae partem complectitur, etiamsi non constet ex tribus angulis planis rectis.
2. Omnes autem anguli plani, ex quibus circumscribitur, simul sumti, aequales sunt tribus rectis.
3. Sicut in figura plana omnes anguli externi, simul sumti, aequales sunt quatuor rectis: ita in corpore solido omnes anguli solidi externi, simul sumti, aequales sunt octo solidis recti.

4. Per angulum externum intelligo curvaturam & inclinationem planorum ad invecem, quam metiri oportet ex angulis planis angulum solidum comprehentibus.
5. Nam illa pars qua aggregatum ex omnibus angulis planis unum angulum solidum facientibus minus est quam quatuor anguli recti planum, designat angulum externum solidum.
6. Si quatuor anguli plani recti ducantur per numerorom angulorum solidorum and & ex producto tollantur 8 anguli recti plani, remanet aggregatum ex omnibus angulis planis qui in superficie talis corporis solidi existunt.

This translates as follows:

1. A solid right angle is one that encloses one eighth of the sphere, even if it does not consist of three plane right angles.
2. But all the plane angles, by which it is circumscribed, taken together are equal to three right angles.
3. Just as in a plane figure all external angles, taken together, are equal to four right angles: so in a solid body all the external solid angles, taken together, are equal to eight solid right angles.
4. By external angle I mean the curvature and inclination of the planes with respect to each other, which it behooves to measure from the plane angles that comprise the solid angle.
5. For that part by which the sum of all the plane angles that constitute a solid angle is smaller than four plane right angles represents the external solid angle. ["planum" should of course read "plani".]
6. If four plane right angles are multiplied by the number of solid angles and from the product are subtracted 8 plane right angles, there remains the sum of all the plane angles that exist on the surface of such a solid body.

A lot is going on in these few sentences, and much of it is not explained.

The first sentence is clear enough. It defines the notion of solid right angle, in terms of the intersection with a sphere (which might as well be the unit sphere), and notes that it doesn't have to be the usual octant. [This becomes generalized later on in the note with the notion of solid angles of equal capacity (i.e., equal area of the parts of the sphere enclosed by the solid angles).]

The second sentence poses a problem. For the special case of a three sided cone it seems to tell us that Descartes was aware of the theorem on the spherical excess (if $\alpha_1, \alpha_2.\alpha_3$ are the angles of a spherical triangle, then the area of the triangle is $\alpha_1 + \alpha_2 + \alpha_3 - \pi$). This theorem had been found by Thomas Harriot in 1603, but apparently remained quite unknown till 1627 or 1629, when it was rediscovered; this is pretty late in the period 1620-1630, during which Descartes presumably wrote his note; on the other hand the subject was certainly in the air.

In addition Descartes clearly has in mind the case of a polyhedral cone of any number of sides. Now the words "anguli plani" seem to suggest the face angles

at the vertex of the solid angle. But those angles do not determine the area in question. As suggested in *[6]*, p.43, one should correct "plani" to "planorum", so that the phrase would translate to "But all the angles between the planes, by which . . .". Moreover these angles would have to understood to be not the internal angles of the spherical polygon formed, but their supplements (π minus angle), the external angles. (The excess formula generalizes to "area = 2π minus the sum of the external angles"; so we have $\pi/2 = 2\pi - 3\pi/2$).
Did Descartes know all this and develop it by himself, but considered it simple enough so no details were necessary?

The third sentence takes a fairly simple fact of plane geometry (by the way it seems that Descartes always thought of convex polygons and polyhedra) and generalizes it to space: The curvature integra of a convex polyhedron is 4π; a very crucial step. There is no indication of proof or even argument (but I don't think he meant to argue just by analogy). It is even quite unclear at this point what "external angle" means (more on this comes in the next sentence, but wait). If he thought (as we think now) of external angle (at a vertex of the body) as determined by the dual cone, formed by the normals to the faces at the vertex (or rather the exterior half-normals), then the argument would be clear enough (the external angles at the various vertices just divide the sphere into a number of pieces).

Then we would have to interpret the first two sentences as indicating a notion of measure of a solid angle (by the area of the part of the sphere cut out by the solid angle); however they don't say that, and, as noted above, only later on is there a (very brief) mention of this idea. Of course he might have taken that sort of thing for granted.

A slightly disturbing comment is that since Euclid's time the external angle of a plane angle is formed not by forming the "dual cone", but by extending one leg of the angle. Of course for a solid angle this doesn't work. Again, the notion of dual cone didn't get published till about 1627, but it must have been in the air.

The fourth sentence is the central piece; it is the one that contains the "Theorema Egregium". It starts with what seems like a definition or at least description of external angle at a vertex, namely as "curvature" (!) or mutual inclination of the planes, and goes on immediately to state that this is to be computed from the angles of the faces of the body at the vertex [so that it is an intrinsic property]. And the next, 5th, sentence gives the formula for the curvature at a vertex: 2π minus the sum of the angles just mentioned.

Again no hint of the argument that leads to the formula. If he had the dual cone in mind and knew the expression for the area of a spherical polygon in terms of the angles (as discussed above for sentence 2), the argument would be evident enough: simple geometry shows that the face angles of the body at the vertex in question are the supplements of the angles of the dual cone.

Is it possible that he took the formula as definition of curvature? For a flat solid angle where all the faces lie in the same plane the sum of the angles is of course 2π; thus he could have taken the difference as measure of the deviation from flatness. But then how would he arrive at the curvatura integra, the fact that the sum over all vertices gives 4π? Could he have worked backwards in some way?

I do think that he saw clearly the central fact, namely that the extrinsic property "curvature", which depends on the position of the surface in space, is actually intrinsic.

(Incidentally, the term "inclination" in sentence 4 is a little dangerous. One should not interpret it as describing the angles between the planes carrying the faces at the vertex; their sum does not determine the curvature and is not even intrinsic.)

Now sentence 6: Adding the formula of sentence 5 over all vertices and using the result of sentence 3 one gets $4\pi = 2\pi v - \sum \gamma$, where v is the number of vertices and the sum goes over all the angles of all the faces. With that one is almost at the Euler formula as we know it.

I will say just a few words about that part of the story. It is easy to get from one formula to the other. Euler stated both equations and emphasized their equivalence. One only has to note that the sum of the angles of any given face is $(s - 2)\pi$, where s is the number of the vertices or of the edges of the face (a variant of the first part of sentence 3), and that the sum of the s-values is twice the number e of edges of the body. By summing in $\sum \gamma$ first over each face, one gets thus $2e\pi - 2f\pi$ (with f the number of faces), and there is Euler's formula, and conversely.

Somewhat surprisingly Descartes does not go on to make this step, although it seems that he knew the result (see below). Instead he goes on to a number of miscellaneous topics, mostly using the formula $\sum \gamma = 2\pi v - 4\pi$. One topic is a short argument showing that there are only the well known five regular solids. (It looks as if he used duality between v and f; as always, one can supply a line of reasoning or several; what he had in mind, remains dark).

Then comes the sentence whose first part is Euler's formula:

Numerus verorum angulorum planorum est $2\phi + 2\alpha - 4$, qui non debet esse major quam $6\alpha - 12$; sed si minor est, excessus erit $+4\alpha - 8 - 2\phi$.

This translates as: The number of true plane angles is $2f + 2v - 4$, which must not be greater than $6v - 12$; but if it is smaller, the excess is $+4v - 8 - 2f$.

The "number of true plane angles'" is equal to $2e$, as we saw above; so there is Euler's formula, except that Descartes uses another quantity instead of e. There is no indication of proof or argument for the claim made. Could he have considered it as too simple or obvious? This very brief statement of what to us seems the main fact is immediately followed by a not very important inequality and the formula for the difference of the two terms involved. One gets the feeling that Descartes did not think the relation had much interest, and one is tempted to say with Lebesgue that he had the Euler formula, but didn't see it (*[8]*, pp.318-319).

Foucher de Careil published the note in 1860 in *[7]*. It was published again in 1906-1909 in *[1]*, and in a new edition 1966 in *[2]*. A number of mathematicians over the years made comments on the note, ranging from brief statements such as: "Descartes appears to have known Euler's formula" to a few pages of discussion. (E. Steinitz, in *[9]*, quotes Descartes's proof (reconstructed, as above) in full.) Surprisingly only quite recently have detailed studies of the note appeared *[5, 6]*;

they contain a lot of historical material, and contain photographic reproductions of Leibniz's manuscript.

In one of the earliest comments, *[3]*, J. Bertrand already draws attention to the analogy between Descartes's idea of the sum of the external angles of a polyhedron and Gauss's notion of total curvature of a surface. He notes that the total curvature of a convex surface is 4π, just like in Descartes's case. He then goes on, curiously enough, to disavow the analogy, and says: The most remarkable fact about surfaces is that the total curvature is invariant under bending of the surface, whereas this makes no sense for polyhedra because they can't be deformed.

The last statement is probably a reference to Cauchy's rigidity theorem for convex polyhedra; but of course polyhedral cones (of more than three sides) are not rigid. Regarding the statement about surfaces, well, he didn't know Cohn-Vossen's rigidity theorem for convex surfaces, and apparently thought that convex surfaces can be deformed. Maybe he was temporarily confused about the distinction between total curvature and curvature at a point, the "most remarkable fact" about the latter being the Theorema Egregium, i.e. invariance of curvature under local bending. That Bertrand made a mistake here and that in fact there is a close analogy between Gaussian curvature and Descartes's notion was recently noted by P. Costabel *[4]*. My main point above has been that the analogy goes further and that Descartes made good use of the Theorema Egregium for polyhedra.

References

[1] Ch. Adam and P. Tannery, eds., *Œuvres de Descartes*, Paris 1906–09, vol. X, pp. 257–276, vol. XI, pp. 690–692.

[2] —, *Œuvres de Descartes, Nouvelle Présentation*, Paris 1966, vol. X, pp. 276 and 687–689, with notes by P. Costabel.

[3] J. Bertrand, in: E. Prouhet, *Remarques sur un passage des oeuvres inédites de Descartes*, C.R. de l'Académie des Sciences, v. 50, Paris, 1860, pp. 779–782.

[4] P. Costabel, La courbure et son apparition chez Descartes, in: *Mélanges Charles Morazé*, Édouard Privat, Toulouse, 1979, pp. 337–344.

[5] P. Costabel, *René Descartes, Exercices pour les Eléménts des Solides*, Presses Universitaires de France, Paris, 1987.

[6] P.J. Federico, *Descartes on Polyhedra. A Study of the De Solidorum Elementis*, New York, Berlin, 1982.

[7] Foucher de Careil, *OEuvres inédites de Descartes, précédées d'une introduction sur la Méthode*, 2 vol., Paris ,1859–60, vol. II, pp. 214–226.

[8] H. Lebesgue, Remarques sur les deux premières démonstrations du théoréme d'Euler relatif aux polyèdres, Bulletin de la Société mathématique de France vol. 52, Paris, 1924, pp. 315–336.

[9] E. Steinitz, *Vorlesungen über die Theorie der Polyeder*, Springer-Verlag, Berlin, 1934, pp. 10–11.

Characters, Characteristic Cycles, and Nilpotent Orbits

Wilfried Schmid[1]
Department of Mathematics
Harvard University
Cambridge, MA 02138, USA

Kari Vilonen[2]
Department of Mathematics
Brandeis University
Waltham, MA 02254, USA

1 Introduction

Two quite different constructions - one algebraic, due to Vogan *[22]*, the other of an analytic nature, by Barbasch-Vogan *[3]* - attach similar invariants to representations of semisimple Lie groups. It had been conjectured that the two types of invariants are related in a definite manner. Our result is an affirmative answer to the conjecture. We connect the algebraic and the analytic invariant via geometry: Kashiwara's Morse theoretic construction of the characteristic cycle of a constructible sheaf *[9]*, and properties of the moment map attached to a Hermitian form (Ness *[17]*).

To be more specific, we let $G_{\mathbb{R}}$ denote a semisimple Lie group with maximal compact subgroup $K_{\mathbb{R}}$; $K_{\mathbb{C}}$ the complexification of $K_{\mathbb{R}}$; $\mathfrak{g}_{\mathbb{R}}$, $\mathfrak{k}_{\mathbb{R}}$ the Lie algebras of $G_{\mathbb{R}}$, $K_{\mathbb{R}}$; and $\mathfrak{g}_{\mathbb{C}}$, $\mathfrak{k}_{\mathbb{C}}$ the complexified Lie algebras. By a representation of $G_{\mathbb{R}}$, we shall mean an admissible representation of finite length on a complete, locally convex, Hausdorff topological vector space. Each such representation π has a character Θ_π, a conjugation invariant distribution on $G_{\mathbb{R}}$. Via the exponential map, Θ_π drops to a conjugation invariant distribution θ_π on the Lie algebra $\mathfrak{g}_{\mathbb{R}}$. To avoid unnecessarily convoluted statements, we suppose that π is irreducible. In that case, both Θ_π and θ_π are eigendistributions for $Z(\mathfrak{g}_{\mathbb{C}})$, the center of the universal enveloping algebra $U(\mathfrak{g}_{\mathbb{C}})$.

Vogan *[22]* puts a "good filtration" on the Harish-Chandra module V_π of the irreducible representation π. The associated graded $gr(V_\pi)$ turns out to be

[1] W.Schmid was partially supported by NSF
[2] K.Vilonen was partially supported by NSA, NSF, and AMS Centennial Fellowship

Geometry, Topology, and Physics for Raoul Bott

Cambridge, MA 02138 USA

a finitely generated module over $gr(U(\mathfrak{g}_\mathbb{C}))/\mathfrak{k}_\mathbb{C}(gr(U(\mathfrak{g}_\mathbb{C}))) \cong Sym(\mathfrak{g}_\mathbb{C}/\mathfrak{k}_\mathbb{C})$, which therefore corresponds to an algebraic cycle $Ass(V_\pi)$ in $(\mathfrak{g}_\mathbb{C}/\mathfrak{k}_\mathbb{C})^*$, the "associated variety" of V_π. While $gr(V_\pi)$ depends on the choice of good filtration, $Ass(V_\pi)$ is a well defined invariant. It is $Ad\,K_\mathbb{C}$-invariant, equidimensional, and lies in the cone $\mathcal{N}^*$ of nilpotents[3] in $\mathfrak{g}_\mathbb{C}^*$. The cone $\mathcal{N}^*$ intersects $(\mathfrak{g}_\mathbb{C}/\mathfrak{k}_\mathbb{C})^*$, the annihilator of $\mathfrak{k}_\mathbb{C}$ in $\mathfrak{g}_\mathbb{C}^*$, in a finite number of $K_\mathbb{C}$-orbits. In effect, $Ass(V_\pi)$ may be thought of as a finite linear combination of $K_\mathbb{C}$-orbits in $\mathcal{N}^* \cap (\mathfrak{g}_\mathbb{C}/\mathfrak{k}_\mathbb{C})^*$, all of the same dimension, with positive integral coefficients.

The Fourier transform $\hat{\theta}_\pi$ of θ_π has an asymptotic expansion about the origin in $i\mathfrak{g}_\mathbb{R}^*$, obtained by letting a scaling parameter go to infinity. This was shown by Barbasch-Vogan *[3]*, who also identify the leading term in the expansion as a real linear combination of invariant measures on $G_\mathbb{R}$-orbits in $\mathcal{N}^* \cap i\mathfrak{g}_\mathbb{R}^*$, again all of the same dimension. Since coadjoint orbits come equipped with canonical symplectic forms, and hence with canonical invariant measures, the Barbasch-Vogan construction assigns a real linear combination of $G_\mathbb{R}$-orbits in $\mathcal{N}^* \cap i\mathfrak{g}_\mathbb{R}^*$ to each irreducible representation π.

Let $\mathfrak{p}_\mathbb{C}$ denote the Ad$K_\mathbb{C}$-invariant complement of $\mathfrak{k}_\mathbb{C}$ in $\mathfrak{g}_\mathbb{C}$, and $\mathcal{N}$ the nilpotent cone in $\mathfrak{g}_\mathbb{C}$. Sekiguchi *[21]*, motivated in part by the results of *[22]* and *[3]*, sets up a natural bijection between nilpotent orbits in the (-1)-eigenspaces of certain commuting involutions on a semisimple Lie algebra, and in particular between the set of $K_\mathbb{C}$-orbits in $\mathcal{N} \cap \mathfrak{p}_\mathbb{C}$ on the one hand, and the set of $G_\mathbb{R}$-orbits in $\mathcal{N} \cap i\mathfrak{g}_\mathbb{R}$ on the other. The latter situation was also treated, on a case by case basis, by Doković *[6]*, who credits Kostant for making the conjecture. Kostant informs us that his conjecture precedes *[22, 3]*, and that he had previously verified it, except for a minor technical gap. Ever since *[21]*, it has been suspected that the algebraic invariant constructed in *[22]* and the analytic invariant appearing in *[3]* become one and the same when the two types of orbits are identified via the Sekiguchi correspondence and the isomorphism $\mathfrak{g}_\mathbb{C} \cong \mathfrak{g}_\mathbb{C}^*$ determined by the Killing form. Explicit statements of this conjecture appear in *[23]*, *[11]*, for example. In some special cases, the conjecture was recently established by Chang *[4, 5]*. One of Chang's intermediate arguments applies in the general case, and we shall use that argument in section 6.

Our result is a proof of the conjecture in general; we shall give an outline of our arguments in this note - full details will appear elsewhere.

2 Preliminaries

We keep the notation of the introduction, but we need to be more precise about the nature of the group $G_\mathbb{R}$: for simplicity we suppose that it is connected and has finite center, though these assumptions can be relaxed considerably by standard techniques *[7]*. We write $G_\mathbb{C}$ for the adjoint group of $\mathfrak{g}_\mathbb{C}$, and X for the flag variety of $G_\mathbb{C}$. Then $G_\mathbb{R}$ and $K_\mathbb{C}$ are finite coverings of their images in $G_\mathbb{C}$; both operate on X with finitely many orbits.

[3] A linear functional on $\mathfrak{g}_\mathbb{C}$ is said to be nilpotent if, under the isomorphism $\mathfrak{g}_\mathbb{C} \cong \mathfrak{g}_\mathbb{C}^*$ determined by the Killing form, it corresponds to a nilpotent element of $\mathfrak{g}_\mathbb{C}$.

We need to recall some basic facts about geometric realizations of $G_{\mathbb{R}}$-representations and Harish-Chandra modules. As in the introduction, by a "representation of $G_{\mathbb{R}}$", we shall mean an admissible representation of finite length, on a complete, locally convex, Hausdorff topological vector space; we tacitly identify representations which are infinitesimally equivalent. We fix some $\lambda \in \mathrm{H}^2(X, \mathbb{C})$ - equivalently, a linear functional on the universal Cartan for $\mathfrak{g}_{\mathbb{C}}$. Harish-Chandra modules can be lifted to representations of $G_{\mathbb{R}}$, uniquely up to infinitesimal equivalence. This gives an isomorphism

$$\{\text{virtual Harish-Chandra modules}\}_\lambda \longrightarrow \{\text{virtual } G_{\mathbb{R}}\text{-representations}\}_\lambda, \quad (2.1)$$

where $\{\ldots\}_\lambda$ refers to the group of virtual Harish-Chandra modules, respectively representations, with infinitesimal character χ_λ, in Harish-Chandra's notation.

Let $D_{K_{\mathbb{C}}}(X)_{-\lambda}$ and $D_{G_{\mathbb{R}}}(X)_{-\lambda}$ denote the equivariant derived categories with twist $-\lambda$, and $\mathcal{O}_X(\lambda)$ the sheaf of holomorphic functions with twist λ - see for example *[19]*, *[20]*; here, as in *[20]*, we incorporate the shift by ρ into the notation. For $\mathcal{F} \in D_{K_{\mathbb{C}}}(X)_{-\lambda}$, the Verdier dual $\mathbb{D}\mathcal{F}$ has twist λ. Hence one can consider the groups $\mathrm{Ext}^p(\mathbb{D}\mathcal{F}, \mathcal{O}_X(\lambda))$ in the algebraic category. As a consequence of the results of Beilinson-Bernstein *[1]*, coupled with the Riemann-Hilbert correspondence (*[8]*, *[15]*), these algebraic Ext-groups are Harish-Chandra modules with infinitesimal character χ_λ, and the map

$$D_{K_{\mathbb{C}}}(X)_{-\lambda} \to \{\text{virtual Harish-Chandra modules}\}_\lambda, \quad (2.2)$$

defined by the formula

$$\mathcal{F} \mapsto \sum (-1)^p \mathrm{Ext}^p(\mathbb{D}\mathcal{F}, \mathcal{O}_X(\lambda)),$$

is a surjective homomorphism.

Analogously, for any $\mathcal{F} \in D_{G_{\mathbb{R}}}(X)_{-\lambda}$, the groups $\mathrm{Ext}^p(\mathbb{D}\mathcal{F}, \mathcal{O}_X(\lambda))$ — this time taken in the analytic category — are representations of $G_{\mathbb{R}}$ *[10, 13]*. Moreover,

$$\mathcal{F} \mapsto \sum (-1)^p \mathrm{Ext}^p(\mathbb{D}\mathcal{F}, \mathcal{O}_X(\lambda))$$

defines a surjective homomorphism

$$D_{G_{\mathbb{R}}}(X)_{-\lambda} \to \{\text{virtual } G_{\mathbb{R}}\text{-representations}\}_\lambda. \quad (2.3)$$

The morphisms (2.1-3) fit into a commutative diagram

$$\begin{array}{ccc} \{\text{virtual H-C-modules}\}_\lambda & \longrightarrow & \{\text{virtual } G_{\mathbb{R}}\text{-representations}\}_\lambda \\ \uparrow & & \uparrow \\ D_{K_{\mathbb{C}}}(X)_{-\lambda} & \longrightarrow & D_{G_{\mathbb{R}}}(X)_{-\lambda}. \end{array} \quad (2.4)$$

In this diagram, the second horizontal arrow

$$\gamma : D_{K_{\mathbb{C}}}(X)_{-\lambda} \to D_{G_{\mathbb{R}}}(X)_{-\lambda} \tag{2.5}$$

is the functor whose existence was conjectured by Kashiwara *[10]* and proved in *[16]*. The commutativity of the diagram (2.4), finally, is already implicit in *[7]*.

3 Characteristic Cycles

In this section X shall denote a real algebraic manifold, or somewhat more generally, a semi-algebraic open subset in some ambient real algebraic manifold. We write $\mathrm{D}(X)$ for the bounded derived category of $\mathbb{C}$-sheaves constructible with respect to a fixed semi-algebraic Whitney stratification $\mathcal{S}$. Kashiwara's characteristic cycle map *[12]* associates a semi-algebraic Lagrangian cycle in the cotangent bundle T^*X to each $\mathcal{F} \in \mathrm{D}(X)$. Strictly speaking, this is a cycle with coefficients in the orientation sheaf of X. It becomes a geometric - i.e., oriented - cycle in T^*X once an orientation is put on X. The point is that the construction makes sense without choice of orientation, and even for non-orientable manifolds X. It will be convenient for our purposes to fix an orientation on X once and for all. Thus the characteristic cycle $\mathrm{CC}(\mathcal{F})$, for $\mathcal{F} \in \mathrm{D}(X)$, becomes an oriented Lagrangian cycle. It is supported on $T^*_{\mathcal{S}}X$, the union of the conormal bundles of the various strata, so

$$\mathrm{CC} : \mathrm{D}(X) \to \mathrm{H}^{inf}_m(T^*_{\mathcal{S}}X, \mathbb{Z}), \tag{3.1}$$

where m is the real dimension of X and H^{inf} homology with infinite support (Borel-Moore homology).

Let us consider an algebraic map $F : X \to Y$ between Whitney stratified spaces $(X, \mathcal{S}_X)$ and $(Y, \mathcal{S}_Y)$, such that F respects the stratifications and is smooth along the strata. Then F induces various maps between the derived categories of the stratified spaces X and Y. One may ask for a description of the effect of these induced maps on characteristic cycles. Kashiwara-Schapira *[12]* do so, in particular, for $\mathrm{CC}(RF_*\mathcal{F})$ when F is proper. We shall use this result, but we shall also need concrete descriptions of $\mathrm{CC}(RF_*\mathcal{F})$ and $\mathrm{CC}(RF_!\mathcal{F})$ when F is not assumed to be proper. We shall now discuss the latter case.

First we suppose that $F : X \to Y$ is an open inclusion. We choose a real analytic exhaustion function $f : X \to \mathbb{R}$, in other words, a function which tends to $+\infty$ as one approaches the boundary of X in Y.

Theorem 3.1: *Under the hypotheses just stated,*

$$\mathrm{CC}(RF_*\mathcal{F}) = \lim_{s \to 0+} \left(\mathrm{CC}(\mathcal{F}) - s\, df\right),$$
$$\mathrm{CC}(RF_!\mathcal{F}) = \lim_{s \to 0+} \left(\mathrm{CC}(\mathcal{F}) + s\, df\right),$$

provided the closure of $\{\mathrm{CC}(\mathcal{F}) - s\, df\} \subset T^*X \times \mathbb{R}$*, respectively* $\{\mathrm{CC}(\mathcal{F}) + s\, df\} \subset T^*X \times \mathbb{R}$*, can be Whitney stratified.*

In slightly different form, this statement already appears in *[20]*, where it is used to prove an explicit Rossmann-type character formula. The meaning of limit in theorem 3.1 can be made explicit as follows. For each generic point $p \in T^*_{\mathcal{S}_Y} Y$ and any normal slice Σ to $T^*_{\mathcal{S}_Y} Y$ in T^*Y through p, the total signed intersection multiplicity between Σ and $\mathrm{CC}(\mathcal{F}) \pm s\,df$, for sufficiently small s, equals the multiplicity of $\mathrm{CC}(Rj_*\mathcal{F})$, respectively $\mathrm{CC}(Rj_!\mathcal{F})$, at p.

The case of a general non-proper map F can be reduced, by passage to the graph of F, to that of the projection

$$p : X \times Y \to Y, \tag{3.2}$$

with possibly non-compact first factor X. Again we introduce a real analytic exhaustion function $f : X \to \mathbb{R}$, but we now suppose that f is generic in the following sense:

$$\begin{gathered}(T^*_{\mathcal{S}_X} X \times T^*_{\mathcal{S}_Y} Y) - s\,df, \text{ respectively } (T^*_{\mathcal{S}_X} X \times T^*_{\mathcal{S}_Y} Y) + s\,df, \\ \text{is transverse to } \; T^*_X X \times T^*Y \; \text{ for small positive values of } s\end{gathered} \tag{3.3}$$

($T^*_X X =$ zero section in T^*X). The existence of such a function f follows from general arguments. In our application, we shall work with a concrete function f which will turn out to satisfy (3.3). As in the statement of theorem 3.1, we also suppose that the closure of $\{\mathrm{CC}(\mathcal{F}) - s\,df\} \subset T^*X \times \mathbb{R}$, respectively $\{\mathrm{CC}(\mathcal{F}) + s\,df\} \subset T^*X \times \mathbb{R}$, can be Whitney stratified.

Theorem 3.2: *For $\mathcal{F} \in \mathrm{D}(X \times Y)$,*

$$\begin{aligned}\mathrm{CC}(Rp_*\mathcal{F}) &= \lim_{s \to 0+} q_* i^*(\mathrm{CC}(\mathcal{F}) - s\,df)), \\ \mathrm{CC}(Rp_!\mathcal{F}) &= \lim_{s \to 0+} q_* i^*(\mathrm{CC}(\mathcal{F}) + s\,df)),\end{aligned}$$

where $q : T^*_X X \times T^*Y \to T^*Y$ is the projection and $i : T^*_X X \times T^*Y \to T^*X \times T^*Y$ the inclusion.

We note that i^* in the statement of the theorem is a Gysin map. Since X and Y are oriented by assumption, we get orientations also on T^*X, T^*Y, and on their product. Thus oriented cycles become co-oriented, and vice versa. By definition, the Gysin map sends co-oriented cycles to co-oriented cycles, hence, in our situation, oriented cycles to oriented cycles.

Theorem 3.2 follows from theorem 3.1 and the formula for $\mathrm{CC}(RF_*\mathcal{F})$ when F is proper *[12, Proposition 9.4.2]*. To prove theorem 3.1 we note that the characteristic cycle can be computed using germs of (stratified) Morse functions and the index theorem *[12, Theorem 9.5.6]*. The crucial point is that when applying the index theorem to $RF_*\mathcal{F}$ for an open inclusion F, any Morse function g can be replaced by the Morse function $g + sf$ when s is sufficiently small .

4 Microlocalization of the Matsuki Correspondence

We return to the setting of section 2. We need to describe the morphism γ in (2.5) explicitly. Let us consider, first of all, the untwisted case, i.e., $\lambda = 0$. Let

$$a : G_{\mathbb{R}} \times X \to X$$

be the action map, composed with inversion in the $G_{\mathbb{R}}$-factor: $a(g,x) = g^{-1}x$,

$$q : G_{\mathbb{R}} \times X \to G_{\mathbb{R}}/K_{\mathbb{R}} \times X$$

the natural quotient map, and

$$p : G_{\mathbb{R}}/K_{\mathbb{R}} \times X \to X$$

the projection onto the second factor. These three maps fit into a diagram

$$X \xleftarrow{a} G_{\mathbb{R}} \times X \xrightarrow{q} G_{\mathbb{R}}/K_{\mathbb{R}} \times X \xrightarrow{p} X. \tag{4.1}$$

For each $\mathcal{F} \in D_{K_{\mathbb{C}}}(X)$, the pullback $a^!(\mathcal{F})$ can be viewed as an object in

$$D_{G_{\mathbb{R}} \times K_{\mathbb{R}}}(G_{\mathbb{R}} \times X) \cong D_{G_{\mathbb{R}}}(G_{\mathbb{R}}/K_{\mathbb{R}} \times X). \tag{4.2}$$

Here we let $G_{\mathbb{R}}$ act diagonally on $G_{\mathbb{R}} \times X$ and $G_{\mathbb{R}}/K_{\mathbb{R}} \times X$; each $k \in K_{\mathbb{R}}$ acts on $G_{\mathbb{R}} \times X$ via right multiplication by k^{-1} on the $G_{\mathbb{R}}$-factor. Since this $K_{\mathbb{R}}$-action is free, we get the equivalence of categories (4.4). In other words, $a^!\mathcal{F} = q^!\tilde{\mathcal{F}}$, for a uniquely determined $\tilde{\mathcal{F}} \in D_{G_{\mathbb{R}}}(G_{\mathbb{R}}/K_{\mathbb{R}} \times X)$. Then

$$\gamma : D_{K_{\mathbb{C}}}(X) \to D_{G_{\mathbb{R}}}(X), \quad \text{with} \quad \gamma(\mathcal{F}) = Rp_!\tilde{\mathcal{F}}, \tag{4.3}$$

defines an equivalence of derived categories *[16]*. This, then, is the map (2.5) in the untwisted case.

The twisted case is essentially the same. Technically, the twisted equivariant derived category is constructed in terms of complexes of sheaves on the enhanced flag variety $\hat{X}$, with a monodromy condition on the right action of the universal Cartan on $\hat{X}$. This right action of the universal Cartan commutes with the left action of $G_{\mathbb{C}}$, so our geometric description of γ carries over without any essential changes; for details see *[16]*.

Elements of the equivariant derived category $D_{K_{\mathbb{C}}}(X)$ are constructible with respect to the orbit stratification. Thus, if we let $T^*_{K_{\mathbb{C}}}X$ denote the union of the conormal bundles of the $K_{\mathbb{C}}$-orbits, the characteristic cycle map (3.1) assigns a cycle $\mathrm{CC}(\mathcal{F}) \in \mathrm{H}^{inf}_m(T^*_{K_{\mathbb{C}}}X, \mathbb{Z})$ to each $\mathcal{F} \in D_{K_{\mathbb{C}}}(X)$; here m denotes the real dimension of the flag variety X. The characteristic cycle construction is local, so CC makes sense equally in the twisted case:

$$\mathrm{CC} : D_{K_{\mathbb{C}}}(X)_{-\lambda} \to \mathrm{H}^{inf}_m(T^*_{K_{\mathbb{C}}}X, \mathbb{Z}).$$

Analogously we get the chacteristic cycle map

$$\mathrm{CC} : D_{G_{\mathbb{R}}}(X)_{-\lambda} \to \mathrm{H}^{inf}_m(T^*_{G_{\mathbb{R}}}X, \mathbb{Z}).$$

On general grounds, there exists a linear map

$$\Phi : \mathrm{H}_m^{inf}(T^*_{K_\mathbb{C}}X, \mathbb{Z}) \to \mathrm{H}_m^{inf}(T^*_{G_\mathbb{R}}X, \mathbb{Z}),$$

which makes the following diagram commutative:

$$\begin{array}{ccc} D_{K_\mathbb{C}}(X)_{-\lambda} & \xrightarrow{\gamma} & D_{G_\mathbb{R}}(X)_{-\lambda} \\ \Big\downarrow \mathrm{cc} & & \Big\downarrow \mathrm{cc} \\ \mathrm{H}_m^{inf}(T^*_{K_\mathbb{C}}X, \mathbb{Z}) & \xrightarrow{\Phi} & \mathrm{H}_m^{inf}(T^*_{G_\mathbb{R}}X, \mathbb{Z}) \end{array} \tag{4.4}$$

We now proceed to describe the map Φ explicitly.

We let $\mu : T^*X \to \mathfrak{g}_\mathbb{C}$ denote the moment map, composed with the identification $\mathfrak{g}_\mathbb{C} \cong \mathfrak{g}_\mathbb{C}^*$ which is induced by the real part of the Killing form. Under this identification, the nilpotent cone $\mathcal{N} \subset \mathfrak{g}_\mathbb{C}$ corresponds to the nilpotent cone $\mathcal{N}^* \subset \mathfrak{g}_\mathbb{C}^*$. The map μ takes values in $\mathcal{N}$:

$$\mu : T^*X \to \mathcal{N}. \tag{4.5}$$

More specifically, if the Borel subalgebra $\mathfrak{b}_x \subset \mathfrak{g}_\mathbb{C}$ corresponds to the point $x \in X$, then

$$\mu : T^*_xX \xrightarrow{\cong} \mathfrak{n}_x =_{\mathrm{def}} [\mathfrak{b}_x, \mathfrak{b}_x]. \tag{4.6}$$

Let $\mathfrak{p}_\mathbb{C}$ denote the Ad$(K_\mathbb{C})$-invariant linear complement of $\mathfrak{k}_\mathbb{C}$ in $\mathfrak{g}_\mathbb{C}$. Then

$$T^*_{K_\mathbb{C}}X = \mu^{-1}(\mathfrak{p}_\mathbb{C}), \ \text{ and } \ T^*_{G_\mathbb{R}}X = \mu^{-1}(i\mathfrak{g}_\mathbb{R}). \tag{4.7}$$

We define a one parameter family of bianalytic maps

$$\begin{gathered} F_s : T^*X \to T^*X, \quad s \in \mathbb{R}^+, \\ F_s(\xi) = \ell(\exp(-s^{-1}\mathrm{Re}\mu(\xi)))^*\xi \quad (\xi \in T^*X), \end{gathered} \tag{4.8}$$

where $\ell(g^{-1}) : X \to X$, for $g \in G_\mathbb{C}$, denotes translation by g^{-1}, and $\ell(g^{-1})^* : T^*_xX \to T^*_{gx}X$ the induced map. For each cycle $C \in \mathrm{H}_m^{inf}(T^*_{K_\mathbb{C}}X, \mathbb{Z})$ and each $s \in \mathbb{R}^+$, $(F_s)_*(C)$ is a Lagrangian cycle in T^*X.

We apply theorem 3.2 to the projection $p : G_\mathbb{R}/K_\mathbb{R} \times X \to X$. The role of the exhaustion function $f : G_\mathbb{R}/K_\mathbb{R} \to \mathbb{R}^+$ will be played by the Killing form, composed with the inverse of

$$\mathfrak{p}_\mathbb{R} \xrightarrow{\exp} G_\mathbb{R} \longrightarrow G_\mathbb{R}/K_\mathbb{R}.$$

With this choice of f, the hypothesis of theorem 3.2 are satisfied. Combining these remarks with (4.3), we get the following result:

Theorem 4.1: *For* $C \in \mathrm{H}_m^{inf}(T^*_{K_\mathbb{C}}X, \mathbb{Z})$,

$$\Phi(C) = \lim_{s \to 0+} (F_s)_*(C).$$

5 Passage to the Nilpotent Variety

Let $\pi : T^*X \to X$ be the projection, and μ the moment map as in (4.6). The adjoint action of $G_{\mathbb{C}}$ stratifies the nilpotent cone $\mathcal{N}$ with strata of even complex dimension; we denote the $2k$-dimensional stratum by $\mathcal{N}_k$. As was first observed by Kostant-Rallis *[14]*, the intersections $\mathcal{N}_k \cap \mathfrak{p}_{\mathbb{C}}$ and $\mathcal{N}_k \cap i\mathfrak{g}_{\mathbb{R}}$ are Lagrangian with respect to the intrinsic symplectic structure on $\mathcal{N}_k$, and thus have real dimension $2k$. We filter the groups $\mathrm{H}_m^{inf}(T^*_{K_{\mathbb{C}}}X, \mathbb{Z})$ and $\mathrm{H}_*^{inf}(\mathcal{N} \cap i\mathfrak{g}_{\mathbb{R}}, \mathbb{C})$ according to support:

$$\begin{aligned} \mathrm{H}_m^{inf}(T^*_{K_{\mathbb{C}}}X, \mathbb{Z})_k &= \text{Image of } \{\mathrm{H}_m^{inf}(T^*_{K_{\mathbb{C}}}X \cap \mu^{-1}\bar{\mathcal{N}}_k, \mathbb{Z}) \to \mathrm{H}_m^{inf}(T^*_{K_{\mathbb{C}}}X, \mathbb{Z})\}, \\ \mathrm{H}_m^{inf}(T^*_{G_{\mathbb{R}}}X, \mathbb{Z})_k &= \text{Image of } \{\mathrm{H}_m^{inf}(T^*_{G_{\mathbb{R}}}X \cap \mu^{-1}\bar{\mathcal{N}}_k, \mathbb{Z}) \to \mathrm{H}_m^{inf}(T^*_{G_{\mathbb{R}}}X, \mathbb{Z})\}, \end{aligned} \tag{5.1}$$

where $\bar{\mathcal{N}}_k$ denotes the closure of $\mathcal{N}_k$.

We recall our earlier identification of the universal Cartan with $\mathrm{H}_2(X, \mathbb{C})$. The exponential e^λ of the twisting parameter $\lambda \in \mathrm{H}^2(X, \mathbb{C})$ is thus a mixed cohomology class on X with complex coefficients. Integration over the fibres of μ defines $\mathbb{Z}$-linear maps

$$\begin{aligned} (\mathrm{gr}\mu_*)_\lambda &: \mathrm{grH}_m^{inf}(T^*_{K_{\mathbb{C}}}X, \mathbb{Z}) \longrightarrow \bigoplus \mathrm{H}_{2k}^{inf}(\mathcal{N}_k \cap \mathfrak{p}_{\mathbb{C}}, \mathbb{C}), \\ (\mathrm{gr}\mu_*)_\lambda &: \mathrm{grH}_m^{inf}(T^*_{G_{\mathbb{R}}}X, \mathbb{Z}) \longrightarrow \bigoplus \mathrm{H}_{2k}^{inf}(\mathcal{N}_k \cap i\mathfrak{g}_{\mathbb{R}}, \mathbb{C}), \end{aligned} \tag{5.2}$$

as follows. For each k, restriction to $\mu^{-1}\mathcal{N}_k$ defines a map

$$\mathrm{gr}_k\mathrm{H}_m^{inf}(T^*_{K_{\mathbb{C}}}X, \mathbb{Z}) \to \mathrm{H}_m^{inf}(T^*_{K_{\mathbb{C}}}X \cap \mu^{-1}\mathcal{N}_k, \mathbb{Z}).$$

We then integrate the cohomology class e^λ over the fibres of μ:

$$\int_\mu \pi^* e^\lambda : \mathrm{H}_m^{inf}(T^*_{K_{\mathbb{C}}}X \cap \mu^{-1}\mathcal{N}_k, \mathbb{Z}) \longrightarrow \mathrm{H}_{2k}^{inf}(\mathcal{N}_k \cap \mathfrak{p}_{\mathbb{C}}, \mathbb{C}).$$

The composition of these two maps gives the first map in (5.2); the $G_{\mathbb{R}}$-case works exactly the same way. Ordinarily it is necessary to specify a sign convention when integrating over the fibres, but this is not an issue in our case since the cohomology class e^λ has non-zero components only in even degrees.

For each k, $\mathcal{N}_k \cap \mathfrak{p}_{\mathbb{C}}$ is the disjoint union of $K_{\mathbb{C}}$-orbits, each of which has real dimension $2k$, as was noted already. Thus, when we orient the orbits, $\bigoplus \mathrm{H}_{2k}^{inf}(\mathcal{N}_k \cap \mathfrak{p}_{\mathbb{C}}, \mathbb{C})$ can be identified with the vector space of formal complex linear combinations of all the $K_{\mathbb{C}}$-orbits in $\mathcal{N} \cap \mathfrak{p}_{\mathbb{C}}$. Similarly, $\bigoplus \mathrm{H}_{2k}^{inf}(\mathcal{N}_k \cap i\mathfrak{g}_{\mathbb{R}}, \mathbb{C})$ becomes the vector space of formal linear combinations of the $G_{\mathbb{R}}$-orbits in $\mathcal{N} \cap i\mathfrak{g}_{\mathbb{R}}$, once theses orbits have been oriented. We pick concrete orientations as follows: Each $K_{\mathbb{C}}$-orbit in $\mathcal{N} \cap \mathfrak{p}_{\mathbb{C}}$ is a complex manifold, and thus becomes naturally oriented; on the $G_{\mathbb{R}}$-side, the intrinsic holomorphic symplectic form σ_k on $\mathcal{N}_k$ is purely imaginary on each $G_{\mathbb{R}}$-orbit, and thus $\frac{\sigma_k}{2\pi i}$ turns the orbits into symplectic - hence oriented - manifolds. We should remark that the orientation of $\mathcal{N}_k \cap i\mathfrak{g}_{\mathbb{R}}$

involves a choice of $i = \sqrt{-1}$, but so does the process of orienting a complex manifold. Caution: In our definition of the moment map (4.5) we used the real part of the Killing form to identify $\mathfrak{g}_{\mathbb{C}} \cong \mathfrak{g}_{\mathbb{C}}^*$, and this identification is anti-holomorphic. On the other hand, to obtain the holomorphic symplectic form σ_k on $\mathcal{N}_k$, we use the holomorphic identification $\mathfrak{g}_{\mathbb{C}} \cong \mathfrak{g}_{\mathbb{C}}^*$ induced by the Killing form itself, not its real part.

For the purpose of our next statement, we reinterpret the maps (5.2) as non-linear maps

$$\begin{aligned}(\mathrm{gr}\mu_*)_\lambda &: \mathrm{H}_m^{inf}(T^*_{K_{\mathbb{C}}}X,\mathbb{Z}) \longrightarrow \bigoplus \mathrm{H}_{2k}^{inf}(\mathcal{N}_k \cap \mathfrak{p}_{\mathbb{C}}, \mathbb{C}),\\ (\mathrm{gr}\mu_*)_\lambda &: \mathrm{H}_m^{inf}(T^*_{G_{\mathbb{R}}}X,\mathbb{Z}) \longrightarrow \bigoplus \mathrm{H}_{2k}^{inf}(\mathcal{N}_k \cap i\mathfrak{g}_{\mathbb{R}}, \mathbb{C}),\end{aligned} \tag{5.3}$$

by projecting cohomology classes in the domains onto their leading graded components. With this convention, theorem 4.1 gives us a commutative diagram

$$\begin{array}{ccc} \mathrm{H}_m^{inf}(T^*_{K_{\mathbb{C}}}X,\mathbb{Z}) & \xrightarrow{\ \Phi\ } & \mathrm{H}_m^{inf}(T^*_{G_{\mathbb{R}}}X,\mathbb{Z}) \\ {\scriptstyle (\mathrm{gr}\mu_*)_\lambda}\big\downarrow & & \big\downarrow{\scriptstyle (\mathrm{gr}\mu_*)_\lambda} \\ \bigoplus \mathrm{H}_{2k}^{inf}(\mathcal{N}_k \cap \mathfrak{p}_{\mathbb{C}}, \mathbb{C}) & \xrightarrow{\ \phi\ } & \bigoplus \mathrm{H}_{2k}^{inf}(\mathcal{N}_k \cap i\mathfrak{g}_{\mathbb{R}}, \mathbb{C}), \end{array} \tag{5.4}$$

with a specific, geometric map ϕ, which we shall now describe. We consider the one parameter family of bianalytic maps

$$\begin{aligned} f_s &: \mathcal{N} \longrightarrow \mathcal{N}, \quad s \in \mathbb{R}^+,\\ f_s(Z) &= \mathrm{Ad}(\exp(s^{-1}\mathrm{Re}Z))Z. \end{aligned} \tag{5.5}$$

The f_s preserve the orbit stratification of $\mathcal{N}$. As a consequence of theorem 4.1,

$$\phi(c) = \lim_{s\to 0+} (f_s)_*(c), \quad c \in \bigoplus \mathrm{H}_{2k}^{inf}(\mathcal{N}_k \cap \mathfrak{p}_{\mathbb{C}}, \mathbb{C}). \tag{5.6}$$

As mentioned in the introduction, Sekiguchi *[21]* sets up a bijection between certain nilpotent classes attached to a commuting pair of involutions in a real semisimple Lie algebra. Specialized to the case of $\mathfrak{g}_{\mathbb{C}}$, viewed as real Lie algebra, and the involutions "complex conjugation with respect to $\mathfrak{g}_{\mathbb{R}}$" and "complex conjugation with respect to $\mathfrak{k}_{\mathbb{R}} \oplus i\mathfrak{p}_{\mathbb{R}}$", his result gives a bijection between the $K_{\mathbb{C}}$-orbits in $\mathcal{N} \cap \mathfrak{p}_{\mathbb{C}}$ on the one hand, and the $G_{\mathbb{R}}$-orbits in $\mathcal{N} \cap i\mathfrak{g}_{\mathbb{R}}$ on the other.

Theorem 5.1: *Let c be a $K_{\mathbb{C}}$-orbit in $\mathcal{N} \cap \mathfrak{p}_{\mathbb{C}}$ of real dimension $2k$, oriented by its complex manifold structure. Then $\phi(c) \in \mathrm{H}_{2k}^{inf}(\mathcal{N}_k \cap i\mathfrak{g}_{\mathbb{R}}, \mathbb{C})$ is the $G_{\mathbb{R}}$-orbit associated to c under the correspondence established in [21], oriented by the symplectic structure induced by $\frac{\sigma_k}{2\pi i}$, where σ_k is the holomorphic symplectic form on $\mathcal{N}_k$.*

Our proof of the theorem, about which we shall say only a few words, works more generally in the setting of affine symmetric spaces, and thus provides a geometric interpretation of the Sekiguchi correspondence. Let $U_{\mathbb{R}}$ be the compact real form of $G_{\mathbb{C}}$ with Lie algebra $\mathfrak{k}_{\mathbb{R}} \oplus i\mathfrak{p}_{\mathbb{R}}$. In each $G_{\mathbb{C}}$-orbit $\mathcal{O} \subset \mathcal{N}$, the $U_{\mathbb{R}}$-moment map as defined by Ness *[17]* picks out a distinguished $U_{\mathbb{R}}$-orbit $\mathcal{O}_U$. The flow (5.5) leaves $\mathcal{O}_U$ invariant. It seems likely that the flow increases distance from $\mathcal{O}_U$, and that would imply the theorem fairly directly. We derive estimates for the moment map which are weaker than the distance increasing property, but still suffice to establish the theorem.

6 The Barbasch-Vogan Conjecture

We assemble the commutative rectangles (2.4), (4.5), and (5.4) into a single commutative diagram, as follows:

$$\begin{array}{ccc}
\{\text{virtual H-C-modules}\}_\lambda & \longrightarrow & \{\text{virtual } G_{\mathbb{R}}\text{-representations}\}_\lambda \\
\alpha \big\uparrow & & \big\uparrow \beta \\
D_{K_{\mathbb{C}}}(X)_{-\lambda} & \xrightarrow{\ \gamma\ } & D_{G_{\mathbb{R}}}(X)_{-\lambda} \\
\mathrm{CC} \big\downarrow & & \big\downarrow \mathrm{CC} \\
\mathrm{H}_m^{inf}(T^*_{K_{\mathbb{C}}}X, \mathbb{Z}) & \xrightarrow{\ \Phi\ } & \mathrm{H}_m^{inf}(T^*_{G_{\mathbb{R}}}X, \mathbb{Z}) \\
(\mathrm{gr}\mu_*)_\lambda \big\downarrow & & \big\downarrow (\mathrm{gr}\mu_*)_\lambda \\
\bigoplus \mathrm{H}_{2k}^{inf}(\mathcal{N}_k \cap \mathfrak{p}_{\mathbb{C}}, \mathbb{C}) & \xrightarrow{\ \phi\ } & \bigoplus \mathrm{H}_{2k}^{inf}(\mathcal{N}_k \cap i\mathfrak{g}_{\mathbb{R}}, \mathbb{C}).
\end{array} \qquad (6.1)$$

The next two results relate the invariants of Vogan and Barbasch-Vogan to the vertical columns in (6.1).

Theorem 6.1: *Let $V \in$ {virtual H-C-modules}$_\lambda$ be irreducible, and $\mathcal{F} \in D_{K_{\mathbb{C}}}(X)_{-\lambda}$ a sheaf such that $\alpha(\mathcal{F}) = V$. Then $(gr\mu_*)_\lambda \circ \mathrm{CC}(\mathcal{F})$ is the associated variety $Ass(V)$ in the sense of Vogan.*

The proof of this theorem is due to Chang *[4]*. It amounts to a considerable refinement of *[2, Theorem 1.9]*. The relevant statement in *[4]*, Corollary 2.5.6, relates the $\mathcal{D}$-module characteristic variety of a Harish-Chandra module V to the associated variety Ass(V). Our characteristic cycle map CC in the diagram (4.4) and elsewhere gives the topological charactersitic cycle. These two characteristic cycle maps do not induce the same orientation; rather, they are related by the twist conj : $T^*X \to T^*X$ given by complex conjugation on the fibres. Our moment map μ becomes holomorphic when it is composed with conj, and thus also affects the orientation of holomorphic cycles. In effect, these two sign changes cancel.

Theorem 6.2: *Let $\pi \in \{virtual\ G_{\mathbb{R}}\text{-}representations\}_\lambda$ be an irreducible representation, and $\mathcal{G} \in D_{G_{\mathbb{R}}}(X)_{-\lambda}$ a sheaf such that $\beta(\mathcal{G}) = \pi$. Then $(gr\mu_*)_\lambda \circ \mathrm{CC}(\mathcal{G})$ is the Barbasch-Vogan invariant of π.*

This is a consequence of our integral formula for the character θ_π *[20]*. In the case of complex groups, the formula is a result of Rossmann *[18]*, who also observed that his integral formula relates the Barbasch-Vogan invariant to the cycle in T^*X over which the integration takes place. We should note that the formula (3.6) in *[20]* is not quite correct as stated, because of our misinterpretation of Rossmann's conventions. Rossmann's form τ_λ, for integral λ, is the negative of the curvature form of $\mathcal{O}_X(\lambda)$, and $\frac{\tau_\lambda}{2\pi i}$ is the Chern form. We therefore need to change the sign in the formula by the factor $(-1)^n$.

At this point, theorems 5.1, 6.1, and 6.2 imply the Barbasch-Vogan conjecture.

References

[1] A. Beilinson and J. Bernstein, Proof of Jantzen's conjecture, preprint.

[2] W. Borho and J.-L. Brylinski, Differential operators on homogenous spaces III, Inventiones Math. **80** (1985), 1–68.

[3] D. Barbasch and D. Vogan, The local structure of characters, Jour. Func. Anal. **37** (1980), 27–55.

[4] J.-T. Chang, Asymptotics and characteristic cycles for representations of complex groups, Compositio Math., to appear.

[5] J.-T. Chang, Characteristic cycles and character contours for Harish-Chandra modules, preprint.

[6] D. Doković, Proof of a conjecture of Kostant, Trans. Amer. Math. Soc. **302** (1987), 577 – 585.

[7] H. Hecht, D. Miličić, W. Schmid and J. Wolf, Localization and standard modules for real semisimple Lie groups I: The duality theorem, Invent. Math. **90** (1987), 297–332.

[8] M. Kashiwara, The Riemann-Hilbert problem for holonomic systems, Publ. R.I.M.S. Kyoto Univeristy **20** (1984), 319 – 365.

[9] M. Kashiwara, Index theorem for constructible sheaves, in: *Systèmes Differentiels et Singularités*, A. Galligo, M. Maisonobe, and Ph. Granger (eds.), Astérisque **130** (1985), 193 – 209.

[10] M. Kashiwara, Open problems, Proceedings of Taniguchi symposium at Katata (1986).

[11] M. Kashiwara, $\mathcal{D}$-modules and representation theory of Lie groups, RIMS preprint 940, Kyoto Univerity (1993).

[12] M. Kashiwara and P. Schapira, *Sheaves on Manifolds*, Springer, 1990.

[13] M. Kashiwara and W. Schmid, Equivariant derived category and representations of semisimple Lie groups, to appear.

[14] B. Kostant and S. Rallis, Orbits and representations associated with symmetric spaces, Amer. J. Math. **93** (1971), 753 – 809.

[15] Z. Mebkhout, Une équivalance de catégories – Une autre équivalence de catégories, Comp. Math. **51** (1984), 55 – 64.

[16] I. Mirković, T. Uzawa, and K. Vilonen, Matsuki correspondence for sheaves, Inventiones Math. **109** (1992), 231–245.

[17] L. Ness, A stratification of the null cone via the moment map, Am. Jour. Math. **106** (1984), 1281 – 1325.

[18] W. Rossmann, Invariant eigendistributions on a semisimple Lie algebra and homology classes on the conormal variety I, II, Jour. Func. Anal. **96** (1991), 130–193.

[19] W. Schmid and K. Vilonen, Character, fixed points and Osborne's conjecture, Contemp. Math. **145** (1993).

[20] W. Schmid and K. Vilonen, Weyl group actions on Lagrangian cycles and Rossmann's formula, announcement, to appear in: the Proceedings of NATO Advanced Research Conference on Mathematical Physics and Group Theory, San Antonio, January 1993, Kluwer Academic Press.

[21] J. Sekiguchi, Remarks on nilpotent orbits of a symmetric pair, Jour. Math. Soc. Japan **39** (1987), 127–138.

[22] D. Vogan, Gelfand-Kirillov dimension for Harish-Chandra modules, Inventiones Math. **48** (1978), 75–98.

[23] D. Vogan, Lectures at Bowdoin College, August 1989.

Mirror Transform and String Theory[1]

C. Vafa
Lyman Laboratory of Physics, Harvard University
Cambridge, MA 02138, USA
and
School of Natural Sciences, Institute for Advanced Study
Olden Lane, Princeton, NJ 08540, USA

Abstract: Some aspects of Mirror symmetry are reviewed, with an emphasis on more recent results extending mirror transform to higher genus Riemann surfaces and its relation to the Kodaira-Spencer theory of gravity[2].

One of the most beautiful aspects of string theory is that strings moving on one manifold may behave identically with strings moving on a different manifold. Any pair of manifolds which behave in this way is called a mirror pair. In this paper I will review some aspects of this phenomenon (see *[1]* for a collection of articles on this subject).

One of the most successful methods in solving difficult questions in mathematical physics has been to transform the problems from their initial setup to a more trivial setup where everything can be understood in simple terms. One of the most classical examples of this is the Fourier transform, which transforms a differential equation to a polynomial equation which is easy to solve. A more recent such example is the twistor transform, which maps the classification of self-dual geometries and self-dual Yang-Mills fields in four dimensions to the question of classification of appropriate complex manifolds and holomorphic vector bundles in six (real) dimensions, which is more manageable. The most recent example of such a transformation is the mirror transform which is the subject of this paper. The idea is that one maps certain difficult questions of interest in algebraic geometry on one manifold to simple questions of variation of Hodge structure on a different (mirror) manifold. As we shall see, however, the status of this transformation is not as clearly understood as the other two cases we mentioned.

We first need to recall some basic aspects of string theory. The main ingredient in a string theory is a two dimensional quantum field theory. An interesting class of 2d QFTs are sigma models, which may be viewed as path integrals over the space of maps ϕ from a Riemann surface Σ^g (of genus g) to a manifold M:

$$\phi : \qquad \Sigma^g \longrightarrow M$$

---

[1] HUTP–94/A005, IASSNS-HEP-94/16

[2] Talk given in the *Geometry and Topology Conference*, April 93, Harvard, in honor of Raoul Bott.

Geometry, Topology, and Physics for Raoul Bott

Cambridge, MA 02138 USA

The integral over the space of maps is written abstractly as

$$\int D\phi e^{-S(\phi)} = F_g(M)$$

where $S(\phi)$ is the energy functional $S(\phi) = \int_{\Sigma^g} |d\phi|^2$ and $F_g(M)$ is called the genus g partition function of strings propagating on M. In most cases of interest in superstrings one considers supersymmetric sigma models which means that in addition to maps from the Riemann surface to M one considers fermionic fields which take their values in the tangent bundle to M and one modifies S in an appropriate way to obtain a supersymmetric theory.

We say that strings *cannot* distinguish between M and M', or $M \equiv M'$ if and only if[3]

$$F_g(M) = F_g(M')\lambda^{g-1} \quad \text{for all} \quad g$$

for some constant λ (playing the role of string coupling constant). The simplest example of mirror symmetry corresponds to choosing M to be a circle of circumference L and M' to be a circle of circumference $1/L$. This case is particularly easy because $F_g(M)$ can be explicitly computed (this is seldom the case in more interesting examples). Identifying the map ϕ from Σ^g to the circle with the coordinate on the circle, we have

$$\phi \sim \phi + L \qquad for \quad M_1$$

$$\phi \sim \phi + \frac{1}{L} \qquad for \quad M_2$$

The space of maps from the Riemann surface to the circle decomposes into infinitely many components depending on how the surface wraps around the circle. To be concrete, let a_i, b_i as i runs from 1 to g denote a canonical basis for $H_1(\Sigma^g)$, and let α^i and β^i be the corresponding harmonic one forms. Then for each set of integers (n_i, m_i) we get a component of the map of the surface to M characterized by the condition that

$$d\phi = (n_i\alpha^i + m_i\beta^i)L + d\tilde{\phi}$$

where $\tilde{\phi}$ is a single valued function on Σ. Expanding the action $S[\phi]$ for this component we get

$$S[\phi] \to L^2 \int (n\alpha + m\beta) \wedge *(n\alpha + m\beta) + S[\tilde{\phi}]$$

Now we can consider the L dependence of F_g. We will have to sum over all disconnected components of such maps. However the fact that for each component S decomposes in the way described above we see that

$$F_g(L) = (\sum_{n,m} exp[-\int (n\alpha + m\beta) \wedge *(n\alpha + m\beta)]) \cdot \int D\tilde{\phi} exp[S[\tilde{\phi}]]$$

[3] I am giving a definition of equivalence which is necessary but has not been proven to be sufficient, though I believe it is also sufficient. The strict definition of equivalence will have to include not just the partition function but all the correlation functions.

Note that the path integral we are left to perform is independent of n, m. Even though the path integral over $\tilde{\phi}$ can be computed easily the important aspect to emphasize in our case is its dependence on L. The only dependence on L comes from the constant maps $\tilde{\phi}$ where it gives a factor of L (the rest of it just gives us $(det'\Delta / \int \sqrt{g})^{-1/2}$ where Δ is the Laplacian on the Riemann surface). We thus immediately deduce that

$$\frac{F_g(L_1)}{F_g(L_2)} = \frac{L_1 \sum_{n,m} exp[-L_1^2 \int (n\alpha + m\beta) \wedge *(n\alpha + m\beta)]}{L_2 \sum_{n',m'} exp[-L_2^2 \int (n'\alpha + m'\beta) \wedge *(n'\alpha + m'\beta)]}$$

Note that if $L_2 = 1/L_1$ then this ratio is equal to $(1/L_1^2)^{g-1}$ by using Poisson resummation on $(n, m) \to (n', m')$. We have thus seen that

$$F_g(L) = F_g(1/L)\lambda^{g-1}$$

where $\lambda = 1/L^2$. This implies that two circles with circumferences which are inverse to each other are mirror pairs. One can easily extend this example to the target being a d-dimensional torus. This mirror symmetry has actually been used to give a model of string cosmology *[2]*. One of the main troubles of early universe cosmology for point particle theories is that the universe would be singular at the time of the big bang. This gives rise to many unphysical things such as infinite temperature and infinitely strong gravitational fields. In the context of strings a toroidal universe will not have these difficulties since it maps a singular universe (a torus with zero size) to an infinite size universe which is manifestly non-singular.

In the context of superstrings one usually considers target spaces being a d-complex dimensional Kähler manifold which admits a Ricci-flat metric. Such Kähler manifolds are known as Calabi-Yau manifolds (Kähler manifolds with trivial canonical line bundle which admit a nowhere vanishing holomorphic d-form). Of particular importance in studying sigma models on Calabi-Yau is the structure of the moduli space. There are two types of deformations to consider: complex deformations and Kähler deformations. The complex deformations of Calabi-Yau relevant for sigma models is the classical one (studied in *[3]*). The dimension of this moduli space is equal to the dimension of $H^1(M, T)$ which for Calabi-Yau manifolds is the Hodge number $h^{1,d-1}$. The Kähler deformations in the context of sigma models are more complicated than in the classical picture and in particular belong to the complexified $H^{1,1}(M)$. The real part plays the usual role of a Kähler class whereas the complex part plays the role of introducing a phase in the measure of the sigma model. The basic idea is that, if b denotes the complex part of the Kähler class, this means that we modify the path integral measure $D\phi$ by

$$D\phi \to D\phi \exp(i \int_\Sigma \phi^*(b))$$

where $\phi^*(b)$ is the pullback of the two form b to the Riemann surface. The reason it is called the complexified Kähler class is that it can be unified with the real part to write the action in a way which automatically includes the above twisting of the path integral measure. Let $\tilde{g}_{i\bar{j}} = g_{i\bar{j}} + ib_{i\bar{j}}$. Let ϕ^i denote the map ϕ in component form. Then the action can be written as

$$S = \int \tilde{g}_{i\bar{j}} \partial\phi^i \bar{\partial}\phi^{\bar{j}} + \tilde{g}^*_{i\bar{j}} \bar{\partial}\phi^i \partial\phi^{\bar{j}}$$

It often is convenient to formally think of $\tilde{g}$ and $\tilde{g}^*$ as independent parameters and take $\tilde{g}^* \to \infty$. In this case the second piece of the action blows up for generic maps and the path integral will be finite only for holomorphic maps, where the second piece vanishes. In this way the asymmetric limit of fixing g and sending $\tilde{g}^* \to \infty$ gives a path-integral which basically measures how many holomorphic maps there are from the Riemann surface to the Kähler manifold[4].

Let us consider a simple example of one dimensional Calabi-Yau manifolds. This is nothing but the elliptic curve. The moduli space of this manifold as far as the complex deformation is concerned is parametrized by one complex parameter τ on the upper half plane. One represents the torus in the standard fashion as a parallelogram with sides 1 and τ imbedded in the complex plane. To get the moduli space we have to recall that τ's differing by the $PSL(2, \mathbf{Z})$ action represent the same torus. As far as the Ricci-flat Kähler metrics are concerned, the area A of the flat torus is the only parameter. However, as discussed above this is complexified to $A - iB$. Let us define the complex parameter ρ parametrizing complexified Kähler deformations by

$$\rho = iA + B$$

Note that since B affects the path-integral only by multiplication by phases it is a periodic variable as far as the moduli space of the theory is concerned. So we have to identify

$$B \sim B + 1$$

which means that

$$\rho \sim \rho + 1$$

There is in fact a larger symmetry to mod out in order to get the moduli space: Let us consider a rectangular torus of size (R_1, R_2) on each side. Setting $B = 0$ this gives $\rho = iR_1R_2$. From the discussion about the duality of the circle, it should be clear that if we consider a rectangular torus with sides $(1/R_1, 1/R_2)$ we get an equivalent theory: This gives $\rho = i/(R_1R_2)$. So we see that in this case we have the following identification

$$\rho \sim \frac{-1}{\rho}$$

It can be shown that this applies to even more general configurations of the torus, and so the moduli space of Kähler deformations, is identified with ρ up to the group generated by these two transformations which magically enough is again $PSL(2, \mathbf{Z})$. We therefore see that τ and ρ play very similar roles in this example, even though one parametrizes the complex structure and the other parametrizes the Kähler structure. In fact, as it turns out, to get the full moduli space there is an

[4]Strictly speaking one needs to consider the topologically twisted sigma model to get this *[4]*.

extra symmetry to mod out and that is the $\tau \leftrightarrow \rho$ exchange. To see this consider the special case of the rectangular torus discussed above. To begin with we have

$$(R_1, R_2) \to \tau = \frac{iR_2}{R_1} \qquad \rho = iR_1R_2$$

We can consider an equivalent theory by taking the duality transformation on the first circle. This leads to

$$(\frac{1}{R_1}, R_2) \to \tau' = iR_2R_1 \qquad \rho' = \frac{iR_2}{R_1}$$

which means that there is an exchange symmetry of τ and ρ. This can be shown to be more generally valid for any torus and leads to the first example of the mirror phenomenon. Below we shall see that this is a special case of the general phenomenon of Calabi-Yau mirrors, where the role of complex and Kähler deformations are exchanged.

Even though the above example is very instructive it is a bit too special and one would like to find more non-trivial examples of the above phenomenon. It turns out there is at least a hint: In distinguishing manifolds we consider topological invariants; for example, the Euler characteristic χ. We conclude that

$$\chi(M_1) \neq \chi(M_2) \to M_1 \neq M_2$$

So we can easily disprove the existence of the isomorphism between M_1 and M_2 if their Euler characteristic is different. Similarly, for string theory we should try to construct *invariants of* 2*d QFT* in order to distinguish them. If two QFT's have different such invariants then they cannot be mirror pairs. These invariants are to be constructed out of objects which canonically make sense in the algebraic formulation of QFT's without any recourse to a target manifold. For supersymmetric QFT's, which are the ones which arise in considering superstrings propagating on Kähler manifolds, there is a basic invariant, which is Witten's index

$$I = Tr(-1)^F$$

where $(-1)^F$ is a mod 2 gradation of the Hilbert space of the QFT (by using the fermion number) and the trace is over the full Hilbert space. $(-1)^F$ is characterized by the fact that it squares to one and it anticommutes with the operator which implements the supersymmetry Q. However this makes $(-1)^F$ well defined only up to an overall sign change

$$(-1)^F \to -(-1)^F$$

We thus see that only $|I|$ is an invariant for the QFT. If two QFT's arising from sigma models on two manifolds have different $|I|$'s we can deduce that they cannot be mirror pairs. In the case of a supersymmetric sigma model on a manifold M it turns out that *[5]*

$$|I| = |\chi(M)|$$

We therefore deduce that if $|\chi(M_1)| \neq |\chi(M_2)|$ then M_1 and M_2 cannot be mirror pairs. This however leaves the door open for two manifolds to be topologically distinct and to have $\chi(M_1) = -\chi(M_2)$, but which are nevertheless mirror pairs; at least the index $|I|$ does not distinguish them. Note that this condition for finding mirror pairs fits very well with the example of the torus considered above because as we saw the torus was its own mirror and on the other hand the Euler characteristic of the torus is zero.

We could try to ask if there are analogs of more refined invariants such as the Betti numbers, or for the case of Calabi-Yau manifolds, which is of most interest to us, the Hodge number $h^{p,q}$, which can distinguish different sigma models. It turns out that $h^{p,q}$ can be defined as the dimension of a canonical subspace of the Hilbert space of the QFT (the ground state subsector with a particular $U(1) \times U(1)$ charge). But there is still an ambiguity. Let d be the complex dimension of the Calabi-Yau manifold. Define

$$\hat{p} = -\frac{d}{2} + p$$

Then there is a canonical $U(1) \times U(1)$ gradation of the Hilbert space of the corresponding QFT. Let $\mathcal{H}^{\hat{p},\hat{q}}$ denote the subspace of the Hilbert space consisting of ground states with $U(1) \times U(1)$ gradation given by $\hat{p}, \hat{q}$. Then

$$h^{p,q} = \dim \mathcal{H}^{\hat{p},\hat{q}}$$

However, again it turns out that there is a Z_2 ambiguity in canonically assigning the $U(1) \times U(1)$ gradation which is obtained by flipping the relative sign of the two $U(1)$'s: $(\hat{p}, \hat{q}) \rightarrow (-\hat{p}, \hat{q})$. This in particular means that the dimension of a given $\mathcal{H}^{\hat{p},\hat{q}}$ could in principle correspond either to $h^{p,q}$ or $h^{d-p,q}$. Just as before the existence of this ambiguity leaves the door open for having mirror pairs (M_1, M_2) for which in particular

$$h^{p,q}(M_1) = h^{d-p,q}(M_2)$$
$$\chi(M_1) = (-1)^d \chi(M_2)$$

It was conjectured based on the existence of simple toroidal examples and the lack of any method to resolve the Z_2 ambiguity in computing $h^{p,q}$ that this kind of mirror pairs always exist *[6]* .

Note that for a Calabi-Yau manifold the dimension of moduli space of complex deformations is given by $h^{d-1,1}$ (this is proven in *[3]*) and the Kähler deformations by $h^{1,1}$. Therefore for mirror pairs the two moduli spaces interchange their roles: The complex deformations for one correspond to Kähler deformations for the other and vice-versa.

The mirror conjecture as stated cannot be possibly true because there exist rigid Calabi-Yau manifolds with no complex deformations. So their mirrors will have no Kähler deformations, i.e., it will not even be a Kähler manifold. However, there is a generalization of the conjecture (involving supermanifolds) which can take care of such cases *[7]*.

An important element in relating ($N = 2$) supersymmetric 2d QFT's (and in particular conformal field theories) with the geometry of Calabi-Yau was Gepner's construction *[8]*. These were CFT's which were constructed by taking the tensor product of the simplest representations of the ($N = 2$) Virasoro algebra (minimal models) which had an unexpected similarity with what one would expect for CFT's coming from sigma models on certain Calabi-Yau manifolds. Subsequently it was discovered that the Landau-Ginzburg description of ($N = 2$) minimal models *[9]* associates an isolated singularity (taken as the superpotential of the Landau-Ginzburg model) to each minimal model. The simplest ones correspond to monomials x^n. Thus Gepner's construction was reinterpreted in this context as taking combinations of these monomials and constructing quasi-homogeneous polynomials to be identified with the defining equation of the associated Calabi-Yau manifold in weighted projective space *[10]* (see also the recent work *[11]*).

From this point on one could use amazing properties of conformal field theories to construct mirror pairs. In particular, if one considers the minimal model given by x^n, the theory has a Z_n symmetry given by multiplying x by an n-th root of unity. It turns out that if we divide out this Z_n symmetry we end up with the original theory again! This is somewhat unexpected as usually one expects that if we divide out a theory by a symmetry we should get a theory with fewer degrees of freedom. This is not the case for conformal theories because dividing by a symmetry gets rid of some states which are not invariant but at the same time introduces new states coming from loops which are closed only up to the group action. So the total degrees of freedom remain constant. That we end up with the same theory in this case can be seen by realizing that this model has an associated circle corresponding to phases of x, and modding out by Z_n changes its radius from R to $1/R$, and thus the previous argument applies to show that we end up with the original theory.

The first non-trivial examples of Calabi-Yau mirrors were constructed in *[12]* using this symmetry. In particular it was shown there that if we consider the quintic given by $\sum_{i=1}^{5} x_i^5 = 0$ in CP^4 and modding out by the maximal subgroup of Z_5 phase multiplications of each monomial consistent with preserving the holomorphic 3-form ($Z_5 \times Z_5 \times Z_5$) one gets the same theory back but now interpreted as a sigma model on a quotient of the quintic three-fold. There was also a large class of examples suggested by computation of the Euler class of Calabi-Yau threefolds (as hypersurfaces in weighted projective space) in which for almost all manifolds searched of Euler class χ there was one with Euler class $-\chi$ *[13]* . There has been further work, giving a conjectured construction for mirror pairs for a very wide class of examples *[14]*.

The existence of mirror pairs as discussed up to now seems like a pure curiosity as far as string theory is concerned. It simply points out that topological invariants are not necessarily good stringy invariants. However, what makes the existence of mirror pairs mathematically interesting is that certain computations of QFT correlation functions in one manifold which have one mathematical interpretation for one manifold have a completely different interpretation for the other. Therefore the existence of mirror symmetry implies in particular that two different mathematical computations on two distinct manifolds are unexpectedly equal. Moreover it turns out that on the one side the computation is simple and on the other side no one knows how to perform it! The mirror pairs thus serve in this context as a way to transform a difficult computation to an easy one.

The difficult computation is what is known as the computation of the quantum cohomology ring *[4]* , which is a deformation of the classical cohomology ring. The deformation parameters are in one to one correpsondence with Kähler deformations of the manifold. In the limit in which we take the size of the manifold to infinity the quantum cohomolgy ring reduces to the classical one. The simplest way to desribe it is to note that the classical cohomology ring can be described by counting the intersection *pionts* of dual cycles representing the cohomology classes. The quantum cohomlogy ring is simply a *blurred* version of classical intersection theory where the intersection *point* is replaced by a rational curve in the manifold. We say three cycles intersect if they all intersect a given rational curve. We weigh each such intersection by the number of points in which they intersect the rational curve and in addition by $\exp(-\int k)$ where k is the (complexified) Kähler class and it is integrated over the rational curve in question. Note that if we let $k \to \infty$ the rational curves are suppressed in this quantum intersection theory and so we are left with the degree zero rational curves which are just points, i.e. we get back the classical intersection theory.

The hard computation is thus a coholomolgy ring deformation which is a function of Kähler moduli. The general structure of mirror pairs discussed above suggests that the mirror computation should involve rings which are functions of complex moduli. It turns out that such objects have already been encountered in the mathematics literature under the general subject of studying variations of Hodge structure. For the particular case of the Calabi-Yau threefold the mirror of the quantum triple intersection defined above can be computed as follows: For each cycle dual to the Kähler class on one manifold there is a direction for the variation of complex structure on the mirror manifold. So, fixing three classes in the previous computation will correspond to fixing three directions for the variation of complex sturcture on the mirror, which we denote by (i, j, k). Let ω be a non-vanishing holomorphic 3-form of the Calabi-Yau manifold. The mirror computation is simply written as

$$C_{ijk} = \int \omega \wedge \partial_i \partial_j \partial_k \omega$$

This is easy to compute and so we have managed to transform a difficult computation of quantum triple intersection on one manifold to classical questions of variations of Hodge structure on the other manifold. The first explicit example was carried out in *[15]* for the quintic three fold and led to the impressive computation of the number of rational curves of arbitrary degree on the quintic.

One can continue this line of correspondence in the context of arbitrary genus Riemann surface. On the difficult side one has to compute the number of holomorphic maps from a genus g Riemann surface to the Calabi-Yau threefold; more precisely, the natural question is to allow the moduli of the Riemann surface to vary and ask how many holomorphic maps there are for some point in the moduli space of the Riemann surface. Actually, for high enough genus one expects not to get isolated holomorphic maps but families of them. In such cases the natural thing to compute is the Euler character of a certain bundle over the moduli space of holomorphic maps from Riemann surfaces of genus g to Calabi-Yau manifolds *[16]*.

This is the hard part of the question and we would like to discuss what is the mirror computation which is expected to be simple? This question has recently been answered *[17]* and we shall now summarize some of the results of this work.

At the level of the sphere the easy side of the computation corresponded to classical computations, which means ordinary integrals which arise even for point particle theories. In other words the sphere computation, which is the tree level for strings, leads to classical computations (ordinary integrals) which in principle could arise from point particle theories (the usual overlap of wave functions). Therefore it is natural to expect that the higher genus computations which are quantum corrections for string theory will be mapped to quantum corrections for the would be point particle theory. This expectation is borne out, and the corresponding point particle theory which quantizes the variation of Hodge structure is called the Kodaira-Spencer theory of gravity *[17]*.

The basic idea of this theory is to start with the classical equations of the theory which is the variation of complex structure of a Calabi-Yau threefold M. According to Kodaira-Spencer theory, the variation of the complex structure can be encoded by saying how $\bar{\partial}$ varies:

$$\bar{\partial} \rightarrow \bar{\partial} + A\partial$$

where A is a section of $TM \times \overline{T}^* M$, which I will denote, counting the vectors with negative weight as opposed to forms by a (-1,1) 'vector-form'. The Kodaira-Spencer equation, which is the consistency condition for the new $\bar{\partial}$ for a finite shift of complex structure holomorphic tangent bundle, is

$$\bar{\partial} A + \frac{1}{2}[A, A] = 0$$

where the commutator takes the commutator of A as a vector and wedges the antiholomorphic forms. To solve this equation we follow *[3]*. We use the notation that when we contract a vector-form with the holomorphic three form Ω we put a prime over the vector-form. Contracting A with Ω gives us a (2,1) form A'. One chooses variations A which in addition respect the three form, which means we choose A such that

$$\partial A' = 0$$

Then the lemma of *[3]* shows that in such a case

$$[A, A]' = \partial(A \wedge A)'$$

By contracting the KS equation with Ω (which is nowhere zero) we get the equivalent equation

$$\bar{\partial} A' + \frac{1}{2}\partial(A \wedge A)' = 0$$

For a physical realization, this has to be the classical equation of an action, which is indeed the case. The action is given by

$$S = \frac{1}{2}\int_M A'\frac{1}{\partial}\bar{\partial}A' + \frac{1}{6}\int_M A' \wedge (A \wedge A)'$$

This action including the condition $\partial A' = 0$ arises naturally in string field theory *[17]* . The action as written above is formal because of $\frac{1}{\partial}$ but it can be made precise sense of *[17]* . Tree level expansions of this theory naturally give rise to perturbative solutions of the KS equation which had been studied before *[3]* . The fact that the tree level computations of this theory leads to the computations which arise in the variations of Hodge structure is not surprising–after all the tree level computations are equivalent to classical equations, which in this case is the KS equation.

One should then go on to compute the quantum corrections for the above action. In such cases one typically encounters divergences which should be regularized. The first such case is the one-loop correction. For the Kodaira-Spencer action the one-loop correction turns out to be a certain particular combination of holomorphic Ray-Singer torsions. Let $\Delta_{p,q}$ be the Laplacian acting on (p,q) forms on M. The one-loop correction turns out to be[5]

$$F_1 = \frac{1}{2}\log \prod_{p,q} \det \Delta_{p,q}^{pq(-1)^{p+q}}$$

of course for such an object to make sense one needs to delete the zero modes and in addition it needs to be regularized. This kind of determinant has been rigorously studied in *[19]* and regularized using the zeta function regularization techniques[6]. In order to compute F_1 one generally studies how F_1 varies as a function of moduli of complex structures of the Calabi-Yau manifold. Formally one argues that F_1 is a sum of a holomorphic and an antiholomorphic function on moduli space. But as is well known there is an anomaly, the Quillen anomaly, which falsifies the formal argument. This anomaly is captured by studying the curvature $\partial\bar{\partial}F_1$. This can be computed either using string techniques *[20]* or the more conventional techniques *[19]* with the result that[7]

$$\partial\bar{\partial}F_1 = \frac{1}{2}\partial\bar{\partial}[\sum_{p,q}(-1)^{p+q}qd_{p,q}] + \frac{1}{2}\int_M Td(T)\sum_p(-1)^p pCh(\wedge^{n-p}T^*)\Big|_{(1,1)\text{-part}}$$

[5]The stringy version of this is $F_1 = \frac{1}{2}\int \frac{d^2\tau}{\tau_2}Tr(-1)^F F_L F_R q^{H_L}\bar{q}^{H_R}$ where the integral is over the moduli space of torus, q is the modular parameter, $H_{L,R}$ denote the left- and right- moving Hamiltonians, and $F_{L,R}$ denote the $U(1) \times U(1)$ left-right fermion number gradiation. This object was defined in *[18]* as a generalization of Ray-Singer torsion to the loop space of Kähler manifolds.

[6]To the best of my knowledge the particular combination of torsions appearing here which leads to a magically simple anomaly formula for the Calabi-Yau manifolds has not been studied in the mathematics literature.

[7]In the stringy computation the $\frac{1}{24}$ in the second term arises in computation as a result of the volume of the moduli space of torus whereas in the index derivation it appears in the expansion of todd class.

$$= \frac{1}{2}\partial\bar{\partial}[\sum_{p,q}(-1)^{p+q} q d_{p,q}] - \frac{\chi(M)}{24} G$$

where $d_{p,q}$ denotes the determinant of harmonic (p, q) forms (in a holomorphic basis), Td, Ch denote the Todd and Chern classes, T and T^* denote the holomorphic tangent and cotangent bundles, $\chi(M)$ denotes the Euler characteristic of M, and G is the Kähler form on moduli space (which is the same as $c_1(\mathcal{L})$, where $\mathcal{L}$ is the line bundle over the moduli space whose section is a holomorphic 3-form). In evaluating the index integral above use has been made of the fact that Calabi-Yau manifolds have vanishing first Chern class. Integrating the anomaly equation to get F_1 is now an easy task with the additional boundary condition of how F_1 behaves near the boundaries of moduli space. It should be noted that F_1 defined above makes sense for CY manifolds of arbitrary dimension.

Let us consider the simplest example of the mirror phenomenon, namely the elliptic curve, and compute F_1 in either language. On the easy side of the computation as we discussed above we are going to get the analytic torsion which is simply

$$F_1(q) = -\log\sqrt{\tau_2}\eta(q)\overline{\eta(q)}$$

where $q = \exp(2\pi i \tau)$ specifies the complex structure of the torus and η is the Dedekind η function given by

$$\eta = q^{\frac{1}{24}} \prod_{n=1}^{\infty}(1 - q^n).$$

Of course the easy way to show this is to study the Quillen anomaly as discussed above: Applying the analysis to this case is rather simple and we end up with the anomaly equation that

$$\partial_\tau \bar{\partial}_{\bar{\tau}} F_1 = \frac{-1}{2(\tau - \bar{\tau})^2}.$$

Modular invariance and regularity in the interior gives rise to the η function as we integrate the above anomaly equation. On the 'difficult' mirror to this, as we mentioned we would expect to be counting the number of holomorphic curves from the torus to the manifold which in this case is again a torus where as discussed before now the role of τ and ρ are exchanged–in other words we now think of τ as parametrizing the complexified Kähler class of the torus. Fixing a point on the torus and mapping it to a particular point on the target torus (dual to the Kähler class) is equivalent to studying $q\partial F_1/\partial q$. Note that in the dictionary for the mirror map q counts the degree of the holomorphic map to the manifold, so the coefficient in front of q^n signifies the number of maps of degree n. However, the existence of the anomaly means that we have an inevitable mixture of q and $\bar{q}$. To simply count the holomorphic curves we should consider an asymmetric limit fixing g but taking $g^* \to \infty$ which concentrates the path integral on holomorphic maps as discussed before. In this case this means keeping τ fixed and sending $\bar{\tau} \to \infty$.

We thus have (expanding the logarithm of the Dedekind η function)

$$q\frac{\partial F_1}{\partial q}\Big|_{\bar{\tau}\to\infty} = \frac{-1}{4\pi i(\tau-\bar{\tau})}\Big|_{\bar{\tau}\to\infty} + \sum_{n,m=1}^{\infty} nq^{nm} - \frac{1}{24} = \sum_{n,m=1}^{\infty} nq^{nm} - \frac{1}{24}$$

Let us check whether this is what we expect by directly counting the number of holomorphic maps from torus to torus: At degree zero, every elliptic curve can be mapped holomorphically to the torus by the constant map. So at degree zero we have a continuous family parametrized by the moduli space of tori. The physics computes (up to sign) roughly the volume of this space, which in this case is $\frac{1}{24}$. The coefficients in front of q^{nm} should signify the number of holomorphic curves which cover the target torus nm times. Let us start with $n = m = 1$. For a fixed target torus, there is clearly a unique torus for which there exists a holomorphic map, namely the torus with exactly the same modulus as the target torus and with the map being the identity map. This explains the coefficient of q being 1. We can consider an nm-fold covering of this torus, by going n times covering in one direction and m times covering in another direction. By tilting the parallelogram representing the torus in one direction we can actually get n inequivalent such tori. One can show by a careful check that up to $SL(2, Z)$ transformation there are no more (in particular the number of holomorphic maps of degree r is equal to $\sum_{n \text{ divides } r} n$) in agreement with the above formula derived by using mirror symmertry.

We can also ask how higher quantum corrections can be computed. As remarked before, higher loop corrections correspond to maps from higher genus Riemann surfaces to the Calabi-Yau manifold. It is at this point that threefolds play a very distinguished role, in that the quantum corrections vanish for all Calabi-Yau n-folds except for $n = 3$. On the 'easy side', which for higher loops are not so easy, one is computing the quantum corrections to the Kodaira-Spencer theory. This will have to be studied very carefully with an eye on regularization of the theory. On the hard side, one is computing the number of holomorphic maps (or the Euler characteristic of an appropriate bundle on the moduli space of holomorphic maps) from genus g Riemann surface to the three fold.

So the question is how one does the computation? The method followed in *[17]* makes use of the fact that the answer is expected to be essentially a holomorphic function of moduli (or more precisely a holomorphic section of an appropriate line bundle on moduli space). If it were exactly holomorphic, to completely fix it one would only need to know its behavior near the boundaries of moduli space which in general would require only finite data. However the holomorphicity is not exactly right at higher genus just as it was not exactly right for genus one, where one encountered Quillen's anomaly. In the genus one case, to get a well defined answer, as mentioned above we have to take the holomorphic derivative of F_1 in the direction of moduli and the Quillen anomaly is the statement that this function is not a holomorphic function of moduli. For higher genus F_g is well defined without having to take any derivatives, because higher genus does not have any isometries, so the anomaly is going to be captured by a statement involving $\bar{\partial} F_g \neq 0$. In order to compute this anomaly, one has in principle two options: Either use the Kodaira-Spencer theory directly and study its anomaly at higher loops (which is the generalization of what has been studied in the case of one loop in *[19]*), or study it directly in string theory where F_g is defined as the integral of a particular measure on the moduli space of Riemann surfaces of genus g. Since the regularization of Kodaira-Spencer theory has not been studied at higher loops the first approach seems difficult at present. It turns out that the second approach, i.e. using string theory techniques, is extremely easy. Even if the regularization of Kodaira-Spencer theory had been studied it is hard to believe that the derivation of the anomaly would have been as simple as the stringy derivation. In the context of string theory one finds that

$$\bar{\partial}_i F_g = \int_{\mathcal{M}_g} \partial \bar{\partial} \rho_{\bar{i}}$$

where $\mathcal{M}_g$ is the moduli space of Riemann surfaces of genus g, and $\rho_{\bar{i}}$ is a well defined form on moduli space. Using this structure the computation reduces to the contributions from the boundary of moduli space of Riemann surfaces. Naturally the objects one encounters there would be related to lower genus computations. The answer one obtains turns out to be

$$\partial_{\bar{i}} F_g = \frac{1}{2} \overline{C}_{\bar{i}\bar{j}\bar{k}} g^{j\bar{j}} g^{k\bar{k}} \left(D_j D_k F_{g-1} + \sum_{r=1}^{g-1} D_j F_r \, D_k F_{g-r} \right)$$

The two boundary contributions on the right hand side come from the two distinct type of degenerations of the Riemann surface (from the handle degeneration and splitting of the surface, respectively). Without going to too much detail let us just note that F_g is a section of $\mathcal{L}^{2-2g}$ where $\mathcal{L}$ is the line bundle of $H^{3,0}(M)$ on the moduli space, and D_i represent covariant derivatives with respect to the natural connection and $g_{i\bar{j}} = \int_M \omega_i \wedge \overline{\omega}_j$ represent the canonical metric on the $(2,1)$ forms ω_i. This anomaly equation can be naturally captured in the master form including all genera at once by considering $Z = \exp \sum_{g=1}^{\infty} \lambda^{2g-2} F_g$ and we have

$$[\partial_{\bar{i}} - \bar{\partial}_{\bar{i}} F_1 - \frac{\lambda^2}{2} \overline{C}_{\bar{i}\bar{j}\bar{k}} g^{j\bar{j}} g^{k\bar{k}} D_j D_k] Z = 0$$

So the basic strategy to compute F_g at higher genus is by induction. Suppose we know F_g up to a given genus. To get the next one, we integrate the above anomaly formula using the lower genus F_g and the behaviour of F_g near the boundaries of moduli space. Using this strategy the case of the quintic threefold was studied in *[17]* with the result indicated in Table 1.

In Table 1 we have included also the genus 0 answer which was computed in *[15]* and the genus 1 result which was computed using the anomaly in *[20]* . As far as the numbers which have been checked mathematically, up to now the genus 0 answer up to degree 3 has been confirmed. For higher genus, the direct computations are even more difficult. The only non-trivial one which has been

Degree	$g = 0$	$g = 1$
n=0	5	50/12
n=1	2875	0
n=2	609250	0
n=3	317206375	609250
n=4	242467530000	3721431625
n=5	229305888887625	12129909700200
n=6	248249742118022000	31147299732677250
n=7	295091050570845659250	71578406022880761750
n=8	375632160937476603550000	154990541752957846986500
n=9	503840510416985243645106250	324064464310279585656399500
...	...	...
large n	$a_0 n^{-3}(\log n)^{-2} e^{2\pi n\alpha}$	$a_1 n^{-1} e^{2\pi n\alpha}$

Degree	$g = 2$	g
n=0	-5/144	$-100 \cdot [c_{g-1}^3]$
n=1	0	
n=2	0	
n=3	0	
n=4	534750	
n=5	75478987900	
n=6	871708139638250	
n=7	5185462556617269625	
n=8	90067364252423675345000	
n=9	325859687147358266010240500	
...	...	...
large n	$a_2 n(\log n)^2 e^{2\pi n\alpha}$	$a_g n^{2g-3}(\log n)^{2g-2} e^{2\pi n\alpha}$

Table 1: # curves of genus g on quintic hypersurface

confirmed is by Stromme et. al. *[21]* (using a formula of Bott) and it is for the $n = 4$ for $g = 1$ (the $n = 3$ for $g = 1$ can be easily argued to be equal to $n = 2$ for $g = 0$ which agrees with the physical computation). For $g > 2$ we have determined only the asymptotic behavior of the number of holomorphic curves (to fix all the numbers we have to have a little bit more precise knowledge about the finite number of coefficients which dominate the divergence behavior of F_g near the boundaries of moduli space of the quintic and fixes the holomorphic ambiguity in integrating the anomaly equation).

Mirror symmetry is a reflection of the rich structure in the loop space of Calabi-Yau manifolds in that they give rise to 2 dimensional conformal theories with $N = 2$ supersymmetry. As is well known these are only very special examples of $N = 2$ superconformal theories, many of which have no target manifold interpretation. We have learned from the existence of mirror symmetry that classical concepts of invariants are to be modified when going to loop space, and in fact we are led to ask more generally what are the invariants of the $N = 2$ conformal field theories which arise for CY sigma models. Having a good understanding of these invariants will be a necessary tool in classifying them. So far only very limited set of invariants of the $N = 2$ theories have been constructed. There could very well be more refined invariants (by invariants we mean aspects of the theory which are unaffected by moving in the moduli space of these theories). In particular, defining invariants over integers which is natural for manifolds has not yet been defined for the $N = 2$ CFT's. This is a challenging and important area to develop.

I would like to thank the hospitality of the Institute for Advanced Study at Princeton where this paper was written.

This research is supported in part by a fellowship from Packard foundation and NSF grants PHY-89-57162 and PHY-92-18167.

References

[1] Contributions to *Essays on Mirror Manifolds*, ed. by S. T. Yau, International Press, 1992.

[2] R. Brandenberger and C. Vafa, Nucl. Phys. **B316** (1989), 391; A. Tseytlin and C. Vafa, Nucl. Phys. **B372** (1992), 443.

[3] G. Tian, in *Mathematical aspects of String theory*, ed. by S. T. Yau, World Scientific, Singapore, 1987 ; A.N. Todorov, Comm. Math. Phys. **126** (1989), 325.

[4] E. Witten, Comm. Math. Phys. **117** (1988), 353; Nucl. Phys. **B340** (1990), 281.

[5] E. Witten, Nucl. Phys. **B202** (1982), 253.

[6] W. Lerche, C. Vafa and N. P. Warner, Nucl. Phys. **B324** (1989), 427; L. Dixon in *Proc. of the 1987 ICTP Summer Workshop in High Energy Physics and Cosmology, Trieste.*

[7] S. Sethi and C. Vafa, in preparation.

[8] D. Gepner, Phys. Lett. *B199* (1987), 380.

[9] C. Vafa and N. Warner, Phys. Lett. **B218** (1989), 51; E. Martinec, Phys. Lett. **B217** (1989), 431.

[10] B. Greene, C. Vafa and N. Warner, Nucl. Phys. **B324** (1989), 371.

[11] E. Witten, Nucl. Phys. **B403** (1993), 159.

[12] B. R. Greene and M. R. Plesser, Nucl. Phys. **B338** (1990), 15.

[13] P. Candelas, M. Lynker, R. Schimmrigk, Nucl. Phys. **B341** (1990), 383.

[14] V.V. Batyrev, Dual polyhedra and mirror symmetry for Calabi-Yau hypersurfaces in toric varieties, J. Alg. Geom. **3**, 493–535; V.V. Batyrev and D. van Straten, Generalized hypergeometric functions and rational curves on Calabi-Yau complete intersections on toric varieties, preprint, 1993; L.A. Borisov, Towards the mirror symmetry for Calabi-Yau complete intersection in Gorenstein toric fano varieties, preprint, 1993 (alg-geom 10000193).

[15] P. Candelas, Xenia C. de la Ossa, Paul S. Green and L. Parkes, Nucl. Phys. **B359** (1991), 21.

[16] E. Witten, Nucl. Phys. **B371** (1992) 191.

[17] M. Bershadsky, S. Cecotti, H. Ooguri and C. Vafa, Harvard University preprint, HUTP-93/A025.

[18] S. Cecotti and C. Vafa, Comm. Math. Phys. **157** (1993), 139.

[19] J.M. Bismut and D.S. Freed, Comm. Math. Phys. **106** (1986), 159 and **107** (1986), 103; J.M. Bismut, H. Gillet and C. Soule, Comm. Math. Phys. **115** (1988), 49, 79, 301; J.M. Bismut and K. Köhler, Higher analytic torsion forms for direct images and anomaly formulas, Univ. de Paris-sud, preprint 91-58.

[20] M. Bershadsky, S. Cecotti, H. Ooguri and C. Vafa, Nucl. Phys. **B405** (1993), 279.

[21] Private communication.

The Verlinde Algebra and the Cohomology of the Grassmannian

Edward Witten [1]
School of Natural Sciences
Institute for Advanced Study
Princeton, NJ 08540

Abstract: The article is devoted to a quantum field theory explanation of the relationship between the Verlinde algebra of the group $U(k)$ at level $N-k$ and the "quantum" cohomology of the Grassmannian of complex k planes in N space. In §2, I explain the relation between the Verlinde algebra and the gauged WZW model of G/G; in §3, I describe the quantum cohomology and its origin in a quantum field theory; and in §4, I present a path integral argument for mapping between them.

1 Introduction

My main goal in these lecture notes will be to elucidate a formula of Doron Gepner *[1]*, which relates two mathematical objects, one rather old and one rather new. Along the way we will consider a few other matters as well.

The old structure is the cohomology ring of the Grassmannian $G(k, N)$ of complex k planes in N space – except that one considers the quantum cohomology (or Floer instanton homology) rather than the classical cohomology. The new structure is the Verlinde algebra, which computes the Hilbert polynomial of the moduli space of vector bundles on a curve. Gepner's formula, as we will consider it here, [2] says that the quantum cohomology ring of $G(k, N)$ coincides with the Verlinde algebra of the group $U(k)$ essentially at level $N-k$.[3]

Gepner discovered his formula by computing the left and right hand side and observing that they were equal. We will seek a more conceptual explanation, by representing the quantum cohomology ring of the Grassmannian in a quantum field theory and reducing that quantum field theory at low energies to another quantum field theory which is known to compute the Verlinde algebra.

[1] Research supported in part by NSF Grant PHY92-45317.

[2] Gepner actually discusses the classical cohomology of the Grassmannian and identifies it with a close cousin of the Verlinde algebra. The refinement of Gepner's formula that we will consider was conjectured by Vafa *[2]* and Intriligator *[3]*.

[3] Actually if one decomposes the Lie algebra of $U(k)$ as $su(k) \times u(1)$, then the level is $(N-k, N)$, that is level $N-k$ for the $su(k)$ factor and level N for the $u(1)$ factor. The source of this subtlety will become clear in §4.6.

Geometry, Topology, and Physics for Raoul Bott

Cambridge, MA 02138 USA

The Verlinde formula appears in several quantum field theories. Of these, the one that is relevant here is the gauged WZW model, of G/G. The shortest and most complete explanation of its relation to the Verlinde formula is due to Gerasimov *[4]*, and I will explain his argument in §2. Part of the charm of the G/G model is that it can be abelianized, that is, reduced to a theory in which the gauge group is the maximal torus of G, extended by the Weyl group *[5]*. The argument is simple in concept and will be summarized in §2.6.

In §3, I explain at a qualitative level how the quantum cohomology of the Grassmannian is represented in a quantum field theory, and some general techniques for studying this field theory and reducing it to a problem in gauge theory. In §4, I describe the arguments in more technical detail. The analysis actually should be adaptable to other manifolds that can be realized as symplectic quotients of linear spaces, such as flag manifolds and toric varieties. (The quantum cohomology of a toric variety has been studied by Batyrev *[6]*; that of a general flag manifold has apparently not yet been studied.)

This paper, despite its length, is based on an idea that can be described very simply. The two dimensional supersymmetric sigma model with target $G(k, N)$ can be described as a $U(k)$ gauge theory (in $\mathcal{N} = 2$ superspace) with N multiplets of chiral superfields in the fundamental representation of $U(k)$. It was studied from this point of view in the case of $k = 1$ (that is $\mathbf{CP}^{N-1}$) many years ago *[7, 8]*, and the generalization to arbitrary k is also familiar *[9, 10]*. At low energy, a suitable $U(k)$ gauge theory with the matter content just stated reduces to the supersymmetric sigma model of the Grassmannian. On the other hand, integrating out the N matter multiplets, one gets an effective action for the $U(k)$ gauge multiplet. Because of a sort of mixing between scalars and vectors, this low energy effective action has no massless particles; this is how the presence of a mass gap has been shown in the past. The novelty in the present paper is simply the observation that the low energy effective action is in fact a gauged WZW model of $U(k)/U(k)$. Under this low energy reduction, the topological correlation functions of the sigma model – which compute the quantum cohomology of $G(k, N)$ – are mapped into correlation functions of the $U(k)/U(k)$ model that can be computed (as we recall in §2) in terms of the Verlinde algebra. This gives the map between the two theories.

The quantum cohomology of the Grassmannian has also been studied – using, more or less, a classical version of the same setup we will follow – by Bertram, Daskalopoulos, and Wentworth *[11]*. And there is a forthcoming mathematical approach to Gepner's formula in work of Braam and Agnihorti. The cohomology of the Grassmannian is closely related to the chiral ring of a certain $\mathcal{N} = 2$ superconformal field theory *[12]* (somewhat misleadingly called a $U(N)/U(k) \times U(N-k)$ coset model); this model probably should be included in the story, but that will not be done here. Some of the phenomena we will study have analogs for real and symplectic Grassmannians, as in *[13]* and the second paper cited in *[1]*; it would be interesting to try to extend the analysis for those cases.

§2 and §3 can be read independently of one another. §4 requires more familiarity with methods of physics than either §2 or §3. Physicists may want to start with §4.

2 The Verlinde Formula And The G/G Model

First of all, the Verlinde algebra counts theta functions, such as the classical theta functions of Jacobi and their generalizations. In modern language, the classical theta functions can be described as follows. Let $\mathcal{T}$ be a complex torus of dimensions g, $\mathcal{L}$ a line bundle defining a principal polarization, and s a positive integer. Then the space of level s theta functions is $H^0(\mathcal{T}, \mathcal{L}^{\otimes s})$. The dimension of this space can be readily determined from the Riemann-Roch theorem.

For example, $\mathcal{T}$ might be the Jacobian $\mathcal{J}$ of a complex Riemann surface Σ, that is, the moduli space of holomorphic line bundles over Σ of some given degree. This example suggests the generalization to the "non-abelian theta functions" of A. Weil. Here one replaces the Jacobian of Σ by the moduli space $\mathcal{R}$ of rank k (stable) holomorphic vector bundles over Σ; now a "non-abelian theta function" at level s is an element of $H^0(\mathcal{R}, \mathcal{L}^{\otimes s})$. Though the Riemann-Roch theorem gives a formula for the dimension of this space, this formula is difficult to use in the non-abelian case as it involves invariants of $\mathcal{R}$ that are not easy to determine directly.

The Verlinde algebra gives on the other hand a practical formula for the dimension of $H^0(\mathcal{R}, \mathcal{L}^{\otimes s})$; the formula was described very explicitly by Raoul Bott in *[14]*. Roughly speaking, the origin of the Verlinde formula in differential geometry is as follows. Via Hodge theory, $\mathcal{R}$ is endowed with a natural Kähler metric. As the dimension of the space of non-abelian theta functions is independent of the complex structure of Σ, one can choose the complex structure to simplify the problem. It is convenient to take Σ to be a nearly degenerate surface consisting of three-holed spheres joined by long tubes. Using the behavior of the differential geometry of $\mathcal{R}$ in this limit, one can write $H^0(\mathcal{R}, \mathcal{L}^{\otimes s})$ as a sum of tensor products of similar spaces for a three-holed sphere with some branching around the holes. The Verlinde algebra encodes the details of this.

The Verlinde algebra arises in several quantum field theories:

(1) It originally arose *[15]* in the WZW model, a conformal field theory (whose Lagrangian we will recall later) that governs maps from a Riemann surface Σ to a compact Lie group G. Mathematically, this model is related to representations of affine Lie algebras, the unitary action of the modular group $SL(2, \mathbf{Z})$ on their characters, etc.

(2) The Verlinde formula is an important ingredient in understanding Chern-Simons gauge theory on a three-manifold. Thus it is relevant to the knot and three-manifold invariants constructed from quantum field theory.

(3) The Verlinde formula also enters in the gauged WZW model, governing a pair (g, A), where A is a connection on a principal G bundle P over a Riemann surface, and g is a section of $P \times_G G$ (where G acts on itself via the adjoint action).

Of these it is the third – which was discovered most recently – that will enter our story. The present section is therefore mainly devoted to an explanation – following Gerasimov *[4]*, who reinterpreted earlier formulas *[16, 17]* – of the gauged WZW model and its relation to non-abelian theta functions.

2.1 Gauge Theory And The Prequantum Line Bundle In Two Dimensions

Let G be a compact Lie group, Σ a closed oriented two-manifold without boundary, and P a principal G bundle over Σ. To achieve some minor simplifications in the exposition, I will suppose G simple, connected, and simply connected. (Notation aside, the only novelty required to treat a general compact Lie group is that more care is required in defining the functional $\Gamma(g, A)$ that appears below; (see *[18]*, §4.)) One consequence of the assumption about G is that P is trivial.

Let $\mathcal{A}$ be the space of connections on P. $\mathcal{A}$ has a natural symplectic structure ω that can be defined with no choice of metric or complex structure on Σ. (This and some other facts that I summarize presently are originally due to Atiyah and Bott *[20]*.) The symplectic structure can be defined by the formula

$$\omega(a_1, a_2) = \frac{1}{2\pi} \int_\Sigma \mathrm{Tr}\, a_1 \wedge a_2, \tag{2.1}$$

where a_1 and a_2 are adjoint-valued one-forms representing tangent vectors to $\mathcal{A}$. Here Tr is an invariant quadratic form on the Lie algebra of G, defined for $G = SU(k)$ to be the trace in the k dimensional representation; in general one can take Tr to be the smallest positive multiple of the trace in the adjoint representation such that the differential form Θ introduced below has periods that are multiples of 2π.

A prequantum line bundle $\mathcal{L}$ over $\mathcal{A}$ is a unitary line bundle with a connection of curvature $-i\omega$. $\mathcal{L}$ exists and is unique up to isomorphism since $\mathcal{A}$ is an affine space. We can take $\mathcal{L}$ to be the trivial bundle with a connection defined by the following formula:

$$\frac{D}{DA_i} = \frac{\delta}{\delta A_i} + \frac{i}{4\pi}\epsilon^{ij} A_j. \tag{2.2}$$

(ϵ^{ij} is the Levi-Civita antisymmetric tensor; when local complex coordinates are introduced, we will take $\epsilon^{z\bar{z}} = -\epsilon^{\bar{z}z} = i$.) The kth power $\mathcal{L}^{\otimes k}$ is therefore the trivial bundle endowed with the connection

$$\frac{D}{DA_i} = \frac{\delta}{\delta A_i} + \frac{ik}{4\pi}\epsilon^{ij} A_j. \tag{2.3}$$

Let $\widehat{G}$ be the group of gauge transformations. If P is trivialized, a gauge transformation is a map $g : \Sigma \to G$, and acts on the connection A by

$$A \to A^g = gAg^{-1} - dg \cdot g^{-1}. \tag{2.4}$$

More invariantly, g is a section of $P \times_G G$, where G acts on itself in the adjoint representation, and the action of g on $\mathcal{A}$ should be written as

$$d_A \to g d_A g^{-1}, \tag{2.5}$$

with d_A the gauge-covariant extension of the exterior derivative. At the Lie algebra level this is

$$A \to A - d_A \alpha, \tag{2.6}$$

where α is a section of $P \times_G \mathbf{g}$, with $\mathbf{g}$ being the Lie algebra of G, on which G acts by the adjoint action.

The action of the gauge group on the space $\mathcal{A}$ of connections lifts to an action on the prequantum line bundle. At the Lie algebra level, the lift is generated by the operators

$$D_i \frac{D}{DA_i} - \frac{ik}{4\pi} \epsilon^{ij} F_{ij}, \tag{2.7}$$

with $F = dA + A \wedge A$ the curvature form. (2.7) means very concretely that the infinitesimal gauge transformation (2.6) is represented on sections of $\mathcal{L}$ by the operator

$$\int_\Sigma \mathrm{Tr}\, \alpha \left(D_i \frac{D}{DA_i} - \frac{ik}{4\pi} F \right). \tag{2.8}$$

Even globally, at the group level, the $\widehat{G}$ action on $\mathcal{L}$ can be described rather explicitly *[21]*. Pick a three manifold B with $\partial B = \Sigma$. Given $g : \Sigma \to G$, extend g to a map (which I will also call g) from B to G. (The extension exists because of our assumption that $\pi_0(G) = \pi_1(G) = 0$. See *[19, 18]* for the definition of Γ without this simplifying assumption.) Define as in *[22]*

$$\Gamma(g) = \frac{1}{12\pi} \int_B \mathrm{Tr}\, g^{-1} dg \wedge g^{-1} dg \wedge g^{-1} dg, \tag{2.9}$$

which is known as the Wess-Zumino anomaly functional *[23]*. This is equivalent to

$$\Gamma(g) = \int_B g^*(\Theta), \tag{2.10}$$

where

$$\Theta = \frac{1}{12\pi} \mathrm{Tr}\, g^{-1} dg \wedge g^{-1} dg \wedge g^{-1} dg \tag{2.11}$$

is a left- and right-invariant closed three-form on G. The periods of Θ are multiples of 2π. This ensures that, regarded as a map to $\mathbf{R}/2\pi\mathbf{Z}$, $\Gamma(g)$ depends only on the restriction of g to Σ. Note that Γ is defined purely in differential topology; no metric or complex structure on Σ is required.

It follows rather directly from the definition of Γ that for $g, h : \Sigma \to G$,

$$\Gamma(gh) = \Gamma(g) + \Gamma(h) - \frac{1}{4\pi} \int_\Sigma \mathrm{Tr}\, g^{-1} dg \wedge dh \cdot h^{-1}. \tag{2.12}$$

A variant of this equation is called the Polyakov-Wiegmann formula *[24]*.

Now, given a connection A on P, set

$$W(g, A) = \Gamma(g) - \frac{1}{4\pi} \int_\Sigma \mathrm{Tr}\, A \wedge g^{-1} dg. \tag{2.13}$$

From (2.12), it follows almost immediately that

$$W(gh, A) = W(g, A^h) + W(h, A). \tag{2.14}$$

From this we can define an action of the gauge group $\widehat{G}$ on the space of functions of A. In fact, setting

$$g^*\chi(A) = \exp(ikW(g, A)) \cdot \chi(A^g), \tag{2.15}$$

we have $(gh)^* = h^*g^*$. Differentiating (2.15) with respect to g at $g = 1$, one sees that this particular lift induces (2.7) at the Lie algebra level; so (2.15) is the desired lifting of the action of the gauge group to an action on the prequantum line bundle $\mathcal{L}$.

2.2 Non-Abelian Theta Functions

So far we have considered Σ simply as a closed, oriented two-manifold without boundary. If one picks a complex structure on Σ, some additional interesting constructions can be made *[20]*. A complex structure on Σ induces a complex structure on the space $\mathcal{A}$ of connections. One simply declares that the $(0, 1)$ part of A is holomorphic and the $(1, 0)$ part is antiholomorphic. If $z, \overline{z}$ are local complex coordinates on Σ, then the connection (2.3) characterizing $\mathcal{L}^{\otimes k}$ can be written

$$\begin{aligned} \frac{D}{DA_z} &= \frac{\delta}{\delta A_z} - \frac{k}{4\pi} A_{\overline{z}} \\ \frac{D}{DA_{\overline{z}}} &= \frac{\delta}{\delta A_{\overline{z}}} + \frac{k}{4\pi} A_z . \end{aligned} \tag{2.16}$$

The complex structure on $\mathcal{A}$ can be described in very down-to-earth terms by saying that a holomorphic function on $\mathcal{A}$ is a function annihilated by $\delta/\delta A_z$. Correspondingly, a holomorphic section of $\mathcal{L}^{\otimes k}$ is a section annihilated by D/DA_z. Even more explicitly, a holomorphic section of $\mathcal{L}^{\otimes k}$ is a function $\chi(A_z, A_{\overline{z}})$ which can be written

$$\chi(A_z, A_{\overline{z}}) = \exp\left(\frac{k}{4\pi}\int_\Sigma \operatorname{Tr} A_z A_{\overline{z}}\right) \cdot \widehat{\chi}(A_{\overline{z}}), \tag{2.17}$$

with $\widehat{\chi}(A_{\overline{z}})$ an ordinary holomorphic function on $\mathcal{A}$.

Once a complex structure is picked on Σ, the connection A determines operators $\overline{\partial}_A$ giving complex structures to vector bundles $P \times_G \mathbf{r}$, with $\mathbf{r}$ a representation of G. The action of gauge transformations on A can be described by the action on the $\overline{\partial}_A$ operators:

$$\overline{\partial}_A \to g \cdot \overline{\partial}_A \cdot g^{-1}. \tag{2.18}$$

Since this formula makes sense for complex g, the $\widehat{G}$ action on $\mathcal{A}$ extends to an action of the complexified gauge group $\widehat{G}_{\mathbf{C}}$ (consisting of maps of Σ to the complexification $G_{\mathbf{C}}$ of G).

Two $\overline{\partial}_A$ operators define equivalent holomorphic bundles if and only if they are related as in (2.18). So the quotient $\mathcal{A}/G_{\mathbf{C}}$ (in case the $G_{\mathbf{C}}$ action is not free, the quotient must be taken in the sense of geometric invariant theory) is the same as the moduli space $\mathcal{R}$ of (stable) holomorphic principal G bundles over Σ.

The formulas used to describe the lift of the $\widehat{G}$ action to $\mathcal{L}$ make sense when g is complex, so we get a lift of the $\widehat{G}_{\mathbf{C}}$ action to $\mathcal{L}$. One defines a line bundle over $\mathcal{R}$ – which we will also call $\mathcal{L}$ – whose sections over an open set $U \subset \mathcal{R}$ are the same as the $\widehat{G}_{\mathbf{C}}$-invariant sections of $\mathcal{L}$ over the inverse image of U in $\mathcal{A}$.

Non-Abelian Theta Functions

The space of non-abelian theta functions, at level k, is $H^0(\mathcal{R}, \mathcal{L}^{\otimes k})$. From what has just been said, this is the same as the $\widehat{G}$-invariant (or equivalently, $\widehat{G}_{\mathbf{C}}$-invariant) subspace $\mathcal{H}^{\widehat{G}}$ of $\mathcal{H} = H^0(\mathcal{A}, \mathcal{L}^{\otimes k})$. We want to determine the dimension of $\mathcal{H}^{\widehat{G}}$.

The strategy, as in *[4]*, will be as follows. We will find a very convenient description of the action of $\widehat{G}$ on $\mathcal{H}$. In fact, for $g \in \widehat{G}$, we will find an explicit integral kernel $K(A, B; g)$ (with $A, B \in \mathcal{A}$) such that for $\chi \in \mathcal{H}$,

$$g^*\chi(A) = \int DB \;\; K(A, B; g)\chi(B). \tag{2.19}$$

(For fixed g, $K(A, B; g)$ is a section of $p_1^*(\mathcal{L}^{\otimes k}) \otimes p_2^*(\mathcal{L}^{\otimes(-k)})$ over $\mathcal{A} \times \mathcal{A}$, with p_1 and p_2 being the two projections to $\mathcal{A}$.) Here DB is the natural symplectic measure, normalized in a way that will be specified in §2.4, on the symplectic manifold $\mathcal{A}$.

Now the projection operator $\Pi : \mathcal{H} \to \mathcal{H}^{\widehat{G}}$ can be written

$$\Pi = \frac{1}{\mathrm{vol}(\widehat{G})} \int_{\widehat{G}} Dg \;\, g^*. \tag{2.20}$$

Here formally Dg is a Haar measure on $\widehat{G}$ and $\mathrm{vol}(\widehat{G})$ is the volume of $\widehat{G}$ computed with the same measure. The dimension of $\mathcal{H}^G = H^0(\mathcal{R}, \mathcal{L}^{\otimes k})$ is the same as $\mathrm{Tr}\,\Pi$, and so can evidently be written

$$\dim H^0(\mathcal{R}, \mathcal{L}^{\otimes k}) = \frac{1}{\mathrm{vol}(\widehat{G})} \int Dg \; DA \quad K(A, A; g). \tag{2.21}$$

It remains to construct a suitable kernel K. This will be done using gauged WZW models.

2.3 Gauged WZW Models

Let Σ be a complex Riemann surface; the complex structure determines the Hodge duality operator $*$ on one-forms. For a map $g : \Sigma \to G$, the WZW functional is

$$I(g) = -\frac{1}{8\pi} \int_\Sigma \mathrm{Tr}\, g^{-1}dg \wedge *g^{-1}dg - i\Gamma(g), \tag{2.22}$$

where $\Gamma(g)$ was defined in (2.9). While $\Gamma(g)$ is defined purely in differential topology, the first term in the definition of $I(g)$ depends on the complex structure of

Σ through the $*$ operator. The quantum field theory with Lagrangian $L(g) = kI(g)$ is conformally invariant and describes level k highest weight representations of the loop group of G and the action of the modular group on their characters.

$G \times G$ acts on G by left and right multiplication ($g \to agb^{-1}$). Let us denote this copy of $G \times G$ as $G_L \times G_R$ with G_L and G_R acting on the left and right respectively. $I(g)$ is invariant under $G_L \times G_R$. We want to pick a subgroup $H \subset G_L \times G_R$ and construct a gauge invariant extension of $I(g)$ with gauge group H. What this means is that we introduce a principal H bundle P, with connection A, and we replace the map $g : \Sigma \to G$ by a section of the bundle $P \times_H G$; here G is understood as the trivial principal G bundle over Σ, and H acts on G via its chosen embedding in $G_L \times G_R$. We want to construct a natural, gauge invariant functional $I(g, A)$ that reduces at $A = 0$ to $I(g)$.

There is no problem in constructing a gauge invariant extension of the first term in (2.22). One simply replaces the exterior derivative by its gauge-covariant extension:

$$-\frac{1}{8\pi}\int_\Sigma \operatorname{Tr} g^{-1} d_A g \wedge * g^{-1} d_A g. \tag{2.23}$$

On the other hand, there is a topological obstruction to constructing a gauge invariant extension of $\Gamma(g)$. The requirement is that the class in $H^3(G, \mathbf{Z})$ that determines the functional Γ (and which in real cohomology is represented by the differential form Θ) should have an extension to the equivariant cohomology group $H^3_H(G, \mathbf{Z})$. This is explained in *[18]*,§4; a quick explanation at the level of de Rham theory (ignoring the torsion in $H^3_H(G, \mathbf{Z})$) is in the appendix of *[17]*.

As explained, for instance, in that appendix, the condition for existence of a gauge invariant extension of $\Gamma(g)$ can be put in the following very explicit form. If T_a, $a = 1 \ldots \dim(H)$ are a basis of the Lie algebra of H, and if the embedding $H \subset G_L \times G_R$ is described at the Lie algebra level by $T_a \to (T_{a,L}, T_{a,R})$, then the requirement is

$$\operatorname{Tr} T_{a,L} T_{b,L} = \operatorname{Tr} T_{a,R} T_{b,R}, \quad \text{for all } a, b. \tag{2.24}$$

A subgroup $H \subset G_L \times G_R$ obeying this condition is said to be anomaly-free. For such an H, the gauge invariant extension of Γ exists and is explicitly

$$\begin{aligned}\Gamma(g, A) = \Gamma(g) &- \tfrac{1}{4\pi} \textstyle\sum_a \int_\Sigma A^a \operatorname{Tr}\left(T_{a,L} dg \cdot g^{-1} + T_{a,R} g^{-1} dg\right) \\ &- \tfrac{1}{8\pi} \textstyle\sum_{a,b} \int A^a \wedge A^b \operatorname{Tr}\left(T_{a,R} g^{-1} T_{b,L} g - T_{b,R} g^{-1} T_{a,L} g\right).\end{aligned} \tag{2.25}$$

Combining these formulas, one gets for anomaly-free H a gauge invariant extension of the WZW functional,

$$I(g, A) = -\frac{1}{8\pi}\int_\Sigma \operatorname{Tr} g^{-1} d_A g \wedge * g^{-1} d_A g - i\Gamma(g, A). \tag{2.26}$$

The quantum field theories with Lagrangians $L(g, A) = kI(g, A)$, k a positive integer, are called G/H models. [4] Note that for given G and H, there may be several G/H models, since there may be several anomaly-free embeddings of H in $G_L \times G_R$.

If H is any subgroup of G, then the diagonal embedding of H in $G_L \times G_R$ is always anomaly free. The model determined by such a diagonal embedding is often called "the" G/H model. If we pick local complex coordinates $z, \bar{z}$ on Σ (which will facilitate a small calculation needed presently) and write the measure $|dz \wedge d\bar{z}|$ as d^2z, then the Lagrangian of the diagonal G/H model is explicitly k times

$$I(g,A) = I(g) - \frac{1}{2\pi}\int_\Sigma d^2z \operatorname{Tr} A_z \partial_{\bar{z}} g \cdot g^{-1}$$
$$+\frac{1}{2\pi}\int_\Sigma d^2z \operatorname{Tr} A_{\bar{z}} g^{-1}\partial_z g - \frac{1}{2\pi}\int_\Sigma d^2z \operatorname{Tr}\left(A_z A_{\bar{z}} - A_z g A_{\bar{z}} g^{-1}\right). \tag{2.27}$$

Our interest will center on the special case of the diagonal G/H model for $H = G$. This is then the G/G model with adjoint action of G on itself. The Lagrangian is k times (2.27), and the partition function at level k is

$$Z_k(G,\Sigma) = \frac{1}{\operatorname{vol}(\widehat{G})}\int Dg \; DA \quad \exp\left(-kI(g,A)\right). \tag{2.28}$$

Our goal is to use (2.21) to show that $Z_k(G,\Sigma)$ coincides with the dimension of the space of non-abelian theta functions at level k.

2.4 The Kernel

One more special case is important: $H = G_L \times G_R$. This is an anomalous subgroup, so there is no gauge invariant G/H Lagrangian and no G/H quantum field theory for this H. We will do something else instead.

Denote the G_L and G_R components of an H connection as A and B. Set

$$I(g,A,B) = I(g) + \frac{1}{2\pi}\int d^2z \operatorname{Tr}\left(A_{\bar{z}} g^{-1}\partial_z g - B_z \partial_{\bar{z}} g \cdot g^{-1} + B_z g A_{\bar{z}} g^{-1}\right.$$
$$\left. - \frac{1}{2}A_z A_{\bar{z}} - \frac{1}{2}B_z B_{\bar{z}}\right). \tag{2.29}$$

This functional is determined by the following: it is not gauge invariant, but its change under a gauge transformation is independent of g and related in a useful way to the geometry of the prequantum line bundle. In fact, under an infinitesimal gauge transformation

$$\delta g = vg - gu, \quad \delta A = -d_A u, \quad \delta B = -d_B v, \tag{2.30}$$

[4] The terminology is somewhat misleading since these models are not the most obvious sigma models with target space G/H; and one is not allowed to use the most obvious H actions on G, such as the left or right actions, which are anomalous. The terminology is used because the models are believed *[26, 32, 16, 17]* to be equivalent to GKO models *[25]*, which were originally described algebraically, and are conventionally called G/H models or coset models. The claimed equivalence to the GKO models implies in particular that the models are conformally invariant at the quantum level.

we have

$$\delta I(g,A,B) = \frac{i}{4\pi}\int_\Sigma \mathrm{Tr}\,(u\,dA - v\,dB)\,. \tag{2.31}$$

(The fact that an extension $I(g,A,B)$ of $I(G)$ exists with these properties has a conceptual explanation noted in the appendix to *[17]*.)

Now, set

$$K(A,B;g) = \exp\left(-kI(g,A,B)\right). \tag{2.32}$$

In its dependence on A, K can be interpreted as a holomorphic section of $\mathcal{L}^{\otimes k}$; this just means that K is independent of A_z except for the exponential factor prescribed in (2.17). Likewise, in its dependence on B, K is an anti-holomorphic section of $\mathcal{L}^{\otimes(-k)}$; this means that it is independent of $B_{\overline{z}}$ except for a similar exponential. In *[17]*, the above facts were used to describe holomorphic factorization of WZW and coset models. Gerasimov's insight *[4]* was that K is actually the kernel representing the action of the gauge group on $\mathcal{H} = H^0(\mathcal{A},\mathcal{L}^{\otimes k})$. This means that for $\chi \in \mathcal{H}$ and $g \in \widehat{G}$,

$$g^*\chi(A) = \int DB \quad K(A,B;g)\chi(B). \tag{2.33}$$

To show this, we first as in (2.17) write the holomorphic section χ as

$$\chi(B) = \exp\left(\frac{k}{4\pi}\int_\Sigma d^2z \mathrm{Tr}\, B_z B_{\overline{z}}\right)\widehat{\chi}(B_{\overline{z}}) \tag{2.34}$$

with $\widehat{\chi}$ an ordinary holomorphic function. The B-dependent factors in the integral on the right hand side of (2.33) are

$$\int DB \quad \exp\left(\frac{k}{2\pi}\int_\Sigma d^2z \mathrm{Tr}\,(B_z B_{\overline{z}} - B_z A_{\overline{z}}{}^g)\right)\cdot\widehat{\chi}(B_{\overline{z}}). \tag{2.35}$$

To perform such an integral, the basic fact is that if $f(\phi)$ is a holomorphic function that grows at infinity more slowly than $\exp(|\phi|^2)$, then

$$\frac{1}{\pi}\int_{\mathbf{C}} |d\phi\wedge d\overline{\phi}|\exp(-\overline{\phi}\phi + a\overline{\phi})f(\phi) = f(a). \tag{2.36}$$

Using this fact and normalizing the symplectic measure on $\mathcal{A}$ so that

$$\int DB \;\exp\left(\frac{k}{2\pi}\int_\Sigma d^2z \mathrm{Tr}\, B_z B_{\overline{z}}\right) = 1 \tag{2.37}$$

(to avoid a determinant that would otherwise arise in using (2.36)), we get simply

$$\int DB \quad \exp\left(\frac{k}{2\pi}\int_\Sigma d^2z \mathrm{Tr}\,(B_z B_{\overline{z}} - B_z A_{\overline{z}}{}^g)\right)\cdot\widehat{\chi}(B_{\overline{z}}) = \widehat{\chi}(A^g_{\overline{z}}). \tag{2.38}$$

The integral in (2.33) thereby becomes

$$\begin{aligned}\int DB \quad K(A,B;g)\chi(B) = \exp\Big(-k\Big(I(g) + \tfrac{1}{2\pi}\int d^2z \mathrm{Tr}\, A_{\overline{z}} g^{-1}dg \\ -\tfrac{1}{4\pi}\int d^2z \mathrm{Tr}\, A_z A_{\overline{z}}\Big)\Big)\cdot\widehat{\chi}(A_{\overline{z}}{}^g).\end{aligned} \tag{2.39}$$

Using (2.34) to reexpress $\widehat{\chi}$ in terms of χ, and using the explicit forms of $I(g)$ and A^g, we get

$$\int DB\ K(A,B;g)\chi(B) = \exp\left(ik\left(\Gamma(g) - \frac{1}{4\pi}\int_\Sigma \mathrm{Tr}\, A \wedge g^{-1}dg\right)\right)\cdot \chi(A^g). \tag{2.40}$$

Using the definition of g^* in (2.15), this indeed coincides with the desired formula (2.33).

As we saw in arriving at (2.21), it follows that the dimension of the space of non-abelian theta functions is

$$\dim H^0(\mathcal{R}, \mathcal{L}^{\otimes k}) = \frac{1}{\mathrm{vol}(\widehat{G})}\int Dg\ DA\ K(A,A;g). \tag{2.41}$$

But the integral on the right is precisely the partition function (2.28) of the G/G model (since $K(A,A;g) = \exp(-kI(g,A))$). So we have arrived at the main goal of this section: identifying the dimension of the space of non-abelian theta functions with the partition function of the G/G model.

Inclusion Of Marked Points

Now we would like to extend the analysis slightly to the case of a Riemann surface Σ with marked points labeled by representations of G. The G/G model in this situation will give a path integral representation of the Verlinde algebra. (This generalization might be omitted on a first reading.)

Suppose one has a represention ρ of a compact Lie group G in a Hilbert space $\mathcal{H}$. Then as in (2.20), the projection operator onto the invariant subspace of $\mathcal{H}$ is

$$\Pi = \frac{1}{\mathrm{vol}(G)}\int_G Dg\ \rho(g), \tag{2.42}$$

with Dg an invariant measure on G and $\mathrm{vol}(G)$ the volume of G computed with that measure. The trace of Π is the multiplicity with which the trivial representation of G appears in $\mathcal{H}$.

Now pick an irreducible representation V of G, that is a vector space V in which G acts irreducibly by $g \to \rho_V(g) \in \mathrm{Aut}(V)$. We want a formula for the multiplicity with which V appears in G. We can reduce to the previous case as follows. Let $\overline{V}$ be the dual or complex conjugate representation of G. The multiplicity with which V appears in $\mathcal{H}$ is the same as the multiplicity with which the trivial representation appears in $\mathcal{H}\otimes\overline{V}$. So we define the projection operator Π_V onto the G-invariant subspace of $\mathcal{H}\otimes\overline{V}$:

$$\Pi_V = \frac{1}{\mathrm{vol}(G)}\int Dg\ \rho(g)\otimes\rho_{\overline{V}}(g). \tag{2.43}$$

The multiplicity with which V appears in $\mathcal{H}$ is

$$\mathrm{mult}(V) = \mathrm{Tr}\,\Pi_V. \tag{2.44}$$

We want to apply this to the case in which G is replaced by the group $\widehat{G}$ of gauge transformations of a principal G bundle $P \to \Sigma$; and $\mathcal{H}$ will be, as above, $H^0(\mathcal{A}, \mathcal{L}^{\otimes k})$. The representations we will use will be the following simple ones. For a point $x \in \Sigma$, let $r_x : \widehat{G} \to G$ be the map of evaluation at x. For any representation $\rho_V : G \to \mathrm{Aut}(V)$ of G, we have the corresponding representation $\rho_{x,V} = \rho_V \circ r_x$ of $\widehat{G}$. Pick now points $x_i \in \Sigma$, labeled by representations V_i, and let $V = \otimes_i V_i$ with $\widehat{G}$ acting by

$$\rho_V = \otimes_i \rho_{x_i,V_i}. \tag{2.45}$$

The conjugate representation is $\rho_{\overline{V}} = \otimes_i \rho_{x_i,\overline{V}_i}$.

We want to find a path integral representation of the multiplicity with which V appears in $\mathcal{H}$, along the lines of (2.44). To this aim we must calculate

$$\mathrm{Tr}\,\left(\rho(g) \otimes \rho_{\overline{V}}(g)\right) = \mathrm{Tr}\,\rho(g) \cdot \mathrm{Tr}\,\rho_{\overline{V}}(g). \tag{2.46}$$

Here the first factor has a path integral expression; in fact,

$$\mathrm{Tr}\,\rho(g) = \int DA\ K(A, A; g), \tag{2.47}$$

with $K(A, B; g)$ the kernel introduced in (2.32). The second factor is simply

$$\mathrm{Tr}\,\rho_{\overline{V}}(g) = \prod_i \mathrm{Tr}_{\overline{V}_i} g(x_i). \tag{2.48}$$

So we get

$$\mathrm{mult}(V) = \dim\left(\mathcal{H} \otimes \overline{V}\right)^{\widehat{G}} = \frac{1}{\mathrm{vol}(\widehat{G})} \int Dg\ DA\ \exp(-kI(g, A)) \cdot \prod_i \mathrm{Tr}_{\overline{V}_i} g(x_i). \tag{2.49}$$

The right hand side is usually called the (unnormalized) correlation function,

$$\left\langle \prod_i \mathrm{Tr}_{\overline{V}_i} g(x_i) \right\rangle \tag{2.50}$$

in the gauged WZW model. (2.50) would be unchanged if all $\overline{V}_i$ are replaced by V_i; the gauged WZW action has a symmetry (coming from an involution of G that exchanges all representations with their complex conjugates) that ensures this.

Relation To The Verlinde Algebra

Now let us relate this to the Verlinde algebra. Let T be the maximal torus of G and G/T the quotient of G by the right action of T. For any irreducible representation V of G, there is a homogeneous line bundle S over G/T such that $H^0(G/T, S)$ is isomorphic to V.

Given marked points $x_1, \ldots, x_s$ on Σ, let $\widehat{\mathcal{A}}$ be the symplectic manifold

$$\widehat{\mathcal{A}} = \mathcal{A} \times \prod_{i=1}^{s} (G/T)_i \tag{2.51}$$

where $(G/T)_i$ is a copy of G/T "sitting" at x_i. This is an informal way to say that the gauge group $\widehat{G}$ (and its complexification $\widehat{G}_{\mathbf{C}}$) acts on $(G/T)_i$ by composition of the evaluation map r_{x_i} with the natural action of G (or $G_{\mathbf{C}}$) on G/T.

If we are given irreducible representations V_i of G, let for each i $\mathcal{S}_i$ be a homogeneous line bundle over $(G/T)_i$ such that $H^0((G/T)_i, \mathcal{S}_i) \cong \overline{V}_i$. Define a homogeneous line bundle $\widehat{\mathcal{L}}$ over $\widehat{\mathcal{A}}$ by

$$\widehat{\mathcal{L}} = \mathcal{L}^{\otimes k} \otimes (\otimes_i \mathcal{S}_i) . \tag{2.52}$$

(In an obvious way, I have identified the line bundles $\mathcal{L}$ and $\mathcal{S}_i$ with their pullbacks to $\widehat{\mathcal{A}}$.) Then

$$H^0(\widehat{\mathcal{A}}, \widehat{\mathcal{L}}) = H^0(\mathcal{A}, \mathcal{L}^{\otimes k}) \otimes (\otimes_i \overline{V}_i) . \tag{2.53}$$

The multiplicity mult(V) of (2.49) is therefore the same as the dimension of the $\widehat{G}$-invariant subspace of $H^0(\widehat{\mathcal{A}}, \widehat{\mathcal{L}})$:

$$\text{mult}(V) = \dim \left(H^0(\widehat{\mathcal{A}}, \widehat{\mathcal{L}})^{\widehat{G}} \right) . \tag{2.54}$$

On the other hand, let $\widehat{\mathcal{R}}$ be the quotient of $\widehat{\mathcal{A}}$ by $\widehat{G}_{\mathbf{C}}$ (the quotient being taken in the sense of geometric invariant theory, using the ample line bundle $\widehat{\mathcal{L}}$). $\widehat{\mathcal{R}}$ is called the moduli space of holomorphic bundles over Σ with parabolic structure, the parabolic structure being a reduction of the structure group to T at the marked points x_i. (By a theorem of Mehta and Seshadri *[28]*, $\widehat{\mathcal{R}}$ coincides with the moduli space of flat connections on $P \to \Sigma - \{x_i\}$ with certain branching about the x_i, up to gauge transformation.) The $\widehat{G}_{\mathbf{C}}$-invariant line bundle $\widehat{\mathcal{L}} \to \widehat{\mathcal{A}}$ descends to a line bundle over $\widehat{\mathcal{R}}$, which we will also call $\widehat{\mathcal{L}}$, whose sections over an open set $U \subset \mathcal{R}$ are $\widehat{G}$-invariant sections of $\widehat{\mathcal{L}}$ over the inverse image of U in $\widehat{A}$. So in particular

$$H^0(\widehat{\mathcal{R}}, \widehat{\mathcal{L}}) = H^0(\widehat{\mathcal{A}}, \widehat{\mathcal{L}})^{\widehat{G}} . \tag{2.55}$$

Both $\widehat{\mathcal{R}}$ and $\widehat{\mathcal{L}}$ depend on the V_i, but I will not indicate this in the notation.

The left hand side of (2.55) is the space of non-abelian theta functions with parabolic structure. If we combine (2.49), (2.50), (2.54), and (2.55), we find that the dimension of this space is naturally written as a correlation function in the gauged WZW model:

$$\dim H^0(\widehat{\mathcal{R}}, \widehat{\mathcal{L}}) = \left\langle \prod_{i=1}^{s} \text{Tr}_{\, V_i}\, g(x_i) \right\rangle . \tag{2.56}$$

The Verlinde Algebra

As a special case of this, the Verlinde algebra is defined as follows. For given "level" k, the loop group of the compact Lie group G has a finite number of isomorphism classes of unitary, integrable representations; their highest weights are a distinguished list of isomorphism classes V_α, $\alpha \in W$ of representations of G. Let X be the $\mathbf{Z}$ module freely generated by the V_α. X has a natural metric given by $g(V_\alpha, V_\beta) = 1$ if $V_\alpha = \overline{V}_\beta$ and otherwise $g(V_\alpha, V_\beta) = 0$. It also has a natural multiplication structure that we will describe presently. X endowed with this structure is called the Verlinde algebra.

Using the metric on X, a multiplication law $V_\alpha \cdot V_\beta = \sum_\gamma N_{\alpha\beta}{}^\gamma V_\gamma$ can be defined by giving a cubic form $N_{\alpha\beta\gamma}$ which is interpreted as $\sum_\delta g_{\gamma\delta} N_{\alpha\beta}{}^\delta$. Such a cubic form is defined as follows.

Take Σ to be a curve of genus zero with three marked points x_i, $i = 1 \ldots 3$, labeled by integrable representations V_{α_i}, $\alpha_i \in W$. The choice of the α_i and of a level k determines a moduli space $\widehat{\mathcal{R}}$ of parabolic bundles with a line bundle $\widehat{\mathcal{L}}$. The structure constants of the Verlinde algebra are

$$N_{\alpha_1,\alpha_2,\alpha_3} = \dim H^0(\widehat{\mathcal{R}}, \widehat{\mathcal{L}}). \tag{2.57}$$

So in other words, from (2.56), the Verlinde structure functions are the genus zero three point functions of the G/G model:

$$N_{\alpha_1,\alpha_2,\alpha_3} = \left\langle \prod_{i=1}^{3} \operatorname{Tr}_{V_i} g(x_i) \right\rangle. \tag{2.58}$$

The basic phenomenon under study in the present paper is a relation between the quantum cohomology of the Grassmannian and the G/G model; the result can be applied to the Verlinde algebra because of (2.58). The special case of a genus zero surface with three marked points is fundamental because the general case can be reduced to this by standard sewing and gluing arguments. In fact, such sewing and gluing arguments, applied to a genus zero curve with four marked points, yield the associativity of the Verlinde algebra.

Higher Cohomology

Obviously, the above discussion has only a physical level of rigor. Among many points that should be clarified I will single out one.

If the V_i are integrable representations at level k, then the higher cohomology $H^i(\widehat{\mathcal{R}}, \widehat{\mathcal{L}})$, $i > 0$ vanishes, and $\dim H^0(\widehat{\mathcal{R}}, \widehat{\mathcal{L}})$ coincides with the Euler characteristic $\chi(\widehat{\mathcal{R}}, \widehat{\mathcal{L}}) = \sum_i (-1)^i \dim H^i(\widehat{\mathcal{R}}, \widehat{\mathcal{L}})$. From comments made to me by R. Bott and G. Segal, it appears that for (2.56) to hold for arbitrary representations V_i (perhaps not integrable), one must replace $\dim H^0(\widehat{\mathcal{R}}, \widehat{\mathcal{L}})$ by $\chi(\widehat{\mathcal{R}}, \widehat{\mathcal{L}})$. A rigorous treatment of the G/G model should show the restriction to integrable representations in deriving (2.56); there may also be a supersymmetric version of the derivation that naturally gives the Euler characteristic and holds for all representations.

2.5 Some Additional Properties

The reader may wish at this stage to turn to §3. However, I will pause here and in §2.6 below to explain a few additional facts that have their own interest and will be needed at a few points in §4.

Topological Field Theory

First of all, the gauged WZW theory of G/H is in general conformally invariant but not topologically invariant. A conformal structure appears in the definition of the Lagrangian. However, for $H = G$ we have evaluated the partition function of the G/H model, and found it to be an integer, independent of the conformal structure of Σ, and equal to the dimension of the space of non-abelian theta functions. This strongly suggests that the G/G model is actually a topological field theory. Let us try to demonstrate that directly.

A conformal structure on Σ can be specified by giving a metric h, uniquely determined up to Weyl scaling. Under a change in h, the change in the G/G Lagrangian is

$$\delta L = \frac{k}{8\pi}\int_\Sigma d^2z\sqrt{h}(h^{z\overline{z}})^2\left(\delta h_{\overline{z}\overline{z}}\mathrm{Tr}\,(g^{-1}D_z g)^2 + \delta h_{zz}\mathrm{Tr}\,(D_{\overline{z}}g\cdot g^{-1})^2\right). \tag{2.59}$$

Though this expression does not vanish identically, it vanishes when the classical equations of motion are obeyed. In fact, under a variation of the connection A, the Lagrangian changes by

$$\delta' L = \frac{k}{2\pi}\int_\Sigma d^2z\sqrt{h}h^{z\overline{z}}\mathrm{Tr}\,\left(\delta A_{\overline{z}}g^{-1}D_z g - \delta A_z D_{\overline{z}}g\cdot g^{-1}\right). \tag{2.60}$$

So the classical Euler-Lagrange equations, asserting the vanishing of $\delta' L$, are

$$0 = g^{-1}D_z g = D_{\overline{z}}g\cdot g^{-1}. \tag{2.61}$$

Since (2.59) vanishes when (2.61) does, the G/G model is classically a topological field theory. Quantum mechanically the analog of using the equations of motion is to make a suitable change of variables in the path integral. In this case, we consider the infinitesimal redefinition of A

$$\begin{aligned}\delta A_z &= \tfrac{1}{4}\delta h_{zz}h^{z\overline{z}}D_{\overline{z}}g\cdot g^{-1}\\ \delta A_{\overline{z}} &= -\tfrac{1}{4}\delta h_{\overline{z}\overline{z}}h^{z\overline{z}}g^{-1}D_z g.\end{aligned} \tag{2.62}$$

(This is a complex change of coordinates that entails an infinitesimal displacement of the integration contour in the complex plane, or more exactly a displacement of the cycle of integration in the complexification of $\mathcal{A}$.) Substituting in (2.61), we see that the Lagrangian $L(g, A)$ is invariant under a change of metric on Σ compensated by the transformation (2.62) of the field variables. The path integral for the partition function

$$\int Dg\ DA\ \exp(-L(A, g)) \tag{2.63}$$

is therefore invariant under the combined change of metric and integration variable, provided the measure DA is invariant. To this effect, we must compute a Jacobian or, at the infinitesimal level, the divergence of the vector field that generates the change of variables (2.62). This is formally

$$\int_\Sigma \left(\frac{\delta}{\delta A_z(x)} \delta A_z(x) + \frac{\delta}{\delta A_{\overline{z}}(x)} \delta A_{\overline{z}}(x). \right) \tag{2.64}$$

This vanishes, as δA_z is independent of A_z and $\delta A_{\overline{z}}$ is independent of $A_{\overline{z}}$. This completes the explanation of why the G/G model is a topological field theory.

Let us note now that the other Euler-Lagrange equation of motion, obtained by varying with respect to g, is

$$D_{\overline{z}}(g^{-1}D_z g) + F_{\overline{z}z} = 0, \tag{2.65}$$

with F the curvature of the connection A. So given (2.61), this implies that

$$F = 0. \tag{2.66}$$

Comparison To The Obvious Topological Field Theory

If the goal were to construct a topological field theory using the fields g, A, the more obvious way to do it would be to take the Lagrangian to be simply

$$L'(g, A) = -ik\Gamma(g, A), \tag{2.67}$$

which manifestly corresponds to a topological field theory, since it is defined without use of any metric or conformal structure. How does this theory compare to the G/G WZW model?

More generally, let us consider the family of theories

$$L_{k'}(g, A) = -\frac{k'}{8\pi}\int_\Sigma \operatorname{Tr} g^{-1}d_A g \wedge *g^{-1}d_A g - ik\Gamma(g, A), \tag{2.68}$$

with positive k'. This coincides with the G/G model at $k' = k$, and with the manifestly topologically invariant model at $k' = 0$. It is straightforward to work out that the classical equations of motion are

$$0 = g^{-1}D_z g - \lambda D_z g \cdot g^{-1} = D_{\overline{z}} g \cdot g^{-1} - \lambda g^{-1} D_{\overline{z}} g, \tag{2.69}$$

with

$$\lambda = \frac{k' - k}{k' + k}. \tag{2.70}$$

For $0 < k' < \infty$, one has

$$-1 < \lambda < 1. \tag{2.71}$$

(2.69) implies

$$d_A g = 0. \tag{2.72}$$

For instance, the first equation in (2.69) is equivalent to

$$\left(1 - \lambda \mathrm{Ad}(g)\right)(D_z g) = 0, \tag{2.73}$$

with $\mathrm{Ad(g)}(x) = gxg^{-1}$. Since $|\mathrm{Ad}(g)| \leq 1$ and $|\lambda| < 1$, (2.73) implies $D_z g = 0$, and similarly (2.69) implies $D_{\overline{z}} g = 0$.

Given that (2.72) follows from the classical equations of motion, the same sort of reasoning as above shows that the Lagrangians $L_{k'}$ describe a family of topological field theories: a change of metric can be compensated by a change of integration variable with trivial Jacobian.

Now, to study the k' dependence, look at

$$\frac{\partial L_{k'}}{\partial k'} = -\frac{1}{8\pi}\int_\Sigma \mathrm{Tr}\,(g^{-1}d_A g \wedge * g^{-1} d_A g). \tag{2.74}$$

By virtue of (2.72), this expression vanishes by the classical equations of motion, so classically the family of theories governed by $L_{k'}$ is constant.

From the above discussion, we know how we should proceed quantum mechanically: we should find a change of integration variable that compensates for the k' dependence of the Lagrangian. Such a change of variable exists because of (2.72); one can take explicitly

$$\delta A_z = -\frac{\delta k'}{k+k'}\left(1 - \lambda \mathrm{Ad}(g^{-1})\right)^{-1}(D_z g \cdot g^{-1}). \tag{2.75}$$

Now, however, a difference arises from our earlier discussion. Because δA_z is a function of A_z, the Jacobian of the transformation in (2.75) is not necessarily 1; the integration measure in the path integral may not be invariant. The change of the integration measure is formally

$$\int_\Sigma \mathrm{Tr}\,\frac{\delta}{\delta A_z(x)}\delta A_z(x). \tag{2.76}$$

Since

$$\frac{\delta}{\delta A_z(x)}\delta A_z(y) \sim \delta^2(x,y), \tag{2.77}$$

this is ill-defined, proportional to $\delta^2(0)$. In any event, since (2.76) is the integral over Σ of a local quantity, any regularization should be of that form. Quantities analogous to (2.76) are regularized (albeit in a slightly *ad hoc* fashion) in *[5]*, in deriving eqn. (6.22). I will not repeat such a calculation here, but I will just explain what general form the answer must have, by asking what is the most general possible perturbation of the G/G model.

The Complete Family Of Theories

Let us simply go back to the gauged WZW model of G/G, and ask what kind of perturbations it has (see also *[29]*). We permit the perturbation of the Lagrangian to be the integral of an arbitrary local functional of g, A, and a metric

h on Σ. In this way we will obtain continuous perturbations of the G/G model, but forbid discrete perturbations (notably changes in k) that cannot be described via the addition of a local functional to the Lagrangian.

Perturbations that vanish by the classical equations of motion are irrelevant, since they can be eliminated by a change of integration variables as described above. (Even if the integration measure is not invariant under the change of variables, changes of variables can be used to eliminate the perturbations that vanish by the equations of motion in favor of other perturbations that do not so vanish.) In classifying perturbations, we therefore can work modulo operators that vanish by the classical equations of motion. Given (2.61) and (2.66), this means that we can discard anything proportional to $d_A g$ or F.

The gauge invariant local operators, modulo operators that vanish by the equations of motion, are generated by operators of the form $U(g)$, with U some function on G that is invariant under conjugation. Since $U(g)$ is a zero-form, to construct from it a perturbation of the Lagrangian, we need also a metric h on Σ, or at least a measure μ, such as the Riemannian measure. The curvature scalar of h will be called R. The most interesting perturbations are

$$Q_U = \int_\Sigma d\mu \; U(g) \tag{2.78}$$

and

$$S_U = \int_\Sigma d^2z\sqrt{h}R \; U(g). \tag{2.79}$$

(2.78) breaks the diffeomorphism invariance of the G/G model down to invariance under the group of diffeomorphisms that preserve the measure μ. The G/G model perturbed as in (2.78) is an interesting family of theories invariant under area-preserving diffeomorphisms (and reducing for $k \to \infty$ to two dimensional Yang-Mills theory, which has the same invariance).

Slightly less obviously, the G/G model perturbed by (2.79) is still a topological field theory. In fact, under an infinitesimal change in h, $\sqrt{h}R$ changes by a total derivative (so that $\int_\Sigma d^2z\sqrt{h}R$ is a topological invariant, a multiple of the Euler characteristic). After integrating by parts, the change in (2.79) under a change in h is

$$\delta S_U \sim \int_\Sigma d^2z\sqrt{h}\left(\delta h_{i'j'} - h_{i'j'}h^{kl}\delta h_{kl}\right) h^{i'i}h^{j'j}D_iD_jU(g). \tag{2.80}$$

This vanishes by the equations of motion, since $d_A(g) = 0$ implies $dU = 0$. Hence one can compensate for δS_U with a redefinition of A (and the Jacobian for the transformation is trivial, since the requisite δA is independent of A).

Other perturbations, such as $\int_\Sigma d^2z\sqrt{h}R^2U(g)$, are less interesting, since (i) they do not possess the large invariances of the theories perturbed by Q_U or S_U; (ii) they vanish as a negative power of t if the metric of Σ is scaled up by $h \to th$, $t >> 1$. The latter property means that in most applications of these systems, such perturbations (if not prevented by (i)) can be conveniently eliminated.

Since the most general perturbation of the G/G model that preserves the diffeomorphism invariance is of the form of S_U, the regularized version of (2.76) must be equivalent to S_U for some U. By the same token, for any k', the G/G model must be equivalent to the $L_{k'}$ model perturbed by some S_U (with a k'-dependent U), and vice-versa. In particular, setting $k' = 0$, the G/G model is equivalent to the manifestly topologically invariant model with Lagrangian $-ik\Gamma(g, A)$, perturbed by some S_U. The requisite U's in these statements can in fact be computed at least heuristically along the lines of the derivation of eqn. (6.22) of *[5]*, but I will not do so here.

Interpretation

Note that the conjugation-invariant function $U(g)$ that entered above can be expressed as a linear combination of the characters $\mathrm{Tr}_V\, g$, as V runs over irreducible representations of G. These are precisely the operators whose correlation functions were interpreted algebro-geometrically in (2.56), so the theories obtained by perturbing the G/G model are all computable in terms of the Verlinde algebra.

2.6 Abelianization

I will now briefly describe another interesting facet of the G/G model, introduced in *[5]*, which apart from its beauty will enter at a judicious moment in §4. [5]

A recurring and significant theme in the theory of compact Lie groups is the reduction to the maximal torus T, extended by the Weyl group W. As explained in *[5]*, the G/G model admits such an reduction to the maximal torus. It is equivalent to the T/T model (that is, the G/H model with both G and H set equal to T) perturbed by S_U, where U is a certain Weyl-invariant function on T and S_U is defined in (2.79).

At the level of precision explained in *[5]*, the abelianization of the model proceeds as follows. Pick a maximal torus $T \subset G$, with Lie algebra $\mathbf{t}$. Impose the "gauge condition" $g \in T$. [6] Decompose the connection as $A = A_0 + A_\perp$, where A_0 is the part of the connection valued in $\mathbf{t}$, and $A_\perp$ is valued in the orthocomplement $\mathbf{t}_\perp$ of $\mathbf{t}$. In this gauge the G/G Lagrangian takes the form

$$L_{G/G}(g, A) = L_{T/T}(g, A_0) - \frac{k}{2\pi}\int_\Sigma d^2z \mathrm{Tr}\left(A_{\perp,z}A_{\perp,\bar{z}} - A_{\perp,z}gA_{\perp,\bar{z}}g^{-1}\right). \tag{2.81}$$

Here

$$L_{T/T}(g, A_0) = kI_{T/T}(g, A_0) \tag{2.82}$$

is the Lagrangian of the T/T model, at level k. The G/G model, in this gauge, differs from the T/T model by the last term in (2.81), which involves $A_\perp$. To

---

[5]A computation reaching a rather similar conclusion is sketched in *[4]*, but unfortunately the fermionic symmetry δ introduced in equations (71)-(74) of that paper does not obey $\delta^2 = 0$, which would be needed to justify the computation. I will therefore concentrate on sketching the argument of *[5]*.

[6]This is not really valid globally as a gauge condition. One must think in terms of integrating over the fibers of the map $G \to T/W$ that maps a group element to its conjugacy class.

reduce the G/G model to something like the T/T model, one must "integrate out" $A_\perp$ to reduce to a description involving g and A_0 only. Happily, the $A_\perp$ integral is Gaussian:

$$\int DA_\perp \exp\left(\frac{k}{2\pi}\int_\Sigma d^2z \mathrm{Tr}\left(A_{\perp,z}A_{\perp,\bar{z}} - A_{\perp,z}gA_{\perp,\bar{z}}g^{-1}\right)\right). \tag{2.83}$$

Such a Gaussian integral formally gives rise to a determinant (as we briefly explain in §3.5 below). In comparing the G/G model to the T/T model, another determinant arises: the Feddeev-Popov determinant comparing the volume of $\widehat{G}$ to the volume of $\widehat{T}$. These two determinants are rather singular but at the same time extremely simple, because the exponent in (2.83) (like the corresponding expression in the Feddeev-Popov determinant) is a local functional without derivatives. In *[5]*, Blau and Thompson calculate these determinants, with a plausible regularization, and argue that the G/G model is equivalent to a T/T model with Lagrangian

$$\widehat{L}_{T/T}(g, A_0) = (k+\rho)I(g, A_0) - \frac{1}{4\pi}\int_\Sigma d^2x\sqrt{h}R \log \det{}_{\mathbf{t}_\perp}(1 - \mathrm{Ad}(g)). \tag{2.84}$$

Here ρ is the dual Coxeter number of G, and $\det_{\mathbf{t}_\perp}(1 - \mathrm{Ad}(g))$ is the determinant of $1 - \mathrm{Ad}(g)$, regarded as an operator on $\mathbf{t}_\perp$. (This well-known Weyl-invariant function enters in the Weyl character formula, where it has a somewhat similar origin, involving a comparison of the volumes of G and T.) In §4.6, we will have occasion to use (2.84) for the case that $G = U(k)$. For that case, if $g = \mathrm{diag}(\sigma_1, \ldots, \sigma_k)$, the eigenvalues of $1 - \mathrm{Ad}(g)$ acting on $\mathbf{t}_\perp$ are the numbers $1 - \sigma_i{\sigma_j}^{-1}$, for $1 \le i, j \le k$, $i \ne j$. Hence the correction term in (2.84) becomes in this case

$$\Delta L = -\frac{1}{4\pi}\int_\Sigma d^2x\sqrt{h}R\left(\sum_{i\ne j}\ln(\sigma_i - \sigma_j) - (k-1)\sum_i \ln\sigma_i\right). \tag{2.85}$$

Given the role of the G/G model in counting non-abelian theta functions, its reduction to a T/T model is a kind of abelianization of the problem of counting such functions. In §7.3 of *[5]*, this is pursued further to obtain a completely explicit count of non-abelian theta functions for $G = SU(2)$. The role of the endpoint contributions in equation (7.14) of that paper still deserves closer study.

3 The Quantum Cohomology Of The Grassmannian

The Grassmannian $G(k, N)$ is the space of all k dimensional subspaces of a fixed N dimensional complex vector space $V \cong \mathbf{C}^N$. If we want to make the dependence on V explicit, we write $G_V(k, N)$.

By associating with a k dimensional subspace of V the $N - k$ dimensional orthogonal subspace of the dual space V^*, we see that $G_V(k, N) \cong G_{V^*}(N - k, N)$. The relation that we will explain here and in §4 between the Verlinde algebra of $U(k)$ at level $(N-k,N)$[7] and the quantum cohomology of $G(k,N)$ therefore implies that the Verlinde algebra of $U(k)$ at level $(N-k,N)$ coincides with that of $U(N-k)$ at level (k,N). This is a surprising fact that had been noted earlier. (For instance, see *[30]* and *[31]*, p. 212, Proposition (10.6.4) for $k \leftrightarrow N-k$ symmetry of loop group representations and *[32, 33]* for such symmetry of the Verlinde algebra.)

One way to describe $G(k,N)$ is as follows. Let B be the space of all linearly independent k-plets $e_1,\dots,e_k \subset V$. A point in B labels a k-plane V with a basis. The group $GL(k,\mathbf{C})$ acts on B by change of basis, $e_i \to \sum_j W_i{}^j e_j,\ \ W \in GL(k,\mathbf{C})$. Since $GL(k,\mathbf{C})$ acts simply transitively on the space of bases of V, upon dividing by $GL(k,\mathbf{C})$ we precisely forget the basis and therefore

$$G(k,N)=B/GL(k,\mathbf{C}). \tag{3.1}$$

B is dense and open in the k-fold product $\mathbf{C}^{kN}=V\times V\times\ldots\times V$ (since the generic k-plet $e_1,\dots,e_k \subset V$ is a basis of V), so $G(k,N)$ is a quotient of a dense open subset of $\mathbf{C}^{kN}$ by $GL(k,\mathbf{C})$. In fact, $G(k,N)$ is the good quotient of $\mathbf{C}^{kN}$ by $GL(k,\mathbf{C})$ that would be constructed in geometric invariant theory.

There is also a symplectic version of this, which will be more relevant in what follows. Pick a Hermitian metric on V so that $V^k=\mathbf{C}^{kN}$ gets a metric and a symplectic structure. In linear coordinates ϕ^{is}, $i=1\ldots k$, $s=1\ldots N$ on $\mathbf{C}^{kN}$, the symplectic form is

$$\omega=i\sum_{i,s} d\phi^{is}\wedge d\overline{\phi}_{is}. \tag{3.2}$$

ω is not invariant under $GL(k,\mathbf{C})$, but it is invariant under a maximal compact subgroup $U(k)\subset GL(k,\mathbf{C})$.

To this symplectic action is associated a "moment map" μ from $\mathbf{C}^{kN}$ to the dual of the Lie algebra of $U(k)$, given by the angular momentum functions that generate $U(k)$ via Poisson brackets. In this case we can take the moment map to be

$$\mu:(e_1,\dots,e_k)\to\{(e_i,e_j)-\delta_{ij}\}. \tag{3.3}$$

In other words, $\mu=0$ precisely if the vectors $e_1,\dots,e_k$ are orthonormal.

Every k-plane has an orthonormal basis, unique up to the action of $U(k)$, so

$$G(k,N)=\mu^{-1}(0)/U(k). \tag{3.4}$$

This is the description of $G(k,N)$ that we will actually use. We will also want to remember one fact: μ is a quadratic function on the real vector space underlying $\mathbf{C}^{kN}$. In components,

$$\mu^i{}_j=\sum_s \phi^{is}\overline{\phi}_{js}-\delta^i{}_j. \tag{3.5}$$

[7]That is, at levels $N-k$ and N for the $su(k)$ and $u(1)$ factors in the Lie algebra of $U(k)$.

3.1 Cohomology

Now we need to discuss the cohomology of $G(k, N)$. We begin with the classical cohomology. Over $G(k, N)$ there is a "tautological" k-plane bundle E (whose fiber over $x \in G(k, N)$ is the k plane in V labeled by x) and a complementary bundle F (of rank $N - k$):

$$0 \to E \to V \cong \mathbf{C}^N \to F \to 0. \tag{3.6}$$

Obvious cohomology classes of $G(k, N)$ come from Chern classes. We set

$$x_i = c_i(E^*), \tag{3.7}$$

where $*$ denotes the dual. (It is conventional to use E^* rather than E, because $\det E^*$ is ample.) This is practically where Chern classes come from, as $G(k, N)$ for $N \to \infty$ is the classifying space of the group $U(k)$. It is known that the x_i generate $H^*(G(k, N))$ with certain relations. The relations come naturally from the existence of the complementary bundle F in (3.6). Let $y_j = c_j(F^*)$, and let $c_t(\cdot) = 1 + tc_1(\cdot) + t^2c_2(\cdot) + \ldots$. Then as a consequence of (3.6),

$$c_t(E^*)c_t(F^*) = 1, \tag{3.8}$$

and $H^*(G(k, N))$ is generated by the x_i, y_j with relations (3.8). If one wishes, these relations can be partially solved to express the y_j in terms of the x_i (or vice-versa).

Now we come to the quantum cohomology, which originally entered in string theory, where *[34]* it enters the theory of the Yukawa couplings (which are related to quark and lepton masses), and in Floer/Gromov theory of symplectic manifolds *[35]*. Additively, the quantum cohomology is the same as the classical one, but the ring structure is different.

Giving a ring structure on $W = H^*(G(k, N))$ is the same as giving the identity $1 \in W$ and a cubic form

$$(\alpha, \beta, \gamma) = \int_{G(k,N)} \alpha \cup \beta \cup \gamma. \tag{3.9}$$

The cubic form determines a metric

$$g(\alpha, \beta) = (\alpha, \beta, 1), \tag{3.10}$$

and given a metric the cubic form $W \times W \times W \to \mathbf{C}$ determines a ring structure $W \times W \to W$.

So I will explain the quantum cohomology ring by describing the quantum cubic form. To this aim, let Σ be a closed oriented two-manifold (which in string theory would be the "world-sheet," analogous to the world-line of a particle). Let $P \in \Sigma$. Let $\mathcal{W} = \text{Maps}(\Sigma, G(k, N))$. Evaluation at P gives a map

$$\mathcal{W} \xrightarrow{\text{ev}(P)} G(k, N), \tag{3.11}$$

by which $\alpha \in H^*(G(k,N))$ pulls back to $\widehat{\alpha}(P) = \mathrm{ev}(P)^*(\alpha) \in H^*(\mathcal{W})$.

Now pick a complex structure on Σ, and let $\mathcal{M} \subset \mathcal{W}$ be the space of holomorphic maps of Σ to $G(k,N)$. We have $\mathcal{M} = \cup_\lambda \mathcal{M}_\lambda$, with $\mathcal{M}_\lambda$ being the connected components of $\mathcal{M}$. In the case of the Grassmannian, the components $\mathcal{M}_\lambda$ are determined by the degree, defined as follows. If $\eta = c_1(E^*)$, which generates $H^2(G(k,N),\mathbf{Z})$, and $\Phi : \Sigma \to G(k,N)$ is such that $\int_\Sigma \Phi^*(\eta) = d$, then Φ is said to be of degree d. Since $\det E^*$ is ample, holomorphic curves only exist for $d \geq 0$.

The quantum cubic form is defined as follows (ignoring analytical details and tacitly assuming that the $\mathcal{M}_\lambda$ are smooth and compact). Let Σ be of genus zero. Let P, Q, R be three points in Σ. Then for $\alpha, \beta, \gamma \in H^*(G(k,N))$, we set

$$\langle \alpha, \beta, \gamma \rangle = \sum_d e^{-dr} \cdot \int_{\mathcal{M}_d} \widehat{\alpha}(P) \cup \widehat{\beta}(Q) \cup \widehat{\gamma}(R), \tag{3.12}$$

with r a real parameter.

In what sense does $\langle \alpha, \beta, \gamma \rangle$ generalize the classical cubic form? One component of $\mathcal{M}$, namely $\mathcal{M}_0$, consists of constant maps $\Sigma \to G(k,N)$. This component is a copy of $G(k,N)$ itself. Under that identification the evaluation maps at P, Q, and R all coincide with the identity, so the contribution of $\mathcal{M}_0$ to $\langle \alpha, \beta, \gamma \rangle$ coincides with the classical cubic form defined as in (3.9). The quantum cubic form differs from the classical one by contributions of the rational curves of higher degree. These contributions are small for $r >> 0$. In practice, for dimensional reasons, for every given α, β, γ of definite dimension, the sum in (3.12) receives a contribution from at most one value of d. (This is in marked contrast to the much-studied case of a Kähler manifold of $c_1 = 0$, where every positive d can contribute to the same correlation function.) Therefore, no information is lost if we set $r = 0$, and that is what we will do in the rest of this section.

It follows from the definition (for any Kähler manifold, not just the Grassmannian) that

$$\langle \alpha, \beta, 1 \rangle = (\alpha, \beta, 1) \tag{3.13}$$

and thus that the classical and quantum metrics coincide. This is equivalent to the statement that rational maps of positive degree do not contribute to $\langle \alpha, \beta, 1 \rangle$. In fact (as $\widehat{1}(R) = 1$), the contribution of a component $\mathcal{M}_\lambda$ of positive degree is

$$\int_{\mathcal{M}_\lambda} \widehat{\alpha}(P) \cup \widehat{\beta}(Q). \tag{3.14}$$

A group $F \cong \mathbf{C}^*$ acts on $\mathbf{CP}^1$ leaving fixed the points P and Q. F acts freely on $\mathcal{M}_\lambda$, if $\mathcal{M}_\lambda$ is a component of rational maps of positive degree. The classes $\widehat{\alpha}(P)$ and $\widehat{\beta}(Q)$ in the cohomology of $\mathcal{M}_\lambda$ are pullbacks from $\mathcal{M}_\lambda / F$. Therefore, on dimensional grounds (3.14) vanishes.

3.2 The Grassmannian

Let us now work out the quantum cohomology ring of the Grassmannian. As a preliminary, we note that the contribution of a moduli space $\mathcal{M}_d$ to the quantum cubic form obeys an obvious dimensional condition: it vanishes unless the sum of the dimensions of α, β, γ equals the (real) dimension of $\mathcal{M}_d$. The component $\mathcal{M}_d$ of genus zero holomorphic curves of degree d in $G(k, N)$ has (according to the Riemann-Roch theorem) complex dimension $\dim_{\mathbf{C}} G(k, N) + dN$. The fact that this depends on d means that the dimensional condition depends on d and therefore that the quantum cohomology ring is not $\mathbf{Z}$-graded. However, the fact that the real dimensions are all equal modulo $2N$ means that the cohomology is $\mathbf{Z}/2N\mathbf{Z}$-graded.

Returning to the relations $c_t(E^*)c_t(F^*) = 1$ that define the cohomology of the Grassmannian, we see that (as the left hand side is *a priori* a polynomial in t of degree N) the classical relations are of dimension $0, 2, 4, \ldots, 2N$. To a classical relation of degree $2k$, the rational curves of degree $d > 0$ will add a correction of degree $2k - 2dN$; this therefore must vanish unless $k = N$ and $d = 1$. Therefore, of the defining relations of the cohomology, the only one subject to a quantum correction is the "top" relation $c_k(E^*)c_{N-k}(F^*) = 0$, and the correction is an element of $H^*(G(k, N))$ of degree 0 and hence simply an integer. So the non-trivial effect of the quantum corrections will be simply to generate a relation of the form

$$c_k(E^*)c_{N-k}(F^*) = a, \tag{3.15}$$

for some $a \in \mathbf{Z}$. Moreover, a is to be computed by examining rational curves in the Grassmannian of degree 1. We will find that $a = (-1)^{N-k}$, so the quantum cohomology ring can be described by the relations

$$c_t(E^*)c_t(F^*) = 1 + (-1)^{N-k}t^N. \tag{3.16}$$

This correction has been described previously *[36]* in the special case of $k = 1$ (complex projective space). Despite its simple form, the correction has a dramatic effect: while the classical cohomology ring is nilpotent (in the sense that every element of positive degree is nilpotent), the quantum cohomology ring is semi-simple. This is evident in its Landau-Ginzburg description *[12, 3, 1]* which we consider presently.

Computation Of a

For X a submanifold of $G(k, N)$, let $[X]$ be its Poincaré dual cohomology class. For instance, for p a point in the Grassmannian, $[p]$ is a top dimensional class, obeying $g(1, [p]) = 1$. (It does not matter here if the metric $g(\, , \,)$ is defined using the classical or quantum cubic form, since we have seen that these determine the same metric.) The definition of the quantum ring structure from the quantum cubic form is such that $a = c_k(E^*)c_{N-k}(F^*)$ can be computed as

$$a = \langle c_k(E^*), c_{N-k}(F^*), [p] \rangle. \tag{3.17}$$

$c_k(E^*)$ equals the Poincaré dual of the zero locus of a generic section of E^*. The dual of the exact sequence (2.6) reads

$$0 \to F^* \to V^* \to E^* \to 0, \tag{3.18}$$

with V^* a fixed N dimensional complex vector space. The image in E^* of any fixed vector $w \in V^*$ gives a holomorphic section $\overline{w}$ of E^*. If as before $e_1, \dots, e_N$ is a basis of V, and w is the linear form that maps $\sum_{i=1}^N r^i e_i$ to r^1, then the restriction of w to $E \subset V$ vanishes precisely if E consists only of vectors with $r^1 = 0$. This is a copy of $G(k, N-1)$ which we will call X_w. Since $\overline{w}$ has only a simple zero along X_w (any E can be perturbed in first order to get one for which $\overline{w} \neq 0$), we have

$$c_k(E^*) = [X_w]. \tag{3.19}$$

For future use, let us note that

$$\int_{G(k,N)} c_k(E^*)^{N-k} = 1. \tag{3.20}$$

Indeed, we can pick $N-k$ holomorphic sections of E^* whose zero sets intersect transversely at a single point. To do so, let w_i for $i = 1, \dots, N-k$ be the linear form on V that maps $\sum_{i=1}^N r^i e_i$ to r^i. Then the $\overline{w}_i$ have the required properties, vanishing precisely for E the k-plane spanned by $e_{N-k+1}, \dots, e_N$.

Now let us compute $c_{N-k}(F^*) = (-1)^{N-k} c_{N-k}(F)$. Under the holomorphic surjection $V \to F$, any vector $v \in V$ projects to a holomorphic section $\overline{v}$ of F. $\overline{v}$ vanishes precisely if $v \in E$; let $Y_v = \{E \in G(k,N) | v \in E\}$. Then $\overline{v}$ has a simple zero along Y_v, so $c_{N-k}(F) = [Y_v]$ and therefore

$$c_{N-k}(F^*) = (-1)^{N-k}[Y_v]. \tag{3.21}$$

Rational curves of degree one in $G(k,N)$ can all be described as follows. Let (s,t) be homogeneous coordinates for $\mathbf{CP}^1$. For $r_1, \dots, r_k$ a set of k linearly independent vectors in the N dimensional vector space V, let $\{r_1, \dots, r_k\}$ be the k-plane that they span. Then a rational curve of degree one in $G(k,N)$ is of the form

$$(s,t) \to \{sr_0 + tr_1, r_2, r_3, \dots, r_k\}, \tag{3.22}$$

with $r_0, \dots, r_k$ being linearly independent vectors in V.

We have to calculate

$$a = (-1)^{N-k} \int_{\mathcal{M}_1} \widehat{[X_w]}(P) \cup \widehat{[Y_v]}(Q) \cup \widehat{[p]}(R). \tag{3.23}$$

Here $\mathcal{M}_1$ is the space of degree 1 rational curves, $w \in V^*$, $v \in V$, and P, Q, R are points in $\mathbf{CP}^1$. If everything is sufficiently generic, a is simply the number of degree one curves that pass through X_e at P, through Y_f at Q, and through p at R.

We choose p to be an arbitrary point in $G(k,N)$ corresponding to a k-plane spanned by vectors $v_1, \dots, v_k$. We take $v = v_0$ to be linearly independent of these, and we pick w to be any linear form that maps v_0 to 1, v_1 to -1, and the v_j of $j > 1$ to 0.

From the explicit description of degree one curves in (3.22), we see that the k-planes represented by points in the image of such a curve are subspaces of a common $k+1$-plane. For a curve that passes through Y_v at Q and through p at

R, this is clearly the $k+1$-plane W spanned by $v_0, v_1, \ldots, v_k$. Requiring that the curve pass also through X_w at P determines the curve uniquely. For instance, if $Q = (1,0)$, $R = (0,1)$, and $P = (1,1)$, then the degree 1 curve must be

$$(s,t) \to \{v_0 s + v_1 t, v_2, \ldots, v_k\}. \tag{3.24}$$

The subvarieties $\widehat{[X_w]}(P)$, $\widehat{[Y_f]}(Q)$, and $\widehat{[p]}(R)$ of $\mathcal{M}_\infty$ meet transversely at that point, so we get finally

$$a = (-1)^{N-k} \tag{3.25}$$

as claimed above.

Landau-Ginzburg Formulation

Write

$$c_t(E^*) = \sum_{i=0}^{k} x_i t^i, \tag{3.26}$$

with $x_i = c_i(E^*)$. Define functions $y_j(x_i),\ j \geq 0$ by

$$\frac{1}{c_t(E^*)} = \sum_{j \geq 0} y_j t^j. \tag{3.27}$$

Classically, the cohomology ring of $G(k,N)$ is described by the relations

$$y_j = 0, \text{ for } N-k+1 \leq j \leq N. \tag{3.28}$$

Let

$$-\log c_t(E^*) = \sum_{r \geq 0} U_r(x_1, \ldots, x_k) t^r. \tag{3.29}$$

So

$$-t^j c_t(E^*)^{-1} = -\frac{\partial}{\partial x_j} \log c_t(E^*) = \sum_{r \geq 0} \frac{\partial U_r}{\partial x_j} t^r. \tag{3.30}$$

Hence if

$$W_0 = (-1)^{N+1} U_{N+1} \tag{3.31}$$

then

$$\frac{\partial W_0}{\partial x_i} = (-1)^N y_{N+1-i}, \text{ for } 1 \leq i \leq k. \tag{3.32}$$

So the defining relations of the classical cohomology take the form

$$dW_0 = 0. \tag{3.33}$$

To obtain in a similar way the quantum cohomology ring, set

$$W = W_0 + (-1)^k x_1. \tag{3.34}$$

The relations $dW = 0$ now give

$$y_{N+1-i} + (-1)^{N-k}\delta_{i,1} = 0. \tag{3.35}$$

Therefore the relation $c_t(E^*) \cdot (\sum_j y_j t^j) = 1$ becomes

$$\left(\sum_{i=0}^{k} x_i t^i\right) \cdot \left(\sum_{j=0}^{N-k} y_j t^j - (-1)^{N-k} t^N + O(t^{N+1})\right) = 1. \tag{3.36}$$

Keeping only the terms of order at most t^N, this becomes

$$\left(\sum_{i=0}^{k} x_i t^i\right) \cdot \left(\sum_{j=0}^{N-k} y_j t^j\right) = 1 + (-1)^{N-k} t^N. \tag{3.37}$$

This coincides with the quantum cohomology ring as described in (3.16). The function W is called the Landau-Ginzburg potential.

If we introduce the roots of the Chern polynomial

$$c_t(E^*) = \prod_{i=1}^{k} (1 + \lambda_i t), \tag{3.38}$$

then W can be written

$$W(\lambda_1, \ldots, \lambda_k) = \frac{1}{k+1} \sum_{j=1}^{k} \left({\lambda_j}^{N+1} + (-1)^k \lambda_j\right). \tag{3.39}$$

Now let us discuss integration. Integration defines a linear functional on the top dimensional cohomology of $G(k, N)$, which is the cohomology in real dimension $2k(N-k)$:

$$f \to I(f) = \int_{G(k,N)} f. \tag{3.40}$$

Since $H^{2k(N-k)}(G(k, N))$ is one dimensional, any two linear functionals on that space are proportional. Such a linear functional can be obtained as follows in the Landau-Ginzburg description. We examine the classical case first. If we consider $x_i = c_i(E^*)$ to be of degree i, then the top dimensional cohomology consists of polynomials f of degree $k(N-k)$ modulo the ideal generated by $\partial W_0 / \partial x_i$, $i = 1 \ldots k$. Consider the linear form on homogeneous polynomials of degree $k(N-k)$ defined by

$$J(f) = (-1)^{k(k-1)/2} \left(\frac{1}{2\pi i}\right)^k \oint dx_1 \ldots dx_k \frac{f}{\prod_{i=1}^{k} \partial W_0 / \partial x_i}. \tag{3.41}$$

The integration contour is a product of circles enclosing the poles in the denominator. $J(f)$ annihilates the ideal generated by dW_0, since if f is divisible by, say, $\partial W/\partial x_i$, then one of the denominators in (3.41) is canceled and one of the contour integrals vanishes.

For f of degree $k(N-k)$, the integral in (3.41) is unaffected if W_0 is replaced by W; this follows from taking the contour integral on (3.41) to be a large contour. The integral can then be evaluated as a simple sum of residues:

$$J(f) = (-1)^{k(k-1)/2} \sum_{dW=0} \frac{f}{\det\left(\frac{\partial^2 W}{\partial x_i \partial x_j}\right)}. \tag{3.42}$$

It is convenient to change variables from the x_i to the λ_a. One has

$$\det\left(\frac{\partial^2 W}{\partial x_i \partial x_j}\right)\bigg|_{dW=0} = \det\left(\frac{\partial^2 W}{\partial \lambda_a \partial \lambda_b}\right) \cdot \det\left(\frac{\partial \lambda_a}{\partial x_i}\right)^2. \tag{3.43}$$

The Jacobian in the change of variables from x_i to λ_a is the Vandermonde determinant:

$$\det\left(\frac{\partial x_i}{\partial \lambda_a}\right)^2 = \prod_{a<b} (\lambda_a - \lambda_b)^2. \tag{3.44}$$

So

$$\begin{aligned} J(f) &= \frac{(-1)^{k(k-1)/2}}{k!} \sum_{dW(\lambda_a)=0} \frac{f \cdot \prod_{a<b}(\lambda_a - \lambda_b)^2}{\prod_a dW/d\lambda_a} \\ &= \frac{(-1)^{k(k-1)/2}}{k!(2\pi i)^k} \oint d\lambda_a \frac{f \cdot \prod_{a<b}(\lambda_a - \lambda_b)^2}{\prod_a dW/d\lambda_a}. \end{aligned} \tag{3.45}$$

The integration contour in each λ_a integral is a circle running counterclockwise around the origin. A factor of $k!$ comes here because, as the map from the λ_a to the x_i is of degree $k!$, each critical point of $W(x_i)$ corresponds to $k!$ critical points of $W(\lambda_a)$.

To verify that $J(f)$ is correctly normalized to coincide with $I(f)$, we set $f = c_k(E^*)^{N-k} = \prod_a \lambda_a^{N-k}$. According to (3.20), $I(f) = 1$. To verify that $J(f) = 1$, we use the contour integral version of (3.45). In the denominator we can replace $dW/d\lambda_a = {\lambda_a}^N + (-1)^k$ by ${\lambda_a}^N - {\lambda_a}^{N-k}$ without changing the behavior on a large contour enough to affect the integral. Then

$$J(f) = \frac{(-1)^{k(k-1)/2}}{k!(2\pi i)^k} \oint d\lambda_1 \dots d\lambda_k \frac{\prod_{a<b}(\lambda_a - \lambda_b)^2}{\prod_c(\lambda_c^k - 1)}. \tag{3.46}$$

The integral is easily done as a sum of residues. The poles are at ${\lambda_a}^k = 1$, for $1 \le a \le k$. Because of the Vandermonde determinant in the numerator, the λ_a must be distinct. Up to a permutation, one must have $\lambda_a = \exp(2\pi i a/k)$; evaluating the residue at this value of the λ_a and including a factor of $k!$ from the sum over permutations, one gets $J(f) = 1$.

3.3 Quantum Field Theory Interpretation

Physicists would never actually begin with the definition that I have given above for the quantum cubic form. Rather, everything begins with considerations on the function space $\mathcal{W} = \text{Maps}(\Sigma, G(k, N))$. Physicists are mainly interested in quantum field theory, which is conveniently formulated in terms of integration over spaces such as $\mathcal{W}$.

For instance, let Σ be a complex Riemann surface with Hodge duality operator $*$, pick a Hermitian metric on $G(k, N)$ (such as the natural $U(k)$-invariant metric), and for a map $\Phi : \Sigma \to G(k, N)$, set

$$L(\Phi) = \int_\Sigma (d\Phi, *d\Phi). \tag{3.47}$$

Then in the "bosonic sigma model with target space $G(k, N)$" we consider integrals such as

$$\int_{\mathcal{W}} D\Phi \, \exp\left(-\frac{L(\Phi)}{\lambda}\right), \tag{3.48}$$

with λ a positive real number. This is not complete pie in the sky. For instance, to make the definition more concrete, one can triangulate Σ and make a finite dimensional approximation to the integral. Then the problem is to adjust λ, while refining the triangulation, so that the given integral (and related ones) converges as the triangulation is infinitely refined.

For a homogeneous space of positive curvature such as the Grassmannian, one knows at a physical level of rigor precisely how to do this: λ must be taken to vanish in inverse proportion to the logarithm of the number of vertices in the triangulation. This is a consequence of a phenomenon known as "asymptotic freedom," which plays a crucial role in the theory of the strong interactions in four dimensions; sigma models with targets such as the Grassmannian were intensively studied in the late 1970's and early 1980's as simple cases of asymptotically free quantum field theories. Asymptotic freedom actually plays an important role in our story, since it leads to the mass gap that will be essential in §4.

Supersymmetric Sigma Models

What we actually want to do is to transfer the integral over the space of holomorphic maps that defined the quantum cohomology ring,

$$\langle \alpha, \beta, \gamma \rangle = \sum_\lambda \int_{\mathcal{M}_\lambda} \widehat{\alpha}(P) \cup \widehat{\beta}(Q) \cup \widehat{\gamma}(R), \tag{3.49}$$

to an integral over the space $\mathcal{W}$ of all maps of Σ to $G(k, N)$. Reversing the usual logic, this is done as follows. The condition that $\Phi : \Sigma \to G(k, N)$ is holomorphic is an equation

$$0 = \left(\overline{\partial}\Phi\right)^{1,0} \tag{3.50}$$

which asserts the vanishing of a section

$$s : \Phi \to (\overline{\partial}\Phi)^{1,0} \tag{3.51}$$

of an infinite dimensional vector bundle Y over $\mathcal{W}$. (Y is the bundle whose fiber at $\Phi \in \mathcal{W}$ is the space of $(0,1)$ forms on Σ with values in $\Phi^*(T^{1,0}G(k,N))$, with $T^{1,0}G(k,N)$ being the $(1,0)$ part of the complexified tangent bundle of $G(k,N)$. The point of the definition is just that $(\overline{\partial}\phi)^{1,0}$ is a vector in Y.)

The space $\mathcal{M}$ of holomorphic maps, being defined by the vanishing of a section $s : \mathcal{W} \to Y$, is Poincaré dual to the Euler class $\chi(Y)$ of Y. So formally we can write

$$\langle \alpha, \beta, \gamma \rangle = \int_{\mathcal{M}} \widehat{\alpha} \cup \widehat{\beta} \cup \widehat{\gamma} = \int_{\mathcal{W}} \widehat{\alpha} \cup \widehat{\beta} \cup \widehat{\gamma} \cup \chi(Y). \tag{3.52}$$

Now, there are any number of ways to write a differential form representing $\chi(Y)$, but one nice way (formulated mathematically by Mathai and Quillen *[37]*) uses a section s and has a nice exponential factor $\exp(-|s|^2/\lambda)$, with $|s|^2$ the norm of s with respect to a metric on Y, and λ a positive real number. For the section indicated in (3.51), the norm with respect to the natural metric is

$$|s|^2 = \int_\Sigma (d\Phi, *d\Phi), \tag{3.53}$$

which is precisely the Lagrangian introduced above for the bosonic sigma model with target space the Grassmannian.

So the long and short of it is that we get a representation

$$\langle \alpha, \beta, \gamma \rangle = \int_{\mathcal{W}\times\ldots} \int D\Phi \ \ldots \ \exp\left(-\frac{1}{\lambda}\int_\Sigma |d\Phi|^2 + \ldots\right) \widehat{\alpha}(P)\widehat{\beta}(Q)\widehat{\gamma}(R), \tag{3.54}$$

much like the bosonic sigma model, but with "fermions," represented by "..." The quantum field theory that appears here is in fact a twisted form of the usual supersymmetric nonlinear sigma model, as I explained in *[38]*; in the twisted model, the fermions can be interpreted in terms of differential forms on the function space $\mathcal{W}$. The relation to the Mathai-Quillen formula was explained by Atiyah and Jeffrey *[39]* in the analogous case of four dimensional Donaldson theory.

It should be fairly obvious that instead of $G(k,N)$ we could use a general Kähler manifold X in the above discussions, at least at the classical level. (If we are willing to give up the interpretation in terms of twisting of a unitary supersymmetric model, we can even consider almost complex manifolds that are not Kähler.) At the quantum level, the situation is more subtle. There are two main branches in the subject. If $c_1(X) = 0$, the supersymmetric sigma model (with a suitable choice of the Kähler metric of X) is conformally invariant; such models provide classical solutions of string theory. On the other hand, if $c_1 > 0$, as in the case of the Grassmannian, one is in a quite different world, with asymptotic freedom and analogs of the mass generation and chiral symmetry breaking seen in the strong interactions.

3.4 Strategy

To try to say something of substance in this situation, we use the realization of $G(k,N)$ as $\mu^{-1}(0)/U(k)$, where μ is the moment map from $\mathbf{C}^{kN}$ to the Lie algebra of $U(k)$. One is tempted to try to lift a map $\Phi : \Sigma \to G(k,N)$ to a map $\widehat{\Phi} : \Sigma \to \mathbf{C}^{kN}$. There is not a natural way to do this, and there may even be a topological obstruction.

So instead we proceed as follows. Let P be a principal $U(k)$ bundle over Σ, A a connection on P, $\widehat{\Phi}$ a section of $P \times_{U(k)} \mathbf{C}^{kN}$, and S a two-form on Σ with values in the adjoint bundle $\mathrm{ad}(P)$. Take

$$\widehat{L}(\widehat{\Phi}, A, S) = \int_\Sigma \left((d_A\widehat{\Phi}, *d_A\widehat{\Phi}) + i(S, \mu \circ \widehat{\Phi}) \right). \tag{3.55}$$

Then classically the theory described by $\widehat{L}(\widehat{\Phi}, A, S)$ is equivalent to the bosonic sigma model with target $G(k,N)$. This can be seen as follows. The Euler-Lagrange equation of S is $\mu \circ \widehat{\Phi} = 0$, so, under the natural projection $P \times_{U(k)} \mathbf{C}^{kN} \to \mathbf{C}^{kN}/U(k)$, $\widehat{\Phi}$ maps to $\Phi : \Sigma \to \mu^{-1}(0)/U(k) = G(k,N)$. The Euler-Lagrange equation for A identifies P and A with the pull-back by $\widehat{\Phi}$ of the tautological principal $U(k)$ bundle and connection over $G(k,N)$. Once these restrictions and identifications are made, $\widehat{L}(\widehat{\Phi}, A, S)$ reduces to the Lagrangian $L(\Phi)$ of the bosonic sigma model of the Grassmannian.

This sort of reasoning is still valid quantum mechanically. For instance, using

$$\int_{-\infty}^{\infty} \frac{dx}{2\pi} e^{ixy} = \delta(y) \tag{3.56}$$

(and the obvious generalization of that formula to several variables) we get the path integral formula

$$\int DS \, \exp\left(-i \int_\Sigma (S, \mu \circ \widehat{\Phi}) \right) = \delta(\mu \circ \widehat{\Phi}). \tag{3.57}$$

So the S integral places on $\widehat{\Phi}$ precisely the restriction that one would guess from the classical Euler-Lagrange equations. From simple properties of Gaussian integrals (which are introduced below), one similarly deduces that, quantum mechanically as classically, the A integral has the effect of identifying P, A with the pull-backs of the tautological objects over the Grassmannian.

Similar reasoning holds after including fermions, so we get for the quantum cubic form a representation of the general kind

$$\begin{aligned}\langle \alpha, \beta, \gamma \rangle = \textstyle\int D\widehat{\Phi}\, DA\, DS \ldots \exp\left(- \int_\Sigma \left((d_A\widehat{\Phi}, *d_A\widehat{\Phi}) + i(S, \mu \circ \widehat{\Phi}) + \ldots \right) \right) \\ \cdot \widetilde{\alpha}(P)\widetilde{\beta}(Q)\widetilde{\gamma}(R).\end{aligned} \tag{3.58}$$

As before, "..." represents terms involving fermions that are not indicated explicitly.

3.5 Reversing The Order Of Integration

In sum, (3.58) will reduce to (3.54) if we integrate over A and S first. To get something interesting, we instead integrate first over $\widehat{\Phi}$. The key point is that $\widehat{\Phi}$ is a section of a bundle over Σ with linear fibers (a $\mathbf{C}^{kN}$ bundle) and that $\widehat{L}$ is quadratic in $\widehat{\Phi}$. Consequently, the $\widehat{\Phi}$ integral is a Gaussian integral.

The basic one dimensional formula

$$\int_{-\infty}^{\infty} \frac{dx}{\sqrt{2\pi}} \exp(-\lambda x^2/2) = \frac{1}{\sqrt{\lambda}} \tag{3.59}$$

has the n dimensional generalization

$$\int_{-\infty}^{\infty} \frac{dx_1 \ldots dx_n}{(2\pi)^{n/2}} \exp\left(-\frac{1}{2}\sum_{i,j} M_{ij}x_i x_j\right) = \frac{1}{\sqrt{\det M}}, \tag{3.60}$$

for any quadratic form M with positive real part; this is demonstrated by picking a coordinate system in which $M = \operatorname{diag}(m_1, \ldots, m_n)$.

In our case the $\widehat{\Phi}$ integral is (apart from terms involving fermions)

$$\int D\Phi \ \exp\left(-\int_\Sigma \left((d_A\widehat{\Phi}, *d_A\widehat{\Phi}) + i(S, \mu \circ \widehat{\Phi})\right)\right). \tag{3.61}$$

This is an infinite dimensional Gaussian integral with M the quadratic form associated with the elliptic differential operator

$$M' = (d_A^* d_A + iS) \otimes 1_N. \tag{3.62}$$

The notation reflects the fact that the $\mathbf{C}^{kN}$ bundle of which $\widehat{\Phi}$ is a section is actually a sum of N copies of a $\mathbf{C}^k$ bundle, and M is the sum of N copies of a quadratic form derived from an operator (namely $d_A^* d_A + iS$) on sections of that $\mathbf{C}^k$ bundle. So the integral over $\widehat{\Phi}$ gives

$$\frac{1}{\sqrt{\det M'}} = \left(\frac{1}{\sqrt{\det(d_A^* d_A + iS)}}\right)^{N/2}. \tag{3.63}$$

The determinant of the elliptic differential operator $d_A^* d_A + iS$ can be conveniently defined using the ζ-function regularization of Ray and Singer.

So modulo fermions we get

$$\langle \alpha, \beta, \gamma \rangle = \int DA\,DS \ldots \left(\frac{1}{\sqrt{\det(d_A^* d_A + iS)}}\right)^{N/2} \cdot \widehat{\alpha}\widehat{\beta}\widehat{\gamma}. \tag{3.64}$$

So we have transformed the problem of computing the quantum cohomology of the Grassmannian to a problem involving integration over the connection A and over S – a problem in quantum gauge theory. This brings us into an entirely different world, that of §2 of this paper.

has a sharp peak at the minimum of the determinant, and "everything" can be calculated in an asymptotic expansion in powers of $1/N$, by expanding around this peak.

For fixed N, it is not true that "everything" can be calculated, but the topological quantities can be, reducing to a saddle point by a more elaborate argument. The essence of the matter is that although the classical Lagrangian $\widehat{L}$ is conformally invariant, the quantum theory is not (because, for instance, with Ray-Singer or any other regularization, the determinant introduced above is not conformally invariant). The topological quantities are however not just conformally invariant but completely independent of the metric of Σ. Scaling up the metric of Σ by a very large real factor, life simplifies because of the basic physical properties of the model – asymptotic freedom and the dynamically generated mass gap. At very large distances (that is, if the metric on Σ is scaled up by a very big factor), the complicated integral over A, S, and fermions in (3.64) reduces to a local and tractable quantum field theory – in fact it reduces to the gauged WZW model (of $U(k)/U(k)$) that was analyzed in §2.

There is a basic principle here: every quantum field theory with a mass gap reduces at very big distances to a topological field theory. Often the topological field theory that so arises is more or less trivial, but in the case of the supersymmetric sigma model of the Grassmannian, it is the gauged WZW model. This large distance reduction of the Grassmannian sigma model to a gauged WZW model, plus the relation explained in §2 between the gauged WZW model and the Verlinde algebra, give the relation between the quantum cohomology of the Grassmannian and the Verlinde algebra.

It is well known that at large distances, massive particles can be neglected and massless particles dominate. Less fully appreciated is that beyond the reach of the propagating fields, a non-trivial dynamics of the vacuum or topological field theory may prevail.

Differential Geometry Of The Moduli Space Of Bundles

A detailed discussion of the reduction of the sigma model to the gauged WZW model will be the subject of §4, but here I will make a few naive remarks. The integrand in (3.64) actually has its maximum for flat connections – with some branching at the points $P, Q, R \in \Sigma$ at with $\widehat{\alpha}, \widehat{\beta}$, and $\widehat{\gamma}$ are inserted. The moduli space of such flat connections is (by a theorem of Mehta and Seshadri *[28]*) the same as the moduli space $\mathcal{R}$ of rank k stable holomorphic vector bundles over Σ, with some parabolic structure at P, Q, R determined by the branching. So the integral gives some differential geometry of $\mathcal{R}$. (In view of (2.66), $\mathcal{R}$ can be interpreted as a space of classical solutions of the gauged WZW model.) In the large N limit, direct analysis of the determinant in (3.64) shows that the differential geometric quantity that appears is the volume of the symplectic manifold $\mathcal{R}$, times $N^{\dim_{\mathbf{C}} \mathcal{R}}$. This is the leading large N behavior of the Riemann-Roch formula for the dimension of the space $H^0(\mathcal{R}, \mathcal{L}^{N-k})$ of non-abelian theta functions. This simple direct argument relates the quantum cohomology of $G(k, N)$ to the dimension of the space of non-abelian theta functions for large N.

This simple direct argument relates the quantum cohomology of $G(k, N)$ to the dimension of the space of non-abelian theta functions for large N.

The only way I know to establish this as an exact relation, not just an asymptotic one for large N, is to reduce the sigma model of $G(k, N)$ to the gauged WZW model as we will do in §4, and then study that model as in §2.

4 From The Grassmannian To The G/G Model

This section is organized as follows. After recalling some background about $\mathcal{N} = 2$ models in two dimensions in §4.1, we construct in §4.2 the sigma model whose target space is the Grassmannian $G(k, N)$. Then we analyze its behavior at long distances in §4.3-6. In §4.7-8, we enter the computational stage and work things out in detail in the simplest non-trivial case.

In the past, the long distance behavior of the Grassmannian sigma model has been analyzed on $\mathbf{R}^2$ *[7]–[10]*; the main results were spontaneously broken chiral symmetry, the existence of a mass gap, and a determination for large N of the spectrum of low-lying states. The novelty here is to examine the long distance behavior more globally, uncovering the relation to the gauged WZW model and thereby (in view of §2) the Verlinde algebra.

I will make a small change in notation in this section. In §2, we considered general compact Lie groups, and (as is conventional in mathematics) we took the Lie algebra to consist of anti-hermitian matrices (so the quadratic form $(a, b) = \mathrm{Tr}\, ab$ is negative definite). The reason that this convention is standard for general Lie groups is that in the case of a real group, whose representations may also all be real, it is unnatural to introduce factors of i and therefore the group generators are naturally anti-hermitian. In this section, the gauge group will be the unitary group $U(k)$, which will arise in a natural complex representation, and I will follow the standard physics convention that the group generators are hermitian matrices; thus $(a, b) = \mathrm{Tr}\, ab$ will be positive definite. The complexification of the Lie algebra of $U(k)$ consists of all $k \times k$ complex matrices; if σ is such a matrix, then $\overline{\sigma}$ will denote its hermitian adjoint.

4.1 Background

We will work in $\mathcal{N} = 2$ superspace in two dimensions, conventions and the basic setup being as explained in *[40, 41]*. The detailed formulas of this subsection are presented mainly for reference, and most readers will want to skim them.

We consider first flat superspace with bosonic coordinates x^m, $m = 0, 1$ (and Lorentz signature $-+$) and fermionic coordinates θ^α, $\overline{\theta}^\alpha$. In a light-cone basis, supersymmetry is realized geometrically by the operators

$$\begin{aligned} Q_\pm &= \frac{\partial}{\partial \theta^\pm} + i\overline{\theta}^\pm \left(\frac{\partial}{\partial x^0} \pm \frac{\partial}{\partial x^1} \right) \\ \overline{Q}_\pm &= -\frac{\partial}{\partial \overline{\theta}^\pm} - i\theta^\pm \left(\frac{\partial}{\partial x^0} \pm \frac{\partial}{\partial x^1} \right). \end{aligned} \tag{4.1}$$

These operators commute with the superspace covariant derivatives

$$
\begin{aligned}
D_\pm &= \frac{\partial}{\partial\theta^\pm} - i\overline{\theta}^\pm\left(\frac{\partial}{\partial x^0} \pm \frac{\partial}{\partial x^1}\right)\\
\overline{D}_\pm &= -\frac{\partial}{\partial\overline{\theta}^\pm} + i\theta^\pm\left(\frac{\partial}{\partial x^0} \pm \frac{\partial}{\partial x^1}\right)
\end{aligned}
\tag{4.2}
$$

which are used in constructing Lagrangians.

To formulate gauge theory, one introduces a gauge field in superspace, replacing the differential operators D_α, $\overline{D}_\alpha$, and $\partial_m = \partial/\partial x^m$ by gauge covariant derivatives $\mathcal{D}_\alpha$, $\overline{\mathcal{D}}_\alpha$, and $\mathcal{D}_m$. On the superspace gauge fields one imposes the very strong constraints

$$
\begin{aligned}
0 = \{\overline{\mathcal{D}}_\alpha, \overline{\mathcal{D}}_\beta\} &= \{\mathcal{D}_\alpha, \mathcal{D}_\beta\}\\
\{\mathcal{D}_\pm, \overline{\mathcal{D}}_\pm\} &= 2i\left(\mathcal{D}_0 \pm \mathcal{D}_1\right).
\end{aligned}
\tag{4.3}
$$

Among other things, these conditions permit the existence of "chiral superfields," superspace fields Φ obeying

$$
\overline{\mathcal{D}}_\alpha\Phi = 0. \tag{4.4}
$$

With the aid of the constraints one can take locally

$$
\begin{aligned}
\mathcal{D}_\alpha &= e^{-V}D_\alpha e^V\\
\overline{\mathcal{D}}_\alpha &= e^V\overline{D}_\alpha e^{-V}
\end{aligned}
\tag{4.5}
$$

where V is a real Lie algebra-valued function on superspace, called a vector superfield. After also fixing some residual gauge invariance, one can go to a "Wess-Zumino gauge," in which

$$
\begin{aligned}
V = {}&\theta^-\overline{\theta}^-(v_0 - v_1) + \theta^+\overline{\theta}^+(v_0 + v_1) - \sqrt{2}\sigma\theta^-\overline{\theta}^+ - \sqrt{2}\overline{\sigma}\theta^+\overline{\theta}^-\\
&+2i\theta^-\theta^+\left(\overline{\theta}^-\overline{\lambda}_- + \overline{\theta}^+\overline{\lambda}_+\right) + 2i\overline{\theta}^+\overline{\theta}^-\left(\theta^+\lambda_+ + \theta^-\lambda_-\right) + 2\theta^-\theta^+\overline{\theta}^+\overline{\theta}^- D.
\end{aligned}
\tag{4.6}
$$

Here v_m is an ordinary two-dimensional gauge field, and the other fields are bose and fermi matter fields. σ is a complex $k \times k$ matrix, and – as V is hermitian – $\overline{\sigma}$ is its hermitian adjoint. We write $F = F_{01} = \partial_0 v_1 - \partial_1 v_0 + [v_0, v_1]$ for the curvature of v. The supersymmetry transformation laws for this multiplet are

$$\begin{aligned}
\delta v_m &= i\bar{\epsilon}\sigma_m\lambda + i\epsilon\sigma_m\bar{\lambda} \\
\delta\sigma &= -i\sqrt{2}\bar{\epsilon}_+\lambda_- - i\sqrt{2}\epsilon_-\bar{\lambda}_+ \\
\delta\bar{\sigma} &= -i\sqrt{2}\epsilon_+\bar{\lambda}_- - i\sqrt{2}\bar{\epsilon}_-\lambda_+ \\
\delta D &= -\bar{\epsilon}_+(D_0 - D_1)\lambda_+ - \bar{\epsilon}_-(D_0 + D_1)\lambda_- \\
&\quad +\epsilon_+(D_0 - D_1)\bar{\lambda}_+ + \epsilon_-(D_0 + D_1)\bar{\lambda}_- \\
&\quad +\sqrt{2}\epsilon_+[\sigma, \bar{\lambda}_-] + \sqrt{2}\epsilon_-[\bar{\sigma}, \bar{\lambda}_+] + \sqrt{2}[\sigma, \lambda_+]\bar{\epsilon}_- + \sqrt{2}[\bar{\sigma}, \lambda_-]\bar{\epsilon}_+ \\
\delta\lambda_+ &= i\epsilon_+ D + \sqrt{2}(D_0 + D_1)\bar{\sigma}\epsilon_- - F_{01}\epsilon_+ - [\sigma, \bar{\sigma}]\epsilon_+ \\
\delta\lambda_- &= i\epsilon_- D + \sqrt{2}(D_0 - D_1)\sigma\epsilon_+ + F_{01}\epsilon_- + [\sigma, \bar{\sigma}]\epsilon_- \\
\delta\bar{\lambda}_+ &= -i\bar{\epsilon}_+ D + \sqrt{2}(D_0 + D_1)\sigma\bar{\epsilon}_- - F_{01}\bar{\epsilon}_+ + [\sigma, \bar{\sigma}]\bar{\epsilon}_+ \\
\delta\bar{\lambda}_- &= -i\bar{\epsilon}_- D + \sqrt{2}(D_0 - D_1)\bar{\sigma}\bar{\epsilon}_+ + F_{01}\bar{\epsilon}_- - [\sigma, \bar{\sigma}]\bar{\epsilon}_-.
\end{aligned} \tag{4.7}$$

The basic gauge invariant field strength is

$$\begin{aligned}
\Sigma = \tfrac{1}{2\sqrt{2}}\{\overline{\mathcal{D}}_+, \mathcal{D}_-\} &= \sigma + i\sqrt{2}\theta^+\bar{\lambda}_+ - i\sqrt{2}\bar{\theta}^-\lambda_- + \sqrt{2}\theta^+\bar{\theta}^- D \\
&\quad -i\bar{\theta}^-\theta^-(D_0 - D_1)\sigma - i\theta^+\bar{\theta}^+(D_0 + D_1)\sigma \\
&\quad +\sqrt{2}\bar{\theta}^-\theta^-\theta^+(D_0 - D_1)\bar{\lambda}_+ - \sqrt{2}\theta^+\bar{\theta}^+\bar{\theta}^-(D_0 + D_1)\lambda_- - i\sqrt{2}\theta^+\bar{\theta}^- F_{01} \\
&\quad -2i\theta^-\bar{\theta}^-\theta^+[\sigma, \bar{\lambda}_-] - 2i\bar{\theta}^-\theta^+\bar{\theta}^+[\sigma, \lambda_+] \\
&\quad -\bar{\theta}^-\theta^-\theta^+\bar{\theta}^+\left((D_0{}^2 - D_1{}^2)\sigma - [\sigma, [\sigma, \bar{\sigma}]]\right) + i\theta^-\bar{\theta}^-\theta^+\bar{\theta}^+[\sigma, \partial_m v^m]
\end{aligned} \tag{4.8}$$

(The last term does not really spoil gauge invariance: the gauge transformations that preserve Wess-Zumino gauge have a certain θ dependence which requires this term to be present.) Σ is a twisted chiral superfield; this means that (by the Bianchi identity together with the constraints)

$$\overline{\mathcal{D}}_+\Sigma = \mathcal{D}_-\Sigma = 0. \tag{4.9}$$

With the aid of Σ, it is straightforward to construct gauge invariant Lagrangians. The standard gauge kinetic energy is

$$\begin{aligned} L_g &= -\tfrac{1}{4e^2}\int d^2x\, d^4\theta \operatorname{Tr}\overline{\Sigma}\Sigma \\ &= \tfrac{1}{e^2}\int d^2x \operatorname{Tr}\left(\tfrac{1}{2}F_{01}^2 + |D_0\sigma|^2 - |D_1\sigma|^2 + i\overline{\lambda}_-(D_0+D_1)\lambda_-\right. \\ &\quad \left. + i\overline{\lambda}_+(D_0-D_1)\lambda_+ + \tfrac{1}{2}D^2 - \tfrac{1}{2}[\sigma,\overline{\sigma}]^2 - \sqrt{2}\lambda_+[\sigma,\overline{\lambda}_-] + \sqrt{2}[\overline{\sigma},\lambda_-]\overline{\lambda}_+\right) \end{aligned} \tag{4.10}$$

One more term constructed from gauge fields only is important. Using the fact that Σ is a twisted chiral superfield, there is an invariant interaction of the form

$$\begin{aligned} L_{D,\theta} &= \tfrac{it}{2\sqrt{2}}\int d^2x\, d\theta^+\, d\overline{\theta}^- \operatorname{Tr}\Sigma|_{\theta^-=\overline{\theta}^+=0} + c.c. \\ &= \int d^2x\left(-r\operatorname{Tr}D + \tfrac{\theta}{2\pi}\operatorname{Tr}F_{01}\right), \end{aligned} \tag{4.11}$$

with

$$t = ir + \frac{\theta}{2\pi}. \tag{4.12}$$

Matter Fields

Chiral superfields are functions Φ on superspace, transforming in some given unitary representation V of the gauge group, and obeying

$$\overline{\mathcal{D}}_\pm\Phi = 0. \tag{4.13}$$

Such a field has an expansion in components

$$\Phi = \phi + \sqrt{2}\theta^\alpha\psi_\alpha + \theta^\alpha\theta_\alpha F. \tag{4.14}$$

The supersymmetry transformation laws for this multiplet are (by dimensional reduction from *[40]*, p. 50)

$$\begin{aligned} \delta\phi &= \sqrt{2}\left(\epsilon_+\psi_- - \epsilon_-\psi_+\right) \\ \delta\psi_+ &= i\sqrt{2}\left(D_0+D_1\right)\phi\overline{\epsilon}_- + \sqrt{2}\epsilon_+F - 2\overline{\sigma}\phi\overline{\epsilon}_+ \\ \delta\psi_- &= -i\sqrt{2}\left(D_0-D_1\right)\phi\overline{\epsilon}_+ + \sqrt{2}\epsilon_-F + 2\sigma\phi\overline{\epsilon}_- \\ \delta F &= -i\sqrt{2}\overline{\epsilon}_+\left(D_0-D_1\right)\psi_+ - i\sqrt{2}\overline{\epsilon}_-\left(D_0+D_1\right)\psi_- \\ &\quad +2\left(\overline{\epsilon}_+\overline{\sigma}\psi_- + \overline{\epsilon}_-\sigma\psi_+\right) + 2i\left(\overline{\epsilon}_-\overline{\lambda}_+ - \overline{\epsilon}_+\overline{\lambda}_-\right)\phi \end{aligned} \tag{4.15}$$

The usual kinetic energy for a multiplet of such chiral superfields is

$$L_{ch} = \tfrac{1}{4}\int d^2x\, d^4\theta\, \overline{\Phi}\Phi = \int d^2x\, \Big(|D_0\phi|^2 - |D_1\phi|^2 + |F|^2 + i\overline{\psi}_+(D_0 - D_1)\psi_+$$
$$+ i\overline{\psi}_-(D_0 + D_1)\psi_- + \overline{\phi}D\phi - \overline{\phi}\{\sigma,\overline{\sigma}\}\phi$$
$$-\sqrt{2}\,\overline{\psi}_+\overline{\sigma}\psi_- - \sqrt{2}\,\overline{\psi}_-\sigma\psi_+ + i\sqrt{2}\,\overline{\psi}_+\overline{\lambda}_-\phi - i\sqrt{2}\,\overline{\psi}_-\overline{\lambda}_+\phi$$
$$+ i\sqrt{2}\,\overline{\phi}\lambda_+\psi_- - i\sqrt{2}\,\overline{\phi}\lambda_-\psi_+\Big). \tag{4.16}$$

The $|D_\alpha\phi|^2$ term in (4.16) is the conventional free kinetic energy corresponding to a sigma model with a flat metric on $V \cong \mathbf{C}^r$. The $\overline{\phi}D\phi$ term is the coupling of D to the moment map, in the sense that if we pick a basis T_a, $a = 1\ldots \dim G$ for the Lie algebra of G, then this term is

$$\int d^2x \sum_a D^a(\overline{\phi}, T_a\phi), \tag{4.17}$$

and the functions $(\overline{\phi}, T_a\phi)$ are the components of the moment map.

A more general Kähler metric on V (and accordingly, a more general form of the moment map) could be obtained by replacing the function $\overline{\Phi}\Phi$ on the left hand side of (4.16) with a more general Kähler potential $K(\Phi,\overline{\Phi})$. These matters are explained in some detail in *[42]*.

4.2 The Model

Now we can construct the actual model of interest. We take the gauge group to be $G = U(k)$. We take kN chiral superfields Φ^{is}, $i = 1\ldots k$, $s = 1\ldots N$, regarded as N copies of the defining k dimensional representation of G. The action of G commutes with a global symmetry group $H \cong U(N)$ which one can think of as the unitary transformations of $\mathbf{C}^N$.

The Lagrangian that we actually wish to study is simply

$$L = L_{gauge} + L_{D,\theta} + L_{ch}. \tag{4.18}$$

The potential energy is determined by the following terms in L:

$$L_{pot} = \frac{1}{2e^2}\mathrm{Tr}\, D^2 - r\mathrm{Tr}\, D + \overline{\phi}D\phi - \frac{1}{2e^2}\mathrm{Tr}\,[\sigma,\overline{\sigma}]^2 - \overline{\phi}\{\sigma,\overline{\sigma}\}\phi. \tag{4.19}$$

Upon integrating out D, the potential energy is

$$V = \frac{e^2}{2}\sum_{i,j=1}^{k}\left(\sum_s \overline{\phi}_{is}\phi^{js} - \delta_i{}^j r\right)^2 + \frac{1}{2e^2}\mathrm{Tr}\,[\sigma,\overline{\sigma}]^2 + \overline{\phi}\{\sigma,\overline{\sigma}\}\phi. \tag{4.20}$$

The space of classical vacua is the space of zeroes of V up to gauge transformation. For V to vanish, ϕ must be non-zero, and therefore σ must vanish. As anticipated in §3, the first term in the potential is the square of the moment map for the action of $U(k)$ on $\mathbf{C}^{kN}$. This term vanishes precisely if the vectors in $\mathbf{C}^N$ represented by the rows of ϕ, divided by $\sqrt{r}$, are orthonormal. The k dimensional subspace $V \subset \mathbf{C}^N$ spanned by the rows of ϕ is gauge invariant, and is the only gauge invariant data determined by ϕ (since any two orthonormal bases of V are related by the action of $U(k)$). Moreover, every k dimensional subspace $V \subset \mathbf{C}^N$ has such an orthonormal basis. So the space of classical vacua is the Grassmannian $G(k, N)$ of k dimensional subspaces of $\mathbf{C}^N$.

Since the condition for vanishing energy is

$$\sum_s \overline{\phi}_{is}\phi^{js} = \delta_i{}^j r, \tag{4.21}$$

the radius of the space of vacua is $\sqrt{r}$ and the Kähler class is proportional to r. Classically, the space of vacua shrinks to a point for $r = 0$; for $r < 0$ the classical energy can no longer vanish and it appears that supersymmetry is spontaneously broken. Quantum mechanically, the situation is rather different and there is a smooth continuation to negative r with unbroken supersymmetry, as discussed (for $k = 1$) in (*[41, §3.2]*); the existence of this continuation will be exploited below.

The choice of a classical vacuum spontaneously breaks the symmetry group $U(N)$ to $U(k) \times U(N-k)$, while leaving supersymmetry unbroken. The oscillations in the vacuum are massless Goldstone bosons at the classical level. Their supersymmetric partners are, of course, also massless classically. Other modes are readily seen to have masses proportional to e. The model therefore reduces at long distances (or equivalently for $e \to \infty$) to the supersymmetric nonlinear sigma model with target space the Grassmannian; we will more briefly call this the Grassmannian sigma model.

At the quantum level, spontaneous breaking of a continuous symmetry such as the $U(N)$ symmetry of this model is not possible in two dimensions. The symmetry must therefore be restored by quantum corrections. Exhibiting this symmetry restoration, and the associated mass gap, was a primary goal of early investigations of the model.

R Symmetries

A right-moving R-symmetry in an $\mathcal{N} = 2$ model in two dimensions is a symmetry under which $\theta^+ \to e^{i\alpha}\theta^+$, $\overline{\theta}^+ \to e^{-i\alpha}\overline{\theta}^+$, while $\theta^-, \overline{\theta}^-$ are invariant. A left-moving R-symmetry obeys the analogous condition with θ^+ and θ^- exchanged.

The Grassmannian sigma model as constructed above has at the classical level a right-moving R-symmetry J_R[8] under which the charges of the various fields are as follows: $(\psi_+, F, \sigma, \lambda\ )$ have charges $(-1, -1, 1, 1)$, their complex conjugates have opposite charge, and other fields have charge zero. Similarly

[8] We will somewhat imprecisely use the symbol J_R to denote either the current or the corresponding charge; and similarly for other currents introduced momentarily.

there is classically a left-moving R-symmetry J_L under which $(\psi_-, F_i, \sigma, \lambda_+)$ have charges $(-1,-1,-1,1)$, their complex conjugates have opposite charges, and other fields are neutral.

At the quantum level, the sum $J_V = J_R + J_L$ is a "vector" symmetry, that is, it transforms left- and right-moving fermions the same way, so it is free of anomaly and generates a $U(1)$ symmetry.

However, the "axial" combination $J_A = J_R - J_L$ is anomalous. The anomaly can be described as follows. Let a be a $U(1)$ connection with first Chern class 1, and embed this in $G = U(K)$ so that the $U(k)$ gauge field is $v = \mathrm{diag}(a, 0, 0, \ldots, 0)$. In such an instanton field, the index of the Dirac operator acting on ψ_+ is N. [9] Similarly the ψ_- index is $-N$. The total anomaly in $J_A = J_R - J_L$ is the difference of these or $2N$. The anomaly in any instanton field would be an integer multiple of this.

So J_A is conserved only modulo $2N$. The only symmetries we can construct from J_A are the discrete transformations $\exp(2\pi i t J_A/2N)$, with $t \in \mathbf{Z}$. This gives a discrete group, isomorphic to $\mathbf{Z}_{2N}$, of chiral symmetries. If unbroken, these symmetries would prevent ψ and λ from gaining a mass. One of the main results of the old literature on this model was that this $\mathbf{Z}_{2N}$ is spontaneously broken down to $\mathbf{Z}_2$, making a mass gap possible; the surviving $\mathbf{Z}_2$ is just the operation $(-1)^F$ that counts fermions modulo two.

The Twisted Model

Any $\mathcal{N} = 2$ supersymmetric theory in two dimensions with an R symmetry can be twisted to obtain a topological field theory. The construction, as explained in *[43]*, which the reader can consult for details, involves adding to the usual stress tensor the derivative of the R-current. As we have just seen, in the case of the Grassmannian there is only one anomaly-free R-symmetry. Consequently, only one twisted topological field theory can be constructed; it is related to the quantum cohomology of the Grassmannian, which was introduced in §3.

In going from the untwisted to the twisted model, the spin of every field decreases (in the convention of *[41]*) by $J_V/2$. For instance, in the untwisted model, ψ_+ and $\overline{\psi}_+$ have spin $1/2$ and $J_V = \mp 1$, so in the twisted model they have respectively spin 1 and 0. More generally the fermi fields that have spin zero in the twisted model are $\overline{\psi}_+, \psi_-, \overline{\lambda}_-, \lambda_+$. If the twisted model is formulated on $\mathbf{S}^2$, the spin zero fields each have one zero mode and the spin one fields have none. Since $\overline{\psi}_+, \psi_-, \overline{\lambda}_-, \lambda_+$ have kN, kN, k^2, k^2 components respectively and have $J_A = 1, 1, -1, -1$, the total J_A value of the zero modes is $kN + kN - k^2 - k^2 = 2k(N-k)$, and this is the anomaly in J_A conservation due to coupling to the curvature of S^2. (Not coincidentally, $2k(N-k)$ is the dimension of $G(k, N)$.) More generally, on a surface of genus g, the spin one fields would have g zero

[9] The index is defined as the number of ψ_+ zero modes minus the number of $\overline{\psi}_+$ zero modes. Recall that ψ_+ transforms as a sum of N copies of the defining k-dimensional representation of $U(k)$; each of these contributes 1 to the index.

modes, so the violation of J_A is

$$\Delta J_A = 2k(N-k)(1-g) = k(N-k)\int_\Sigma d^2x\sqrt{h}\frac{R}{2\pi}. \qquad (4.22)$$

Here I have written the Euler characteristic of Σ, which of course equals $2(1-g)$, as the familiar curvature integral.

Fermionic Symmetry Of The Twisted Theory

The untwisted theory, formulated on a flat world-sheet, possesses fermionic symmetries, that were described in detail in equations (4.7), (4.15). After twisting, the fermionic parameters ϵ_+ and $\overline{\epsilon}_-$ in the transformation laws have spin zero; let Q_- and $\overline{Q}_+$ be the symmetries generated by those transformations, and let $Q = Q_- + \overline{Q}_+$. By a standard calculation, ${Q_-}^2 = {\overline{Q}_+}^2 = \{Q_-, \overline{Q}_+\} = 0$ and in particular $Q^2 = 0$. Moreover, the stress tensor can be written as $T = \{Q, \Lambda\}$ for some Λ. It follows that if we restrict ourselves to operators that are annihilated by Q (or more exactly to cohomology classes of such operators), the theory can be interpreted as a topological field theory. Each cohomology class of Q-invariant operators has representatives annihilated by both Q_- and $\overline{Q}_+$.

The relevant observables are easily found. The transformation laws of the topological theory are found from the microscopic transformation laws (4.7), (4.15) by setting $\epsilon_- = \overline{\epsilon}_+ = 0$ and keeping $\epsilon_+, \overline{\epsilon}_-$. By inspection of the transformation laws, σ (but not $\overline{\sigma}$) is invariant, so that any gauge invariant holomorphic function of σ is a suitable vertex operator in the topological theory. Such functions are linear combinations of characters, so the basic operators constructed this way are

$$O_V(x) = \mathrm{Tr}_V\,\sigma(x), \qquad (4.23)$$

with V an irreducible representation of $G = U(k)$ and Tr_V the trace in that representation.

Actually, these are the only relevant operators. In fact, even before twisting, the model (for $r >> 0$) is equivalent at long distances, as we saw above, to a sigma model with target the Grassmannian $G(k, N)$. Consequently, the twisted model is simply the standard A model of $G(k, N)$ (the A model for any Kähler target is explained in detail in *[43]*) so the cohomology classes of observables are in one-to-one correspondence with the de Rham cohomology of $G(k, N)$.

Indeed, upon integrating out the massive fields, $\sigma(x)$ turns into a bilinear expression in massless fermions tangent to $G(k, N)$, with values in the adjoint representation of $U(k)$. It is easy to calculate this explicitly in the weak coupling, low energy limit. To this aim, we need only evaluate a tree diagram, and we can ignore the kinetic energy of the massive σ field. The relevant part of the Lagrangian is simply

$$-\int d^2x \left(\sum_{ijs}\overline{\phi}_{is}\{\sigma,\overline{\sigma}\}^i{}_j\phi^{js} + \sqrt{2}\sum_{ijs}\overline{\psi}_{+is}\overline{\sigma}^i{}_j\psi_-{}^{js}\right). \qquad (4.24)$$

Because of the overall $U(N)$ invariance, it suffices to work out the effective operator representing σ in the low energy theory at one particular point on $G(k,N)$. We take this to be the point represented by $\phi^{is} = \sqrt{r}\delta^{is}$ for $1 \leq s \leq k$, $\phi^{is} = 0$ for $s > k$. With this choice, (4.24) becomes

$$-\int d^2x \left(2r\mathrm{Tr}\,\overline{\sigma}\sigma + \sqrt{2}\sum_{ijs} \overline{\psi}_{+is}\overline{\sigma}^i{}_j\psi_-{}^{js} \right). \tag{4.25}$$

The quickest way to evaluate the tree diagram is simply to impose the equation of motion of $\overline{\sigma}$; this gives

$$\sigma^j{}_i = -\frac{1}{r\sqrt{2}}\sum_s \overline{\psi}_{+is}\psi_-^{js}. \tag{4.26}$$

In the interpretation of the low energy theory in terms of differential forms on $G(k,N)$, $\psi_-/\sqrt{r}$ and $\overline{\psi}_+/\sqrt{r}$ are $(1,0)$ and $(0,1)$ forms. (ψ_- and $\overline{\psi}_+$ have been normalized to have canonical kinetic energies; their natural normalization as differential forms involves dividing by $\sqrt{r}$.) So σ is represented in the low energy theory by a $(1,1)$ form or more exactly by the operator in the $G(k,N)$ model determined by this $(1,1)$ form.

The chosen vacuum $\phi^{is} = \sqrt{r}\delta^{is}$ is invariant up to a gauge transformation under a subgroup $U(k) \times U(N-k)$ of $U(N)$. For σ to be a $U(N)$-invariant form in the adjoint representation of the gauge group, it must transform in the adjoint representation of the unbroken $U(k)$ (since the unbroken symmetry is a mixture of this with a gauge transformation) and be invariant under $U(N-k)$. The $U(N)$ action can then be used to extend σ in a unique way to an invariant $(1,1)$ form on $G(k,N)$. It is evident that the right hand side of (4.26) has the required properties.

Conversely, the right hand side of (4.26) is the only bilinear expression in $\overline{\psi}_+$ and ψ_- with the claimed properties, so any adjoint-valued $U(N)$-invariant $(1,1)$ form would be a multiple of σ. Such a form is the curvature of the tautological $U(k)$ bundle E^* with its natural connection. So up to a constant, which I will not verify directly (it can be absorbed in the constant later called c), σ coincides in the low energy theory with the tautological curvature. Hence classical expressions $O_V = \mathrm{Tr}_V\sigma$ coincide with the corresponding polynomials in Chern classes on $G(k,N)$, and as quantum operators in the twisted theory, the O_V coincide with the elements of the quantum cohomology determined by those classes. The fact that the tautological classes generate the cohomology of $G(k,N)$ ensures that the O_V span the space of observables of the twisted theory.

There is a more conceptual approach to identifying σ with the tautological curvature which I will indicate very briefly. Let $\lambda_- = \eta_0 - \eta_1$, $\overline{\lambda}_+ = \eta_0 + \eta_1$. Restrict to the diagonal fermionic symmetry with $\epsilon_+ = \overline{\epsilon}_- = \epsilon$. Then a key part of the symmetry algebra is

$$\begin{aligned} \delta v_m &= 2i\epsilon\eta_m \\ \delta\eta_m &= \sqrt{2}\epsilon D_m\sigma \\ \delta\sigma &= 0. \end{aligned} \tag{4.27}$$

This multiplet describes the equivariant cohomology of the gauge group acting on the space $\mathcal{A}$ of connections. The interpretation of σ as the curvature of the tautological bundle over the quotient is standard in equivariant cohomology. This interpretation holds independent of any specific Lagrangian model; the salient feature of the particular model we are considering is that the connection v_m is identified via the low energy equations of motion with the pullback of the tautological connection on $E^* \to G(k, N)$.

Instantons

A correlation function

$$\left\langle \prod_i O_{V_i}(x_i) \right\rangle \tag{4.28}$$

on a Riemann surface Σ can be computed as follows. The O_{V_i} determine cohomology classes of $G(k, N)$ as we have just seen; pick Poincaré dual cycles H_i. Let d be the non-negative integer, if any, such that the moduli space of holomorphic maps $\Phi : \Sigma \to G(k, N)$ obeying

$$\Phi(x_i) \in H_i \tag{4.29}$$

has virtual dimension zero. The correlation function (4.28) is zero if such a d does not exist; otherwise it is

$$\left\langle \prod_i O_{V_i}(x_i) \right\rangle = \exp(-dr) \cdot N_{\{H_i\}} \tag{4.30}$$

with $N_{\{H_i\}}$ the "number" of holomorphic maps $\Phi : \Sigma \to G(k, N)$ that obey (4.29). (In general, in defining this number, one must make a suitable perturbation of the equation to avoid possible degeneracies; that is why I have put the word "number" in quotes.) (4.30) follows from the standard description of the A model, as explained in *[43]*.

More microscopically, to see the appearance of instantons, one can begin with the transformation laws (4.7), (4.15). The calculation of the correlation function in (4.28) can be localized, by a standard argument, on the fixed points of $Q_-, \overline{Q}_+$. An analysis as in *[41]*, pp. 184-8, identifies those fixed points (for $r >> 0$) with the holomorphic maps of Σ to the Grassmannian. Those holomorphic maps appear in precisely the form in which they were studied by Bertram, Daskalopoulos, and Wentworth *[11]*.

4.3 Some Renormalization Factors

Before analyzing the quantum theory, I want to first point out a few details involving renormalization.

Any topological field theory in two dimensions could be modified by the addition of a term

$$\Delta L = a \int_{\Sigma} d^2x \sqrt{h} \frac{R}{2\pi} \tag{4.31}$$

without affecting the topological invariance. The affect of this is merely to multiply a genus g amplitude by a factor of $\exp(a(2-2g))$.

An important role in the analysis will be played by the Kähler parameter r. For instance, r enters in the basic formula (4.30) expressing correlation functions in terms of instantons. However, as we see in (4.30), the r dependence of a degree d instanton contribution is known *a priori*. Moreover, because of the $\mathbf{Z}$ grading of the classical cohomology, every given correlation function in genus g receives a contribution at most only from one known value of d. Therefore, there is no material loss in setting r to 0, and we will do that eventually in §4.7.

Another normalization question involves the possibility of multiplying an operator of degree w by a factor $\exp(uw)$ with some constant u. One can show by keeping track of the classical $\mathbf{Z}$ grading that this can be absorbed in adding constants to r and a. This normalization question will arise below because we will find that the field σ of the Grassmannian sigma model has a macroscopic interpretation as

$$\sigma = cg, \tag{4.32}$$

where c is a constant that we will determine only approximately and g is the elementary field of a gauged WZW model.

In practice, in our computations we will not try to determine the precise values of a and c. At the end, when we enter the computational stage. we will identify the values of these parameters by checking a couple of special cases of the formulas.

4.4 Quantum Properties Of The Model

We come finally to the point of the present paper – the calculation mapping the Grassmannian sigma model onto the G/G model, and thence the Verlinde algebra. The calculations themselves are not new *[7] –[10]*, and I will therefore present them rather briefly. What is new is the result that we will get by considering these computations in a global context.

We begin with the expression for D in terms of matter fields that is obtained by varying the potential energy term (4.19) with respect to D:

$$-\frac{1}{e^2} D^i{}_j = \sum_{s=1}^{N} \phi^{is} \overline{\phi}_{js} - r\delta^i_s. \tag{4.33}$$

At the classical level, for $r >> 0$, vanishing of the $D^i{}_j$ – which is needed to set the energy to zero – requires that the ϕ^{is} should have non-zero vacuum expectation values. This in turn spontaneously breaks the global $U(N)$ symmetry (and ensures the existence of massless Goldstone bosons and the absence of a mass gap). Such spontaneous breaking of a continuous symmetry is, however, impossible in two dimensions.

The resolution of this conundrum has long been known. Quantum mechanically the operator $O^i{}_j = \sum_{s=1}^N \phi^{is}\overline{\phi}_{js}$ can have an expectation value even if the ϕ^{is} do not. If this expectation value can equal $r\delta^i{}_j$, then the $D^i{}_j$ can vanish without spontaneous breaking of the $U(N)$ symmetry.

To investigate this phenomenon, let us compute the expectation value of $O^i{}_j$. We will first do this in a naive approximation, treating the ϕ's as free fields with the mass term that we can read off from the classical Lagrangian. Then we will discuss the conditions for validity of the approximation. We will do the calculation on Euclidean $\mathbf{R}^2$, making the standard Wick rotations from the Lorentz signature Lagrangian given above.

The mass term for the ϕ field in the Lagrangian is $\sum_{i,j,s}\overline{\phi}_{is}\{\sigma,\overline{\sigma}\}^i{}_j\phi^{is}$, with $\{\cdot,\cdot\}$ the anticommutator. Treating the ϕ's as free fields with that mass term, the expectation value of $O^i{}_j$ is simply

$$\langle O\rangle = N\int\frac{d^2k}{(2\pi)^2}\frac{1}{k^2+\{\sigma,\overline{\sigma}\}}. \tag{4.34}$$

The factor of N comes from summing over s.

The integral in (4.34) is logarithmically divergent. The divergence can be regularized by subtracting a similar integral with $\{\sigma,\overline{\sigma}\}$ replaced by a multiple of the identity, say $2\mu^2$, with μ an arbitrary "subtraction point." The subtraction can be interpreted as an additive renormalization of r. After this regularization, the integral can be evaluated, and one gets

$$\langle O\rangle = -\frac{N}{4\pi}\ln\left(\{\sigma,\overline{\sigma}\}/2\mu^2\right). \tag{4.35}$$

The condition for D to vanish in this approximation is hence that

$$-\frac{N}{4\pi}\ln\left(\{\sigma,\overline{\sigma}\}/2\mu^2\right) - r = 0, \tag{4.36}$$

or

$$\{\sigma,\overline{\sigma}\} = 2\mu^2\exp\left(-4\pi r/N\right). \tag{4.37}$$

This is however only a necessary condition for vanishing of the energy. Another condition comes from the presence in the classical Lagrangian of a term proportional to $\mathrm{Tr}\,[\sigma,\overline{\sigma}]^2$. This term gives a contribution to the energy that vanishes precisely when $[\sigma,\overline{\sigma}] = 0$, so in seeking to describe the vacuum, we may assume that σ and $\overline{\sigma}$ commute and therefore rewrite (4.37) in the form

$$\sigma\overline{\sigma} = \mu^2\exp\left(-4\pi r/N\right). \tag{4.38}$$

This means that

$$\sigma = cg \tag{4.39}$$

with g a unitary matrix – $g\bar{g} = 1$ – and c the constant

$$c = \mu \exp\left(-2\pi r/N\right). \tag{4.40}$$

Thus, we have obtained a kind of sigma model with a field g taking values in the unitary group.

The vacuum expectation value of σ that we have just found gives a positive mass squared to the ϕ^{is}, so that they will have zero vacuum expectation value, restoring the $U(N)$ symmetry. However, the discrete chiral symmetry (conservation of J_A modulo $2N$) is spontaneously broken in this process. Indeed, since σ has $J_A = 2$, the vacuum expectation value of σ breaks $\mathbf{Z}_{2N}$ down to $\mathbf{Z}_2$. (For instance, this is discussed in detail for $k = 1$ on p. 310 of *[8]*.) As the broken symmetry is discrete, this does not produce Goldstone bosons and is compatible with the existence of a mass gap. In fact, the broken symmetry helps in getting a mass gap, since most of the fermions obtain masses at tree level from the vacuum expectation value of σ.

Validity Of The Approximation

Before proceeding to unravel further subtleties, let us discuss the conditions for validity of the approximation.

The traditional region of validity of the above approximation, as in *[7, 8]*, is k fixed, $N \to \infty$, with r and $1/e^2$ of order N. In this limit, the corrections to the approximation (of treating the ϕ's as free fields with a σ-dependent mass) are of order $1/N$. The above computation is part of the beginning of a systematic expansion of all physical observables in powers of $1/N$. Many important features of the theory involve properties that are stable under perturbation – like whether there is a mass gap, what symmetries are spontaneously broken, and certain aspects of the topological sector. For addressing such questions, the $1/N$ expansion is good enough for fixed k and sufficiently big N.

That is not enough for us, because we wish to relate the Verlinde algebra to the cohomology of the Grassmannian for all k and N. Happily, there is another region of validity of the approximation. At the classical level, the matrix O is positive definite, and accordingly for $r < 0$ it would be impossible for the energy to vanish. Quantum mechanically, because of the subtraction that was needed in the above computation, O is not positive definite. Accordingly, zero energy is possible also for negative r at the quantum level; indeed, the solution (4.38) makes sense for either sign of r. (The continuation of the model to negative r was discussed in (*[41, §3.2]*) for the case $k = 1$.)

I claim that for any k and N, the computation leading to (4.38) is a valid approximation for $r << 0$. The reason for this is that the approximate vacuum state given by this computation has exponentially *large* σ for $r \to \infty$. This gives an exponentially large mass to the $\mathbf{\Phi}$ multiplet, so ϕ and ψ loops can be ignored except perhaps for renormalization effects involving diagrams with poor ultraviolet convergence. In this super-renormalizable theory, the only such diagram is the one loop diagram whose evaluation leads to (4.38).

The quantum cohomology of the Grassmannian involves, naively, the behavior for $r >> 0$. However, because the first Chern class of $G(k, N)$ is positive, every topological correlation function of the twisted theory (of operators of definite dimension or J_A) receives a contribution only from one value of the instanton number and hence depends on r as $\exp(-dr)$ with a known constant d that appeared in (4.30). The behavior for $r << 0$ therefore determines the behavior for $r >> 0$. Consequently, the fact that our approximation is valid for the theory continued to $r << 0$ means that it is good enough for studying the topological sector of the twisted theory. We will now explore the implications.

4.5 The Mass Gap And The WZW Action

Because σ was determined to be an arbitrary unitary matrix (times a fixed constant), it appears at first sight that the model has a continuous vacuum degeneracy and therefore massless particles, at least in this approximation. This can hardly be correct because the massless σ particles, subject to the constraint (4.38), do not furnish a representation of $\mathcal{N} = 2$ supersymmetry. (The ϕ fields are massive in this approximation, as we have noted, and so cannot help.) This puzzle was resolved in the old literature in the context of the $1/N$ expansion; the resolution involves giving a mass to σ by mixing with the $U(k)$ gauge field v. (See, for instance, p. 308 of *[8]* for $k = 1$ and the discussion of the $\phi_5 - \lambda$ propagator on pp. 165-6 of *[9]* for general k.) I will not present the detailed computations here, as they are standard; I will merely summarize them and focus on the interpretation, which is all that is new.

The $1/N$ expansion amounts to integrating out the chiral superfields Φ^{is} to obtain an effective action for the gauge multiplet. The $\sigma - v$ mixing comes from the one loop diagram of Figure (4.1). The non-vanishing contribution is the one in which the particles running around the loop are fermions. However, perhaps even more fundamental is the one-loop diagram with external sigma fields only and internal fermions, shown in Figure (4.2).

The fermions ψ^{is}, for $s = 1 \ldots N$, form N copies of the fundamental representation of $U(k)$. Let us suppress the s index and look at a single multiplet ψ^i. The key point is that, looking back to the Lagrangian (4.16), the fermions

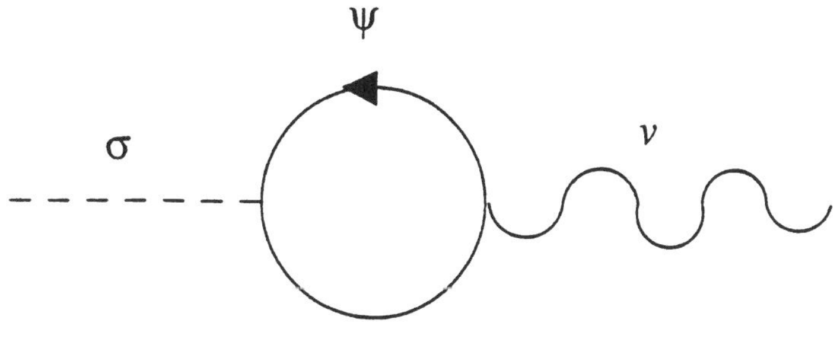

Figure 4.1: The one-loop diagram describing $\sigma - v$ mixing

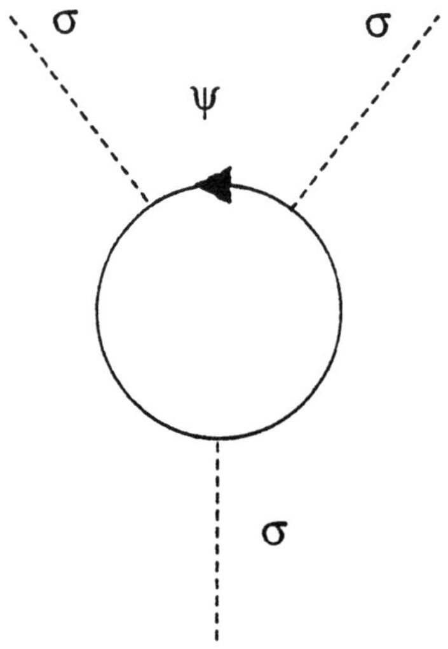

Figure 4.2: The one-loop diagram generating the Wess-Zumino coupling for σ.

receive their mass from a coupling $-\sqrt{2}\,\overline{\psi}_{-i}\sigma^i{}_j\psi_+{}^j$ – *c.c.* This coupling breaks the $U(k)_L \times U(k)_R$ chiral symmetry of the fermion kinetic energy down to a diagonal $U(k)$. Therefore, when we integrate out the fermions to get an effective action for σ, we are dealing with the standard problem of integrating out massive fermions that receive their mass from spontaneous chiral symmetry breaking. It is precisely in connection with this problem that the anomalous Wess-Zumino interaction, defined in (2.9), was originally discovered. The long wavelength limit of the effective interaction obtained by integrating out σ is therefore – allowing for the N multiplets – precisely

$$L_{eff}(\sigma) = N\Gamma(\sigma). \tag{4.41}$$

The form of this interaction is completely determined by $U(k)_L \times U(k)_R$ invariance and the chiral anomaly. (Let me warn the reader that the $U(k)_L \times U(k)_R$ symmetry just invoked is explicitly broken by interactions, such as the gauge couplings, that do not contribute to $L_{eff}(\sigma)$ in leading order in $1/N$. The corrections to the leading large N behavior are the subject of the next sub-section.)

Now we include the gauge fields and Feynman diagrams such as that of Figure (4.1). Such diagrams must extend (4.41) to a gauge invariant effective action $L_{eff}(\sigma, v)$. The minimal choice, in some sense, is the gauge invariant extension of the Wess-Zumino action that was defined in (2.25):

$$L_{eff}(\sigma, v) = N\Gamma(\sigma, v). \tag{4.42}$$

Is this minimal form correct? Apart from terms that vanish by the equations of motion and terms of higher dimension that can be ignored at long distances, a non-minimal gauge invariant term (on a flat world sheet) would have to be of the form

$$\int_\Sigma \mathrm{Tr}\, FW(\sigma), \tag{4.43}$$

with $F = dv + v \wedge v$ and W a function of σ that transforms in the adjoint representation. In the next sub-section, we will show that such a term is not generated, even by corrections to the $1/N$ expansion. We will also discuss the role of the terms that vanish by the equations of motion, and some curvature-dependent terms.

Synthesis

If we take the Lagrangian $N\Gamma(\sigma, v)$ by itself, it describes a level N gauged WZW model of $U(k)/U(k)$. This sort of model was analyzed in §2, and as we know from §2.5, it describes a topological field theory. There are no propagating modes at all, massless or massive. If one adds conventional kinetic energy for σ and v (such terms are certainly present in our underlying Lagrangian), one has propagating modes but massive ones. Indeed the conventional kinetic energy is irrelevant in the infrared and the large distance behavior is that of the gauged WZW model.

Thus, the Grassmannian sigma model – even if one does not restrict *a priori* to its topological sector – reduces at long distances to a topological field theory. In fact, any theory with a mass gap will do this, since at distances at which the massive particles can be neglected, all that survives is dynamics of the vacuum or topological field theory.

In the case of the Grassmannian sigma model, there was an underlying topological sector, described in §4.2, and visible from the classical Lagrangian before any analysis of its quantum properties. The basic observable in this topological sector was the σ field that appears in (4.42) (but now restricted to $\overline{\sigma}\sigma = \text{constant}$). Thus the topological sector, defined microscopically, passes over at large distances to the gauged WZW model governing the σ field. This is the mapping from the quantum cohomology of the Grassmannian to the gauged WZW model (and thence the Verlinde algebra) that is the main goal of this paper.

In what follows, we will analyze the corrections to the $1/N$ expansion and eventually pin down the details of the mapping from the quantum cohomology to the gauged WZW model.

Search For Manifest Supersymmetry

It would be attractive to find an extension of (4.42), including the fermi partners of σ and v, with manifest $\mathcal{N} = 2$ supersymmetry. Of course, the one-loop effective action from which (4.42) was defined is such an extension, but it would be nice to find, for instance, a compact description in $\mathcal{N} = 2$ superspace of a local interaction describing the long-wavelength part of the one-loop effective action. I have been unable to do this and leave it as an interesting question.

However, let us truncate to the abelian case in which σ, v, and their fermionic partners are diagonal matrices (for instance $\sigma = \mathrm{diag}(\sigma_1, \ldots, \sigma_k)$). In this case, the field strength is similarly diagonal (say $\Sigma = \mathrm{diag}(\Sigma_1, \ldots, \Sigma_k)$). With this truncation, it is possible to find an explicit, local superspace interaction that describes all of the anomalous interactions. This interaction (which in the abelian case was discussed in *[41]*, §3.2), is

$$L_{\mathcal{N}=2} = \frac{1}{\sqrt{2}} \sum_{i=1}^{k} \int d^2x\, d\theta^+\, d\bar{\theta}^- \left(\frac{it\Sigma_i}{2} - \frac{N}{2\pi} \Sigma_i \ln(\Sigma_i/\mu) \right) + c.c. \qquad (4.44)$$

The ease of writing this interaction in the diagonal case and the difficulty of describing its full non-abelian generalization may be related to the utility of abelianization in the next subsection.

4.6 Corrections

Now we turn to analyzing the corrections to this approximation. We can ignore operators of dimension higher than two, which are irrelevant at long distances. Terms of dimension less than two, such as (2.78), cannot arise as they would violate the underlying $\mathcal{N} = 2$ supersymmetry or (as a consequence) the topological invariance of the twisted sector. Also, we can ignore terms that vanish by the v equations of motion and so can be eliminated, as in §2.5, by a redefinition of v. Such terms (which are of the form $\int_\Sigma \sigma^*(B)$, with B an adjoint-invariant two-form on $U(k)$) would play no role in our subsequent analysis.

We are left with three issues to consider:

(1) First, there might be corrections to the discrete data, the "level" of the effective WZW model. In the one loop approximation, we found the level to be N; however, corrections of relative order $1/N$ could shift this by a constant. Actually, this has to be formulated more precisely because the Lie algebra of $U(k)$, which we will call $u(k)$, is not simple; it can be split as $su(k) \oplus u(1)$, where $su(k)$ consists of the traceless $k \times k$ hermitian matrices and $u(1)$ is the center of $u(k)$. In general, one could have a gauged WZW model for $U(k)$ of level (N_1, N_2), by which I mean that the Lagrangian would be determined by the quadratic form $(\cdot, \cdot)$ on $u(k)$ such that $(a, b) = N_1 \mathrm{Tr}\, ab$ for $a, b \in su(k)$, and $(a, b) = N_2 \mathrm{Tr}\, ab$ for $a, b \in u(1)$. Thus, the first correction to the $1/N$ approximation might lead to $(N_1, N_2) = (N + u, N + v)$ where u, v are integers (perhaps depending on k); higher order corrections in $1/N$ must vanish as they could not be integral for all N.

(2) Second, the low energy effective action might contain a term

$$\int_\Sigma \mathrm{Tr}\, FW(\sigma), \qquad (4.45)$$

with as above F the $u(k)$ curvature, and W a function of σ that transforms in the adjoint representation. Though this term vanishes by the equations of motion of the low-energy gauged WZW model and so could be eliminated even from the quantum theory by a field redefinition (as described in §2.5), it could still play a role that will be explained later.

Note that a constant term in W (that is, a multiple of the identity) could be absorbed in an additive renormalization of t; we will not try to determine such a renormalization, and all of our statements about W will hold modulo an additive constant.

(3) Finally, we need to know whether, when the twisted theory is formulated on a curved world-sheet, the effective Lagrangian contains a term

$$\Delta L = \int_\Sigma d^2x\sqrt{h}\frac{R}{4\pi}U(\sigma), \tag{4.46}$$

with R the world-sheet curvature and $U(\sigma)$ a function invariant under conjugation. As we have discussed in §2.5, any continuous deformation of the gauged WZW model that preserves the topological invariance and cannot be eliminated by a change of variables is of this form. (By contrast, the deformations considered above in (1) are discrete, not continuous, and the deformations in (2) can be removed by a change of variable.)

Now, here are the answers that I will claim for these three questions:

(A1) I will claim that the level of the effective gauged WZW model is really $(N-k, N)$. The correction can be thought of as a $1/N$ correction that comes from integrating out the $U(k)$ gauge multiplet (the $u(1)$ level is not shifted since the gauge multiplet is neutral under $u(1)$).

(A2) I will claim that $W=0$, in other words that no term of the form (4.45) is generated.

(A3) I will claim that a term of the form (4.46) is generated, with

$$U = (N-k)\ln\det\sigma + \text{constant}. \tag{4.47}$$

This might be regarded as the minimal possibility compatible with the anomaly formula (4.22).

Abelianization

Now I will explain how I will do the calculation. A $1/N$ expansion will not suffice, since we do not want to be limited to sufficiently large N. Instead, we will study the theory in the alternative regime of $r << 0$.

To identify the corrections to the effective action of the three types discussed above, it suffices to work in the region of field space in which σ is a diagonal matrix with distinct eigenvalues $\sigma_1, \dots, \sigma_k$. Moreover, we impose the condition of unbroken supersymmetry (or vanishing vacuum energy); in the approximation of (4.38) – which is valid for $r << 0$ – the condition is

$$\overline{\sigma}_i\sigma_i = \mu^2\exp(-4\pi r/N), \text{ for } i = 1, \dots, k. \tag{4.48}$$

The distinct values of the σ_i break $U(k)$ to a diagonal subgroup $U(1)^k$. Calculations are relatively easy because the chiral superfields and the "off-diagonal" part of the gauge multiplet have large masses, of order $\overline{\sigma}\sigma$, which can be read off from the classical Lagrangian. The fields which remain massless in this approximation (and actually get masses at one loop, smaller by a factor of e^2) are the diagonal part of the gauge multiplet. The effective action for the massless modes, including the one loop correction, has already been written with manifest $\mathcal{N} = 2$ supersymmetry in (4.44). This is a kind of gauged WZW model of $U(1)^k$. So in this regime, we get a kind of abelianization of the Grassmannian sigma model.

This should not come as a complete surprise, since as we have recalled in §2.6, the gauged WZW model has a precisely analogous abelianization. Now, we will have to be careful in using abelianization to compute the effects of types (1), (2), and (3), because in going from the gauged WZW model to its abelianization, precisely analogous terms are generated. These were computed in *[5]* and described in §2.6, and are as follows.

(B1) The shift in level in going from the gauged WZW model of $U(k)$ to its abelianization is $(k, 0)$. (There is obviously no shift of the $u(1)$ level under abelianization since $u(1)$ is already abelian.)

(B2) No term of the form (4.45) is generated.

(B3) The term of the form (4.46) that is generated in abelianizing the gauged WZW model was presented in equation (2.85).

Now in verifying claims (A1), (A2), and (A3), we will integrate out from the Grassmannian sigma model the fields that, in the abelianized regime, have tree level masses; thus we will get the precise abelianized theory that is equivalent to the Grassmannian sigma model. Then we will interpret the result as a sum of two contributions: the terms claimed in (A1), (A2), and (A3) which describe how to go from the topological sector of the Grassmannian sigma model to an equivalent gauged WZW model; and the terms (B1), (B2), and (B3), which arise in abelianization of the gauged WZW model.

So claims (A1), (A2), and (A3) are equivalent to the following claims, which are the ones that we will actually check:

(C1) After abelianization, there is no shift in the level of the Grassmannian sigma model from the naive result (N, N). We interpret this to mean that the topological sector of the Grassmannian sigma model is equivalent to a gauged WZW model of $U(k)/U(k)$ at level $(N - k, N)$, and the level of that model is shifted by $(k, 0)$ upon abelianization.

(C2) There will be no induced term of the type (4.45), in abelianizing either of the two models or in comparing them.

(C3) The induced term of type (4.46) in abelianization of the Grassmannian model will be the sum of (4.47) and the contribution (2.85) that arises in abelianizing the gauged WZW model. The sum of these is simply

$$\widetilde{U}(\sigma) = (N-1)\ln\det\sigma - \sum_{i\neq j}\ln(\sigma_i - \sigma_j). \qquad (4.49)$$

The Calculation

Now I will explain the calculation justifying (C1), (C2), and (C3). In discussing (C1) and (C2), world-sheet curvature is irrelevant, and we can work on a flat $\mathbf{R}^2$.

(C1) and (C2) can be taken together and deduced from the following principle. Suppose that a $u(1)$ gauge field v, with field strength $f = dv$, is coupled to a Dirac fermion χ, of charge q. Let χ have a mass term

$$L_{mass} = -\int_{\Sigma} d^2x \left(\overline{\chi}_- m\chi_+ + \overline{\chi}_+\overline{m}\chi_-\right). \tag{4.50}$$

We want to integrate out χ to get an effective action for v. The dependence of the effective action on the phase of m comes only from the chiral anomaly and is

$$L_{\text{eff}} = \ldots + q\int_{\Sigma}\frac{f}{(2\pi)}\text{Im}\log m. \tag{4.51}$$

Now we look at the Grassmannian sigma model in the abelianized regime of $r << 0$, σ_i large (obeying (4.48)) and distinct. The chiral superfields Φ^{js} and the off-diagonal part of the gauge multiplet have bare masses of order $|\sigma_i|$. They can be integrated out in a one loop approximation; higher order corrections would be of order e^2 and irrelevant. Integrating out massive bosons does not give terms relevant to (C1) or (C2), while the contributions of fermions can be deduced from (4.51).

To do so explicitly, let v_i, $i = 1\ldots k$, be the diagonal components of the gauge field. First we work out the contributions of chiral superfields. Each v_i is coupled to N chiral superfields Φ^{is}, $s = 1\ldots N$, of charge 1. The fermi elements of these superfields have mass $\sqrt{2}\sigma_i$ (by inspection of (4.16)), so their contribution is

$$N\sum_{i=1}^{k}\int_{\Sigma}\frac{dv_i}{2\pi}\text{Im}\ln\sigma_i. \tag{4.52}$$

This is the level N gauged WZW action computed of equation (4.42), specialized to the case that only the diagonal components of v and σ are non-zero.

Now, we come to the off-diagonal part of the gauge fields. Again, the relevant contribution comes from the phases of the masses of the off-diagonal fermions $\lambda^i{}_j$, $i \neq j$. Since (4.52) coincides with the level N gauged WZW action, the claims (C1) and (C2) amount to the assertion that no additional contribution will come from integrating out the $\lambda^i{}_j$.

By inspection of (4.10), the mass of $\lambda^i{}_j$ is $\sqrt{2}(\sigma_i - \sigma_j)$. The gauge field v_i interacts with the $\lambda^i{}_j$, $j \neq i$, of charge 1, and with the $\lambda^m{}_i$, $m \neq i$, of charge -1. Their contribution adds up to

$$\sum_{i,j}\int_{\Sigma}\frac{dv_i}{2\pi}\left(\text{Im}\ln(\sigma_i - \sigma_j) - \text{Im}\ln(\sigma_j - \sigma_i)\right). \tag{4.53}$$

This is zero, or more exactly, it is independent of the σ_i. Consequently it can be interpreted as a constant term in W or an additive renormalization of t; as noted in the paragraph following (4.45), we will not keep track of such effects. As for the diagonal components of λ, they are neutral and do not couple to the v_i.

It remains to discuss (C3). To this aim, we can take Σ to be a Riemann surface of genus zero and take the σ_i to be constants. As $\int_\Sigma d^2x\sqrt{h}R/4\pi = 1$ in genus zero, the claim (C3) amounts to the assertion that the path integral $\int D\Phi_i \dots e^{-L}$ is a constant multiple of

$$(\det\sigma)^{-(N-1)} \prod_{i\neq j}(\sigma_i - \sigma_j). \tag{4.54}$$

To verify this, we first integrate out the massive fields in the same one loop approximation as above. As before, the boson determinant is real and depends only on $|\sigma_i|$, while the fermion determinant has a phase that can be extracted from the chiral anomaly. Ordinarily, there is no chiral anomaly for fermions in a gravitational field in two dimensions, but the twisting to produce the topological theory involves a modification of the fermion kinetic energy that introduces such an anomaly.

We could proceed as above, starting with the anomaly formula analogous to (4.51). For the sake of variety, however, let us note that the anomaly can be captured by the path integral over the zero modes of the fermion kinetic energy. For instance, the fermions ψ^{is} from the chiral multiplets have components $\overline{\psi}_+, \psi_-$ of spin zero and other components of spin one. The zero modes of the fermion kinetic energy are the constant modes of $\overline{\psi}_+, \psi_-$, and the path integral over those modes is

$$\int d\overline{\psi}_{+is} d\psi_-{}^{jt} \exp\left(\sum_{is} \overline{\psi}_{+is}\overline{\sigma}_i\psi_-{}^{is}\right) = \det\overline{\sigma}^N = \text{constant}\cdot\det\sigma^{-N}, \tag{4.55}$$

where we have used the fact that $\sigma\overline{\sigma} = \text{constant}$. Similarly, for the off-diagonal λ fields, the zero modes of the kinetic energy are the constant modes of $\overline{\lambda}_-$, λ_+, and the path integral over those modes is

$$\prod_{i\neq j} \int d\overline{\lambda}_-{}^i_j\, d\lambda_+{}^j{}_i \exp\left(\sqrt{2}(\sigma_i - \sigma_j)\overline{\lambda}_-{}^i{}_j\lambda_+{}^j_i\right) = \text{const}\cdot \prod_{i\neq j}\left(\sigma_i - \sigma_j\right). \tag{4.56}$$

Comparing (4.55) and (4.56) with the claim made in (4.49) concerning (C3), we see that we are missing precisely one factor of $\det\sigma$. This must come from the remaining integral over the diagonal components of the gauge field. Indeed, though the diagonal fermions $\lambda^i{}_i$ are massless at tree level, they receive at the one loop level a mass term with the form

$$\sum_{i=1}^{k}\left(\lambda_+{}^i{}_i\overline{\sigma}_i{}^{-1}\overline{\lambda}_-{}^i{}_i + \text{c.c.}\right) \tag{4.57}$$

This term can be straightforwardly calculated or can be read off from the $\mathcal{N} = 2$ extension (4.44) of the bosonic anomalous interactions. Integrating out the diagonal fermions therefore gives (up to a constant) a factor of

$$(\det\overline{\sigma})^{-1} = \text{constant}\cdot(\det\sigma). \tag{4.58}$$

This is the last factor needed for (C3).

This factor could in a more general way be predicted as follows. Just because the diagonal theory (including the $\lambda^i{}_i$) is a product of k sub-theories, the phase it produces must be of the form

$$\prod_{i=1}^{k} F(\sigma_i) \tag{4.59}$$

for some function F. Given this factorized form, to agree with the anomaly formula (4.22) it must be that $F(\sigma) = \sigma$.

5 The Verlinde Algebra And The Grassmannian

In this subsection, we will put the pieces together and write down the precise connection between the Verlinde algebra and the quantum cohomology of the Grassmannian.

First of all, we consider the map from the cohomology ring of the Grassmannian to the Verlinde algebra. The quantum cohomology ring of $G(k, N)$ is generated by operators of the form

$$O_V = \mathrm{Tr}_V\sigma, \tag{4.60}$$

with V an irreducible representation of $U(k)$. Thinking of σ is the curvature of the natural connection on the tautological rank k bundle over $G(k, N)$, $\mathrm{Tr}_V\sigma$ can be interpreted as a characteristic class of that bundle and hence a cohomology class of $G(k, N)$.

On the other hand, working at long distances, σ becomes a unitary matrix (up to a constant that we will eventually pin down) and then $\mathrm{Tr}_V\sigma$ can be interpreted as an operator in the effective gauged WZW model of $U(k)/U(k)$. We worked out in (2.56), (2.58) the interpretation of this operator: it is the element of the Verlinde algebra determined by the representation V. Of course, from what we have said in §4.6, the Verlinde algebra in question is the one for the group $U(k)$ at level $(N - k, N)$.

So we have gotten the precise mapping from the cohomology of the Grassmannian to the Verlinde algebra. A couple of points should be clarified:

(1) It is essential that in mapping from the Grassmannian sigma model to the gauged WZW model, there is no correction of type (4.45). Such a term, since it vanishes by the equations of motion in the gauged WZW model, could be transformed away by a redefinition of σ and v. But the resulting redefinition of σ would cause the operator O_V to mix with similar operators for other representations. Thus, were terms with the structure (4.45) to appear, their precise form would enter in determining the map from the cohomology of $G(k, N)$ to the Verlinde algebra.

By contrast, corrections to the gauged WZW action that vanish by the v equations of motion and so can be removed by redefinition of v are immaterial, since $O_V(\sigma)$ is independent of v. As noted at the beginning of §4.3, we have made no claim that corrections that vanish by the v equations of motion are not generated or have any particular structure.

(2) In the rest of this paper, we will set the Kähler parameter r to zero; as explained in §4.3, this involves no essential loss of information. Two other constants discussed in §4.3 also enter. One is the constant c in the relation $\sigma = c \cdot$ unitary matrix, which we evaluated only approximately in (4.40). The other is the additive renormalization constant called a in (4.31), which is unknown since we did not attempt to determine the constant in (4.47). For the time being, we will set c and a to 1 and 0; eventually we will verify that this is correct (for $r = 0$) by checking special cases of the formulas.

Correlation Functions And The Metric

Now, let us determine precisely how the correlation functions and the metric in the Grassmannian sigma model compare to those in the gauged WZW model. The essential point that goes beyond what we have just said above is that one must include the correction term of (4.46), (4.47). By topological invariance, σ can be treated as a constant, so the correction factor in the path integral is

$$\exp(-\Delta L) = \exp\left(-\int_\Sigma d^2x\sqrt{h}\frac{R}{4\pi}(N-k)\ln\det\sigma\right) = (\det\sigma)^{(g-1)(N-k)}. \tag{4.61}$$

The point or points at which $\det\sigma$ is inserted are immaterial. (It will turn out that $\det\sigma$ is an invertible element of the Verlinde algebra or quantum cohomology.)

So if $\langle\ \ \rangle_{G(k,N)}$ denotes an expectation value of the path integral of the Grassmannian sigma model, and $\langle\ \ \rangle_{WZW}$ denotes a path integral in the gauged WZW model, then the relation between these symbols in genus g is

$$\left\langle\prod_{i=1}^{s}\mathrm{Tr}_{V_i}(\sigma_i)\right\rangle_{G(k,N)} = \left\langle\prod_{i=1}^{s}\mathrm{Tr}_{V_i}(\sigma_i)\cdot(\det\sigma)^{(g-1)(N-k)}\right\rangle_{WZW}. \tag{4.62}$$

Henceforth we will abbreviate $\mathrm{Tr}_{V_i}(\sigma_i)$ as V_i. From (4.62) one can see that the natural metric on the cohomology of $G(k,N)$ (given by Poincaré duality) does not coincide with the natural metric on the Verlinde algebra. Let us call these metrics (the sigma model and Verlinde metrics) g_σ and g_V, respectively. We recall that the metric is defined by a two point function in genus 0, so

$$\begin{aligned} g_\sigma(V_1,V_2) &= \langle V_1V_2\rangle_{G(k,n)} \\ g_V(V_1,V_2) &= \langle V_1V_2\rangle_{WZW} = \langle V_1V_2\cdot(\det\sigma)^{(N-k)}\rangle_{G(k,N)}, \end{aligned} \tag{4.63}$$

with the correlation functions being in genus zero.

Now let us compare the ring structure on the cohomology of the Grassmannian to the ring structure of the gauged WZW model. We recall that in either of the two theories, the ring structure is introduced by interpreting the genus zero three point function in terms of a binary operation, say $V_1, V_2 \to V_1 \cdot V_2$, according to the following formula:

$$\langle V_1 V_2 V_3 \rangle = g(V_1 \cdot V_2, V_3). \tag{4.64}$$

The relation between the genus zero three point functions of the two theories is from (4.62)

$$\langle V_1 V_2 V_3 \rangle_{WZW} = \langle V_1 V_2 V_3 (\det \sigma)^{(N-k)} \rangle_{G(k,N)}. \tag{4.65}$$

In particular the three point functions of the $G(k, N)$ and gauged WZW models do not coincide. However, they differ by the same factor of $(\det \sigma)^{(N-k)}$ that enters in the relation between the metrics. This means in fact, upon putting together the last few formulas, that the multiplication laws are the same in the two theories.

So finally, our natural map from the quantum cohomology of the Grassmannian to the Verlinde algebra is a ring homomorphism – justifying terminology that was used above.

Non-Abelian Theta Functions

In the title of this paper and in much of the writing of it, I have emphasized the Verlinde algebra, which determines the dimension of the space of non-abelian theta functions. However, the above gives directly a formula for the dimension of the space $\mathcal{H}$ of non-abelian theta functions without having to detour via the Verlinde algebra. Of course, we will count non-abelian theta functions for the group $U(k)$ at level $(N-k, N)$; because the $U(1)$ theory is well understood, there is no essential difficulty in generalizing to other levels.

Let $\langle 1 \rangle^g$ denote the partition function in genus g. Then the dimension of $\mathcal{H}$ on a Riemann surface Σ of genus g is

$$\dim \mathcal{H} = \langle 1 \rangle^g_{WZW} = \left\langle (\det \sigma)^{-(g-1)(N-k)} \right\rangle_{G(k,N)} = \left\langle (\det \sigma)^{k(g-1)} \right\rangle_{G(k,N)} \tag{4.66}$$

In the last expression, I have used the fact that $(\det \sigma)^N = 1$, as one can deduce from the Landau-Ginzburg description of the quantum cohomology (we do this below for $k = 2$). The right hand side of (4.66) can be evaluated by counting holomorphic maps of Σ to $G(k, N)$ obeying certain conditions.

5.1 Getting Down To Earth

In this section, we will make everything completely explicit in the cases of $k = 1$ and $k = 2$. (Some of the issues are discussed by Gepner for general k in the last paper in *[1]*.)

First we dispose of $G(1, N)$, that is, $\mathbf{CP}^{N-1}$. Over $\mathbf{CP}^{N-1}$ there is a tautological principal $U(1)$ bundle P. Let W be the standard representation of $U(1)$, of "charge one." Associated to P in the representation W is a line bundle $\mathcal{L}$. The representation W determines an operator $\mathrm{Tr}_W \sigma = \sigma$ which we will call x. In the sigma model, interpreting σ as the curvature of the natural connection on $\mathcal{L}$, $x = c_1(\mathcal{L})$.

The cohomology ring of $\mathbf{CP}^{N-1}$ is generated by x and classically is $\mathbf{C}[x]/x^{N-1}$. But from *[36]* or the $k = 1$ case of (3.16), the quantum cohomology ring is

$$R = \mathbf{C}[x]/(x^N - 1). \tag{4.67}$$

The metric on the cohomology determined by Poincaré duality is

$$g_\sigma(x^k, x^l) = \delta_{k+l,N-1}, \tag{4.68}$$

where in view of (4.67), k and l are evaluated modulo N.

On the other hand, in the gauged WZW model, $x = \mathrm{Tr}_W\sigma$ should be identified with the element of the Verlinde algebra for $U(1)$ at level N determined by the representation W. The structure of this algebra is well known. It is generated by W with the relation $W^N = 1$, just as in (4.67), and the metric is

$$g_V(W^k, W^l) = \delta_{k+l,0}, \tag{4.69}$$

with again k and l taken modulo N. (4.68) and (4.67) are related in the fashion predicted by (4.63). Indeed since for either metric $g(a,b) = g(ab,1)$, (4.63) is equivalent to

$$g_V(x^k, 1) = g_\sigma(x^k, x^{N-1}). \tag{4.70}$$

That disposes of $k = 1$. Obviously, we cannot expect the non-abelian case $k = 2$ to be as trivial as that.

The Verlinde Algebra For $k = 2$

First we describe explicitly (but not in a fully self-contained fashion) the Verlinde algebra of the group $U(2)$ at the desired level $(N-2, N)$.

We have an exact sequence

$$1 \to Z_2 \to SU(2) \times U(1) \xrightarrow{f} U(2) \to 1. \tag{4.71}$$

Here the map f is as follows: we identify $SU(2)$ with the 2×2 unitary matrices of determinant 1, $U(1)$ with 2×2 unitary matrices that are multiples of the identity, and for $x \in SU(2)$, $y \in U(1)$, let $f(x,y) = xy$.

The gauged WZW action of $U(2)$ at level $(N-2, N)$ restricts, if one takes the fields to lie in $SU(2)$, to the $SU(2)$ action at level $N-2$; but if one takes the fields to be in $U(1)$, it restricts to the $U(1)$ gauged WZW action at level $2N$. (A factor of two arises simply because the trace of the identity matrix in the fundamental representation of $U(2)$ is 2.) Therefore, we will proceed by comparing the Verlinde algebra of $U(2)$ at level $(N-2, N)$ to that of $SU(2) \times U(1)$ at level $(N-2, 2N)$.

The $SU(2)$ Verlinde algebra was described explicitly in *[14]*. If V_1 is the two dimensional representation of $SU(2)$, and V_n is its nth symmetric tensor power, then the Verlinde algebra of $SU(2)$ is the usual representation ring of $SU(2)$, subject to the relation

$$V_{N-1} = 0. \tag{4.72}$$

The representation ring of $SU(2)$, subject to this relation, is spanned additively by $V_0, \ldots, V_{N-2}$. The multiplication law can be described explicitly as

$$V_i \times V_j = \sum_t N_{ijt} V_t, \tag{4.73}$$

where N_{ijt} is 1 if the following relations and their cyclic permutations are obeyed:

$$i + j \geq t, \quad N - 2 - i + j \geq N - 2 - t, \quad 2(N-2) - i - j \geq t. \tag{4.74}$$

Otherwise $N_{ijt} = 0$. The metric on the Verlinde algebra is

$$g(V_s, V_t) = \delta_{s,t}. \tag{4.75}$$

The $U(1)$ Verlinde algebra has already been introduced above. It is generated by the charge one representation W, and the defining relation at level $2N$ is

$$W^{2N} = 1. \tag{4.76}$$

The metric is

$$g(W^u, W^v) = \delta_{u+v,0}. \tag{4.77}$$

The $SU(2) \times U(1)$ Verlinde algebra at level $(N-2, 2N)$, is therefore spanned additively by the elements $V_i W^j$, for $i = 0, \ldots, N-2$, $j = 0, \ldots, 2N-1$, corresponding to the representation $V_i \otimes W^{\otimes j}$. The multiplication law and metric are products of the multiplication law and metric of $SU(2)$ and $U(1)$.

Now we want to proceed to $U(2) = (SU(2) \times U(1))/\mathbf{Z}_2$. Dividing by $\mathbf{Z}_2$ halves the volume of the group manifold (if one uses a fixed Haar measure in an obvious sense). The Verlinde algebra is defined on a certain space of conformal blocks which can be constructed by quantizing an appropriate phase space $\mathcal{M}$ – for instance, the phase space of the gauged WZW model. When the volume of the group is halved, the volume of $\mathcal{M}$ is[10] divided by $2^2 = 4$; therefore, in the semiclassical limit of large N, the Verlinde algebra of $U(2)$ at level $(N-2, N)$ will have one fourth the dimension of that of $SU(2) \times U(1)$ at level $(N-2, 2N)$.

One factor of two is obvious. Among all $SU(2) \times U(1)$ representations, we must restrict to those that are representations of $U(2)$. This means keeping only $V_i W^j$ with $i + j$ even. The second factor of two is less obvious. One must impose the equivalence relation

$$V_i W^j = V_{N-2-i} W^{j+N}. \tag{4.78}$$

I refer the interested reader to *[44]* for an explanation (in the analogous case of $SO(3) = SU(2)/\mathbf{Z}_2$) of such matters.

Note that if we set $\tau(V_i W^j) = V_{N-2-i} W^{j+N}$, then the Verlinde algebra of $SU(2) \times U(1)$ at level $(N-2, N)$ obeys $\tau(a)b = a\tau(b) = \tau(ab)$. This ensures

[10] $\mathcal{M}$ is the moduli space of flat connections in genus one and consists of pairs of commuting elements of the gauge group G, divided by the Weyl group. Because one has a pair of elements of G, the volume of $\mathcal{M}$ is decreased by a factor of n^2 if one divides G by a group of order n.

that the Verlinde algebra of $SU(2) \times U(1)$ induces a natural algebra structure on the quotient by the relations (4.78). This quotient algebra, restricted to $i + j$ even (a τ-invariant condition) is the Verlinde algebra of $U(2)$ at level $(N-2, N)$. The metric is

$$g_V(V_i W^s, V_j W^t) = \delta_{i,j}\delta_{s+t,0} + \delta_{i,N-2-j}\delta_{s+t-N,0}. \tag{4.79}$$

A complete but redundant set of relations for the $U(2)$ Verlinde algebra would be the relations

$$V_{N-1} = 0, \quad W^{2N} = 1 \tag{4.80}$$

inherited from the $SU(2)$ and $U(1)$ algebras, along with (4.78). A special case of (4.78) is

$$V_{N-2}W^N = 1. \tag{4.81}$$

It will become clear presently that (4.80) and (4.81) suffice to characterize the Verlinde algebra.

Representations And Characters

Two representations of the group $U(2)$ will play a distinguished role. The first is the standard two dimensional representation $\mathcal{V}_1$. Under restriction to $SU(2) \subset U(2)$, $\mathcal{V}_1$ restricts to the standard two dimensional representation V_1 of $SU(2)$, and the scalars in $U(2)$ act with charge 1. So $\mathcal{V}_1$ pulls back to the representation $V_1 \otimes W$ of $SU(2) \times U(1)$.

The other important representation is $\eta = \wedge^2 \mathcal{V}_1$. $SU(2)$ acts trivially on η, and the scalars in $U(2)$ act with charge 2, so η pulls back to the representation W^2 of $SU(2) \times U(1)$.

So the elements of the $U(2)$ Verlinde algebra determined by $\mathcal{V}_1$ and η are just $V_1 W$ and W^2. What elements in the quantum cohomology do these same representations determine? Over $G(2, N)$, there is a tautological complex two-plane bundle E^* with curvature matrix represented by the quantum field σ. The Chern classes of E^* are

$$\begin{aligned} c_1(E^*) &= \operatorname{Tr}_{\mathcal{V}_1}\sigma \\ c_2(E^*) &= \operatorname{Tr}_{\eta}\sigma. \end{aligned} \tag{4.82}$$

So under the natural mapping from the quantum cohomology of $G(2, N)$ to the Verlinde algebra of $U(2)$, $c_1(E^*)$ and $c_2(E^*)$ correspond to the representations $\mathcal{V}_1$ and η.

Let us now briefly discuss the classical representation ring of $U(2)$; the Verlinde algebra is a quotient of this, as we have described. Consider the maximal abelian subgroup of $U(2)$ of matrices of the form

$$\sigma = \begin{pmatrix} \lambda_1 & 0 \\ 0 & \lambda_2 \end{pmatrix}, \tag{4.83}$$

with $|\lambda_1| = |\lambda_2| = 1$. The character of a representation R of $U(2)$ is $\operatorname{Tr}_R \sigma$

regarded as a function of the λ_i. For instance, the characters of $\mathcal{V}_1$ and η are

$$\begin{aligned} \mathrm{Tr}_{\mathcal{V}_1}\sigma &= \lambda_1 + \lambda_2 \\ \mathrm{Tr}_{\eta}\sigma &= \lambda_1\lambda_2. \end{aligned} \tag{4.84}$$

If $\mathcal{V}_n$ is the nth symmetric tensor power of $\mathcal{V}_1$, then its character is

$$\mathrm{Tr}_{\mathcal{V}_n}\sigma = \frac{{\lambda_1}^{n+1} - {\lambda_2}^{n+1}}{\lambda_1 - \lambda_2}. \tag{4.85}$$

Any irreducible representation of $U(2)$ is of the form $\mathcal{V}_s\eta^t$ for some integers s, t (with $s \geq 0$). The equivalence relation (4.78) becomes

$$\mathcal{V}_s\eta^t \leftrightarrow \mathcal{V}_{N-2-s}\eta^{s+t+1}. \tag{4.86}$$

In general, the map from a representation of $U(2)$ to its character is an isomorphism between the ring of representations of $U(2)$ and the ring of Laurent polynomials in λ_1 and λ_2 that are invariant under the Weyl group, which acts by $\lambda_1 \leftrightarrow \lambda_2$.

As we have sketched above, the Verlinde algebra of $U(2)$ is a quotient of the classical representation ring of $U(2)$ by a certain ideal. Under the isomorphism between the representation ring and the character ring, the generators of this ideal can be identified with certain Laurent polynomials in the λ's. For instance, the relations in (4.80) and (4.81) become

$$\begin{aligned} \frac{{\lambda_1}^N - {\lambda_2}^N}{\lambda_1 - \lambda_2} &= 0 \\ (\lambda_1\lambda_2)^N &= 1 \\ \lambda_1\lambda_2\frac{{\lambda_1}^{N-1} - {\lambda_2}^{N-1}}{\lambda_1 - \lambda_2} &= 1. \end{aligned} \tag{4.87}$$

The second relation in (4.87) has the following implication. The classical representation ring of $U(2)$ is a ring of *Laurent* polynomials in the λ's, including negative powers. But by multiplying by a suitable power of $1 = (\lambda_1\lambda_2)^N$, one can clear the denominators and regard the Verlinde algebra as the quotient of the ring of Weyl-invariant polynomials (not Laurent polynomials) in the λ's by a certain ideal $\mathcal{I}$. We will learn that $\mathcal{I}$ is in fact generated by the first and third relations in (4..87).

If we multiply the first relation in (4.87) by $\lambda_1 + \lambda_2$ and subtract the third, we learn that

$$\frac{{\lambda_1}^{N+1} - {\lambda_2}^{N+1}}{\lambda_1 - \lambda_2} + 1 = 0. \tag{4.88}$$

Cohomology Ring Of $G(2, N)$

Now return to the sigma model interpretation of σ as the curvature of the tautological two-plane bundle E over $G(2, N)$. If we introduce the roots $\widetilde{\lambda}_1, \widetilde{\lambda}_2$ of the Chern polynomial, then the Chern classes of E are

$$\begin{aligned} c_1(E^*) &= \widetilde{\lambda}_1 + \widetilde{\lambda}_2 \\ c_2(E^*) &= \widetilde{\lambda}_1 \widetilde{\lambda}_2. \end{aligned} \tag{4.89}$$

We observed earlier that under the map from cohomology of $G(2,N)$ to the Verlinde algebra, $c_1(E^*)$ and $c_2(E^*)$ correspond to $\mathcal{V}_1$ and η. If in turn we identify the Verlinde algebra as a quotient of the character ring, $\mathcal{V}_1$ and η are identified with their characters, which were given in (4.84). As (4.84) and (4.89) coincide, the identification between these rings can be interpreted as $\lambda_i \leftrightarrow \widetilde{\lambda}_i$. Henceforth, we make this identification and drop the tildes.

The quantum comology ring of $G(2,N)$ is the ring of polynomials in $c_1(E^*)$ and $c_2(E^*)$ modulo an ideal $\mathcal{J}$. As explained in §3.2, $\mathcal{J}$ can be described as follows. Let

$$W(\lambda_1, \lambda_2) = \frac{1}{N+1}\left(\lambda_1{}^{N+1} + \lambda_2{}^{N+1}\right) + (\lambda_1 + \lambda_2). \tag{4.90}$$

Since it is Weyl-invariant, W can be regarded as a polynomial in $c_1 = \lambda_1 + \lambda_2$ and $c_2 = \lambda_1\lambda_2$. The ideal $\mathcal{J}$ is generated by the relations

$$0 = dW = \frac{\partial W}{\partial c_1} dc_1 + \frac{\partial W}{\partial c_2} dc_2. \tag{4.91}$$

Since

$$\begin{aligned} d\lambda_1 &= \frac{\lambda_1 dc_1 - dc_2}{\lambda_1 - \lambda_2} \\ d\lambda_2 &= \frac{-\lambda_2 dc_1 + dc_2}{\lambda_1 - \lambda_2}, \end{aligned} \tag{4.92}$$

we have

$$\begin{aligned} dW = \left(\lambda_1{}^N + 1\right) d\lambda_1 + \left(\lambda_2{}^N + 1\right) d\lambda_2 = dc_1 \left(\tfrac{\lambda_1{}^{N+1} - \lambda_2{}^{N+1}}{\lambda_1 - \lambda_2} + 1\right) \\ -dc_2 \left(\tfrac{\lambda_1{}^N - \lambda_2{}^N}{\lambda_1 - \lambda_2}\right). \end{aligned} \tag{4.93}$$

The quantum cohomology of $G(2,N)$ is therefore defined by the relations

$$0 = \frac{\lambda_1{}^N - \lambda_2{}^N}{\lambda_1 - \lambda_2} = \frac{\lambda_1{}^{N+1} - \lambda_2{}^{N+1}}{\lambda_1 - \lambda_2} + 1. \tag{4.94}$$

If we compare this to (4.88) and to the first equation in (4.87), we see that these relations hold in the Verlinde algebra, and consequently the Verlinde algebra is a quotient of the quantum cohomology of $G(2,N)$.

To show that the two algebras coincide (and that all additional relations we found earlier for the Verlinde algebra are consequences of (4.93)), it suffices to compare the dimensions of the two algebras. From the description of the

Verlinde algebra as being spanned by the elements $V_i W^j$, with $0 \leq i \leq N-2$, $0 \leq j \leq 2N-1$, with a two-fold restriction and a two-fold identification, its dimension is $N(N-1)/2$. On the other hand, with c_1 and c_2 considered to be of degree 1 and 2, respectively, the potential $W(c_1, c_2)$ is homogeneous of degree $N+1$; it follows by a simple counting that the polynomial ring in the c_j modulo the ideal $dW = 0$ has dimension $N(N-1)/2$. This completes the explicit verification of the equivalence between these rings.

Moreover, we can now dispose of the constant c in the relation $\sigma = cg$. This constant corresponds to a possible constant in the relation $\lambda_i \leftrightarrow \widetilde{\lambda}_i$. The relations above such as (4.94) are not invariant under rescaling of the λ's, and so the agreement with the Verlinde algebra would be ruined if we modified the value of c. A similar argument holds for $k > 2$.

The Metric

It remains to show that the metric on the Verlinde algebra and the metric on the quantum cohomology of $G(2, N)$ are related in the expected fashion.

Since either metric obeys $g(a, b) = g(ab, 1)$, to verify (4.63), it suffices to show that

$$g_V(a, 1) = g_\sigma(a, (\det \sigma)^{N-2}). \tag{4.95}$$

We already know the Verlinde metric:

$$g_V(\mathcal{V}_s \eta^t, 1) = \delta_{s,0}\delta_{t,0} + \delta_{N-2-s,0}\delta_{t-1,0}. \tag{4.96}$$

Now let us compute the metric on the cohomology of $G(2, N)$. Since there are no instanton corrections to the metric, we need only compute the classical metric on the cohomology. According to (3.45), that metric is

$$g_\sigma(a, b) = -\frac{1}{2} \sum_{dW(\lambda)=0} \frac{ab(\lambda_1 - \lambda_2)^2}{N^2 {\lambda_1}^{N-1} {\lambda_2}^{N-1}} = -\frac{1}{2} \sum_{{\lambda_1}^N = {\lambda_2}^N = -1} \frac{ab(\lambda_1 - \lambda_2)^2}{N^2 {\lambda_1}^{N-1} {\lambda_2}^{N-1}}. \tag{4.97}$$

The symmetric polynomials in the λ's of degree $2(N-2)$ (corresponding to the top dimensional cohomology of $G(2, N)$) are of the form

$$f_r = \frac{{\lambda_1}^{2r+1} - {\lambda_2}^{2r+1}}{\lambda_1 - \lambda_2} (\lambda_1 \lambda_2)^{N-2-r}, \quad \text{with } 0 \leq r \leq N-2. \tag{4.98}$$

A simple calculation gives

$$\begin{aligned} g_\sigma(f_r, 1) &= -\tfrac{1}{2N^2} \cdot \\ &\sum_{{\lambda_1}^N = {\lambda_2}^N = -1} \left(\left(\frac{\lambda_1}{\lambda_2}\right)^{1+r} - \left(\frac{\lambda_1}{\lambda_2}\right)^{r} - \left(\frac{\lambda_1}{\lambda_2}\right)^{-r} + \left(\frac{\lambda_1}{\lambda_2}\right)^{-1-r} \right) \\ &= \delta_{r,0}. \end{aligned} \tag{4.99}$$

This reproduces the first term on the right of (4.96) up to the shift predicted in (4.95). To interpret the second term on the right of (4.96), note that while classically for homogeneous f, $g_\sigma(f, 1)$ is non-zero unless f is of degree $2(N-2)$, the quantum cohomology is only graded modulo N (in complex dimension), so we can also consider the case that f is of degree $N-4$; by a calculation similar to the above, this reproduces the second term in (4.96).

Moreover, we can now dispose of the renormalization constant a of equation (4.31). Inclusion of this term would rescale the metric by a factor of e^{2a}; the agreement between the two metrics means that the above formulas are normalized correctly, at $r = 0$. Though we have made this check on the value of a (and a similar, earlier check for c) only for $k = 2$, the arguments are similar for any k.

Acknowledgements

I benefited from discussions with G. Segal at an early stage of this work.

References

[1] D. Gepner, Fusion Rings And Geometry, Commun. Math. Phys. **141** (1991) 381; D. Gepner and A. Schwimmer, Symplectic Fusion Rings And Their Metric, Nucl. Phys. **B380** (1992) 147; D. Gepner, Foundations Of Rational Conformal Field Theory, I, (Cal Tech preprint, 1992).

[2] C. Vafa, Topological Mirrors And Quantum Rings, in *Essays On Mirror Manifolds*, ed. S.-T. Yau (International Press, 1992).

[3] K. Intriligator, Fusion Residues, Mod. Phys. Lett. **A6** (1991) 3543.

[4] A. Gerasimov, Localization In GWZW And Verlinde Formula, hepth@xxx/9305090.

[5] M. Blau and G. Thompson, Derivation Of The Verlinde Formula From Chern-Simons Theory And The G/G Model, Nucl. Phys. **B408** (1993) 345.

[6] V. Batyrev, Quantum Cohomology Rings Of Toric Manifolds, Journées de Géometrie algébrique d'Orsay Asterisque **218** (1993), 9.

[7] A. D'Adda, M. Luscher, and P. DiVecchia, A $1/N$ Expandable Series Of Nonlinear Sigma Models With Instantons, Nucl. Phys. **B146** (1978) 63, Topology And Higher Symmetries Of The Two Dimensional Nonlinear Sigma Model, Phys. Report **49** (1979) 239.

[8] E. Witten, Instantons, The Quark Model, And The $1/N$ Expansion, Nucl. Phys. **B149** (1979) 285.

[9] E. Abdalla, M. Forger, and A. Lima Santos, Non-Local Charges For Nonlinear Sigma Models On Grassmann Manifolds, Nucl. Phys. **B256** (1985) 145.

[10] S. Cecotti and C. Vafa, On Classification Of $N = 2$ Supersymmetric Theories, Commun. Math. Phys. **158**, 569–644.

[11] A. Bertram, G. Daskalopoulos, and R. Wentworth, Gromov Invariants For Holomorphic Maps From Riemann Surfaces To Grassmannians, preprint (April, 1993).

[12] W. Lerche, C. Vafa, and N. Warner, Chiral Rings In $\mathcal{N} = 2$ Superconformal Theories, Nucl. Phys. **B324** (1989) 427.

[13] M. Bourdeau, E. J. Mlawer, H. Riggs, and H. J. Schnitzer, Topological Landau-Ginzburg Matter From $SP(N)_K$ Fusion Rings, Mod. Phys. Lett. **A7** (1992) 689.

[14] R. Bott, On E. Verlinde's Formula In The Context Of Stable Bundles, in *Topological Methods In Quantum Field Theories*, ed. W. Nahm et. al. (World Scientific, 1991).

[15] E. Verlinde, Fusion Rules And Modular Transformations In 2-D Conformal Field Theory, Nucl. Phys. **B300** (360) 1988.

[16] K. Gawędzki and A. Kupiainen, A G/H Conformal Field Theory From Gauged WZW Models, Phys. Lett. **215B** (1988) 119, Coset Construction From Functional Integrals, Nucl. Phys. **B320(FS)** (1989) 649; K. Gawędzki, Constructive Conformal Field Theory, in *Functional Integration, Geometry, And Strings*, eds. Z. Hava and J. Sobczyk (Birkhauser, 1989).

[17] E. Witten, On Holomorphic Factorization Of WZW And Coset Models, Commun. Math. Phys. **114** (1992) 189.

[18] S. Axelrod, Geometric Quantization Of Chern-Simons Gauge Theory, Ph.D. Thesis, Princeton University (1991).

[19] K. Gawędzki, Topological Actions In Two-Dimensional Quantum Field Theories, in *Non-perturbative Quantum Field Theory,* ed. G. 't Hooft (Plenum Press, 1988); G. Felder, K. Gawędzki, and A. Kupiainen, Spectra of Wess-Zumino-Witten Models With Arbitrary Simple Groups, Commun. Math. Phys. **117** (1988) 127.

[20] M. F. Atiyah and R. Bott, The Yang-Mills Equations Over Riemann Surfaces, Philos. Trans. R. Soc. London. **A308** (1982) 523.

[21] T. R. Ramadas, I. M. Singer, and J. Weitsman, Some Comments On Chern-Simons Gauge Theory, Commun. Math. Phys. **126** (1989) 409.

[22] E. Witten, Global Aspects Of Current Algebra, Nucl. Phys. **B223** (1983) 422, Non-Abelian Bosonization In Two Dimensions, Commun. Math. Phys. **92** (1984) 455.

[23] J. Wess and B. Zumino, Consequences Of Anomalous Ward Identities, Phys. Lett. **37B** (1971) 95.

[24] A. M. Polyakov and P. B. Wiegmann, Theory Of Non-Abelian Goldstone Bosons In Two Dimensions, Phys. Lett. **B131** (1983) 121.

[25] P. Goddard, A. Kent, and D. Olive, Virasoro Algebras And Coset Space Models, Phys. Lett. **152B** (1985) 88.

[26] A. Altschuler, K. Bardakci, and E. Rabinovici, String Models With $c < 1$ Components, Nucl. Phys. **B299** (1988) 157, A. Altschuler, K. Bardakci, and E. Rabinovici, A Construction Of the $c < 1$ Modular Invariant Partition Functions, Commun. Math. Phys. **118** (1988) 241.

[27] D. Karabali and H. J. Schnitzer, BRST Quantization Of The Gauged WZW Action And Coset Conformal Field Theories, Nucl. Phys. **B329** (1990) 649; D. Karabali, Q-Han Park, H. J. Schnitzer, and Zhu Yang, A GKO Construction Based On A Path Integral Formulation Of Gauged Wess-Zumino-Witten Actions, Phys. Lett. **B216** (1989) 307.

[28] V. B. Mehta and C. S. Seshadri, Moduli Of Vector Bundles On Curves With Parabolic Structures, Math. Ann. **248** (1980) 205.

[29] S. Elitzur, A. Forge, and E. Rabinovici, On Effective Theories Of Topological Strings, Nucl. Phys. **B388** (1992) 131.

[30] I. B. Frenkel, Representations of affine Lie algebras, Hecke modular forms and Korteweg-de Vries equations, in *Lie Algebras And Related Topics*, Lecture Notes In Mathematics Vol. 933 (Springer-Verlag, 1982), p. 71; Representations Of Kac-Moody Algebras And Dual Resonance Models, in *Lectures In Applied Mathematics*, vol. 21 (American Mathematical Society, 1985), p. 325.

[31] A. Pressley and G. Segal, *Loop Groups* (Oxford University Press, 1986).

[32] E. J. Mlawer, S. G. Naculich, H. A. Riggs, and H. J. Schnitzer, Group Level Duality of WZW Fusion Coefficients And Chern-Simons Link Observables, Nucl. Phys. **B352** (1991) 863; S. G. Naculich, H. A. Riggs, and H. J. Schnitzer, Group Level Duality In WZW Models And Chern-Simons Theory, Phys. Lett. **B246** (1990) 417.

[33] T. Nakanishi and A. Tsuchiya, Level-Rank Duality Of WZW Models In Conformal Field Theory, Commun. Math. Phys. **144** (1992) 351.

[34] M. Dine, N. Seiberg, X.-G. Wen, and E. Witten, Non-Perturbative Effects On The String World Sheet, I, II Nucl. Phys. **B278** (1986) 769, **B289** (1987) 319.

[35] M. Gromov, Pseudo-Holomorphic Curves In Symplectic Manifolds, Invent. Math. **82** (1985) 307.

[36] E. Witten, Two Dimensional Gravity And Intersection Theory On Moduli Space, Surveys in Differential Geometry **1** (1991) 243.

[37] V. Mathai and D. Quillen, Superconnections, Thom Classes, and Equivariant Differential Forms, Topology **25** (1986) 85.

[38] E. Witten, Topological Sigma Models, Commun. Math. Phys. **118** (1988) 411.

[39] M. F. Atiyah and L. Jeffrey, Topological Lagrangians And Cohomology, J. Geom. Phys. **7** (1990) 119.

[40] J. Wess and J. Bagger, *Supersymmetry And Supergravity*, second edition (Princeton University Press, Princeton, 1992).

[41] E. Witten, Phases Of $\mathcal{N} = 2$ Models In Two Dimensions, Nucl. Phys. **B403** (1993) 159 .

[42] N. J. Hitchin, A. Karlhede, U. Lindstrom, and M. Roçek, Hyperkahler Metrics And Supersymmetry, Commun. Math. Phys. **108** (1987) 535.

[43] E. Witten, Mirror Manifolds And Topological Field Theory, in *Essays On Mirror Manifolds*, ed. S.-T. Yau (International Press, 1992).

[44] G. Moore and N. Seiberg, Taming The Conformal Zoo, Phys. Lett. **B220** (1989) 422.

Geometry, Topology, and Physics for Raoul Bott

Cambridge, MA 02138 USA

Reflections on Geometry and Physics

Michael Atiyah
Trinity College
Cambridge CB2 1TQ
England

1 Philosophical reflection. In this lecture I will discuss in general terms what has been happening to the theoretical physics/mathematics frontier over the past 15 years.

Specifically I refer to the geometric and topological aspects of quantum field theory which have now spread in a variety of directions. New terms such as quantum groups, quantum geometry, quantum cohomology are appearing. These indicate the scope and significance of the interaction, but it is premature in my view to try to force everything into a particular mould. Time will tell what the significant aspects really are and then the right title to adopt will be clearer.

However, there has been unease expressed in certain quarters, most recently by Jaffe and Quinn, about the doubtful mixture that is emerging. Not tied closely to experimental physics nor to rigorous mathematics, standards are endangered and warning signs should be erected! Like a ship exploring unchartered seas, with inadequate maps and faulty compasses, catching glimpses of beautiful tropical islands: mirages or reality?

We can distinguish perhaps four different types of reaction by mathematicians towards these developments

(A) Take the heuristic results "discovered" by physicists and try to give rigorous proofs by other methods. Here the emphasis is on ignoring the physics background and only paying attention to mathematical results that emerge from physics. Like Ramanujan who intuited marvelous formulae that defied mathematical proof so physicists are viewed in the same light. The task of the mathematician is to start from scratch and aim to prove these marvelous "intuited truths". This is, of course, the minimalist reaction: the mathematician reacting in his own terms to an externally posed problem. We cannot ignore such challenges and we would all agree that a rigorous proof is a desirable objective.

(B) The second approach is try to understand the physics involved and enter into a dialogue with physicists concerned. This has great potential benefits since we mathematicians can get behind the scenes and see something of the stage machinery. This may provide clues for possible proofs, it may enable us to generalize the story and it may help us to see unexpected links with other areas. We may also be able to assist the physicists in their task, by pointing out relevant bits of mathematics or suggesting new points of view. This dialogue has, in fact, been developing widely in recent years, so that a whole new generation of mathematicians and physicists have begun to speak a common language. The worry of Jaffe and others is that this is a kind of pidgin English, with little grammar and no literary merit. But in its own terms it has been a remarkable success. The "results" keep growing in scope and depth and inevitably attract the incoming generation. The question is: where is it all leading?

(C) Following on from (B), one natural road for mathematicians to take is to try to develop the physics on a rigorous basis so as to give formal justification to the conclusions. This is the traditional role of the "mathematical physicist", of whom Jaffe is a fine exponent, and who have made rich contributions in the past. While undeniably the "right" approach for a respectable mathematician, it is sometimes too slow to keep up with the action. Depending on the maturity of the physical theory and the technical difficulties involved, the gap between what is mathematically provable and what is of current interest to theoretical physicists can be immense. Moreover, proofs are not always constructed from the bottom up. They may start from the top, or from the side, and only emerge after many hesitant steps and experimentation. Moreover, the right framework has to be established before rigorous work can begin, just as an architectural plan is necessary before the builders move in.

(D) Finally, and most ambitious of all, we may try to understand the deeper meaning of the physics-mathematics connection. Rather than view mathematics as a tool to establish physical theories, or physics as a way of pointing to mathematical truths, we can try to dig more deeply into the relation between them. This may lead us into the perennial problem of deciding whether mathematical results are invented or discovered. This investigation may only have philosophical or theoretical interest but it could lead to better understanding and even to new insights and genuine progress.

These four approaches are not, of course, mutually exclusive but many people will only dip their toes into this whole area and are happy to stick with (A). A sizeable community goes as far as (B), while (C) and (D) are definitely minority tastes. I do not disguise my attraction to (D) and this lecture will try to develop my ideas in that direction.

2 A Survey. Having set the philosophical scene, and raised some questions, I want to spend some time surveying briefly some of the new ideas and results in mathematics that have emerged from the interaction with theoretical physics.

2.1 Index Theory. The index theorem for the Dirac operator on compact Riemannean manifolds has turned out to be of great interest and relevance in gauge theories, since it measures the difference between left-handed and right-handed spinors or other physical entities. Various new proofs emerged naturally from the physicist's viewpoint. In particular, supersymmetry, an algebraic formalism that is increasingly used to bring fermions and bosons onto an equal footing, has led to useful simplifications. Moreover, a whole range of generalizations. including the study of the dependence of the Dirac operator on background gauge potentials, have been suggested by the physics. These have subsequently been given rigorous proofs by Bismut and others.

2.2 Elliptic Genera. Quantum field theory (for one space dimension) led Witten to introduce an appropriate Dirac operator on loop spaces. This has shed light on elliptic genera: these are generating functions for an appropriate sequence of Dirac operators coupled to bundles associated to the tangent bundle. It turns out that they are modular forms and the physics gives a natural interpretation of modularity as a consequence of (2-dimensional) relativistic invariance. Moreover a conjectured rigidity theorem (for compact group actions) also followed naturally from the physics and was eventually given rigorous proofs by Bott and Taubes.

2.3 Topological Quantum Field Theories. A number of extremely interesting topological theories, including Jones' work on knot invariants and Donaldson's work on 4-manifolds, have been given quantum field theory formulations by Witten. This has provided a unifying framework and has also led to generalizations of the original work. Thus the Jones invariants of knots in S^3 have now been extended to knots in general closed 3-manifolds.

Theories of this type, in dimension 2, have led to very precise and new information about the moduli space of flat unitary bundles on Riemann surfaces.

2.4 Conformal Field Theory. The representation theory of certain infinite-dimensional algebras, related to the circle, has a globalization over Riemann surfaces. The original circle here appears as the boundary of a "puncture" on the surface. Such "conformal field theories" are reasonably precise algebraic objects which connect representation theory to topology, via the topological Jones theory of (2.3).

2.5 Quantum Cohomology. Quantum field theory leads to a natural deformation of the ordinary cohomology ring of a manifold. This may briefly be referred to as "quantum cohomology". For example, for the complex projective line P_1 the ordinary cohomology is generated by $x \in H^2(P_1)$ with $x^2 = 0$, while the quantum cohomology has $x^2 = \beta$, where β is a real number (a parameter of the theory so that $\beta \to 0$ is the classical limit).

These "quantum cohomologies" are of considerable mathematical interest. For projective spaces and more generally Grassmannians they are related to the "Verlinde algebra" which plays a key role in conformal field theory and related topics. For 3-dimensional Calabi-Yau manifolds it contains information about the numbers of rational curves. This information is consistent with known results but does not yet have a rigorous mathematical proof. The physicist's "proof" involves the intriguing notion of dual or mirror manifolds, a pair of Calabi-Yau manifolds which are supposed to yield the same quantum field theory, but in dual ways.

2.6 2-dimensional gravity. The examples so far all fall within the class of gauge theories for forces other than gravity. However, there have been interesting developments related to gravity in 2-dimensions. These are closely involved with the moduli spaces of Riemann surfaces. In particular, triangulations of these moduli spaces link up with the combinatorial techniques of Feynman diagrams. The most exciting developments in this direction are due to Kontsevich and they also link up with the topology of 3-manifolds.

2.7 "Twisted" theories. Witten has shown how many physical quantum field theories can be "twisted" to yield topological theories. The twisting involves changing the spin of various fields. Certain correlation functions of the physical theory can be identified with some of the correlation functions of the twisted topological theory. This link has potentially important consequences. For example, Witten has suggested that the presence of a mass gap for $N = 2$ supersymmetric Yang-Mills theory in 4-dimensions may be related to conjectural properties of the Donaldson polynomials (which are derived from the topological Yang-Mills theory).

3 Interpretation. All these examples of fruitful interaction between quantum field theory and topology indicate that something substantial and widespread is involved. How should we interpret all this, what does it imply for "real" physics and how are we to deal with its mathematical aspects?

Perhaps it is helpful if we recall the role of symmetry (and group theory) in physics. Over the years symmetry has come to be recognized as a crucial guiding principle in large parts of fundamental physics. Starting with finite symmetries (as in crystals) and then moving on to continuous symmetries of compact groups, quantum physicists eventually introduced Hilbert space representations of non-compact Lie groups. This introduces extra analytical difficulties and, at first, there was no systematic mathematical theory to build on. However, mathematicians such as Gelfand and Harish-Chandra soon moved in to establish a base and develop an elaborate theory. Infinite-dimensional representations are now regarded as a vital part of many branches of mathematics including those like Number Theory, which are far removed from Physics.

I suggest that we are now seeing a similar, but more elaborate story involving the impact of Topology on Quantum Theory. Early topological ideas go back to Dirac (and even to Maxwell) but have only played a major part in the past decade or two. Again we are essentially dealing with infinite-dimensional phenomena (quantum fields) and it is the topology of these infinite-dimensions that is making itself felt. Topology and Symmetry have close analogies and relations, but Topology is inherently broader and more complex. For this reason we should not be surprised if Quantum Topology is a difficult subject which will take many years to mature.

Topology and Group Theory have something in common in their relation to Physics. Both interact, in principle, via Analysis but for many purposes the Analysis can be by-passed and replaced by Algebra. This is why so much of the Physics literature is filled with formulae. In the absence of a fully-fledged theory able to handle all the difficult analysis, physicists work formally and heuristically with algebraic formulae.

It is clear that the presence of symmetry in a physical situation imposes strong constraints and these can be exploited algebraically. What is the corresponding impact of topology? As Witten has explained, topology tends to provide information about low-energy states. For example, Hodge's theory of harmonic forms shows that the zero-energy states (for differential forms) correspond to the cohomology. It is worth noting that no significant topology enters for scalar fields, but in the super-symmetric version, when differential forms are brought in, the topological consequences become very significant. Corresponding statements can be made when we pass from quantum mechanics to quantum field theory. Interesting topology usually requires many nonscalar fields and frequently involves super-symmetry.

Symmetry and topology can play complementary roles with topology helping to determine the ground-states and symmetry then telling us how to build up higher states.

Now let me return to the general question of the "meaning" of all this "quantum topology". It would be hard to deny, in the face of all accumulated evidence, that the physicists who dabble with topology and quantum field theory are really on to something. How should we mathematicians respond, giving that a great deal rests on heuristic calculations and physical insight? Physicists will say that they are trying to develop quantum field theories which will explain all elementary particles and, if they are ambitious, also gravity. They are experimenting with a wide variety of models, many of which are "toy" models in the sense that they are grossly over-simplified in order to make them tractable. Given the extreme difficulty of the "real" physical theories, it is not unreasonable to focus on easier ones where one can make progress and gain insight - science always progresses in this way, although the mark of good scientists is to play with the right toys. A simplified model may be one in lower dimensions or having additional symmetries which lead perhaps to exact solutions. What the physicists typically extract is a lot of algebraic information (they like formulae!) and a toy model usually bristles with formal algebra.

This is the conventional, and acceptable, explanation of the physicists. What should be the reaction of the mathematician? Here we find a marked contrast, depending on the mathematician's background. Analysts, particularly those who have been trying to provide a rigorous basis for quantum field theory, dislike the algebraic superstructure which they think skirts the issue and hides the analytical difficulties. They would prefer to concentrate on the simplest possible theory algebraically so as to face up to analytical difficulties. Other mathematicians, coming from algebra, geometry or topology, are attracted by the superstructure and recognize there numerous features linked to their own experience. They are more than happy to follow the physicists in postponing any consideration of real analysis and concentrating on the formal structure. The hope, and ultimate justification, is that the formal apparatus may in the end lead the way to producing a rigorous theory. Perhaps the analysis will prove more tractable when approached the right way. It is already clear that a more formally complicated theory may turn out to be essentially simpler and better behaved than an apparently elementary theory, For example, in 4-dimensions, it is now recognized that Yang-Mills theory is better behaved than a scalar ϕ^4 theory.

Now let me turn to a more difficult question. What are to make of the striking results in 3 and 4-dimensional geometry that have emerged from field theory ideas? Specifically I have in mind the Jones invariants of knots and Donaldson's profound results on 4-manifolds. Is the relation with physics an accident which will in due course be eliminated, and replaced by more conventional mathematical techniques, or is the physics here to stay?

My own view is that the quantum standpoint is essential and that we are dealing with aspects of geometry or topology which are best understood in terms of quantum physics. For example, the fact that the 4-dimensional phenomena unearthed by Donaldson do not occur in other dimensions is surely an indication of their depth, and an indication that other conventional mathematical techniques will be inadequate to explain them. In 3-dimensions the work of Vassiliev, based on conventional homology (of a function space), has cast new light on the Jones invariants but it has not yet displaced the quantum approach. It provides an alternative avenue with different merits.

My conclusion is that, as in earlier episodes, mathematicians will absorb and abstract the essential quantum theory ideas, and develop an appropriate branch of mathematics. Because of the complexity and depth of the theories, especially if gravity is to be included, this may take time and may develop into an imposing edifice.

References

[1] A. Jaffe, F. Quinn, *Theoretical Mathematics*, Toward a cultural synthesis of Mathematics and Theoretical Physics. Bull. Amer. Math. Soc. (1993).

Geometry, Topology, and Physics for Raoul Bott

Cambridge, MA 02138 USA

Homology cobordism and the simplest perturbative Chern-Simons 3-manifold invariant

CLIFFORD HENRY TAUBES

1 Introduction. Witten predicted [**18**] that certain products of a certain 2-form could be integrated over products of a compact, oriented 3-manifold to give differential invariants of the 3-manifold. These predicted invariants were first constructed by Axelrod and Singer [**2, 3**] in the case where the 3-manifold has the rational homology of S^3. (A similar prediction in [**18**] for computing Jones' knot invariants had been partially realized by Bar Natan [**4**].) Subsequently, Kontsevich [**9**] gave an alternative realization of Witten's proposed invariants, with the same constraint on the homology of the 3-manifold. (Presumably, the invariants of Axelrod/Singer and of Kontsevich are the same, but the author has not seen a proof that such is the case.) Note that the invariants of Axelrod/Singer and Kontsevich have only been calculated for the 3-sphere (where they vanish).

The Axelrod/Singer and Kontsevich invariants are formally related to the 3-manifold invariants of Reshitikin and Turaev [**14**]. (The relationship here is presumed analogous to that between Jones, HOMFLY and other knot invariants and the knot invariants of Vassiliev [**16**], [**17**]; see [**6**], [**5**], [**10**].) There is no theorem at present which describes the precise relationship between these various 3-manifold invariants. Such a theorem would be useful in light of the fact that the invariants of Reshitikin and Turaev can be explicitly computed; they have been computed in closed form for lens spaces [**8**] and Seifert fibered 3-manifolds [**13**].

This is the first of two articles focusing solely on the simplest of the invariants of Kontsevich, an invariant, I_2, which assigns a number to a 3-manifold M (as constrained above) by integrating the cube of a certain real valued 2-form over $M \times M$. Of particular concern here is the value of I_2 on the 3-manifold boundaries of a 4-dimensional spin cobordism which has the rational homology of S^3. The results in this article, together with those in the sequel [**15**], prove that $I_2(M) = I_2(M')$ when M and M' are the boundary components of an oriented, spin 4-manifold W for which:

(1.1)

1. The intersection form on W's second homology (mod torsion) is conjugate to a direct sum of metabolic pairs.
2. The inclusions of M and M' into W induce injections of $H_1(\cdot;\mathbb{Z}/2)$.

(A metabolic pair is a symmetric, 2×2 matrix with zero's on the diagonal.)

In particular, the preceding result implies that $I_2(M) = 0$ when M has the integral homology of S^3. These results are restated and proved in [**15**].

This article makes a large step on the way to (1.1); the main theorem here, Theorem 2.9, states (in part) that $I_2(M) = I_2(M')$ when M and M' are the boundaries of an oriented, spin 4-manifold W for which the inclusions of M and M' into W induce

(1.2)

1) Isomorphisms on $H_p(\cdot;\mathbb{Q})$ for $p = 0, ..., 4$.
2) Injective maps on $H_1(\cdot;\mathbb{Z}/2)$.

In the course of proving Theorem 2.9, $I_2(S^3)$ is shown to vanish. Thus, even without the sequel **[15]**, the main theorem here can be used, in principle, to show that I_2 vanishes for certain 3-manifolds. (It is possible that $I_2 \equiv 0$ for all M!)

The author hopes that the constructions in this article will prove useful in studying the full set of invariants of Axelrod/Singer and Kontsevich, and this accounts, in part, for the length of the presentation. (The constructions here play a crucial role in **[15]**.)

Before beginning the story, the author wishes to thank Robion Kirby and Paul Melvin for their comments concerning this work, and also for their encouragement and support. A debt is owed as well to Dror Bar-Natan for sharing his knowledge of knot invariants.

This article is organized as follows: The definition of I_2 and the main theorem (Theorem 2.9) are given in the next section. The remaining sections (3-11) are occupied with constructions that are needed for the main theorem's proof. Section 3 is a digression to present certain facts from Morse theory. Section 4 studies the homological constraints which arise in the proof. Sections 5-10 contain the construction of a solution to the homological constraints. The final aspects of the proof of the main theorem are provided in Section 11.

2 The definition and properties of $I_2(M)$. The purpose of this section is to give a definition of Kontsevich's invariant, $I_2(\cdot)$, for compact, oriented 3-manifolds that have the rational homology of S^3. This section also contains the paper's main theorem about the equality of I_2 for a pair of 3- manifolds which occur as the boundary components of a certain kind of 4-dimensional cobordism.

a) Topological considerations.

Let M be a compact, oriented 3-manifold with the rational homology of S^3. Fix a point $p_0 \in M$. Let $\Delta \subset M \times M$ denote the diagonal. Define the subspace

$$\Sigma \equiv \Delta \cup (p_0 \times M) \cup (M \times p_0). \tag{2.1}$$

Lemma 2.1 describes the cohomology of $M \times M - \Sigma$. Before reading Lemma 2.1, be forewarned that a regular neighborhood of Σ in $M \times M$ is a neighborhood of Σ which strongly deformation retracts (rel Σ) onto Σ. It is an exercise to show that such neighborhoods exist. Also, in Lemma 2.1, cohomology is computed with real ($\mathbb{R}$) coefficients.

LEMMA 2.1. *Let Σ be as defined in (2.1). Then*

1) $H^2((M \times M) - \Sigma) \approx \mathbb{R}$.

2) Let $N \subset M \times M$ be a regular neighborhood of Σ. Then, restriction gives an isomorphism $H^2((M \times M) - \Sigma) \approx H^2(N - \Sigma)$.

3) Let $i : \mathbb{R}^3 \to N$ be an embedding which intersects $\Sigma - (p_0, p_0)$ transversely in a single point, $i(0)$. Then $i^* : H^2(N - \Sigma) \approx H^2(\mathbb{R}^3 - 0)$ is an isomorphism.

4) $H^1((M \times M) - \Sigma) \approx H^1(N - \Sigma) \approx 0$.

Proof. For the first assertion, use Meyer-Vietoris to prove that $(M \times M) - ((p_0 \times M) \cup (M \times p_0))$ has the rational homology of $\mathbb{R}^6$. Then, use Meyer-Vietoris again to compute the cohomology of the remainder when Δ is deleted. In fact, this calculation with Meyer-Vietoris shows that $M \times M - \Sigma$ has the rational cohomology of $S^3 \times S^2$.

Prove the second assertion using the Meyer-Vietoris exact sequence for the cover of $M \times M$ by N and $M \times M \cup \Sigma$. (The Kunneth formula gives $H^2(M \times M) = 0$, while restriction injects $H^3(M \times M)$ into $H^3(\Sigma)$.)

The third and fourth assertions are left as exercises with Meyer-Vietoris. □

The cohomology of $(M \times M) - \Sigma$ with rational coefficients is isomorphic to its DeRham cohomology.

b) An invariant.

Let C denote the set of pairs (N, φ) where N is a regular neighborhood of Σ, and where $\varphi : N \to \mathbb{R}^3$ is a smooth map with the property that $\varphi^{-1}(0) = \Sigma$. Define an equivalence relation on C as follows: Say that (N_0, φ_0) and (N_1, φ_1) are equivalent if there is a regular neighborhood $N_2 \subset N_0 \cap N_1$ and a smooth map

$$\Phi : [0,1] \times N_2 \to \mathbb{R}^3 \tag{2.2}$$

which obeys $\Phi(0, \cdot) = \varphi(0)$ and $\Phi(1, \cdot) = \varphi_1$ and $\Phi^{-1}(0) = [0,1] \times \Sigma$.

Let c denote the set of equivalence classes in C.

Now, change gears somewhat and pick a smooth, closed 2- form, μ, on $\mathbb{R}^3 - 0$ whose integral over the standard unit 2-sphere is equal to 1. For example,

$$\mu = (4\pi)^{-1}|x|^{-3}(x_1\, dx_2\, dx_3 + x_2\, dx_3\, dx_1 + x_3\, dx_1 dx_2) \tag{2.3}$$

Let $\varphi \in C$. According to Lemma 2.1, there exists a smooth, closed 2-form on $M \times M - \Sigma$ which agrees with $\varphi^*\mu$ on N. Fix such a form and call it ω_φ.

PROPOSITION 2.2. *Let* $(\mathrm{N}, \varphi) \in C$ *and choose* ω_φ *as described above. Then the following integral converges:*

$$I_2 \equiv \int_{M \times M - \Sigma} \omega_\varphi \wedge \omega_\varphi \wedge \omega_\varphi \tag{2.4}$$

Furthermore, I_2 *is independent of the choice of* ω_φ *to extend* $\varphi^* \mu$*, and it is independent of the choice of* μ*. Infact,* I_2 *depends only on the equivalence class of* (N, φ) *in* c *.*

Proof. The integral converges because the integrand has compact support on $M \times M - N$. Indeed, $\omega_\varphi \wedge \omega_\varphi$ vanishes on N because ω_φ on N is the pull back of a form on S^2.

Now, suppose that (N_0, φ_0) and (N_1, φ_1) define the same equivalence class in c . Suppose that μ_0 and μ_1 are different choices for μ in Proposition 1.2. Suppose that ω_0 and ω_1 are closed 2-forms on $M \times M - \Sigma$ which extend $\varphi_0^* \mu_0$ and $\varphi_1^* \mu_1$ from N_0 and N, respectively.

Let $N_2 \subset N_0 \cap N_1$ and $\Phi : [0,1] \times N_2 \to \mathbb{R}^3 - 0$ give the equivalence between (N_0, φ_0) and (N_1, φ_1). There are no obstructions to finding a 2-form μ on $[0,1] \times (\mathbb{R}^3 - 0)$ which is closed and restricts to $0 \times (\mathbb{R}^3 - 0)$ as μ_0 and to $1 \times (\mathbb{R}^3 - 0)$ as μ_1. Meanwhile, Lemma 2.1 insures that there are no obstructions to extending $\Phi^* \mu$ to $[0,1] \times (M \times M - \Sigma)$ as a closed 2-form ω.

With ω defined, compute

$$0 \equiv \int_{[0,1] \times (M \times M - \Sigma)} d(\omega \wedge \omega \wedge \omega) \tag{2.5}$$

using Stokes' theorem to express 0 (i.e. Equation (2.5)) as a sum of three terms. (Note that the integrand in (2.5) is compactly supported away from $[0,1] \times \Sigma$ since ω is pulled back from a 4- dimensional manifold on $[0,1] \times N_2$.) The three terms alluded to above are as follows: The first term is the contribution to Stokes' theorem from $\{1\} \times (M \times M - \Sigma)$; it is the integral in (2.4) as computed using the data with subscript "$_1$".

The second term is the contribution to Stokes theorem from $\{0\} \times (M \times M - \Sigma)$; it is the integral in (2.4) as computed using the data with subscript "$_0$".

To write down the third term which contributes to the Stokes' theorem computation of (2.5), one must first fix $N \subset N_2$, a smooth, oriented, codimension 1 submanifold that separates $\Sigma \subset M \times M$ from $M \times M - N_2$. With N understood, here is the third contribution to (2.5):

$$\delta I_2 \equiv \int_{[0,1] \times N} \omega \wedge \omega \wedge \omega \tag{2.6}$$

Note that (2.6) is zero because ω on $[0,1] \times N$ equals $\Phi^* \mu$, the pull-back of a 2-form on a 4-dimensional manifold.

Thus, the number I_2 is the same, whether computed using the data with subscript "$_0$" or with subscript "$_1$". □

c) Singular framings.

The previous subsection introduced the set c of equivalence classes of pairs (N, φ), where N is a regular neighborhood of Σ, and where $\varphi : N \to \mathbb{R}^3$ has $\Sigma = \varphi^{-1}(0)$. The purpose of this subsection is to describe a fiducial set of such classes. However, a preliminary, digression is required to define the notion of a singular framing of T^*M. The digression has four parts.

Part 1 of the digression introduces the standard framing of $T^*\mathbb{R}^3$, $dx = (dx_1, dx_2, dx_3)$. Part 1 also introduces the framing $\underline{\delta}$ of $T^*(\mathbb{R}^3 - 0)$ which is given at x $\in \mathbb{R}^3$ by

$$\underline{\delta} \equiv -dx + 2 \mid x \mid^{-2} < x,\, dx > x, \tag{2.7}$$

where $< x, dx > \equiv \Sigma_{i=1}^3 x_i dx_i$.

Part 2 of the digression makes the remark that a framing (such as $\underline{\delta}$) can be changed to a different framing using a matrix in $GL(3, \mathbb{R})$. Indeed, if $g \equiv (g_{ij})$ is a such a matrix, and if $\zeta \equiv (\zeta_1, \zeta_2, \zeta_3)$ is a framing, then $g\,\zeta$ is the framing given by $(g\,\zeta)_i = \Sigma_{j=1}^3\, g_{ij}\,\zeta_j$.

Part 3 of the digression defines the notion of a singular framing:

DEFINITION 2.3. *A singular frame, ζ, for T^*M is an oriented trivialization of $T^*(M - p_0)$ which has the following property: Let $\varphi : \mathbb{R}^3 \to M$ be an orientation preserving embedding (coordinate system) with $\varphi(0) = p_0$. There should exist an element $g \in GL(3, \mathbb{R})$, with positive determinant, and such that*

$$\lim_{r \to 0} \sup_{|x| = r} \mid \varphi^*\zeta - g\,\underline{\delta} \mid (x) = 0. \tag{2.8}$$

(Note: Let ζ be a frame for $T^*(M - p_0)$. Suppose that φ and φ' are two coordinates systems as in Definition 2.3 and that there exists g which makes (2.8) true for φ. It is an exercise to show that there will exist g' which makes (2.8) true for φ'.)

Roughly, a singular frame for T^*M is a frame which looks like $\underline{\delta}$ in some coordinate system centered at p_0.

Part 4 of the digression defines a homotopy class of singular framing of T^*M. Measure the distance between singular frames using the C^0 norm on sections of $T^*(M - p_0)$. Then, give the set of C^0 norms the induced metric topology. A homotopy class of singular frames is just a path component of the space of singular frames.

LEMMA 2.4. *The set c of homotopy classes of singular frames for T^*M is naturally a principal bundle over a point for the abelian group $\pi_0 \equiv (Maps(M;\, SO(3))$.*

Proof. Since $\pi_2(SO(3)) = 0$, the set of homotopy classes of singular frames is in $1 - 1$ correspondence with the set of homotopy classes of honest framings of T^*M. Fix a frame of T^*M and then the space of framings of T^*M can be identified with the space of maps from M into $SO(3)$. Thus, the set of homotopy classes of singular frames is in $1-1$ correspondence with the group π_0.

The action of π_0 on the set of homotopy classes of singular frames comes about as follows: Let $\zeta \equiv (\zeta_1, \zeta_2, \zeta_3)$ be a singular frame, and let $g \equiv (g_{ij})_{i,j=1}^3$ be a map from M to $SO(3)$. Then $g\,\zeta$ is the frame whose i'th component is given by $\Sigma_{j=1}^3\, g_{ij}\,\zeta_j$. It is left to the reader to check that the aforementioned action is free and transitive.

By the way, the group π_0 is naturally isomorphic to an abelian extension of $H^1(M;\mathbb{Z}/2)$ by $\mathbb{Z}$. The projection,

$$w : \pi_0 \to H^1(M;\mathbb{Z}/2), \tag{2.9}$$

of a homotopy class $[g]$ of map $g : M \to SO(3)$ is the cohomology class of the pull-back by g of the generator of $H^1(SO(3);\mathbb{Z}/2)$. Any two maps which have the same pull-back of said generator differ by a map which lifts to a map from M to S^3. The homotopy class of such maps to S^3 is classified by assigning to a map its degree, an integer.

There is also a homomorphism,

$$p_\infty : \pi_0 \to \mathbb{Z}, \tag{2.10}$$

which is defined on a class $[g]$ by taking the generator $H^3(SO(3);\mathbb{Z})$ and evaluating its pull-back on M's fundamental class. Note that classes in π_0 which lift to map M into S^3 are sent by p_∞ into $2\,\mathbb{Z}$.

End the digression. □

PROPOSITION 2.5. *A homotopy class, $[\zeta]$, of singular frames for T^*M canonically defines an equivalence class, $c_\zeta \equiv c$ of pairs (N,φ), where N is a regular neighborhood of Σ and where $\varphi : N \to \mathbb{R}^3$ is a smooth map with $\varphi^{-1}(0) = \Sigma$. Furthermore, when $(N,\varphi) \in c_\zeta$, then φ^* is an isomorphism between $H^2(\mathbb{R}^3 - 0)$ and $H^2(N - \Sigma)$.*

Propositions 2.2 and 2.5 define a map, I_2, from the set, c, of equivalence classes of canonical frames to $\mathbb{R}$.

PROPOSITION 2.6. *The map $I_2 : c \to \mathbb{R}$ is equivariant under the action of π_0 on c when the π_0 action on $\mathbb{R}$ is defined by sending a pair $([g], r) \in \pi_0 \times \mathbb{R}$ to $2^{-1} p_\infty[g] + r$.*

The propositions in this subsection are proved below.

d) A canonical singular frame.

Note that I_2 above is not a numerical invariant of M as it is apriori defined on the quotient, $\underset{\sim}{c}$, of the set of homotopy classes of singular frames by the action of the kernel of the homomophism p_∞. However, (as suggested by Kevin Walker

[??]) one can apply an observation of Atiyah to produce a canonical element in c, and then apply I_2 to this canonical element to produce the numerical invariant, $I_2(M)$.

The definition of $I_2(M)$ requires the following four part digression: Part 1 of the digression recalls the observation of Atiyah [**1**] that a compact, oriented 3-manifold has a canonical homotopy class of framing of $T^*M \oplus T^*M$. Atiyah calls this 2-frame the *canonical* 2-frame. It will be denoted here by A.

A frame Ξ for $T^*M \oplus T^*M$ which defines A is characterized by two conditions. The first condition requires that Ξ differ from a product frame (ζ, ζ) by a map from M into $SO(6)$ which lifts to a map M into Spin(6).

To describe the second characterizing condition, remark first that when M is the boundary of a compact, oriented 4-manifold, X, with boundary, then the frame Ξ defines a framing of $(T^*X \oplus T^*X)|_M$ since $T^*X|_M \approx T^*M \oplus \underline{\epsilon}$ with $\underline{\epsilon}$ being the trivial line bundle. Remark second that a framing Ξ of $(T^*X \oplus T^*X)|_M$ has a relative first Pontrjagin number $p_1(T^*X \oplus T^*X, \Xi)$. (This number is defined to be the first Pontrjagin number of an $\mathbb{R}^6$-bundle over the space X/M obtained by crushing M to a point. The bundle in question is trivial near M, and is isomorphic to $T^*X \oplus T^*X$ away from M, with Ξ defining the isomorphism.)

With $p_1(T^*X \oplus T^*X, \Xi)$ understood, here is the 2nd condition that characterizes Atiyah's canonical 2-frame: For any X as above,

$$p_1(T^*X \oplus T^*X, \Xi) = 6 \ \text{signature}(X) \tag{2.11}$$

(Note that the Hirzebruch signature theorem insures that when (2.11) holds for one X as above, it holds for all such X.)

Part 2 of the digression serves as a reminder that every compact, oriented 3-manifold is the boundary of some compact, oriented, spin 4-manifold with boundary [**18**]. Furthermore, the signature mod(8) of such a bounding 4-manifold is an invariant of M.

Part 3 of the digression remarks that any map from a 3- manifold into Spin(6) deforms into Spin(3) and has a natural degree. Part 4 of the digression describes the relationship between singular framings for T^*M and framings of $T^*M \oplus T^*M$:

PROPOSITION 2.7. *Let M be a compact, oriented 3-manifold with the rational homology of S^3.*

1) *A homotopy class of singular frame $[\zeta]$ for T^*M naturally defines a pair, $([\zeta_-], [\zeta_+])$, of homotopy classes of honest frames for T^*M. Here, $\zeta_+ = g\,\zeta_-$ where $g : M \to SO(3)$ is a map with $p_\infty(g) = 2$ and which lifts to S^3.*
2) *The assignment $[\zeta] \to ([\zeta_-], [\zeta_+])$, above, induces a natural, injective map θ from $\underline{c}$ into the set of homotopy classes of framings of $T^*M \oplus T^*M$.*
3) *If M bounds a compact, oriented, spin 4-manifold whose signature is zero mod(4), then Atiyah's canonical 2-frame is in the image of θ.*

4) *In general, there is a map g from M to Spin(6) with non-negative degree and such that $g\,A$ is in the image of θ.*

This proposition is also proved below.

End the digression and consider the following definition of $I_2(M)$:

DEFINITION 2.8. *Let M be a compact, oriented 3-manifold which has the rational homology of S^3. Let A denote the canonical 2-frame of Atiyah.*

a) *If M bounds a spin 4-manifold with signature 0 mod(4), then define $c_M \in \underline{c}$ to be $\theta^{-1}(A)$.*

b) *In general, define $c_M \in c$ so that $\theta(c_M) = g\,A$, where g is a map from M to Spin(6) whose degree is non-negative and minimal among the set of all g such that $g\,A \in$ Image(θ).*

c) *Define $I_2(M)$ to be the value of Proposition 2.6's homomorphism I_2 on c_M.*

e) $I_2(\cdot)$ and cobordisms.

The main purpose of this article is to prove that $I_2(M)$ is an invariant of a certain type of cobordism. A precise statement requires a two part digression.

For Part 1 of the digression, consider a pair, M_0 and M_1, of compact, oriented 3-manifolds. A 4-manifold with boundary, W, will be called an oriented, rational homology, spin cobordism between M_0 and M_1 when the following requirements are met: First, W is oriented and spin. Second, W's boundary is the disjoint union of M_0 and M_1. Third, let $i_{0,1} : M_{0,1} \to W$ denote the inclusions as boundary components. These inclusions, plus the given orientation of W, orient M_0 and M_1. This boundary orientation of M_1 should agree with its given orientation, but the boundary orientation of M_0 should disagree with M_0's given orientation. Fourth, the inclusions of M_0 and M_1 into W should induce isomorphisms on the rational homology.

End Part 1 of the digression and start Part 2. Let W be a compact, oriented, spin 4-manifold with boundary and let M be a component of ∂W. Let $K(M;W)$ denote the cokernel of the restriction induced homomorphism $H^1(W;\mathbb{Z}/2) \to H^1(M;\mathbb{Z}/2)$. The purpose of Part 2 of the digression is to define a homomorphism

$$l_W : c :\to K(M;W). \tag{2.12}$$

The definition of l_W takes five steps: Step 1 remarks that T^*W is isomorphic to the trivial bundle because W is spin and not a compact manifold. In particular, T^*W has a frame, ζ. For Step 2, let ψ be a singular frame for T^*M and let ψ' be any honest frame for T^*M which agrees with ψ on the compliment of a ball around p_0. Note that ψ' defines a frame for $T^*W|_M \approx T^*M \oplus \underline{\epsilon}$ by adding an appropriately oriented frame, ϵ, for the trivial real line bundle ϵ. Step 3 observes that there is a unique map $g : M \to SO(4)$ which is characterized by the equation $(\psi', \epsilon) = g\,(\zeta|_M)$. Step 4 observes that pull-back by g defines an element $l(\psi', g) \in H^1(M;\mathbb{Z}/2)$. Finally, Step 5 observes that $l(\psi', g)$ depends only on the pair (ψ, g), and that the image of $l(\psi', g)$ in $K(M; W)$ depends only on the homotopy class of the singular frame f and on the 4-manifold W. This image is l_W.

End the digression. Here is the main theorem in this article:

THEOREM 2.9. *Let M_0 and M_1 be compact, oriented 3-manifolds with the rational homology of S^3. Let W be an oriented, rational homology, spin cobordism between M_0 and M_1.*

1) *If the inclusions of both M_0 and M_1 into W induce injective maps on $H_1(\cdot; \mathbb{Z}/2)$, then $I_2(M_0) = I_2(M_1)$.*
2) *More generally, if both the canonical homotopy class of singular frames for M_0 and for M_1 (as defined in Proposition 2.8) are represented by $c \in c$ with $l_W(c) = 0$, then $I_2(M_0) = I_2(M_1)$.*
3) *$I_2(S^3) = 0$ and so $I_2(M) = 0$ if M and S^3 are cobordant by a spin cobordism with the rational homology of S^3.*

The proof of Theorem 2.9 occupies Sections 3-11 of this article, but see Subsections 2i for the proof that $I_2(S^3) = 0$ and see Subsection 2k for an outline of the strategy for the proof of the rest of the theorem.

f) Proof of Proposition 2.5.

The proof of Proposition 2.5 uses the Pontrjagin-Thom construction; and here is a short digression to outline how it works: Let Y be a smooth manifold and let $Z \subset Y$ be a smooth submanifold of codimension p with trivial normal bundle $N \subset TY|_Z$. Note that a trivialization of Z's co-normal bundle, N^*, defines a unique homotopy class of maps from a regular neighborhood of Z in Y to R^p which have Z as the inverse image of 0.

Indeed, a trivialization, ζ, of N^* defines a (fiberwise) linear map $\underline{\varphi}_\zeta : N \to \mathbb{R}^p$ which has zero as a regular value, and which has the zero section as the inverse image of 0.

Now, call a map $e : N \to Y$ an *exponential* map if it has the following two properties:

1) e's restriction to the zero section, $Z_0 \subset N$, is the identity.
2) e's differential along Z_0 induces the canonical identification between

$$(TN|_{Z_0})/TZ_0 \quad \text{and} \quad (TY|_Z)/TZ \equiv N. \tag{2.13}$$

It is not hard to show that exponential maps exist.

Because of (2.13), an exponential map defines a diffeomorphism between a neighborhood of the zero section in N with a neighborhood of Z in Y. (Use the inverse function theorem.) Furthermore, any two exponential maps are homotopic through exponential maps. (The zero'th and first order terms in the Taylor's expansion off the zero section of an exponential map are fixed completely by (2.12).)

Fix an exponential map $e : N \to Y$. As remarked, e defines a diffeomorphism between a regular neighborhood, N, of Z in Y with a regular neighborhood of the zero section in N. Thus

$$\varphi_{\zeta,e} \equiv \underline{\varphi}_{\zeta}\, e^{-1} : N \to \mathbb{R}^p \tag{2.14}$$

is a map with Z as the inverse image of zero.

Because any two exponential maps are homotopic through exponential maps, one can conclude that the coframe ζ, all by itself, defines an equivalence class, c_ζ, of pair (N, φ) where N is a regular neighborhood of Z in Y, and where $\varphi : N \to \mathbb{R}^p$ is a smooth map with $\varphi^{-1}(0) = Z$. (The equivalence relation between such pairs is analgous to the equivalence relation in Proposition 2.2.)

By the way, this construction has an inverse. Let $Z \subset Y$ be an embedded submanifold, and suppose there is a neighborhood $N \subset Y$ of Z and a map $\varphi : N \to \mathbb{R}^p$ with 0 a regular value and with $Z = \varphi^{-1}(0)$. Let $dx \equiv (dx_i)_{i=0}^{p}$ denote the standard basis for $T^*\mathbb{R}^p$. Then $(\varphi^* dx)|_Z$ defines a framing for Z's conormal bundle in Y.

End the digression.

The plan of proof of Proposition 2.5 is as follows: Step 1 presents an essentially canonical map, Φ_0, from a neighborhood of (p_0, p_0) to $\mathbb{R}$ with the correct inverse of 0. Step 2 finds canonical conormal framings for $p_0 \times (M - p_0)$ and for $(M - p_0) \times p_0$ to be used with the Pontrjagin-Thom construction to extend Φ_0 along neighborhoods of $p_o \times M$ and $M \times p_0$. Step 3: The given singular framing of T^*M will then be used to define a conormal framing for $\Delta - (p_0, p_0)$ and, finally, the Pontrjagin-Thom construction will be used to turn this conormal framing into an extension of Φ_0 to a neighborhood of Δ.

Step 1: Choose an oriented frame for $T^*M|_{p_0}$ and fix an exponential map from $T^*M|_{p_0}$ into M. This identifies a neighborhood of p_0 in M with a ball of some non-zero radius in $\mathbb{R}^3$. To avoid complicated notation, assume the radius is 1. This identification will be made implicitly in what follows. Use $x \equiv (x_1, x_2, x_3)$ to denote the Euclidean coordinates in $\mathbb{R}^3$.

Note that the coordinate system so defined depends on a choice of frame for $T^*M|_{p_0}$ and also on an exponential map. As described above, the dependence on the choice of exponential map is inconsequential if one is interested only in a homotopy class of map from a neighborhood of Σ to $\mathbb{R}^3$. The choice of frame for $T^*M|_{p_0}$ is also inconsequential because two such frames differ by the action of an element in the component of the identity matrix in $GL(3;\mathbb{R})$.

The preceding coordinate system identifies a neighborhood, $N_0 \times N_0$, of (p_0, p_0) with a neighborhood of $(0,0)$ in $\mathbb{R}^3 \times \mathbb{R}^3$. Make this identification implicit in what follows.

A point in $\mathbb{R}^3 \times \mathbb{R}^3$ will be written as (x, y). With this understood, $p_0 \times M$ intersects $N_0 \times N_0$ as the subspace where $x \equiv 0$, while $M \times p_0$ intersects $N_0 \times N_0$ as the subspace where $y = 0$. The diagonal Δ intersects $N_0 \times N_0$ as the subspace where $x - y = 0$.

Define Φ_0 on $N_0 \times N_0$ by the formula

$$\Phi_0(x, y) \equiv |x|^2 y - |y|^2 x \tag{2.15}$$

(Dror Bar-Natan described Φ_0 to the author.) Here are the salient features of Φ_0:

LEMMA 2.10. *The map Φ_0 has the following properties:*
1) $\mid \Phi_0(x,y) \mid = \mid x \mid \mid y \mid \mid x - y \mid$.
2) Φ_0 *is a submersion on* $\mathbb{R}^3 \times \mathbb{R}^3 - (0,0)$.
3) $d\Phi_0 \mid_{x=0} = - \mid y \mid^2 dx$.
4) $d\Phi_0 \mid_{y=0} = \mid x \mid^2 dy$.
5) Let $w \equiv x + y$ *and let* $u = x - y$. *Then*

$$d\Phi_0 \mid_{u=0} = 2 \mid w \mid^2 (-du + 2 \mid w \mid^{-2} \langle w, du\rangle\, w), \tag{2.16}$$

where $\langle w, du\rangle \equiv \Sigma_{i=1}^3 w_i\, du_i$.

Proof. All remarks are exercises with multivariable calculus. □

Step 2: The coordinate system on N_0 identifies $N_0 \times M$ with $B \times M$, where $B \subset \mathbb{R}^3$ is a neighborhood of the origin. Map $N_0 \times M$ to B by first using the preceding identification and then projecting onto B. Call the preceding map φ. Note that 0 is a regular value of φ and that $p_0 \times M = \varphi^{-1}(0)$. Thus, $-\varphi^* dx$ defines a frame for the conormal bundle of $p_0 \times M$ in $M \times M$.

Now, restrict attention to $N_0 \times N_0 \subset N_0 \times M$. The map Φ_0 of Lemma 2.10 maps $p_0 \times N_0$ to zero, and is a submersion along $p_0 \times N_0$ except at $p_0 \times p_0$. Note that $\Phi_0^* dx$ and $-\varphi^* dx$ differ along $p_0 \times N_0$ only by multiplication of a scalar function. (This is Assertion 3 of Lemma 2.10.) In the coordinates of Lemma 2.10, $\Phi_0^* dx = - \mid y \mid^2\, dx$ while $\varphi^* dx = -dx$. Note also that the former frame is defined for $|y| < 1$, while the latter is defined for $\mid y \mid > 0$. Take a favorite, positive function h on $[0, \infty)$ which obeys

(2.17)

1) $h(t) = t$ for $t < 1/2$
2) $h(t) = 1$ for $t \geq 3/4$.

Use $-h(|y|^2)\, dx$ to interpolate between $\Phi_0^* dx$ where $\mid y \mid < 1/2$ and between $\varphi^* dx$ where $\mid y \mid > 1$ and so construct a coframe for $p_0 \times M$ on the compliment of $p_0 \times p_0$ which agrees with $\Phi_0^* dx$ on $p_0 \times N_0$ and with $-\varphi^* dx$ on $p_0 \times (M - N_0)$. Use the Pontrjagin-Thom construction (as described above) to extend this coframe to a smooth map from a neighborhood of $p_0 \times M$ in $M \times M$ to $\mathbb{R}^3$ which agrees with Φ_0 on a neighborhood of $p_0 \times p_0$.

Given that the coordinate system about p_0 is fixed, the above extension will be canonical up to homotopy.

As for $M \times p_0$, introduce the switch map,

$$\Theta : M \times M \to M \times M \tag{2.18}$$

which interchanges the factors. Obviously, the switch map interchanges $p_0 \times M$ with $M \times p_0$. Use the switch map to extend Φ_0 along a neighborhood of $M \times p_0$. □

Step 3: To extend the map Φ_0 to a neighborhood of Δ, the strategy will be to find a coframe, ζ, for the normal bundle of $\Delta - p_0$ which agrees with $\Phi_0^* dx$ on $(\Delta - p_0) \cap N_0 \times N_0$. Given such a frame ζ, one can copy the arguments from Step 2, above, to extend Φ_0 along Δ. As in Step 2, the extension will be unique up to homotopy.

Now, Δ is canonically diffeomorphic to M, while the conormal bundle of Δ in $M \times M$ is canonically isomorphic to T^*M. (The isomorphism here is given by $\pi_R^* - \pi_L^*$, where $\pi_{R,L} : M \times M \to M$ are the projections onto the right (R) and left (L) factors.) So, the coframe ζ is a frame for $T^*(M - p_0)$ with a prescribed form near p_0. The constraint on ζ near p_0 comes from the requirement that ζ agree with $\Phi_0^* dx$. The latter is given in (2.16). Thus, up to the scalar factor, ζ should be a singular frame for T^*M in the sense of Definition 2.3, but with the matrix g in Definition 2.3 equal to the identity.

If $g \equiv g_\zeta$ in Definition 2.3 is not the identity for the given frame ζ, then replace ζ by $g_\zeta\, \zeta$; this last frame will obey (2.8) with g the identity matrix.

Thus, a singular frame for T^*M can be used to build a map from a neighborhood of Σ to $\mathbb{R}^3$ with all of the requisite properties. That said map is unique up to homotopy follows from two already mentioned facts: First, positive determinant matrices in $GL(3, \mathbb{R})$ form the path component of the identity. Second, the map in (2.14) is insensitive (modulo homotopies) to the choice of exponential map.

g) Proof of Proposition 2.6.

To consider the behavior of I_2 when π_0 acts on c, remember that the action of π_0 is generated as follows: Let ζ be a singular frame for T^*M. Let $g : M \to SO(3)$ be a smooth map which equals the identity near to p_0. Then $[g] \in \pi_0$ acts on the class $[\zeta]$ of the frame ζ to give the class of the singular frame $g\,\zeta$.

To compare the value of Φ_2 on $[\zeta]$ and on $[g\,\zeta]$, construct the map $\varphi_\zeta : N \to \mathbb{R}^3$ as described in Proposition 2.5. One could construct $\varphi_{g\,\zeta}$ too, but here is a shortcut to this map: Introduce the neighborhood $N_0 \subset N$ as before. Then, define a smooth map ψ from N_0 to $SO(3) \times \mathbb{R}^3$ as follows:

(2.19)

1) If $p \in (B_0 \times M) \cup (M \times B_0)$, then set $\psi(p) \equiv (1, \varphi_\zeta(p))$.
2) If $p \in N_\Delta$, then set $\psi(p) \equiv (g(p), \varphi_\zeta(p))$.

Let $m : SO(3) \times \mathbb{R}^3 \to \mathbb{R}^3$ denote the map which describes the group action. Let $p : SO(3) \times \mathbb{R}^3 \to \mathbb{R}^3$ denote the projection. With m and with ψ understood, one can take $\varphi_{g\,\zeta}$ as follows:

$$\varphi_{g\,\zeta} = m \circ \psi \tag{2.20}$$

Equation (2.20) can be exploited with the help of the following observation: On $SO(3) \times (\mathbb{R}^3 - 0)$, the 2-forms $m^* \omega_0$ and $p^* \omega_0$ are cohomologous, so find a 1-form α_0 on $SO(3) \times (\mathbb{R}^3 - 0)$ such that $m^*\omega_0 = p^*\omega_0 + d\alpha_0$. Make α_0 restrict to 0 on $1 \times (\mathbb{R}^3 - 0)$. With α_0 understood, notice that $\varphi^*_{g\,\zeta}\omega_0 = \varphi^*_\zeta\omega_0 + \psi^* d\alpha_0$ on $N_0 - \Sigma$.

Let ω and ω' be closed 2-forms on $M \times M - \Sigma$ which extend $\varphi^*_\zeta\omega_0$ and $\varphi^*_{g\,\zeta}\omega_0$, respectively. There is no obstruction to extending α from $N_0 - \Sigma$ as a smooth 1-form α on $M \times M - \Sigma$ such that $\omega' = \omega + d\alpha$ (See Lemma 1.1.) Then, $\Phi_2[g\,\zeta] - \Phi_2[\zeta]$ is given by

$$\Phi_2[g\,\zeta] - \Phi_2[\zeta] = \int_{M\times M-\Sigma} (3\,d\alpha \wedge \omega \wedge \omega + 3\,d\alpha \wedge d\alpha \wedge \omega + d\alpha \wedge d\alpha \wedge d\alpha) \tag{2.21}$$

Here is how to evaluate the right side of (2.21): Let $S^2 \subset \mathbb{R}^3$ denote a ball of small radius about 0 and which is in the image of φ_ζ on $N_\Delta - \pi^{-1}(B_0)$ and which is transversal to φ_ζ on $N_\Delta - \pi^{-1}(B_0)$. (One can assume, with no loss of generality, that such a 2-sphere exists.) Look carefully at the proof of Proposition 2.5. Introduce S to denote $\varphi_\zeta^{-1}(S^2)$. This is a two sphere bundle over $N_\Delta - \pi^{-1}(B_0)$. Because α_0 vanishes on $B_0 \times M$ and on $M \times B_0$, Stokes theorem equates the right hand side of (2.21) with

$$\Phi_2[g\,\zeta] - \Phi[\zeta] = \int_S \varphi^*(\alpha_0 \wedge d\alpha_0 \wedge p^*\omega_0). \tag{2.22}$$

(One uses here the fact that $\omega_0 \wedge \omega_0 = 0$ and that $2\,d\alpha_0 \wedge p^*\omega_0$ is equal to $-d\alpha_0 \wedge d\alpha_0$.)

Introduce $i : SO(3) \to SO(3) \times S^2$ to be the inclusion as a fiber of the projection, p, to S^2. Then, the integral in (2.22) can be evaluated by pushing the integrand forward to $SO(3) \times S^2$, and then by doing the S^2 integration first. The result is

$$\Phi_2[g\,\zeta] - \Phi_2[\zeta] = p_\infty[g] \int_{SO(3)} i^*(\alpha_0) \wedge di^*(\alpha_0). \tag{2.23}$$

The factor of $p_\infty[g]$ in (2.23) is due to the fact that g^* maps the fundamental class of M to $p_\infty[g]$ times the fundamental class of $SO(3)$.

Now, the integral in (2.23) is simply a Hopf invariant, and can be computed to be equal to 1/2. (Restrict $m : SO(3) \times S^2 \to S^2$ to $m_0 : SO(3) \times n \to S^2$, where n is the north pole. Then $di^*\alpha_0 = m_0^*\omega_0$. But, m_0 defines a principal $SO(2)$ bundle with Euler class equal to 2, so $2\,i^*\alpha_0$ will integrate to 1 over the fiber of m_0, while ω_0 integrates to 1 over the base. Thus, the value of the integral in (2.23) is equal to 1/2.) So, $\Phi_2[g\,\zeta] = 2^{-1}\,p_\infty[g] + \Phi_2[\zeta]$ as claimed. □

h) Proof of Proposition 2.7.

A singular frame, ζ, for T^*M defines a 2-frame for TM as follows: Choose a coordinate system around p_0 and introduce the matrix g as in (2.8). Then $\zeta' \equiv g\,\zeta$ obeys (2.8) with g replaced by the identity. Thus, on a small ball about p_0, ζ' differs from the frame $\underline{\delta}$ of (2.7) by a small amount, and so there is a canonical homotopy of ζ' so that it agrees with $\underline{\delta}$ on a small ball, B, about p_0.

Now, $\pi_2(SO(3))$ vanishes, so there is no obstruction to deforming $\underline{\delta}$ inside B so that the result, $\underline{\delta}'$, agrees with $\underline{\delta}$ near B's boundary, and agrees with the constant framing dx on an even smaller ball, $B' \subset B$. Such a deformation would change ζ' to an honest framing of TM. Unfortunately, $\pi_3(SO(3)) \approx \mathbb{Z}$, so a topologist might argue that there is no canonical way to get an honest framing from ζ'.

However, the frame $\underline{\delta}$ is rather special (Kevin Walker pointed this out to the author); restrict it to the 2-sphere boundary $S^2 \subset B$. Compare $\underline{\delta}$ with the constant framing to define a map from $S^2 \to SO(3)$. Lift this map to $S^3 = SU(2)$ and one has the inclusion of S^2 in S^3 as the equator which is invariant under multiplication by ± 1 on S^3. And, there are precisely two canonical deformations to a point of this equatorial embedding of S^2 in S^3. Indeed, consider taking the family of two spheres of decreasing latitude starting from $\pi/2$ and going to 0. Or, take the family of increasing latitude, starting from $\pi/2$ and going to π.

Thus, there are two canonical ways to obtain an honest framing from the singular framing ζ. Denote these two honest framings by $\zeta_\pm$. By construction, $\zeta_+ = g\,\zeta_-$ where $g : M \to SO(3)$ has degree 2 and lifts to map M into S^3. This exhibition of $\zeta_\pm$ completes the proof of the first assertion of Proposition 2.7.

To prove Assertion 2 of Proposition 2.7, take $\zeta_\pm$ above and produce the 2-framing $\Theta_\zeta \equiv \zeta_+ \oplus \zeta_-$. (Note that $\zeta_+ \oplus \zeta_-$ is homotopic to $\zeta_- \oplus \zeta_+$ as a framing of $TM \oplus TM$.) Clearly, if ζ_0 and ζ_1 define the same equivalence class of singular framing for T^*M, then Θ_{ζ_0} and Θ_{ζ_1} will be homotopic as framings of $TM \oplus TM$ and so define the same 2-framing of TM.

To prove that the map Θ is injective on $\underline{c}$, consider first a 4- manifold X which bounds M. Let Ξ, Ξ' be frames for $T^*M \oplus T^*M$. Then Ξ is homotopic to Ξ' only if $p_1(TX \oplus TX; \Xi) = p_1(TX \oplus TX; \Xi')$.

With the preceding understood, let ζ and ζ' be a pair of singular frames for T^*M. Then $\zeta = g\,\zeta'$ where $g :\to SO(3)$. A direct computation reveals that

$$p_1(TX \oplus TX; \Theta_{g\zeta}) = p_1(TX \oplus TX; \Theta_\zeta) + 4\,p_\infty[g]. \tag{2.24}$$

This shows that the map Θ is injective and completes the proof of Assertion 2 of Proposition 2.7.

To prove Assertion 3 of Proposition 2.7, consider a compact, oriented 4-manifold, X, with boundary M. Suppose that X is connected, spin, and that X's signature is even. Define

$$\chi_0(X) \equiv \sum_{i=1}^{3}(-1)^i \dim(H^i(X;\mathbb{R})). \tag{2.25}$$

Now χ_0 is even because M is a rational homology sphere, and because the signature of X is assumed even. With this understood, then one can assume, with no loss of generality that $\chi_0 = 0$. (Indeed, if χ_0 is initially positive, connect sum with $\chi_0/2$ copies of $S^1 \times S^3$ to obtain a compact, oriented, spin 4-manifold with $\chi_0(\cdot) = 0$. If $\chi_0 < 0$ initially, connect sum with $\chi_0/^\frown$ copies of $S^2 \times S^2$ to obtain a compact, oriented, spin 4-manifold with $\chi_0(\cdot) = 0$.)

Choose an embedding of the closed unit 4-ball in $B \subset X$ and let X_0 denote the compliment of this 4-ball. This X_0 has two boundary components, one M and the other S^3. Since $\chi_0(X) = 0$, the boundary splittings, $T^*X_0 \mid_M = T^*M \oplus \underline{\epsilon}$ and $T^*X_0 \mid_{S^3} = T * S^3 \oplus \underline{\epsilon}$, extend over X_0 as a splitting $T^*X_0 = V \oplus \underline{\epsilon}$. Here, V is an oriented 3-plane bundle over X_0. Furthermore, because X_0 is a spin manifold, V is a spin 3- plane bundle, and so trivial; $V \approx \oplus_3 \underline{\epsilon}$.

Let h be a singular frame for T^*S^3, and construct the frames $h_\pm$ for T^*S^3 as instructed above. The fact that V is trivial implies that the frame h_+ for $V \mid_{S^3}$ extends over X_0; so does h_-. Restrict these frames to $M \subset \partial X_0$ to give frames, $\zeta_\pm$, for $T^*M^3 = V \mid_{M^3}$. Notice that $\zeta_+ = g\,\zeta_-$ for some $g : M \to SO(3)$ which lifts to a map into S^3.

The point of the preceding is this: The extendability of $h_\pm$ over X_0 implies $p_1(TB \oplus TB, \Theta_h) = p_1(TZ \oplus TZ, \Theta_\zeta)$. Now, for S^3, one can compute rather explicitly that $p_1(TB \oplus TB; \Theta_h) = 0 \bmod 8$. Indeed, one can compute with the following choice for $h_\pm$: Take h_+ to be the Lie group framing given by the left-invariant 1-forms, and take h_- to be the Lie group framing given by the right invariant 1-forms. For $h_\pm$ as above, $p_1(TB \oplus TB; \Theta_h) = 0$.

Since, $p_1(TZ \oplus TZ; \Theta_\zeta) = 0 \bmod (8)$, Atiyah's canonical framing will be realized by Θ_ζ if Z's signature is 0 mod(4).

Assertion 4 of Proposition 2.7 follows from Proposition 1 in [1]. □

i) $I_2(S^3)$.

The purpose of this subsection is to provide a proof of the assertion in Theorem 2.9 that $I_2(S^3) = 0$. (Dror Bar-Natan taught the author this proof.) This fact is needed later and so, for future reference, is stated as

LEMMA 2.11. $I_2(S^3) = 0$.

Proof. The strategy has two parts: Part 1 extends the map Φ_0 in (2.15) to define a map, Φ_1, from $S^3 \times S^3$ to $\mathbb{R}^3$ with inverse image Σ_{S^3}. Given that such an extension exists, take φ in Proposition 2.2 to be the restriction of Φ_1 to a regular neighborhood of Σ_{S^3} and use $\omega_\varphi \equiv \Phi_1^*\mu$ in (2.4). The integrand is then zero so the integral in (2.4) is zero.

Part 2 observes that $d\Phi_1 |_{S^3\times p_0}$ and $d\Phi_1 |_{p_0\times S^3}$ define linear maps from the respective normal bundles of $S^3\times p_0$ and $p_0\times S^3$ to $\mathbb{R}^3$. Part 2 also observes that $d\Phi_1|_\Delta$ defines a linear map from the normal bundle of $\Delta\equiv\Delta_{S^3}$ into $\mathbb{R}^3$. With these facts understood, the fact that $I_2(s^3)=0$ is established by demonstrating the following two points:

1) $d\Phi_1|_{S^3\times p_0}$ and $d\Phi_1|_{p_0\times S^3}$ define respective normal bundle framings of $(S^3-p_0)\times p_0$ and $p_0\times(S^3-p_0)$ which are homotopic (rel neighborhoods of $p_0\times p_0$) to the normal bundle framings which are respectively induced by the projections, $\pi_{R,L}$, on the right and left factors of S^3 in $S^3\times S^3$.
2) $d\Phi_1|_\Delta$ defines a normal bundle framing of $\Delta_{S^3}-(p_0,p_0)$ which is gives Definition 2.8's canonical singular framing of T^*S^3.

Part 1: To exhibit the map Φ_1, agree first to use stereographic projection from the north and south poles of S^3 to cover S^3 by two coordinate patches, U_s and U_n. (So, the north pole is the origin in U_n and the south pole is the origin in U_s). Agree to take the point p_0 to equal the south pole.

Define Φ_1 by

(2.26)

1) On $U_s\times U_s:\Phi_1(x,y)\equiv(|x|^2+1)^{-1}(|y|^2+1)^{-1}(|x|^2y-|y|^2x)$.
2) On $U_n\times U_s:\Phi_1(x,y)\equiv(|x|^2+1)^{-1}(|y|^2+1)^{-1}(y+|y|^2x)$.
3) On $U_s\times U_n:\Phi_1(x,y)\equiv(|x|^2+1)^{-1}(|y|^2+1)^{-1}(-|x|^2y-x)$.
4) On $U_n\times U_n:\Phi_1(x,y)\equiv(|x|^2+1)^{-1}(|y|^2+1)^{-1}(x-y)$.

It is left to the reader to verify that Φ_1 is consistently defined, and that $\Phi_1^{-1}(0)=\Sigma_{S^3}$.

Part 2: Using line 1 or line 2 of (2.26), one finds $d\Phi_1|_{S^3\times p_0}$ to be proportional to the pull-back by π_R of the constant frame on $\mathbb{R}^3$. Likewise, using line 1 or line 3 of (2.26), one finds that $d\Phi_1 |_{p_0\times S^3}$ is proportional to the pull-back by π_L of the constant frame.

As for $d\Phi_1 |_\Delta$, compute on $U_n\times U_n$ to find that $d\Phi_1 |_\Delta$ is equal to $(\pi_R^*-\pi_L^*)(-(1+|x|^2)^{-2}dx)$. Remember that $S^3-p_0=U_n$ and thus $-(1+|x|^2)^{-2}dx$ is a framing for T^*U_n which extends as a singular framing for T^*S^3. The image of this framing under the map Θ of Proposition 2.7 is the pair (ζ_-,ζ_+) where $\zeta_\pm$ are the Lie group framings given by the left and right invariant 1-forms on S^3 as $SU(2)$.

Let $B\subset\mathbb{R}^4$ denote the unit 4-ball. It is left as an exercise to check that (ζ_-,ζ_+) stabilizes to give a framing of $(TB\oplus TB)|_B$ which extends over B. □

k) Outline of Theorem 2.9's proof.

Here is a simplified outline for a proof of Theorem 2.9: One would like to find a smooth, oriented, 7-dimensional manifold Z whose boundary is the disjoint union of $M_1 \times M_1$ and $M_0 \times M_0$, with the latter oriented in reverse. This Z should contain an oriented, dimension 4 subvariety, Σ_Z, (a union of submanifolds) whose boundary is the union of Σ_{M_1} and Σ_{M_0}, with the latter oriented in reverse. (Here, Σ_M is defined in (2.1) as the union of submanifolds. Each submanifold has its fundamental class; these are oriented so that the inclusion $\Sigma_M \to M \times M$ sends $[\Delta_M] - [M \times p_0] - [p_0 \times M]$ (a linear combination of fundamental classes) to zero in $H_3(M \times M; \mathbb{R})$.)

Given a singular frame for M_0 and an appropriate singular frame for M_0, one would like to find a smooth 2-form ω_Z on $Z - \Sigma_Z$ and then use Stokes' theorem to compare the integrals in (2.4) for M_0 and for M_1.

Unfortunately, this simplified strategy has not been realized. However, there is a modification of this strategy which can be carried out. Here are the steps:

Step 1: Find a smooth, oriented manifold Z with boundary, and suppose that ∂Z is the disjoint union of $M_1 \times M_2$ and $M_0 \times M_0$ (with the latter oriented in reverse) plus some number of copies of $S^3 \times S^3$. Label these extra boundary components by a finite set, crit.

Step 2: Inside Z, find an oriented, dimension 4 subvariety Σ_Z with $\partial \Sigma_Z = \Sigma_{M_0} \cup \Sigma M_0 \cup_{p \in \text{crit}} (\Sigma_{S^3})_p$.

Step 3: Make sure that there is a closed 2-form, ω_Z, on $Z - \Sigma_Z$ with the choice of a singular frame for M_0 and an "appropriate" singular frame for M_1. (See Step 4, below, for the definition of "appropriate".) Require the following:

(2.27)

1) The 2-form ω_Z should restrict to $M_0 \times M_0 - \Sigma_{M_0}$ as the 2-form used in (2.4) to compute I_2 for M_0. It's restriction to $M_1 \times M_1 - \Sigma_{M_1}$ should give the 2-form used in (2.4) for the computation of I_2 for M_1.
2) The 2-form ω_Z should restrict to each copy of $S^3 \times S^3$ as the 2- form used for (2.4) with the singular frame that gives $I_2(S^3)$ (see Lemma 2.11).
3) The wedge product $\omega_Z \wedge \omega_Z$ should vanish near Σ_Z.

Step 4: If the canonical frame (Definition 2.8) was chosen for M_0, check that the "appropriate" frame for M_1 is M_1's canonical frame.

Step 5: Given that ω_Z exists as prescribed above, use Stokes' theorem to prove that the values of I_2 in (2.4) for M_0 and for M_1 agree:

$$(2.28) \qquad 0 \equiv \int_Z d(\omega_Z \wedge \omega_Z \wedge \omega_Z) = \int_{M_1 \times M_1} \omega_Z \wedge \omega_Z \wedge \omega_Z$$
$$- \int_{M_0 \times M_0} \omega_Z \wedge \omega_Z \wedge \omega_Z \quad + \int_{\partial \underline{N}_Z} \omega_Z \wedge \omega_Z \wedge \omega_Z \,.$$

Here, $\underline{N}_Z$ is the closure of a regular neighborhood of Σ_Z in Z (with smooth boundary) on which $\omega_Z \wedge \omega_Z = 0$. Thus, the last term in (2.28) vanishes.

Section 3 begins the proof of Theorem 2.9 with the definition of the space Z. Later sections construct Σ_Z and carry out the remaining steps of the proof. The construction of ω_Z is completed in Section 11.

3 Morse theory. This subsection reviews various Morse theoretic constructions which will be used in subsequent sections. The final two subsections define and explore the space Z for Section 2k's first step in the proof of Theorem 2.9.

Suppose here that M_0 and M_1 are both compact, connected, oriented 3-manifolds whose rational homology is the same as that of S^3. Let W be an oriented, spin cobordism between M_0 and M_1. Assume that W is connected. By surgery on embedded circles, one can modify W so that $H^1(W;\mathbb{R}) = H^3(W;\mathbb{R}) = 0$. This will be assumed as well.

a) Good Morse functions.

A function $f : W \to [0,1]$ will be called a good Morse function if the criteria below are met:

(3.1)

1) $M_0 = f^{-1}(0)$ and $M_1 = f^{-1}(1)$.
2) $df\,|_{M_0}$ and $df\,|_{M_1}$ are never zero.
3) The critical points of f are non-degenerate.
4) There are no critical points of index 0 or 4.
5) Let c be an index i critical point. Then $|\, f(c) - i/4 \,| < 1/100$.
6) If c, c' are distinct critical points of f, then $f(c) \neq f(c')$.

It is shown in [**12**] (for example) that W has *good Morse functions*.

When $f : W \to [0,1]$ is a good Morse function, use $\mathrm{crit}(f) \subset W$ to denote the set of critical points of f. Use $\mathrm{crit}_k(f) \subset \mathrm{crit}(f)$ to denote the set of critical points of index k.

Let $p \in \mathrm{crit}_k(f)$. There is an almost canonical coordinate system on a neighborhood of p. This coordinate system is an embedding $\psi_p : \mathbb{R}^4 \to W$ with the following properties:

1) $\psi_p(0) = p$.
2) There exists $\delta > 0$ such that $\psi_p^* f$ restricts to the radius δ ball about p as the function

$$\psi_p^* f = -x_1^2 \cdots - x_k^2 + x_{k+1}^2 + \cdots + x_4^2. \tag{3.2}$$

(See, e.g. [**12**].) These coordinates will be called *Morse coordinates.* The image of $\mathbb{R}^4$ under ψ_p^* will be called U_p.

b) Pseudo-gradient vector fields.

Aside from a Morse function and Morse coordinates, the standard machinery for Morse theory requires the choosing of a *pseudo-gradient vector field* for f. This is a vector field, v, on W with the property that $v(f) > 0$ on $W - \mathrm{crit}(f)$. Also, require of v that it have the following form near $p \in \mathrm{crit}_k(f)$: The pushforward by $(\psi_p)^{-1}$ of v should restrict to a small ball about the origin in $\mathbb{R}^4$ to equal

$$-x_1\partial_1 \cdots - x_k\partial_k + x_{k+1}\partial_{k+1} + \cdots + x_4\partial_4. \tag{3.3}$$

A pseudo-gradient vector field for f will often be called a *pseudo- gradient*, for short.

A gradient flow line of a pseudo-gradient v is a map γ, of a closed interval, I, into W with the following properties:

(3.4)

1) $I = [a, b]$ with $-\infty \leq a < b \leq +\infty$.
2) If $a = -\infty$, then $\gamma(a) \in \text{crit}(f)$; and if $b = +\infty$, then $\gamma(b) \in \text{crit}(f)$.
3) If $a > -\infty$, then $\gamma(a) \in M_0$; and if $b < +\infty$, then $\gamma(b) \in M_1$.
4) $\gamma_*(\partial_t) = v \mid_{\gamma(t)}$ for all $t \in I$

(Here, ∂_t differentiates the coordinate t to give 1.)

If γ is a gradient flow line of a pseudo-gradient, v, say that γ begins at $\gamma(a)$ and ends at $\gamma(b)$.

There is a great deal of flexibility in the choice of a pseudo- gradient. And, there are specific constraints which can be imposed on a pseudo-gradient which simplify some subsequent constructions.

c) The Morse complex.

With the help of good Morse function f and an appropriate pseudo gradient, v, one can define a finite dimensional complex whose homology is naturally isomorphic to the relative homology $H^*(W, M_0; \mathbb{Z})$. (See, e.g. [**12**].) The complex is written

$$0 \longrightarrow C_3 \xrightarrow{\partial} C_2 \xrightarrow{\partial} C_1 \longrightarrow 0 \tag{3.5}$$

To describe the $\{C_k\}$ in (3.5), it is necessary to first digress to review the construction of ascending and descending disks: As described in [**12**], one can use v to define, for each $p \in \text{crit}_k(f)$, a pair of open subsets, $B_{p-} \subset \text{int}(W)$ and $B_{p+} \subset \text{int}(W)$, which are embedded disks of dimension k and $4-k$, respectively. Here, B_{p+} is the ascending disk from p, and B_{p-} is the descending disk from p.

As a subset, B_{p-} is the union $\{p\}$ with the set of points of $\text{int}(W) - \text{crit}(f)$ which lie on gradient flow lines which end at p. And, B_{p+} is the union of $\{p\}$ with the set of points in $W - \text{crit}(f)$ which lie on gradient flow lines which start at p.

Note that (3.4) implies that

(3.6)

1) $\psi_p(B_{p-} \cap U_p) = \{(x_1, \cdots, x_4) \in \mathbb{R}^4 : x_{k+1} = \cdots = x_4 = 0\}$,
2) $\psi_p(B_{p+} \cap U_p) = \{(x_1, \cdots, x_4) \in \mathbb{R}^4 : x_1 = \cdots = x_k = 0\}$.

These disks intersect at one point, p, and there transversally. Otherwise,

$$f \mid (B_{p-} - p) < f(p) < f \mid (B_{p+} - p) \tag{3.7}$$

As W is assumed oriented, an orientation for B_{p-} orients B_{p+} so that their intersection number, $[B_{p-}] \cdot [B_{p+}]$, is equal to $\{1\}$.

End the digression.

One defines C_k in (3.5) from the free $\mathbb{Z}$-module, $\underline{C_k}$, on the set of pairs

$$\{(p, \epsilon) : p \in \text{crit}_k(f) \text{ and } \epsilon \text{ is an orientation for } B_{p-}\} \tag{3.8}$$

Indeed, set $C_k \equiv \underline{C_k}/\sim$, with the equivalence relation $(p, \epsilon) \sim -(p, -\epsilon)$. (The C_k for different choices of pseudo-gradient are canonically isomorphic.)

To define the operator ∂ in (3.5), it is necessary to make a two part digression.

Part 1 of the digression introduces some constraints on the pseudo-gradient v. These are described next.

DEFINITION 3.1. *Let f be a Morse function on W. A pseudo-gradient v will be called* good *if the following criteria are met:*

1) *If $p, q \in \text{crit}_k(f)$ and if $p \neq q$, then $B_{p+} \cap B_{q-} = \emptyset$.*
2) *If $p \in \text{crit}_k(f)$ and $q \in \text{crit}_{k+1}(f)$ then B_{p+} intersects B_{q-} transversely. Furthermore, $B_{p+} \cap B_{q-}$ is a finite union of gradient flow lines, the closure of each starts at p and ends at q.*
3) *If $p \in \text{crit}_1(f)$ and $q \in \text{crit}_3(f)$, then B_{p+} intersects B_{q-} transversally.*
4) *If $p_0 \in M_0$ and $p_1 \in M_1$ have been apriori specified, then require that p_0 start a gradient flow line with end at p_1.*

(By the way, because of their definitions in terms of v's flow lines, descending disks from distinct critical points do not intersect, and likewise, ascending disks.)

See [**12**] for a proof that good pseudo-gradients exist. Henceforth, assume that v is a good pseudo-gradient.

Part 2 of the digression considers the intersections of ascending and descending disks. Start the discussion with the introduction of $M_{k,k-1} \equiv f^{-1}(4^{-1}k - 1/8)$. Due to (3.1), one can conclude that df is nowhere zero along $M_{k,k-1}$, so this subspace is an embedded submanifold of W. Furthermore, $M_{k,k-1}$ is naturally oriented by using df to trivialize its normal bundle.

Because of (1) in Definition 3.1, each B_{p-} intersects $M_{k,k-1}$ in its interior as a $(k-1)$ sphere S_{p^-} which is oriented (by df) when B_{p-} is oriented. Likewise, B_{p+} intersects $M_{k+1,k}$ in a sphere, S_{p^+}, of dimension $3-k$ which is oriented when B_{p-} is oriented.

Note that Definition 3.1 implies (in part) the following assertion: If $p \in \text{crit}_k(f)$, then S_{p^-} has transversal intersection in $M_{k,k-1}$ with any S_{q^+} from $q \in \text{crit}_{k-1}(f)$. With the preceding understood, use $[S_{p^-}] \cdot [S_{q^+}] \in \mathbb{Z}$ to denote the algebraic intersection number of S_{p^-} with S_{q^+} in $M_{k,k-1}$.

End the digression.

Here is the definition of the boundary map ∂ in (3.5):

$$\partial(p, \epsilon) \equiv \sum_{q \in C_{k-1}} ([S_{p^-}] \cdot [S_{q^+}])\, (q, \epsilon_q). \tag{3.9}$$

See [**12**] for a proof that (3.5) with ∂ as in (3.9) is a chain complex whose homology is isomorphic to $H*(W, M_0; \mathbb{Z})$.

Note that there is a dual complex to (3.5),

$$0 \longrightarrow C_3^* \xrightarrow{\partial^*} C_2^* \xrightarrow{\partial^*} C_1^* \longrightarrow 0 \tag{3.10}$$

which is defined using $-f$ and $-v$ when (3.5) is defined from the pair f and v. The homology of (3.10) computes $H^*(W, M_1; \mathbb{Z})$. Poincare' duality identifies $H^*(W, M_1; \mathbb{Z})$ with $H^{4-*}(W, M_0; \mathbb{Z})$, hence the duality between (3.10) and (3.5).

d) Factoring the cobordism.

The purpose of this subsection is to indicate how to factor the cobordism W into two simpler cobordisms. The following proposition summarizes:

PROPOSITION 3.2. *Let M_0, M_1 be compact, oriented 3-manifolds with the rational homology of S^3. Suppose that there is an oriented, spin cobordism, W', between M_0 and M_1. Then there exists an oriented, spin cobordism, W, between M_0 and M_1 which decomposes as $W = W_1 \cup W_1 \cup W_3$, where*

1) *$\partial W_1 = -M_0 \cup M_0', \partial W_2 = -M_0' \cup M_1'$, and $\partial W_3 = -M_1' \cup M_1$, where M_0' and M_1' are compact, oriented 3-manifolds with the rational homology of S^3.*
2) *$W_{1,2,3}$ are oriented, spin manifolds.*
3) *Both W_1 and W_3 have the rational homology of S^3. Meanwhile, W_2 has vanishing first and third Betti numbers.*
4) *W_1 and W_3 have a good Morse functions with no index 3 critical points. Meanwhile, W_2 has a good Morse function without index 1 and index 3 critical points.*
5) *If W' has the rational homology of S^3, then W above can be assumed to have the rational homology of S^3. And, one can assume that $M_0' = M_1'$ and that W_2 is the product cobordism.*
6) *Let $l_{W'}$ and l_W be as given in (2.12). Suppose that c_{M_0} or c_{M_1} (as in Definition 2.8) is represented by c in $ker(l_{W'})$. Then $l_W(c) = 0$ too.*
7) *The intersection forms of W and W' are conjugate by an element of $Gl(\cdot, \mathbb{Z})$.*

In particular, Assertions 5 and 6 of the preceding proposition allow one to prove Theorem 2.9's statements concerning 4- dimensional spin cobordisms with the rational homology of S^3 between a pair of 3-manifolds with the rational homology of S^3 by restricting to the following special case:

Special Case: Let M_0, M_1 be compact, oriented 3-manifolds with the rational homology of S^3. Let W be an oriented, spin cobordism between M_0 and M_1. Assume that W has the rational homology of S^3 and assume that W has a good Morse function f with no index 3 critical points.

(3.11)

The remainder of this subsection proves Proposition 3.2.

Proof of Proposition 3.2. First of all, let W' be the original spin cobordism between M_0 and M_1. Then, surgery on W' will produce an oriented, spin cobordism W which has vanishing first and third Betti numbers. The surgery removes tubular neighborhoods of embedded circles and replace them with copies of $B^2 \times S^2$. (Here, B^2 is the unit ball in $\mathbb{R}^2$.)

Given such W, find a good Morse function f on W and a good pseudo-gradient, v; and then define the complex in (3.5). Label the critical points of index 1 as $\{a_1, \cdots, a_r\}$, label those of index 2 as $\{b_1, \cdots, b_{r+s+t}\}$, and label the index 3 critical points as $\{c_1, \cdots, c_t\}$. Here, $s = \dim(H_2(W;\mathbb{R}))$. (Remember that W has, by assumption, vanishing rational homology in dimensions 1 and 3.) Fix orientations for the descending disks from all of these critical points. With this understood, this set of critical points defines a basis for the complex $\{C_k\}$ in (3.5).

Now, it is convenient to relable the basis for C_2 as follows: Since the map $\partial \oplus \partial^* : C_2 \to C_1 \oplus C_3$ is a surjection, one can relable the critical points $\{b_\alpha\}$ so that

$$\partial : \operatorname{Span}\{b_i\}^r{}_{i=1} \to C_1, \tag{3.12}$$

$$\partial^* : \operatorname{Span}\{b_{r+s+i}\}_{i=1}^t \to C_3,$$

are both isomorphisms over $\mathbb{Q}$. At the same time, one can require that the projection of Span $\{b^s_{r+i}\}_{i=1}$ onto $C_2/(\partial^* C_1 \oplus \partial C_3)$ is an isomorphism.

With (3.12) understood, one can use 4.1 in **[12]** to find a new good Morse function f_{new} which has the following three properties: First, f_{new} agrees with f outside small neighborhoods of the points in $\text{crit}_2(f)$. Second, f_{new} has the same critical points and pseudo- gradient as f. Third, there exists small $\epsilon > 0$ such that

(3.13)

1) $f_{\text{new}}(\{b_1, \cdots, b_r\}) \in (1/2 - 2\epsilon, 1/2 - \epsilon)$,
2) $f_{\text{new}}(\{b_{r+1}, \cdots, b_{r+s}\}) \in (1/2 - \epsilon, 1/2 + \epsilon)$,
3) $f_{\text{new}}(\{b_{r+s+1} \cdots, b_{r+s+t}\}) \in (1/2 + \epsilon, 1/2 + 2\epsilon)$

Note that (3.12), (3.13) indicate that $W = W_1 \cup W_2 \cup W_3$, where

(3.14)

1) $W_1 \equiv f^{-1}([0, 1/2 - \epsilon])$
2) $W_2 \equiv f^{-1}([1/2 - \epsilon, 1/2 + \epsilon])$
3) $W_3 \equiv f^{-1}([1/2 + \epsilon, 1])$

The boundaries of $W_{1,2,3}$ are compact, oriented 3-manifolds with the rational homology of S^3. This is guaranteed by (3.12). Meanwhile, the inclusion of any boundary component of $W_{1,3}$ into $W_{1,3}$ induces an isomorphism of rational homology. This is not true for W_2; this W_2 has the zero first and third Betti numbers, but the second Betti number of W_2 is equal to s.

Note that the function f_{new} can be used as a Morse function on $W_{1,2,3}$. On W_1, it has no critical points of index 3, on W_3 it has no critical points of index 1, while on W_2, it has only critical points of index 2.

The preceding remarks prove Assertions 1-5. To prove Assertion 6, suppose, for the sake of argument that c_{M_0} is represented by c in $\ker(l_{W'})$. Let ξ be a singular frame for T^*M_0 in the class c and let ξ' be a smooth frame for T^*M_0 which agrees with ξ on the complement of a ball about p_0. Write $T^*W' \mid_{M_0} \approx T^*M_0 \oplus \underline{\mathbb{R}}$, where $\underline{\mathbb{R}}$, is the trivial bundle, spanned by $df \mid_{M_0}$. With this understood, ξ' extends to a frame (ξ', df) for $T^*W' \mid_{M_0}$. Note that $l_{W'}(c)$ is the obstruction to extending this frame over W'. Likewise, $l_W(c)$ is the obstruction to extending (ζ', df) over W. With this understood, remark that Assertion 6 will be proved by demonstrating that (ξ', df) extends over W if it extends over W'. This demonstration requires four steps.

Step 1: Fix a frame e' for T^*W' which extends (ξ', df).

Step 2: Let $\sigma \subset W'$ be an oriented, embedded circle whose fundamental class is a generator of $H_1(W';\mathbb{Z})/\text{Torsion}$. Suppose a surgery is done on W' to kill the class generated by σ. Such a surgery will replace a tubular neighborhood of σ in W' with $B^2 \times S^3$. Because $\pi_2(SO(3)) = 0$, all framings of $T^*(B^2 \times S^2)$ are mutually homotopic. A framing of $T^*(B^2 \times S^2)$ restricts to the boundary where it can be written as $h\,e'$, where $h \equiv h(e')$ is a map from $S^1 \times S^2$ to $SO(4)$. If h lifts to $SU(2) \times SU(2)$, then the frame (σ', df) will also extend over the manifold which is obtained from W' by surgery on σ.

Step 3: With this last point understood, suppose that h does not lift as required. The strategy is to abandon e' and find a new extension, e'' for (σ', df) so that the resulting $h(e'')$ does lift to $SU(2) \times SU(2)$.

Step 4: To construct e'', let $s : \sigma \to S^1$ be a degree 1 map. Since the restriction map $H^1(W';\mathbb{Z}) \to H^1(\sigma;\mathbb{Z})$ is surjective, the map s extends as a smooth map from W' to S^1. Since M_0 has vanishing first cohomology, the map s can be taken to map M_0 to point, $1 \in S^1$. Let $j : S^1 \to SO(4)$ be a map which generates $\pi_1(SO(4))$ and which takes 1 to the identity matrix. The composition $k \equiv j \circ s$ maps W' to $SO(4)$ and maps M_0 to the identity. Thus, $e'' \equiv k\,e'$ defines an extension of (ξ', df) over W', and $h(k\,e') = h(e)\,k^{-1}$ will lift to map $S^1 \times S^2$ into $SU(2) \times SU(2)$.

Thus, Assertion 6 follows from this last remark with Step 3.

As for Assertion 7, it is directly a consequence of the fact that W is obtained from W' by surgery on a set of circle generators of $H^1(W';\mathbb{Z})/\text{Torsion}$. □

e) A basis theorem for the Special Case.

Assume here that W is a cobordism which satisfies the assumptions of (3.11). In particular, W has the rational homology of S^3, and also W has a good Morse function f with only index 1 and index 2 critical points. Fix a good pseudo-gradient v for f.

Introduce the complex in (3.5) for W. This is a 2-step complex, and the boundary map $\partial : C_2 \to C_1$ is an isomorphism over the rationals. Let $\{a_1, \cdots, a_r\}$ label the index 1 critical points and let $\{b_1, \cdots, b_r\}$ label the index 2 critical points. Orient the descending disks from these critical points so that these sets of critical points can be considered as a basis for $C_{1,2}$, respectively.

With the basis for $C_{1,2}$ chosen as above, the boundary maps in (3.5) are simply integer valued matrices. That is, $\partial b_i = \Sigma_j S_{i,,j}\, a_j$, where $S \equiv \{S_{i,,j}\}$ is an integer valued, $r \times r$ matrix.

Here is a useful observation: The precise form of the matrix S is determined by the choice of good pseudo-gradient v. With this fact understood, one can ask whether there is a choice of peudo- gradient for f which gives a "nice" matrix S. The answer to this question is given by Milnor's basis theorem (Theorem 7.6 in [**12**]):

PROPOSITION 3.3. *Let W be a cobordism which satisfies (3.11). Then W has a good Morse function, f, with no index 3 critical points and with the following additional properties: There exists a labeling, $\{a_1, \cdots, a_r\}$ and $\{b_1, \cdots b_r\}$, for the respective index 1 and index 2 critical points of f. And, there exists a good pseudo- gradient for f and a choice of orientations for the descending disks from f's critical points. And, this data is such that*

(3.15)

1) *For all $i \in \{1, \cdots, r\}$,*
2) *$\partial b_i = \Sigma_j S_{i,j}\, a_j$, where $S \equiv \{S_{i,j}\}$ is an upper triangular, integral matrix with positive entries along the diagonal.*
3) *For all $i \in \{1, \cdots, r-1\}$, one has $f(a_i) > f(a_{i+1})$ and $f(b_i) > f(b_{i+1})$.*

The remainder of this subsection is occupied with proving this proposition.

Proof of Proposition 3.3. Start with a good pseudo-gradient, v, for f. Fix orientations for the descending disks so that the boundary operator in (3.5) can be represented as a matrix, T, so that $\partial b_i = \Sigma_j T_{i,j}\, a_j$. Note that the matrix T is integral and invertible over the rationals.

Now, a fundamental result in algebra (see, e.g. [**11**]) states that there exists a unimodular, integral matrix V such that $V\,T \equiv T'$ has only zeros below the diagonal. Let $\underline{b} \equiv \{\underline{b}_i \equiv \Sigma_j V_{i,j}\, b_j\}$. This is a new basis for C_2, and $\partial \underline{b} = V\,T\,a = T'\,a$.

With V and T' understood, appeal to Theorem 7.6 in [**12**] to find a pseudo-gradient for f, v', for which the resulting descending disks represent the basis b for C_2. For this pseudo-gradient, the boundary operator in (3.5) is given by the matrix T'. By changing the orientations of the descending disks if necessary, one can change the signs of the diagonal elements of T' so that they are all positive. Call the resulting matrix S.

The given arrangement of the critical values of f can be insured by making an appropriate, small perturbation. □

By the way, if the boundary $\partial C_2 \to C_1$ is an isomorphism over Z, then the matrix S in Propostion 3.3 can be taken to be a diagonal matrix.

As a last remark, note that the matrix for the adjoint complex, $\partial^* : C_1 \to C_2$, will be the transpose of the matrix S in Proposition 3.3. This matrix, S^T, will be lower triangular. □

f) Morse theory on $W \times W$

The manifold $W \times W$ is a manifold with boundaries and corners. Here it is:

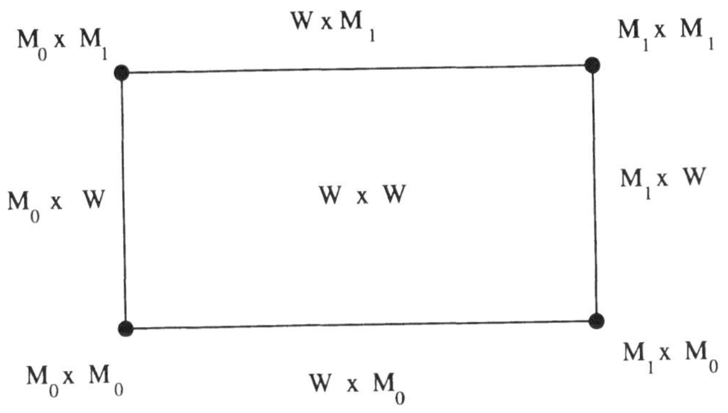

(3.16)

The reader is invited to formalize a "manifold with boundaries and corners", but the picture above should be self explanatory.

The good Morse function f on W can be used to illuminate (3.16) near the corners. To do so, one must note first that Properties 1 and 2 in (3.1) make it possible to use the pseudo gradient to give W its product structure near ∂W. To be precise, there is a diffeomorphism,

$$\lambda_0 : f^{-1}([0,1/8)) \to M_0 \times [0,1/8) \tag{3.17}$$

which restricts to $f^{-1}(0)$ as the identity and which has $\lambda_0^* f$ given by projection to $[0,1/8)$. There is a corresponding

$$\lambda_1 : f^{-1}((7/8,1]) \to M_1 \times (7/8,1]. \tag{3.18}$$

Using (3.17), a neighborhood of $M_0 \times M_0$ in $W \times W$ is mapped by $\lambda_0 \times \lambda_0$ to

$$(M_0 \times [0,1/8)) \times (M_0 \times [0,1/8)) \approx M_0 \times M_0 \times [0,1/8) \times [0,1/8). \tag{3.19}$$

Of course, $\lambda_0 \times \lambda_1, \lambda_1 \times \lambda_0$ and $\lambda_1 \times \lambda_1$ give similar structure to the other corners of $W \times W$.

With a good Morse function, f, chosen for W, introduce the function $F : W \times W \to [-1,1]$ which sends (x, y) to

$$F(x,y) \equiv f(y) - f(x). \tag{3.20}$$

This is a function with properties that are listed in the next lemma. The lemma's statement uses the following notation: First, introduce the projections, $\pi_L : W \times W \to W$ and $\pi_R : W \times W \to W$ which send (x, y) to x and to y, respectively. Second, when v is a vector field on W, introduce the vector fields v_L and v_R on $W \times W$ which are defined so that

(3.21)

1) $d\pi_L\, v_L = v$ and $d\pi_R\, v_L = 0$;
2) $d\pi_L\, v_R = 0$ and $d\pi_R\, v_R = v$.

LEMMA 3.4. *Let f be a good Morse function for W and let v be a good pseudo-gradient for f. Then, the function F of (3.20) has only non- degenerate critical points. Furthermore:*

1) *$crit_n(f) = \cup_k(crit_{4+k-n}(f) \times crit_k(f))$.*
2) *The vector field $v_R - v_L$ is a pseudo-gradient for F which obeys 1- 3 of Definition 3.1.*
3) *The pseudo-gradient $v_R - v_L$ gives the following descending and ascending disks for $(p, q) \in crit_{4+k-n}(f) \times crit_k(f) \subset crit_n(f)$:*

$$B_{(p,q)-} = B_{p+} \times B_{q-}, \tag{3.22}$$

$$B_{(p,q)+} = B_{p-} \times B_{q+}.$$

4) *The pseudo-gradient $v_R - v_L$ is nowhere tangent to a boundary or a corner in (3.16).*

Proof. The proofs of these assertions are left as exercises. But, for Assertion 3, note for example that near $M_0 \times M_0$, $(\lambda_0 \times \lambda_0)^{-1}$ (Of (3.19)) pulls back F to send the point $((x, t), (y, s))$ in $(M_0 \times [0, 1/8)) \times (M_0 \times [0, 1/8))$ to

$$((\lambda_0 \times \lambda_0)^{-1})^* F((x,t),(y,s)) = s - t. \tag{3.23}$$

Note, by the way, that (3.22) indicates how to orient $B_{(p,q)-}$ given orientations for B_{p-} and B_{q+}. And, with orientations to the descending disks $\{B_{(p,q)-} : (p,q) \in \text{crit}(F)\}$, one can consider the analog to the chain complex C in (3.5) as constructed for $W \times W$ using the function F and the pseudo-gradient $v_R - v_L$. The following lemma describes the homology of this complex. □

LEMMA 3.5. *The analog of the chain complex C in (3.5) as constructed for $W \times W$ using F and the pseudo-gradient $v_R - v_L$ gives a chain complex, C^F, which is canonically isomorphic to $C^* \otimes C$, where C^* is the complex in (3.10). The homology of the complex C^F is canonically isomorphic to $H*(W \times W, (W \times M_0) \cup (M_1 \times W); \mathbb{Z})$.*

Notice that the relative homology above is that of the square in (3.16) relative to the union of its bottom and right sides.

Proof. This follows from Lemma 3.4 and (3.22). □

g) The space Z.

As outlined in Section 2k, the first step to proving Theorem 2.9 is to construct an oriented, 7-dimensional manifold Z whose boundary is the disjoint union of $M_0 \times M_0, M_1 \times M_1$ and some number of copies of $S^3 \times S^3$. The purpose of this subsection is to construct such a Z using the cobordism W and a good Morse function f on W. To begin, construct F from f as in (3.16). Use F to define

$$\underline{Z} \equiv F^{-1}(0) = \{(x, y) \in W \times W : f(x) = f(y)\}. \tag{3.24}$$

This subspace $\underline{Z}$ plays a central role in subsequent parts of the story, and the purpose of this subsection is to describe some of $\underline{Z}$'s properties.

To begin, note that both $M_0 \times M_0$ and $M_1 \times M_1$ lie in $\underline{Z}$ since f is constant on M_0 and also on M_1. Near these corners, $\underline{Z}$ is a manifold with boundary given by the disjoint union of $M_0 \times M_0$ and $M_1 \times M_1$. See (3.19).

Unfortunately, $\underline{Z}$ is not a manifold everywhere unless f has no critical points. This is because 0 is not a regular value of the function F. Fortunately, the singularities of $\underline{Z}$ are not hard to describe; they occur at the points of crit$(F) \cap \underline{Z}$, that is, points of the form $(p, p) \subset W \times W$ where $p \in$ crit(f). (Remember that the critical points of f are assumed to have distinct critical values.) Furthermore, the neighborhoods of these critical points are relatively easy to describe. The picture is given in the following lemma. The lemma introduces the notion of a *cone* on a manifold N. This is the space which is obtained by taking $[0, 1) \times N$ and crushing $\{0\} \times N$ to a point.

LEMMA 3.6. *Let f be a good Morse function on W. Let $p \in crit_k(f)$. Then, a neighborhood of (p, p) in $\underline{Z}$ is naturally isomorphic to the cone on $S^3 \times S^3$. In fact with ψ_p and $U_p = \psi_p(\mathbb{R}^4)$ as in (3.2), then the map $(\psi_p \times \psi_p)^{-1}$maps $\underline{Z} \cap (U_p \times U_p)$ to a subset of $\mathbb{R}^4 \times \mathbb{R}^4$ which intersects a ball neighborhood of $(0, 0)$ as the set of (x, y) which obey*

$$y_1^2 + \cdots + y_k^2 + x_{k+1}^2 + \cdots + x_4^2 = x_1^2 + \cdots + x_k^2 + y_{k+1}^2 + \cdots + y_4^2. \tag{3.25}$$

Warning: *As indicated by (3.25), the cone on $S^3 \times S^3$ here is not induced by the obvious product structure on $W \times W$. The product structure which induces (3.25) is the product structure in*

$$B_{(p,p)-} \times B_{(p,p)+}, \tag{3.26}$$

with $B_{(p,p)\pm}$ as in (3.22).

Proof. Equation (3.25) is an immediate consequence of (3.2). □

The manifold (with boundary) Z in Section 2k will be found inside $\underline{Z}$; it is obtained by excising from $\underline{Z}$ a small ball about each of the singular points (p,p) for $p \in \mathrm{crit}(f)$. More precisely, one fixes some small $r > 0$. Then, the intersection of Z with $U_p \times U_p$ is mapped by $\psi_p \times \psi_p$ to the set of (x,y) which obey

$$|x|^2 + |y|^2 \geq r. \tag{3.27}$$

With (3.27) understood, $\partial Z \cap U_p \times U_p$ is mapped by $\psi_p \times \psi_p$ to the set of (x,y) which obey

(3.28)

1) $y_1^2 + \cdots + y_k^2 + x_{k+1}^2 + \cdots + x_4^2 = r$,
2) $x_1^2 + \cdots + x_k^2 + y_{k+1}^2 + \cdots + y_4^2 = r$.

As the precise value of r here is immaterial (as long as r is small), the precise value will not be specified.

There is an alternative approach to defining Z. Here, Z is a "blow up" of $\underline{Z}$ at the points of the form $(p,p) \in \mathrm{crit}(f)$. In this case, Z maps to $\underline{Z}$ by a map π. Each point in $\underline{Z} - \{(p,p) \in \mathrm{crit}(F)\}$ has a single point in its inverse image. But, the inverse image of any point $(p,p) \in \mathrm{crit}(F)$ is the corresponding $S^3 \times S^3 \subset \partial Z$. This blow up corresponds to resolving the cone point in $\underline{N} \equiv ([0,1) \times N)/(\{0\} \times N)$ with the tautological projection $\pi : [0,1) \times N \to N$.

h) Properties of Z.

With Z now defined, here are its salient features:
A manifold: Z is a manifold with boundary,

$$\partial Z = (M_0 \times M_0) \cup (M_1 \times M_1) \cup_{p \in \mathrm{crit}(f)} (S^3 \times S^3)_p. \tag{3.29}$$

Orientation: The manifold $\mathrm{int}(Z)$ has a natural orientation. Indeed, $W \times W$ has a natural orientation. Then, $\mathrm{int}(Z) \subset F^{-1}(0)$ is open, and $dF \neq 0$ on $\mathrm{int}(Z)$, so the 1 form dF trivializes the normal bundle to $\mathrm{int}(Z) \subset W \times W$. This serves to orient Z. The induced orientation on $M_1 \times M_1 \subset \partial Z$ agrees with its canonical orientation, but the induced orientation on $M_0 \times M_0 \subset \partial Z$ disagrees with the canonical orientation.

To orient $(S^3 \times S^3)_p$, use the inclusion of $W \approx \Delta_W \subset W \times W$ to orient Δ_W and hence Δ_Z. The boundary of Δ_Z intersects $(S^3 \times S^3)_p$ as $\Delta_{S^3} (\equiv (\Delta_{S^3})_p)$ Give $(\Delta_{S^3})_p$ the induced orientation from Δ_Z. Then, orient the left factor of S^3 in $(S^3 \times S^3)_p$ so that the composition of $\pi_L : \Delta_{S^3} \to S^3$ and then the inclusion $S^3 \to (S^3 \times \text{ point}) \subset (S^3 \times S^3)_p$ is orientation preserving. Orient the right factor analogously and use the product orientation to orient $(S^3 \times S^3)_p$. (Remark that the induced orientation on $(S^3 \times S^3)_p$ as a boundary component of Z agrees with this orientation if index p is odd, and it disagrees if index p is 2.)

Homology: The rational homology in dimensions 0-3 of Z is of some concern in subsequent sections. Consider

LEMMA 3.7. *Suppose that W has the rational homology of S^3. Then the rational homology of Z is as follows:*

1) $H_0(Z) \approx \mathbb{R}$.

2) $H_1(Z) \approx H_2(Z) \approx 0$.

3) There is a surjection

$$0 \leftarrow H_3(Z) \leftarrow L_- \oplus L_+ \oplus H_3(\partial Z). \tag{3.30}$$

Here L_- is freely generated over $\mathbb{R}$ by

$$\underline{L}_- \equiv \{B^4_{(p,q)-} \cap Z : (p,q) \in crit_4(F) \text{ and } F(p,q) > 0\}, \tag{3.31}$$

while L_+ is freely generated over $\mathbb{R}$ by

$$\underline{L}_+ \equiv \{B^4_{(p,q)+} \cap Z : (p,q) \in crit_4(F) \text{ and } F(p,q) < 0\}, \tag{3.32}$$

Note that the intersections which define $\underline{L}_\pm$ in (3.31), (3.32) are all embedded 3-spheres.

Also note that the inclusion of Z in $W \times W$ gives an isomorphism on π_1 and π_2.

Proof. Note first that $H_{0,1,2}(Z)$ and $H_{0,1,2}(\underline{Z})$ agree, and that

$$H_3(Z) \approx H_3(\underline{Z}) \oplus_{p \in \text{crit}(f)} H_3((S_3 \times S_3)_p) \tag{3.33}$$

This follows using Meyer-Vietoris for the cover of $\underline{Z}$ by the union of Z and the cones on the $(S^3 \times S^3)_p$ in ((3.25).

Next, pick $\epsilon > 0$, but small so that F has only critical points of the form (p,p) in $F^{-1}((-\epsilon, \epsilon))$. Let $V \equiv F^{-1}((-\epsilon, \epsilon))$ observe that V strongly deformation retracts into $\underline{Z}$. Thus, $H_i(V) \approx H_i(\underline{Z})$.

To compute $H_i(V)$, observe that $W \times W$ can be constructed from V by a sequence

$$V \equiv V_3 \subset V_4 \subset V_5 \subset V_6 \equiv W \times W, \tag{3.34}$$

where V_{k+1} is obtained from V_k by the attachment of disjoint handles, $(B^k \times B^{8-k})$'s, on disjointly embedded $(S^{k-1} \times B^{8-k})$'s in the boundary of V_k.

To be precise, V_4 contains all of F's index 4-critical points,

$$V_4 \equiv F^{-1}([-1/8, 1/8]); \tag{3.35}$$

and V_5 contains all index $3, 4$, and 5 critical points,

$$V_5 \equiv F^{-1}([-3/8, 3/8]). \tag{3.36}$$

The attaching 3-spheres for the handles that change V_3 to V_4 are given by (3.31), (3.32). Meanwhile, the attaching 4-spheres for the handles that change V_4 to V_5 are

$$\begin{aligned} &\{B_{(p,q)-} \cap F^{-1}(1/8)\}_{(p,q)\in \mathrm{crit}_5(F)} \\ &\qquad \cup \{B_{(p,q)+} \cap F^{-1}(-1/8)\}_{(p,q)\in \mathrm{crit}_3(F)}. \end{aligned} \tag{3.37}$$

The 5-spheres for the attachments that change V_5 to V_6 should be obvious.

The resulting Meyer-Vietoris sequences from (3.34) read, in part,

$$\begin{aligned} &H_3(\underline{L}_+) \oplus H_3(\underline{L}_-) \to H_3(V_3) \to H_3(V_4) \to 0, \\ &H_3(V_4) \approx H_3(V_5) \approx H_3(V_6). \end{aligned} \tag{3.38}$$

The third assertion in Lemma 3.7 follows from (3.38) and (3.33). The other assertions follow by Meyer-Vietoris from (3.34)-(3.37). □

4 Homological constraints. In this section, M_0 and M_1 will both be oriented, 3- dimensional manifolds with the rational homology of S^3. And, W will be an oriented, connected, spin cobordism between M_0 and M_1. Let $f : W \to [0, 1]$ be a good Morse function. Use f to construct the the space Z as described in Sections $3g$ and $3h$. The proof of Theorem 2.9 is a five step affair which is outlined in Section $2k$. The manifold (with boundary) Z of Sections $3g, h$ realizes the first step in the proof. The next step in the proof is to construct a subvariety $\Sigma_Z \subset Z$ with various properties as outlined in Steps 2 and 3 of Section $2k$. The purpose of this section is to reformulate some of these requirements in a purely homological way.

a) The homology of Σ_M and $M \times M$.

In order to understand the homological constraints on Σ_Z, it proves useful to digress first with a homological interpretation of some of the constructions in Section 2. Return then to the milieu of Section 2 where M is a compact, oriented 3-manifold with the rational homology of S^3 and where $\Sigma_M \subset M \times M$ is defined by (2.1).

The inclusion $\Sigma_M \subset M \times M$ induces a surjective homomorphism on the respective rational homology groups in dimension 3, with a one dimensional kernel.

$$\sigma_M \equiv [\Delta_M] - [p_0 \times M] - [M \times p_0] \tag{4.1}$$

This σ_M bounds (rationally) in $M \times M$, and a bounding cyle defines a class,

$$\rho_M \in H_4(M \times M, \Sigma_M). \tag{4.2}$$

(Here, $H.(X,Y)$ for a space X and subspace $Y \subset X$denotes the relative homology with rational coefficients.) The Poincare dual of ρ_M is the generator of $H_2(M \times M - \Sigma_M)$ which figures so prominently in Section 1. End the digression.

b) Homological constraints on Σ_Z from ω_Z.

Return to the bordism milieu of the introduction. The subvariety Σ_Σ should have a physical boundary (as a cycle, for example) which is given by

$$\partial\Sigma_Z = \Sigma_{M_1} \cup \Sigma_{M_2} \cup_{p \in \text{crit}(f)} (\Sigma_{S^3})_p, \tag{4.3}$$

where $(\Sigma_{S^3})_p$ is the obvious Σ_{S^3} in the boundary component $(S^3 \times S^3)_p$ of Z. Finding Σ_Z to satisfy (4.3) would satisfy Step 2 in Section 1h.

However, there are certain cohomological constraints on a solution to (4.3) which must be satisfied before it can solve the constraints which are implicit in Step 3 of Section 2k, and in particular, Parts 1 and 2 of (2.27). These are expressed by the following lemma:

LEMMA 4.1. *Let $\Sigma_Z \subset Z$ be a subvariety which obeys (4.3). Then, there is a closed 2-form, ω_Z, on $Z - \Sigma_Z$ which restricts to any component $Y \subset \partial Z - \partial\Sigma_Z$ to generate $H^2(Y)$ if and only if $H_4(\Sigma_Z, \partial\Sigma_Z)$ contains a class σ_Z which obeys:*

1) The image of σ_Z in $H_4(Z, \partial Z)$ is zero.

2) $\partial\sigma_Z$ in $H_3(\partial\Sigma_Z)$ obeys

$$\partial\sigma_Z = \sigma_{M_1} - \sigma_{M_0} + \sum_{p \in crit(f)} (\sigma_{S^3})_p. \tag{4.4}$$

(The absence of signs in the last term in (4.4) stems from the convention of Section 3h for orienting the right and left factors of S^3 in the boundary component $(S^3 \times S^3)_p \subset \partial Z$.)

The third constraint in (2.27) is the most difficult of all to satisfy. The strategy for satisfying the third constraint in (2.27) has two parts, one homological and the other geometric. For both parts, fix $N_Z \subset Z$, a regular neighborhood of Σ_Z. The homological issue is to characterize a closed 2-form on $N_Z - \Sigma_Z$ which is the restriction from $Z - \Sigma_Z$ of a closed 2-form ω_Z from Lemma 4.1. The geometric issue is to find such an ω which obeys $\omega \wedge \omega = 0$.

The following lemma resolves the homological issue:

LEMMA 4.2. *Suppose that Conditions 1 and 2 of Lemma 4.1 are obeyed. Let $N_Z \subset Z$ be a regular neighborhood of Σ_Z. A closed 2-form, ω, on $N_Z - \Sigma_Z$ is the restriction to $N_Z - \Sigma_Z$ of a closed 2-form ω_Z on $Z - \Sigma_Z$ as described in Lemma 4.1 if the following occur:*

1) *The connecting homomorphism from $H^2(N_Z - \Sigma_Z)$ to $H^3_{comp}(N_Z)$ sends ω to a multiple of the Poincare dual of $\sigma_Z \in H_4(N_Z, N_Z \cap \partial Z)$.*
2) *The restriction homomorphism $H^2(Z) \to H^2(\Sigma_Z)$ is surjective.*

This lemma is proved below. The last subsection in this section discusses the strategy for finding an appropriate ω near Σ_Z with $\omega \wedge \omega = 0$.

Proof of Lemma 4.1. To prove necessity, start with the observation that the cohomology class in $H^2(Z - \Sigma_Z)$ of the 2-form in question has Poincare dual

$$\rho_Z \in H_5(Z, \Sigma_Z \cup \partial Z). \tag{4.5}$$

The requirements in (2.27.1) and (2.27.2) concerning the restriction of ω_Z to ∂Z imply the following homological condition on $\partial\rho_Z$

$$\partial\rho_Z = \rho_{M_1} - \rho_{M_2} + \sum_{p\in\mathrm{crit}(f)} (\rho_{S^3})_p - \sigma_Z, \tag{4.6}$$

where

$$\sigma_Z \in H^4(\Sigma_Z, \partial\Sigma_Z) \tag{4.7}$$

is a class which obeys (4.4) (so that $\partial^2\rho_Z$ will vanish).

To prove the sufficiency assertion of the lemma, start with ρ_Z as described. Represent σ_Z as a cycle on Σ_Z. By assumption, one has $\sigma_Z - \tau = \partial\rho_Z$, where τ is a 4-cycle on ∂_Z, and where ρ_Z is a 5-cycle on Z. Note that $\partial\tau$ is equal to the right side of (4.6) also. Thus,

$$\tau - (\rho_{M_1} - \rho_{M_2} + \sum_{p\in\mathrm{crit}(f)} (\rho_{S^3})_p) \tag{4.8}$$

has zero boundary, and so defines a class in $H_4(\partial Z)$. However, this group is zero ($H_4(\partial Z) \approx H^2(\partial Z) = 0$ (see Section 2). Thus, (4.6) holds for some 5-cycle

ρ_Z on Z. The Poincare dual of ρ_Z is a class in $H^2(Z - \Sigma_Z)$ with the required properties. □

Proof of Lemma 4.2. The question of extending a closed 2-form on $N_Z - \Sigma_Z$ over $Z - \Sigma_Z$ is described by part of the Meyer-Vietoris sequence for the cover of Z by $(Z - \Sigma_Z) \cup N_Z$. The relevent part is:

$$H^2(Z) \to H^2(Z - \Sigma_Z) \oplus H^2(\Sigma_Z) \to H^2(N_Z - \Sigma_Z) \to H^3(Z) \tag{4.9}$$

The last arrow in (4.9) factors through the inclusion induced map $H^3_{\text{comp}}(N_Z) \to H^3(Z)$. So, if the image of ω in $H^3_{\text{comp}}(N_Z)$ is Poincare dual to a multiple of σ_Z as a class in $H_4(N_Z, N_Z \cap \partial Z)$, then the image of ω in $H^3(Z)$ is zero if the image of σ_Z in $H_4(Z, \partial Z)$ is zero. This is the first condition in Lemma 4.1. Thus, under Condition 1 of Lemma 4.2, the class ω maps to zero in $H^3(Z)$.

When Condition 2 of Lemma 4.2 holds, then ω must be in the image of the restriction homomorphism from $H^2(Z - \Sigma_Z)$ because of the exactness of (4.9). □

c) Satisfying Lemma 4.1's constraints.

The second constraint in Lemma 4.1 will be satisfied by construction; as it is essentially a restatement of (4.3) with orientations taken into account. The first constraint in Lemma 4.1 is more subtle. Here is a strategy for finding a solution: The variety Σ_Z will be constructed from a union of varieties,

$$\Sigma_Z = \Delta_Z \cup E_L \cup E_R \cup E_- \cup E_+. \tag{4.10}$$

Each variety on the right side of (4.10) will carry a fundamental class. (Here, a variety is a union of embedded submanifolds. If the constituent submanifolds are oriented, then the variety has a fundamental class which is the sum (in the relevent homology group) of the fundamental classes of the constituent submanifolds.) And, for a particular integer $N > 0$, the class σ_Z will be given as

$$\sigma_Z = [\Delta_Z] - [E_L] - [E_R] - N^{-1}[E_-] - N^{-1}[E_+]. \tag{4.11}$$

In (4.10), (4.11), Δ_Z and $E_{L,R}$ are honest submanifolds; these will be defined in subsequent subsections. Meanwhile, $E_\pm$ will be honest varieties unless $N = 1$ in (4.11). The construction of $E_\pm$ is quite lengthy and starts in the next section with the completion in Section 10. But, see subsections $4f$, g below.

With (4.11) understood, the first constraint of Lemma 4.1 will be solved with the help of Lemma 4.3, below. (The statement of this lemma reintroduces $\underline{L}_\pm$ from (3.31), (3.32).)

LEMMA 4.3. *Suppose that W has the rational homology of S^3. Let $V \subset Z$ be a union of dimension 4 submanifold with boundary such that $\partial V \subset \partial Z$. Suppose* *that each component of V carries a fundamental class. Then* $[V] \in H_4(Z, \partial Z; \mathbb{R})$ *vanishes if:*

1) $[\partial V] = 0$ *in* $H_3(\partial Z; \mathbb{R})$.

2) V has zero intersection number with any component $x \subset (\underline{L}_- \cup \underline{L}_+)$.

(The intersection number of V with an embedded, 3-dimensional submanifold of Z is defined to be the sum of the intersection numbers of the components of V.)

Proof. Poincare duality equates $H_4(Z, \partial Z)$ with $H^3(Z)$. Intersection theory makes this explicit, as the intersection pairing between $H_4(Z, \partial Z)$ and $H_3(Z)$ becomes, under Poincare duality, the dual pairing between $H^3(Z)$ and $H_3(Z)$. Now, use this fact with Assertion 3 of Lemma 3.7. □

d) The subspace Δ_Z

Let $\Delta_W \subset W \times W$ denote the diagonal. Clearly, $\Delta_W \subset Z$. Let Δ_Z denote the intersection of Δ_W with $Z \subset \underline{Z}$. (Alternately, if Z is thought of as the blow up of $\underline{Z}$, then Δ_Z can be defined as the inverse image of Δ_W under this blow up.) Note that Δ_Z is a submanifold with boundary in W, and

$$\partial \Delta_Z = \Delta_{M_0} \cup \Delta_{M_1} \cup_{p \in \mathrm{crit}(f)} (\Delta_{S^3})_p. \tag{4.12}$$

The orientation of W defines an orientation for Δ_W and thus for Δ_Z. The orientation of $(\Delta_{S^3})_p$ is induced from the orientation of Δ_Z in Section $3h$ as a boundary component of Δ_Z. With this understood, one has:

LEMMA 4.4. *Let* $[\Delta_Z] \in H_4(Z, \partial Z)$ *denote the fundamental class of* Δ_Z. *Then*

$$\partial[\Delta_Z] = -[\Delta_{M_0}] + [\Delta_{M_1}] + \sum_{p \in crit(f)} [(\Delta_{S^3})_p]. \tag{4.13}$$

as a class in $H_3(\partial Z)$.

Proof. This is left as an exercise.

As a final remark, note that

$$\Delta_Z \cap (L_- \cup L_+) = \emptyset. \tag{4.14}$$

This is a consequence of Condition 1 in Definition 3.1. □

e) The submanifolds $E_{R,L}$.

By assumption (see 4 of Definition 3.1), there is a gradient flow line for the pseudo-gradient v which starts at p_0 and which ends at p_0. Let γ denote this line. Define

(4.15)

1) $E_R \equiv (\gamma \times W) \cap Z$,
2) $E_L \equiv (W \times \gamma) \cap Z$.

Here are the properties of these spaces:

LEMMA 4.5. *Both E_R and E_L are embedded submanifolds (with boundary) of Z. Also,*

1) $\partial E_R = (p_0 \times M_0) \cup (p_1 \times M_1)$.
2) $\partial E_L = (M_0 \times p_0) \cup (M_1 \times p_1)$.
3) *Let π_L and π_R denote the respective right and left factor projections from $W \times W$ to W. Then $\pi_R : E_R \to W$ and $\pi_L : E_L \to W$ are both diffeomorphisms.*
4) $E_R \cap \Delta_Z = E_L \cap \Delta_Z = E_R \cap E_L = (\gamma \times \gamma) \cap \Delta_Z$. *Furthermore, this subspace $(\gamma \times \gamma) \cap \Delta_Z$ has a neighborhood $U \subset Z$ with a diffeomorphism (of manifold with boundary) $\psi_U : U \approx [0,1] \times \mathbb{R}^3 \times \mathbb{R}^3$ which obeys*
   (a) $\psi_U((\gamma \times \gamma) \cap \Delta_Z) = [0,1] \times (0,0)$.
   (b) $\psi_U(E_R) = [0,1] \times \{0\} \times \mathbb{R}^3$.
   (c) $\psi_U(E_L) = [0,1] \times \mathbb{R}^3 \times \{0\}$.
   (d) $\psi_U(\Delta_Z) = [0,1] \times \Delta_{\mathbb{R}^3}$.
   (e) $\psi_U(M_0 \times M_0) = \{0\} \times \mathbb{R}^3 \times \mathbb{R}^3$.
   (f) $\psi_U(M_1 \times M_1) = \{1\} \times \mathbb{R}^3 \times \mathbb{R}^3$.
   (g) *The interchange map $(z, z') \to (z', z)$ on Z is mapped by ψ_U to $(t, x, y) \to (t, y, x)$.*
5) *Both E_R and E_L have empty intersection with the components of $\underline{L}_- \cup \underline{L}_+$ of (3.30), (3.31).*
6) *Orient E_R and E_L by π_L and π_R, respectively. Then*

$$\partial[E_R] = -[p_0 \times M_0] + [p_1 \times M_1], \tag{4.16}$$

$$\partial[E_L] = -[M_0 \times p_0] + [M_1 \times p_1].$$

The remainder of this subsection is occupied with the proof of this lemma.

Proof. Since γ is a flow line of v, it has a parametrization

$$\underline{\gamma} : [0,1] \to W \tag{4.17}$$

with $(\underline{\gamma}^* f)(t) = t$. This implies that the function F of (3.16) restricts without critical points to $\gamma \times W$ and to $W \times \gamma$; therefore, both E_R and E_L are submanifolds of Z.

Assertions 1 and 2 of the lemma follow because γ is assumed to miss crit(f).

To prove the third assertion, use $\underline{\gamma}$ to view E_R as the graph of f in $[0,1] \times W$, where π_R restricts as the obvious projection to W. The proof of Assertion 3 for E_L is analogous.

To prove Assertion 4, note that γ, being embedded, has a neighborhood $U_\gamma \subset W$ with a diffeomorphism $\psi_\gamma : U_\gamma \to [0,1] \times \mathbb{R}^3$ which obeys $f(\psi_\gamma^{-1}(t,x)) = t$ and $\psi_\gamma(\underline{\gamma}(t)) = (t,0)$. (Use the implicit function theorem to construct such a ψ_γ .) Take $U = U_\gamma \times U_\gamma$ and take $\psi_U = \psi_\gamma \times \psi_\gamma$. The verification of (a)-(g) follow immediately. □

f) The varieties $E_\pm$.

With Δ_Z and $E_{R,L}$ defined in the preceding section, the solution σ_Z of (4.11) to Lemma 4.1's constraints is missing still $[E_-]$ and $[E_+]$. Indeed, the class $[\Delta_Z] - [E_R] - [E_L]$ is a class in $H_4(Z, \partial Z)$ whose boundary is equal to

$$-\sigma_{M_0} + \sigma_{M_1} + \sum_{p \in \text{crit}(f)} [\Delta_{S^3}]_p, \tag{4.18}$$

which is only a part of the right side of (4.4).

As remarked earlier, the construction of $E_\pm$ is quite lengthy. To simplify matters, the decomposition given in Proposition 3.2 will be invoked to break the discussion into two parts so that the cobordism W can be assumed to obey the conditions of (3.11). That is, W will be assumed to have the rational homology of S^3 and W has a good Morse function with no index 3 critical points.

The construction of $E_\pm$ for W given by (3.11) is started in the next section with a digression to describe certain constructions on such W. The construction of $E_\pm$ for (3.11) is completed in Section 9.

With W understood to be given by (3.11), here is a rough description of $E_\pm$: Fix a good Morse function $f : W \to [0,1]$ with no index 3 critical points. Let $a_1, \cdots, a_r$ and $b_1, \cdots, b_r$ label the index 1 and index 2 critical points of f.

Now fix a good pseudo-gradient, v, for f, and fix orientations from the descending disks from crit(f) such that the conclusions of Proposition 3.3 hold. That is, with the orientations implicit, the points $a_1, \cdots, a_r$ and $b_1, \cdots, b_r$ define a basis for C_1 and C_2, respectively. And, with respect to this basis, the boundary map in (3.5), $\partial : C_1 \to C_2$, is represented by an upper triangular matrix, S, with positive entries on the diagonal.

A pair $E_\pm$, of subvarieties (with boundary) of Z will be constructed with $\partial E_\pm \subset \partial Z$. The variety E_- is obtained as the intersection with Z of a subvariety of $W \times W$; this subvariety is constructed by performing various surgeries on multiple copies of products of the ascending disks from points in crit$_1(f)$ with the descending disks from the points in crit$_2(f)$. Meanwhile, the variety E_+ is obtained as the intersection with Z of a different subvariety of $W \times W$. In this case, the subvariety is constructed by surgery on multiple copies of the product of the descending disks from $\mathrm{crit}_2(f)$ with the ascending disks from $\mathrm{crit}_1(f)$.

The varieties $E_\pm$ will be naturally oriented and seen to define classes $[E_\pm] \subset H_4(Z, \partial Z)$. The boundaries of these classes are

$$\partial[E_-] = N \sum_{p \in \mathrm{crit}(f)} [S^3]_{p^-}, \tag{4.19}$$

$$\partial[E_+] = N \sum_{p \in \mathrm{crit}(f)} [S^3]_{p^+},$$

where (4.19) has introduced the following shorthand: When p is a critical point of f, use $[S^3]_{p^-}$ to denote $[S^3 \times \text{point}] \in H_3((S^3 \times S^3)_p)$, and use $[S^3]_{p^+}$ to denote $[\text{point} \times S^3] \in H_3((S^3 \times S^3)_p),$. Here, the orientations on $(S^3 \times \text{point})$ and $(\text{point} \times S^3)$ are defined in Section $3h$. (The diagonal in $(S^3 \times S^3)_p$ is oriented as a component of the boundary of Δ_Z and then the right and left factors of S^3 in $(S^3 \times S^3)_p$ are oriented by using the canonical identification of S^3 with Δ_{S^3}.)

The $[E_\pm]$, of (4.19) will be constructed to have zero intersection pairing with the classes in $L_\pm$ of (3.30). This will insure that σ_Z of (4.11) satisfies both requirements of Lemma 4.1. (See Lemma 4.3.)

g) Constraints from $\omega_Z \wedge \omega_Z = 0$.

With Σ_Z in (4.10) constructed so that both requirements of Lemma 4.1 are satisfied, there is a 2-form on $Z - \Sigma_Z$ which is a candidate for the form ω_Z in Step 3 of Theorem 2.9's proof.

The issue then arises as to whether σ_Z can be found to satisfy the conditions in (2.27). The construction of a closed 2-form which satisfies the conditions of (2.27) is carried out in Section 10. However, to motivate some of the intervening contortions, here is a rough summary of the difficulties:

Remark 1: As long as $E_\pm$ in (4.10) have empty intersection with $M_0 \times M_0$ and with $M_1 \times M_1$, then there is no obstruction to finding ω_Z which obeys (2.27.1). (See Lemma 2.1)

Remark 2: The remaining requirements of (2.27) are harder to satisfy. In particular, the second requirement in (2.27) will require that for each $p \in \mathrm{crit}(f)$,

(4.20)

1) $E_- \cap (S^3 \times S^3)_p = S^3 \times x_p$
2) $E_+ \cap (S^3 \times S^3)_p = x_p \times S^3.$

This requirement and (4.19) are incompatible unless $N = 1$ or unless $E_\pm$ are singular. Together, (4.19), (4.20) force the use *subvarieties* for $E_\pm$ instead of submanifolds.

Given (4.20), the second constraint in (2.27) can also be satisfied. (See Lemma 2.1 again.)

Remark 3: The first condition of Lemma 4.2 is not easy to satisfy with a 2-form ω which obeys $\omega \wedge \omega = 0$. In the case where $N = 1$ in (4.11) (so $E_\pm$ are manifolds) the strategy will be to find a regular neighborhood $N_Z \subset Z$ of Σ_Z and a map

$$\varphi_Z : N_Z \to \mathbb{R}^3 \tag{4.21}$$

which obeys $\varphi_Z^{-1}(0) = \Sigma_Z$ and which pulls back the generator of $H^3_{\text{comp}}(\mathbb{R}^3)$ to a non-zero multiple of the Poincare dual in $H^3_{\text{comp}}(N_Z)$ to $\sigma_Z \in H_4(N_Z, N_Z \cap \partial Z)$. In this case,

$$\omega_Z \equiv \varphi_Z^{-1}(\mu) \tag{4.22}$$

with μ as in (2.3).

In the case where $N > 1$ in (4.11), the preceding strategy will be modified. When $N > 1$ in (4.11), then φ_Z, as in (4.21), will be defined only in a neighborhood of $\Delta_Z \cup E_L \cup E_R \subset \Sigma_Z$, and ω_Z will be defined near $\Delta_Z \cup E_L \cup E_R$ by (4.22). But, near the remainder of Σ_Z (i.e. near most of $E_\pm$), the form ω_Z will be defined somewhat differently. (The basic difference being that ω_Z will be defined *locally* as the pull-back of a closed 2-form from a space of dimension less than 4. However, the space in question will not always be S^2. In some places, the space will be the compliment in S^3 of $N + 1$ distinct points.)

This strategy for constructing a closed, square zero solution to Condition 1 of Lemma 4.2 requires $\Delta_Z, E_{L,R}$ and the constituent submanifolds of $E_\pm$ to have trivial normal bundles in Z. (See Remark 4, below.) The success of this strategy also requires that the mutual intersections of $Z, E_{L,R}$ and $E_\pm$ have a canonical form. (See Remark 5, below.)

Remark 4: The normal bundle of Δ_Z in Z is trivial if and only if Z is a *spin* manifold. Indeed, $H^4(\Delta_Z) = 0$ and therefore, an oriented 3-plane bundle over Z is classified by its 2nd Stieffel-Whitney class. Since Z has trivial normal bundle in $W \times W$, the Stieffel-Whitney classes of the normal bundle to Δ_Z in Z are the same as those of the normal bundle to Δ_Z in $W \times W$. The latter are the restrictions of the Stieffel-Whitney classes of the normal bundle of Δ_W in $W \times W$. And, this last normal bundle is naturally isomorphic to the tangent bundle of W. Finally remember that W is said to be spin if the 2nd Stieffel-Whitney class of its tangent bundle vanishes.

The normal bundles to $E_{L,R}$ are trivial, since they are isomorphic to the normal bundle to the path γ in W.

A constituent submanifold of E_+ (or E_-) has a normal bundle in Z. If care is taken in constructing $E_\pm$, then these normal bundles will be trivial too.

Remark 5: The construction of a square zero ω to satisfy the first condition of Lemma 4.2 seems to require that $E_\pm$ do not intersect each other or Δ_Z and $E_{L,R}$ in a complicated way. Infact, the $E_\pm$ that are finally constructed will have empty intersections with $E_{L,R}$, while

$$E_- \cap E_+ = E_- \cap \Delta_Z = E_+ \cap \Delta_Z = \cup_{i=1}^r v_i, \tag{4.23}$$

where $\{\epsilon_i\}_{i=1}^r$ are disjoint, embedded paths in Δ_Z. In fact, fix the label $i \in \{1, \cdots, r\}$ and let $a \equiv a_i$ and $b \equiv b_i$ be the i'th pair of index 1 and index 2 critical points of the Morse function f as described by Proposition 3.3. Then, v_i will start at the point $(x_a, x_a) \in (S^3 \times S^3)_a$ and will end at the point $(x_b, x_b) \in (S^3 \times S^3)_b$.

Assertion 4 of Lemma 4.5 describes the intersections amongst $E_{L,R}$ and Δ_Z. Assertion 4 of Lemma 4.5 and (4.23) (with some conditions on normal bundle framings) insure that the intersection of $E_\pm$ with Δ_Z has the appropriate form for the construction of a square zero ω to satisfy the first condition of Lemma 4.2.

Remark 6: The second condition in Lemma 4.2 will be satisfied by taking care to construct $E_\pm$ to have vanishing H^2. Note that Z, Δ_Z and $E_{L,R}$ all have vanishing H^2. (See Lemmas 3.7, 4.4 and 4.5, respectively).

Care must also be taken to insure that $E_\pm$ do not intersect each other or Δ_Z and $E_{L,R}$ in a complicated way. Infact,

LEMMA 4.6. *Suppose that Σ_Z is given by (4.10) with Δ_Z and $E_{L,R}$ as described in Sections* 4d *and* 4e, *respectively. Suppose that $E_\pm \subset Z$ are varieties which have empty intersection with $E_{L,R}$ and which intersect each other and Δ_Z as described in (4.23). Suppose, in addition, that $H^2(E_\pm; \mathbb{Q}) = 0$. Then $H^2(\Sigma_Z; \mathbb{Q}) = 0$ and the homomorphism $H^2(Z; \mathbb{Q}) \to H^2(\Sigma_Z; \mathbb{Q})$ of Lemma 4.2 is surjective by default.*

Proof. Because the intersections of $\Delta_Z, E_{L,R}$ and $E_\pm$ with each other are a union of line segments (which have vanishing H^1), the Meyer-Vietoris exact sequence shows that $H^2(\Sigma_Z)$ is isomorphic to the direct sum of $H^2(\cdot)$ for $(\cdot) = \Delta_Z, E_{L,R}$ and $E_\pm$. By assumption $H^2(E_\pm) = 0$. Meanwhile, $H^2(\Delta_Z) \approx 0$, since Δ_Z is the compliment in $\Delta_W \approx W$ of a finite union of disjoint (open) 4-balls. And, $H^2(E_{L,R}) \approx 0$ since $E_{L.R} \approx W$. □

Remark 7: In summary, the construction of $E_\pm$ for the case of (3.11) will proceed with care taken with:

(4.24)

1) Normal bundle framings.
2) Intersections with $\Delta_Z, E_{L,R}$ and with each other.
3) Keeping $H^2(E_\pm)$ equal to zero.

5 Disk intersections for the Special Case. The construction of $E_\pm$ for W given by (3.11) starts in this section with a digression to describe certain constructions on such W. The constructions here serve to modify the ascending disks from index 1 critical points and also descending disks from index 2 critical points.

With W understood to be given by (3.11), begin the discussion by fixing a good Morse function $f : W \to [0,1]$ as described by Proposition 3.3. As in Proposition 3.3, let $a_1, \cdots, a_r$ label f's index 1 critical points and $b_1, \cdots, b_r$ label the index 2 critical points.

Fix a good pseudo-gradient, v, for f, and fix orientations from the descending disks from crit(f) so that the conclusions of Proposition 3.3 hold. That is, with the orientations implicit, the points $a_1, \cdots, a_r$ and $b_1, \cdots, b_r$ define a basis for C_1 and C_2, respectively. And, with respect to this basis, the boundary map in (3.5), $\partial : C_1 \to C_2$, is represented by an upper triangular matrix, S, with positive entries on the diagonal.

The matrix S gives a certain amount of algebraic information about the intersections of the descending disks from crit$_2(f)$ and the ascending disks from crit$_1(f)$. That is, the intersection of the descending disk from b_i and the ascending disk from a_j is a discrete set of flow lines which start at a_j and end at b_i. Each such flow line carries a sign, ± 1. And, the matrix element $S_{i,j}$ computes the sum of these ± 1's. In particular, Proposition 3.3 insures that the algebraic intersection number of the descending disk from b_i and the ascending disk from a_j is zero if $i > j$.

However, even when $i > j$, the point a_j may lie in the closure of the descending disk from b_i. This is an unpleasant fact which must be circumvented in order to facilitate certain constructions in the subsequent subsections. The purpose of this subsection is to modify the descending and ascending disks so as to make this eventuality irrelevent. The expense here is to replace the disk with a more complicated submanifold of W.

a) Past and future.

The purpose of this subsection is to introduce some terminology which will arise in the modification constructions below. To begin, focus on a subset $U \subset W$. Define the past of U, written past(U), as follows: A point$x \in$ past(U) if there is a gradient flow line $\gamma : [a, b] \to W$ and times $t, t' \in (a, b]$ with $t' \geq t$ and with

(5.1)

1) $\gamma(t) = x$,
2) $\gamma(t') \in U$.

Define the future of U, written fut(U), as the subset of points x in W for which there is a gradient flow line which obeys (5.1) but where $t, t' \in [a, b)$ and $t' \geq t$. Note that past$(U)\cap$ fut$(U) = U$.

For example, if $p \in W$ is not a critical point, then past(p) is the set of points which are hit before p on the gradient flow line through p. However, if $p \in$ crit(f), then past$(p) = B_{p-}$.

b) Tubing descending disks from crit$_2(f)$.

This subsection begins the modification process; it describes a construction, *tubing*, which modifies the descending disk from an index 2 critical point b_i so that the closure of the modified submanifold is disjoint from any index 1 critical point a_j for $j < i$.

To make the tubing construction, focus first on some index 2 critical point $b = b_i$ and a particular index 1 critical point $a = a_j$ for $j < i$. A descending disk from the index 2 critical point b will intersect a neighborhood, U_a of a in a finite set of components. Each of these components contains the intersection of U_a with a gradient flow line which starts at a and ends at b. To be precise, let $V \subset B_{b-} \cap U_a$ be a component. After a small isotopy, one can find Morse coordinates for U_a so that

$$\psi_a(V) = \{(x_1, x_2, x_3, x_4) : x_2 \geq 0 \text{ and } x_3 = x_4 = 0\}. \tag{5.2}$$

With (5.2) understood, the flow line between a and b which lies in V is given in the Morse coordinates by intersecting $\psi_a(V)$ with the ray $\{(x_1, x_2, x_3, x_4) : x_2 > 0 \text{ and } x_1 = x_3 = x_4 = 0\}$.

To consider the full intersection of B_{b-} with a neighborhood of a in W, it is convenient to first intersect $\psi_a(B_{a+})$ with a small radius sphere in $\mathbb{R}^4$ about the origin. Call the result S_+; in Morse coordinates, this S_+ is a small radius 2-sphere in the 3-plane where $x_1 = 0$. The intersection $B_{b-} \cap S_+$ is transverse, and is a finite number of points, $B_{b-} \cap S_+ = \{e_\alpha\}$. Because $j < i$ and the matrix S is upper triangular, the 2-sphere S_+ has zero algebraic intersection number with B_{b-}. This means that the points $\{e_\alpha\}$ of B_{b-}'s intersection with S_+ can be paired so that each pair contains one point with positive intersection number and one with negative intersection number. Write this pairing as

$$\{e_\alpha\} = \{\{e_1, e_2\}, \cdots, \{e_{2n-1}, e_{2n}\}\}. \tag{5.3}$$

Since S_+ is a 2-sphere, the two points of any pair can be connected by a path in S_+. These paths can be drawn so that paths coming from distinct pairs in $B_{b-} \cap S_+$ do not intersect. The paths should also be drawn to avoid intersections of S_+ with any other descending sphere from $\text{crit}_2(f)$. Let $\{\zeta_\mu\}_{\mu=1}^n$ be the set of paths just defined.

The value of f on S_+ is some constant, $f_0 > f(a)$. Then, introduce $M \equiv f^{-1}(f_0) \cap U_a$. This will be a smooth 3-manifold given by

$$\psi_a(M_+) = \{(x_1, \cdots, x_4 : -x_1^2 + x_2^2 + x_3^2 + x_4^2 = f_0\}. \tag{5.4}$$

With M_+ understood, thicken each $\zeta \in \{\zeta_\mu\}$ to a thin ribbon in M_+; call this ribbon $\underline{\zeta} \approx I \times I$, where $I \equiv [0, 1]$. (The ribbon should be thin so that it's only intersection with a descending disk from $\text{crit}_2(f)$ is with $\partial\underline{\zeta}$.) Thus parameterized, $I \times \{1/2\} = \zeta$, while $\partial I \times I$ is embedded in $B_{b-} \cap M_+$. To be explicit, parameterize as $\zeta(\tau) \in S_+$ for $\tau \in [0, 1]$. Then, to a first approximation, $\underline{\zeta}$ should be parameterized by (τ, τ') as

$$\begin{aligned} &1) \quad x_1 = f_0^{1/2}\, \epsilon\, (2\tau' - 1), \\ &2) \quad (x_2, x_3, x_4) = (1 + \epsilon^2\, (2\tau' - 1)^2)^{1/2}\, \zeta(\tau)\} \end{aligned} \tag{5.5}$$

for small $\epsilon > 0$.

Let η_{0-} denote the past of $\underline{\zeta}(0, 0)$; it is part of a gradient flow line which starts on M_0. Let η_{0+} denote the past of $\underline{\zeta}(0, 1)$, part of another gradient flow line starting on M_0.

The union $\eta_0 \equiv \eta_{0-} \cup \zeta(0,\cdot) \cup \eta_{0+}$ is a piece-wise smooth curve in B_{b-}. Here is a picture of η_0 and past(η_0):

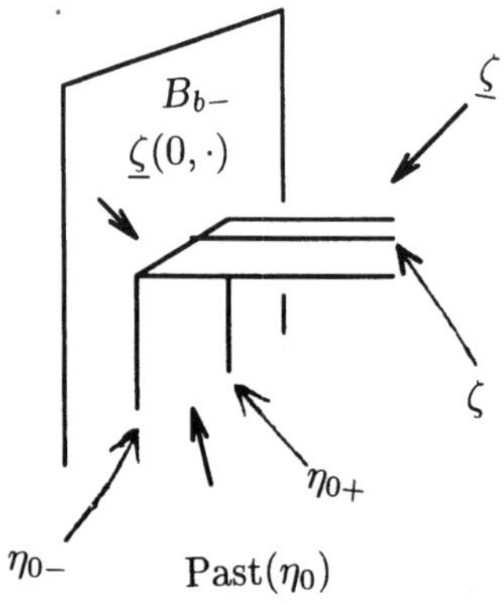

(5.6)

Let η_1 - denote the past of $\underline{\zeta}(1,0)$ and let η_{1+} denote the past of $\underline{\zeta}(1,1)$. Set $\eta_1 \equiv \eta_{1-} \cup \underline{\zeta}(1,\cdot) \cup \eta_{1+}$. This is a piece-wise smooth curve in B_{b-}.

With the preceding understood, here is a surgery on B_{b-}: Delete from B_{b-} the set past$(\eta_0)\cup$ past(η_1) to get a manifold with piecewise smooth boundary $\eta_0 \cup \eta_1$, and then glue on to this boundary image $(\underline{\zeta})\cup$ past$(\underline{\zeta}(\cdot,0))\cup$ past$(\underline{\zeta}(\cdot,1)$. Call the resulting space B'_{b-}. See the following picture:

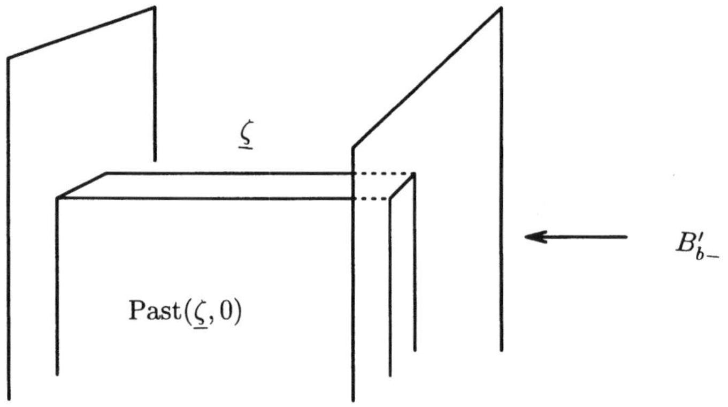

(5.7)

The surgery just described is the *tubing* construction on a cancelling pair of intersection points of B_{b-} with S_+

Effect this tubing construction for all the pairs in (5.3) which comprise B_{b-}'s intersection with S_+. Because the surgeries are constructed using gradient flow lines, the surgeries from different pairs in (5.3) do not interfere (or intersect) each other.

After all n surgeries are performed, the result is a piecewise smooth submanifold of W whose closure misses the critical point a_j. This submanifold can be smoothed after a small perturbation and will henceforth will be assumed smooth.

Effect the same tubing construction for all pairs of intersection points for all a_j with $j < i$. Use B_{1b-} to denote the result of doing this surgery. (Note: Because the surgeries are defined using gradient flow lines, the surgeries which come from different index 1 critical points do not interfere nor intersect with each other.)

Finally, effect this same tubing construction for all b_i in $\text{crit}_2(f)$. Note that this can be done so that the resulting set of submanifolds $\{B_{1b-} : b = b_i\}_{i=1}^{r}$ are disjoint in W. (The point here is that the paths ζ and the ribbons $\underline{\zeta}$ in (5.5) have only the two boundary points of ζ as intersection points with descending disks from $\text{crit}_2(f)$. The rest of the tubing construction uses only gradient flow lines–and so won't create intersections with descending disks.)

c) Normal bundles.

Let $b = b_i \in \text{crit}_2(f)$. The submanifold $B_{b-} \subset W$ is oriented as the negative disk from the degree 2 critical point $b = b_i$. As an oriented submanifold of W, B_{b-} has a canonical trivialization of its normal bundle (up to homotopy). Simply flow the trivialization of the normal bundle of B_{b-} at b along B_{b-} using the pseudo-gradient v.

The preceding subsection described the construction of a submanifold B_{1b-} from B_{b-} by doing surgery on embedded arcs in B_{b-} with endpoints on $B_{b-} \cap M_0$. The resulting 2-dimensional submanifold can be seen to be orientable, and it inherits a canonical orientation from B_{b-}. (Note that each surgery that is performed on B_{b-} is constrained to lie in a 3-dimensional ball in W. One dimension of this ball is the pseudo-gradient flow direction, the other two dimensions can be parameterized by the ribbon coordinates on $\underline{\zeta}$ in (5.5).)

As B_{1b-} is not closed in W, the normal bundle to B_{1b-} will be a trivial bundle, and the claim is that there is an essentially canonical trivialization up to homotopy. The point is that in constructing B_{1b-} from B_{b-} one does a large number, say N, of essentially identical, non-interfering surgeries. So, one need only check that the canonical normal trivialization of B_{b-} extends over any one of these surgeries to give a normal trivialization of the postoperative manifold which agrees with the normal trivialization of B_{b-} away from the area of surgery. That such is the case is easy to check, since each individual surgery can be performed inside a 3- dimensional ball inside of W.

d) Tubing ascending disks from $\mathrm{crit}_1(f)$.

Let $a \equiv a_i \subset \mathrm{crit}_1(f)$. The closure of the ascending disk from a will typically intersect many of the points in $\mathrm{crit}_2(f)$. The purpose of this subsection is to modifies the ascending disk so that the closure of the resulting submanifold of W is disjoint from $\{b_j\}_{j>i}$. This modification procedure will also be called *tubing*.

To begin the tubing construction, focus attention first on an index 2 critical point $b \equiv b_j$ with $j > i$. Introduce the Morse coordinates, ϕ_b of (3.2), on a neighborhood of b. Let $S_- \subset \mathbb{R}^4$ denote a small radius circle in the (x_1, x_2)-plane, centered at the origin. This is a small radius circle in the inverse image by ϕ_b of the descending disk from b.

The submanifold B_{a+} intersects S_- transversely in a finite set. The algebraic intersection number here is zero because the transpose of the matrix S in Proposition 3.3 is lower triangular. Thus, the intersection points can be paired off so that the one member of each pair is a positive intersection point and the other member is negative. However, care must be taken in making this pairing.

The pairing process is an inductive one, where each induction step pairs off some non-zero number of points until all points have been paired. To describe the pairing process in an induction step, remark that the n'th induction step will pair up a non-zero number of points and leave a subset, $Y_{n+1} \subset S_-$ of points which have not yet been paired. Here, Y_{n+1} contains an even number of points, half are points of positive intersection and half are points of negative intersection. By definition, $Y_1 \equiv B_{a+} \cap S_-$.

Being points on the circle, the points of Y_{n+1} have a cyclic ordering. Break this ordering by choosing one point, e_1, as the fiducial point. With e_1 chosen, there is a decomposition of Y_{n+1} into subsets of consecutive points,

$$Y_{n+1} = \Lambda_1 \cup \Lambda_2 \cup \cdots \cup \Lambda_J, \tag{5.8}$$

where the points in Λ_j can be described as follows:

(5.9)

1) The signs for all points in Λ_j agree, and for $j > 1$, disagree with those in Λ_{j-1}.
2) In clockwise order, Λ_1 starts with e_1.
3) The points in Λ_j follow clockwise those in Λ_{j-1}.

The $(n+1)$'st round of pairings is obtained by assigning the last point in Λ_j to the first point in Λ_{j+1}; doing this for $j = 1, \cdots, J-1$. See the following diagram:

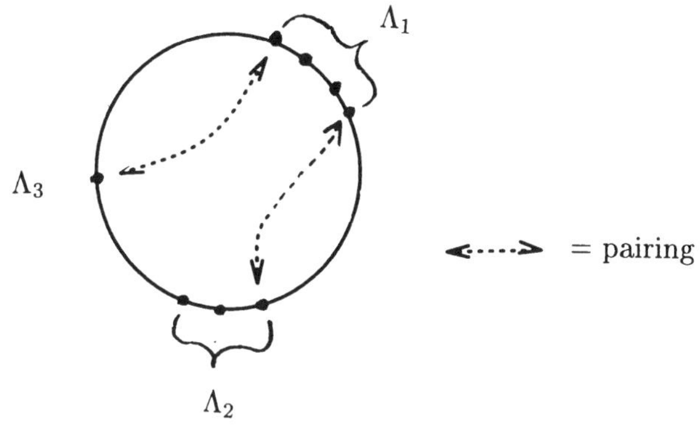

(5.10)

Let $P_{n+1} \subset Y_{n+1}$ denote the set of points just paired. This P_{n+1} contains at least two points (i.e., one pair) unless Y_{n+1} is already the empty set. If $Y_{n+1} \neq \emptyset$, then $Y_{n+2} \equiv Y_{n+1} - P_{n+1}$ contains strictly fewer points then Y_{n+1}. Thus, iterating the pairing process as just described will eventually pair up all the points of $B_{a+} \cap S_-$.

With the pairing process complete, turn to the tubing process. Start the discussion by considering the set P_1 of points which were paired on the first round. By construction, P_1 decomposes as a union of nearest neighbor pairs, with a pair, $\{e, e'\}$, in the decomposition composed of one point with positive intersection number and one with negative intersection number. Write $\{e, e'\}$ with e' clockwise from e on S_-. Let ζ denote a (closed) interval of S_- which sits clockwise between e and e'.

The value of the function f on S_- is some constant, f_0. One can assume, with no loss of generality, that $M_- \equiv f^{-1}(f_0)$ is a smooth submanifold of W near the critical point b. Indeed,

$$\psi_b(M_-) = \{(x_1, \cdots, x_4 : -x_1^2 - x_2^2 + x_3^2 + x_4^2 = f_0\}. \tag{5.11}$$

With M_- understood, thicken ζ to an embedded $I \times D^2$ inside M_-. Here, $I = [0,1]$ and D^2 is the standard 2-disk. The embedding sends $I \times \{0\}$ to ζ with $\{0\} \times \{0\}$ going to e and with $\{1\} \times \{0\}$ going to e'. Meanwhile, the embedding should embed $\partial I \times D^2$ into $M_- \cap B_{a+}$ as neighborhoods of $\{e, e'\}$ in M_-. Only $I \times \{0\}$ should intersect S_- and only $\partial I \times D^2$ should intersect B_{a+}.

Agree to identify $I \times D^2$ with its image inside M_-. See the following picture:

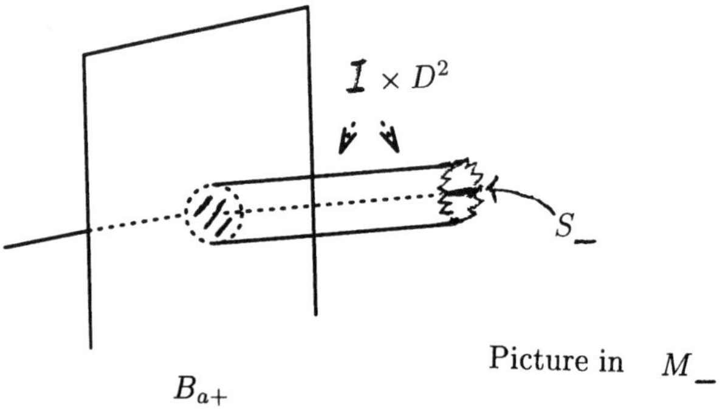

(5.12)

With $I \times D^2$ understood, perform the following surgery on B_{a+}: Delete $\text{fut}(\{0\} \times D^2) \cup \text{fut}(\{1\} \times D^2)$ from B_{a+}. The closure of the resulting space has a new "boundary" which is piecewise smooth, being the union

$$\text{(5.13)} \quad \text{fut}(\{0\} \times \partial D^2) \cup \text{fut}(\{1\} \times \partial D^2) \cup (\{0\} \times D^2) \cup (\{1\} \times D^2).$$

With this deletion complete, glue onto the boundary above the set $\text{fut}(I \times \partial D^2) \cup (I \times D_2)$. The result will be a piece-wise smooth manifold (in its interior) which can be smoothed to give a smooth manifold (in its interior) whose closure has two less intersections with S_-. Let B'_{a+} denote the resulting smoothed manifold.

See the following diagram:

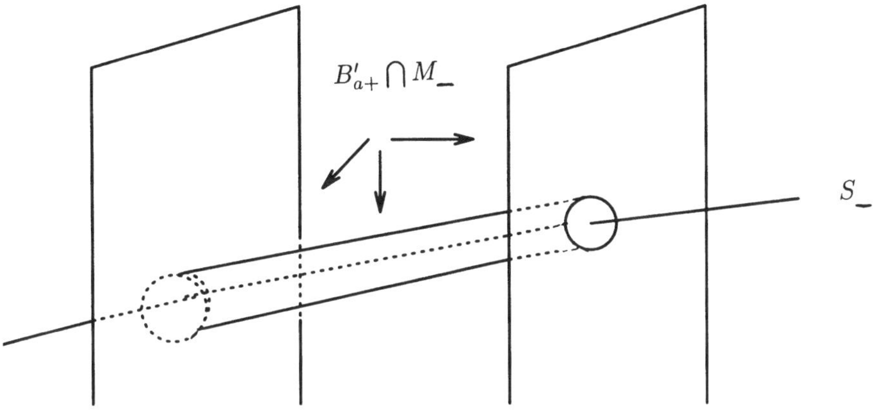

(5.14)

Notice that the B'_{a+} intersects S_- in the set Y_2. And, the subset P_1 of points paired on the second round consists, by construction, of nearest neighbor pairs on S_-. One can therefore repeat the preceding tubing construction to obtain a sub-manifold $B''_{a+} \subset W$ whose intersection with S_- is precisely the set Y_3. Clearly, an iteration of the tubing construction (as described above) will result, finally, with a sub-manifold of W with no intersections with S_-. See the following diagram:

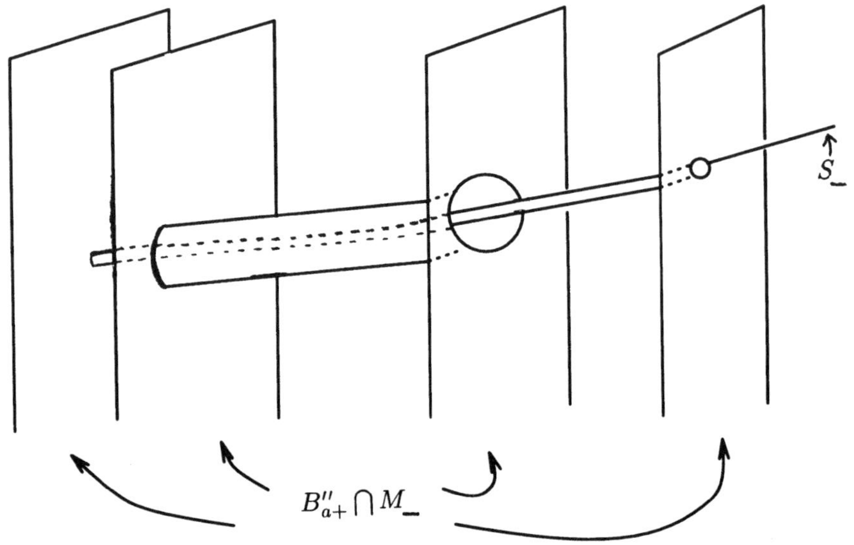

(5.15)

One can make the same tubing constructions simultaneously at all of the points in $\{b_{i+1}, \cdots, b_r\}$. This is because tubings from distinct critical points will not interfere with each other. (They are defined by gradient flow lines so do not intersect each other in $W-$ crit(f).) Use B_{1a+} to denote the submanifold of W which results from doing these tubing surgeries near all of the points in the set $\{b_{i+1}, \cdots, b_r\}$.

Complete this tubing construction for each $a \in \text{crit}_1(f)$. The resulting set $\{B_{1a+} : a \in \text{crit}_2(f)\}$ may contain pairs which mutually intersect, but notice that these intersections will occur only in small ball-neighborhoods of the points of $\text{crit}_2(f)$.

e) Normal framings for B_{1a+}

The purpose of this subsection is to point out that the submanifolds $\{B_{1a+}\}$ all come with a canonical normal bundle framing. Indeed, B_{a+} is oriented and the tubing constructions do not destroy the orientability of B_{1a+} and it is left to the reader to check that B_{1a+} inherits a natural orientation from B_{a+}. An orientation for B_{1a+} induces one on the normal bundle of B_{1a+}. And, since $B_{1a+} \subset W$ has codimension 1, the act of orienting the normal bundle of B_{1a+} gives that bundle a trivialization.

f) The flow line γ.

There is a flow line, γ, which starts at $p_0 \in M_0$ and which ends at $p_1 \in M_1$. (See 4 of Definition 3.1.) One can assume that all B_{1a+} and B_{1b-} constructed here are disjoint from γ. Here is why: The flow γ must avoid all of the critical points of f. Since W is compact, there is an open subset of $U \subset W$ which contains $\mathrm{crit}(f)$ and is such that $\mathrm{past}(U)$ and $\mathrm{fut}(U)$ are disjoint from γ. Given such a set, one can assume without loss of generality that

$$(5.16)\qquad \begin{array}{ll} 1) & B_{1a+} \subset \mathrm{fut}(U) \text{ for all } a \in \mathrm{crit}_1(f). \\ 2) & B_{1b-} \subset \mathrm{past}(U) \text{ for all } b \in \mathrm{crit}_2(f). \end{array}$$

6 The first pass at $E_\pm$. This section will construct submanifolds $E_{1\pm} \subset Z$ which plug into (4.10), (4.11) to solve the constraints of Lemma 4.1 in the case when W is described by (3.11). (So W has the homology of S^3 and W has a good Morse function with no index 3 critical points.)

a) The submanifolds $Y_{i,j\pm}$.

As in the preceding section, fix a good Morse function f on W with no index 3 critical points, and label the index 1 and index 2 critical points of f as $\{a_1, \cdots, a_r\}$ and $\{b_1, \cdots, b_r\}$, respectively. For each $a \in \mathrm{crit}_1(f)$, construct the submanifold $B_{1a+} \subset W$ as directed in the preceding section. Also, for each $b \in \mathrm{crit}_2(f)$, construct the submanifold $B_{1b-} \subset W$ as directed in said section.

When $1 \le i \le j \le r$, set $a \equiv a_i$ and set $b \equiv b_j$. Now define

$$(6.1)\qquad \begin{array}{ll} 1) & Y_{i,j-} \equiv Z \cap (B_{1a+} \times B_{1b-}) \\ 2) & Y_{i,j+} \equiv Z \cap (B_{1b-} \times B_{1a+}) \end{array}$$

These subspaces will be used to construct the varieties $E_\pm$ of (4.10). (See also (6.3) and Definition 6.4, below.)

Lemmas 6.1 and 6.2, below describes some of the salient features of $Y_{i,j\pm}$.

LEMMA 6.1. *Let $1 \le i \le j \le r$, and set $a \equiv a_i$ and $b \equiv b_j$.*

1) *After (arbitrarily) small isotopies of B_{1a+} and B_{1b-}, the former the identity near a and the latter the identity near b, the subspaces $Y_{i,j\pm} \subset Z$ will be closed, embedded, dimension 4 submanifolds (with boundary).*
2) *$\partial Y_{i,j\pm} = \emptyset$ if $i \neq j$.*
3) *$\partial Y_{i,i\pm} \subset (S^3 \times S^3)_a \cup (S^3 \times S^3)_b$.*
4) *$\partial Y_{i,i-} \cap (S^3 \times S^3)_a$ is the disjoint union of embedded 3-spheres, each isotopic to ($S^3\times$ point). Likewise, $\partial Y_{i,i-} \cap (S^3 \times S^3)_b$ is the disjoint union of embedded 3-spheres, each isotopic to ($S^3\times$ point).*
5) *$\partial Y_{i,i+} \cap (S^3 \times S^3)_a$ is the disjoint union of embedded 3-spheres, each isotopic to (point $\times S^3$). Meanwhile, $\partial Y_{i,i+} \cap (S^3 \times S^3)_b$ is the disjoint union of embedded 3-spheres, each isotopic to (point $\times$ S^3).*
6) *All $Y_{i,j\pm}$ are orientable.*
7) *The product normal framings of B_{1a+} and B_{1b-} in W induce a framing of the normal bundles to $Y_{i,j\pm}$ in Z.*
8) *All $Y_{i,j\pm}$ have $H^2(Y_{i,j\pm}) = 0$.*

9) *All $Y_{i,j\pm}$ are disjoint from $E_{R,L}$ in (4.15).*

(Remark that Assertions 8 and 9 are not needed until one reaches the part of Theorem 2.9's proof where (2.27.3) must be verified.)

The proof of this lemma is deferred to Subsection 6c.

Since all the $Y_{i,j\pm}$ are orientable, one can consider their intersection numbers with the generators of $H_3(Z)$. These intersection numbers are computed in Lemma 6.2, below.

Lemma 6.2 uses the following notations and conventions: Let p and p' be critical points of f with the same index, and for which $f(p) > f(p')$. Consider $L_{(p,p')+} \equiv (B_{p-} \times B_{p'+}) \cap Z \in \underline{L}_+$. This is the boundary of the subset of $B_{p-} \times B_{p'+}$ where $F \leq 0$. (The latter is a manifold with boundary.) Orient $B_{p-} \times B_{p'+}$ with the product orientation and then agree henceforth to orient $L_{(p,p')+}$ with the induced orientation as the boundary of the subset where $F \leq 0$.

Consider now the 3-sphere $L_{(p',p)-} \equiv (B_{p'+} \times B_{p-}) \cap Z \in \underline{L}_-$. This is the boundary of the subset of $B_{p'+} \times B_{p-}$ where $F \geq 0$. (The latter is a manifold with boundary.) Orient $B_{p'+} \times B_{p-}$ with the product orientation and then agree henceforth to orient $\underline{L}_{(p',p)-}$ with the induced orientation as the boundary of the subset where $F \geq 0$.

LEMMA 6.2. *Add the following to the conclusions of Lemma 6.1: The submanifolds $\{Y_{i,j\pm}\}$ have transversal intersections with the 3- spheres in $\underline{L}_\pm$ and $Y_{i,j-} \cap \underline{L}_- = \emptyset$ and $Y_{i,j+} \cap \underline{L}_+ = \emptyset$, where $\underline{L}_\pm$ are given by (3.31), (3.32). Furthermore, the $\{Y_{i,j\pm}\}$ can be oriented so that*

1) *The intersection of $Y_{i,j-}$ with $L_{(p,p')+} \in \underline{L}_+$ is empty unless $p = a_i$ or $p' = b_j$. If $p = a_i$ and $p' = a_k$, then the intersection number is $-S_{j,k}$. If $p = b_k$ and $p' = b_j$, then this intersection number is $S_{k,i}$.*
2) *The intersection of $Y_{i,j+}$ with $L_{(p',p)-} \in \underline{L}_-$ is empty unless $p' = b_j$ or $p = a_i$. If $p' = b_j$ and $p = b_k$, then the intersection number is $S_{k,i}$. If $p = a_i$ and $p' = a_k$, then the intersection number is $S_{j,k}$.*
3)

$$\partial[Y_{i,i-}] = (S_{i,i})\,([S^3]_{a-} + [S^3]_{b-}), \tag{6.2}$$

$$\partial[Y_{i,i+}] = (S_{i,i})\,([S^3]_{a+} + [S^3]_{b+}).$$

Here, $S_{i,i} > 0$ is given in (3.15). (For $p = a$ or b, the classes $[S^3]_{p\pm}$ are defined subsequent to (4.19).)

The proof of this lemma is deferred to Subsection 6d, below.

b) The construction of $[E_\pm]$.

With the orientations of Lemma 6.2, the submanifolds $\{Y_{i,j\pm}\}$ of Lemma 6.1 will define homology classes in $H_4(Z, \partial Z)$ and linear combinations of these classes will produce classes $[E_\pm]$ which fit into (4.11) to solve the constraints of Lemma 4.1. To be precise here, introduce the matrix S of (3.15) and the integer valued matrix $T \equiv \det(S)\, S^{-1}$. Note that $T \equiv (T_{i,j})$ is upper triangular (when $i > j$, then $T_{i,j} = 0$) with $T_{i,i} = \det(S)/S_{i,i}$.

With T understood, introduce

$$(6.3) \qquad [E_{1-}] \equiv \sum_{i,j} T_{i,j}\, [Y_{i,j-}] \text{ and } [E_{1+}] \equiv \sum_{i,j} T_{j,i}\, [Y_{i,j+}].$$

(In (6.3), the sums are over all pairs i, j with $1 \le i \le j \le r$.) Here are the salient features of these classes:

LEMMA 6.3. *Define the classes* $[E_{1\pm}]$ *by (6.3). Then*

$$(6.4) \qquad \partial[E_{1-}] = det(S) \sum_{p\in\ crit(f)} [S^3]_{p-}.$$

$$\partial[E_{1+}] = det(S) \sum_{p\in\ crit(f)} [S^3]_{p+}.$$

Furthermore, $[E_{1\pm}]$ *have zero intersection pairing with the classes which are generated by the 3-speres in* $\underline{L}_{\pm}$ *of (3.31) and (3.32).*

It follows from this lemma that Lemma 4.1 is satisfied if the classes $[E_{\pm}]$ *in (4.11) are set equal to* $[E_{1\pm}]$ *from (6.3). In this case, (4.11)'s integer* N *must equal* det(S). *(In later constructions, it proves convenient to take* $[E_{\pm}]$ *in (4.11) to be some multiple of* $[E_{1\pm}]$ *from (6.3).)*

Proof. Consider first the properties of $[E_{1-}]$. It follows from Assertion 1 of Lemma 6.1 and Assertion 1 of Lemma 6.2 that $\partial[E_{1-}]$ obeys (6.4). This is because the boundary annihilates all terms in (6.3) save those for which $i = j$. Then, (6.4) follows from (6.2) and the fact that $T_{i,i} = \det(S)/S_{i,i}$.

According to Assertion 2 of Lemma 6.2, $[E_{1-}]$ is represented by the fundamental classes of submanifolds with empty intersection with the classes from $\underline{L}_{-}$. To study the intersection pairing between $[E_{1-}]$ and a class from $\underline{L}_{+}$, fix integers m and n with $1 \le m < n \le r$. Let $a \equiv a_m$ and let $a' \equiv a_n$. Consider the pairing between $[E_{1-}]$ and the class of $L_{(a,a')+}$. Using Assertion 3 of Lemma 6.2, one finds that this number is equal to

$$(6.5) \qquad \sum_k T_{m,k}\, S_{k,n},$$

which is zero because $m \neq n$ and T is proportional to S^{-1}.

Next, let $b \equiv b_m$ and let $b' \equiv b_n$ and consider the pairing between $[E_{1-}]$ and the class of $L_{(b,b')_+}$. Using Assertion 3 of Lemma 6.2 again, one finds that this pairing is equal to

$$(6.6) \qquad \sum k S_{m,k}\, T_{k,n},$$

which is also zero, because $m \neq n$ and $T = \det(S)\, S^{-1}$.

Thus Lemma 6.3 is proved for $[E_{1-}]$. The proof for $[E_{1+}]$ is analogous and is left to the reader. □

c) Proof of Lemma 6.1

Fix i and j such that $1 \leq i \leq j \leq r$ and let $a \equiv a_i$ and $b \equiv b_j$. For Assertion 1's proof, note that $Y_{i,j-} \cap \operatorname{int}(Z)$ will be a submanifold of $\operatorname{int}(Z)$ if F's restriction to $B_{1a+} \times B_{1b-}$ has zero as a regular value. This will follow if f's restriction to B_{1a+} has disjoint critical values from its restriction to B_{1b-}. With an arbitrarily small isotopy, of B_{1b-} near $f^{-1}(f(a))$, one can insure that $f(a)$ is not a critical value of f on B_{1b-}. Likewise, an arbitrarily small isotopy of B_{1a+} near where $f = f(b)$ will insure that $f(b)$ is not a critical value of f on B_{1a+}. With this understood, a small isotopy of B_{1a+} which is the identity near a will insure that the critical values of f on B_{1a+} are disjoint from those of f on B_{1b-}.

Argue as follows to prove that $Y_{i,i-}$ is closed: The closure of B_{1b-} in W adds only the descending disks from $\{a_k\}_{k\geq j}$. However, $f(B_{1a+}) \geq f(a_i) > f(\{a_k\}_{k>i})$ (see Assertion 2 of Proposition 3.2). Therefore, where $(\pi_L^* f) \leq 3/8$, the closure of $(B_{1a+} \times B_{1b-}) \cap Z$ adds nothing unless $i = j$, and then, only the point (a,a) is added.

Likewise, B_{1a+} is not closed in W, but its closure adds only ascending disks from $\{b_k\}_{k\leq i}$. (By construction the closure of B_{1a+} misses $\{b_k\}_{k>i}$.) However, $f(B_{1b-}) \leq f(b_j) < f(\{b_k\}_{k<j})$ because of Assertion 2 in Proposition 3.2. Therefore, where $(\pi_L^* f) \geq 3/8$, the closure of $(B_{1a+} \times B_{1b-}) \cap Z$ adds nothing except when $i = j$, in which case only the point (b,b) is added.

The preceding proves that $(B_{1a+} \times B_{1b-}) \cap Z$ is closed.

A similar argument proves that $Y_{i,j+}$ is closed.

Note that the preceding argument proves Assertions 2 and 3 also.

To prove Assertion 4, consider first the neighborhood U_a of a in W as described by the Morse coordinates (3.2). The submanifold B_{1b-} intersects this ball in at least $S_{i,i}$ components; and a typical component, say V, has the following form: There is a unit vector v with coordinates $(0, v_2, v_3, v_4)$ and

$$V = \{(x_1, t v_1, t v_2, t v_3) : x_1 \in \mathbb{R} \text{ and } t > 0.\}. \tag{6.7}$$

(The unit vectors (i.e. v) are distinct for distinct components of $B_{1b-} \cap U_a$.)

Equation (3.6) describes B_{1a+} near a since near a, it is identical to B_{a+}.

Consider next the neighborhood $U_a \times U_a$ of (a,a) in $W \times W$, and use the coordinates of (3.25). One sees that near (a,a), each component of $B_{1a+} \times B_{1b-}$ has the form $B_{1a+} \times V$, where $V \subset B_{1b-}$ is given by (6.7). Thus, the intersection $B_{1a+} \times V$ with Z near (a,a) is given by the set of points $((x_1, x_2, x_3, x_4), (y_1, y_2, y_3, y_4)) \in \mathbb{R}^4 \times \mathbb{R}^4$ where

(6.8)

1) $x_1 = 0$

2) $(y_2, y_3, y_4) = t\,v$ for $t > 0$,
3) $t^2 = y_1^2 + x_2^2 + x_3^2 + x_4^2$.

Note that this set intersects $(S^3 \times S^3)_a \subset \partial Z$ as $S^3 \times p_v$, where

$$p_v = (0, r\,v_2, r\,v_3, r\,v_4). \tag{6.9}$$

Equations (6.7) and (6.8) establish the first part of Assertion 3 concerning the intersection of $Y_{i,j-}$ with $(S^3 \times S^3)_a$.

An analogous argument shows that the intersection of $Y_{i,j-}$ with $(S^3 \times S^3)_b$ has the following form: The coordinate chart U_b describes a neighborhood of b in W. A component, V, of the intersection of B_{1a+} with U_b is given as

$$\{(x_1, x_2, x_3, x_4) \in \mathbb{R}^4 : (x_1, x_2) = t\,v, \text{ for } t > 0 \text{ and } v \in \mathbb{R}^2 \text{ with } |\,v\,| = 1\}. \tag{6.10}$$

Use $U_b \times U_b$ to describe a neighborhood of (b, b) in W. The intersection of $B_{1a+} \times B_{1b-}$ with this neighborhood will be a union of components, each of the form $V \times B_{1b-}$ with V as above. With this understood, $V \times B_{1b-}$ intersects Z as the set of points in $\mathbb{R}^4 \times \mathbb{R}^4$ of the form $((x_1, x_2, x_3, x_4), (y_1, y_2, y_3, y_4))$ where

(6.11)

1) $(x_1, x_2) = t\,v$ for $t > 0$ and $|\,v\,| = 1$
2) $y_3 = y_4 = 0$
3) $t^2 = y_1^2 + y_2^2 + x_3^2 + x_4^2$.

The preceding equation demonstrates that $V \times B_{1b-}$ intersects $(S^3 \times S^3)_b \subset \partial Z$ as $S^3 \times p_v$, where $p_v = (r\,v_1, r\,v_2, 0, 0)$. The proof of Assertion 5 of the Lemma 6.1 follows essentially the same arguments which prove Assertion 4. The details for Assertion 5's proof are omitted.

Consider now the proof of Assertion 6: Both B_{1a+} and B_{1b-} are orientable (as described in the previous section), and so their product is orientable. Then, the restriction of dF to the product trivializes the normal bundle of $Y_{i,j-}$ in $B_{1a+} \times B_{1b-}$ and similarly that of $Y_{i,j+}$ in $B_{1b-} \times B_{1a+}$.

To prove Assertion 7, remark that both B_{1a+} and B_{1b-} were constructed with canonically trivial normal bundles. Thus, their product has a canonical (up to homotopy) trivialization of its normal bundle in $W \times W$. With this understood, remember that Z is cut out of $W \times W$ as part $F^{-1}(0)$, while $Y_{i,j-}$ is cut out of $B_{1a+} \times B_{1b-}$ as part of $F^{-1}(0)$, so the trivialization of $(B_{1a+} \times B_{1b-})$'s normal bundle in $W \times W$ defines, upon restriction to $F^{-1}(0)$, a trivialization of the normal bundle to $Y_{i,j-}$ in Z.

Once again, the argument for $Y_{i,j+}$ is analogous and omitted.

To prove Assertion 8, first remember that B_{1a+} and B_{1b-} are constructed from B_{a+} and B_{b-}, respectively by surgery. The surgery on B_{b-} occurs near where $f = 1/4$, while the surgery on B_{1a+} occurs near $f = 1/2$. This implies that $Y_{i,j-}$ can be seen as the result of a surgery on the 4-sphere which is the intersection of the descending disk from F's index 5 critical point (a,b) with $F^{-1}(1/8)$. The surgery is on embedded $S^0 \times B^4$'s in said 4-sphere. The number of these surgeries is the combined total of the surgeries which make B_{1a+} from B_{a+} and B_{1b-} from B_{b-}. Each such surgery increases the rank of $H_1(\cdot; Z)$ by one, but leaves $H_2(\cdot; Z) = 0$.

Assertion 9 follows from (4.15) and (5.16). □

d) Proof of Lemma 6.2.

Consider first that the 3-spheres in $\underline{L}_{\pm}$ do not come near the critical points (p,p) of F. This follows from Proposition 3.2. Therefore, an (arbitrarily) small isotopy of B_{1a+} or of B_{1b-} will result in transversal intersections between $Y_{i,j\pm}$ and any of the spheres in $\underline{L}_{\pm}$.

Remark next that the intersection of $Y_{i,j-}$ with some $L_{(p',p)-}$ is non-empty only if $B_{1a+} \cap B_{p'+} \neq \emptyset$ and also $B_{1b-} \cap B_{p-} \neq \emptyset$. The former is empty if p and p' have index 2, while the latter are empty if p and p' have index 1.

To prove Assertion 1, one should consider orienting $Y_{i,j-}$ as follows: Orient $B_{1a+} \times B_{1b-}$ with its product orientation. Then, note that $Y_{i,j-}$ is a codimension zero part of the boundary of the subset of $B_{1a+} \times B_{1b-}$ where $F \geq 0$. Give $Y_{i,j-}$ the induced boundary orientation. Use o to denote said orientation. With the orientation o, the intersection number between $Y_{i,j-}$ and some $L_{(p,p')+} \in \underline{L}_+$ is equal to the coefficient in front of (p,p') in the expression for the $\partial(a,b)$ in the complex C^F of Lemma 3.5. (Note that $B_{1a+} \times B_{1b-}$ is homologous to the descending 5-disk from (a,b).) The computation of this coefficient is straightforward and leads to Assertion 1. (The fact that the intersection in question is empty unless $p = a_i$ or $p' = b_j$ follows from the fact that when a and a' are index 1 critical points of f, then $B_{1a+} \cap B_{a'-} = \emptyset$ unless $a = a'$. Likewise, when b and b' are index 2 critical points of f, then $B_{1b-} \cap B_{b'+} = \emptyset$ unless $b = b'$.)

The proof of Assertion 2 is analogous. Here, the orientation o for $Y_{i,j+}$ is defined by considering $Y_{i,j+}$ as a codimension zero part of the boundary of the subset of $B_{1b-} \times B_{1a+}$ where $F \leq 0$.

Consider now the proof of Assertion 3. There is a proof along the lines of the proof of Assertion 1, but a direct proof is had by the following argument: Let $a \equiv a_i$ and $b \equiv b_i$. An intersection point, q, of $(B_{a+} \cap M_{3/8})$ with $(B_{b-} \cap M_{3/8})$ corresponds to one boundary component of $(B_{a+} \times B_{b-}) \cap Z$ in $(S^3 \times S^3)_a$ and, likewise, to one boundary component in $(S^3 \times S^3)_b$. (And vice-versa.) The orientation of these boundary components relative to the given orientations of $(S^3)_{a-}$ and to $(S^3)_{b-}$ will be found equal, but opposite the local intersection number at q of $(B_{a+} \cap M_{3/8})$ with $(B_{b-} \cap M_{3/8})$ in $M_{3/8}$.

Step 1: This step compares the local intersection number at q with the orientation of the corresponding boundary component of $(B_{a+} \cap B_{b-}) \cap Z$ in $(S^3 \times S^3)_a$. To begin, take the Morse coordinates near a of (3.2) so that $B_{a+} = \{x \equiv (x_1, x_2, x_3, x_4) : x_1 = 0\}$. Orient B_{a+} by $\partial_2\partial_3\partial_4 \in \Lambda^3 TB_{a+}$. A neighborhood, $U \subset M_{3/8}$ of $M_{3/8}$'s intersection with B_{a+} is isotopic to $\{x : -x_1^2 + x_2^2 + x_3^2 + x_4^2 = R^2\}$ for some $R > r$. This U is oriented at $q \in (0, R, 0, 0) \in U$ by $-\partial_1\partial_3\partial_4$.

Now q lies in B_{a+}, but suppose that q is also a point of intersection B_{b-} and B_{a+}. Suppose further that the local intersection number at q of $(B_{a+} \cap M_{3/8})$ with $(B_{b-} \cap M_{3/8})$ is equal to $\epsilon = \pm 1$. Without loss of generality, B_{b-} can be assumed to intersect a neighborhood of U as $\{x : x_2 > 0 \text{ and } x_3 = x_4 = 0\}$. To obtain the correct intersection number at q, it is necessary to orient B_{b-} using $-\epsilon\, \partial_1 \partial_2$. (Note that $df = dx_2$ at q, so $(B_{a+} \cap M_{3/8})$ is oriented near q by $\partial_3 \partial_4$, while $(B_{b-} \cap M_{3/8})$ at q is oriented by $-\epsilon\, \partial_1$. Then, their intersection at q has local orientation $-\epsilon \partial_1 \partial_3 \partial_4$ which agrees or disagrees with the orientation $-\partial_1 \partial_3 \partial_4$ of $M_{3/8}$ depending on whether $\epsilon = \pm 1$.)

With the preceding understood, it follows that $B_{a+} \times B_{b-}$ is oriented near (q, q) by $-\epsilon \partial_2 \partial_3 \partial_4 \partial_1' \partial_2'$; here the prime indicates a vector field from the second factor of W in $W \times W$, while the absence of a prime indicates a vector field from the first factor of W. Now, at the point (q, q), the 1-form $dF = dx_2' - dx_2$; this implies that $\epsilon\, (\partial_2 + \partial_2') \partial_3 \partial_4 \partial_1'$ orients $(B_{a+} \times B_{b-}) \cap Z$ in $(S^3 \times S^3)_a$ near (q, q). Finally, the boundary of the component of $(B_{a+} \times B_{b-}) \cap Z$ in $(S^3 \times S^3)_a$ which corresponds to the point q is oriented by contracting this last frame with $-dx_2' - dx_2$ which yields $-\epsilon \partial_3 \partial_4 \partial_1'$. This disagrees with the orientation of $(S^3)_{a-}$ when $\epsilon = +1$ and it agrees with said orientation when $\epsilon = -1$.

Step 2: This step compares the local intersection number at q with the orientation of the corresponding component in $(S^3 \times S^3)_b$ of the boundary of $(B_{a+} \times B_{b-}) \cap Z$. To begin, take the Morse coordinates of (3.2) around b. Then, $B_{b-} = \{x : x_3 = x_4 = 0\}$. Orient B_{b-} by $\partial_1 \partial_2$. A neighborhood, U, of the point q in $M_{3/8}$ is isotopic to the subset given in Morse coordinates as $\{x : -x_1^2 - x_2^2 + x_3^2 + x_4^2 = -R^2\}$. Here $R >> r$ and q is the point $(0, R, 0, 0)$. The orientation of $M_{3/8}$ is determined from the fact that df at q is $-dx_2$. Thus, $\partial_1 \partial_3 \partial_4$ orients $M_{3/8}$.

Meanwhile, a neighborhood of q in B_{a+} can be assumed given by the set $\{x : x_1 = 0 \text{ and} x_2 > 0\}$. This part of B_{a+} is oriented by $\epsilon\, \partial_2 \partial_3 \partial_4$. (Thus, $(B_{a+} \cap M_{3/8})$ is oriented at q by $-\epsilon\, \partial_3 \partial_4$ while $-\partial_1$ orients $(B_{b-} \cap M_{3/8})$ at q. Their intersection gives $\epsilon \partial_1 \partial_3 \partial_4$ for the orientation of $M_{3/8}$ as it should.) The orientation for $B_{a+} \times B_{b-}$ near (q, q) is thus given by $\epsilon\, \partial_2 \partial_3 \partial_4 \partial_1' \partial_2'$.

The 1-form dF at (q, q) is given by $-dx_2' + dx_2$, and this means that $\epsilon\, (\partial_2 + \partial_2') \partial_3 \partial_4 \partial_1'$ orients the part near (q, q) of $(B_{a+} \cap Bb-) \cap Z$. With this last point understood, it follows that $-\epsilon\, \partial_3 \partial_4 \partial_1'$ orients the part of $\partial((B_{a+} \cap B_{b-}) \cap Z)$ in $(S^3 \times S^3)_b$ which corresponds to q. Note that this orientation disagrees with the given orientation of $(S^3)_{b-}$ when $\epsilon = +1$, but it agrees when $\epsilon = -1$. In particular, note that this anti-correlation with the local intersection number at q is the same as that for components of $\partial((B_{a+} \times B_{b-}) \cap Z)$ in $(S^3 \times S^3)_a$.

It follows from the preceding calculations that (6.2) holds if the orientation $-o$ is used on $\{Y_{i,j-}\}$.

A similar argument shows that the second line of (6.2) is correct if the $\{Y_{i,j+}\}$ are also oriented with $-o$. The details here are left to the reader.

e) Push-offs.

The next task is to provide a representative of each $[E_{1\pm}]$ as the fundamental class of a smoothly embedded submanifold (with boundary), $E_{1\pm} \subset Z$. Here,

$\partial E_{1\pm} \subset \partial Z$.

The construction of $E_{1\pm}$ requires the introduction of a procedure, called *push-off*, for making copies of embedded submanifolds. The following digression described the push-off procedure.

Start the digression by considering the following abstract situation: Let X be a compact manifold with boundary, and let $Y \subset X$ be a compact submanifold with boundary, which intersects ∂X transversally as ∂Y. Let $N_Y \to Y$ denote the normal bundle to Y in X. (Note that N_Y restricts to ∂Y as the latter's normal bundle in ∂X.) Suppose that N_Y admits a section, s, which never vanishes. Let $e : N_Y \to X$ be an exponential map which maps $N_Y \mid_{\partial Y}$ to ∂X. (See (2.13).)

Together, e and s and a real number $\lambda \neq 0$ define a map,

$$e(\lambda\, s(\cdot)) : Y \to X, \tag{6.12}$$

whose image is disjoint from Y. If λ has small absolute value, then the image, Y', of (6.12) will be an embedding of Y into X, where $\partial Y'$ is an embedding of ∂Y into ∂X. This image, Y', is called a *push-off* of Y. Here are some properties of the push-off:

(6.13)

($\cdot$) Y' is disjoint from Y, but smoothly isotopic to Y. (The obvious isotopy is to consider $\lambda \to 0$ in (6.12). This isotopy will isotope $\partial Y'$ to ∂Y in ∂X.)

($\cdot$) If Y comes with some apriori orientation, then Y' has a canonically induced orientation which makes $[Y] = [Y']$ in $H^*(X, \partial X)$.

($\cdot$) Let $V \subset X$ be a submanifold which intersects Y transversally. Then V will also intersect Y' transversally if λ in 6.12) has sufficiently small absolute value.

($\cdot$) Let $V \subset X$ be a closed submanifold with empty intersection with Y. Then $V \cap Y' = \emptyset$ if λ in (6.12) has sufficiently small absolute value.

($\cdot$) If Y has a framed normal bundle, then this framing naturally induces a framing of the normal bundle to Y'.

Note also that one can define any finite number of disjoint push-offs of Y by using different values of λ in (6.12). Alternately, one can use different sections $\{s_1, \cdot\}$ of N_Y with fixed λ as long as the $\{s_j\}$ are no-where vanishing and no two are anywhere equal.

In the sequel, assume the following conventions:

(6.14)

($\cdot$) Any pair of distinct push-offs of the same submanifold are mutually disjoint.

($\cdot$) Suppose that the normal bundle N_Y is trivial, and that an apriori trivialization has been specified. (Call it the *canonical* trivialization.) In this case, agree that all push-offs of Y will be defined by using for s in (6.12) a constant linear combination of basis vectors for the canonical trivialization.

($\cdot$) When the precise choice of exponential map or parameter λ or section s in (6.12) are irrelevent to subsequent discussions, their presence will not be explicitly noted. (But, keep the preceding convention on the section s when the normal bundle to Y has been trivialized.)

(The last two conventions in (6.14) allow one to speak of a push-off of Y with-out cluttering the conversation with a list of irrelevent (but necessary) choices.)

End the digression.

f) $E_{1\pm}$ as submanifolds.

The purpose of this last subsection is to define $[E_{1\pm}]$ of (6.3) as the fundamental class of a closed, embedded submanifold (with boundary) of Z. Consider first $[E_{1-}]$.

This $[E_{1-}]$ is a sum of fundamental classes of the $\{Y_{i,j-}\}$. The first observation is that each $Y_{i,j-} \cap Y_{n,m-} = \emptyset$ unless $m = j$. This is because the various $\{B_{1b-}\}_{b\in \text{ crit}_2(f)}$ are mutually disjoint. There may be non-empty intersections between $Y_{i,j-}$ and $Y_{k,j-}$ when $i \neq k$. These can be avoided if the following convention is used: Remember that each B_{1b-} has trivial normal bundle in W. And, remember that said normal bundle has a canonical trivialization up to homotopy. For each $b \in \text{ crit}_2(f)$, choose a trivialization of the normal bundle of B_{1b-} which is in the canonical homotopy class. Then, fix j and when $i < j$, define $Y_{i,j-}$ as in (6.1) but where B_{1b-} is replaced by a push-off copy. For each such i, use a *different* push-off copy. This will make $Y_{i,j-}$ disjoint form $Y_{k,j-}$ when $i \neq k$.

Now, generalize this process of separating the $\{Y_{i,j-}\}$ as follows: Reintroduce the matrix $T = (T_{i,j})$ which appears in (6.3). For each pair (i,j) with $1 \leq i \leq j \leq r$, let $a \equiv a_i$ and $b \equiv b_j$. Take $\mid T_{i,j} \mid$ distinct push-off copies of B_{1b-} and use these in 6.1) to define $\mid T_{i,j} \mid$ distinct push-off copies of $Y_{i,j-}$. It will prove convenient to require that all such push-off copies are disjoint from the flow line γ of Part 4 in Definition 3.1. (One can make all such copies in past(U), where $U \subset W$is an open subset which contains $\text{crit}(f)$ and whose past and future are disjoint from γ . See (5.16).)

Since the various $\{B_{1b-}\}_{b\in \text{ crit}_2(f)}$ are mutually disjoint, one can make all of these push-offs so that each copy of $Y_{i,j-}$ is disjoint from each copy of $Y_{k,l-}$ when $(i,j) \neq (k,l)$.

With the preceding understood, consider:

PROPOSITION 6.4. *Define $E_{1-} \subset Z$ as an oriented submanifold of Z (with boundary) as the union over all pairs (i,j) (with $1 \leq i \leq j \leq r$)) of the $\mid T_{i,j} \mid$ push-offs of $Y_{i,j-}$ as defined above. Take these copies with the following orientations: Orient the copies of $Y_{i,j-}$ as in Lemma 6.2 if $T_{i,j} > 0$; and oriented them in reverse if $T_{i,j} < 0$. Define $E_{1+} \subset Z$ as a submanifold to be the image of E_{1-} under the switch map on $W \times W$ which sends (x,y) to (y,x).*

(This map preserves Z.) Then these oriented submanifolds can be assumed to have the following properties:

1) *The fundamental classes of $E_{1\pm}$ obey (6.3).*
2) *$E_{1\pm}$ intersect ∂Z transversely in $\partial E_{1\pm}$.*
3) *$E_{1\pm}$ have empty intersection with $M_0 \times M_0$ and $M_1 \times M_1$.*
4) *$E_{1\pm}$ have trivial normal bundles in Z, and said normal bundles have canonical trivializations up to homotopy.*
5) *$H^2(E_{1\pm}; \mathbb{Z}) = 0$.*
6) *$E_{1\pm}$ have empty intersection with $E_{R,L}$ of (4.15).*

The proof of this proposition is left to the reader.

7 The second pass at $E_\pm$. Assume here that W obeys the constraints of (3.11). If $E_\pm$ in (4.10) is $E_{1\pm}$ of Proposition 6.4, then the resulting Σ_Z satisfies Steps 1 and 2 plus Part 1 of Step 3 in Section 2k's outline of the proof of Theorem 2.9. However, the completion of Step 3 requires modifications of $E_{1\pm}$. The problem is that $E_{1\pm}$ intersect the various $(S^3 \times S^3)_p \subset \partial Z$ too many times, and they intersect each other too many times, and they intersect Δ_Z too many times.

The change of $E_{1\pm}$ into $E_\pm$ is a multi-step process which begins in this section and ends in Section 10. Then, Section 11 constructs a 2-form ω_Z to satisfy (2.27). This section starts the process by modifying $E_{1\pm}$ to make a submanifold, $E_{2\pm}$, with simpler intersections with the $(S^3 \times S^3)_p \subset \partial Z$.

a) The submanifold E'_{1-}.

To begin the modification process, fix $i \in \{1, \cdots, r\}$ and, as usual, set $a \equiv a_i$ and $b \equiv b_i$. Make 2det(S) additional push-off copies of B_{1b-}. Make these copies so that they are disjoint from all other push-off copies of $\{B_{1b'-} : b' \in \text{crit}_2(f)\}$ which have so far been constructed. Use these 2det(S) push-off copies of B_{1b-} to make $2\det(S)$ copies of $B_{1a+} \times B_{1b-}$ and then $2\det(S)$ copies of $Y_{i,i-}$ as describe in (6.1). Orient the first det(S) of these $Y_{i,i-}$ canonically, and orient the remaining det(S) of these copies opposite to their canonical orientation. The first det(S) copies of $Y_{i,i-}$ (the ones with the canonical orientation) will be called the special $Y_{i,i-}$.

Define E'_{1-} to be the union of Proposition 6.4's E_{1-} with the (oriented) submanifold which is comprised of the union of the preceding 2det(S) copies of oriented $Y_{i,i-}$. Notice that this E'_{1-} still obeys the conclusions of Lemma 6.3 and Proposition 6.4.

b) Tubing near (a, a)

Consider now the intersection of E'_{1-} with $(S^3 \times S^3)_a$: As described in (6.8), (6.9), this intersection is given as

$$S^3 \times \Lambda'_a, \tag{7.1}$$

where $\Lambda'_a \subset S^3$ is a finite set of points. Each point in Λ'_a comes with a sign (± 1), and there are $\det(S)$ more plus signs than minus signs. This means that the set Λ'_a can be decomposed as $\Lambda_a \cup T_a$, where the points in T_a can be paired so that the signs in each pair add to zero.

It proves useful to take some care in defining the set Λ_a. Here is how: To begin, note that the intersection of $B_{b-} \cap f^{-1}(3/8)$ with $B_{a+} \cap f^{-1}(3/8)$ is transversal, and has intersection number $S_{i,i}$. Pick a point in this intersection where the local intersection number is positive. Such a point lies on a gradient flow line, $\mu(\equiv \mu_i)$ which starts at a and ends at b. The intersection of $\mu \times \mu$ with $(S^3 \times S^3)_a$ is a point, $p_a \times p_a$, where $p_a \in S^3$. With p_a singled out, note that the intersection of any push-off copy of $Y_{i,i}$ with $(S^3 \times S^3)_a$ contains a unique $(S^3 \times p'_a)$ where p'_a is the push-off of p_a. (There is a canonical isotopy between the push-off copy and the original (shrink λ to zero in (6.12), and under this isotopy, p'_a moves to p_a.) In particular, each of the $\det(S)$ special copies of $Y_{i,i}$ defines such a point p'_a, and these $\det(S)$ points are the points that comprise Λ_a.

As remarked above, the points in T_a can be paired up so the signs of each pair sum to zero,

(7.2) $$T_a = U_\alpha \{e_\alpha, e'_\alpha\}$$

For each pair $\{e_\alpha, e'_\alpha\}$ in (7.2), E_1- induces orientations on $S^3 \times e_\alpha$ and $S^3 \times e'_\alpha$, and these orientations are opposite.

Now, for each pair, $\{e, e'\}$ on (7.2)'s right side, embed $[0,1]$ into S^3 to have boundary $\{e, e'\}$. (Do this in such a way that the embedded intervals from distinct pairs do not intersect.) The associated $S^3 \times [0,1]$ has boundary $(S^3 \times e) \cup (S^3 \times e')$ and the orientations here agree with those which are induced by E_{1-}. Hence, $S^3 \times [0,1] \subset (S^3 \times S^3)_a$ can be surgered to E_{1-} along their common boundaries, $(S^3 \times e) \cup (S^3 \times e')$. The result is a topological embedding in Z of a smooth, oriented manifold with boundary, where the boundary embeds (smoothly) in ∂Z. (The embedding has "corners", these being the components of $S^3 \times \{e_\alpha, e'_\alpha\}$ where the surgery took place.) The point is that this new manifold has two less boundary components then E'_{1-}. Here is a picture:

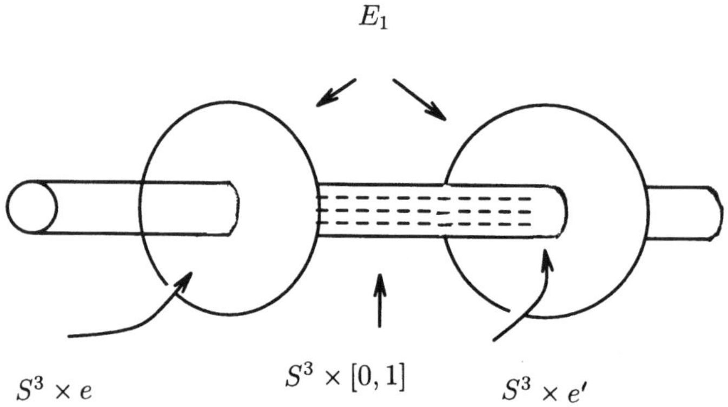

(7.3)

Make the preceding construction for each pair on the right side of (7.2). The result is a topological embedding of a surgered E'_{1-}. (The "corners" of the embedding are the components of $S^3 \times T_a$.) The embedding of this surgered E'_{1-} intersects $(S^3 \times S^3)_a$ in $S^3 \times \Lambda_a$ (where it is the same as E'_{1-}) and also in a copy of $S^3 \times [0,1]$ for each pair on the right in (7.2). (The copies of $S^3 \times [0,1]$ for distinct pairs will not intersect if one takes care to insure that the embedded $[0,1]$'s from different pairs do not intersect.)

Now note that the copies of $S^3 \times [0,1]$ can be isotoped normally off $(S^3 \times S^3)_a$ (push radially outward in the coordinates of Lemma 3.6 so that the resulting embedding of the surgered E'_{1-} intersects $(S^3 \times S^3)_a$ in $S^3 \times \Lambda_a$. And, note that all of the "corners" in the resulting embedding can be readily smoothed so that the result is an embedded submanifold of Z. The following diagram illustrates:

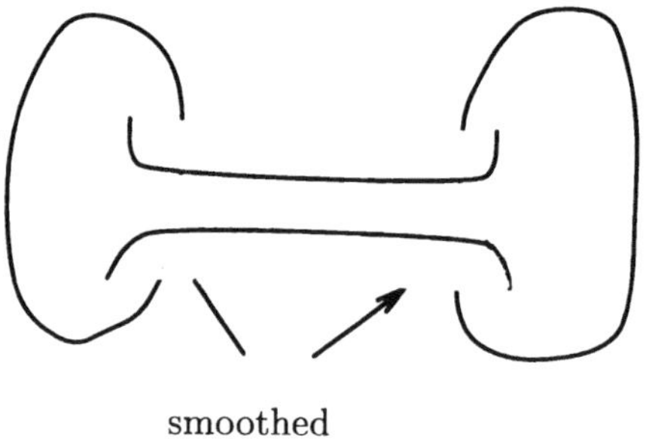

(7.4)

The preceding construction can be done at all $a \in \mathrm{crit}_1(f)$. The result is a submanifold, $E''_{1-} \subset W$. Note that E''_{1-} has a minimal number of intersections with any $(S^3 \times S^3)_a \subset \partial Z$ as its intersection is equal $S^3 \times \Lambda_a$, a set of $\det(S)$ push-off copies of $S^3 \times p_a$. Note also that E''_{1-} agrees with E'_{1-} away from $\{(S^3 \times S^3)_a\}_{a \in \mathrm{crit}_1(f)}$.

c) Tubing near (b, b)

Let $b \equiv b_i \in \operatorname{crit}_1(f)$. Consider now the intersection of E''_{1-} with $(S^3 \times S^3)_b$: As described in (6.8), 6.9), this intersection is given as

$$S^3 \times \Lambda'_b, \tag{7.5}$$

where $\Lambda'_b \subset S^3$ is a finite set of points. Each point in Λ'_b comes with a sign (± 1), and there are $\det(S)$ more plus signs than minus signs. This means that the set Λ'_b can be decomposed as $\Lambda_b \cup T_b$, where the points in T_b can be paired so that the signs in each pair add to zero.

It proves useful to take some care in defining the set Λ_b. Here is how: The flow line $\mu(\equiv \mu_i)$ which starts at a and ends at b. The intersection of $\mu \times \mu$ with $(S^3 \times S^3)_b$ is a point, $p_b \times p_b$, where $p_b \in S^3$. With p_b singled out, note that the intersection of any push-off copy of $Y_{i,i}$ with $(S^3 \times S^3)_b$ contains a unique $(S^3 \times p'_b)$ where p'_b is the push-off of p_b. (There is a canonical isotopy between the push- off copy and the original (shrink λ to zero in (6.12), and under this isotopy, p'_b moves to p_b.) In particular, each of the $\det(S)$ special copies of $Y_{i,i}$ defines such a point p'_b, and these $\det(S)$ points are the points that comprise Λ_b.

As remarked above, the points in T_b can be paired up so the signs of each pair sum to zero,

$$T_b = U_\alpha \{e_\alpha, e'_\alpha\} \tag{7.6}$$

For each pair $\{e_\alpha, e'_\alpha\}$ in (7.5), E''_{1-} induces orientations on $S^3 \times e_\alpha$ and $S^3 \times e'_\alpha$, and these orientations are opposite. With this understood, one can repeat the tubing construction as described in the previous subsection (see (7.4), (7.5)) to surger E''_{1-} near (b, b) and then isotope the result to obtain an embedded submanifold of Z which intersects $(S^3 \times S^3)_b$ in $S^3 \times \Lambda_b$. Furthermore, this last construction can be done simultaneously near all (b, b) for $b \in \operatorname{crit}_2(f)$. Use E_{2-} to denote the resulting submanifold of Z.

d) The intersection of $E_{2\pm}$ **and** $E_{R,L}$**.**

The next four subsections describe various properties of $E_{2\pm}$. The purpose of this subsection is to prove

LEMMA 7.1. *The submanifolds $E_{2\pm} \subset Z$ can be constructed as described above so that they do not intersect $E_{L,R}$ of (4.15).*

Proof. Let $U \subset W$ be an open neighborhood of $\operatorname{crit}_1(f)$ and let $U' \subset W$ be an open neighborhood of $\operatorname{crit}_2(f)$. Then $E_{2\pm}$ can be made (as described above) so that they are supported in Z's intersection with $(\operatorname{fut}(U) \times \operatorname{past}(U'))$. The latter set is disjoint from $E_{R,L}$ if U and U' are not too big; this is because the flow line γ misses f's critical points. □

e) The intersection of $E_{2\pm}$ with Δ_Z:

Fix $i \in \{1, \cdots, r\}$ and let $a \equiv a_i$ and $b \equiv b_i$. By construction, E_{2-} intersects $(S^3 \times S^3)_a$ in $S^3 \times \Lambda_a$. It intersects $(S^3 \times S^3)_b$ in $S^3 \times \Lambda_b$. Here, Λ_a and Λ_b are sets of $\det(S)$ points.

Now, there is a natural way to pair the points in Λ_a with those in Λ_b and here it is: When $p \in \Lambda_a$ and $p' \in \Lambda_b$ are partners, then (p,p) and (p',p') are the endpoints of a transversal component of $E_{2-} \cap \Delta_z$ which is an embedding of $[0,1]$.

Such a pairing exists for the following reasons: If $p \in \Lambda_a$, then $S^3 \times p$ is a component of the intersection of a push-off copy of $Y_{i,i-}$ with $(S^3 \times S^3)_a$. By design, there exists a unique $p' \equiv p'(p) \in \Lambda_b$ for which $S^3 \times p'$ is a component of the intersection of *the same* push- off copy with $(S^3 \times S^3)_b$. This is another definition of the pairing between Λ_a and Λ_b.

To finish the story, remark that the afore-mentioned push-off copy of $Y_{i,i-}$ is $(B_{1a+} \times B'_{1b-}) \cap Z$, where B'_{1b-} is a push-off copy of B_{1b-}. And, both p and p' lie on a push-off copy, $\mu' \subset B_{1a+} \cap B'_{1b-}$, of a chosen flow line, $\mu (\equiv \mu_i)$, which starts at a and ends at b. Finally, $(\mu' \times \mu')$ intersects Z transversally in Δ_Z and $(\mu' \times \mu') \cap Z$ is an embedded interval in Δ_Z and a transversal component of $E_{2-} \cap \Delta_Z$.)

With the preceding understood, one sees that

$$E_{2-} \cap \Delta_Z = (\cup_{i=1}^{r} \Gamma_i) \cup C, \tag{7.7}$$

where Γ_i is the union of $\det(S)$ push-offs (in Δ_Z) of $(\mu_i \times \mu_i) \cap \Delta_Z$, and where $C \subset \operatorname{int}(\Delta_Z)$ is compact. Infact, after an (arbitrarilly small) isotopy of the push-offs of the $\{B_{1b-} : b \in \operatorname{crit}_2(f)\}$ (with support away from $\operatorname{crit}(f)$), one can arrange for the intersection in (7.7) to be transversal. In this case, C is a disjoint union of embedded circles in $\operatorname{int}(\Delta_Z)$.

f) Normal framings.

Consider now the normal bundle to E_{2-}. Of particular interest in subsequent sections is the fact (see Lemma 7.2, below) that E_{2-} has trivial normal bundle. Also of interest is the behavior of a framing of this normal bundle on ∂E_{2-} and along the components of $\{\Gamma_i\}$ from (7.7).

Two digressions are required before Lemma 7.2: The first digression defines the notion of a *product* framing of the normal bundle of a submanifold in Z: This is a framing of the normal bundle with the property that each basis vector is annihilated by the differential of either π_L or π_R. The same definition works to define the product framing of a submanifold of $W \times W$.

A second digression is required to set the stage for a discussion of the normal framing near ∂E_{2-} and $\{\Gamma_i\}$. To start, consider $i \in \{1, \cdots, r\}$. As usual, let $a \equiv a_i$ and $b \equiv b_i$. Let $\mu \subset \Gamma_i$ be a component and define p, p' by requiring $(p,p) \equiv \mu \cap (S^3 \times S^3)_a$ and $(p',p') \equiv \mu \cap (S^3 \times S^3)_b$. Associate to μ the subset of E_{2-}

$$(S^3 \times p) \cup (S^3 \times p') \cup \mu. \tag{7.8}$$

Note the following: Let $\mu' \subset \Gamma_i$ be any other component. Then, E_{2-} near the μ' version of (7.8) is naturally defined as a push-off of E_{2-} near the μ version of (7.8). (Near the μ-version of (7.8), E_{2-} is a push-off copy of an open neighborhood of $(B_{1a+} \times B_{1b-}) \cap Z$. And, near the μ'-version, E_{2-} is a different push-off copy of the same open neighborhood. Infact, each of these push-off copies is constructed as $B_{1a+}\times$ (push-off copy of B_{1b-}).

These last observations give a natural method of comparing a given normal framing of E_{2-} along the μ and μ' versions of (7.8). See (6.13).

End the second digression.

LEMMA 7.2. *The submanifold E_{2-} has trivial normal bundle in Z. Furthermore, the normal bundle to E_{2-} has a framing with the following properties: Let $i \in \{1, \cdots, r\}$ and let $\mu \in \Gamma_i$. Then the frame is a product frame along (7.8) and it restricts as a constant frame along $S^3 \times p$ and $S^3 \times p'$. Furthermore, let $\mu' \subset \Gamma_i$ be a different component. Then the push-off which identifies E_{2-} near the μ and μ' versions of (7.8) will identify the restriction of the frame to the μ and μ' versions of (7.8).*

Proof. Because E_{2-} is constructed by surgering E_{1-} and the latter is a union of (6.1)'s $\{Y_{i,j-}\}$, the proof starts with a description of the normal bundles to (6.1)'s $\{Y_{i,j\pm}\}$. To begin, consider $i, j \in \{1, \cdots, r\}$ such that $i \leq j$. Let $a \equiv a_i$ and $b \equiv b_j$. Then $B_{1a+} \times B_{1b-} \subset W \times W$ has trivial normal bundle with a natural product framing. This implies that $Y_{i,j-}$ in (6.1) has a natural product framing of its normal bundle in Z. (See Lemma 6.1.)

Consider now $i = j$ and the induced normal framing of a component of $\partial Y_{i,i-}$. □

LEMMA 7.3. *Let c denote either a_i or b_i. Let $S^3 \times p$ be a component of $\partial Y_{i,i-} \cap (S^3 \times S^3)_c$. Then the product normal framing of $Y_{i,i-}$ in Z induces a product normal framing of $S^3 \times p$ in $(S^3 \times S^3)_c$ and this induced normal framing is homotopic through product framings to the constant normal framing as defined by choosing a fixed basis for $TS^3 \mid_p$ and using the projection π_R to write the normal bundle in question as $S^3 \times TS^3 \mid_p$.*

Proof. Consider first the case $c = a_i$. Here, p is described by (6.9). Think of the vector $v \equiv (v_2, v_3, v_4)$ as a point in the unit 2-sphere about the origin in the 3-plane spanned by the coordinates (y_2, y_3, y_4). With this understood, then (6.9) implies that a product normal frame to $Y_{i,i-}$ restricts to $(S^3 \times p) \subset \partial Y_{i,i-}$ to have the form $\partial_{x_1}, e_2, e_3)$, where the vectors $e_{2,3} \in TS^2 \mid_v$, and where ∂x_1 is tangent to the x_1 axis. In particular, this is a normal frame for $S^3 \times p$ in $(S^3 \times S^3)_c$. Furthermore, it is homotopic through product frames to the trivial frame because $\pi_3(SO(2)) = 1$. (In fact, the vectors $e_{2,3}$ depend only on the y_1 coordinate.)

Next, consider the case where $c = b_i$: Here, $S^3 \times p$ is described in (6.11). Think of the vector $v \equiv (v_1, v_2)$ in (6.11) as a point in the unit circle in the plane $x_3 = x_4 = 0$. Then, a product normal frame from $Y_{i,i-}$ restricts to $(S^3 \times p) \subset \partial Y_{i,i-}$ to have the form $(e_1, \partial_{y_3}, \partial_{y_4})$, where $e_1 \in TS^2 \mid_v$ and where $\partial_{y_{3,4}}$ are tangent to the y_3 and y_4 coordinate axis, respectively. This frame is evidently homotopic through product frames to the constant frame; simply homotope e_1 to a constant length vector.

End the digression. □

To complete the proof of Lemma 7.2, remember that E_{2-} was constructed from E_{1-} by taking a pair, $S^3 \times e$ and $S^3 \times e'$, in the same boundary component and gluing to them a boundary $S^3 \times I$. Here I is an embedded interval in S^3 with boundary $\{e, e'\}$. According to Lemma 7.3, the induced normal framing on any boundary component is homotopic to the constant framing; and so there is no obstruction to connecting the normal framing on $S^3 \times e$ to the normal framing on $S^3 \times e'$ over the interval $S^3 \times I$. The following diagram illustrates the procedure:

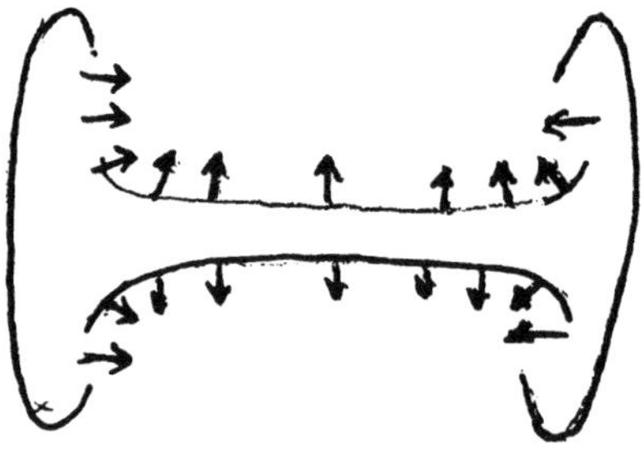

(7.9)

The aforementioned argument shows that E_{2-} has a framing for its normal bundle. But, the argument above also shows that there is a framing for the normal bundle of E_{2-} which agrees with Lemma 7.3's product framing for E_{1-} near (7.8) for any $i \in \{1, \cdots, r\}$ and any $\mu \in \Gamma_i$. (Remember that near (7.8), E_{2-} and E_{1-} agree.) This last observation plus Lemma 7.3 imply the final two statements of Lemma 7.2.

g) Further properties.

Define E_{2-} as above. Then, define $E_{2+} \subset Z$ to be the image of E_{2-} under the switch map which sends $(x, y) \subset Z$ to (y, x). The following proposition lists the salient features of $E_{2\pm}$:

PROPOSITION 7.4. *Define $E_{2\pm}$ as above. These submanifolds can be constructed and oriented so that the following hold:*

1) *The fundamental classes of $E_{2\pm}$ obey (6.3).*
2) *$E_{2\pm}$ intersect ∂Z transversely in $\partial E_{2\pm}$*

3) $E_{2\pm}$ *have empty intersection with* $M_0 \times M_0$ *and* $M_1 \times M_1$.

4) *If* $p \in$ *crit*(f), *the intersection of* E_{2-} *with* $(S^3 \times S^3)_p$ *is* $S^3 \times \Lambda_p$ *where* $\Lambda_p \subset S^3$ *is a set of det(S) points. Similarly, the intersection of* E_{2+} *with* $(S^3 \times S^3)_p$ *is* $\Lambda_p \times S^3$.

5) *The normal bundle of* E_{2-} *are described by Lemma 7.2 and the normal bundle of* E_{2+} *is described by Lemma 7.2 if (7.8) is replaced by its switched version,* $(p \times S^3) \cup (p' \times S^3) \cup \mu$.

6) $H^2(E_{2\pm}; \mathbb{Z}) = 0$.

7) $E_{2\pm}$ *have empty intersection with* $E_{R,L}$ *of (4.15).*

Proof. The only assertion which is not already proved is Assertion 6. To prove Assertion 6 for E_{2-}, remark first that E_{1-} has vanishing H^2. (See Proposition 6.4.) Then, note that E_{2-} is constructed from E_{1-} by gluing various copies of $S^3 \times I$ onto boundary $(S^3 \times S^0)$'s. This sort of surgery will decrease H^0 or increase H^1, but it can not change H^2. □

8 The third pass at $E_\pm$. The submanifolds $E_{2\pm}$ of the preceding section intersect the diagonal as described in (7.7), with $C \subset \operatorname{int}(\Delta_Z)$ being a finite union of embedded circles. The purpose of this subsection is to modify some number of like oriented, push-off copies of $E_{2\pm}$ so that the result, $E_{3\pm}$, intersects Δ_Z as in (7.7) but with $C = \emptyset$. To be precise, consider:

PROPOSITION 8.1. *There are oriented submanifolds (with boundary)* $E_{3\pm} \subset Z$ *with the following properties:*

1) E_{3+} *is the image of* E_{3-} *under the switch map on* Z *sending* (x, y) *to* (y, x).

2) *The fundamental classes* $[E_{3\pm}]$ *are equal to* $N\,[E_{1\pm}]$ *for some integer* $N \geq 1$. *Here,* $[E_{1\pm}]$ *are described by (6.3) and Lemma 6.3.*

3) $E_{3\pm}$ *have empty intersection with* $M_0 \times M_0$ *and* $M_1 \times M_1$.

4) $E_{3\pm}$ *have empty intersection with* $E_{L,R}$ *of (4.15).*

5) *If* $p \in$ *crit*(f), *then the intersection of* E_{3-} *with* $(S^3 \times S^3)_p$ *has the form* $S^3 \times \Lambda_p$, *where* Λ_p *is a set of* N *points. Similarly, the intersection of* E_{3+} *with* $(S^3 \times S^3)_p$ *is* $\Lambda_p \times S^3$.

6) $E_{3-} \cap \Delta_Z = \cup_{i=1}^r \Gamma_i$, *where* $\Gamma_i \subset \Delta_Z$ *is as follows: There is a flow line* μ_i *which starts at* a_i *and ends at* b_i. *With the canonical identification of* Δ_W *with* W *understood,* Γ_i *is the union of* N *like oriented, disjoint, push-off copies of a closed interval,* $I \subset \mu_i$. *And, each of these* N *push-offs of* I *starts in* $(\Lambda_a \times \Lambda_a) \cap \Delta_Z$ *and ends in* $(\Lambda_b \times \Lambda_b) \cap \Delta_Z$.

7) *Both* $E_{3\pm}$ *have trivial normal bundles in* Z. *The normal bundle of* E_{3-} *has a framing,* ζ, *which restricts to a product normal framing on a neighborhood of* $(\cup_{i=1}^r \Gamma_i) \cup \{S^3 \times \Lambda_p\}_{p\in \, crit(f)}$. *Furthermore, this framing* ζ *restricts to* $\{S^3 \times \Lambda_p\}_{p\in \, crit(f)}$ *as a constant framing. The normal bundle to* E_{3+} *in* Z *is described by applying the switch map to the preceding.*

8) $H^2(E_{3\pm};\mathbb{Q}) = 0$.

(Compare with Proposition 7.4.)

The rest of this section is devoted to the construction of E_{3-}. The first subsection below ($8a$) introduces some of the basic tools. Subsections $8b - 8e$ apply the tools from $8a$ to the proof of Proposition 8.1. The final subsections, $8f - 8h$, contain the proofs of three propositions that are stated in $8a$.

a) Deleting circles.

In comparing Propositions 8.1 and 7.4, one sees that the essential difference between E_{2-} and E_{3-} is that the intersection of both are described by a form of (7.7), but that $E_{3-} \cap \Delta_Z$ has no compact components. With this understood, remark that E_{3-} will be constructed from some number of like oriented, disjoint, push-off copies of E_{2-} by surgery, with the point of the surgery to eliminate the unwanted compact components of the intersection with Δ_Z. Of course, this must be done so as not to destroy any of desired properties of E_{2-}–i.e., Assertions 2-5 and 7, 8 of Proposition 8.1.

In abstraction, the problem is to remove circles which are components of the transversal intersection between two four dimensional submanifolds inside a seven dimensional submanifold. Here is the model:

(8.1)

MODEL: Let X be a connected, oriented 7-manifold, and let $A, B \subset X$ be oriented, dimension 4 submanifolds which intersect transversally. Let $O \subset X$ be an open set and let $\sigma \equiv (A \cap B) \cap O$. Suppose that σ is compact; a disjoint union of oriented, embedded circles.

Given the model, here are the problems:

PROBLEM 1: Find an oriented, dimension 4 submanifold $A' \subset X$ with the following properties:

1) $A' \cap (B \cap O) = \emptyset$.
2) $A - (A \cap O) = A' - (A' \cap O)$.
3) $[A] = [A']$ in $H_4(X, X - O)$.

PROBLEM 2: Find A' as in Problem 1 with $H^2(A';\mathbb{Q}) = 0$.

PROBLEM 3: Assuming that $A - (A \cap O)$ has trivial normal bundle, find A' solving Problems 1 and 2 with trivial normal bundle. And, given, apriori, a frame ζ for A' as normal bundle over $A - (A \cap O)$, extend ζ over A' as a normal bundle framing.

(8.2)

These three problems will arise a number of times in the subsequent two sections and will be solved under various assumptions on A, B and O.

The solution to Problem 1 begins with the following basic surgery result:

PROPOSITION 8.2. *Let X, A, B, and O be as described in (8.1) and in Problem 1 of (8.2). If the class, $[\sigma]$, of σ is zero in $H_1(B \cap O; \mathbb{Z})$, then there is a solution to Problem 1.*

Problem 2 can be solved when extra conditions are added:

PROPOSITION 8.3. *Let X, A, B, and O be as in (8.1) and Problems 1 and 2 of (8.2). Assume that*

a) $[\sigma] = 0$ *in* $H_1(B \cap O; \mathbb{Z})$.

b) The map $H^3_{comp}(A; \mathbb{Q}) \to H^3(A; \mathbb{Q})$ *is injective. And, assume either*

c) $H_1(\sigma; \mathbb{Q}) \to H_1(A; \mathbb{Q})$ *is injective, or else assume*

d) $B \cap O$ *is connected and* $[\sigma] \neq 0$ *in* $H_1(A; \mathbb{Q})$*, Then, there exists* $A' \subset X$ *which solves Problems 1 and is such that* $H^2(A'; \mathbb{Q}) \approx H^2(A; \mathbb{Q})$*. Thus, Problem 2 is solved by* A' *if* $H^2(A; \mathbb{Q}) = 0$.

Remark that Condition *b* of this proposition will be true automatically if $A \cap O$ is the interior of a manifold with boundary, $\underline{A}$, whose boundary, $\partial \underline{A}$, obeys $H^2(\partial \underline{A}; \mathbb{Q}) = 0$.

To solve Problem 3 of (8.2), it is necessary to digress first to define a $\mathbb{Z}/2$ valued invariant for homologically trivial, normally framed circles in an oriented 4-manifold with even intersection form. (This is invariant is well known to 4-manifold topologists.)

To start the digression, let B denote the oriented 4-manifold. To say that B's intersection form is even is to say that the self- intersection number of any embedded, orientable surface in B is an even number. (Note that B need not be compact.)

Let $\sigma \subset B$ be the finite union of disjointly embedded, oriented circles which represents the trivial element in $H_1(B; \mathbb{Z})$. The invariant in question, $\chi_{B,\sigma}(\cdot)$, assigns ± 1 to the various homotopy classes of framings of the normal bundle to σ in B. (If σ is a single circle, then there are precisely two normal framings up to homotopy since $\pi_1(SO(3)) \approx \mathbb{Z}/2$.)

To calculate $\chi_{B,\sigma}$, first choose an oriented surface with boundary, $R \subset B$, such that $\partial R = \sigma$. An oriented frame $\zeta \equiv (e_1, e_2, e_3)$ for the normal bundle to σ in B will be called an *adapted* frame when the vector e_3 is the inward pointing normal vector to R along ∂R.

LEMMA 8.4. *Let B, σ and R be as described above. Let ζ be an oriented, normal frame for $\sigma \subset B$. Then ζ is homotopic to an adapted frame.*

Proof. On a component, C, of σ, two normal frames differ by a map from S^1 to $SO(3)$. With this understood, note that $\pi_1(SO(3)) \approx \mathbb{Z}/2$, so there are two homotopy classes of normal frames along C. Two normal frames for which e_3 is the inward normal to R differ by a map from S^1 to $SO(3)$ which factors through a map from S^1 to $SO(2) \subset SO(3)$. With the preceding understood, the lemma follows because the induced homomorphism from $\pi_1(SO(2))$ to $\pi_1(SO(3))$ is surjective. □

The important feature of an adapted normal frame is that an adapted normal frame allows one to make an unambiguous definition of the mod(2) self-intersection number, $(R \cdot R)_2$, of R. Here is how: Take a section of R's normal bundle in B which agrees with e_1 on ∂R. Perturb the section away from ∂R so that it has transverse intersection with the zero section. Then, count the number of such intersection points mod(2).

One can also define $R \cdot R \in \mathbb{Z}$ by counting intersections with sign, but only the mod(2) intersection number is required for the definition of $\chi_{B,\sigma}$.

LEMMA 8.5. *If two adapted frames are homotopic in the space of* all *normal frames for σ, then the corresponding values of $(R \cdot R)_2$ agree.*

Proof. Adapted, normal frames to a given component $C \subset \sigma$ can be found which differ by a degree one map to $SO(2)$ and are such that the corresponding push-offs of R are identical save for a small open set near a point in C. With this understood, one need only check the lemma for the case where R is a planar 2-disk in $\mathbb{R}^4$. See, e.g Section 1.3 of [**7**]. □

It follows from Lemmas 8.4 and 8.5 that the surface R defines a map, $\chi_{B,\sigma}(\cdot)$, from the set of homotopy classes of normal frames of $\sigma \subset B$ to $\mathbb{Z}/2$. By definition, $\chi_{B,\sigma}(\zeta)$ assigns to ζ the number $(R \cdot R)_2$ that is computed by using an adapted frame which is homotopic to ζ.

Consider the dependence of $\chi_{B,\sigma}(\cdot)$ on the surface R:

LEMMA 8.6. *Suppose that B has even intersection pairing in its second homology. Then $\chi_{B,\sigma}(\cdot)$ is the same for all surfaces R bounding σ.*

Proof. Let ζ be a framing of the normal bundle to σ in B. Let $R_{1,2} \subset B$ be a pair of surfaces which bound σ. The task is to show that $R_1 \cdot R_1 = R_2 \cdot R_2$ mod(2).

One can assume, with no loss of generality, that ζ is adapted to R_1. Since $\pi_1(S^2) \approx 0$, the surface R_2 can be isotoped, with σ fixed, so that e_3 is the outward pointing normal vector to R_2. With this understood, R_1 and R_2 can be joined together along σ to obtain a C^1 immersion of a closed, oriented surface, R, in B. (The lack of smoothness occurs across σ.) The surface R may not be embedded because R_1 and R_2, though individually immersed, may intersect each other. Any way, with a small isotopy of R_1 (away from ∂R_1), the intersections of R_1 with R_2 can be made transverse.

An embedded surface in B has a well defined self-intersection number. An immersed surface has a well defined intersection number also. In this case,

$$R \cdot R = R_1 \cdot R_1 + R_2 \cdot R_2 - 2(R_1 \cdot R_2) \tag{8.3}$$

The number in (8.3) is the intersection number for the embedded surface that is obtained by resolving all of the double points of R.

Given that (8.3) is the self intersection number of an embedded surface in B, the assumptions in Lemma 8.5 require that (8.3) be an even number. Thus $R_1 \cdot R_1 = R_2 \cdot R_2 \bmod(2)$ as required.

(Here is how to resolve a double point of an immersed surface: In local coordinates the transveral intersection of the two sheets of the surface is described by the zeros in $\mathbb{C}^2 = \mathbb{R}^4$ of the equation

$$z_1 z_2 = 0. \tag{8.4}$$

The resolution of the intersection point replaces the solution to (8.4) with the solution to the equation $z_1 z_2 = \epsilon$. Here, $\epsilon \in \mathbb{C}$ is small but not zero.) □

With the invariant $\chi_{B,\sigma}(\cdot)$ of Lemma 8.6 understood, end the digression.

Here is a solution to (8.2)'s third problem:

PROPOSITION 8.7. *Let X, A, B, and O be as in (8.1) and Problems 1 and 2 of (8.2). Assume that Conditions a, b and either c or c' of Proposition 8.3 hold. Suppose that B has even intersection form and that $A \subset X$ has a trivial normal bundle. Let ζ be a given frame for A's normal bundle in X.*

1) *The restriction of ζ to $\sigma \equiv (A \cap B) \cap O$ defines a normal frame, $\zeta_\sigma \equiv e\,|_\sigma$, to σ in B.*
2) *If $\chi_{B,\sigma}(\zeta_\sigma) = 0$, then there is a solution, $A' \subset X$, to Problem 1 such that the normal frame ζ over $A - (A \cap O)$ extends over A'.*
3) *Thus, if $H^2(A; \mathbb{Q}) = 0$, then A' solves Problems 1-3 of (8.2).*

The proofs for Propositions 8.2, 8.3 and 8.7 are given in Subsections 8f-8h.

b) The proof of Proposition 8.1.

Let E'_{2-} denote the disjoint union of some number $N \geq 1$ disjoint copies of E_{2-}. The goal is to apply Propositions 8.2, 8.3 and 8.7 to remove the compact (circle) components, C, of the intersection of E'_{2-} with Δ_Z. With this goal understood, Proposition 8.2, 8.3 and 8.7 will be considered with the following identifications: Take

$$X = \operatorname{int}(Z), \quad A = \operatorname{int}(E'_{2-}), \quad B = \operatorname{int}(\Delta_Z). \tag{8.5}$$

Take O to be the compliment in $\operatorname{int}(Z)$ of the closure of a regular neighborhood of

$$\partial Z \cup (\cup_{i=1}^{r} \underline{\mu}_i) \cup E_R \cup E_L. \tag{8.6}$$

Here, $\underline{\mu}_i \equiv (\mu_i \times \mu_i) \cap \Delta_Z$ with μ_i as in Section 7b. This regular neighborhood should contain $\{\Gamma_i\}$ in (7.7) of $E_{2-} \cap \Delta_Z$, and it should also contain the push-off copies of $\{\Gamma_i\}$which comprise the interval components of $E'_{2-} \cap \Delta_Z$. Needless to say, O should contain the compact components of $E'_{2-} \cap \text{ int}(Z)$.

With this choice of X, A, B and O, the assertions of Proposition 8.1 will follow from Proposition 7.4 if the hypothesis of Propositions 8.2, 8.3 and 8.7 can be verified for a suitable N. (Remember that E'_{2-} is comprised of N push-off copies of E_{2-}.) Note: With regard to Proposition 8.7, the normal framing, ζ, of any push- off copy of $E_{2-} \subset E'_{2-}$ should be the normal framing of E_{2-} which is described by Lemma 7.2.

Subsections 8c-8e verify that there exists $N \geq 1$ for which the hypothesis of these three propositions are satisfied. □

c) Removing circles in $E'_{2-} \cap \Delta_Z$.

The purpose of this subsection is to verify that there exists an integer $N_1 \geq 1$ which is such that the hypothesis of Proposition 8.1 can be verified when E'_{2-} is any multiple of N_1 push-off copies of E_{2-}.

The discussion begins with a digression to study the first homology of $B \cap O$. (Equations (8.5) and (8.6) define B and O.) The projection π_L (or π_R) identifies B with $\text{int}(W)$. This projection identifies $B \cap O$ with the compliment in W of a regular neighborhood of $\partial W \cup \text{ crit}(f) \cup (\cup_{i=1}^r \mu_i) \cup \gamma$, where γ is the flow line in 4 of Definition 3.1.

Now consider $\sigma \subset B \cap O$, a finite union of embedded, oriented circles. After a small isotopy, the circles in σ can be arranged to have empty intersection with the descending disks from $\text{crit}_2(f)$. With this isotopy understood, the pseudo-gradient flow will isotope the circles in σ so that the resulting circles, σ_1, lies in the open submanifold $W_3 \equiv \{x \in W : 3/4 < f(x) < 1\}$. That is, $f(\sigma_1)$ is larger than any critical value of f.

The pseudo-gradient flow defines a diffeomorphism between W_3 and $M_1 \times (3/4, 1)$. By assumption, M_1 is a rational homology sphere, which means that the homology class, $[\sigma_1]$, of σ_1 is zero in $H_1(W_3 - W_3 \cap \gamma; \mathbb{Q})$. (Note that $W_3 \cap \gamma = p_1 \times (3/4, 1)$.) Alternately, one can conclude that

$$N_1\,[\sigma_1] = 0 \in H_1(W_3 - W_3 \cap \gamma, \mathbb{Z}) \tag{8.7}$$

for some integer $N_1 \geq 1$. This means that N_1 push-off copies of σ_1 bounds an embedded surface in $W_3 - W_3 \cap \gamma$. (Orient all N_1 push-off copies of σ_1 identically.)

Thus, N_1 push-off copies of σ will bound an embedded surface in $B \cap O$.

End the digression.

To verify Proposition 8.2's hypothesis for E'_{2-}, consider the discussion of the preceding subsection where σ is equal to C in (7.7). This choice of σ determines the integer N_1 in (8.7). If $N_1 = 1$ in (8.7), then Proposition 8.1 can be directly applied to $A \equiv E_{2-}$ so that the result, A', intersects $\Delta_Z \subset Z$ as described by (7.7) but with $C = \emptyset$.

However, the case $N_1 > 1$ in (8.7) can not be ruled out. In the case that $N_1 > 1$, let $m \geq 1$ and let E'_{2-} denote the disjoint union $m\,N_1$ disjoint, push-off copies of E_{2-}, all oriented as E_{2-}. (Use the normal framing of Section $7f$ when making these push-offs.)

With E'_{2-} understood, observe that

$$E'_{2-} \cap \Delta_Z = (\cup_{i=1}^{r} \Gamma_i) \cup C', \tag{8.8}$$

where C' in (8.8) is, by design, $m\,N_1$ disjoint, push-off copies of C from (7.7). In (8.8), each Γ_i is the union of $m\,N_1 \det(S)$ push-offs (in Δ_Z) of $\mu_i \equiv \Delta_Z \cap (\mu_i \times \mu_i)$.

By construction, the homology class of C' in $H_1(B \cap O; \mathbb{Z})$ is zero. (Because $[C'] = m\,N_1\,[C]$ and the class of C is N_1-torsion.)

With the preceding understood, then Proposition 8.1 can be applied with X, A, B and O as described by (8.5) and (8.6) so long as the number N is a multiple of N_1 in (8.7).

d) Constraining H^2.

Proposition 8.2 constructs a submanifold $A' \subset Z$ from some number $N \geq 1$ push-off copies of E_{2-}. (Here, N must be a multiple of N_1 from (8.7).) This A' is constructed so that it misses $E_{L,R}$ and a form like E_{2-} near ∂Z. And, the intersection of A' with Δ_Z is the union $\cup_{i=1}^{r} \Gamma_i$, where Γ_i is the union of $N \det(S)$ push-off copies of the path $\underline{\mu}_i$. If Proposition 8.1's E_{3-} is this A', then A' will have to have vanishing 2nd cohomology. That is, A' must be a solution to Problem 2 in (8.2).

Proposition 8.3 will be used to solve Problem 2 in the case at hand; this is the subject of the present subsection.

The task here is to verify that the conditions of Proposition 8.2 can be met for $A \equiv E'_{2-} \cap X$ with E'_{2-} some number, N, of like- oriented, push-off copies of E_{2-}. (Note that Assertion 6 of Proposition 7.4 asserts that $H^2(A; \mathbb{Q}) = 0$.) Taking Conditions $a - c$ in order, remark that the previous subsection has established that Condition a is satisfied when N is divisable by a certain integer (N_1 of (8.7)).

Condition b is satisfied because of Assertion 4 of Proposition 7.4. That is, A has closure a manifold with boundary, and the boundary is a number of copies of S^3. Since $H^2(S^3) = 0$, the required injectivity holds.

Condition c is established by the following lemma:

LEMMA 8.8. *Let C denote the union of the compact components of $E_{2-} \cap \Delta_Z$. The inclusion of C into E_{2-} induces a monomorphism from $H_1(C; \mathbb{Z})$ into $H_1(E_{2-}; \mathbb{Z})$.*

Proof. Remark that E_{2-} is obtained via ambient surgery (in Z) on various embedded $(S^0 \times B^4)$'s in disjoint unions of $\{Y_{i,j-} : i \leq j \in \{1, \cdots, r\}\}$ (see (7.3)).

As remarked earlier, $Y_{i,j-}$ can be viewed as the result of ambient surgery (in $F^{-1}(1/8)$) on various embedded $(S^0 \times B^4)$'s in the 4-sphere which is the intersection of the descending 5-disk from (a_i, b_j) with $F^{-1}(1/8)$. For a given $S^0 \times B^4$, the $S^0 \times \{0\}$ is a pair of algebraically cancelling intersections of said descending 4-sphere with the ascending 4-disk from some (a_i, a_k) or (b_k, b_j) in $\text{crit}_4(F)$.

Thus, E_{2-} is obtained from a disjoint union of embedded 4- spheres in $F^{-1}(1/8)$ by ambient surgery on embedded $(S^0 \times B^4)$'s. It follows from the preceding that $H_1(E'_{2-})$ is a summand of some number of $\mathbb{Z}$'s. And, it follows that a union, σ, of oriented, embedded circles in E_{2-} injects its first homology into $H_1(E_{2-}; \mathbb{Q})$ if:

(8.9)

1) An added $I \times S^3$ which intersects σ has intersection number ± 1 with $\{\text{point}\} \times S^3$.
2) Each component of σ intersects at least one $I \times S^3$.

In the present circumstances, σ is the union of the compact components of $E'_{2-} \cap \Delta_Z$. To understand σ, remember that E_{2-} is constructed from E_{1-} by ambient surgery. The reader can check that this surgery is disjoint from any compact components of $E_{1-} \cap \Delta_Z$. Indeed, the surgery from E_{1-} to E_{2-} takes place on push-off copies of $\{Y_{i,i-}\}_{i=1}^r$, but the compact components of $E_{1-} \cap \Delta_Z$ are the components of the various push-offs of $\cup_{i<j}(Y_{i,j-} \cap \Delta_Z)$.

Thus, the compact components of $E_{2-} \cap \Delta_Z$are of two types: A Type 1 component is a compact component from $E_{1-} \cap \Delta_Z$. And, a Type 2 component is created by the surgery which changed E_{1-} into E_{2-}. (The latter are made in the surgery on the various push-off copies of $\{Y_{i,i-}\}_{i=1}^r$.)

To understand the Type 1 components, use π_L or π_R to identify Δ_W with W and this intersection is identitified with $B_{1a+} \cap B_{1b-}$. (Here, $a \equiv a_i$ and $b \equiv b_j$.) To see the latter, start with $B_{a+} \cap B_{b-}$. This is a disjoint union of flow lines which start at a and end at b. The surgery which changes B_{b-} to B_{1b}-effects the intersection with B_{a+}. The effect is to surger the flow lines near a.

See the following picture:

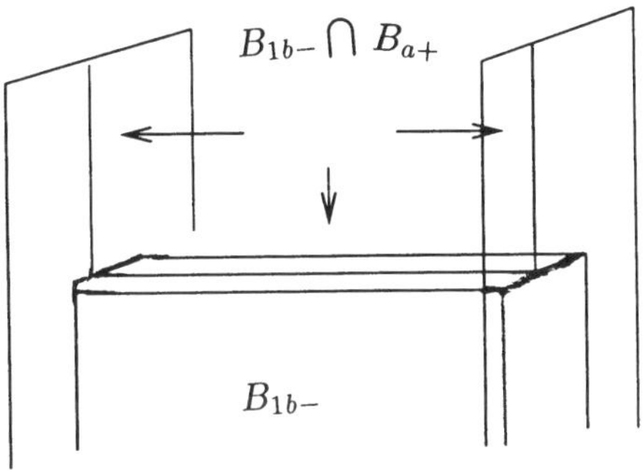

(8.10)

A similar picture occurs near b when B_{a+} is surgered to produce B_{1a+}. The resulting intersection $B_{1a+} \cap B_{1b-}$ differs from $B_{a+} \cap B_{b-}$ in that the ends of the flow lines in the latter have been tied together near a and near b to produce a compact intersection with some number, $n_{i,j}$, of components. (This $n_{i,j}$ is at least one, but no more than half of the number of components in $B_{a+} \cap B_{b-}$.) See the following (very schematic) picture:

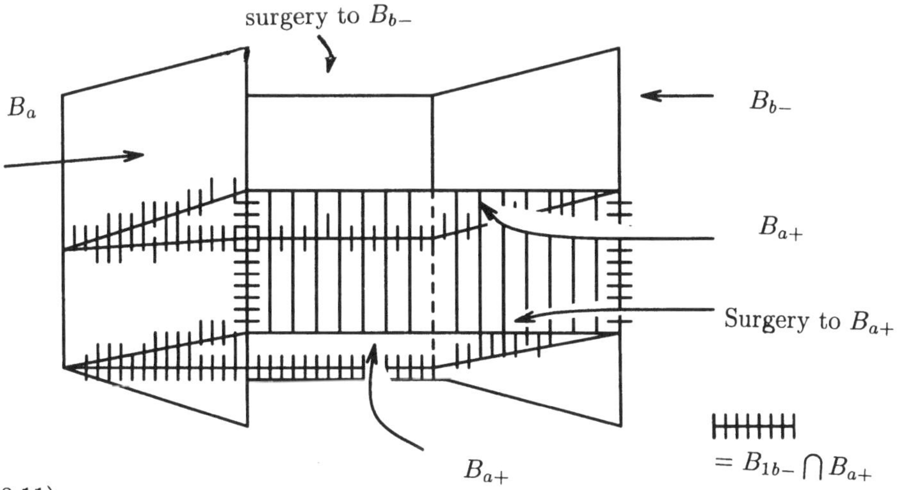

(8.11)

The effect of the preceding picture for the intersection with Δ_Z of one push-off copy of $Y_{i,j-}$ is as follows: Each surgery ties together an end of one flow line (for f's pseudo-gradient) in Δ_Z with the nearby end of a second flow line in Δ_Z the tie being across the associated $I \times S^3$. (Here, the canonical identification of Δ_W with W is taken implicitly.) See below:

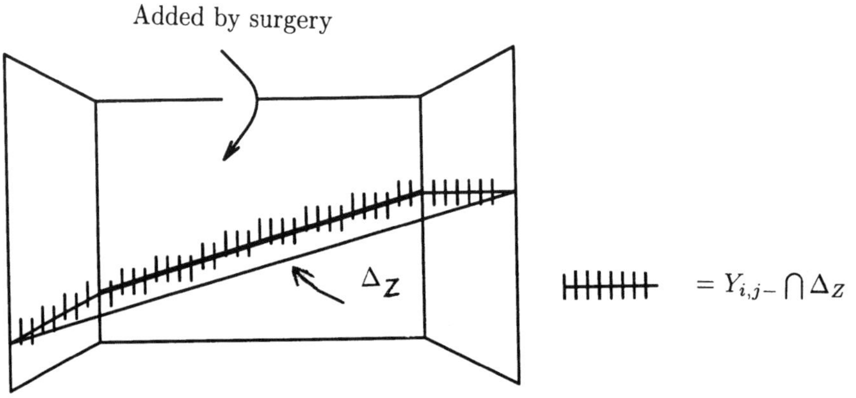

(8.12)

Thus, each copy of $Y_{i,j-}$ in E_{2-} (for $i < j$) produces $n_{i,j}$ components in σ. And, (8.9) is satisfied for each such copy.

As (8.9) is obeyed for each copy of $Y_{i,j-}$, and as the surgeries on the different copies of $Y_{i,j}$ are independent, it follows that (8.9) is satisfied by the set of all Type 1 components. That is, (8.9) is satisfied by the union of the compact components of $E_{1-} \cap \Delta_Z$.

Consider now the Type 2 components. To understand these components, remember that E_{2-} was constructed from E_{1-} by ambient surgery on various push-off copies of $\{Y_{i,i-}\}$. The surgeries do not connect a push-off of $Y_{i,i-}$ with one of $Y_{j,j-}$ if $j \neq i$.

Fix $i \in \{1, \cdots, r\}$. The union of push-off copies of a given $Y_{i,i}$ intersects Δ_Z in the union of the corresponding push-offs of the set of flow lines which is $B_{a+} \cap B_{b-}$. (Use $a \equiv a_i$ and $b \equiv b_i$). As far as these copies of $Y_{i,i-}$ are concerned, E_{2-} is constructed from them by surgery on embedded $(S^0 \times B^4)$'s. The result surgery changes the afore mentioned intersection with Δ_Z; each such surgery near (a, a) ties the ends near (a, a) of two of flow line copies across the added $I \times S^3$. There is a similar effect near (b, b). See (8.12).

It follows from the preceding picture that (8.9) holds for all of the Type 1 flow lines also, and since the added $(I \times S^3)$'s which effect Type 1 flow lines are disjoint from Type 2 flow lines, the lemma is established in total. □

e) Prescribed framing.

The purpose of this subsection is to establish that the conditions of Proposition 8.7 can be met for X, A, B and O of (8.5) and (8.6) if the number N (of copies of E_{2-} in E'_{2-}) is an even multiple of the integer N_1 in (8.7). Here, the framing ζ of the normal bundle in Z to each push-off copy of $E_{2-} \subset E'_{2-}$ is described by Lemma 7.2.

To begin, observe first that B has vanishing rational homology in dimension 2, so the condition on B's intersection form is trivially satisfied. Also, as A is some number of push-off copies of E_{2-}, it has trivial normal bundle in Z.

Next, remember that an integer $N_1 > 1$ has already been found which has the following properties: If $N \geq 1$ is divisible by N_1 and if A is taken to be N push-off copies of E_{2-} (all like oriented), then σ is the boundary of an embedded, oriented surface $R \subset B$.

With the frame σ as described above, let $\zeta_\sigma \equiv \zeta\mid_\sigma$. The final question is the value of $\chi_{B,\sigma}(\zeta_\sigma)$. Here is the answer: If N is an *even* multiple of N_1, then $\chi_{B,\sigma}(\zeta_\sigma) = 0$. This assertion follows from the following lemma:

LEMMA 8.9. *Let X be an oriented 4-manifold with even intersection form. Let $\sigma_1, \sigma_2 \subset X$ be compact, oriented, embedded 1-manifolds which are disjoint. Let ζ_1 and ζ_2 be normal frames for $\sigma_{1,2}$, respectively. Let $\sigma = \sigma_1 \cup \sigma_2$ and let e by the normal frame for σ which is given by $\zeta\mid_{\sigma_{1,2}} \equiv \zeta_{1,2}$. Then $\chi_{B,\sigma}(\zeta) = \chi_{B,\sigma_1}(\zeta_1) + \chi_{B,\sigma_2}(\zeta_2)$.*

Proof. By assumption, σ_1 bounds an embedded, oriented surface, $R_1 \subset X$. Likewise, σ_2 bounds a similar surface, R_2. If R_1 and R_2 are in general position, then $R_2 \cap \sigma_2 = \emptyset$ and vice- versa. Meanwhile, R_1 will intersect R_2 transversally in a finite set of points. Resolve the double points in $R_1 \cup R_2$ (as in (8.4)) to obtain a compact, oriented, embedded surface, $R \subset X$, with boundary σ.

Now, no generality is lost by assuming that $\zeta_{1,2}$ are adapted frames for $R_{1,2}$, respectively. In this case, ζ will be an adapted frame for R. Then, $R \cdot R$ is given by (8.3), and the lemma follows. □

f) Proof of Proposition 8.2.

The proof starts with a digression for some constructions on a neighborhood of B in X. To start the digression, define $N \to B$ to be the normal bundle to B in X. Fix an exponential map,

$$e : N \to X \tag{8.13}$$

which maps $N\mid_\sigma$ into A. Put a smooth fiber metric on N with the property that e embeds the set

$$N' \equiv \{v \in N : \mid v \mid < 2\} \tag{8.14}$$

onto a neighborhood $N \subset X$ of B. Agree now to identify N with N' using e.

Introduce the 2-sphere bundle $S \to B$,

$$S \equiv \{v \in N : \mid v \mid = 1\}. \tag{8.15}$$

Identify S with its image by e in X. This S is the boundary of a tubular neighborhood of B in X,

$$T \equiv \{v \in N : |v| \leq 1\}. \tag{8.16}$$

End the digression.

The proof proper of Proposition 8.2 starts by remarking that σ, by assumption, is the boundary of a smooth, oriented, embedded surface (with boundary), $R \subset B \cap O$. Find such an R for which $\text{int}(R)$ has no compact components. Because σ has codimension 3 in B, one can require that $\text{int}(R) \cap \sigma = \emptyset$. (The local model for R near σ is given by taking σ to be the line $x_1 = x_2 = x_3 = 0$ in $\mathbb{R}^4$, and R the half plane $x_1 = x_2 = 0$ with $x_3 \geq 0$.)

Let $S_R \equiv S|_R$. This is a smooth oriented 4-manifold (with boundary) which is embedded in X. The boundary of this 4-manifold is $S_\sigma \equiv S|_\sigma$. Note that $S_\sigma \subset A$ is the boundary of $T_\sigma \equiv T|_\sigma$, the embedded image of $\sigma \times B^3$ onto a tubular neighborhood of σ in A.

With the preceding understood, introduce the following surgery on A:

$$A_0' \equiv (A - \text{int}(T_\sigma)) \cup S_R. \tag{8.17}$$

Note that $A_0' \subset X$ is a C^0 embedding of a smooth, oriented manifold; the embedding has a corner at S_σ where S_R and $A - \text{int}(T_\sigma)$ overlap. See the following picture:

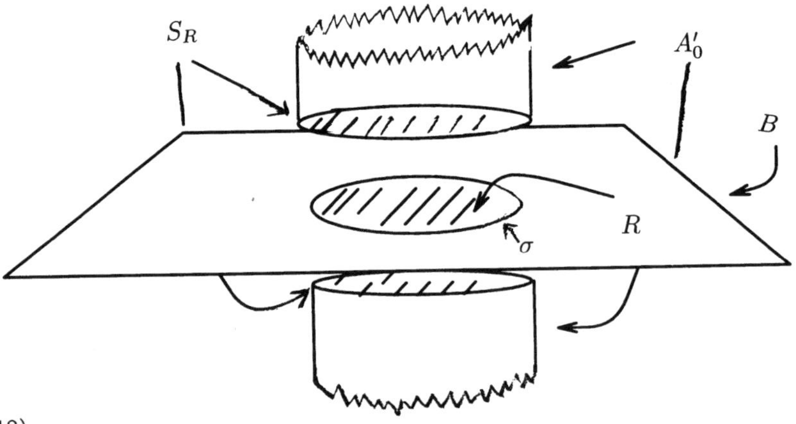

(8.18)

A neighborhood of this corner of A_0' is embedded in T_R. Smooth A_0' in T_R along the corner, and one obtains a smoothly submanifold, $A' \subset X$ which solves Problem 1 in (8.2). Indeed, the first two requirements are met by construction. As for the third, remark that $[T|_R]$ defines a 5-dimensional cycle in O whose boundary is $[A'] - [A]$. □

g) Proof of Proposition 8.3.

Consider first the proof of the proposition under the Assumption $a - c$. Let A' be as described above (see (8.17)). Then, $H^2(A')$ can be computed using the following homology exact sequences for the pairs (A, T_σ) and (A', S_R):

(8.19)

1) $H^1(A) \to H^1(T_\sigma) \to H^2(A, T_\sigma) \to H^2(A) \to H^2(T_\sigma)$.
2) $H^1(A') \to H^1(S_R) \to H^2(A', S_R) \to H^2(A') \to H^2(S_R) \to H^3(A', S_R)$.

(Use rational coefficients please.)

In Sequence 1, the first arrow is surjective because of Assumption c. And, $H^2(T_\sigma) \approx 0$ because T_σ is a tubular neighborhood of a disjoint union of circles. Thus

$$H^2(A, T_\sigma) \approx H^2(A). \tag{8.20}$$

To analyze the second sequence of (8.19), note that its first arrow is surjective. This is because R is path connected, thus forcing S_R to be a topologically trivial 2-sphere bundle. For the same reason, $H^2(S_R) \approx H^2(\partial T_\sigma)$. Meanwhile, excision identifies $H^*(A', S_R) \approx H^*(A, T_\sigma)$. Thus, with (8.20), the second sequence in (8.19) implies

$$0 \to H^2(A) \to H^2(A') \to H^2(\partial T_\sigma) \to H^2(A, T_\sigma). \tag{8.21}$$

Poincare' duality plus Assumptions b and c of Proposition 8.3 imply that the last arrow in (8.21) is surjective, thus establishing an isomorphism between $H^2(A)$ and $H^2(A')$.

Now consider Proposition 8.3 with Assumptions a, b and c'. Remark here that when σ is connected (i.e. just one circle), then Assumption c' implies Assumption c. With this fact understood, here is the task ahead: Under Assumptions a, b and c', find an ambient (in O) surgery on A so that the result, $\underline{A}$, has the following properties:

(8.22)

1) $[\underline{A}] = [A]$ in $H_4(X, X - O; \mathbb{Z})$.
2) $H^2(\underline{A}) \approx H^2(A)$.
3) $H^3_{comp}(\underline{A}) \to H^3(\underline{A})$ is injective.
4) $\underline{A}$ intersects B inside O in a single compact component which is not trivial in $H_1(A, \mathbb{Q})$ but which bounds in $B \cap O$.

A solution to (8.22) will validate Proposition 8.3.

Here is an algorythm for constructing $\underline{A}$: To begin choose a pair of components, $C_{1,2} \subset \sigma$. Fix $p_1 \in C_1$ and $p_2 \in C_2$. By assumption, $B \cap O$ is path connected, so there is a path in B (a smoothly embedded interval), τ, which starts at p_1 and ends at p_2. Make sure that int(τ) has empty intersection with A. Also, arrange τ so that it is not tangent to A along its boundary. The choice of p_1 as the starting point and p_2 as the ending point orients τ.

Let $V \to \tau$ denote the normal bundle to τ in B. This is an oriented 3-plane bundle over τ. Note that $TC_1 \mid_{p_1} \subset V \mid_{p_1} \approx \mathbb{R}^3$ and also $TC_2 \subset V \mid_{p_2} \approx R^3$ are oriented lines. As S^2 is path connected, there is an oriented, dimension 1 sub-bundle $V_0 \subset V$ whose restriction to p_1 is TC_1 and whose restriction to p_2 is the line TC_2, but oriented in reverse.

With V_0 understood, remark that $V_0 \oplus N \to \tau$ is an oriented 4-plane sub-bundle of the normal bundle to τ in O. Also note that this bundle restricts to p_1 and as $TA \mid_{p_1}$, and it restricts to p_2 as the 4- plane $TA \mid_{p_2}$, but with its orientation reversed.

The normal bundle to τ in X is isomorphic to $V \oplus N$. Fix an exponential map $e_\tau : V \oplus N \to X$ which restricts to N as e in (8.13), which restricts to map V into B, and which maps $V_0 \mid_{p_{1,2}}$ into $C_{1,2}$, respectively. (Thus, $e_\tau \mid_V$ is an exponential map for τ in B.) Put a fiber metric on the bundle $V_0 \oplus N$ such that e_τ embeds the subspace of vectors v with norm less than 2. Let $S \subset V_0 \oplus N$ denote the radius 1 sphere bundle, and identify S with its embedded image under e_τ.

Let $T \subset V_0 \oplus N$ denote the radius one, 3-ball bundle. Identify T with its embedded image under e_τ. Note that $T \mid_{p_1}$ is a tubular neighborhood in A of p_1, while $T \mid_{p_2}$ is a tubular neighborhood of p_2 in A.

Now define the following surgery on A:

$$A_0 \equiv (A - \text{int}(T \mid_{p_1} \cup T \mid_{p_2})) \cup S. \tag{8.23}$$

This is a C^0 embedding of a smooth manifold, A_1, in X. Here, the embedding is smooth away from $S \mid_{p_1} \cup S \mid_{p_2}$, where there is a corner. Near this corner, A_0 is embedded in $V_0 \oplus N$. Smooth out the corner in $V_0 \oplus N$ and the result is an embedding of A_1 into X:

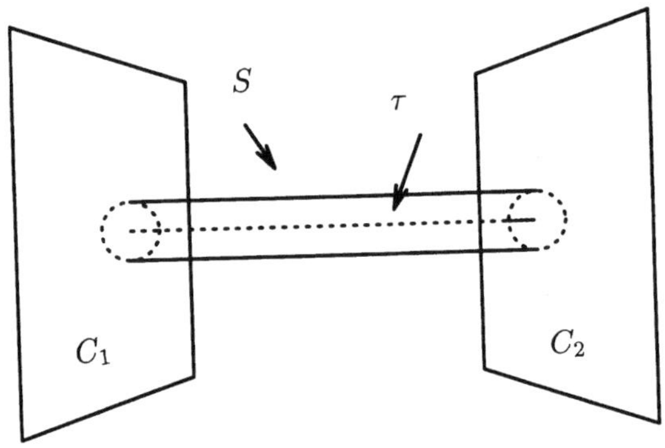

(8.24)

With A_1 understood, consider its properties with respect to (8.22): First, the homology classes of A_1 and A agree are equal in $H_4(X, X - O)$. This is because A_1 and A_0 define the same class and the 5-manifold T defines a cycle in O with boundary $[A_0] - [A]$.

Second, $H^2(A_1) = 0$, because A_1 has been obtained from A by surgery on an embedded $S^0 \times B^4$ (i.e. $T\,|_{p_1} \cup T\,|_{p_2}$).

Third, $H^3_{\text{comp}}(A_1) \to H^3(A_1)$ is injective. Both are either equal to their A counter-parts, or are obtained from their A counter parts by the addition of 1 generator which is dual to the 3-sphere $S\,|_{p_1}$. (Prove this with Meyer-Vietoris.)

Fourth, the intersection of A_1 with B in O has one less component then that of A with B in O. This is because the surgery from A to A_1 has surgered C_1 to C_2 by removing an embedded $S^0 \times B^1$ from $C_1 \cup C_2$ (i.e. $(T \cap V_0)\,|_{p_1} \cup (T \cap V_0)_{p_1}$); the missing $S^0 \times B^1$ is replaced by $S \cap V_0$. (See (8.24).)

Note that the homology class in $H_1(B \cap O)$ of $\sigma_1 \equiv A_1 \cap B \cap O$ is the same as that of $\sigma \equiv A \cap B \cap O$. Indeed, σ_1 defines the same homology class as $\sigma_0 \equiv A_0 \cap B \cap O$, and $T \cap V_0$ as a 2-cycle in $B \cap O$ has boundary $[\sigma_0] - [\sigma_1]$.

If σ had only two components to start with, then set $\underline{A} \equiv A_1$ and stop, because (8.22) has just been verified for this $\underline{A}$. If σ had more than two components, iterate the preceding procedure by renaming $A_1 \equiv A$ and $\sigma_1 \equiv \sigma$. The iteration stops with $\underline{A}$ which obeys (8.22). □

h) Proof of Proposition 8.7.

The construction of A' from A for solving Problem 1 (and Problem 2) of (8.2) via Propositions 8.2 and 8.3 is described in the preceding two subsections. The construction involves two types of ambient surgeries in X. The first type of surgery gives the smoothing, A_1, of A_0 in (8.23). The second type of surgery gives the smoothing, A', of the A'_0 in (8.17), (8.18).

Type 1: Extending frames for surgery on $S^0 \times B^4$.

Consider the smoothing, A_1, of A_0 in (8.23). Let $U \subset A$ be an open neighborhood of $T\,|_{p_1} \cup T\,|_{p_2}$. □

LEMMA 8.10. *Let ζ be a normal frame for A in X. Then there is a normal frame for A_1 in X which agrees with ζ on $A - U$.*

Type 1: Extending frames over surgery on $S^1 \times B^3$.

The assumption here is that A' is obtained from A by smoothing the surgery A'_0 in (8.17). (See (8.18) too.) Let ζ be a normal frame for A in X, and let $U \subset A$ be a neighborhood of T_σ.

LEMMA 8.11. *The normal frame ζ on $A - U$ extends as a normal frame over the smoothing, A', of (8.17) if and only if $\zeta\,|_\sigma$ is homotopic to an adapted frame for which $(R \cdot R)_{\,mod(2)} = 0$.*

If B has even intersection pairing, then according to Lemma 8.6, the $\mathbb{Z}/2$ number $(R \cdot R)_{\text{mod}(2)}$ is the invariant $\chi_{B,\sigma}(\zeta\,|_\sigma)$. Thus, Lemmas 8.10 and 8.11 with the constructions in the two preceding subsections prove Proposition 8.7.

The remainder of this section is occupied with the proofs of the preceding two lemmas.

Proof of Lemma 8.10. The strategy is to first define a normal frame, ζ_1, for int(S) in X. Having done so, the final step proves that there are no obstructions to connecting ζ_1 to ζ on $A - U$.

To construct ζ_1, first fix a normal frame, (e_1, e_2, e_3), for the normal bundle (V) of τ in B. One can arrange such a frame so that e_3 is tangent to the sub-line V_0. Then, (e_1, e_2) orient V/V_0.

Use the exponential map $e_\tau : V \oplus N \to X$ to identify a neighborhood of the zero section of $V \oplus N$ with a neighborhood of τ in X. With this identification understood, then the normal bundle to int(S) in X is spanned by $\zeta_1 \equiv (e_1, e_2, e_3')$, where $e_3' \in T(V_0 \oplus N) \mid_S$ restricts to the fiber over $x \in \tau$ as the inward pointing normal vector to $S \mid_x$ in $(V_0 \oplus N) \mid_x$.

With ζ_1 understood, consider connecting ζ_1 to ζ near ∂S. To make such a connection, introduce the tangent vector, v to τ. Let $\underline{v}$ denote a lift of v to $V \oplus N$.

At p_1 or p_1, the triple (e_1, e_2, v) defines a normal frame for A in X. The normal frame ζ for A in X can be homotoped inside U so that it agrees with (e_1, e_2, v) at p_1 and p_2 and equals $(e_1, e_2, \underline{v})$ on a neighborhood, U', of $T \mid_{p_{1,2}}$ with compact support in U. Thus, 2/3 of the normal frame ζ, i.e. (e_1, e_2), have been extended over A_1.

The compliment in the normal bundle to A_1 of the 2-plane span of (e_1, e_2) is an oriented line bundle which is framed by $\underline{v}$ on $U - U'$ and by e_3' on int(S). There is no obstruction to framing this compliment by a frame which agrees with $\underline{v}$ on $U - U'$ and with e_3' on the compliment in S of an apriori specified neighborhood of ∂S. □

Proof of Lemma 8.11. The strategy for extending ζ on A - U as a normal frame for A' will be to construct a normal frame, ζ_1, for S_R in X, and then consider whether ζ and ζ_1 can be joined.

To construct ζ_1, introduce the normal 2-plane bundle, V, to R in B. Since R is oriented, V is an oriented bundle and so trivial because int(R) has no compact components. Let e_1, e_2 be a frame for V.

Note that $V \oplus N \to R$ is the normal bundle to R in X. Choose an exponential map $e_R : V \oplus N \to X$ which restricts to N as e in (8.13). Use e_R to identify a neighborhood of the zero section in the bundle $V \oplus N$ with a neighborhood of R in X.

Let e_3 be the inward pointing normal vector to $S_R \subset T_R \subset N \mid_R$. Then the triple $\zeta_1 \equiv (e_1, e_2, e_3)$ span the normal bundle to S_R in X.

With the normal frame for S_R understood, consider its extension to a normal frame for A on the compliment in U of a neighborhood U' of T_σ. For this purpose, introduce v to denote the inward pointing normal vector field to R along σ. Lift v to a vector field $\underline{v}$ on $(V \oplus N)_\sigma$. Since A near σ is identified by (8.13) with a neighborhood of the zero section of $N \mid_\sigma$, it follows that a normal frame for A near σ is given by the triple $(e_1 e_2, \underline{v})$. Furthermore, there is no obstruction to joining this frame on the compliment of a neighborhood U'

of T_σ with the frame $\zeta_1 \equiv (e_1, e_2, e_3)$ on the interior of S_R. (The pair (e_1, e_2) define 2/3 of the extension, and $\underline{v}$ and e_3 define the same orientation for the complimentary line.)

With the preceding understood, then one can conclude that the normal frame ζ extends from $A-U$ to A' if the restriction of ζ to σ is homotopic to the normal frame $(e_1, e_2, \underline{v})$. Now the latter frame is an adapted frame and, by construction, $R \cdot R = 0$ for $(e_1, e_2, \underline{v})$. Thus, $\zeta \mid_\sigma$ is homotopic to (e_1, e_2, σ) if and only if $\zeta \mid_\sigma$ is homotopic to an adapted frame for which the corresponding $R \cdot R$ is even. (See Lemma 8.5.) □

9 The fouth pass at $E_\pm$. The submanifold E_{3-} of the preceding section intersects Δ_Z as described by Assertion 6 of Proposition 8.1. Let $\Theta : Z \to Z$ denote the switch map which sends (x, y) to (y, x). Since $E_{3+} = \Theta(E_{3-})$, the intersections of E_{3-} with Δ_Z are also intersections of E_{3+} with Δ_Z. Unfortunately, there may be compact components to $E_{3-} \cap E_{3+}$ which occur in $Z - \Delta_Z$. Such extra components are troublesome and must be eliminated, and their elimination is the goal of this section.

As will be seen, surgery on $E_{3\pm}$ will result in oriented submanifolds (with boundary) $E_{4\pm} \subset Z$ which have the following properties:

PROPOSITION 9.1. *There are oriented, embedded submanifolds (with boundary) $E_{4\pm} \subset Z$ with the following properties:*

1) *There is an open neighborhood $U \subset Z$ of $\Delta_Z \cup \partial Z$ such that $E_{4+} \cap U$ and $E_{4-} \cap U$ are images of each other under the switch map on Z.*
2) *The fundamental classes $[E_{4\pm}]$ are equal to $N\,[E_{1\pm}]$ for some integer $N \geq 1$. Here, $[E_{1\pm}]$ are described by (6.3) and Lemma 6.3.*
3) *$E_{4\pm}$ have empty intersection with $M_0 \times M_0$ and $M_1 \times M_1$.*
4) *$E_{4\pm}$ have empty intersection with $E_{L,R}$ of (4.15).*
5) *If $p \in$ crit(f), then the intersection of E_{4-} with $(S^3 \times S^3)_p$ has the form $S^3 \times \Lambda_p$, where Λ_p is a set of N points. Similarly, the intersection of E_{4+} with $(S^3 \times S^3)_p$ is $\Lambda_p \times S^3$.*
6) *$E_{4-} \cap \Delta_Z = \cup_{i=1}^r \Gamma_i$, where $\Gamma_i \subset \Delta_Z$ is as follows: There is a flow line μ_i which starts at a_i and ends at b_i. With the canonical identification of Δ_W with W understood, Γ_i is the union of N like oriented, disjoint, push-off copies of a closed interval, $I \subset \mu_i$. And, each of these N push-offs of I starts in $(\Lambda_a \times \Lambda_a) \cap \Delta_Z$ and ends in $(\Lambda_b \times \Lambda_b) \cap \Delta_Z$. Likewise, $E_{4+} \cap \Delta_Z = \cup_{i=1}^r \Gamma_i$.*
7) *$E_{4-} \cap E_{4+} = \cup_{i=1}^r \Gamma_i'$, where $\Gamma_i' \subset Z$ is the union of Γ_i with $N-1$ like oriented, push-off copies of Γ_i in $Z - \Delta_Z$.*
8) *Both $E_{4\pm}$ have trivial normal bundles in Z. The normal bundle of E_{4-} has a framing, ζ, which restricts to a product normal framing on a neighborhood of $(\cup_{i=1}^r \Gamma_i) \cup \{S^3 \times \Lambda_p\}_{p \in \mathit{crit}(f)}$. Furthermore, this framing ζ restricts to $\{S^3 \times \Lambda_p\}_{p \in \mathit{crit}(f)}$ as a constant framing. The normal bundle*

to E_{4+} in Z has a framing which restricts to $E_{4+} \cap U$ as the image of ζ under the switch map.

9) $H^2(E_{4\pm};\mathbb{Q}) = 0$.

(Compare with Proposition 8.1. The only essential change is in $E_{4-} \cap E_{4+}$. But note that the integer N, the points $\{\Lambda_p\}$ and the line segments $\{\Gamma_i\}$ which appear here may be different from those which appear in Proposition 8.1.)

The rest of this section is devoted to the construction of $E_{4\pm}$.

a) $E_{3-} \cap E_{3+}$ on $Z - \Delta_Z$.

Isotope E_{3+} in $Z - (\Delta_Z \cup \partial Z)$ so that its intersection on $Z - \Delta_Z$ with E_{3-} is transversal. (Still use E_{3+} to denote the after isotopy submanifold.) This intersection is now a finite union of disjoint, embedded, oriented circles, σ, with $N-1$ like oriented, push-off copies of $\{\Gamma_i\}$ in $Z - \Delta_Z$. (These push-offs of $\{\Gamma_i\}$ are disjoint from σ.) There is a natural inclination to remove the circle components, σ, by apply the techniques from the previous section (Propositions 8.2, 8.3 and 8.7). However, either

$$[\sigma] = 0 \in H_1(E_{3\pm};\mathbb{Q}) \tag{9.1}$$

or not; but $[\sigma]$ is never non-trivial in one and trivial in the other. (Because, before perturbing E_{3+}, one was the image of the other under the switch map.)

If (9.1) holds, then one can not (directly) apply Proposition 8.2 to remove the intersection σ. If (9.1) does not apply, then Proposition 8.2 can be applied, but not (directly) Proposition 8.3. So, whether or not (9.1) holds, some preliminary work must be done before the techniques from Section 8 can be employed.

Consider the case where (9.1) holds.

LEMMA 9.2. *Let $O \equiv Z - (\Delta_Z \cup \partial Z \cup E_L \cup E_R)$. There is an ambient surgery (in O) of E_{3-} which results in an oriented submanifold (with boundary) $E'_{3-} \subset Z$ with the following properties:*

1) *Assertions 2-8 of Proposition 8.1 hold when E'_{3-} is substituted for E_{3-}.*
2) *There is a tubular neighborhood, $U_\Delta \subset Z$, of Δ_Z which intersects $E'_{3-} \cap E'_{3+}$ as $\cup_{i=1}^{r} \Gamma'_i$, where Γ'_i is the union of Γ_i with $N-1$ disjoint, like oriented, push-off copies of Γ_i.*
3) *$E'_{3-} \cap E'_{3+}$ intersects $Z - U_\Delta$ as a disjoint union, σ', of oriented circles which obey*

(9.2)

(a) $[\sigma'] \neq 0 \in H_1(E'_{3-};\mathbb{Q})$.

(b) $[\sigma'] = 0 \in H_{-1}(E_{3+};\mathbb{Q})$.

(Note that there is no E'_{3+}; the symmetry under the switch map will be broken here.)

The proof of this lemma will be given shortly.

Consider the case where (9.1) is false. The goal here is to modify E_{3+} as described in the following lemma:

LEMMA 9.3. *Let $O \equiv Z - (\partial Z \cup \Delta_Z \cup E_L \cup E_R)$. There is an ambient surgery (in O) of some number $n \geq 1$ of like-oriented push-offs of E_{3+} which results in an oriented, embedded submanifold, (with boundary) $E'_{3+} \subset Z$ with the following properties:*

1) *Let E'_{3-} denote the union of n like-oriented, push-off copies of E_{3-}. The Assertions 2-8 of Proposition 8.1 hold when $E'_{3\pm}$ are substituted for $E_{3\pm}$.*
2) *There is a tubular neighborhood, $U_\Delta \subset Z$, of Δ_Z which intersects $E'_{3-} \cap E'_{3+}$ is $\cup_{i=1}^{r} \Gamma'_i$, where Γ'_i is the union of Γ_i with $N-1$ disjoint, like oriented, push-off copies of Γ_i.*
3) *$E'_{3-} \cap E'_{3+}$ intersects $Z - U_\Delta$ as a disjoint union, σ', of oriented circles which obey*

(9.3)

(a) $[\sigma'] \neq 0 \in H_1(E'_{3-}; \mathbb{Q})$.
(b) $[\sigma'] = 0 \in H_{-1}(E_{3+}; \mathbb{Q})$.

This lemma will also be proved shortly.

Proof of Proposition 9.1.

When (9.1) is true, take E'_{3-} from Lemma 9.2 and set $E'_{3+} \equiv E_{3+}$. When (9.1) is false, take $E'_{3\pm}$ from Lemma 9.3. Use σ' to denote the intersection in $O \equiv Z - (\partial Z \cup U_\Delta \cup E_L \cup E_R)$ between $E'_{3\pm}$. Use (9.2b) or (9.3b) in the respective cases, to find an integer $N_3 \geq 1$ with the property that $N_3\,[\sigma'] = 0 \in H_1(E_{3+}; \mathbb{Z})$.

Take $2\,N_3$ push-off copies of E'_{3-}, all with the same orientation as E_{3-}, and let A denote the interior of the resulting union. Take $2\,N_3$ push-off copies of E'_{3+}, all oriented as E'_{3+}, and let B_0 denote the interior of the resulting union. Let $X \equiv int(Z)$ and let O be as before. Now, Proposition 8.2 can be invoked using B_0 for B, but not Proposition 8.3 because Assumption c has not been shown to hold, and because Assumption c' will be false because B_0 will not be connected. However, there is a surgery which remedies this problem: □

LEMMA 9.4. *Let B_0, A, X and O be as described above. Then, there is an ambient surgery in $O - A$ on some finite number of embedded $(S^0 \times B^4)$'s in B_0 such that the result, B is path connected. This surgery does not change either $H_1(\cdot; \mathbb{Z})$ or $H_2(\cdot; \mathbb{Z})$. Finally, if ζ is a normal frame for B_0 in X, and if $U \subset B_0$ is a neighborhood of the $(S^0 \times B^4)$'s, then $\zeta\,|_{B^0 - U}$ extends smoothly over B as a normal frame for B in X.*

The proof of this lemma is given below.

With Lemma 9.4 understood, Propositions 8.2, 8.3 and 8.7 can be applied using X, O, A and B as described above. Use E_{4-} to denote the closure in Z of the promised solution, A', to Problems 1-3 in (8.2). Relable E_{4+} to denote the closure in Z of B. The pair $E_{4\pm}$ will satisfy the requirements of Proposition 9.1.

b) Making $[\sigma'] \neq 0 \in H_1(E'_{3-};\mathbb{Q})$.

This subsection is concerned with the construction of E'_{3-} of Lemma 9.2.

This construction requires a preliminary digression to introduce another surgery tool. The digression concerns the abstract model of (8.1). Proposition 9.5, below, summarizes the digression.

The statement of Proposition 9.5 requires the following remarks to set the stage: When $S \to A$ is an embedded 2-sphere, let $v_S \to S$ denote the normal bundle to S in A. Suppose that $P \subset X$ is an embedded 3-dimensional ball with boundary S. (The local model here takes S to be the plane $x_3 = \cdots = x_7 = 0$ in $\mathbb{R}^7$ and then P is the half plane $x_3 \geq 0, x_4 = \cdots = x_7 = 0$.). Use N_P to denote the normal bundle to P in X. The natural inclusion

$$0 \to v_S \to N_P\mid_S \tag{9.4}$$

plays an important role in Proposition 9.5.

PROPOSITION 9.5. *Let A, B, O, X and σ be as described in (8.1). Assume that:*

a) $H^3_{comp}(A;\mathbb{Q}) \to H^3(A;\mathbb{Q})$ is injective.

b) $[\sigma] = 0 \in H_1(A;\mathbb{Q})$. Suppose that there exists an embedded 2-sphere $S \subset (A \cap O) - \sigma$ which bounds an embedded 3-ball $P \subset O$, and:

c) The fundamental class of S is homologically trivial in $H_2(A;\mathbb{Q})$.

d) $int(P) \cap A = \emptyset$.

e) P intersects B transversally.

f) With respect to some orientation on $P, [P \cap B] \neq 0 \in H_0(P;\mathbb{Z})$.

g) $(N_P\mid_S)/v_S \to S$ is a trivial 2-plane bundle.
Let $U \subset X$ be an open neighborhood of P. There exists an embedded, oriented, 4-dimensional submanifold $A' \subset X$ with the following properties:

1) A' intersects B transversally in σ', and $[\sigma'] \neq 0 \in H_1(A';\mathbb{Q})$.

2) $H^2(A';\mathbb{Q}) = H^2(A,\mathbb{Q})$.

3) $H^3_{comp}(A';\mathbb{Q}) \to H^3(A';\mathbb{Q})$ is injective.

4) $A' = A$ on $X - U$ and $[A] = [A']$ in $H_4(X, X - O)$.

5) $[\sigma'] = [\sigma] \in H_1(B;\mathbb{Q})$. Furthermore,

6) If A has trivial normal bundle in X, then so does A'. And, if ζ is a frame for the normal bundle to A in X, then $\zeta\mid_{A-U}$ extends over A' as a frame for the normal bundle to A' in X.

This proposition will be proved in the next subsection. Consider now its application to (9.2).

Proof of Lemma 9.2. The lemma will be proved by applying Proposition 9.5. For this purpose, take X to be int(Z) and then define $O \equiv Z - (\partial Z \cup$

$U_\Delta \cup E_L \cup E_R$). Take $A \equiv \operatorname{int}(E_{3-})$ and, likewise, take $B \equiv \operatorname{int}(E_{3+})$. Given that the assumptions of Proposition 9.5 hold when (9.1) is true, one should take E'_{3-} to be the closure in Z of the submanifold A' of Proposition 9.5. As for the validity of the assumptions of Proposition 9.5, remark that Assumptions $a - b$ are satisfied by construction; see Proposition 8.1. (Assumption b holds since E_{3-} is a manifold with boundary whose boundary is a union of 3-spheres; and $H^2(S^3) = 0$.)

The remaining assumptions of Proposition 9.5 will be verified with the exhibition of a 2-sphere $S \subset A \cap O$ with the requisite properties. To find the appropriate 2-sphere, it is important to remember that E_{3-} was constructed from E'_{2-}. Here is a brief summary: The compact components of $E'_{2-} \cap \Delta_Z$ bound a connected, oriented, embedded surface (with boundary) $R \subset \Delta_Z$. Let $N \to \Delta_Z$ be the normal bundle. A fiber metric was chosen for N, and an exponential map $e : N \to Z$ was chosen so that e mapped the radius 2 ball fiber bundle in N diffeomorphically onto its image in Z. (This ball bundle in N was identified using e with a neighborhood of Δ_Z in Z.) Also, e was constrained to map $N\mid_{\partial R}$ into E'_{2-}.

Next, the radius 1 sphere bundle, $S_R \subset N\mid_R$ was introduced, as well as the radius 1 ball bundle, $T_R \subset N\mid_R$. Finally, E_{3-} was defined to be the result of smoothing the corner in the surgery

$$(9.5)\qquad (E'_{2-} - T_{\partial R}) \cup S_R.$$

By the way, no generality is lost by assuming that the surgery in (9.5) occured. Indeed, if $E_{2-} = E_{3-}$ and no such surgery occurs (in which case $E_{2-} \cap \operatorname{int}(\Delta_Z)$ has no compact components), then there are isotopies of $E_{1\pm}$ so that the resulting submanifolds obey the conclusions of Proposition 9.1. Or, if $E_{2-} = E_{3-}$, then one could add two push-off copies of $Y_{1,1-}$ to E'_{1-} of Section 7a, one oriented positively and the other oriented negatively. Then, the tubing construction of E_{2-} will insure that $E_{2-} \cap \operatorname{int}(\Delta_Z)$ has a compact component.

With (9.5) understood, remark that E_{3+} is obtained from $\Theta(E_{3-})$ by an isotopy. Let $R' \subset R$ denote the complement of a (small) collar of ∂R. Near R', E_{3-} is $S_R\mid_{R'}$ and this coincides with $\Theta(E_{3-})$ near R'. (Thus, $\Theta(E_{3-})$ and E_{3-} do not intersect transverally.) Choose the isotopy to obtain E_{3+} from $\Theta(E_{3-})$ so that E_{3+} near R' is a sphere bundle $S_{R+} \subset N\mid_{R'}$ of radius *greater* than 1.

Pick a point $x \in R'$ and let

$$(9.6)\qquad S_x \equiv S_R\mid_{R'} \subset E_{3-}.$$

This is a homologically trivial, embedded 2-sphere in E_{3-}. This S_x bounds the 3-ball $T_x \subset N\mid_x$ which is the unit ball in the fiber of N at x. Notice that this 3-ball has empty intersection with E_{3+} since the latter intersects the fiber at x in a sphere of radius larger than 1.

However, the ball T_x intersects Δ_Z transversally in a single point, namely x. This intersection with Δ_Z will now be traded for N transversal intersections with E_{3+}. (This is the same N as in Proposition 8.1.) The technique used here is called "connect summing with a transverse sphere" (see, e.g. Chapter 1 of [**7**]). This technique proceeds as follows: Fix $p \in \operatorname{crit}(f)$ and then fix a point $q \in S^3 - \Lambda_p$. Observe that the sphere $S^3 \times q \subset (S^3 \times S^3)_p$ intersects Δ_Z transversally once (at $q \times q$), and it intersects E_{3+} transversally N times, at $q \times \Lambda_p$. Orient $S^3 \times q$ and all N intersections with E_{3+} will have the same sign. (See Assertion 5 of Proposition 8.1.) Because $q \notin \Lambda_p$, this $S^3 \times q$ will have empty intersection with E_{3-}.

Take this $S^3 \times q \subset (S^3 \times S^3)_p$ and push it off ∂Z so that it is an embedded submanifold, $Y \subset \operatorname{int}(Z)$. Push if off only slightly, so that Y still intersects E_{3+} in the N points of the push-off of $q \times \Lambda_p$, and so that Y intersects Δ_Z in the push-off of $q \times q$. Also, do not let Y intersect E_{3-}.

Remark that N is an oriented vector bundle, and thus T_x is an oriented 3-ball. Orient Y so that its intersection number with Δ_Z is the opposite of that for $T_x \cap \Delta_Z$. Now, one can "tube" Y to T_x to obtain a new 3-ball, $P \subset Z$ with the following properties:

(9.7)

1) $\partial P = S_x$,
2) $P \cap (\Delta_Z \cup \partial Z \cup E_L \cup E_R) = \emptyset$,
3) $\operatorname{int}(P) \cap E_{3-} = \emptyset$,
4) $P \cap E_{3+}$ is N distinct points, all homologically the same in $H_0(P)$.

Abstractly, P is the connect sum of Y and T_x. Realize this connect sum in Z by choosing a path, τ, between x and $x' \equiv Y \cap \Delta_Z$ in $\operatorname{int}(\Delta_Z)$. Make sure that τ avoids the paths $\{\Gamma_i\}_{i=1}^{r}$ of Assertion 6 in Proposition 8.1. Modify the exponential map $e : N \to Z$ so that e maps $N|_{x'}$ into Y. Let $S_\tau \subset N|_\tau$ denote the radius 1/8 sphere bundle. Let $T_x^0 \subset N|_x$ and $T_{x'} \subset N|_{x'}$ denote the radius balls of radius 1/8. Then Y is obtained by smoothing the corners of the surgery

$$((Y - T_{x'}^0) \cup (T_x - T_x^0)) \cup S_\tau. \tag{9.8}$$

Here is a picture:

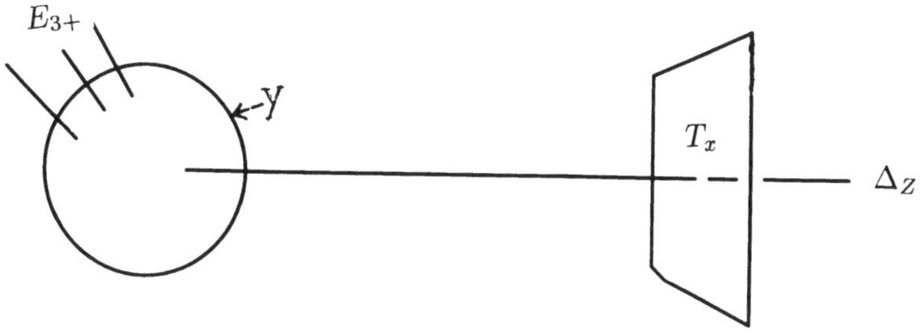

(9.9)

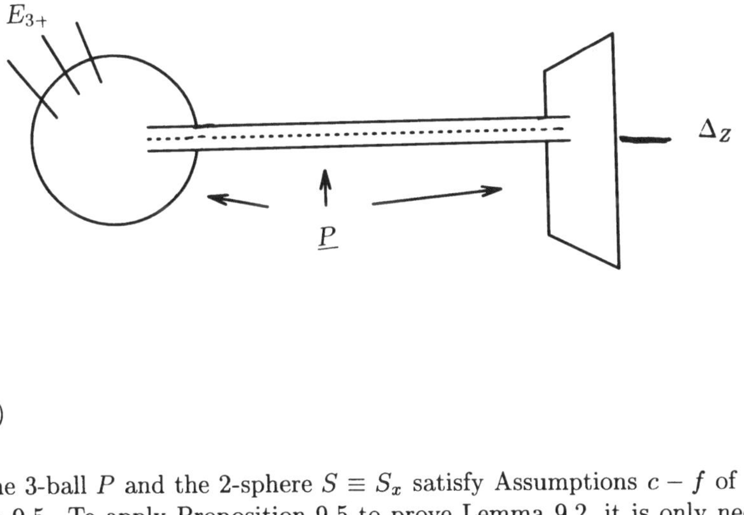

(9.10)

The 3-ball P and the 2-sphere $S \equiv S_x$ satisfy Assumptions $c - f$ of Proposition 9.5. To apply Proposition 9.5 to prove Lemma 9.2, it is only necessary to check that Assumption f of Proposition 9.5 is satisfied. For this purpose, let $\pi : S_x \to x$ denote the projection. Then, the normal bundle to S_x in A is isomorphic to $\pi^* TR \mid_x$. Meanwhile, the normal bundle to P in X along S_x is the same as the normal bundle to T_x in X along S_x which is isomorphic to $\pi^* T\Delta_Z \mid_x$. Thus, the quotient $(N_P \mid_S)/\upsilon_S \approx \pi^*((T\Delta_Z \mid_x)/(TR \mid_x))$, which is trivial, as required. □

c) Proof of Proposition 9.5.

Since $[S]$ is homologically trivial in A, $v_S \to S$ is a trivial 2- plane bundle. Since P is a 3-ball, so $N_P \to P$ is a trivial 4-plane bundle. And, because $(N_P\,|_S)/v_S$ is a trivial bundle, there are no obstructions to extending v_S over P as a 2-plane subbundle of N_P. Use $v \to P$ to denote this 2-plane bundle.

Fix an exponentional map $e_P : N_P \to X$ with the following properties: First, e_P should restrict to v_S as a map into A. And, e_p should restrict to map $N_P\,|_{P\cap B}$ into B. Fix a fiber metric on N_p with the property that e_p embeds the interior of the unit ball bundle in N_p onto a neighborhood of P in X. Then, use e_p to implicitly identify the unit ball bundle in N_P with its e_p-image.

With the preceding understood, let $s_0 \subset v$ denote the 1-sphere bundle of radius 1/4. Let $t_0 \subset v$ denote the ball bundle of radius 1/4. (Thus, $\partial(t_0\mid_{\mathrm{int}(P)}) = s_0\mid_{\mathrm{int}(P)}$.)

Introduce the surgery

$$A_0 \equiv (A - t_0\,|_S) \cup s_0, \tag{9.11}$$

which is a C^0 embedding of a smooth, oriented 4-manifold, A' into X. This embedding is smooth away from the corner at $s_0\,|_S$ and it can be smoothed inside v to produce a smooth embedding of A' into X. Note that A' can be arranged to agree with A on the compliment of any apriori specified open neighborhood $U \subset X$ of P.

With A' understood, consider the various assertions of Proposition 9.5: To prove Assertion 1, consider that

$$\sigma' \equiv (A' \cap B) \cap O = s_0\,|_{P\cap B} \cup \quad \sigma. \tag{9.12}$$

This is a transversal intersection because P intersects B transversaly. The fact that $[\sigma'] \neq 0$ in $H_1(A';\mathbb{Q})$ follows via Meyer- Vietoris and the fact that $[P\cap B] \neq 0$ in $H_0(P)$. (Use the Meyer- Vietoris sequences for the decompositions $A = (A - s_0\,|_S) \cup (t_0\,|_S)$ and also $A' = (A - s_0\,|_S) \cup s_0$.)

Meyer-Vietoris also proves Assertion 2, namely $H^2(A';\mathbb{Q}) = H^2(A;\mathbb{Q})$. (Use the same sequences as above.)

Assertion 3 is true because one can interpret t_0 as a cycle, and this cycle obeys $\partial t_0 = [A] - [A']$.

Assertion 4 is true because the circles in $s_0\,|_{P\cap B}$ bound the discs $t_0\,|_{P\cap B}$.

To prove Assertion 5, the strategy will be to find a framing of the normal bundle to $\mathrm{int}(s_0)$ in X which is compatible with the framing ζ on $A - S$. To begin, introduce the notation $\underline{v}$ to denote the vector field along $s_0 \subset t_0$ which points radially inward on each 2- ball fiber of $t_0 \to P$. With $\underline{v}$ understood, the normal bundle to s_0 in X is isomorphic to $(N_P/v)\oplus$ Span$(\underline{v})$. The bundle $N_P/v \to P$ is a trivial 2-plane bundle (because P is a ball), and so it has a global frame, (e_1, e_2). Thus, $(e_1, e_2, \underline{v})$ is a frame for the normal bundle to $\mathrm{int}(s_0)$ in X.

Now consider the normal bundle to A along S. For this purpose, let e_3 denote the inward pointing tangent bundle to P along S. (So e_3 spans the normal bundle to S in P). Then, along S, the normal bundle to A in X is isomorphic to $(N_P/v)\,|_S \oplus \mathrm{Span}(e_3)$. Thus, $N_P\,|_S \oplus \mathrm{Span}(e_3) \to S$ is isomorphic to the normal bundle of S in X. Let $e_S : N_P\,|_S \oplus \mathrm{Span}(e_3) \to X$ be an exponential map which maps v_S into A and which maps $v\,|_S \quad \oplus \mathrm{Span}(e_3)$ into P. Use e_S to identify a neighborhood of S in X with a neighborhood of the zero section of the bundle $N_P\,|_S \oplus \mathrm{Span}(e_3) \to S$.

Because $\pi_2(SO(3)) = 0$, there are no obstructions to homotoping the given normal frame ζ in a neighborhood of S so that the restriction of ζ to said neighborhood is$(e_1\,|_S, e_2\,|_S, e_3)$. Thus, two thirds of the frame ζ can be extended over A' from $A - S$. As usual, there is no obstruction to homotoping $\underline{v}$ near $s_0\,|_S$ to equal e_3 on the compliment of a neighborhood of S in A. □

d) Making $[\sigma] = 0$ in $H_1(E_{3+})$.

This subsection is concerned with the construction of E'_{3+} of Lemma 9.3. The construction requires a preliminary digression to introduce a modified version of Proposition 8.2. Here is the scenario: As in Proposition 8.2, X is a smooth, oriented 7-manifold and A, B are oriented, 4-dimensional submanifolds of X. Let $O \subset X$ be an open set which contains a component, σ, of $A \cap B$.

PROPOSITION 9.6. *Let X, O, A, B and σ be as described above. Suppose that*

a) $[\sigma] \neq 0$ *in* $H_1(B;\mathbb{Q})$.

b) $[\sigma] = 0$ *in* $H_1(O;\mathbb{Q})$.

c) $H^3_{comp}(B;\mathbb{Q}) \to H^3(B;\mathbb{Q})$ *is injective.*

d) *B has trivial normal bundle in X with a given normal framing ζ.*

e) *O is path connected.*

Then there exists $n \neq 1$ and there exists an oriented, dimension 4 submanifold $B' \subset X$ which obeys:

1) *Let B_0 denote the disjoint union of n distinct, like oriented push- off copies of B. Then $B' = B_0$ in $X - O$.*

2) *B' intersects A transversally, and $\sigma' \equiv (B' \cap A) \cap O$ is compact.*

3) $[\sigma'] = 0$ *in* $H_1(B';\mathbb{Z})$.

4) $[\sigma'] = n\,[\sigma]$ *in* $H_1(A;\mathbb{Q})$.

5) $H^2(B';\mathbb{Q}) = H^2(B_0;\mathbb{Q})$.

6) $[B'] = n\,[B]$ *in* $H_4(X, X - O;\mathbb{Z})$.

7) *B' has trivial normal bundle in X and the push-off normal framing, ζ, of B_0 extends from $B_0 - B_0 \cap O$ to a smooth normal framing of B' in X.*

This proposition is proved below. Consider its application first.

Proof of Lemma 9.3. The strategy is to apply Proposition 9.6. For this purpose, set $X \equiv \mathrm{int}(Z)$ and $O \equiv Z - (\partial Z \cup U_\Delta \cup E_L \cup E_R)$. Set $A \equiv \mathrm{int}(E_{3-})$ and set $B \equiv \mathrm{int}(E_{3+})$. Assumption a of Proposition 9.6 holds under the assumption that (9.1) is false. Assumption b of Proposition 9.6 holds for the following reason: The vanishing of $H_1(Z;\mathbb{Q})$ is guaranteed by Lemma 3.7. The vanishing of $H_1(O;\mathbb{Q})$ then follows using Meyer-Vietoris. (Remember that Δ_Z and $E_{L,R}$ are codimension 3 in Z.) Assumption c of Proposition 9.6 holds because B is the interior of a manifold (E_{3+}) with boundary a disjoint union of 3-spheres. (Remember that $H^2(S^3) = 0$.)

Thus, the assumptions of Proposition 9.6 hold when (9.1) is false. Take E'_{3+} in Lemma 9.3 to be equal to the closure in Z of the submanifold B' of Proposition 9.6. □

Proof of Proposition 9.6. By assumption, there exists an integer n such that $n\,[\sigma] = 0 \in H_1(O;\mathbb{Z})$. Use the given framing of B's normal bundle in X to push-off n parallel copies of B, all with the same orientation as B. Let B_0 denote the union of these n disjoint submanifolds. Make these push-offs close to the original, so that $\sigma_0 \equiv (B_0 \cap A) \cap O$ will be equal to n like oriented, push-off copies of σ. Then $[\sigma_0] = 0$ in $H_1(O;\mathbb{Z})$. This means that σ_0 is the boundary of an oriented, embedded surface with boundary, $R \subset O$. Since A and B have codimension 3 in O, one can arrange R so that $\operatorname{int}(R) \cap A = \emptyset$ and $\operatorname{int}(R) \cap B = \emptyset$. One can also arrange R to be connected because O is connected.

Here is the local picture of R near a component of σ_0: Let C be a component of σ_0. Let v_C denote the normal bundle to C in R, an oriented line bundle. Then, the normal bundle to C in O splits as the direct sum $v_C \oplus L\,|_C \oplus TB_0\,|_C$, where $L \to \sigma_0$ is an oriented 2-plane bundle, and so trivial. (Thus, the normal bundle for B in X along σ_0 splits as $v_C \oplus L\,|_C$.) Choose a framing, ζ_B, for $TB_0\,|_C$ and also choose a framing (e_1, e_2) for L. If a frame, ζ, has been given for B's normal bundle in X, then choose (e_1, e_2) so that with the addition of an oriented frame, e_3, for v_C, the triple (e_1, e_2, e_3) defines an adapted frame to C which is homotopic to $\zeta\,|_C$. (See Lemma 8.4).

Choose an exponential map, e_C, from the normal bundle along C into X which maps the positive axis in v into R and which maps $TB_0\,|_C$ into B_0. Use this exponential map and the given framings to identify an open neighborhood of C in X with one of $C \times (0,0)$ in the triple product $C \times \mathbb{R}^3 \times \mathbb{R}^3$.

Then R near C is identified with the subspace of points (t,x,y)in $C \times R^3 \times R^3$ where $x_1 = x_3 = y = 0$ and $x_1 \geq 0$. Meanwhile, B_0 near C is identified with the subspace of points (t,x,y) with $x = 0$. And, A near C is identified with the subspace of points (t,x,y) where

$$x = \underline{A}(t)\,y + O(|\,y\,|^2), \tag{9.13}$$

where $\underline{A} : C \to Gl(3,\mathbb{R})$.

Because R is connected and has non-trivial boundary, the normal bundle $N_R \to R$ to R in X is isomorphic to the trivial 5- plane bundle. On σ_0, this bundle has a splits as $N_R \approx L \oplus TB_0\,|_{\sigma_0}$. □

LEMMA 9.7. *The subbundle $L \subset N_R\,|_{\sigma_0}$ extends over R as an oriented,*

2-plane subbundle, L_R, of N_R. Furthermore, given a framing (e_1, e_2) for L, there exists an extension $L_R \to R$ of L over which the framing (e_1, e_2) extends.

Proof. Choose a trivialization for N_R to identify this bundle with $R \times \mathbb{R}^5$. Then $L \subset N_R \mid_{\sigma_0}$ is defined by a map from σ_0 into the space of oriented 2-planes in $\mathbb{R}^5$. The space of such oriented 2-planes deformation retracts onto $SO(5)/(SO(3) \times SO(2))$; thus, L is defined by a map, $\eta : \sigma_0 \to SO(5)/(SO(3) \times SO(2))$. Note that the choice of an oriented frame (e_1, e_2) for L defines a lift of η to a map $\underline{\eta} : \sigma_0 \to SO(5)/SO(3)$.

With the preceding understood, the question here is whether or not this map η can be extended over R. The answer is that $\underline{\eta}$ can be extended because $SO(5)/SO(3)$ is simply connected. □

With Lemma 9.7 understood, pick an extension, $L_R \to R$ of L over R over which the given frame (e_1, e_2) for L extends. Introduce $V_R \approx N_R/L_R$ and choose a splitting $N_R \approx L_R \oplus V_R$ with the property that $V_R \mid_{\sigma_0}$ agrees with $TB \mid_{\sigma_0}$.

Fix an exponential map $e_R : N_R \to X$ with the property that e_R on $N_R \mid_{\sigma_0}$ agrees with the restriction of $e\sigma_0$ to $L \oplus TB_0 \mid_{\sigma_0}$. Choose a fiber metric on N_R so that e_R embeds the interior of the radius 2 ball bundle onto an open neighborhood of $\operatorname{int}(R)$ in X.

Let $\epsilon > 0$ and let $T_R \subset V_R$ denote the radius ϵ ball bundle. Let $S_R \subset T_R$ denote the sphere bundle of radius ϵ. Since V_R is oriented, both T_R and S_R are isomorphic to trivial fiber bundles.

With all of the above understood, define the surgery

$$B_1 \equiv (B_0 - \operatorname{int}(T_R \mid_{\sigma_0})) \cup S_R. \tag{9.14}$$

This B_1 is a C^0 embedding of a smooth, oriented manifold into X. The embedding fails to be smooth at the corner, $S_R \mid_{\sigma_0}$. Smooth B_1 inside V_R on a neighborhood of this corner to produce a smoothly embedded, oriented submanifold, $B' \subset X$.

The claim now is that B', as described above, will satisfy Assertions 1-7 of Proposition 9.6: Assertion 1 is true by construction. To prove Assertion 2, use (9.13) to see that σ' is a push-off of σ_0. Indeed, for small ϵ, the copy in σ' of the component C in (9.13) is given, to order ϵ^2, as the set of (t, x, y) with

$$x = (\epsilon(\sum_{1 \leq j \leq 3} \mid \underline{A}^{-1})_{j,1} \mid^2)^{-1/2}, 0, 0) \tag{9.15}$$

and $y = \underline{A}^{-1}(x_1, 0, 0)$. Here $\underline{A}_{i,j}^{-1}$ are the components of the inverse to $\underline{A}$ in (9.13). Note that this identification of σ' confirms Assertion 4 as well.

To prove Assertion 3 of Proposition 9.6, observe first that (9.15) identifies σ' as a section of S_R over a push-off into $\operatorname{int}(R)$ of σ_0. Thus, σ' bounds in B' if the section in question, s, extends as a section over R of S_R. Now, S_R is a trivial 2-sphere bundle so isomorphic to $R \times S^2$. Such an isomorphism identifies the section s, with a map, also called s, from σ to S^2. Since $\pi_1(S^2) = 0$, any such map extends over R.

To prove Assertion 5, invoke the Meyer-Vietoris exact sequences for the decomposition $B_0 = (B_0 - T_R\,|_{\sigma_0}) \cup T_R\,|_{\sigma_0}$ and for the decomposition of B_1 in (9.14). (Note that B_1 and B' are homeomorphic.)

To prove Assertion 6, remark that B' and B_1 are C^0-isotopic by an isotopy with support in O, so $[B'] = [B_1]$ in $H_4(X, X - O; \mathbb{Z})$. Meanwhile, T_R, as a 5-cycle in O satisfies $\partial T_R = [B_1] - [B_0]$.

Finally, consider Assertion 7. To begin, remark that the normal bundle to S_R in X is isomorphic to $L_R \oplus \tau$, where $\tau \to S_R$ is the trivial line bundle which is S_R's normal bundle in V_R. Now, by construction, L_R has a frame, (e_1', e_2') which extends (e_1, e_2).

Meanwhile, the normal bundle, N_B, to B in X splits upon restriction to σ_0 as $L \oplus v_{\sigma_0}$, where v_{σ_0} is the normal bundle to σ_0 in R, an oriented line. And, the frame ζ is homotopic in a neighborhood of σ_0 so that the result restricts to σ_0 as (e_1, e_2, e_3), where (e_1, e_2) frame L and where e_3 is the inward pointing normal vector to σ_0 in R. Thus, two thirds of the normal frame ζ can extended from the compliment of a neighborhood of σ_0 in B_0 over B'. As usual, there are no obstructions to extending the remaining third of ζ over B'.

e) Proof of Lemma 9.4.

If there are some number $q > 1$ components of B_0, label the components of B_0 as $\{B_{0,\alpha}\}_{\alpha=1}^q$. Pick $y_\alpha \in B_\alpha \cap (O - A)$ for each index $\alpha > 1$. Also, choose $q - 1$ distinct points $\{x_\alpha\}_{\alpha \geq 2} \subset B_{0,1}$. Since O is path connected, so $O - A$ will be path connected; and so one can find, for each $\alpha \geq 2$, a path ρ_α (an embedding of $[0, 1]$) which starts at y_α and ends at x_α. Choose the set $\{\rho_\alpha\}_{\alpha \geq 2}$ to be distinct. Now, for each $\alpha \geq 2$, mimick the surgery in (8.23) to make an ambient connect sum of $B_{0,\alpha}$ with $B_{0,1}$. Since the $\{\rho_\alpha\}$ are distinct, these connect sums can be made with out interfering with each other. Use B to denote the result, after smoothing near the corners. The verification that B does the job is left to the reader as an exercise. (For the framing issue, see Lemma 8.10.) □

10 The last pass at $E_\pm$. It is the purpose of this section to explain how to make $E_\pm$ from $E_{4\pm}$ of the preceding section. The metamorphasis from $E_{4\pm}$ to $E_\pm$ will be called *melding*.

This melding operation only changes $E_{4\pm}$ in a neighborhood of $\Delta_Z \cup \partial Z$, and the neighborhood in question can be as small as desired. In particular, in this neighborhood, $E_{4\pm}$ should be the image of E_{4-} under the switch map $\Theta : Z \to Z$ which sends (x, y) to (y, x). (See Assertion 1 of Proposition 9.1). Furthermore, in this neighborhood, E_{4-} (hence, E_{4+}) should consist, locally, of N parallel push-off copies (see Assertions 5 and 6 of Proposition 9.1).

The effect of the melding will be to push all of these parallel copies together on some smaller neighborhood of $\Delta_Z \cup \partial Z$. The cost of the melding is that $E_\pm$ will not be a manifold (unless $N = 1$ in Proposition 9.1).

a) E_{4-} near $\Delta_Z \cup \partial Z$.

Consider E_{4-}. The following construction will be done r times, once for each pair in $\{(a_i, b_i)\}_{i=1}^r$. These r versions can be done simultaneously, so fix attention on one index i, and simplify notation by setting $a \equiv a_i$ and $b \equiv b_i$.

The set $\Gamma_i \subset E_{4-} \cap \Delta_Z$ is a set of N embedded intervals which connect the N components of $S^3 \times \Lambda_a = E_{4-} \cap (S^3 \times S^3)_a$ with the N components of $S^3 \times \Lambda_b = E_{4-} \cap (S^3 \times S^3)_b$. Near

$$(S^3 \times \Lambda_a) \cup \Gamma_i \cup (S^3 \times \Lambda_b), \qquad (10.1)$$

E_{4-} consists of N components (sheets), $\{Y_\alpha\}_{a=1}^N$. Here is a picture:

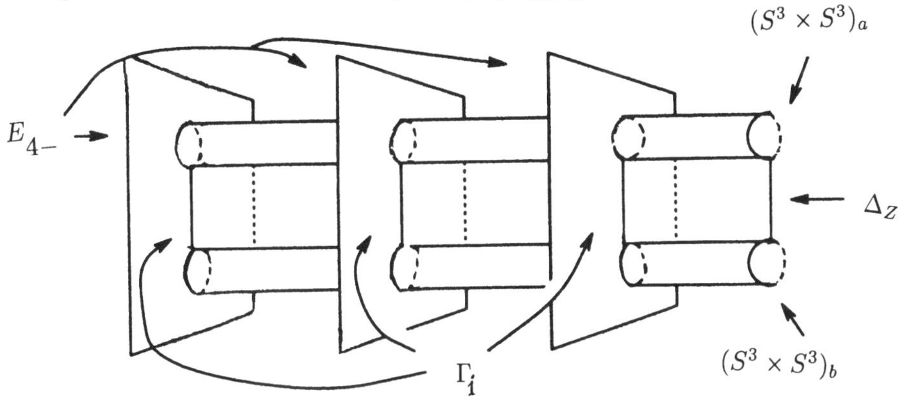

(10.2)

The sheets $\{Y_\alpha\}_{\alpha \geq 2}$ are push-off copies of Y_1. As for Y_1, it is an embedded image in Z of the compliment in the open, unit 4-ball of the interiors of a pair of disjoint 4-balls, $B_\pm$, of radius 1/8, respectively centered at $(\pm 1/4, 0, 0, 0)$. Note that the boundary of B_- is mapped to $S^3 \times p_a \subset S^3 \times \Lambda_a$, and the boundary of B_+ is mapped to the corresponding $S^3 \times p_b \subset S^3 \times \Lambda_b$. Furthermore, $Y_1 \cap \Delta_Z$ is the segment of the x_1 axis between $(\pm 1/8, 0, 0, 0) \in \partial B\pm$. Here is a picture of Y_1:

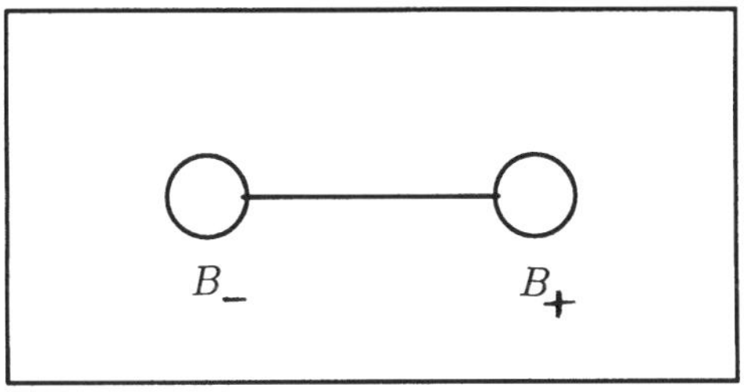

(10.3)

As remarked, the $\{Y_\alpha\}_{\alpha\geq 2}$ in (10.2) are push-off copies of Y_1. To be precise here, remember that Y_1 has a framing, $\zeta \equiv (e_1, e_2, e_3)$, to its normal bundle, N_{Y_1}, which is a product framing that restricts to both $(S^3 \times p_a)$ and $(S^3 \times p_b)$ as a constant framing. (See Assertion 8 of Propositio 9.1.) Fix an exponential map,

$$e : N_{Y_1} \to B, \tag{10.4}$$

which maps N_{Y_1}'s restriction to $(S^3 \times p_a)$ into $(S^3 \times S^3)_a$, and which likewise maps N_{Y_1}'s restriction to $(S^3 \times p_b)$ into $(S^3 \times S^3)_b$. Fix a metric on N_{Y_1} which makes the frame ζ orthonormal, and fix $\epsilon > 0$ such that (10.4) embeds the interior of the radius 2ϵ ball bundle onto a neighborhood of Y_1 in Z. Use e in (10.4) to identify the interior of this ball bundle with its image in Z.

With the preceding understood, the copy Y_α of Y_1 can be taken as the image of the section $s_\alpha : Y_1 \to N_{Y_1}$ that is given by

$$s_\alpha(x) = (\alpha - 1)\, N^{-1}\, \epsilon\, e_1. \tag{10.5}$$

b) The meld.

With the preceding picture E_{4-} near (10.1) understood, here is the meld: Fix a function $\beta : [0,1] \to [0,1]$ which has the following properties:

(10.6)

1) $\beta \equiv 1$ on $[5/8, 1]$.
2) $\beta \equiv 0$ on $[0, 1/2]$.
3) β is nondecreasing.

As described in (10.3), identify Y_1 with a subset of the unit ball about the origin in $\mathbb{R}^4$. Use x to denote the Euclidean coordinate in $\mathbb{R}^4$, and, by restriction, a point in Y_1. By the way, note that the assignment to $x \in Y_1$ of the number $\beta(|\, x\,|)$ defines a smooth function on Y_1 which vanishes in a neighborhood of

$$B_- \cup \{(x_1, 0, 0, 0) : -1/8 \leq x_1 \leq 1/8\} \cup B_+. \tag{10.7}$$

For $\alpha \geq 2$, define the deformation, Y'_α, of Y_α as follows: Y'_α is the image in N_Y of the section s'_α which sends $x \in Y_1$ to

$$s'_\alpha(x) \equiv (\alpha - 1)\, N^{-1}\, \beta(|\, x\,|)\, \epsilon\, e_1. \tag{10.8}$$

Notice that Y'_α agrees with Y_α on the compliment of a regular neighborhood of (10.1) in Z; but that Y'_α coincides with Y_1 on a smaller neighborhood of

(10.1). Here is a picture for $\alpha \geq 1$:

(10.9)

In (10.9), the shaded region marks where Y'_α and Y_1 intersect.
Here is a picture of all the $\{Y'_\alpha\}_{\alpha \geq 1}$:

(10.10)

Use E_- to denote the result of applying the preceding meld operation to E_{4-} in a neighborhood of (10.1) for each $i \in \{1, \cdots, r\}$.

As for E_+, remember that E_{4+} coincided with $\Theta(E_{4-})$ near each of the r versions of (10.2). This neighborhood can be assumed to include the regions that are depicted in (10.2). With this understood, set $E_+ \equiv E_{4+}$ outside of the Θ-image of the regions in (10.2), but inside the Θ-image of each region in (10.2), declare

$$E_+ \equiv \Theta(E_-). \tag{10.11}$$

c) Properties of $E_\pm$.

The following proposition describes some of the salient features of $E_\pm$:

PROPOSITION 10.1. *Construct $E_\pm \subset Z$ as described above. Then:*

1) *There is an open neighborhood $U \subset Z$ of $\Delta_Z \cup \partial Z$ such that $E_+ \cap U$ and $E_- \cap U$ are images of each other under the switch map on Z.*
2) *The fundamental classes $[E_\pm]$ are equal to $N[E_{1\pm}]$ for some integer $N \geq 1$. Here, $[E_{1\pm}]$ are described by (6.3) and Lemma 6.3.*
3) *$E_\pm$ have empty intersection with $M_0 \times M_0$ and $M_1 \times M_1$.*
4) *$E_\pm$ have empty intersection with $E_{L,R}$ of (4.15).*

5) *If* $p \in$ *crit*(f), *then the intersection of* E_- *with* $(S^3 \times S^3)_p$ *has the form* $S^3 \times x_p$, *where* x_p *is a single point. Similarly, the intersection of* E_+ *with* $(S^3 \times S^3)_p$ *is* $x_p \times S^3$.

6) $E_\pm \cap \Delta_Z = \cup_{i=1}^r v_i$, *where* $v_i \subset \Delta_Z$ *is as follows: There is a flow line* μ_i *which starts at* a_i *and ends at* b_i. *With the canonical identification of* Δ_W *with* W *understood,* v_i *is a closed interval in a push-off copy of* μ_i. *And,* v_i *starts at* $(x_a, x_a) \in (S^3 \times S^3)_a$ *and* v_i *ends at* $(x_b, x_b) \in (S^3 \times S^3)_b$. *Here,* $a \equiv a_i$ *and* $b \equiv b_i$.

7) $E_- \cap E_+ = \cup_{i=1}^r v_i$.

8) $H^2(E_\pm; \mathbb{Q}) = 0$.

The proof is straightforward and left to the reader. (See Proposition 9.1. Also, use Meyer-Vietoris to compute $H^2(E_\pm)$.)

Here is a picture:

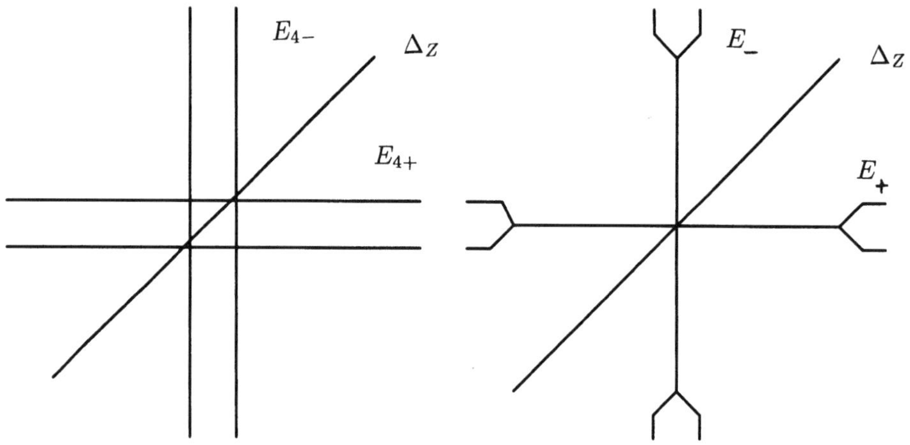

Before melding After melding

(10.12)

11 Completing the proof. The purpose of this last section is to complete the proof of Theorem 2.9. The strategy here will be as follows: Suppose that M_0 and M_1 are cobordant via a spin 4-manifold, W, with the rational homology of S^3. Factor the cobordism as in Assertion 5 of Proposition 3.2 into two pieces, $W_1 \cap W_3$. Both W_1 and W_3 are given by (3.11). Here, W_1 is a cobordism from M_0 to a rational homology sphere M, while W_3 is a cobordism from M to M_1. In both cases, the manifold with boundary, $Z.$ ($\equiv Z_{1,3}$), has been defined, and Sections $4d$, $4e$ and 10 describe the variety $\Sigma_{Z\cdot} \subset Z.$ (The latter require a choice of base point $p \in M$.) Let $Z \equiv Z_1 \cup Z_3$ and $\Sigma_Z \equiv \Sigma_{Z_1} \cup \Sigma_{Z_3}$, where the common boundary components in both cases are identified (these being $M \times M$ in the former and Σ_M in the latter).

With Z and Σ_Z understood, the proof plan from Section $2k$ will be complete with the completions of Steps 3 and 4 in Section $2k$. These steps are considered below.

Completing Step 3 requires the construction of a 2-form ω_Z on $Z - \Sigma_Z$ which satisfies (2.27). This 2-form will be constructed first on the compliment of Σ_Z in a regular neighborhood $N_Z \subset Z$ of Σ_Z. The extension to $Z - \Sigma_Z$ will be made by appeal to Lemma 4.2.

Step 4 of Theorem 2.9's proof (from Section $2k$) will be completed during the construction of ω_Z.

Let $N_{Z_1} \equiv N_Z \cap Z_1$ and define N_{Z_3} analogously. The 2-form ω_Z on $N_Z - \Sigma_Z$ will be constructed first on $N_{Z_1} - \Sigma_{Z_1}$ and second on $N_{Z_3} - \Sigma_{Z_3}$. The case of $N_{Z_1} - \Sigma_{Z_1}$ is considered in Subsections $11a-h$ and that of $N_{Z_3} - \Sigma_3$ is considered in Subsection $11i$. These two constructions are matched in Subsection $11j$ where Step 4 of Section $2k$ is verified. Section $11k$ completes the proof of Theorem 2.9 with a discussion of the conditions in Lemma 4.2.

To avoid cumbersome notation, the subscript "$_1$" will be dropped in Subsections $a - h$. Thus, in these sections, Z will denote Z_1, Σ_Z will denote Σ_{Z_1}, etc.

a) Preliminary remarks.

Construction of ω_Z on $N_Z - \Sigma_Z$ is accomplished in two steps. The first step defines ω_Z near $\Delta_Z \cup E_L \cup E_R$ by using (4.22), but where φ_Z is a map which is defined only on a regular neighborhood, N', in Z of $\Delta_Z \cup E_L \cup E_R$. This φ_Z has $\varphi_Z^{-1}(0) = \Sigma_Z \cap N'$. The second step constructs ω_Z on $N_Z - N'$.

The construction of φ_Z occupies Subsections $b - f$, below. The construction of ω_Z on $N_Z - N'$ occupies Subsections g and h.

b) Near $E_L \cap E_R$.

The purpose of this subsection is to construct φ_Z near $E_L \cap E_R$. To be precise, a neighborhood $U \subset Z$ of $E_L \cap E_R$ will be described with a map

$$\varphi_U : U \to \mathbb{R}^3 \tag{11.1}$$

which obeys

$$\varphi_U^{-1}(0) = (E_L \cup E_R \cup \Delta_Z) \cap U. \tag{11.2}$$

Then, $\varphi_Z \mid U$ will be declared equal to φ_U.

A digression on framings begins the construction of φ_U. To start the digression, fix a frame for $TM \mid_p$. This frame can be thought of as a frame for the normal bundle to p in TM. Use the pseudo- gradient flow to extend this frame as a normal framing to the flow line $\gamma \subset Z$. This normal framing to γ induces a framing of $TM_0 \mid_{p_0}$. End the digression.

Parameterize that flow line γ so that $f(\gamma(t)) = t$. Let $N_\gamma \to \gamma$ denote the normal bundle to γ in W and select an exponential map $e_\gamma : N_\gamma \to W$ which maps $N_\gamma \mid_{p_0, p}$ into M_0, M, respectively. Require that

$$f \circ e_{\gamma(t)} \equiv t. \tag{11.3}$$

Use this exponential map and the normal framing with the afore- mentioned parameterization of γ to define a diffeomorphism, ψ_γ, from a neighborhood, $O_\gamma \subset W$ of γ onto $[0,1] \times B$, where $B \subset \mathbb{R}^3$ is a ball-neighborhood of the origin.

Using the preceding identification, build $\psi_U \equiv (u_\gamma \times u_\gamma) \mid_Z$ of the neighborhood $U \equiv Z \cap (O_\gamma \times O_\gamma)$ of $E_L \cap E_R$ with $[0,1] \times B \times B$. This U and ψ_U obey the conclusions of Assertion 4 in Lemma 4.5 except that $B \subset \mathbb{R}^3$ should everywhere replace $\mathbb{R}^3$ and $B \times B$ should everywhere replace $\mathbb{R}^3 \times \mathbb{R}^3$.

With the preceding understood, define φ_Z on U to be the composition of the map ψ_U with the map from $[0,1] \times \mathbb{R}^3 \times \mathbb{R}^3$ which sends (t, x, y) to $\Phi_0(x, y)$ with Φ_0 given by (2.15).

Note that $\varphi_Z \mid_U$ agrees with Proposition 2.5's map φ when restricted to a neighborhood of $p_0 \times p_0$ in $M_0 \times M_0$, or to a neighborhood of $p \times p$ in $M \times M$.

Note also, for reference below, that the map $\varphi_Z \mid_U$ is invariant under the switch map $\Theta : Z \to Z$ which sends (x, y) to (y, x).

c) Near $E_L \cup E_R$.

The purpose of this subsection is to construct φ_Z near $E_L \cup E_R$. To begin, remark that the normal bundle $N_R \to E_R$ to E_R in Z is naturally isomorphic to $\pi_L^* N_\gamma$. Thus, said normal bundle has a natural framing.

Take the dual to this natural framing to frame the dual bundle, N_R^* and choose an exponential map $e_R : N_R \to Z$. The framing of N_R^* and e_R together define a map, φ_R, from a neighborhood of E_R in Z into $\mathbb{R}^3$ which has E_R as the inverse image of zero. (See (2.14).) Choose this exponential map so that it sends $N_R \mid_{M_0 \times M_0}$ into $M_0 \times M_0$ and likewise sends $N_R \mid_{M \times M}$ into M. (The exponential map $e_\gamma : N_\gamma \to W$ of the preceding subsection induces such an exponential map in a natural way.)

On $E_R \cap U$, the differentials of the maps φ_R and φ_U are scalar multiples of each other, and so there is a homotopy of φ_R near U which has it agree with $\varphi_Z \mid_U$ on U and so extend $\varphi_Z \mid_U$ to map a neighborhood of E_R in Z to $\mathbb{R}^3$ with the correct inverse image of zero. See the Step 2 of the proof of Proposition 2.5 for the details.

With $\varphi_Z \mid_U$ now extended over E_R, extend it further over $E_R \cup E_L$ by using the switch map $\Theta : Z \to Z$. Use $\varphi_Z \mid_{RL}$ to denote this extended map.

Note that $\varphi_Z \mid_{R,L}$ can be made so that its restriction to a neighborhood of $(p_0 \times M_0) \cup (M_0 \times p_0)$ in $M_0 \times M_0$ agrees with the map φ for M_0 in Proposition 2.5. Likewise, its restriction to a neighborhood of $(p \times M) \cup (M \times p)$ in $M \times M$ can be arranged to agree with the analogous φ for M

d) Near $E_\pm \cap \Delta_Z$.

The intersections between $E_\pm$ and between these varieties and Δ_Z form a set of disjoint line segments, $\{v_i\}_{i=1}^r$. (Note that $E_\pm$ are manifolds near these line segments.) The purpose of this subsection is to define the map φ_Z near each v_i.

To start, fix $i \in \{1, \cdots, r\}$. Let $a \equiv a_i$ and $b \equiv b_i$. Note that v_i has end points $(x_a, x_a) \subset (S^3 \times S^3)_a$ and $(x_b, x_b) \subset (S^3 \times S^3)_b$. Also, the identification, using π_L or π_R, of Δ_Z with a subset of W identifies v_i with a subinterval in a pseudo-gradient flow line which starts at a and ends at b.

Remember that E_- is the result of melding E_{4-} of Proposition 9.1. This means, in particular, that $(S^3 \times S^3)_a \cup v_i \cup (S^3 \times S^3)_b$ has a neighborhood, $U_i \subset Z$, such that $E_- \cap U_i$ is the same point set as a component, Y, of $E_{4-} \cap \Delta_Z$. Meanwhile, $E_+ \cap U_i = \Theta(Y)$.

Remark next that E_{4-} has, according to Assertion 8 of Proposition 9.1, a special normal framing, ζ. And, E_{4+} has a special normal frame, ζ', which restricts to U_i as the image of ζ under the switch map Θ. The pair of frames (ζ, ζ') restrict to v_i to frame the normal bundle $N_i \to v_i$ of v_i in Z.

Notice that Θ fixes Δ_Z and the differential of Θ (denoted Θ^*) acts on N_i and interchanges Span(ζ) with Span$\zeta')$.

Fix an exponential map $e : N_i \to v_i$ with the following properties:

(11.4)

1) e :Span$(\zeta) \to E_-$.
2) e :Span$(\zeta') \to E_+$.
3) At $(x_a, x_a), e$ maps N_i into $(S^3 \times S^3)_a$.
4) At $(x_b, x_b), e$ maps N_i into $(S^3 \times S^3)_b$.
5) $\Theta \circ e = e \circ \Theta^*$.

Together, the map e and the frames (ζ, ζ') define a map

$$\psi : v_i \times \mathbb{R}^3 \times \mathbb{R}^3 \to U_i \tag{11.5}$$

with the following properties:

(11.6)

1) There is an open ball $B \subset \mathbb{R}^3$ about 0 and ψ embedds $v_i \times B \times B$.
2) ψ is the identity on $v_i \times (0,0)$.
3) $\psi^{-1}(E_-) = \{(t, x, 0) \in v_i \times \mathbb{R}^3 \times \mathbb{R}^3\}$.
4) $\psi^{-1}(E_+) = \{(t, 0, y) \in v_i \times \mathbb{R}^3 \times \mathbb{R}^3\}$.
5) $\psi^{-1}(\Delta_Z) = \{(t, x, x) \in v_i \times \mathbb{R}^3 \times \mathbb{R}^3\}$.
6) $\psi((x_a, x_a) \times \mathbb{R}^3 \times \mathbb{R}^3) \subset (S^3 \times S^3)_a$.
7) $\psi((x_b, x_b) \times \mathbb{R}^3 \times \mathbb{R}^3) \subset (S^3 \times S^3)_b$.
8) $\psi(t, x, y) = \Theta(\psi(t, y, x))$.

Given the preceding, define the map φ_Z on a neighborhood of v_i in Z by declaring that

$$(\varphi_Z \circ \psi)(t, x, y) \equiv \Phi_0(x, y), \tag{11.7}$$

where Φ_0 is given in (2.15).

e) Near $(S^3 \times S^3)_p$.

The next step is to define the map φ_Z near Z's boundary components $\{(S^3 \times S^3)_p\}_{p\in \text{crit}(f)}$. So, fix $i \in \{1, \cdots, r\}$ and let p denote either a_i or b_i.

A neighborhood, $V_p \subset Z$, of $(S^3 \times S^3)_p$ is diffeomorphic to the product $(S^3 \times S^3)_p \times [0, 1)$ as a manifold with boundary. Furthermore, there is no difficulty in finding such a diffeomorphism so that

(11.8)

1) $E_- \cap V_p = (S^3 \times x_p) \times [0, 1)$.
2) $E_+ \cap V_p = (x_p \times S^3) \times [0, 1)$.
3) $\Delta_Z \cap V_p = \Delta_{S^3} \times [0, 1)$.
4) The switch map acts by $\Theta(x, y, t) = (y, x, t)$.

In V_p, the variety E_- is smooth and it agrees with a component, Y, of $E_{4-} \cap V_p$. Also, $E_+ \cap V_p = \Theta(E_- \cap V_p)$. Also, E_{4-} has the normal frame, ζ, which restricts to Y as a constant frame. And, E_{4+} has the normal frame ζ' which restricts to $\Theta(Y)$ as $\Theta * \zeta$.

Use these constant frames to define frames for the dual bundles to the normal bundles of Y and $\Theta(Y)$ in Z. Then, use these frames for the conormal bundles to extend $\varphi_Z \mid U_i$ of (11.7) to a neighborhood of $E_\pm \cap V_p$ by mimicking Step 2 in the proof of Proposition 2.5. (Exploit the product structure on V_p in (11.8).)

Meanwhile, T^*S^3 has its singular framing which gives (see Proposition 2.7, Definition 2.8 and Lemma 2.11) the canonical homotopy class of singular framing for which the value of $I_2(S^3)$ is zero. As in Step 3 of Proposition 2.5's proof, use this framing to obtain a singular framing of the normal bundle to $\Delta_{S^3} \times [0, 1)$ in Z.

(Remember that $(S^3 \times S^3)_p$ has two obvious projections to S^3, these are given by the product structure in (3.26) and are denoted $\pi_\pm$. To be explicit, introduce the coordinate sytem ψ_p of (3.2) and introduce $U_p \equiv \psi_p(\mathbb{R}^4)$. Note that $U_p \times U_p$ is a neighborhood of (p, p) in $W \times W$. With this understood, $\pi_\pm$ are the restrictions to $(S^3 \times S^3)_p$ of the maps from $U_p \times U_p$ to $\mathbb{R}^4$ which are given by

$$(11.9) \qquad \pi_-(x, y) \equiv (y_1, x_2, x_3, x_4) \text{ and } \pi_+(x, y) \equiv (x_1, y_2, y_3, y_4)$$

when $p \in \text{crit}_1(f)$; and by

$$(11.10) \qquad \pi_-(x, y) \equiv (y_1, , y_2, x_3, x_4) \text{ and } \pi_+(x, y) \equiv (x_1, x_2, y_3, y_4)$$

when $p \in \text{crit}_2(f)$. Then, the map $\pi_+^* - \pi_-^*$ identifies T^*S^3 with the normal bundle to Δ_{S^3} in $(S^3 \times S^3)_p$.)

As in Step 3 of the proof of Proposition 2.5, use the induced framing of the normal bundle to $\Delta_S^3 \times [0,1)$ in Z and the product structure of V_p as described in (11.8) to extend $\varphi_Z \mid U_i$ over a neighborhood of the intersection of $\Delta_Z \cap V_p$.

Note that this extension of φ_Z to a neighborhood of $\Sigma_Z \cap V_p$ has the property that its restriction to $(S^3 \times S^3)_p$ is equal to a map φ in Proposition 2.5 (for S^3) that gives $I_2 = 0$. See Lemma 2.11.

f) Near Δ_Z.

At this point, the map φ_Z is defined on a neighborhood of $E_{L,R}$ and on a neighborhood of each $\{(S^3 \times S^3)_p\}_{p \in \text{crit}(f)}$ and on a neighborhood of each $\{v_i\}_{i=1}^r$. The purpose of this subsection is to extend φ_Z to a neighborhood of Δ_Z.

As a preamble, remark that φ_Z has already been defined near some parts of Δ_Z, specifically, near $\{v_i\}_{i=1}^r$ and near γ and near Δ_{S^3} in each $\{(S^3 \times S^3)_p\}_{p \in \text{crit}(f)}$. The extension over a full neighborhood of Δ_Z is obtained by mimicking the constructions of Step 3 in the proof of Proposition 2.5.

To be more precise about this strategy, it is necessary to first introduce $N_Z^* \to \Delta_Z$, the dual to the normal bundle to Δ_Z in Z. A particular framing over $\Delta_Z - (\gamma \cup (\cup_{i=1}^r v_i))$ will be chosen for N_Z^*, and an exponential map, $e : N_Z \to Z$ as well. These give a map φ_Δ, as in (2.14), from a neighborhood in Z of $\Delta_Z - (\gamma \cup (\cup_{i=1}^r v_i))$ to $\mathbb{R}^3$. The singularities near $\gamma \cup (\cup_{i=1}^r v_i)$ of the chosen framing will be constrained so that φ_Δ can be readily homotoped near γ and $(\cup_{i=1}^r v_i)$ and each $\{\Delta_S^3 \subset (S^3 \times S^3)_p\}_{p=1}^r$ to match up with φ_Z where the latter is already specified.

The framing χ for N_Z^* over $\Delta_Z - (\gamma \cup (\cup_{i=1}^r v_i))$ will have the following form near γ: Write $\chi \equiv (e_1, e_2, e_3)$ and use the coordinates near γ that are given by the map ψ_U of Section 11a, above. Write $x = v + u$ and write $y = v - u$ so that $u \equiv 0$ signifies Δ_Z and $u = v = 0$ signifies γ. Then,

$$\chi \mid_{(t,v,v)} = 2\langle v, du\rangle\, v - \mid v \mid^2\, du, \tag{11.11}$$

where $\langle v, du\rangle \equiv \Sigma_{j=1}^3 v_j\, du_j$.

The singularity of χ near each v_i will have the form of (11.11) when the coordinates of (11.5), (11.6) are used.

Also, χ is constrained on the diagonal Δ_{S^3} in each $(S^3 \times S^3)_p$ so that the inverse of the natural identification $(\pi_+^* - \pi_-^*)$ between T^*S^3 and the conormal bundle of $\Delta_{S^3} \subset (S^3 \times S^3)_p$ sends χ to a singular coframe which gives S^3's canonical singular frame.

With the preceding understood, agree now to further constrain χ along Δ_{M_0} as follows: The inverse of the natural identification $(\pi_R^* - \pi_L^*)$ between T^*M_0 and the conormal bundle to $\Delta_{M_0} \subset M_0 \times M_0$ should send χ to a singular frame for T^*M_0. (See Definition 2.3).

This χ, as constrained above, will extend over the rest of Δ_Z when the following condition is met:

LEMMA 11.1. *Let*

(11.12) $T \equiv (\gamma \cup (\cup_{i=1}^{r} v_i) \cup (\cup_{p\in\ crit(f)}(\Delta_{S^3})_p) \cup \Delta_{M_0}) \subset \Delta_Z,$

and let χ be a frame for $N_Z^* \to (\Delta_Z - (\gamma \cup (\cup_{i=1}^{r} v_i))$ which is defined on a neighborhood of T as described above. If the homotopy class of T^*M_0's singular frame $(\pi_R^* - \pi_L^*)^{-1}(\chi \mid \Delta_{M_0})$ is in $ker(l_W)$ (see (2.12), then the frame χ extends over $\Delta_Z - T$ as a frame for N_Z^*.

Given that Lemma 11.1 is true, the extension of the map φ_Z near Δ_Z is obtained by using (2.14) and a singular frame χ as described above for which $(\pi_R^* - \pi_L^*)^{-1}(\chi \mid \Delta_{M_0})$ gives Proposition 2.7's canonical homotopy class of singular frame for T^*M_0. (Theorem 2.9 assumes that the canonical singular frame for M_0 is annihilated by the homomorphism $l_{W_1 \cup W_3}$; and this implies that this homotopy class is also annihilated by l_{W_1}.) The construction of φ_Z near Δ_Z using χ can be made with a straightforward appropriation of the arguments in Step 3 of the proof of Proposition 2.5. These final details are left to the reader.

Proof of Lemma 11.1. The singular framing χ has been defined near the point $p \in \Delta_M$ by (11.11). First, choose any extension, χ', of χ over the remainder of Δ_M so that $(\pi_R^* - \pi_L^*)^{-1}(\chi' \mid \Delta_{M_0})$ is a singular frame for T^*M as described in Definition 2.3.

Because W is a spin manifold, the bundle $N_Z^* \to \Delta_Z$ is a trivial bundle, so it has a framing, h. If $h_{1,2}$ are a pair of framings of N_Z^*, then $h_1 = g\, h_2$, where $g : \Delta_Z \to SO(3)$.

Let $U_0 \subset \Delta_Z$ be a regular neighborhood of $\Delta_{M_0} \cup \gamma \cup \Delta_M$. This U_0 can be taken so that the boundary of its closure in $\Delta_Z$is a submanifold which is diffeomorphic to the connect sum of M_0 with M. One can also take U_0 so that the extension χ' is defined on the boundary of its closure.

Fix $i \in \{1, \cdots, r\}$ and set $a \equiv a_i$ and $b \equiv b_i$. Let $U_i \subset \Delta_Z$ be a regular neighborhood of $(\Delta_{S^3})_a \cup v_i \cup (\Delta_{S^3})_b$. This U_i can be taken so that the boundary of its closure in Δ_Z is a submanifold which is diffeomorphic to S^3. One can also take U_i so that χ (and so χ') is defined over the boundary of its closure.

Let $X \equiv \Delta_Z - (U_0 \cup (\cup_i U_i))$. By construction, X is a smooth manifold with boundary, and χ' is defined over ∂X. Let h be a frame for N_Z^* over Δ_Z. Then $\chi' = g\,(h\,|_{\partial X})$ where $g : \partial X \to SO(3)$. Extending χ' over int(X) is the same as extending g. Obstruction theory shows that the map g will extend if:

(11.13)

1) $g_* : H_1(\partial X; \mathbb{Z}/2) \to H_1(SO(3); \mathbb{Z}/2)$ annihilates the kernel of the inclusion induced homomorphism $i^* : H_1(\partial X; \mathbb{Z}/2) \to H_1(X; \mathbb{Z}/2)$.
2) $g_* : H_3(\partial X; \mathbb{Z}) \to H_3(SO(3); \mathbb{Z})$ annihilates the kernel of the inclusion induced homomorphism $i^* : H_3(\partial X; \mathbb{Z}) \to H_3(X; \mathbb{Z})$.

These two conditions can be satisfied for some extension χ' of χ provided that the restriction of χ to $\Delta_{M_0} - p_0$ differs from h by a map $g_0 : (M_0 - p_0) \to SO(3)$ for which

$$g_{0*} : H_1(M_0; \mathbb{Z}/2) \to H_1(SO(3); \mathbb{Z}/2) \tag{11.14}$$

annihilates the kernel of $i_* : H_1(M_0; \mathbb{Z}/2) \to H_1(W; \mathbb{Z}/2)$. That is, if the invariant $l_W(\cdot)$ of (2.12) vanishes on the homotopy class of T^*M_0's singular frame $(\pi_R^* - \pi_L^*)^{-1}(\chi \mid \Delta_{M_0})$. This proves the lemma. □

g) ω_Z Near E_-.

The map φ_Z has now been defined near all of Σ_Z save for the compliment in $E_\pm$ of a neighborhood of Δ_Z. Let $N' \subset Z$ denote a regular neighborhood of $\Delta_Z \cup E_L \cup E_R$ over which φ_Z is defined. With this understood, define $\omega_Z \equiv \varphi_Z^* \mu$, where μ is the 2-form of (2.3). The task in this subsection and the next is to extend ω_Z over the rest of Σ_Z.

In order to accomplish this task, it proves useful to focus first on E_-. There are three distinguished regions of E_-. Regions 1 and 2 each consist of r components. Each such component is labeled by $i \in \{1, \cdots, r\}$. To describe the i'th component of Region 1 or 2, it proves convenient to return to the notation and coordinates that are used in Section 10 to describe the meld region in E_- near v_i and $(S^3 \times S^3)_a$ and $(S^3 \times S^3)_b$ for $a \equiv a_i$ and $b \equiv b_i$. In particular, return to (10.7) - (10.10).

With the preceding coordinates understood, the i'th component of *Region 1* is defined to be the compliment in the interior of the ball of radius 15/32 in $\mathbb{R}^4$ of the balls $B_\pm$ of radius 1/8 and center $\pm 1/4$, respectively.

The i'th component of *Region 2* is the transition region between the fully melded part of E_- and the part of E_- which agrees with E_{4-}. Here is a 5 step definition of the i'th component of Region 2: *Step 1*: Introduce the annulus $A \subset \mathbb{R}^4$ which is given as the compliment of the radius 13/32 ball about the origin in the ball of radius 1 about the origin. This A is identified as an open subset of the sheet Y_1 of E_{4-}. (In Section 10, Y_1 is identified with the subset of $\mathbb{R}^4$ that is the compliment interior of the radius 1 ball of the balls $B_\pm$.) *Step 2*: In $\mathbb{R}^4$, intoduce the rays $\{r_\alpha\}_{\alpha=0}^N$ by

$$r_0 \equiv \{(t, y) : y = 0 \text{ and } t \le 0\}. \tag{11.15}$$

$$\{r_\alpha \equiv \{(t, y) : y = N^{-1}(\alpha - 1)\,\epsilon\, e_1 \text{ and } t \ge 0\}.$$

Here, e_1 is the unit vector along the first axis in $\mathbb{R}^3$. *Step 3*: Take the function β of (10.6) and define a map ψ from $\mathbb{R}^4 \times \mathbb{R}^3$ into $\mathbb{R}^4$ by setting

$$\psi(x, y) = (\beta(\mid x \mid), y) \tag{11.16}$$

with β as in (10.6). Here, $\mathbb{R}^4$ is written as $\mathbb{R} \times \mathbb{R}^3$. *Step 4*: Introduce the projection

$$\pi_1 : \mathbb{R}^4 \times \mathbb{R}^3 \to \mathbb{R}^4. \tag{11.17}$$

Step 5: The i'th component of Region 2 is given as

$$\psi^{-1}(\cup_{\alpha=0}^{N} r_\alpha) \cap \pi_1^{-1}(A) \tag{11.18}$$

Region 3 contains the compliment in E_- of Regions 1 and 2. And, Region 3 intersects the i'th component of Region 2 in

$$\psi^{-1}(\cup_{\alpha=0}^{N} r_\alpha) \cap \pi_1^{-1}(A') \tag{11.19}$$

where A' is the compliment of the radius 7/8 ball in the interior of the radius 1 ball.

$\cdot\omega_Z$ in Region 3

With the preceding understood, Here is a four step definition of ω_Z in Region 3: *Step 1*: Note that E_- in Region 3 ($\equiv R_3$) agrees with E_{4-} which has framed normal bundle. Use the framing, ζ, from Assertion 8 of Proposition 9.1. *Step 2*: Choose an exponentional map to map said normal bundle into Z. *Step 3*: Define a map, φ, from a neighborhood of R_3 into $\mathbb{R}^3$ by using (2.14). The map φ will have $\varphi^{-1}(0) = R_3$. *Step 4*: Define ω_Z on Region 3 to be

$$\omega_Z \equiv N^{-1}\,\varphi^{-1}(\mu), \tag{11.20}$$

with μ as in (2.3) and with N as in Assertion 2 of Proposition 10.1.

ω_Z in Region 1

The definition of φ_Z in Region 1 is almost as simple: In each component of Region 1, E_- coincides with a sheet of E_{4-} and so has framed normal bundle. Use the framing ζ again. Again, pick an exponential map for the normal bundle, and define a map φ'_Z using (2.14).

This φ'_Z will not necessarily match up where φ_Z has already been defined, i.e. near each v_i and near each $(S^3 \times x_p) \subset (S^3 \times S^3)_p$. However, the differentials of φ_Z and φ'_Z differ at most by a scalar multiple along E_- where both are defined, so it is a straightforward proceedure to modify φ'_Z to match up with φ_Z where the two disagree. See the argument in Step 2 of the proof of Proposition 2.5.

With this matching complete, define ω_Z in Region 1 by

$$\omega_Z \equiv \varphi_Z^{-1}(\mu), \tag{11.21}$$

where μ is as specified in (2.3).

ω_Z in Region 2

Fix $i \in \{1, \cdots, r\}$ and consider the definition of ω_Z near the i'th component of Region 2, i.e. near (11.18). The 2-form ω_Z will be defined near (11.18) in $\mathbb{R}^4 \times \mathbb{R}^3$ as the pull-back via the map $\psi / \mid \psi \mid$ of a 2-form on $S^3 - \cup_{\alpha=0}^{N}(S^3 \cap r_\alpha)$. This 2-form μ_N, is given as follows: First, let $\{p_\alpha \equiv S^3 \cap r_\alpha\}_{\alpha=0}^{N}$. Then, employ:

LEMMA 11.2. *Let $N \geq 1$ be given as well as $N+1$ distinct points $\{p_\alpha\}_{\alpha=0}^{N} \subset S^3$. For each $\alpha \in \{0, \cdots, N\}$, let $B_\alpha \subset S^3$ be an embedded 3- ball with $B_\alpha \cap (\cup_{\alpha'} p_{\alpha'}) = p_\alpha$. Orient ∂B_α by the normal directed towards p_α. Let ω_α be a closed 2-form on $B_\alpha - p_\alpha$ with the following property:*
1) $\int_{\partial B_0} \omega_0 = 1$.
2) $\int_{\partial B_\alpha} \omega_\alpha = 1/N$ *if* $\alpha \geq 1$.
Then, there is a closed 2-form, μ_N, on $S^3 - \cup_{\alpha=0}^{N} p_\alpha$ which, for all α, restricts to $B_\alpha - p_\alpha$ as ω_α.

Proof. Use Meyer-Vietoris.

The lemma gives μ_N once suitable $\{\omega_\alpha\}_{\alpha=0}^{N}$ are specified. These should be chosen so that $(\psi / \mid \psi \mid)^* \mu_N$ agrees with ω_Z where Region 2 overlaps Regions 1 and 3. (The overlap with Region 1 determines ω_0 and the N components of the overlap with Region 3 determines $\{\omega_\alpha\}_{\alpha \geq 1}$. In this regard, remember that E_- in Region 2 is made by modifying the amounts of push-off of N sheets, $\{Y_\alpha\}_{\alpha=1}^{N}$, of E_{4-}. These push-offs are all parallel and in the e_3 direction with respect to the frame ζ of E_{4-}. On the sheet Y_α, the frame ζ is the push-off copy of the frame ζ on Y_1. Thus, the frame ζ on each sheet of (11.21) agrees with the constant frame (e_1, e_2, e_3) for $\mathbb{R}^3$.) The details here are left to the reader. □

h) ω_Z near E_+.

Where E_+ and E_{4+} differ, $E_+ = \Theta(E_-)$. In fact, E_+ can be divided into three regions, which are Region 3 and the image by Θ of Regions 1 and 2 in E_-. On Region 3, E_+ and E_{4+} agree. With this understood, the form ω_Z should be defined near the Θ image of the E_- regions 1 and 2 by pull-back using the map Θ. The definition of ω_Z on Region 3 of E_+ mimics the definition of ω_Z on Region 3 of E_- and the details are left to the reader.

i) The form ω_Z on Z_3.

Remember (from this section's introduction) that the original cobordism, W, between M_0 and M_1 was split in half as $W_1 \cup W_3$, and this resulted in a corresponding split of $Z = Z_1 \cup Z_3$. The preceding subsections defined ω_Z on $N_{Z_1} - Z_1$, and it is the purpose of this subsection to describe ω_Z on $N_{Z_3} - Z_3$.

But for an obvious change of notation, the construction of ω_Z on $N_{Z_3} - Z_3$ repeats the constructions of the previous subsections (a- h). (The notation change replaces $W \equiv W_1$ by $W \equiv W_3$ and (M_0, p_0) by (M_1, p_1).)

j) Continuity of ω_Z.

Let $Z \equiv Z_3 \cup Z_3$ as in the introduction to this section. Likewise, let $W \equiv W_1 \cup W_3$. Let $\Sigma_Z \equiv \Sigma_{Z_1} \cup \Sigma_{Z_3}$ and let N_Z be the corresponding union of N_{Z_1} and N_{Z_3}.

The continuity of ω_Z cross $(N_Z - \Sigma_Z) \cap (M \times M)$ is a concern because ω_Z has been separately constructed on $N_{Z_1} - \Sigma_{Z_1}$ and $N_{Z_3} - \Sigma_{Z_3}$. Let ω_{Z_3} denote the restriction of ω_Z to $N_{Z_1} - \Sigma_{Z_1}$ and let ω_{Z_3} denote the analogous restriction to $N_{Z_3} - \Sigma_{Z_3}$. The constructions in Subsections 11b, and 11c insure that ω_{Z_1} and ω_{Z_3} match up nicely near $p \times M$ and $M \times p$ in $M \times M$. At issue is the match between ω_{Z_1} and ω_{Z_3} near $\Delta_M \subset M \times M$.

Subsection f describes ω_{Z_1} and ω_{Z_3} near Δ_M. The form ω_{Z_1} is constructed with the help of a singular frame for M which is an extension over W_1 (see Lemma 11.1) of the canonical singular frame for M_0. Likewise, ω_{Z_3} is constructed near Δ_M with the help of a singular frame for M which is an extension over W_3 of the canonical singular frame for M_1. At issue is whether these two singular frames can be chosen to agree. The purpose of this subsection is to prove that these frames can be assumed equal under the assumptions of Theorem 2.9.

To begin, consider W_1 and remark that the singular frame in question which defines ω_{Z_1} near Δ_M comes from a frame χ_1 for the conormal bundle $N^*_{Z_1}$ for $\Delta_{Z_1} - (\gamma_1 \cup (\cap_{i=1}^{r_1} v_{1i}))$ as a submanifold of Z_1. Here, the notation is from Lemma 11.1 except that subscripts "1" now appear to signify subsets of Z_1. In particular, χ_1 has the prescribed singularity of (11.11) along $\gamma_1 \cup (\cup_{i=1}^{r_1} v_{1i})$.

Likewise, ω_{Z_3} is defined with the help of a frame, χ_3, for the conormal bundle $N^*_{Z_3}$ of $\Delta_{Z_3} - (\gamma_3 \cup (\cap_{i=1}^{r_3} v_{3i}))$ with the prescribed singularity of (11.11) along $\gamma_3 \cup (\cup_{i=1} r_3 v_{3i})$.

Theorem 2.9 assumes that (2.12)'s homomorphism l_W annihilates the canonical singular frame for M_0. This implies that the proof of Lemma 11.1 can be repeated with minor notational changes to prove that χ_1 extends over

$$\Delta_Z - ((\gamma_1 \cup (\cup_{i=1}^{r_1} v_{1i})) \cup (\gamma_3 \cup (\cup_{i=1}^{r_3} v_{3i}))). \tag{11.22}$$

as a frame for the conormal bundle N^*_Z for Δ_Z in Z. Likewise, Theorem 2.9 assumes that l_W annihilates the canonical singular frame for M_1; and so χ_3 extends over (11.22) also.

With the preceding understood, then the following lemma implies that χ_1 and χ_3 can be chosen to agree. Thus, the lemma below resolves the continuity issue.

LEMMA 11.3. *Let $\chi \equiv \chi_1$ or χ_3. If $(\pi^*_R - \pi^*_L)^{-1}(\chi \mid \Delta_{M_0})$ gives Definition 2.8's canonical homotopy class of singular frame for T^*M_0, then Definition 2.8's canonical homotopy class of singular frame for T^*M_1 is given by $(\pi^*_R - \pi^*_L)^{-1}(\chi \mid \Delta_{M_1})$.*

Proof. To start the argument consider the following three remarks: *Remark 1* is that the dual, N^*_W, to the normal bundle of $\Delta_W \subset W \times W$ is canonically isomorphic to T^*W. The isomorphism is $\pi_L^* - \pi_R^* : T^*W \to T^*(W \times W) \mid_\Delta$. (The image of the preceding map annihilates $T\Delta_W$.)

Remark 2 is that the restriction of N_W^* to Δ_Z has a natural line subbundle, namely $\mathrm{Span}(dF) \subset N_W$, where F is the function in (3.20). The quotient bundle is naturally isomorphic to N_Z^*. Thus, N_W^* splits over Δ_Z as

$$N_W^* \approx N_Z^* \oplus \mathrm{Span}(dF) \tag{11.23}$$

Remark 3 is the observation that, as in Proposition 2.7, there are, up to homotopy, two canonical, honest framings, (χ_+, χ_-), of $N_Z^* \to \Delta_Z$ which can be obtained from the singular framing χ. (Copy the construction of $\zeta_\pm$ in the proof of Proposition 2.7 and use the fact that the singularity of χ along any of the paths $\gamma_1 \cup \gamma_3$ or $\{v_{1i}\}_{i=1}^{r_1}$ or $\{v_{3i}\}_{i=1}^{r_3}$ are independent of the parameter along that path. See (11.11).) □

Given the following three remarks, it follows that χ with dF gives a framing of $N_W^* \oplus N_W^*$ over Δ_Z, namely $((\chi_+, dF), (\chi_-, dF))$.

LEMMA 11.4. *The framing $((\chi_+, dF), (\chi_-, dF))$ extends to a framing of $N_W^* \oplus N_W^*$ over Δ_W.*

Accept this lemma and here is how to finish the proof of Lemma 11.1's second assertion: It follows from Lemma 11.4 that the framing for $N_W^* \oplus N_W^*$ over Δ_{M_0} given by $((\chi_{1+}, dF), (\chi_{1-}, dF))$ extends over all of Δ_W to give a framing, $((\chi'_{1+}, dF), (\chi'_{1-}, dF))$, for $N_W^* \oplus N_W^*$ over Δ_{M_1}.

Suppose first that (χ_{1+}, χ_{1-}) gives Atiyah's canonical framing on the nose. (See, e.g. Assertion 3 of Proposition 2.7.) Then, because W has index zero, (χ'_{1+}, χ'_{1-}) must give Atiyah's canonical framing for M_1.

Now, suppose that $(\chi_{1+}, \chi_{1-}) = g\,A$, where g is a map from M_0 to Spin(6) with minimal, non-negative degree, and where A is a frame which gives Atiyah's canonical frame. There is no obstruction here to extending g over Δ_W as a map to Spin(6). With this extension understood, it follows that $g^{-1}(\chi'_{1+}, \chi'_{1-})$ must give Atiyah's canonical frame for M_1. (This is because W has signature zero.)

By definition, there exists a map h from M_1 of minimal non- negative degree such that $h\,g^{-1}(\chi'_{1+}, \chi'_{1-}) = (\chi_{3+}, \chi_{3-})$. This implies that degree$(h\,g^{-1}) \leq 0$. (Note that the degrees of g's restrictions to M_0 and to M_1 must agree.)

Now, reverse the roles of M_0 and M_1 and also χ_1 and χ_3 in the preceding argument. The inevitable conclusion has degree$(g\,h^{-1}) \leq 0$. However, $\deg(h\,g^{-1}) = -\deg(g\,h^{-1})$, so g and h must have the same degree. This implies the lemma in the general case.

Proof. The framing $((\chi_+, dF), (\chi_-, dF))$ is defined over each $\{(\Delta_{S^3})_p\}_{p \in \mathrm{crit}(f)}$ and the task is to extend this frame over the open 4-ball components in $\Delta_W - \Delta_Z$ which are bounded by the 3- spheres in $\{(\Delta_{S^3})_p\}_{p \in \mathrm{crit}(f)}$. For this purpose, fix $p \in \mathrm{crit}(f)$ and introduce the coordinate sytem ψ_p of (3.2) and let $U_p \equiv \psi_p(\mathbb{R}^4)$. Note that $U_p \times U_p$ is a neighborhood of (p, p) in $W \times W$. The projections $\pi_\pm$ of (11.9), (11.10) define a second product structure on $U_p \times U_p$.

Let τ denote the 1-form on $\mathbb{R}^4$ which is the exterior derivative of the square of the distance function from the origin. Then, note from (3.25) that

$$(\pi_+^* - \pi_-^*)\tau = dF\,|_\Delta \tag{11.24}$$

Thus, the inverse of $(\pi_+^* - \pi_-^*)$ over $\Delta_{\mathbb{R}^4}$ is a map which identifies N_W^* with $T^*\Delta_{\mathbb{R}^4}$. This map identifies the frame $((\chi_+, dF), (\chi_-, dF))$ with a frame for $(T^*\mathbb{R}^4 \oplus T^*\mathbb{R}^4)\,|_{S^3}$ which gives Atiyah's canonical 2-frame (see Definition 2.8 and Lemma 2.11). Atiyah's canonical 2- frame for S^3 extends over the 4-ball, and so the image of this extension under the map $(\pi_+^* - \pi_-^*)$ extends $((\chi_+, dF), (\chi_-, dF))$. □

k) Verification of Lemma 4.2.

Define ω_Z on $N_Z - \Delta_Z$ as above. If ω_Z can be shown to extend to $Z - \Sigma_Z$ as a closed form, then (2.28) completes the proof of Theorem 2.9. The extension will be made by appealing to Lemma 4.2.

As previously remarked, the second condition of Lemma 4.2 follows from Proposition 10.1 so only Condition 1 of Lemma 4.2 is at issue.

To verify Condition 1 of Lemma 4.2, remark first that the image of the class of ω_Z in $H^3_{\text{comp}}(N_Z)$ is represented by a closed 3- form, η, which is obtained as follows: Let $\rho : N_Z \to \mathbb{R}$ be a smooth function which has compact support and which takes the value 1 near Σ_Z. Then

$$\eta \equiv -d\rho \wedge \omega_Z. \tag{11.25}$$

Clearly, this form integrates to zero over any closed 3-cycle in N_Z. But, to decide whether η is Poincare dual to σ_Z, one must compute the integral of η over cycles which represent certain classes in $H_3(Z, Z - N_Z)$.

To identify the relevant cycles, remark that

$$H_3(Z, Z - NZ) \approx H_3(Z_1, Z_1 - N_{Z_1}) \oplus H_3(Z_3, Z_3 - N_{Z_3}). \tag{11.26}$$

It is therefore permissable to concentrate on each factor in (11.26) separately. Since the arguments are the same for either factor in (11.26), the notation will be simplified starting in the next paragraph by using Z to denote *either* Z_1 or Z_3.

To begin the verification of Condition 1 of Lemma 4.2, remark that it is convenient to replace σ_Z by a homologous cycle. For this purpose, remark that an isotopy of $E_{4\pm}$ pushes this space into N_Z if it is not there already. With $E_{4\pm}$ in N_Z, note that the class σ_Z in $H_4(N_Z, N_Z - \Sigma_Z)$ is the same as $\sigma' \equiv [\Delta_Z] - [E_L] - [E_R] - N^{-1}([E_{4-}] - [E_{4+}])$. The latter is obviously a sum of classes which are represented as the fundamental classes of oriented submanifolds of N_Z which have trivial normal bundles. As such, there is an unambiguous intersection pairing between σ' and classes in $H_3(Z, Z - N_Z)$.

With the preceding understood, note that the Poincare' dual of σ' is characterized by the following fact: Its integral over a cycle in $H_3(Z, Z - N_Z)$ is the same as the intersection number of said cycle with σ'.

Since the Poincare dual of σ' is equal to the Poincare' dual of σ_Z, it is sufficient to consider the integral of the 3-form η over cycles with boundary in $Z - N_Z$ and compare the value of said integral with the intersection number of the cycle with σ'. This last task is straightforward because one can consider each of the constituent submanifolds (i.e., $\Delta_Z, E_{L,R}$ and $E_{4\pm}$) separately. The task is left to the reader as an exercise.

References

[1] M. F. Atiyah, *On framings of 3-manifolds*, Topology 29, 1- 7 (1990).

[2] S. Axelrod and I. M. Singer, *Chern-Simons perturbation theory*, Proc. XXth DGM Conference (New York, 1991), S. Catto and A. Tocha, eds., World Scientific, 1992, 3-45.

[3] S. Axelrod and I. M. Singer, *Chern-Simons perturbation theory II*, Jour. Diff. Geom. 39 (1994) 173-213.

[4] D. Bar-Natan, *Perturbative aspects of the Chern-Simons topological quantum field theory*, Ph.D thesis, Princeton Univ., June 1991.

[5] D. Bar-Natan, *On the Vassiliev knot invariants*, Topology, to appear.

[6] J. S. Birman and X-S Lin, *Knot polynomials and Vassilievs' invarients*, Invent. Math. III (1993) 153-287.

[7] M. F. Freedman and F. Quinn, *Topology of 4-dimensional manifolds*, Princeton University Press, Princeton 1990.

[8] L. Jeffrey, *Chern-Simons invariants of lens spaces and torus bundles, and the semi-classical approximation*, Commun. Math. Phys. 147 (1992) 563-604.

[9] M. Kontsevich, *Feynman diagrams and low dimensional topology*, Max-Planck-Institute (Bonn) Preprint.

[10] X-S. Lin, *Vertex models, quantum groups and Vassiliev's knot invariant*, Columbia Univ. Preprint, 1991.

[11] C. C. MacDuffe, *The theory of matrices*, Springer, Berlin 1933.

[12] J. Milnor, *Lectures on the h-cobordism theorem, notes by L. Siebenmann and J. Sondau*, Princeton University Press, Princeton 1965.

[13] L. Rozansky, *Large k asymptotics of Witten's Invariant of Seifert Manifolds*, University of Texas preprint 1993.

[14] N. Yu. Reshitikin and V. G. Turaev, *Invariants of 3-manifolds via link polynomials and quantum groups*, Invent. Math. 103 (1991) 547-597.

[15] C.H. Taubes, *Homology cobordisms and the simplest perturbative Chern-Simons 3-manifold invariant*, Harvard University preprint 1993.

[16] V. A. Vassiliev, *Cohomology of knot spaces*, in: *Theory of singularities and its applications*, V. I. Arnold, ed. American Math. Soc., Providence,

1990.

[17] V. A. Vassiliev, *Compliments of discriminants of smooth maps: Topology and applications*, Trans. of Math. Mono. 98, Amer. Math. Soc., Providence, 1992.

[18] E. Witten, *Quantum field theory and the Jones polynomial*, Commun. Math. Phys. 117 (1988), 351-399.

Made in United States
Orlando, FL
05 November 2026

90230602R00302